KU-455-676

A Contextual History of Mathematics

to Euler

Ronald Calinger

The Catholic University of America

with the assistance of

Joseph E. Brown   and   Thomas R. West

Rensselaer Polytechnic Institute   *The Catholic University of America*

Prentice Hall, Upper Saddle River, NJ 07458

Library of Congress Cataloging-in-Publication Data

Calinger, Ronald.
    A contextual history of mathematics : to Euler / Ronald
Calinger.
        p. cm.
    Includes bibliographical references and index.
    ISBN 0-02-318285-7
    1. Mathematics--History. I. Title.
QA21.C188   1999
    510'.9--dc21   98-53293
                                                       CIP

Executive Editor: George Lobell
Editorial Assistant: Gale Epps
Editor-in-Chief: Jerome Grant
AVP of Production and Manufacturing: David W. Riccardi
Senior Managing Editor: Linda Mihatov Behrens
Executive Managing Editor: Kathleen Schiaparelli
Editorial/Production Supervision: Barbara Mack, Elaine W. Wetterau
Manufacturing Buyer: Alan Fischer
Manufacturing Manager: Trudy Pisciotti
Marketing Manager: Melody Marcus
Marketing Assistant: Amy Lysik
Art Director: Jayne Conte
Composition: MacroTeX Services
Cover Designer: Bruce Kenselaar
Cover Credit: "Goblet," Kungl. Husgeradskammaren/The Royal
    Collections, Stockholm. (SS10). Photographer Sven Nilsson.

ⓒ 1999 by Prentice-Hall, Inc.
Upper Saddle River, NJ 07458

All rights reserved. No part of this book may
be reproduced, in any form or by any means,
without permission in writing from the publisher.

Printed in the United States of America
10 9 8 7 6 5 4 3 2 1

HAMPSHIRE COUNTY LIBRARY

510.9   0023182857

C003833930

ISBN 0-02-318285-7

PRENTICE-HALL INTERNATIONAL (UK) LIMITED, LONDON
PRENTICE-HALL OF AUSTRALIA PTY. LIMITED, SYDNEY
PRENTICE-HALL CANADA INC., TORONTO
PRENTICE-HALL HISPANOAMERICANA, S.A., MEXICO
PRENTICE-HALL OF INDIA PRIVATE LIMITED, NEW DELHI
PRENTICE-HALL OF JAPAN, INC., TOKYO
PRENTICE-HALL (SINGAPORE) PTE. LTD.
EDITORA PRENTICE-HALL DO BRASIL, LTDA., RIO DE JANEIRO

Contents

Preface

Today mathematicians offer two principal interpretations of their discipline. One group defines it as abstract, universal, independent of time and setting and revealing itself in discoveries that evolve within the rules of its inherent logic; the other holds it to be an intellectual product of culture, society, and the quirks of mathematicians. Operating from one or the other of these perspectives, scholars differ on the analysis of primary technical and related nonmathematical writings and the complex processes of transmission. The ensuing prologue will suggest something of the scope, nature, and sources of mathematics, as well as the evolution of the critical history of the discipline. Drawing on studies from anthropology, ethnography, linguistics, philosophy, psychology, religion, and sociology that intersect with mathematics, this book examines the lives and contributions of extraordinary figures, such as Pythagoras, Archimedes, Hypatia, al-Khayyām, Bhāskara, Fermat, and Leibniz, including their discourse with other practioners in schools, academies, and the rest of the republic of letters. It seeks to illuminate the power, beauty, and variety of mathematics by investigating major problems, guiding methods, and applications. It explores the mediation, fruitful and injurious, of mathematical developments by their cultural, intellectual, and social contexts. This book spans the growth of mathematics from the tallying sticks of the late Age of Stone through classical mathematics up to 1727, when Isaac Newton died and Leonhard Euler commenced his prodigious accomplishments in St. Petersburg and Berlin.[1]

Why a new general history of mathematics? The principal answer comes from a survey of the *Mathematical Reviews*, the abstracts of *Historia Mathematica*, and the annual *Isis Critical Bibliography* that reveals a sharply increasing number of articles and books on historical studies of mathematics. Their detailed findings and interpretations indicate the need for a new synthesis and, where possible, more nearly complete accounts of past achievements. Prominent among recent scholars in the field are Lorraine Daston, Ivor Grattan-Guinness, Wilbur Knorr, Thomas Kuhn, Joseph Needham, Otto Neugebauer, Clifford Truesdell, B. L. van der Waerden, and André Weil. The results of sifting through much of this expanding mass of scholarship have been integrated into this text.

This contextual history is distinctive from current general histories of mathematics in several ways that are intended to assist in the improvement of research, learning, and pedagogy. A brief synopsis of the political and social fabric of each era and culture aims to convey a sense of time and the major circumstances and intellectual demands that mathematicians encountered. Maps delineating geographical settings and cultural and political illustrations—paintings, statuary, and edifices—are joined with the text to contribute to a sense of the unfolding of history. The contextual treatment and extensive endnotes and bibliographies are two features that largely separate this book from the comprehensive history of mathematics by Carl Boyer and the more internal works by David Burton, Howard Eves, and Victor Katz. It not only addresses the ancient Mediterranean together with medieval and early modern Europe but also gives larger treatments of traditional China, India, medieval Islam, and the Maya than do most general histories. It seeks as well to treat in greater depth mathematical problems, connections between fields of mathematics, and relations with the natural sciences and other fields of learning. It raises competing interpretations of topics and themes to suggest the richness and complexity of history and that no single mechanical pathway suffices for the subject.

Longstanding in history is a division between generalists and specialists. Careful negotiation is required between an entanglement in details and resort to generalizations that do injustice to the technical content of mathematics.[2] Preparing this history also demanded attentiveness to the remappings, redivisions, and renarrations that every new crop of historians achieves. This book, presenting both established and promising emerging theses, must be provisional in conclusions.

Rather than including problem sets in the text, this study takes advantage of modern technologies. Professor Mary Ann McLoughlin has prepared a web site including problem sets for each chapter, several course syllabi, maps, a clinic for faculty and student questions, and chapter guides and test banks. The URL is www.strose.edu/facstaff/math/maryann.html. The web site, responding to faculty using this book as a text, will allow problems and open-ended exercises to be updated regularly and new resource materials to be added.

Although this book is written for readers having the equivalent of introductory college level courses in mathematics and might be profitably read by many, it primarily addresses four audiences: university students majoring in mathematics, mathematics education, the exact sciences, engineering, or the history of mathematics; mathematicians wishing to read more about the history of their subject; historians of science and intellectual historians; and readers probing the intersections of mathematics with other fields, such as the work of the ancient Pythagoreans, the role of mathematics in early Christianity, mathematical requirements for the traditional Chinese bureaucracy, and the seventeenth-century debate between faith and reason. This book ends with the inventions of calculus, which begin modern mathematics. For students in calculus courses, it can serve as a supplement, offering more information about its roots and the work, including errors, of its inventors, thus better revealing the human face of mathematics.

Acknowledgments

Among many mathematicians, historians, and students who have worked to improve this book, I am indebted above all to Joe Brown, who wrote Chapters 10 and 12 on mathematics in medieval Europe and made trenchant comments on the entire manuscript. At Rensselaer, Marion Quiroga typed the first draft and she and Roderick Brumbaugh carefully reviewed the grammar and soundness of arguments. Edith Luchins and Mary Ann McLoughlin examined the problems and proofs in the first seven chapters. Subsequently, at Catholic University Thomas West has helped revise the entire manuscript with a keen editing eye that demands clarity and accessibility in writing without compromising on vocabulary and concepts. At our weekly lunches we have laughed, disputed, and discussed the book. A number of mathematical reviewers, working initially with Robert Pirtle at Macmillan and especially now with George Lobell at Prentice Hall, have identified infelicitous statements, obscure passages, and errors in problems and proofs. Reviewers who are known to me include Professors Donald Anderson of the University of Tennessee-Knoxville and William R. Hintzman of San Diego State University. I am grateful for the time and effort shown in their comments, including robust opposition to a few of my interpretations. All obscurities or errors that remain in the text are mine alone.

For their prompt responses and valuable assistance in locating and acquiring sources and illustrations, I am also grateful to the staff of Mullen Library at Catholic University, James C. Miller of the Engineering and Physical Sciences Library at the University of Maryland-College Park, and Ronald Brashear of the Dibner Library, which is part of the Smithsonian Institution Libraries.

Others deserve special mention. My students at Rensselaer and Catholic University have demanded precise, crisp explanations in the history of mathematics that have been important to the reworking of this manuscript. Prentice Hall asked that I put the manuscript in a scientific computer program that would produce page proofs. This could not have been done without the patient assistance of Lynda Wever. Working on this book keeps bringing to my mind the academic preparation from Carl Gustavson at Ohio University, Robert Colodny at the University of Pittsburgh, and Allen Debus and Saunders Mac Lane at the University of Chicago, especially our many discussions. For more than a decade the writing of this book has been an integral part of the lives of my wife Betty and our children John and Anne. Their forbearance of my obsessiveness is deeply appreciated. Watching John act in performances at school, particularly his rendering of Richard III, and Anne play on her soccer teams has been energizing.

Ronald Calinger
calinger@cua.edu

Prologue

Rich, diverse, and vibrant, mathematics is a rapidly expanding field. From the dozen constituent disciplines listed in the *Jahrbuch Über die Fortschritte der Mathematik* in 1868, it has grown into more than sixty in the current classification of research papers in the *Mathematical Reviews*. Convention portions the field into four main branches: algebra, analysis, geometry, and applied mathematics. Of these, geometry, with its visual power, its projection of a universality not attainable in sensory experience, and its logical rigor, dominated mathematics in the West from the fourth century B.C. to the eighteenth century.[3] Also prominent through the centuries was mathematical astronomy. Differential and integral calculus that developed into analysis did not emerge until the late seventeenth century.[4] In modern specialization as the four principal disciplines and their first branches are being split, the new disciplines may have a hybrid character, overlapping one another and blurring points of connection. They thus require ever more finely tuned definitions. Number theory, for example, belongs to algebra and employs analysis, geometry, and probability. Some modern branches also strain traditional classification. Mathematics now includes computer science, differential equations, game theory, logic and foundations, probability and stochastic processes, set theory, statistics, general topology, and statistical physics.

The Nature and Sources of Mathematics

Since antiquity, mathematics has been roughly divided into pure and applied mathematics.[5] The older term covering much of what is now classified as applied is mixed, which obscures the boundaries with pure subjects. Pure mathematics is contemplative, addressing mathematical objects, such as sets. Its research produces significant findings published in refereed journals that have demanding standards. Since at least the eighteenth century, mathematics has been widely perceived in the world of learning as the highest form of thought,[6] with pure mathematics at its zenith, self-contained, Olympian in its separateness from physical objects.[7] Applied mathematics crosses disciplines, and its boundaries with fields such as physics and economics are permeable. Eugene Wigner defines three processes that act in concert when the conjunction is most effective: mathematical models disclose possibilities that might otherwise escape notice; the conjunction generates mathematical research; and applications need not be confined by the stringent criteria for validity, determined by pure reason, in the companion discipline.[8]

Between pure and applied mathematics, a transfer of simple information and a dialectic of ideas and methods have always existed. Creation of differential and integral calculus independently by Isaac Newton and Gottfried Leibniz, together with its ordering and extension by Leonhard Euler, have greatly accelerated these transfers. The mathematical sciences are now powerful instruments for gaining exact knowledge about natural phenomena and exerting some mastery over nature, and their application to daily human conduct increases. The framing discourse of advanced engineering and most natural sciences is heavily mathematical and often at a level beyond calculus. The models into which scientific researchers organize their ideas and observations are set forth in a formal language of statements that links mathematical objects with the behavior of physical phenomena, or operations of nature. Scientists and engineers test their empirical consequences by exacting quantitative measurements. Rarer, but of greater interest, is the discovery of some novel correlation amounting to a law of nature and of the mathematical relations associated with it.

In modern times, the connections of mathematics with commerce, trade, the natural sciences, military science, engineering, and the visual arts have expanded. Mathematics deeply informs as well such activities as decision making, social planning, and communication. The current fecund role of mathematics in human affairs shows clearly in game theory and computers. Game theory, which mathematically analyzes conflict and cooperation, treats, as analogues to strategic choices in competitive games, certain decisions made in economics, business, politics, and the military. The minimax theorem, whose object is to determine the point that will yield a set of optimal outcomes for all the players in a game, is now a cornerstone for economics and decision making. Digital computers have become essential to commerce, banking, industrial production, quality control, architecture, communications, education, medical monitoring, record keeping, and quantitative analysis in the social sciences.[9] The realm of mathematics extends to investigations of harmony and beauty in theoretical aesthetics, music, and the visual and performing arts.[10]

Discourse in academe between pure and applied mathematics is intense and occasionally hostile. Yet the very categories *pure* and *applied* have been questioned.[11] If there is pure mathematics, opponents ask, what mathematics is impure? That deep curiosity about problems in nature along with a sustained quest of exact solutions can motivate not only speculative thinkers but tinkering inventors increases the confusion. Attempting a demarcation that will better identify the shared methods and differing objects studied, Jean Dieudonné urges that the term *pure mathematics* be replaced by *mathematics* and *applied mathematics* by *application of mathematics*.[12]

In the continuing attempt to locate the root of mathematical creativity, twentieth-century scholars seem to depend largely on identifying which mathematicians do original research that pushes back the frontiers of mathematical knowledge. Scholars from Jules-Henri Poincaré to Alfred Adler have divided the general body of mathematicians into two groups: a tiny minority of authentic creators conduct original studies, while most mathematicians are essentially artisans who confirm.[13] The neatness and simplicity of such a dualistic model is attractive, and a historian might see some rhetorical advantage in splitting the mathematical community as clearly as, say, C. P. Snow has modern high culture into two strains, the humanistic and the scientific. But Schrödinger's atom model may give a more accurate metaphor for representing the accomplishment of the mathematical community. Circling its nucleus are fuzzy, smeared particles acting like waves in orbits at varying distances from the nucleus. Instead of a sharp dichotomy between creators and developers, there appears to be a central nuclear pair doing both. Five peerless geniuses—Archimedes, Newton, Euler, Gauss, and Riemann—occupy that center. But the nucleus is incomplete without the more numerous *clarissimi*, to use Gauss's term. The atom's integrity and that of mathematics require that two shells be added. Filling one shell is a larger third group that advances sound research in mathematics and improves its teaching; the outer shell of connoisseurs supply trenchant perspectives on the development of mathematics. Together with a grasp of mathematics, connoisseurship would entail a good knowledge of authors and schools, skill in historical methods, and the ability to convey clearly to the literate public the nature and achievement of mathematics.[14] Stripped of its shells, the atomic like mathematical community in this analogy is ineffective.

Aside from the qualifications for mathematical creativity, what does doing good mathematics require? Despite the inextricable affiliation of mathematics with logic and the manipulation of symbols, Poincaré has declared that talent in these by itself is insufficient for the task. Otherwise, far more people would be doing it. At the heart of good mathematics Poincaré places an aesthetic sensibility that delicately sorts out hidden harmonies, relations, and useful combinations of ideas. In *A Mathematician's Apology*, originally published in 1940, G. H. Hardy also makes aesthetics central to mathematics.[15] The astrophysicist Subrahmanyan Chandrasekhar finds this epistemic and esthetic connection across the sciences.[16] Mortimer Adler's *Six Great Ideas*, published in 1981, differentiates between beauty as a taste providing pleasure to the general public and an admirable beauty transcending taste and restricted to a small circle possessing a special sensibility to it.[17] Poincaré's belief that this aptitude is innate and limited to a few great thinkers has been challenged. Freud's location of the unconscious within the psyche, Jonathan Lear asserts, has made creativity in all fields "more democratically available."[18] The concept of a guiding beauty in mathematics also has critics, who find it vague and mystical.[19]

On the larger, more methodological question, debated since antiquity, of whether mathematical truths are discovered or invented, Platonist realism argues for discovery. It insists that mathematics places students in the company of preexistent, eternally stable, ideal Forms. Mathematical Forms of purely abstract arithmetic numbers and geometric figures are less general than the supreme Forms of truth, goodness, and the beauty of metaphysics. But they are closely related: their study is a necessary gateway for understanding those supreme Forms. Recognizing that the physical world is an imperfect copy, Platonists pursue by rigorous logic the truths of the ideal realm, located at a site below the Forms. Platonist absolutists hold that this does not negate their preexistence. Logic, like a telescope, is merely an instrument of discovery: it does not create the galaxies of number and shape that it brings into view. Through the pure mathematics of Nicolas Bourbaki and other major theorists such as Paul Erdös, the Platonic and other internal interpretations have maintained a strong presence in the twentieth century.[20] A Bourbakiist, André Weil holds that none but research mathematicians are properly prepared to explain mathematical ideas and their evolution.[21] Only they can grasp the timeless facts behind mathematical statements.[22] Weil sees cultural, historical, linguistic, and philosophical issues as irrelevant.

In *Mathematics: Form and Function*, Saunders Mac Lane insists that recent versions of Platonism—the ontological and realist as well as the methodological and epistemological—address not the nature of mathematics but its processes.[23] In the actual practice of mathematics, he observes, these varieties disintegrate. Mac Lane conceives of mathematics as protean, not to be captured by uncompromisingly stable Forms.[24] The increasingly abundant objects inhabiting the universe of mathematics, which are generally contrived to enhance our understanding of counting and motion, include numbers, matrices, differential manifolds, lattices, topological spaces, and functions located in a multilayered conceptual structure.[25] More general objects, for example, replace numbers on a second level of abstraction. In a third level, these objects are conjoined into spaces, which are the basis of functions in a fourth. That mathematics contains many levels of abstraction in structures attenuates the Platonist position, for logic then is not simply a tool of discovery but an essential framework in the edifice of mathematics itself. After Évariste Galois developed his theory of the solvability of equations by concentrating on the structures of fields and groups, algebraists investigated others: fields, rings, and so forth. In his two-part *Moderne Algebre*, published with Springer in 1930 and 1931, Bartel van der Waerden, with the use of Emil Artin's lectures, made a neat presentation of Galois theory that replaced earlier largely obscure accounts. Building on lectures by Emmy Noether on ideal theory, he also formulated new conceptual and structural insights.[26] The work of Noether immensely influenced the evolution of modern algebra. Hermann Weyl praised her "mighty imagination" and the "prophetic lapidary manner" of her statements.[27]

Beside that structural assessment can be put the claim that mathematics is an empirical science of space and number. Among adherents of this persuasion, George Polyá has elucidated the substantive role of guesswork in making discoveries in mathematics and Imre Lakatos, its presence in mathematical arguments.[28] Although the empiricists are a minority, Philip Kitcher detects a recent movement toward that position.[29] An empirical element is added in massive computations that may involve trial and error. To facilitate such operations, mathematicians devise and perfect algorithms. Since the seventeenth-century invention of the telescope, microscope, and pendulum and spring clocks, the steady attainment of greater precision in the study of nature has helped improve algorithms as well as theory. This search for precision, which includes ever closer approximation techniques, is characteristic of mathematics. Carrying out extensive computations reliably, quickly, cheaply, and with little manual effort has required creating logarithms and designing instruments like the abacus and the high speed electronic computer. The experimental character of number theory is shown partly in the devising of theorems in number theory through computations with modern computers. Utilizing the computer, mathematicians also probe established methods for solving equations, determining when they are reliable, erratic, or idiosyncratic.[30]

Whatever the rank of empirical efforts, mathematics exhibits abstraction as another primary characteristic. Abstraction is, of course, a property of all natural sciences and mental activity. Mathematical abstraction addresses partly quantitative relations in algebraic equations and formulas, as well as in spatial forms in geometric diagrams. Throughout it involves symbols, such as numerals and for unknowns letters. "By relieving the brain of all unnecessary work, a good notation sets it free to concentrate on more advanced problems," wrote Alfred North Whitehead, "and, in effect, increases the mental power of the race."[31] Languages themselves, their words and phrases stretched toward generalization, distort physical reality.[32] Felicitous symbols, which make possible modern mathematical discourse, may also contain a hermetic power, awakening creativity and innovation.

Formalization that possesses logical rigor and produces compelling conclusions about mathematical objects and physical phenomena has been another major feature of mathematics. Guided by sober reason and canons of the axiomatic method, reflective mathematicians from Euclid to Weierstrass to the present have developed higher standards of logical rigor. Working with abstract objects independent of physical facts, mathematical theorists may among the sciences come the closest to logical certitude. While most mathematicians are neither philosophers nor deeply engaged in the methodological investigations of philosophy, mathematics has appealed widely to scholars since 1600 as the exemplar for reaching relatively true statements.

At the turn of the twentieth century, grand foundational projects labored to fuse appropriately mathematics and logic. Following upon the work of Gottlob Frege, Bertrand Russell and Whitehead in *Principia Mathematica*, appearing in three volumes between 1910 and 1913, offered a program for reducing to logic alone all proofs and terms of mathematics once certain primitives are defined. Beginning in *Grundlagen der Geometrie*, published in 1899, David Hilbert developed axiomatics into his early version of a strict formalism.[33] He held the definition of relational structures to be the fundamental object of mathematical research. Before logic and all scientific thinking can proceed, a representation of these is necessary. Stringently manipulating symbols, Hilbert derived formulas and proofs by finitistic methods in systems free of contradiction and, thus, consistent. Finitary mathematics is only a small portion of mathematics, but Hilbert thought it possessed the necessary epistemological base. After 1918, he developed his later version of formalism, discussing it in his two volumes of *Grundlagen der Mathematik*, the first published from 1934 and the second in 1939. In radically abstracting the meanings of logical terms, it suggests Kant's distinction between real and ideal propositions.[34] But Hilbert's radical abstraction in formalism gave rise to the common misconception that he supported a popularized version of formalism that considered mathematics simply a formal game played with meaningless symbols. No less a figure than Hermann Weyl conveyed this view.[35]

Even as the strict logicist and formalist schools were being set forth and refined, the proper place and role of reason came under question. In his doctoral dissertation *On the Foundations of Mathematics*, completed in 1907, Luitzen E. J. Brouwer began to build the fundamental parts of mathematics on intuitions or "units of perception." In the Kantian sense, these are forms "inherent in human reason," which might be restricted to the whole numbers.[36] Mental operations as Brouwer understood them are not passive but shape the universe of mathematics. Reversing the direction of Frege and Russell, he stressed algorithms. Discussions of an arbitrary function and set had already raised questions about the application of the two core principles of classical logic: contradiction and the law of the excluded middle. Brouwer acknowledged the importance of eliminating contradiction, but since the law of the excluded middle—*A* or not *A*, no other alternatives being possible—applies only to finite sets, he rejected proofs based on it.[37] His method for banishing contradiction was to require that mathematical objects be explicitly constructed in proofs. Brouwer's belief that consistency alone is inadequate for justifying formal systems was confirmed in 1931, when Kurt Gödel proved that a system combining Giuseppe Peano's axioms of natural numbers and the logicist program that Russell's *Principia Mathematica* advanced lacks the desired completeness. It contains a proposition that is formally undecidable, a proposition q for which neither q nor not q can be proved in the system: a finding whose complete consequences remain to be assimilated.

Although Gödel showed that for certain arithmetic propositions consistency is not demonstrable, Hilbert's strict formalism achieved another goal stated at the end of his Paris problems, an increase in uniformity within modern mathematics. Even when Georg Cantor's theory of infinite sets, including the concept of the power or equivalence of sets of numbers, gave rise to paradoxes, Hilbert refused to leave the mathematical paradise that Cantor had created. To avoid paradoxes, twentieth-century mathematicians such as Ernst Zermelo and Abraham Fraenkel have employed only infinite sets from Cantor's theory that seem unlikely to lead to contradictions, and they have restricted fundamental concepts, relations, and operations to those granted in their axiom system for set theory. Their admissible sets have sufficed to construct practically all of classical mathematical analysis.

In his Paris lecture to the International Congress of Mathematicians in 1900, Hilbert asserted that mathematics relies on an abundance of problems—difficult, significant, and unsolved.[38] The branches of mathematics cannot flourish without these. Problem solving tests the investigator's ability to devise new methods and it widens horizons. Mathematics, George Polyá maintains, "is the art of problem solving."[39] Another modern definition makes it the primary task of mathematics to seek regularities and patterns in behavior, motion, number, or shape, or even in the substrata of chaos.[40] Here too, as throughout good mathematics, problems remain paramount. An upsurge of problems from the physical world, such as those associated with Copernican and Keplerian astronomy and navigation in the late sixteenth century and the seventeenth, influences the financial support and direction of research of many mathematicians. Invigorating problems do not arise uniformly across mathematical disciplines, but surge and shift. A lack of problems in a branch foreshadows its extinction.[41]

To solve problems, mathematicians have to posit theorems and find ways to prove them formally.[42] Proving a theorem involves proceeding by logical argument to deduce it from the fundamental properties of the concepts that occur within it. When a direct attack fails to solve a problem, mathematicians seek more specialized assaults on simpler and easier preliminary problems. Mathematicians may also identify a problem as belonging to a chain and follow across its unknown terrain to a solution. To the tribunal for evaluating the significance of a theorem and its proofs, Hardy's *A Mathematician's Apology* brings an essentially aesthetic element: how much these possess of clarity, simplicity, and order. To that list of virtues, mathematicians add the amount of knowledge that the theorem summarizes.

The rigor in exposition of the polished stage of formal proofs is generally interpreted as being at the opposite pole from the jumbled processes of creativity and discovery, which lack their precise order, linear sequence, and dispassionate coldness.[43] Since antiquity, the removal or concealment of the array of paths followed in solving problems has been a prevalent style in mathematical works. Occasionally, leading mathematicians, such as John Wallis in *Treatise of Algebra Both Historical and Practical*, published in 1685, have criticized it; and in 1954 Jacques Barzun's *Teacher in America* related the frustration among algebra students ignorant of how proofs and the subject developed.[44]

Clearly, no single procedure guides all discovery. Among the techniques and methods employed are effective algorithms, analogies from different fields, experiment, informed guesswork, coherent intuition, and imagination.[45] Intuition may fit not with Frege's judgment of it as superficial knowledge, but with Kant's concept of the mind's ability to impose structure on experience.[46] That a free play of the imagination is critical to discovery supports Georg Cantor's thesis that the essence of mathematics is its freedom. False starts, frustration, and reversals appear in the course leading to discovery. If a problem is sufficiently difficult to resist solution by sustained and deep analysis, Poincaré contends, the search for a solution goes to the subconscious level, even in cases when the researcher has consciously abandoned the problem. He and Jacques Hadamard describe moments that scientists speak of as "aha" or eureka moments: revelations that come suddenly and apparently randomly after subconscious cogitation over days, weeks, or months. For fifteen years William Rowan Hamilton had failed to extend the representation of complex numbers from the binary or ordered pairs of real numbers to ordered triples. He could not mulitiply the triples. In 1843 Hamilton experienced a sudden breakthrough: he could carry out all arithmetical operations if he employed ordered quadruplets or quaternions instead of triples and dropped the commutative law of multiplication. Hamilton described this discovery as a "flash of inspiration." It is a famous example of a eureka moment.[47] Another is Poincaré's recognition that the transformations in his definition of Fuchsian functions are those of non-Euclidean geometries.[48]

Historiographic Issues

Historiography is the basis for writing history. It identifies the critical methods of gathering, examining, and analyzing credible documents, records, and other cultural artifacts, as well as the interpretations and styles employed in the imaginative reconstruction and synthesis of historical materials into expositions and narratives. It grounds the historian's investigations of discourses, mentalities, and intellectual structures. To recognize what continues and what changes through different cultural contexts, to identify invariance in problems usually found at points of convergence between fields, to deny privilege to any one method of scholarship, to address the boundaries and limits of research: all this demands an historiographic taxonomy. During the last decade, the flood of scholarly studies in the history of mathematics listed in the *Mathematical Reviews*, the abstracts of *Historia Mathematica*, and the annual critical bibliographies of *Isis* alone provides a glimpse into a swiftly developing field, wrestling with a variety of historiographic methods and debating requisite competencies.

From at least the mid-twentieth century, each of the two principal interpretations of mathematics has fostered an historical style. Scholars who conceive mathematics to be a world apart, as abstract, universal, independent of time and setting, and the highest in theory of the natural sciences are called internal or mathematical historians.[49] This view goes neatly, though not necessarily, with the Platonist conception of an ideal realm of mathematical truths. Mathematical historians have graduate degrees in mathematics, conduct research in this field, and consider the history of mathematics to be primarily the history of mathematical ideas and objects. Their work has occasionally been taken to fit with the generally discredited Whig theory of history, in which the past is anticipation and programmed preparation leading linearly to present knowledge. As early as 1931 Herbert Butterfield criticized the Whig interpretation for tending to portray the past only in a way to ratify, if not glorify, the present.[50] Pure mathematics offers an abiding, not an evolving standard, and, unlike Whig theory, it is free of warts. Most external or cultural historians, having graduate degrees in history, philosophy, or sociology, are formally prepared in a skill crucial to historical research: the ability to decode the language, symbols, and cultural and intellectual framework of primary sources.[51] Littlewood's thesis challenged the view that mathematics is the product of quirks of mathematicians by holding that a mathematician is more the creation of the field than its creator.[52] Externalists appear more open to including an exploration of past errors, distortions, and pathologies, and most reject the notion that the boundaries between internalist and externalist are fixed and unchangeable. Both groups conceive of the history of mathematics as a technical subject, having its own styles and high standards, while internal historians properly lead in demanding that these be more explicit.[53]

Within the history of science, Gaston Bachelard's *New Scientific Spirit*, published in 1934,[54] challenged the Enlightenment notion of the permanence of reason and the rationality of science, and from the 1940s Georges Canguilhem explored errors, discontinuities, and the reshaping of scientific foundations.[55] Their writings inspired the postmodern critique of modern culture that is associated above all with Michel Foucault. From the 1960s the work of postmodernists underlay the shift of emphasis in the history of science from transcendent ideas to distinctive practices, skills, and activities that constitute normality in a field. Prominent authors in this direction are Paul Forman, Peter Galison, John Heilbron, Thomas P. Hughes, Daniel Kevles, and Frederick Lawrence.[56] Among others, Charles Gillispie and Richard Westfall were to transform contextualism from its positivistic form to the pursuit of a deeper understanding of the social nature of science. Today, within the history of science the nervousness previously felt about what Steven Shapin cites as the "beast of externalism" has dissipated. The debate between internal and external history of science, he asserts, has become obsolete,[57] while Thomas Gieryn eschews it as an "archaic conceptual dichotomy."[58] But in the history of mathematics, contentious lines still exist. David Rowe has documented the quarrel between André Weil and Sabetai Unguru over the origins of geometric algebra. The Bourbaki legacy and Weil's statements, Rowe believes, have reinforced an unfavorable opinion of external history and hardened divisions between internal and external scholars.[59] Joan Richards argues that reconciling these camps is the foremost problem facing the history of mathematics.[60]

Although the two principal interpretations of mathematics cannot cover the entire range of experience in mathematics, these complementary views provide crucial guidance to the history of mathematics that must go beyond the polarized quarrel between internalists and externalists. Resolving the dispute may demand a modern Eudoxus to relate internal and external scholarship intricately, as he did rationals and irrationals in antiquity. Noel Swerdlow identifies a tension within Otto Neugebauer's work, stemming from his skillful probing of internal and cultural perspectives in his complex analytical studies of the exact sciences from ancient Babylon to the European Renaissance.[61] Similar tensions are evident in the research of Alexandre Koyré on early modern science and Marshall Clagett on Archimedes and ancient Egypt. Henk Bos and Michael Mahoney investigate interactions between mathematical ideas and techniques with contextual forces that shape motivation and meaning in canons of intelligible knowledge.[62] Ivor Grattan-Guinness and Karen Parshall also persuasively argue for a cross-fertilizing balance between internal and external history.[63]

The early historiography of mathematics, which may conveniently be dated from Eudemus of Rhodes in Athens during the fourth century B.C., was uncritical. Eudemus' teacher Aristotle opened writings on the sciences with historical surveys and dialectics. In writing his *History of Geometry*, now lost, Eudemus followed that practice. In his *Commentary on the First Book of Euclid's Elements*, the Neoplatonist Proclus in the fifth century A.D. preserved its only known section, which was given the title "Eudemian Summary" or "Catalogue of Geometers." Since its first sentence on Pythagoras exactly copies another from Iamblichus' *Life of Pythagoras*, placeable at about A.D. 400, W. Burkert concludes that the original summary lacked information on Pythagoras.[64] Dominic O'Meara suggests that the addition of material from Iamblichus was part of a rework to offer a Platonic explanation of the findings of Pythagoras.[65] Two other ancient biographers touched on the history of mathematics: Plutarch, author of *Parallel Lives* in the second century A.D., and Diogenes Laërtius, who in the third century wrote *Lives of Philosophers*. None of these men achieved a critical Thucydidean level of history. Like Herodotus, they accepted stories without checking them.

During the efflorescence of mathematics in medieval Islam, al-Nadim and Salid al-Andalusi, along with a few other scholars, introduced historical comments into their mathematical treatises.[66] Analysis of these writings should throw light on the nature and scope of medieval Islamic achievement and on the transmission of its mathematics to the Renaissance Latin West and the perspective assumed by geometers there. Roshdi Rashed points out that it will also work against remnants of the notion advanced by Paul Tannery, Pierre Duhem, and others nearly a century ago that science is only a western phenomenon.[67] The analysis of these treatises, however, remains to be written.

In the West, stirrings of a critical history of mathematics first occur in the seventeenth century in Johannes Kepler's prefaces to mathematical works[68] and in Bernard Fontenelle's early eighteenth-century *Éloges* of the Paris Academy of Sciences. Fontenelle sought to assign proper credit for mathematical discoveries and to identify the sources used. The most famous case he reviewed was the quarrel between Newtonians and Leibnizians over who merited priority for inventing calculus. Fontenelle's colleague Pierre de Montmort planned a general history of mathematics, thinking that he might uncover powerful methods that the classical Greeks had concealed. Montmort died in 1719 without completing his project. Fontenelle forged ahead with it. He considered the history of mathematics most apt to represent the history of the human mind and believed that human cognition approaches a limit asymptotically—a view that Pierre-Simon de Laplace later espoused. Still, as late as the early eighteenth century, histories of mathematics were chronologies marred by errors, legends, and jumbled names, dates, and titles. The *Historia Matheseos Universae* in 1742 by Johann Heilbronner is an example.

A critical history of mathematics soon appeared during Europe's mature Enlightenment. In 1758, Jean Étienne Montucla (1725–1799), a geometer, completed his comprehensive survey *Histoire des mathématiques*. Montucla was influenced by Francis Bacon and Henri de Montmor and was active in the literary circle of the publisher Charles Jombert, which included d'Alembert, Diderot, and Joseph-Jérôme Lefrançais de Lalande. He was attentive to the standards in Voltaire's new intellectual history *Siècle de Louis XIV*, published in 1751, including the secularizing of historical causation. Montucla displays a mastery of ancient sources and a meticulous sifting of fact from legend. This first critical history of mathematics, replacing chronological organization with the narrative that requires active agents, assumes that humankind makes steady advances in mathematics. This agrees with the spirit of the era. French savants, formulating the concept of the progress of the human mind, were referring to mathematics as an exemplary case. Turgot writes that only mathematics has made sure progress "from the first steps" and d'Alembert's *Preliminary Dis-*

course of 1751 maintains that "geometry, astronomy, and mechanics ... by their nature always ... [perfect] themselves."[69] Turgot adjudged his *Histoire* "excellent" and Lagrange considered it "unique in its genre."[70] The revised and enlarged *Histoire* was published in four volumes from 1799 to 1802.

Neither Charles Bossut's outline history of mathematics for the 1784 *Encyclopédie* nor Abraham Kästner's four-volume *Geschichte der Mathematik*, published from 1796 to 1799, of the old, dry, bio-bibliographic variety was to provide the model that Montucla supplied. His first major successor was Moritz Benedikt Cantor, whose *Vorlesungen über Geschichte der Mathematik* appeared in four volumes in several editions from 1880 to 1908.[71] Having more research available, Cantor improved upon Montucla, who had occasionally relied on analogy. Accuracy and a resolute description of primary sources are the objectives of Cantor's text. In the journal *Bibliotheca Mathematica*, Gustav Enestrom corrected Cantor's *Vorlesungen* and added new findings.[72]

Up to the early twentieth century, the history of mathematics occupied only a small place in the history of science. William Whewell's influential *History of the Inductive Sciences*, appearing in three volumes from 1837 to 1857, seems a major reason for this. Whewell denied that the mathematization of nature had been the essential and most distinctive feature of early modern science. But from the mid-1920s, precisely that claim was being put forth. E. A. Burtt, E. J. Dijksterhuis, Alexandre Koyré, and Anneliese Maier were among its supporters. Burtt's *The Metaphysical Foundations of Modern Physical Science*, published in 1924, examines the mathematical world view underlying our mental processes that arose in the sixteenth and seventeenth centuries. The work radically distinguishes between primary qualities, absolute, objective, and mathematical, and the relative, subjective, and fluctuating secondary qualities. The ideas of these four became integral to the history of science. Under the leadership of George Sarton, founder of the journal *Isis*,[73] the history of science developed by mid-century from an episodic affair depending on personal enthusiasm to an organized field of learning having a secure financial base.

Continuity was then a central theme in tracing scientific progress. Dijksterhuis was an important contributor to this viewpoint. Anneliese Maier, like Pierre Duhem earlier, stimulated research on the mathematical sciences in the Latin middle ages, a field that had been thought nonexistent. In refining Duhem's work, she and Ernest Moody found similarities and differences between fourteenth-century impetus theory and the work of Galileo. Alistair Crombie traces some of Galileo's sources to the late middle ages and William Wallace, to Paduan scholastics and lectures at the Collegio Romano.[74] Charles Gillispie's elegant *The Edge of Objectivity*, published in 1960, contending that science advances by the progressive elimination of human wish,[75] may also be placed among the works that define the element of continuity in science. But from 1960 almost to the present, the dominant interpretation of the history of the sciences has been as a sequence of turning points and revolutions, superceding earlier knowledge.

Whewell had investigated cycles of science and found epochs of fundamental change. These epochs, his work maintains, depend upon the ideas and performance of men like Kepler and Newton, impelled by gifted intellect and moral purpose. After Arnold Toynbee adopted a cyclic interpretation of history, Burtt, Dijksterhuis, Koyré, and Maier led in offering fresh and differing interpretations in pioneering what is now termed the scientific revolution, which led them to considering an era of radical transformations. Burtt set its time span at 140 years from Copernicus to Newton, while Koyré placed its crux in the seventeenth century. The term "scientific revolution" first gained wide currency when Herbert Butterfield used it in *The Origins of Modern Science, 1300–1800*, appearing in 1949. In 1957, Marshall Clagett organized and gained National Science Foundation support for an influential ten-day symposium at the University of Wisconsin on critical problems in the history of science. There Koyré described developments before the seventeenth century not as direct precursors but as reasons why early modern science had not emerged earlier. He had studied under David Hilbert and Edmund Husserl at Göttingen, and he was deeply influenced by Hegelian thought and Emile Myerson's Paris discussions of epistemology, as well as by the neo-Kantian Ernst Cassirer. Speaking of mutations of the human intellect, Koyré endeavored to show that the construction of ideas in the exact sciences depend largely on philosophical preconceptions and are mediated by the political, religious, and social contexts.[76]

At mid-century Lynn Thorndike, Frances Yates, and Walter Pagel widened the identification of influences on sixteenth- and seventeenth-century scientific rationality by adding investigations of the fields of magic, hermeticism, and alchemy[77]—studies ably extended by Allen Debus and Betty Jo Teeter Dobbs. Dirk Struik explored Marxist concerns with class issues, economic determinants, and technics,[78] while from the 1930s to the 1980s Robert K. Merton was pioneering the sociology of science,[79] posing what is known today as Merton's thesis on Puritanism as a major force in the rise of science in seventeenth-century England. Building on the work of Emile Durkheim and Max Weber and using statistical analysis and documentary searches, Merton concentrated on the social, cultural, and economic dynamics influencing the direction and tempo of growth in the sciences and technology.[80] Although he agreed with George Sarton, a member of his doctoral committee, that to an extent extrinsic factors condition even mathematical discovery, he argued that modern scientific discoveries and inventions are relatively autonomous and belong to the internal history of science.[81] The older Merton and his partisans added a consideration of the development of scientific institutions in strengthening the independence of modern science. Before the sciences achieved an institutional base in the seventeenth century, their activities and methods in England were not clearly separable from the interests, for example, of religion. As institutional specialization proceeds, the sciences gain autonomy and internal evolution becomes more important.

In *The Structure of Scientific Revolutions*, published in 1962, Thomas Kuhn consolidated the concept of revolutions in the sciences. The sciences, the study argues, do not develop cumulatively and do not possess an infallible empirical base. Karl Popper's idea that the sciences progress through falsification and refutation applies only to logic and theoretical mathematics.[82] Instead, Kuhn proposes that the same sciences of different eras, of Aristotle and of Newton, for example, are incommensurable. Kuhn defines two states of the sciences: normal science or a dominant scientific paradigm and revolutionary science involving a paradigm shift. Anomalies and particularly serious "counter instances" grow over time in normal science. Normative language may mask them initially but, as they grow, they produce a breakdown or crisis in normal puzzle solving that causes scientists to lose faith in it.[83] Falsification of test statements alone is not sufficient. A new paradigm evolves that alters the language, theories, and research techniques of science. New convictions of a social psychological nature based in communities and metaphysical beliefs in conceptual schemes support this. Although Kuhn's revolutionary thesis was challenged,[84] his *Structure* joined with *The Copernican Revolution*, published in 1957, reoriented the historiography of science.[85]

Having labored to separate religion and metaphysics from science, logical empiricists viewed Kuhn's theory as a dangerous relativism. Imre Lakatos believes that critical reason alone can explain fundamental scientific changes. Lakatos distinguishes between an early naive stage of falsification, depending on trial and error, and a sophisticated methodological form, in which researchers severely test a theory, particularly at its borders, and make speculations that contribute to progress in knowledge. Scientific theories, he argues, are fallible, but most are not completely false and provide a basis for growth.[86] Attacking logical empiricism, Paul Feyerabend denies that Lakatos has reestablished critical reason as the final arbiter in the sciences. Feyerabend claims that new research programs must often struggle and to succeed have to add auxiliary or ad hoc hypotheses. Copernican theory, for example, required auxiliary hypotheses for deriving important empirical consequences, and Galileo had to assume the reliability of the early telescope. Sometimes even a degenerative research program can be transformed. Critical discussions alone, Feyerabend contends, need not produce new theories. Feyerabend's *Against Method*, published in 1975, argues that scientists think and interpret the world in different ways and that their thinking may occasionally be anarchic but not nihilistic.[87]

As Gertrude Himmelfarb observes, historians are extremely wary of the term revolution,[88] whose pantheon includes the industrial, French, and Glorious. In the late twentieth century, the political connotations of that metaphor and the associated general limiting of it to sudden, dramatic, and seemingly irreversible upheavals occurring over less than a decade, but altering the course of centuries, have made many historians sparing in its use. Employing the metaphor requires a clear definition.

In the debate among historians of mathematics over whether Kuhn's theory applies to their field,[89] Clifford Truesdell and Michael Crowe deny that a revolution has yet taken place within mathematics, while Herbert Mehrtens believes the term "revolution" can be an adequate metaphor depending on whether a study attains precision conceptually and historically.[90] Crowe requires an irrevocable discard of previously accepted content within mathematics. In the preface to L'Hôpital's *Analyse des infiniment petits*, published in 1696, Fontenelle had applied to Descartes' mathematical achievement the metaphor of revolt. Montucla's *Histoire* calls it a "happy revolution" and Auguste Comte's *Course de philosophies* of 1835, a general revolution.[91] In this century, Yvon Belaval challenged Comte's view, noting that the Cartesian method is finite and rejects the infinitary methods soon to be articulated by Leibniz.[92] Acknowledging as revolutionary Descartes' invention of analytical geometry, I. Bernard Cohen, Thomas Hawkins, and Michael Mahoney disagree with Crowe.[93] Hawkins observes that the Cartesian revolution does not reject ancient geometry, instead introducing powerful new methods. Michael Crowe's ten laws concerning patterns of change, which he now considers mistaken and a hindrance to further historical studies, helped prompt by reaction Donald Gillies's book *Revolutions in Mathematics*.[94] There Joseph Dauben accepts as revolutionary whatever establishes a new vocabulary and grammar and displaces an old by a new authority.[95] Two cases that he cites are Cauchy's introduction of rigor in analysis and Georg Cantor's creation of transfinite set theory. On similar terms, L. Kruger argues that such change has occurred in probability.[96] In her study of the vibrating string and the St. Petersburg problems, Judith Grabiner finds the collapse of older traditions and practices.[97] But Leo Corry denies that these cases are revolutionary in Kuhn's meaning: producing, in addition to seminal breakthroughs, new or distorted languages and criteria for mathematical truths.[98]

While agreeing on the significance of what happened in the phenomenon known as the scientific revolution, many historians of science have grown skeptical about the metaphor itself. H. Floris Cohen believes that it remains a vital historiographic tool and simply needs a more careful balance between the internal and external components of the scientific revolution.[99] But David Lindberg and John Schuster observe that some scholars consider the notion of the scientific revolution as only a heuristic,[100] and Steven Shapin and Roy Porter find lacking in science the abrupt, cataclysmic change that would justify it.[101]

Perhaps most damaging to Kuhn's thesis is the argument that its model of crises and paradigm shifts does not address the full complexity of the thought and dynamics involved in major scientific change. Christia Mercer's replacing by a more subtle and complex model the usual caricature of monolithic Aristotelianism in early modern science fits in this vein.[102] Peter Dear and Paolo Mancosu demonstrate that Kuhn's model does not address variations in support for innovation from within established traditions. Mancosu particularly cites late-sixteenth-century astronomy, and Dear examines Marin Mersenne's work on scholastic Aristotelianism. Mersenne did not attempt to replace but to preserve it. Perceived as part of humanistic pedagogical innovation, the Aristotelian syllogism appears to be strengthened with mathematical reasoning to balance Ciceronian probabilism.[103]

Enhanced by the founding of two important new journals, *Archives for History of Exact Sciences* by Clifford Truesdell in the 1960s and *Historia Math-*

ematica by Kenneth May in 1974, the history of mathematics has developed steadily. As specialization increases, attention, beginning with the work of René Taton and Morris Kline, has shifted from the unity of mathematics, as Carl Boyer presented, it to its diversity.

Mainly internal history remains prevalent, as a reading of Truesdell's *Archives* and May's *Historia Mathematica*, along with *Archives internationales d'histoire des sciences, Centaurus*, and the annual *Isis Critical Bibliography* will reveal. So will a review of the thematic works of Henk Bos, Lenore Feigenbaum, and Eberhard Knobloch on early calculus, Harold Edwards and André Weil on number theory, Craig Fraser and Herman Goldstine on the calculus of variations, Thomas Hawkins and Helena Pycior on modern algebras, E. S. Kennedy and Roshdi Rashed on medieval Arabic mathematics, Wilbur Knorr and Joan Richards on geometry, David Pingree on ancient and medieval India, and Clifford Truesdell on eighteenth-century rational mechanics. But a complementary newer and more finely developed contextualism is growing that explores the complexity of intellectual interactions, their unpredictable consequences, and the partly social nature of scientific knowledge. Among recent histories of probability and statistics, not only the technical studies by Anders Hald and Stephen Stigler, but also Lorraine Daston's exploration of their connections with theology and legal sentencing have received high praise from reviewers.[104] So have Ian Hacking's and Theodore Porter's investigations of the influence of determinism and other elements of philosophy.[105]

To illuminate precisely the creative work of the individual, biographers plumb the intellectual and institutional setting of the subject's thought; examine the formative and adult experience, including mediating historical circumstances; and analyze in depth the contribution of the mathematician under study, all pursued through a mixture of methodologies. Confusing past with present categories within which mathematics has proceeded must be avoided along with an injudicious use of modern symbols that imposes present interpretations on historical materials. A steady flow of skillful biographies in recent years testifies to the vigor of this genre. Among these are Mario Biagoli's *Galileo Courtier: The Practice of Science in the Culture of Absolutism*, Joseph Dauben's *Georg Cantor*, Pierre Dugac's *Richard Dedekind*, Charles Coulston Gillispie's *Pierre-Simon Laplace*, Thomas Hankins' *William Rowan Hamilton*, Ann Hibner Koblitz's *Sofia Kovaleskaia*, Jesper Lützen's *Joseph Liouville*, Michael Mahoney's *The Mathematical Career of Pierre de Fermat*, and Richard Westfall's *Never at Rest: A Biography of Isaac Newton*. A trend is toward group biography and the study of communities of interest that craft mathematical development.

Primary sources, published and unpublished, are indispensable to the history of mathematics as to any other historical study. Prominent among those edited and published since 1960 are Derek T. Whiteside's eight-volume collection of Isaac Newton's mathematical papers, five recent volumes of Gottfried Leibniz's *Sämtliche Schriften und Briefe*, including Hartmut Hecht's 1991 book on Leibniz's extension of abstract mathematics to physics,[106] and Charles

Blanc, Emil A. Fellmann, and Adolph P. Youschkevitch's volumes in Leonhard Euler's eighty-nine volume *Opera omnia*.[107]

New institutional studies of mathematics are also improving our knowledge of its evolution. Kurt Biermann has examined the development of mathematics at the University of Berlin from 1810 to 1920; and Ivor Grattan-Guinness, that at Paris from 1800 to 1840. Before researchers recently uncovered hundreds of commercial arithmetics in Italian trading cities, investigations of late medieval and Renaissance mathematics dealt chiefly with schools and universities. G. Arrighi and Warren van Egmond have catalogued these arithmetics, while Cynthia Hay and Frank Swetz have edited books concerning them.[108] Jens Høyrup studies the scribal schools as cultivators of mathematics in ancient Babylon, and Gert Schubring describes the effort of mathematics to obtain autonomy with the support of educational reforms in nineteenth-century Prussia. Joseph Needham has pioneered in Chinese mathematics, while Roshdi Rashed and J. Lennart Berggren examine medieval Islam and Michael Crossley the Maya. Marcia Ascher and Ubiratan d'Ambrosio add the field of ethnomathematics, and contributions of women are now more closely studied.[109]

The history of science, Peter Burke of Cambridge observes, has now moved from the margins of history into the main current,[110] and Michel Serres proclaims that the importance of science and technology today will make the history of science the cornerstone of modern history.[111] This means that histories of science and mathematics must consider their growing audiences and be competent in applicable general historiographic methods. Among the continuing self-criticism that historiography has undergone during the last seventy-five years is Lewis Namier and R. H. Tawney's attack in the 1930s on the traditional paradigm of Leopold von Ranke (d. 1886) that history is objective, a narrative of events revealed in the systematic study of primary archival sources.[112] Lucien Febvre and Fernand Braudel's school that formed around the *Annales* journal demanded deeper studies of structures, such as institutions and modes of thought, especially by means of social and economic history. Michel Foucault in *Order of Things*, his so-called *Discours de la méthode* published in 1966, and in *Archaeology of Knowledge*, released three years later, parted company with causality and continuity,[113] dismissing origins as chimera and studying chance, accidents, and sudden ruptures in what he called epistemes, the layers of basic grids underlying knowledge. He spoke of contingencies rather than necessities and shifted emphasis to micropolitics and cultural history. In *Les mots et les choses* of 1966, Foucault wrote that the individual author or man is unimportant for intellectual history, a recent invention destined to disappear like the drawing of a face in the sand.

From the 1970s, postmodernists, anthropologist Clifford Geertz, and contextualists have been prominent in historiographic debates. Academic postmodernist or ultrastructuralist critics, such as Jacques Derrida, Bruno Latour, and Georg Gadamer, have found that language is not a stable referent for study of the past, but part of the reality of a given time,[114] while Geertz admonished historians not to confuse the linguistic and symbolic expressions of other societies with ours.[115] Nicholas Jardine argues that historical narration should mix aesthetic, anthropological, historical, and sociological categories from the past and present day. A successful combination will have thick descriptions, minimal confusion of past and present terms, and maximal interpretive power from present-day terms.[116] Natalie Zemon Davis has warned against the rigid boundaries among categories and ideal types espoused by literary critics and sociologists during the last decade.[117] By the late 1970s research in intellectual history had coalesced around the evolving contextual methodology of Quentin Skinner, J. G. Pocock, and George Stocking.[118] These scholars aim to reconstruct the past, if not exactly, by salvaging meanings from documents studied in their historical context, not from absolute meanings of their words. Their position is akin to that of the nineteenth-century philosopher Wilhelm Dilthey, who held that the self cannot be understood in isolation, severed from its practical interactions with the world. Each view has critics. Tian Yu Cao, for example, praises postmodernism, which remains ill defined to the present, for its historiographic criticisms but rejects it for not offering an alternative rigorous cognitive standard to logocentrism and disallowing causal explanations.[119] Mark Lilla spurns Derrida's depiction of justice as lying outside of experience, nature, or reason and equating it with deconstruction.[120]

Part I

Before the Advent of Civilization

1

Origins of Number and Culture

God Himself made the whole numbers —
Everything else is the work of men.
— Leopold Kronecker

The whole of mathematics stems essentially from the study of numbers, space, time, and motion.[121] Among these, number systems have been not only profound sources of ideas, results, and theories in the development of mathematics but indispensable to commerce, economics, science, and engineering—the underpinnings of civilizations. These systems principally begin with the *counting numbers* or *positive integers*, denoted conventionally by the decimal symbols 1, 2, 3, ..., 9, 10, ... and known technically as the *natural numbers*. In conveying messages within the human brain, counting is far more fundamental than language. An embedded feature of the brain, it is measured by electrical pulses and packets of chemical transmitters. Because *number* is not only a science itself but also the language of mathematics and the exact sciences, it had to be developed in some form before either could evolve. In recent centuries, paradoxes and simple conundrums about the abstract nature of number have also served to revitalize mathematics.[122] Our story must therefore begin with the concept of number and its lengthy, complex evolution.

What the term *number* means today is generally not well understood. When a schoolchild who is learning to count asks an adult "what is a number?", the usual answer is not a formal definition but a concrete illustration.[123] The adult may point to fingers or objects or recite some rhymes about counting—counting being the most familiar way of encountering numbers. Such explanations will likely make the child expect numbers to have a physical existence. Yet they do not. Numbers may be defined as *abstract concepts of quantity or, more specifically, aggregates of units*. Their modern formal definition holds further that if x is a positive integer, so is its successor, $x + 1$; they are thus extendable and open ended. Numbers are vocalized, as in the English names *one, two, three, ...* These sounded words are also expressed in more compact and visually powerful written symbols, 1, 2, 3, ..., which the English language terms *numerals*. Numerals, which then constitute a kind of alphabet, are often employed for tagging physical items, as, for example, the page numbers and endnotes in this book.

Numbers themselves are thus intimately connected to everyday sense experience through numerals that fill at least two roles. Numerals may be used to list, label, and tally physical entities and to compare collections of them. Numerals also make it possible to manipulate objects conveniently by addition, subtraction, multiplication, and division or any other operation built on these four fundamental ones.

The theory of numbers is the branch of mathematics that mainly studies the properties of natural numbers or, more generally, rational integers among themselves and their analogies in other systems. Rational integers include negative as well as positive whole numbers: ..., −3, −2, −1, 0, 1, 2, ... According to most of modern mathematics, numbers form their own world, which exists in its own right. During the last two centuries, mathematicians have greatly broadened that world by extending the concept of number from natural to complex numbers and beyond these to transfinite, p-adic, and surreal numbers. Today mathematicians generally rank the theory of numbers above the other branches of mathematics in its aesthetic appeal, deriving, from its boundedness that allows only limited application, its ability to ask serious questions simply without recourse to lengthy definitions, and the difficulty of the challenge posed in solving these questions.[124] Hidden within their simplicity is complexity. Solving them often involves making connections with many different fields, ranging from Galois theory to algebraic geometry, and thus presumes a mastery of large chunks of mathematics. Although number theory is profoundly abstract and only partly empirical, the world of numbers remains clearly linked to the physical world across the bridge of numerals and measurement, which associates a definite quantity with a certain physical entity. This bridge is extremely important for scientific research. On it, information flows in both directions between the physical universe and the abstract realm of mathematics.

Recognition of periodicity or recurrence among phenomena initiates our study of nature. Conspicuous regularity of general recurrences in the physical world—day and night, lunar phases, the seasons, positions of celestial bodies, breathing, pulse, heartbeat—opens up the possibility of measurement. It also gave rise to the idea of exactness. In its turn, the quest for exact measurement has allowed for a deepened understanding of natural phenomena. Expressing natural behavior in numbers makes it possible to describe nature in a network of quantitative laws, generally referred to as the laws of nature. Examples are Galileo's law of freely falling bodies, $s = \frac{1}{2}gt^2$, where s = distance traveled, g = the acceleration due to gravity, and t = time, and Boyle's law, which describes the compressibility of gases at constant temperature: $pv = k$, where p = pressure, v = volume, and k is a constant.

As the ubiquitous substitution of illustration for definition in the response to the schoolchild's question suggests, the abstract concept of number is not easily achieved.[125] And so it was in history. From a notion of number as something concrete and perceptible, lodged in external, physical objects, the advance to a purely symbolic understanding happened over a long, tortuous course with many false starts. Increasing anthropological evidence confirms a hunch that the notion of number began with primitive counting long before the appearance of civilization and progressed with an intuitive search for a comprehensive and efficient system of graphical representation of concrete numbers. The primitive notion of quantity and its use for crude measurement came of practical necessities: trimming the height of a detached branch to match the shoulder height of a hunter, for example, gave a rough unit of distance measurement. If necessity is the mother of invention, ritual was its nurse from at least shortly before the emergence of eastern Mediterranean civilizations. Among other attendants have been magic, comparisons of forms and shapes, and the element of play. Not until the second millennium B.C. in the river civilizations of Mesopotamia and Egypt would the first glimpses appear into the abstraction of number.

1.1   Concrete Number Before Civilization

Customarily historians distinguish between the period of human activity before civilization and the time after its advent. The latter period, often called the historical, starts when writing first appeared, or at any rate when writing left records that have survived. In the absence of writing, records of the earlier period, loosely termed prehistoric, are carried by utensils, tools, weapons, carvings, paintings, and jewelry generally known as artifacts. These demonstrate the great technological, political, and social accomplishments before civilization.

In recent years some scholars have questioned whether number had its origins in the period before civilization. They have based their argument largely on the conception that now prevails of number as abstract and symbolic. The German mathematician Carl Gauss's monumental *Disquisitiones Arithmeticae*, published in 1801, asserts that number is purely a psychical reality—a free creation of the mind. By this definition, which most present-day mathematicians accept, the evolution of number must have begun in historical times. But mathematics did not, like Minerva, emerge without an initial stage of growth. Our knowledge of cognitive psychology reinforces the evidence that before its appearance as Gauss has defined it, number had its origin in a concrete as opposed to symbolic phase.

Those who speak of a phase of concrete numbers have disagreed over how and when they first appeared. Early in this century, the record of precivilization was too fragmentary for a solidly founded investigation of the beginnings of concrete number. In *How Natives Think*, published in 1923, French anthropologist Lucien Lévy-Bruhl (d. 1939) denies that primitive peoples knew the category of number, since their languages lacked progressive, conventional individual numerical terms and failed to disjoin numbers from numbered objects in adjectival numerical words.[126] In his view, similar items were grouped indiscriminately together under a term equivalent to the English "many." In the decades since Lévy-Bruhl, archaeologists have accumulated enough information from artifacts of precivilization hunters, peasants, and craftsmen to reassess the early stages

of concrete number. Using these richer materials and new techniques for understanding languages of preliterate peoples, Vere Gordon Childe (d. 1957), the
widely read Australian pioneer of twentieth-century archaeology, claimed that
linguistic considerations do not support Lèvy-Bruhl's conclusions.[127] Childe's
knowledge of many European languages was extraordinary, but the data he
synthesized are now superceded, and his Marxist interpretation of prehistory,
viewing human cultures as social constructs, was contentious and flawed in
its determinism.[128] Lévy-Bruhl's position on number may be incorrect, but
the findings of Childe, who lacks sufficiently rich and diverse sources, are not
definitive.

Recent biological studies suggest that the human brain communicates in numerical codes. New noninvasive brain imaging techniques, such as PET scans
and magnetoencephalography, are confirming this type of chemical transmission.[129] They are also identifying regions of the integrated system of the brain
that are engaged in calculation. While the British mathematical physicist Roger
Penrose claims that humans have an intuitive grasp of mathematics, cognitive
neuropsychologist Stanislas Dehaene is weaving together a rich study of biological, cultural, and psychological evidence to indicate that the mind and
brain create a number sense, which culture and associated technologies, such
as symbolic numerals, profoundly shapes.[130]

Archaeological and psychological data from tightly controlled experiments
and meticulous observations in a sense expand upon this brain research evidence by suggesting that the origin of a numerical or quantitative sense is
at least coeval with that of human beings. This position gains support from
George Murdock's convincing anthropological claim in 1968 that numerals are
one of seventy-two items that he had discovered to occur in every culture known
to ethnography or history.[131] Against it must be set R. M. Dixon's determination that some Australian aborigine languages lack numerical systems.[132] The
psychological evidence is more compelling. It reveals that some nonhuman animals may have a number sense and that some birds, such as the crow, raven,
and gray parrot, are able to distinguish between one, two, three, and four seeds
or between boxes with two, three, four, five, or six spots on a lid, while pigeons
can differentiate seven items.[133] These experiments support natural history
anecdotes dating to the eighteenth century of numerate animals. One story has
a nobleman trying to kill a crow, but the crow flies out of range when he sees
the hunter. The hunter builds a blind, but the bird waits until he comes out
and departs. The hunter brings a friend into the blind, but the crow waits for
them both to go. The same pattern follows for three, four, and five hunters,
even when they leave at different times. Then six hunters enter the blind. After five leave, the crow returns and the sixth hunter kills it.[134] Some birds and
animals, it seems, have a limited sense of quantity, the ability to distinguish
between some and more, few and many, little and too much, small and great,
and far and near. If nothing else, the need for basic Darwinian survival skills
would suggest this. Surely precivilized people were that sophisticated. These
basic senses of magnitude and spatial distance together with the concept of

progressive differentiation form the basis of the modern procedure of counting. Although they lacked a concept of progressive differentiation, precivilization human beings did develop over the earliest culture periods a primitive counting by way of various tallying techniques.

The Age of Stone, the first of two major periods into which the human experience can be roughly divided (the other is the Age of Metals), is practically identical with the era before civilization. Beginning before 600,000 B.C. and lasting in parts of Europe and the Near East until the fourth millennium B.C., it comprises over ninety-nine percent of the time of human existence on this planet.[135] It may be subdivided into three stages: the Paleolithic, Mesolithic, and Neolithic, each name characterizing the type of stone tools and weapons prevailing at the time. In each stage, number apparently went through some degree of development. The duration of each period varies among the different geographical zones of Earth.

The Paleolithic (from the Greek, "old stone") Age lasted according to conservative estimates more than a half-million years. Anthropologist Richard Leakey proposes a far greater time span. His Kenyan findings include crude knifelike tools thought to be more than two million years old. The Paleolithic Age ended about twelve thousand years ago in Europe and the Middle East. Its characteristic tool, the all-purpose handax, was made by chipping pieces of flint or large flaking stone. The relatively brief Mesolithic Age lasted in the stone regions from about the tenth to the mid-eighth millennium B.C. Sharp, slender stone chips from the ax were now utilized as knives and chisels. During the Neolithic (from the Greek, "new stone") Age, from about the mid-eighth to the fourth millennium B.C. in Europe and the Near East, implements of finely ground and polished stone supplanted chipped stone tools. The human hand with its flexible thumb directed by the developed cortex of the brain produced these tools and the remaining material culture.

Tallying was humankind's initial detectable step in numeration. Several tallying techniques emerged as early as the Paleolithic Age. Paleolithic peoples copied the plurality of things with tallying sticks, which were strips of wood, bone, or ivory variously incised along one edge: the word *tally* comes from the French *tailler*, to cut. Other tallies were kept by knots tied in threads, which were perhaps sinews or thongs of rawhide; by pebble (calculi) counting; and by marks on cave walls (for example, I, II, III, ...) or later on pottery. These methods of tallying or copying reveal that their creators had the power of mind to relate thing to thing, to let one thing, such as a cut or a knot, correspond to another, say, a bird or an implement. Such methods imply another important psychological element, the intuition of exact matching or one-to-one correspondence, the basis for the modern notion of *cardinal number*. Cardinal numbers answer the question of how many items are in a collection.

During the Paleolithic Age, progress in achieving a numerical sense was extremely slow. Tallying techniques can be traced with certainty only from its last one hundred thousand years, when human beings were evolving from Neanderthals to the subspecies *homo sapiens sapiens* (thinking thinking man) with a brain triple or quadruple the size of that of their predecessors. That transition was completed during the last glaciation some forty thousand years

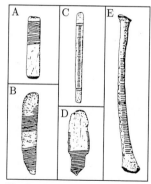

Notched bones from the Upper Paleolithic.

Tally sticks used by French bakers.

Thirteenth-century English tally sticks.

ago. Tallying techniques came into wider use during the subsequent three hundred centuries of the late Paleolithic period, when human beings greatly improved their tools and weapons, began mechanical invention with spears and bows, and created the visual arts by carving bracelets and other objects and by painting pictures on cave walls. In these endeavors, people improved their metrical sense, which also aided protomathematics, that is, mathematics resting purely on perceptual conditions. We surmise that the need to write down the results of tallying helped lead to ideograms or symbols for numbers as group life grew more regular and more highly organized.

Tallying sticks were not limited to the precivilization period, but persisted in use to modern times. The famous Napoleonic Law Code of 1804 refers to the use of tally sticks as contracts, while Charles Dickens notes in his speech on administrative reform that the British Exchequer used double tally sticks until 1834.[136] Burning of long accumulated hazelwood tallies overheated Parliament's furnaces that year, producing a fire that destroyed the House of Lords' building.

During the Mesolithic Age ending about the mid-eighth millennium B.C. in the Near East or perhaps earlier in the late Paleolithic Period, people invented patterned concrete numbers having cycles. This advanced protomathematics, for it implied the intuition of *ordinal numbers*, which indicate not only how many items, but also in what order items occur, for example, in ascending order 1st, 2nd, 3rd, 4th, ... The earliest form of patterned concrete counting is referred to as "two-counting." Among others, primitive tribes in Africa, South America, Alaska, and Australia have used the system of two-counting.[137] For numerals it had some sort of marks for *one* and *two* and either a further mark or word like *many* for *more than two*.[138]

To the modern mind, a progression from cardinal to ordinal number may seem strange. In constructing units of measure, modern people often proceed from smaller to larger in ordered, building-block steps. This suggests that cardinal number presupposes ordinal number. But the perceptual origins of knowledge were probably not identical with the conceptual.[139] Early human perception and imagination closely bound to tangible nature apparently often assigned greater prominence to totalities than to the separate things comprising them. This prioritizing probably carried over into numeration. The Paleolithic tallying techniques suggest a concern with a collection, rather than component parts. It is therefore plausible that the intuition of cardinal number, which defines the total number of items in a group instead of arranging them analytically and separately in a place order, occurred first in the era before civilization.

To communicate the recognition of "oneness," "twoness," and "quantity beyond," primitive tribes widely employed descriptive or adjectival numerical words: one-dog, two-trees, three (or more)-skins. Adjectival numerical words, which were closely connected with an item, are known technically as numeral classifiers.[140] This practice seems characteristic of peoples deeply concerned with context and insistent on joining important attributes to objects, including that of measurement. Vestiges of adjectival numerical words appear in the early Egyptian, Arabic, Hebrew, Sanskrit, Greek, and Gothic languages, which had singular, dual, and plural adjectival forms. Peoples who did not carry numeral classification beyond small numbers apparently thought only in very small numbers. But other peoples, the ancient Chinese and Japanese, many in Southeast Asia and Oceania, a few in Africa, the Maya in America, the Maori of New Zealand, and the tribes of British Columbia, had even more adjectives in their languages to handle 4, 5, 6, . . . , or 10 items.[141] Anthropological studies of these associations in different languages reveal how early peoples categorized or classified items. Apparently, these early peoples sharply distinguished the classifier from the modified root word, as in one-cat and two-cats, which suggests that they understood the classifier as a separate noun number. Linguistically, however, number remained a property of the *enumerability* of things.

The linguistic situation seemingly changed slowly, as adjectives came to be used as nouns, or what historians call *vocal numbers*. This was a profound development. The numbers *one, two, three,* or more as nouns stood for independent entities in language as well as in the mind, objective realities clearly separated from the object being counted. Number words as nouns probably appeared only after the concrete number system had expanded beyond two-counting and after crude graphical marks, such as I and II, had long been employed. As the vocal numbers and subsequent written numerals gained acceptance, the many adjectival forms gradually disappeared.

By the end of the Mesolithic Age, two-counting was possibly being displaced by counting in crude groups of five (quinary), ten (decimal), and twenty (vigesimal). Group 10 refers to a system of number words and visual numbers denoting 1 to 10 and some principle of combination operating thereafter. Visual numbers refer here to pointing to fingers, arms, shoulders, toes, ankles, or knees to indicate a number. Mesolithic human beings thus likely learned to count by collections or groups, a step beyond the tally stick with notches counting by ones. Lacking any direct information about the thought processes of the time and knowing little about what communal activities underlay numeration, historians can only conjecture as to what generated the particular numbers and what pressures produced these number developments in the first place. But in some regards the answers seem clear.

Philological researches have shown that early concrete numbers and their bases were strikingly anthropomorphic and thus almost universally uniform. The notion *one* seemingly involved individuality, wholeness, or unity probably reflecting self-consciousness. The concept *two* apparently reflected paired parts of the body: two eyes, two ears, two arms. Twoness contained the germ of two fundamental concepts in our thought: *polarity*, for example, the opposition between what the right and what the left hand does; and *duality*, juxtaposition of the sky above, for instance, and the earth below.[142] The fundamental duality was male and female; for the ancient Chinese this was the *yin-yang* principle upon which they founded their science. *Three* came from observing such triads as male, female, and issue. It was a symbol of the dynamic. *Four* apparently derived from the number of cardinal directions—originally seen as front, back, left, and right of the individual, but later as east, west, north, and south. Of all anatomical considerations, the ten fingers of human beings left a permanent and perhaps the greatest imprint on the use of number.

By the late precivilization period, human beings probably signaled numbers by holding up fingers or pointing to body parts, as much later Papuans, New Guinea Elemas, Torres Strait Islanders, and various peoples of Africa, Oceania, and America did. Most likely, finger and body numbers were widely used because they transcended language differences. They are still used at auctions and in stock exchange signals.

Duodecimal finger-counting method.
This method is used in India, Indochina, Pakistan, Afghanistan,
Iran, Turkey, Iraq, Syria, and Egypt.

Finger numbers served as a basis for vocal numbers. In many languages the names of the fingers on a hand (thumb excluded) were also those for the numbers 1 through 4. A "hand" was 5, for example, in Sanskrit *pantcha*, in

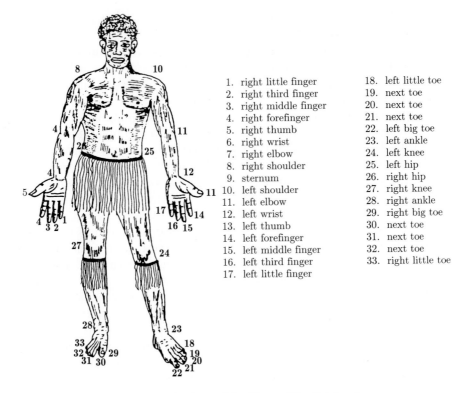

1. right little finger	18. left little toe
2. right third finger	19. next toe
3. right middle finger	20. next toe
4. right forefinger	21. next toe
5. right thumb	22. left big toe
6. right wrist	23. left ankle
7. right elbow	24. left knee
8. right shoulder	25. left hip
9. sternum	26. right hip
10. left shoulder	27. right knee
11. left elbow	28. right ankle
12. left wrist	29. right big toe
13. left thumb	30. next toe
14. left forefinger	31. next toe
15. left middle finger	32. next toe
16. left third finger	33. right little toe
17. left little finger	

Counting methods used by Torres Strait islanders

Persian *pentcha*, and in Russian *piat*.[143] In several languages, 10 was "two hands" or sometimes "man"; 20 was "the fingers of hands and feet."

The early search for numerals in number words, particularly in base 10, clearly left its mark on our speech—one of the chief ways that number has influenced our cultural and intellectual growth. Speech is a condition of culture, and culture a condition of speech.[144] The same material that builds the larger culture also builds its component speech: for example, early number words were drawn from the practice of tallying that was correlated to the anatomy of fingers and toes. Speech and a larger brain are the sources of our formulating and expressing ideas and thus aid cultural growth. Speech is also the primary vehicle for transmitting our culture. Counting in base 10 is an enduring legacy of our remote ancestors. This imparts an additional meaning to the the maxim of Protagoras (fifth century B.C.): "Man is the measure of all things."

Although vocal numbers were an advance over finger numbering, they too were deficient in not having a recordable format and in their awkwardness for purposes of calculation. While they could express quite large numbers and sufficed for memorization and oral communications, simple arithmetic could hardly develop from them. Only the invention of ideograms, the written abstract, as distinct from pictorial, symbols, would advance the conceptual status of number and facilitate the transition from protomathematics to mathematics.

Philological research and investigations of hand-gesture numerals and vocal numbers have given rise to different interpretations of the basis of concrete numbers. Philosopher Ernst Cassirer (d. 1945) averred that the human physiology is the source of all numerical differentiation: body numerals are extended intuitively to cover the whole world.[145] A. Seidenberg disagreed, noting that the human physiology provides only a latter-day potential for numbers, whose genesis is a spontaneous creation. Even understanding most hand-gesture numerals depends on learning and the generation of numbers on environmental circumstances.[146] More recently, Noam Chomsky and James Hurford have maintained that the number faculty is an innate human property and number a linguistic universal.[147] Even so, that capacity remained latent until the last stage of human evolution to *Homo sapiens sapiens*. Without strong social and psychological pressures, neither the germ of number sense that had emerged ex nihilo nor its parallel, the innate ability to master intricate number systems possessing "discrete infinity," would have been made fruitful. Language as a result of these pressures became modified so as to express a new, generalized mensuration. A combination is seemingly essential: a genetically gifted few vigorously exercise their higher degree of brain complexity with parallel support from external circumstances. To this point, Chomsky and Hurford agree. Chomsky, however, contends that the precivilization innovator needed environmental conditions encouraging some improvement in numeration, while Hurford stresses the autonomous power in the process of the mind that chooses alternatives.

Today, agreement is general that the human body provided the means for primitive number systems but not the stimulus for the expansion of those systems that occurred during the late Paleolithic and the Mesolithic Age. Instead, social needs and psychological pressures arising from simple trade and barter, from measuring supplies or distances, from counting warriors, animals, stone implements, or weapons probably supplied the impetus for more sophisticated numbering.[148] Most likely a numerical sense progressed in parallel with advances in material culture. But utilitarian motives alone do not seem to account adequately for late precivilization numeration whose roots seem intertwined with those of ancient number mysticism. Material advance also fostered more elaborate religious ritual linking the mysterious realms of language and number. A. Seidenberg points to religious ritual that seeks to convert the creative power of words into the creative power of numbers, to establish certain numbers of priests to ensure the sacral validity of ceremonies, and to develop myths embedded in magical-mystical enumeration,[149] but his work is faulted for lacking a sufficient basis in anthropology.

The development of both the numerical sense and material culture obviously differed in pace from one precivilization people to another. Growth throughout the Paleolithic and Mesolithic Ages is periodized in millennia rather than centuries, decades, or years. Until almost the end of the Age of Stone, the need to subsist on meager resources kept material culture at an almost static level:

Paleolithic and Mesolithic peoples, who were hunters, fishers, and gatherers constantly on the move and vulnerable to attacks by wild animals and natural calamities, could make little progress in either culture or social organization. Thomas Hobbes appropriately describes their lives as "solitary, poor, nasty, brutish, and short."

1.2   The Neolithic Revolution: The Age of Agriculture

The Mesolithic Age, although it had improved stone implements and weapons, inaugurated no new cultural stage; it simply mapped a cultural past onto a geological present era. The same was not true of its successor, the Neolithic Age. Environmental circumstances still chiefly influenced the evolution and applications of concrete number systems, but environment itself came somewhat under the control of systematic farming, and human existence changed profoundly.

Anthropologists generally believe that women were essential to the emergence of farming. Before and during the Mesolithic Age, it appears, labor was divided by sex. Smaller in stature and less swift afoot than their male counterparts, and constrained by childbearing, Mesolithic women made clothing, gathered wild grains, nuts, and berries, and wove baskets, in part to store dry foods. In picking such foods, these women must have found how to plant seeds and grow food from them.

The slow transition away from migratory hunter-gatherer culture to agriculture and its permanent settlements began first in the Near East some nine thousand years ago and in the British isles, India, and northern China perhaps slightly later. The transition was not a simple process and the reasons for it remain unclear. The previously prominent theory that climatic changes reduced herd sizes and forced hunter-gatherers to shift to agriculture does not seem correct. Investigators now suggest a combination of botanical and human factors.

Among these was the appearance of hybrid strains of wheat with higher yields. Anthropologists have found that when early Near Eastern farmers planted a certain species of wheat, wild grasses nearby fertilized it to produce a richer hybrid. This high-yield cereal allowed a surplus to last beyond the growing season with sufficient seed for the following year's planting. Precivilized farmers also grew sorghum and millet. An improvement in cereal crops and the domestication of more and larger animals—cattle, sheep, pigs, and possibly goats—provided a greater and more dependable food supply than hunting-gathering, although the supply remained precarious because insects or plant diseases could destroy entire crops. In the absence of census figures, anthropologists surmise that better-fed populations grew sharply, creating pressure for more agriculture.

The Neolithic achievement of systematic agriculture profoundly altered ways of life, mentality, values, and social organization. Only three later revolutions have produced comparable turnabouts: the philosophical in China, India, and classical Greece; the religious, stretching from 1200 B.C. to about A.D. 600, ushering in monotheism; and the scientific, dating from the fifteenth century. Farmsteads and permanent villages now replaced roving bands as the fundamental unit of human society. Arduous and regular labor in the fields may have moderated the hunters' boldness and violence into habits of foresight and restraint.

By about 4000 B.C., Neolithic people were settling in substantial numbers on the floodplains of great river valleys, notably the Tigris and Euphrates of Mesopotamia (modern Iraq), the Nile of Egypt, and later the Yellow River of northern China, the Indus of Pakistan, and the Ganges of India. Except for China, which grew rice and millet, all relied primarily on wheat. From dotted patterns of villages, a few urban centers emerged. We do not know the precise reasons why people chose to live in early urban centers with their crowding, lack of waste disposal, danger of contagion, and in time organized warfare. In addition to humans being social animals, reliable food supplies, protection from enemies and floods within fortified city walls, and the availability of amusements are possible reasons.

The prehistory of protomathematical science in each of these river valleys remains unclear and debatable. Did protomathematics develop independently in each area, deriving in all these cases from a compound of local environmental circumstances and human ingenuity, as one camp holds? Or do the three protomathematics show a common origin, as a second camp contends? Of course, no direct documentary evidence exists. Recently, A. Seidenberg and B. L. van der Waerden have offered a tentative hypothesis that a common ground did exist. They base their idea on anthropological records and detailed comparative studies hinting at historical connections between ancient Babylon, Vedic India, and Han China in the calculation of Pythagorean triples, while ancient Babylon, Egypt, Greece, and China had similar methods of computing plane areas and volumes of solids.[150] Seidenberg and van der Waerden surmise that at the source of the differing protomathematics in the ancient civilizations of the great river valleys was an oral tradition in crude arithmetic and geometry applied by its creators to architecture, horizon astronomy, and ritual. Seidenberg and van der Waerden argue that the Beaker people were the originators of this oral protomathematics.

Named after their pottery, the Beakers were a ruling stratum across Europe. They had constructed megaliths in the Iberian peninsula by the fourth millennium B.C. and by about 2800 B.C. in the British Isles, especially Stonehenge I. The Egyptians later drew on similar skills in constructing pyramids. Stonehenge I's postholes are oriented for observation of the rising and setting of fixed stars and possibly for prediction of eclipses. Like the ancient Egyptians and Greeks, the Beakers had rope stretchers who used knotted cords to survey land and design and lay out temples. For example, they fixed cords around one stick to draw circles and apparently around two sticks to trace ellipses. From their study of megaliths and a modified carbon-14 dating of organic material in archaeological excavations,[151] Seidenberg and van der Waerden argue that Beaker protomathematics spread from mid-central Europe to the British Isles to the eastern Mediterranean between 3000 and 2500 B.C. But the lines of transmission are unknown. Among the critics of this idea are scholars who believe that it simply refines discredited classical diffusion theory.

If a Beaker transmission did carry to Mesopotamia by 2500 B.C., it encountered a civilization that had long been evolving. In the riverbank urban communities of the Near East, so it is believed, the older Neolithic social structure based on rule by consensus changed radically during the fourth millennium B.C. Geography and climate had a powerful influence on that political economy. Substantial agriculture was possible there only through irrigation. Near Eastern urban centers thus had increasingly to control flood water with canals, dikes, and ditches. This required coordinated programs involving hundreds or thousands of people, men and women alike. Successful programs producing great surpluses of food demanded strict management of water supplies, seeds, and labor. A despotic managerial elite arose. Surplus grain became a basic article of commerce, which thereby expanded. The first urban community with these features dates from about 3600 B.C. in Sumer in the lower or southernmost Tigris-Euphrates Valley at the head of the Persian gulf.

By 3000 B.C. a rigid social structure had evolved in Sumerian cities. Their social pyramid had a broad human base of peasant commoners and slaves, most of whom lived at the subsistence level. Agricultural surplus permitted the emergence of a small artisan order, roughly a class, usually of low social status. Above it was a powerful temple priesthood, which owned large portions of the city's land. Urban activities centered on an impressive temple, or divine household, which was a storehouse for grain and precious metals. Temple and palace scribes directed agricultural and civil programs, while merchants expanded trade. A hereditary nobility was higher in the pyramid. A king or *lugal*, who exercised absolute political power, was at the top. At first, the king was an elected war leader, but kingship became hereditary during the frequent wars.

1.3   Writing and Metrology in Ancient Sumer

Agricultural, royal, commercial, and religious exigencies seem to have led Sumerian scribes, merchants, and priests to develop two tools crucial to later mathematical progress, writing and an improved art of measurement, or metrology.

Careful records of water distribution, astronomical phenomena, and the erratic behavior of the Euphrates may have been the initial subject of Sumerian writing. The growth and concentration of population, food supplies, and trade made for more extensive and more complex records. Palace and temple scribes kept inventories of animals, slaves, and crops for kings and temples, apportioned taxes, distributed commodities, and recorded complicated business transactions. Writing was also needed to transmit government acts and laws and to develop religious texts. In Sumer and subsequent civilized society, writing facilitated communication and permitted records of past experience to be compiled. These records in turn fostered new ideas.

The Sumerians did not invent writing. A crude writing had long existed in the form of a complicated system of pictographs, simple drawings of a fish, bird, or man, impressed on soft, moist clay tablets, which were then baked. The earlier method was to bake them in the sun; later kilns came into use. Clay was used for writing, because it was plentiful in Mesopotamia and could be easily imprinted and baked to form permanent records.

From 3600 to about 2300 B.C., the Sumerians made fundamental advances in writing. They transformed it from pictographs to a conventionalized script consisting of hundreds of syllabic signs, or ideograms, associated with sounds. Phonetic combination of these sounds conveyed ideas. Because Sumerian script was written with a stylus, a reed cut at an angle, whose base left a triangular or wedge-shaped symbol, the script that evolved is known as cuneiform (Latin *cuneus*, a "wedge," and *forma*, "a shape"). The sharp stylus edge made a vertical stroke. Sumerian progress in writing prepared the way for the Canaanite invention of the alphabet in the second millennium B.C., with twenty-nine symbols representing consonants alone.

The Sumerians meanwhile were also founding a system of scribal schools, which taught writing and were centers of culture. These schools flourished by 2500 B.C. Most students were sons of the wealthy. Discipline was harsh. Students were taught grammar and lists of words, learned how to prepare clay tablets, and solved mathematical problems. Although Sumerian schools stressed business affairs and administration, the practical issues with which scribes had to deal, the curriculum included such topics as literature, linguistics, religion, and botany.

Advances in measurement accompanied sophistication in writing and a permanent medium for it. Apparently, they mainly responded to problems posed by large-scale agriculture, fortifications, and architectural projects. Constructing canals and dikes and parceling land after floods as populations grew required greater accuracy in measurement. Temples and palaces with well-spaced, sizable pillars of the same height and shape, along with properly placed doors and matching windows, demanded a symmetry that in turn necessitated better metrical techniques.

The intuition of space and time is doubtless primal. But to be of use to mathematics and the natural sciences, it must be discovered and examined in enough detail to deepen the understanding of these concepts. Primarily from their pursuit of increasingly precise measurement, palace, temple, and school scribes advanced the awareness of space and time. The complex psychological and analytical sense of extension began to be articulated simply in perceptions of displacement among points, fostered by surveys and large construction projects. The similarly sophisticated intellectual experience of duration had equally plain beginnings in the marking off of the time ordering of two or more events. Agriculture, which demanded accurate chronology, must have prompted an effort to improve early calendars and thus to refine the notion of time.

Sumerians were perhaps the first to realize clearly that success in farming depends on knowing when to plant. Only a reliable calendar could provide this information. Observing cycles of the moon, the most conspicuous marker of the passage of time after the diurnal cycle of day and night, set the Sumerians on the track to a calendar. But this information alone did not produce one. Sumerian astronomers had to observe, measure, and correlate movements of both the sun and the moon before they produced a reliable calendar. Their first luni-solar calendars were roughly $12 \times 29\frac{1}{2} = 354$ days in length, that is, twelve times the number of days in a lunar or synodical month (lunation), the period from new moon to new moon. Consequently, they were about eleven days short of one solar year. About every three years a thirteenth month had to be added to correct the calendar. Sumerian luni-solar calendars generally began with the spring equinox, which was near the time of "Coming Forth" (of seedlings).

The calendar was a remarkable scientific achievement. With it, palace and temple scribes could foretell the seasons accurately. This capability probably strengthened the position of temple scribes in the community. With a seeming mastery over the seasons, could they not interpret omens from the gods? And in their halting beginnings in defining the nature of space and time, Sumerian scribes started the exploration of two enduring concepts of science.

Part II

Antiquity: From Protomathematics to Theoretical Mathematics

2

The Dawning of Mathematics in the Ancient Near East

> The invention of . . . place value notation is undoubtedly
> one of the most fertile inventions of humanity.
> — Otto Neugebauer

Although mathematics clearly did not emanate solely from the ancient Near East, the principal historical origins of mathematics may largely be traced to the urban civilizations of Mesopotamia and Egypt.[152] During the course of ancient Mesopotamian civilization from ca. 3500 B.C. to the time of Christ, three successive peoples—Sumerians, Akkadians, and Babylonians—especially contributed to the origins of arithmetic and algebra. Their scribes and merchants developed a deft number system consisting of a sexagesimal (60) base, number ideograms, and, by 1700 B.C., place-value notation, which facilitated extensive and diverse computations. With their drawings and writing medium confined to moist clay tablets, Mesopotamian scribes made only minor contributions to geometry. By comparison, the Egyptians had a limited number system and algebra but founded more of practical geometry, in part establishing rules to measure the circle, to compute granary volumes, and to determine land boundaries when replacing markers destroyed by annual Nile floods.

Increasingly difficult practical needs, conjoined to a school-and-bureaucracy complex, motivated these impressive accomplishments. The needs in the new Mesopotamian and Egyptian civilizations arose from intensive agriculture, astral religion, monumental building programs, warfare, governmental taxation, and extensive trade and commerce. Palace and priest scribes and teachers in scribal schools responded to these needs. To the extent that they were successful, scribes were to flourish and enjoy prestige. They had to attempt to harness river waters to provide large-scale irrigation and control flooding, an especially difficult task on the erratic Tigris-Euphrates. Assigning fields to farmers and reassigning them after floods required surveying. Palace and priest scribes made astronomical observations and kept accurate records of movements of celestial bodies to find evidence of divine actions, set religious holidays, and prepare calendars for agricultural planning. The buildings they planned included huge adobe granaries, ziggurats in Mesopotamia, and stone pyramids in

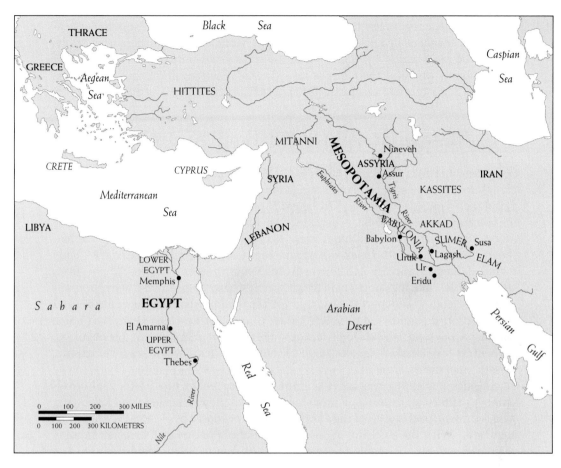

The Ancient Near East

Egypt. Palace and priest scribes also had to determine and apportion taxes annually, while growing economies required basic arithmetic daily for billings and calculating wages and interest on loans. Long-range trade involved different number systems.

Although practical needs principally motivated this protomathematical work, palace and temple scribes and schools supplied the thinking. Allowing scribes time to contemplate problems of organized agriculture, commerce, and government was a major innovation of the first Near Eastern civilizations. Already in antiquity, Aristotle in his *Metaphysics* attributed the origins of mathematical science in Egypt to the leisure of the priestly caste.

To carry out their tasks well, scribes and scribal mathematicians had to develop more sophisticated protomathematics, a degree of speculation going beyond the utilitarian concerns that were its original motive. Written documents suggest that the play element, mathematics enjoyed for its own sake, influenced their pursuit of problems in a higher mathematics. Thus, sometimes a game mentality addressing the challenge of resolving such problems underlies their work. Among at least one group of the more creative scribal mathematicians and teachers in ancient Mesopotamia, virtuosity in mathematics seems to have been a source of pride.[153] But the school-and-bureaucracy administrators probably opposed their forays into theory. The transition to a continuous tradition of theoretical mathematics awaited the classical Greeks.

2.1   Mesopotamian Civilization: An Introduction

Sometime between 3500 and 3000 B.C., the earliest civilization in the West came into existence in Sumer in southern Mesopotamia (from the Greek, the "land between the rivers"). At least three elements—cities, writings, and metallurgy—were apparently necessary to form civilization. Civilization rose among Sumerian urban communities, perhaps twelve in number, which had grown from dense patterns of villages into cities of several thousand.[154] They included Lagash and Ur in the south and Nippur in the north. These cities were complex, having large-scale irrigation systems, an administrative bureaucracy, law codes, and postal services. Surplus from organized agriculture and construction demands for temples, palaces, and fortifications permitted some people to have specialized functions; among them were architects, blacksmiths, carpenters, clothmakers, cooks, dancers, engineers, leather workers, merchants, priests, and tavern owners.

Besides possessing writing, the Sumerians enjoyed a rapid technical development. Sumerian technology included a scratch plow, wheeled vehicles, boats, wheel-thrown pottery, weaving, and impressive adobe brick towers known as ziggurats. With the smelting of copper and, shortly afterward, bronze (an alloy of copper and tin) for tools and weapons, the Age of Stone yielded to an Age of Metals in Sumer.

Our knowledge of ancient Mesopotamian political and cultural history is sketchy but growing. Its sources are archaeological artifacts and documents gathered since the seventeenth century. The documents—cuneiform clay tablets that number more than 400,000—have been found through excavations of the ruins of such cities as Nippur, which yielded 50,000, Nineveh 22,000, and Susa and Ebla thousands more. In 1849–50, Nineveh was called Kuyunijk when its palace library was unearthed, and so its tablets have the inventory letter K in the British Museum. These 400,000 tablets comprise only a small portion of the total number written in ancient Mesopotamia and thus offer a scant glimpse into that society. Added to the problem of gathering and preserving these tablets have been those of deciphering and evaluating them. While even laypersons can detect symbols for numbers in the tablets, deciphering nonnumerical symbols of their forgotten languages with their variety of dialects and modifications over three millennia has proved difficult.

The work of deciphering is relatively recent. The key came in 1846, upon Henry Rawlinson's complete translation of the monumental account of Darius I's crushing of a revolt in Persia in 516 B.C. This account had been carved four hundred feet above ground level on the Behistun Cliff in Iran for passing caravans to see. It was written in three languages—Elamitic (Old Persian), Babylonian, and Persian. Bilingual inscriptions in Sumerian and Akkadian, a Semitic language similar to Babylonian, permitted extrapolations and further deciphering. Rawlinson had cracked the cuneiform enigma and essentially opened for study the records of ancient Mesopotamia. Still, the task of translating and analyzing the content of cuneiform tablets with mathematical content, which the late Otto Neugebauer and François Thureau–Dangin pioneered after 1925, remains far from complete.

Despite such lacunae in the literary record, an outline history of ancient Mesopotamia has emerged. That land experienced successive invasions and intermittent warfare that diversified and invigorated its culture. Civilizations from their origins had mixtures of peoples, not the racial purity of heinous myth. Geography made cities of the river plain open to attack not only from neighboring cities in quarrels over frontiers, water, and food supplies, but also from other invaders: bordering mountain pastoral people to the north, Semites from the west, and Indo-Aryan nomads from Iran to the east. Over a nearly three thousand-year period, waves of invaders battled for hegemony in Mesopotamia.

By 3000 B.C., raids had begun on cities. In Sumer, rule by consensus swiftly gave way to kingship. Around 2340, Sumer fell before the Akkadian chieftain Sargon ("True King"). Two centuries later the city of Ur revolted and reasserted Sumerian supremacy. About 2000 Semitic invaders swept aside Sumerian rule. A century later Semitic Amorites established their capital at Babylon, which dominated regional river commerce. The Amorite dynasty attained its highest political and cultural consolidation under Hammurabi (ca. 1792–ca. 1750 B.C.), known for his harsh but severely logical law code, which repeated the injunction "an eye for an eye and a tooth for a tooth."

During the remainder of the second and first millennia B.C., the center of power in Mesopotamia continued to shift. By 1550, Hammurabi's dynasty had fallen. Outfitted with war chariots, Kassites from Iran overwhelmed the Babylonian foot soldiers. By the ninth century, iron-weaponed Assyrians, known for their cruelty, became prominent. Their last major king, Assurbanipal (668–ca. 626), built Nineveh into a great city. Afterward, a new line of Babylonian kings, particularly Nebuchadnezzar (605–562 B.C.), established the Chaldean Empire. Nebuchadnezzar built the great ziggurat in Babylon known for its hanging gardens and in 586 carried many Hebrews off into Babylonian captivity. Mesopotamia was next conquered by the Persians under Cyrus in 538 and by Alexander the Great in 330 B.C. The period from the Alexandrine conquest to the birth of Christ is known as the Seleucid, after the Greek general Seleucus who controlled the region after Alexander's death.

Under successive military invasions that could have destroyed a culture, that of ancient Mesopotamia not only maintained substantial continuity but also spread to bordering regions by merchant routes. Barbarian invaders borrowed extensively from the older civilizations of the plain and generally respected their traditions. Besides military invasions, an extensive long-distance trade contributed to a great diversity of products and technical knowledge. After 3000 B.C. the Tigris-Euphrates Valley became a crossroad of trade routes. Importation of metals, wood, and other raw materials from Syria, Cyprus, Asia Minor, and more distant regions became vital to Mesopotamian life. In this long-distance trade, merchants exchanged traditional lore as well as merchandise and monies. They had to itemize large numbers of goods, to agree on media of exchange and their value, and to try to establish standard weights and measures.

Cultural patterns changed only slowly and grudgingly. Before Alexander, invaders were gradually assimilated into the older culture and enriched it. Within Mesopotamia, different fields of learning, from botany to linguistics, occasionally flourished, especially under the Sumerians, Akkadians, and Babylonians. Protomathematics was to benefit conspicuously from this continuity of knowledge, practice, and traditions.

Among the larger forces fostering the growth of protomathematics were imperialism and long-distance trade. In extent of impact, both were new to human society. They enlarged the store of knowledge, providing new schools and palace and temple scribes with a large base of information. The stronger influences on the growth of protomathematics, however, were practical necessities related to organized agriculture, commerce, and religion.

When Mesopotamian cities became the heart of a large land empire after 3000 B.C., commercial activities grew in complexity and scale. As its population approached one million by 2000, moreover, Mesopotamian society required closer stewardship of grain supplies and more accurate bookkeeping for the distribution of goods.[155] Harvests stored in granaries had to be apportioned among increasing numbers of peasants, artisans, priests, and officials. Daily business transactions, such as pricing goods, along with the making of wills called for numerical tables. Canals, dams, and irrigation projects involved calculation, as did the construction of granaries and other buildings. Calendar reform, essential both to agriculture and to the religious practice of divination, meant keeping meticulous astronomical records.

Palace and temple scribes skillfully handled great quantities of accounts and other records. Among these careful and punctilious compilers of data were keen observers. The refinement of two humble skills—observation and manual adeptness—probably sharpened habits of thought in general. The ancient Mesopotamian achievement sprang from this precision, as well as from the antiquity and continuity of their records. Astronomers by the Seleucid period were attaining a standard error of less than six minutes a year for celestial positions, an accuracy not equaled even by the ancient Greeks.[156]

The chief source of information about proto- and early mathematics in ancient Mesopotamia is a small group of roughly four hundred numerical tablets and tablet fragments. These table and problem tablets and, to a lesser extent, selected commercial and astronomical tablets disclose the pattern of mathematical development and its substance. Paleographic and linguistic data reveal that all the purely mathematical texts date from one of two widely separated periods: about two-thirds derive from the Old Babylonian or Hammurabic period (ca. 1900–1550 B.C.) and one-third from the Seleucid period. Commercial tablets show that the Sumerians and Akkadians progressed in creating mathematics, although just how far is uncertain. Undoubtedly they laid the groundwork for its flowering in Old Babylon. More than a subsequent millennium of relative stagnation was followed by a brief resurgence.

2.2   Mathematical Achievements in Ancient Mesopotamia

Numeration

Within this social and cultural setting, ancient Mesopotamian scribes and teachers invented a great deal of elementary arithmetic, producing work of the highest caliber in number and numerical operations. They had four types of numbers: cardinal, ordinal, integers, and fractions. The crude beginnings of cardinals and ordinals stem from the late Paleolithic Age and the Mesolithic. By the Old Babylonian Period, Mesopotamian numeration had three striking and yet simple features: an indigenous sexagesimal system (base 60), number ideograms, and place-value notation. These did not appear at once. It took several centuries and three different peoples to produce them.

By 2350 B.C. the Sumerians apparently had developed the sexagesimal system. According to the early twentieth-century German Assyriologist Kewitsch, the sexagesimal system may have been a compromise built naturally on contact between two peoples using different number bases, base 6 and the more common base 10. Alternating these bases results in units of 1, 10, 60, 600, 3600, ...[157] The Sumerians had signs for each of these units. French Assyriologist Thureau-Dangin believes that 6 is an unlikely base, but could be a variant derived from one-half of base 12.

Although our society generally follows the decimal form, we know the sexagesimal system today because our units of time and angle measure are based on it. In time units, 60 seconds = 1 minute and 60 minutes = 1 hour. In angle measure, 60 seconds = 1 minute, while 60 minutes = 1 degree; a circle contains 360 degrees, the angle completely around a point. The sexagesimal system has persisted chiefly as a result of its use in astronomy. Its acceptance by the ancient Greek astronomer Hipparchus (d. ca. 125 B.C.) and the Hellenistic astronomer Claudius Ptolemy (d. ca. A.D. 178) for his *Almagest* (*Mathematikos Syntaxeos*) fixed it firmly in western astronomy.

The sexagesimal system was a significant advance in protomathematics. With its use the Sumerians progressed from counting, which is formlessly additive, to grouping. While primitive people had vague groupings based primarily upon two- or finger-counting, the Sumerians were the first to use grouping consistently in commerce, trade, astronomical records, and the measurement of time. Thereafter, all civilized people of antiquity accepted the method of ordering numbers in groups either of sixties, thirty-six hundreds, and further powers of sixty or, more usually, of tens, hundreds, thousands, and so on.

If base 60 was not an expedient compromise or result of cultural contact, what was the motivation for consciously choosing it?

Theon of Alexandria assigned to the Sumerians a good mathematical reason for selecting it over the more common base 10, which they also knew. They had devised the sexagesimal system, he said, to avoid fractions as much as possible.

Fractions had first appeared in the Bronze Age in the search for greater precision, but Sumerian scribes were uncomfortable with divisions that do not work out exactly. Like other ancients, they found fractions difficult and troublesome. Moderns are no happier:

> Multiplication is vexation,
> Division is as bad;
> The Rule of Three perplexes me,
> And Fractions drive me mad.

Theon may be right that the sexagesimal system was the Sumerian response to reaching smaller units of measure by division. Avoiding complicated fractions—a feature important to their monies—requires finding a number base with many divisors. Since 10 is divisible only by 2 and 5, it is a poor choice. Performing divisions into small whole numbers can shortly give 60 as a more suitable base. It is divisible by 2, 3, 4, 5, 6, 10, 12, 15, 20, and 30. This means that ten important sexagesimal unit fractions can be expressed exactly without any remainder. In our customary numbers, rendered with a 0 symbol and a semicolon as a sexagesimal point, they are:

$$1/2 = 0;30 \qquad 1/5 = 0;12 \qquad 1/12 = 0;5 \qquad 1/20 = 0;3$$
$$1/3 = 0;20 \qquad 1/6 = 0;10 \qquad 1/15 = 0;4 \qquad 1/30 = 0;2$$
$$1/4 = 0;15 \qquad 1/10 = 0;6$$

Moritz Cantor and other scholars have not been persuaded that uneasiness with fractions led the Sumerians to adopt base 60. They have speculated that the Sumerian luni-solar calendar of roughly 360 days was a more likely source than commerce: $60 \times 6 = 360$. At best, this argument is dubious. The Sumerian luni-solar calendar was initially closer to 354 than to 360 days, and it was later refined toward 365 days. Current information suggests that base 60 came first, while the notion of a 360-day calendar occurred later. Perhaps the use of the sexagesimal system in compiling astronomical records and calendar preparations suggested a 360-day year, but nature did not prove amenable to this numerical convention.

Another striking feature of ancient Mesopotamian numeration was number ideograms (abstract symbols). Since neither the Sumerians nor the ancient Egyptians, each of whom had a continuity of one language, produced number ideograms, these abstractions apparently required a critical meeting of languages. That their initial appearance came after the Akkadian conquest suggests that they were intended to provide a symbolism usable in each of two languages of basically distinct grammars.

The Akkadians, who did not match the far advanced Sumerian culture, adopted the Sumerian language for official or sacred purposes, much as Latin was adopted in medieval Europe. Akkadian scribes initially used both languages in their texts, writing them in parallel, and compiled lexicons to give equivalents. This lexicographic work, along with the commercial need for consistent, more uniform numbers, produced less varied forms of words and new numerals. For some unknown reason the scribes chose to represent numbers by the older Sumerian script, which had hundreds of pictorial signs and syllabic representations that today are called phonograms. The more fossilized the Sumerian language became, the more Sumerian signs were used for numerals. In time these signs lost their original concrete sense and became abstract. Thus, the number ideograms emerged free of concrete notions, such as "pointing to" or "nodding the head."

Developing abstract number symbols was but a first phase in Akkadian progress in numeration. Faced with new domestic problems brought on by a growing population and consolidating military and commercial expansion, Akkadian scribes advanced computation considerably. A moral or legal factor, a program to curb embezzlement of goods and silver by dishonest scribes and priests, also brought pressure to improve computation. Adding numerical symbols permitted more sophisticated computations. By 2100 B.C. the Akkadians had invented the abacus and developed a crude arithmetic. They could add and subtract and, with difficulty, divide and multiply. For extensive and widespread computations, having dozens of symbols proved unwieldy. A small group of numerals, which through repetition can represent all numbers, was plainly required for effective and economical computation. This new stage in numerical progress was reached by the successors of the Akkadians. In the Amorite period the Babylonians invented place-value or positional notation, the third and most striking feature of ancient Mesopotamian numeration.

Two tablets listing squares of numbers that were discovered by English geologist William K. Loftus in 1854 first confirmed this invention. These tablets have the expected numbers 1, 4, 9, $\ldots$, up to 7^2 or 49, followed not by 64 but by 14 for 8^2, 121 for 9^2, and so on.

This shows the ambiguity of Babylonian numbers and the necessity to read them in context. They did not have a mark, such as a comma, to separate powers of 60. In this case, the 1 in 14 stands for 1×60, and so $14 = 1 \times 60 + 4 = 64$, and $121 = 1 \times 60 + 21 = 81$. The Babylonians had discovered that the question "How many?" can be enormously simplified if the same number stands for different values depending on its position in a row of numbers. Place-value notation eliminated the need to invent new digits for numbers, no matter how large. With it, a small number of symbols suffices to express any real number to any given degree of accuracy.

The prodigious power of number as a language was revealed in this momentous grammatical breakthrough: position in the expression governs value. American mathematician Oswald Veblen was to call it a singular achievement of the human intellect, and Otto Neugebauer, "undoubtedly one of the most fertile inventions of humanity," comparable to "the invention of the alphabet as contrasted to the use of thousands of picture signs."[158]

What exactly does place value entail? Consider the decimal (base-10) numeral 3742. The value of each digit depends on its position relative to the decimal point. At each further digit leftward from the decimal point, the position values increase by factors of 10:

First place	Multiple of	10^0, or 1	Units
Second place	Multiple of	10^1, or 10	Tens
Third place	Multiple of	10^2, or 100	Hundreds
Fourth place	Multiple of	10^3, or 1000	Thousands

The decimal numeral 3,742 has 3 in the thousand's place, 7 in the hundred's place, 4 in the ten's place, and 2 in the unit's place. Its total value is obtained by the operations of multiplication and addition:

$$3,742 = (3 \times 1000) + (7 \times 100) + (4 \times 10) + (2 \times 1).$$

Place-value or positional arrangement can apply to any number base. In base 8, for example, 3,742 becomes $(3 \times 8^3) + (7 \times 8^2) + (4 \times 8^1) + (2 \times 8^0)$, which equals (in decimal form) 2,018. More generally, if 3,742 is a numeral in number base a, where a is any number greater than 7 here, since it is customary only to include digits up to $a-1$, then 3,742 equals $(3 \times a^3) + (7 \times a^2) + (4 \times a^1) + (2 \times a^0)$. This generality of application is essential for a conceptual understanding of it.

Presumably the highly developed commerce already employing the Akkadian numeral system facilitated the adoption in ancient Babylon of place-value notation in base 60. In monetary transactions, Babylonian scribes and merchants recorded larger and smaller units or weights of silver by simple juxtaposition of numerals all in a row, which denoted units of different value. This corresponds to our giving relative monetary value in the United States as a ratio of dollars to cents: saying "a dollar ten" for $1.10. As Otto Neugebauer has observed, when the Old Babylonians extended this device from monetary units of weights to numbers, they had place-value notation.[159] It perhaps occurred by happenstance rather than design. Its utility, simplicity, and beauty may have made it appealing to Babylonian scribal mathematicians and merchants. They set forth a sexagesimal place-value system because 60 was a monetary base.

Although the Babylonians attained an impressive working knowledge of place-value notation, they appear to have understood it simply as a clever and perhaps as an elegant device, that expedited transactions in the marketplace and allowed them to solve and perform extensive computations. Extant cuneiform texts contain no evidence that they applied it to any number base other than 60. We thus assume that they lacked a conceptual understanding of it. This does not detract, however, from the magnitude of their achievement.

Place-value notation in the hands of Babylonian scribes had particular strengths. Again, their probable reason for choosing the Sumerian 60 as their number base was its greater facility with fractions. As their commercial texts show, they chose it over such others as the Akkadian 10 and bases 2, 12, and 24 known from people with whom they traded. A number system covering only positive integers (they never considered negative numbers) would have been weak and limited. Their place-value system made Babylonian scribes capable of expressing equally well all positive rational numbers, that is, all positive integers and all proper fractions, such as 1/4, 1/8, 3/4, 8/9, ... Today a rational number is defined as any number that can be expressed as a ratio of two integers (m/n), providing $n \neq 0$. Among the ancients, the Babylonians alone had a numerical system that treated positive integers and fractions alike.

In selecting basic number symbols (digits), Babylonian scribes improved upon their predecessors' work, but their preoccupation with economy came at the cost of full flexibility in place-value notation. A repertoire of symbols adequate for a flexible notation had evolved slowly from Sumerian number tags and pictographs through the many Akkadian cuneiform phonograms. Yet to speed up the scribe's impressing his stylus on moist clay, the Babylonians confined themselves to two. In cuneiform (wedge shaped) notation a wedge stands for unit or 1, ❯, and crescent for 10, ◁. The triangular base of the stylus provided the basis for the wedge shape for unit. "Bills of lading" are providing good information on the evolution of this numeral system, but the origins of the crescent symbol for 10 are not yet known with certainty. It may represent a wedge on its side, or it may have portrayed two hands pressed together with the thumbs extended. Babylonian scribes wrote the numbers from 1 to 59 as repetitions and combinations of these two symbols: In place value, ❯ could stand for 60 as well as 1 and the combination process would begin again. Having a sexagesimal base but only two wedge-shaped symbols to represent all natural numbers forced the Babylonians to construct a repetitive rather than a digital system. A repetitive system uses symbol iteration to obtain digits, as the Babylonian numerals 2 through 9 or their Roman numeral counterparts illustrate, rather than self-standing symbols for each, as in the Indo-Arabic numerals 1 through 9. On correcting the Akkadian overabundance of digits, the Babylonians allowed too few for a digital numeral system.

| 60 | 70 | 80 | 120 | 130 |

Besides lacking sufficient digits, the Babylonians had no separator—a sexagesimal semicolon corresponding to our decimal point—distinguishing whole numbers from fractions. Place value alone is ambiguous and not absolute when units and subunits are indistinguishable, a serious conceptual shortcoming from a modern standpoint. So the actual value of a Babylonian numeral could not

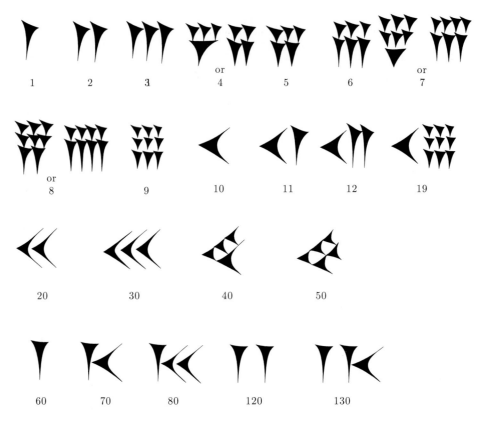

Seleucid cursive form

be distinguished immediately from the symbols alone but had to be determined by adjusting orders of magnitude to the context. For example, the Babylonian numeral 2,15 (with the Neugebauer convention of a comma to separate powers of 60) might equal

(a) $0;2,15 = 2/60^1 + 15/60^2 = 1/30 + 1/240 = 3/80$ or

(b) $2;15 = 2 \times 60^0 + 15/60^1 = 2 \ 1/4$, or

(c) $2,15; = (2 \times 60^1) + (15 \times 60^0) = 135$, and so on.

As this example illustrates, the power of 60 was suppressed. The Babylonians worked with what we today call a "floating sexagesimal point."

There is no evidence that the ambiguity of their place-value system troubled the Babylonians or that they endeavored consciously to make it absolute. That would require either a sexagesimal equivalent of the decimal point and a zero symbol, or more digits, or both. The addition of digits would have contradicted the scribes' bent for economy. The reason for their failure to devise a sexagesimal separator lies, we believe, in their understanding of place-value notation as a computational device and not as a concept of why the computation was valid. Proposing a sexagesimal separator to act like the decimal point presumes conceptual understanding.

The search for a zero concept, as distinct from a fixed zero symbol, had a more satisfactory resolution. Early Babylonians often left empty places or gaps indicating a space that changed the value of the next digit. A place-value system has to specify which power of the base each digit represents. In the modern decimal form, 37 obviously differs from 307 or 370. In these last two numbers, the zero acts as a placeholder, an item the Old Babylonians lacked. Since neither of their two original wedge-shaped numerals could express "nothingness" or "emptiness," these could not serve in their original form as placeholders or space fillers. Leaving a gap, however, gave rise to confusion, because different scribes left different amounts of space, some of which were imperceptible to scribal copyists. The practice of leaving gaps, moreover, was not uniformly followed. Not until the Seleucid period, over a millennium later, did the ancient Mesopotamians have two small oblique wedges or triangles, ⟋⟋ or ▲, to act as medial placeholders and an omicron to act as a medial zero. The omicron o came from the first letter of the Greek word for "nothing". A medial zero is a zero as appears in 205 or 4,906.

Seleucid Mesopotamians still needed a zero for the extreme right position in numbers like 350 or 9,400, a seemingly simple extension of their use of omicron. But they had trouble with the concept of "lack" or "nothingness" to designate a quantity of "something" and so never applied their zero symbol to the extreme right place. Another millennium was to elapse before Indian sages began to develop an operational zero.

Although limited by ambiguous notation and lack of an operational zero, the Babylonian place-value numerical system was a remarkable achievement. It was the most effective basis for computation in antiquity. With it the Babylonians could add, subtract, multiply, or divide positive rational numbers with facility, as their problem and table texts demonstrate. Their sexagesimal place-value system, which included integers and fractions alike, was the exact counterpart of our decimal system for arithmetical operations. Thus, a decimal such as 3.26 is considered 326×0.01. It can then be treated as if it were an integer.

The Babylonian sophistication in handling numbers went well beyond that of other ancient peoples in the West—Middle and New Kingdom Egyptians, classical and Hellenistic Greeks, and Romans—who had advanced cultures. Except briefly for the Old Kingdom Egyptians and perhaps a precursor of the Maya in Mesoamerica, no other ancient people discovered place-value notation, which shows that it was not inevitably connected to the extensive use of number. These other ancient peoples treated fractions separately and, like the Akkadians, had a surfeit of numerals. The Egyptians initially had pictorial symbols, while the twenty-seven characters of the Ionic alphabet eventually constituted Greek numerals. Roman numerals, more familiar to us because of their retention, for example in cornerstone dating, used the Roman alphabet selectively as follows:

I	V	X	L	C	D	M
1	5	10	50	100	500	1000

Like the Egyptian and Greek symbols, these allowed addition and subtraction, but were difficult for multiplication.[160] Multiplying 100×5 in Roman numerals will reveal the tip of the problem: $C \times V = D$. The calculator has to operate by special cases, rather than within a general system. This requires substantial memorization and mnemonic devices or extensive tables. The Babylonian place-value numerical system would not be superseded until around A.D. 800, over two millennia after the Hammurabic period, when Indians produced the flexible and simple decimal system, which had both a digital (not repetitive) and a positional feature.

Arithmetic and Rhetorical Algebra

By 1700 B.C., table and problem texts had become traditional and pervasive throughout Babylonia. What scholars refer to today as table texts deal with arithmetical problems, while problem texts involve formulating and solving algebraic or geometric problems. In preparing both, Old Babylonian scribes displayed extensive and sometimes sophisticated skill. Using their superior place-value notation, they computed directly, or by methods of approximation, or with algorithms.

Table texts demonstrate just how much of arithmetic the Babylonians created. Since they rarely generalized results, they did not invent arithmetic as a theoretical construct and only employed its computational component. In table texts, they capably added, subtracted, multiplied, and divided. Showing an eye for structure and economy in tabulations, their multiplication tables in base 60, which should go from 1×2 to 60×60, list products only from 1 to 20, followed by products for 30, 40, and 50. Adding two numbers from the separate portions of these tables provides economic calculation to determine any of its $(59)^2$ products that remain to have a complete table. In numerous tables, scribes computed squares and cubes of numbers from 1 to 50 and extracted square and cube roots.

Babylonian scribes seem to have restricted knowledge of their methods of solution to an oral tradition in their schools. Modern scholars have been able, however, to reconstruct several of them. One characteristic method of solution reduces division to reading appropriate tables of reciprocals. Babylonian scribes discovered that dividing by an integer a is the same as multiplying by its reciprocal, $1/a$, for example, $n/5 = n \times 1/5$, or in the sexagesimal system $n/5 = n \times; 12$. In modern notation, reciprocal tables usually give two columns with a number to the left and its sexagesimal reciprocal to the right. The product of the pair is 60. A selection from these tables follows:

4	15
5	12
6	10
.	.
8	7;30

9	6;40
10	6
.	.
12	5
.	.

Tables generally gave only reciprocals expressible in finite sexagesimal fractions and avoided sexagesimal irregulars, 7, 11, 13, 14, 17, and so forth. Because these numbers do not exactly divide into 60 or its powers, they give nonterminating answers. For instance, in the sexagesimal

$$1/7 = 0; 8, 34, 17, 8, 34, 17, \ldots ,$$

where the block 8,34,17 repeats itself infinitely. Ancient Mesopotamian scribes disregarded the prohibition on occasion as a Sumerian text from ca. 2500 B.C. shows. It gives the value of $1/7$ as 0;8,34,17,8. A later text suggests that scribes employed a method of approximation (or linear interpolation) to compute sexagesimal irregulars. Knowing values of $1/8$ and $1/6$, it approximated $1/7$ as intermediate between them and gave these upper and lower bounds:

$$8, 34, 16, 59 < 1/7 < 8, 34, 18.$$

Again, the correct solution infinitely repeats 8,34,17. In this particular problem, which the Sumerians and Old Babylonians worked out, lies the initial step toward a mathematical analysis of infinite arithmetical processes and the concept of number in general. Although linear interpolation became commonplace among Babylonian scribes, they did not grasp the profound implications of problems such as determining $1/7$.

The Babylonians calculated square roots better than any other people of antiquity, except for some Hellenistic Greeks. Their roots of nonsquare numbers include the approximation $17/12$ for $\sqrt{2}$. Yale Babylonian Collection tablet 7289, which dates from ca. 1600 B.C., gives in modern notation this remarkably good approximation:

$$\sqrt{2} = 1; 24, 51, 10.$$

This means that $\sqrt{2} = 1 + 24/60 + 51/60^2 + 10/60^3 = 1 + 0.4 + 0.014167 + 0.000046$. The result is 1.414213 instead of the more accurate 1.414214, a difference of $1/10^6$.

To obtain such precision the Babylonians probably calculated square roots by a procedure equivalent to the algorithm now known as Archytas's or Heron's or Newton's method, which parallels the approximation technique for obtaining irregular reciprocals. Computers often use this algorithm today. It is both simple and effective. Given a number $n = ab$, then $\sqrt{n}$ is approximately $(a+b)/2$. The closer that a is to b, the better the approximation. Through successive approximations, the technique provides an ever more accurate value. If the first approximation is a, the next ones are

$$a_1 = (a + b)/2 \quad \text{or, since } b = n/a, \; a_1 = (a + n/a)/2$$
$$a_2 = (a_1 + n/a_1)/2$$
$$a_3 = (a_2 + n/a_2)/2, \text{ and so on.}$$

This series of successive approximations is an iterative process. The subscripts list their order, so that a_1 is the second, a_2 the third, and so forth. The superiority of Babylonian computation rests not simply on numeration, but also on skill in developing such algorithmic procedures.

The Babylonians computed $\sqrt{2}$ at least in part to determine the diagonal of a square of side 1. This indicates, and other texts confirm,[161] that the Babylonians knew the Pythagorean theorem over a millennium before the Greek geometer Pythagoras (d. ca. 480 B.C.) lived, although they did not prove it. They knew that in a right or 90° triangle the sum cf the squares of the lengths of the two shorter sides or legs $(a^2 + b^2)$ equals the square of the length of the hypotenuse (c^2).

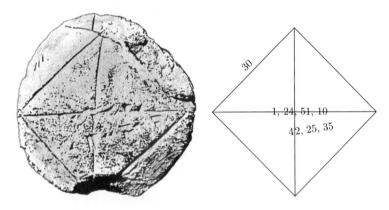

Tablet showing a square having side 30 with its diagonal 42,25,35
and their ratio. This tablet is in the Yale Babylonian Collection.

Scores of problem tablets reveal that by 2000 B.C. Mesopotamian elementary arithmetic had evolved into the beginnings of a rhetorical or prose algebra. Akkadian and later Old Babylonian scribes set and solved in verbal form what are now classified as algebraic problems. In general, these clay tablets give sequences of related problems with computations and answers. They often end with the comment "Such is the procedure." Although these tablets give no general rules, their consistent treatment of problems and concluding comment suggest a nascent theoretical approach. The arrangement and materials of these tablets also indicate occasional formative intellectual exercises with abstract probes of numerical relations.

By definition, rhetorical algebra has a serious shortcoming. Algebra, which generalizes arithmetic, today expresses relationships between numbers symbolically. Mesopotamian problem texts reveal only the beginning steps toward algebraic symbols. From their arithmetic, ancient Mesopotamian scribes had abbreviated forms for the two operations "to multiply" and "to find the reciprocal of" as well as for "plus," "minus," "equals," and "square." Some of these may have stemmed from the crude Akkadian arithmetic. Babylonian scribal mathematicians represented unknowns by the short words *asa* (area), *sag* (breadth), and *us* (flank). These were probably gathered from measuring terms from the dead Sumerian and Akkadian languages. If the scribes had possessed an alphabet containing a small number of letters to work with, they might have been able neatly and economically to represent unknowns and numerical operations. Couched in the language of ordinary discourse, their algebra could not advance far.

Despite symbolic shortcomings, the Old Babylonians proficiently handled an array of algebraic problems. They solved not only linear equations, common in problems of pay and inheritance, but also rectangular or quadratic equations and selected higher degree equations. This put them far beyond Egyptians of the time, whose mastery advanced no further than linear equations. Problems of how to subdivide a field into rectangles generated an early form of quadratic equations. Land areas were computed from the perimeter of rectangles or squares on the sides of right triangles, and army officers estimated numbers of enemy troops from their camp perimeter. Such land area computations may have fostered the mistaken notion in antiquity that the area of a plane figure is always proportional to the size of its perimeter.

In speculating as to how Old Babylonian scribes could solve hundreds of quadratic equations in verbal form, Otto Neugebauer has maintained that, when necessary, they could reduce them to what the Greeks later called the normal form. Thus, they reduced $ax^2 + bx = c$ to the normal form by keeping $y = ax$ fixed and substituting $y = ax$ to obtain $y^2 + by = ac$, which shows flexibility in their algebra.

The solution for this Old Babylonian problem demonstrates that the scribes had the equivalent of the modern formula for solving quadratic equations:

> I have added the area and two-thirds of the side of my square and it is 0;35. What is the side of my square?

In modern notation, this becomes

$$x^2 + \frac{2}{3}x = 35/60.$$

The verbal instructions for its solution are ingenious.

> First multiply the coefficients of the x^2 and x terms. This gives 1 (or 60) $\times \frac{2}{3} = 0;40$. Half of this is 0;20. Square this amount to get 0;6,40 and add 0;35 to obtain 0;41,40. Find its square root, or 0;50. Now subtract the amount 0;20, which was previously multiplied by itself, to give 0;30.

In modern notation, this is

$$x = \sqrt{(0;20)^2 + 0;35} - 0;20 = \sqrt{0;6,40 + 0;35} - 0;20$$
$$= \sqrt{0;41,40} - 0;20 \qquad = 0;50 - 0;20 = 0;30.$$

The Old Babylonians thus had the equivalent of the formula

$$x = \sqrt{(a/2)^2 + b} - a/2$$

to solve the quadratic equation $x^2 + ax = b$. This procedure was probably derived empirically.

Old Babylonian scribes also solved the system

$$(1)\ xy = b \quad \text{and} \quad (2)\ x + y = c \text{ or } x - y = c$$

by assuming from equation (2) that x exceeds $c/2$ by a certain amount w, that is, $x = c/2 + w$ and similarly y lacks being $c/2$ by w, that is $y = c/2 - w$. By substitution, $xy = b$ becomes

$$(c/2 + w)(c/2 - w) = b \Rightarrow c^2/4 - w^2 = b \Rightarrow c^2/4 - b = w^2,$$

a quadratic in w, and $w = \sqrt{(c/2)^2 - b}$. By this substitution,

$$x = c/2 + \sqrt{(c/2)^2 - b} \text{ and } y = c/2 - \sqrt{(c/2)^2 - b}.$$

Babylonian scribes also solved quadratic equations by completing the square.[162] Compound interest problems prompted work on higher degree equations. Scribes had ways of solving selected cubic (third degree) and at least one quartic (fourth degree) equation. When possible, they reduced cubic equations to the Babylonian normal form $(n^3 + n^2 = a)$. One Berlin Museum clay tablet that suggests this approach gives not only squares and cubes of the numbers $n = 1, 2, 3, \ldots, 20, 30, 40,$ and 50, but also the sums of $n^3 + n^2$.

Plimpton 322: Right Triangular Triples

With their arithmetical orientation, Babylonian scribal mathematicians wondered what combination of three positive numbers satisfies what is now known as the Pythagorean relationship. Today, numbers that do so are called Pythagorean triples. The now famous and controversial algebraic clay tablet with catalogue number 322 of the G. A. Plimpton Collection at Columbia University gives important results. In deciphering and analyzing this clay tablet, originally thought to be simply a "commercial account," Otto Neugebauer and Abraham Sachs would demonstrate in 1945 that Babylonian scribes obtained far-reaching insights into the Pythagorean relationship partly from a number theoretical search for Pythagorean triples. Initially that search was probably ancillary to an effort to measure land areas of squares on the sides of a right triangle.

Plimpton 322, which dates from the Old Babylonian Period, is the right-hand portion of what is thought to have been a substantially larger tablet. The extant portion, which measures $13 \times 9 \times 3$ centimeters, has fifteen horizontal rows of numbers arranged in four columns. Here are portions of the four columns with the sexagesimal number and, in brackets, the decimal equivalent:

Column I	Column II	Column III	Column IV
1,59,0,15	1,59 [119]	2,49 [169]	1
1,56,56,58, . . .	56,7 [3367]	3,12,1 [11521] (4825)	2
1,55,7,41, . . .	1,16,41 [4601]	1,50,49 [6649]	3
1,53,10,29, . . .	3,31,49 [12709]	5,9,1 [18541]	4
. . .	. . .	. . .	. . .
1,38,33,36, . . .	9,1 [541] (481)	12,49 [769]	9
. . .	. . .	. . .	. . .
1,33,45	45	1,15	11
. . .	. . .	. . .	. . .
1,23,13,46,49	56	53 (106)	15

Column IV simply lists the line numbers consecutively. The heading for Column II is the word *sag*, roughly "solving width," likely meaning computing the shorter leg. Column III is headed *siliptum*, "solving diagonal," referring probably to computing the hypotenuse of a right triangle. The square of the column III number minus the square of the column II number thus gives another perfect square, that is, III^2 − II^2 = I^2. In the first row

$$(169)^2 - (119)^2 = (120)^2.$$

Its root is the longer leg or *us*, which is translated as "flank."

Except for a few minor scribal errors, with corrections indicated in the table in parentheses, the middle two columns belong to sets of Pythagorean triples. In line 9 the scribal copyist wrongly transcribed the sexagesimal 9,1 or 541 instead of 8,1 or 481, while in line 15 the middle-right column contains 53 instead of 1,46, which is double 53.

How did Old Babylonian scribes derive this table with numbers c^2 (col. III) = a^2 (col. II) + b^2 (col. I), or $c^2 - a^2 = b^2$? How generalized was the procedure to obtain the values given in the table? The magnitude of some numbers indicates that this table was neither pure guesswork nor the result of trial and error. Consulting a "flank" column (b) missing from the facing page suggests an answer. The illuminating column is found elsewhere in Plimpton 322.

Line	b	Line	b	Line	b
1	2,0	6	6,0	11	1,0
2	57,36	7	45,0	12	40,0
. . .	. . .	. . .	. . .	. . .	. . .
5	1,12	10	1,48,0	15	1,30

Values in column I in the previous four-column table correspond very closely to triples derived from the squares of ratios (c/b) in right triangles, or what trigonometry now calls the secant squared. The values decrease regularly a degree at a time. Replacing the first comma in each number in column I with a semicolon thus reflects a decrease, in each row from top to bottom, of a degree in ratios that fall precisely between angles from 45° (row 1) to 31° (row 15). If the angle is 45°, then $a = b$ and $c^2 = 2b^2$, and $(c/b)^2 = 2$. On comparison, column I, row 1, on the facing page has the sexagesimal 1;59,0,15, which is almost 2 and exactly $c^2/b^2 = 28,561/14,400$. Moreover, c/a is about $17/12$ (in decimal numbers), a Babylonian approximation of $\sqrt{2}$, and with $a = 1,59$ (sexagesimal) and $b = 2,0$, $a/b = 1,59/2,0 =$ almost 1 or the diagonal of a square, hence a 45° angle. Plimpton 322 is thus a systematized collection of Pythagorean triples. Another table of Pythagorean triples based on decreases by degree with smaller angles probably existed.

Another fascinating feature of Plimpton 322 is that, except for lines 11 and 15, all Pythagorean multiples given in it are relatively prime. Primitive or relatively prime Pythagorean triplets have no common integral factor other than 1. For example, (3, 4, 5) and (5, 12, 13) are primitive Pythagorean triples, while (6, 8, 10) is not. Not until much later in Hellenistic times was it known that all primitive Pythagorean triples may be obtained in this way:

$$a = p^2 - q^2, \quad b = 2pq, \quad \text{and} \quad c = p^2 + q^2,$$

where p and q are relatively prime integers, not both odd, and $p > q$. The three equalities converge into $a^2 + b^2 = c^2$.

Presumably, Plimpton 322 compilers began with two regular sexagesimal integers p and q, both relatively prime and differing in parity, and $p > q$. By Neugebauer's definition, regular sexagesimal integers p and q have to be multiples of powers of 2, 3, and 5, the three primes in base 60 of the sexagesimal number system. In row 1, for example, $p = 2^2(3)$ and $q = 5$, and in row 2, $p = 2^6$ and $q = 3^3$.

Although we do not know how, Old Babylonian scribes may have worked with an equivalent of $(c/b)^2 - (a/b)^2 = 1$. If in modern notation $x = c/b$ and $y = a/b$, then $x^2 - y^2 = 1$, which can be factored as $(x+y)(x-y) = 1$. Modern mathematicians see that when x and y are rational and the product of the two quantities is equal to 1, these quantities are reciprocals. This may be written as

$$x + y = p/q \quad \text{and} \quad x - y = q/p.$$

Adding the two gives the sum $c/b = x = \frac{1}{2}(pq' + p'q)$, where p' and q' are the reciprocals of p and q. Plimpton 322 scribes employed this sum and $a/b = y = \frac{1}{2}(pq' - p'q)$ to generate the right triangular triples. Remarkably, numbers p and q here are not only regular numbers but nearly all regular numbers in "standard tables" of reciprocals.

Scribal reductions of results to primitive Pythagorean triples, except in the case of lines 11 and 15, support the view that the most fruitful results in Plimpton 322 lie in the origins of arithmetic. Even line 11 may have retained (60,45,75) as more common in base 60 than (4,3,5). While Plimpton 322 suggests a vague awareness of prime numbers, no evidence exists that Old Babylonian scribes recognized their significance. It was classical Greek geometers who made the first notable progress in the study of prime numbers.

Mixed Algebra and Geometry

Accomplished in arithmetic and rhetorical algebra, whose rectangular equations were based on geometry, the Sumerians, Akkadians, and Old Babylonians seem otherwise to have contributed little to geometry. Without the notion of proof, they did not have demonstrative geometry. Neither did they conceive of geometry as a separate discipline. Rather, they had a mixed algebra that was treated equally with other forms of numerical relations among physical objects. They readily converted into algebraic problems questions that had arisen from additions and divisions of agricultural field areas and bricks for construction, and they equated the results with numbers. Babylonian mixed algebra, it should be noted, differed from the later geometric algebra found in al-Khayyām. B. L. van der Waerden discerns in Book II of Euclid the embryo of geometric algebra, which he believes synthesizes Babylonian mixed or mensurational algebra with pre-Euclidean geometric traditions. But Sabetai Unguru denies that geometric algebra existed before the Christian era. Chapter 5 will address that debate.

While clay tablets, to which ancient Mesopotamian scribes were confined, lend themselves well to the abstract notation of arithmetic and algebra, they are not suited for producing consistently well-drawn geometric figures or long treatises. Some scholars have conjectured that since they developed town plans and built large buildings, Babylonian scribes probably drew many blueprints in the sand, using it much as we do chalkboards. Even if these drawings were made, their impermanence would keep them from having a long-range impact.

But the Babylonians had at least a rudimentary geometrical knowledge. They had rules to find areas of plane figures such as triangles, rectangles, trapezoids, and regular polygons, including squares, pentagons, hexagons, and heptagons; to find volumes of simple solids, among them the frustum of a square pyramid; and to compute approximate areas of circles. They utilized the concept of similarity of triangles and the proportionality of corresponding sides in similar triangles. Figures in a Baghdad tablet suggest that areas of four right triangles compare essentially as squares on corresponding sides. Babylonian scribes also found that the altitude of an isosceles triangle bisects the base. In examining the regular hexagon, they approximated $\sqrt{3}$ to be 1;45 (sexagesimal) or 1.75 (decimal). For computing the area of a circle, Mesopotamian scribes generally applied one of these rules: $a = 3r^2$, where $r =$ the radius, or $A = c^2/12$, where $c =$ the circumference. This gives $\pi = 3$ (the same rough approximation suggested in the Hebrew Bible: I Kings 7:23 and II Chronicles 4:2). Tablets unearthed at Susa in 1936 on ratios of perimeters of regular hexagons and heptagons to their circumscribing circles, however, show that the Babylonians had a better approximation of π as 3;7,30 or $3\frac{1}{8}$.

Neither of these values approaches their accuracy in arithmetical computations; nor are they as good as the best Egyptian approximations. The actual value of π is $3.141592\ldots$. Modern scholars often use the degree of precision by ancient and medieval peoples in the approximation of π to indicate their geometrical prowess. This measure may be necessary but is hardly sufficient to establish sophistication in geometry.

Motivation, Discontinuity, and Diffusion

From Paleolithic times environmental needs had been the chief stimulus to protomathematics. But ancient Mesopotamian problem texts cover too wide a range of problems to make it credible that they were undertaken solely in response to practical necessities. Some problem texts, moreover, pursue a higher mathematics that exceeds the practical. Others are simply puzzles and diversions that apparently served no other purpose than recreation. These particular texts reveal that at least a small group of scribal mathematicians within the school-and-bureaucracy component of society now responded to a degree of curiosity, challenge, a sense of beauty and harmony in number and form, and play—mathematics for its own sake. Among these creative scribal mathematicians, virtuosity in mathematics was a source of pride.[163] At times administrators in the school-and-bureaucracy complex apparently opposed this theoretical work. A continuous tradition of the pursuit of mathematics for its own sake was not to begin until the ancient Greek era after 600 B.C.

As noted earlier, table and problem texts disclose a major stagnation from about 1550 to around 300 B.C. No new arithmetic or algebraic techniques developed then. Historians and mathematicians have wondered why Mesopotamian mathematics halted new achievements just when it was flourishing.

Among the most likely causes of the stasis, political and economic factors were probably paramount. The necessary conditions for research in a land that was successively Sumer, Akkad, and Babylon had been a stable politics allowing a few to pursue persistent and accurate research, mainly in schools, and the increasingly complex practical mathematical needs of commerce, agriculture, and architecture leading to more sophisticated work in early mathematics. These soon disappeared. The turbulence that the more militaristic societies of the Hittites, Kassites, Assyrians, and their successors brought to the region after the Hammurabic dynasty was most likely not conducive to encouraging new research. These cultures evidently supported learning less than had their predecessors. The very success of Old Babylonian mathematics, effective for the practical problems of that society and perhaps those immediately following it, may also have obviated new research and pressures for innovation.

Their early mathematics, moreover, acquired a new status that deterred innovation. Initially, religion had acted as a patron of practical protoscientific studies and elementary mathematics. From Sumerian times a delicate balance existed between religion and this early learning. At the end of the Old Babylonian period, the balance shifted to the detriment of its nascent mathematics. Astronomy, which was essential to the ancient Mesopotamians for calendric reform and prompted extensive numerical calculations, apparently degenerated into astrology as a handmaid of religion. This assured the continued prestige of both, but gave no new creative impulse. To paraphrase a statement from the German mathematician Karl Weierstrass for a later mathematics, this germinal mathematics had advanced as the creative servant of society, protoscience, and religion, but would not do so if largely confined within religion.

The shifting from astronomy to astrology probably stemmed partly from the animistic and numerically oriented religion of the region. Ancient Mesopotamians associated their gods with natural phenomena and assigned numbers to each. The head of the Sumerian gods, the sky-god An, was represented by 60, the perfect number in Sumerian protomathematics. The Mesopotamian cosmos centered on seven vagabond stars, identified with the gods, whose nightly position shifted against the background of the fixed stars. They were the Moon (Nannar = 30), Sun (Utu = 20), and five planets: Nebo (Mercury), Ishtar (Venus), Nergal (Mars), Marduk (Jupiter), and Ninib (Saturn). (Our planets are still called by the names of gods, although not the Mesopotamian ones.)

Shortly after 1700 B.C., according to the *Gilgamesh Epic*,[164] Babylonian stargazers described a belt across the celestial equator. The mapping of this equatorial belt was originally a calendric device to partition the agricultural year. But during the troubles associated with the downfall of the Hammurabic dynasty, it came to be seen more as a theater for operations of the gods and, centuries later, as a model for judicial astrology and divining the fate of human beings (horoscope casting). This development thus precipitated a shifting from astronomy to astrology. Around 1000 B.C. Mesopotamian tablets relate a mass of omens. Two centuries later Assyrian astronomical tablets give primacy to them.[165] Neugebauer traces to the fourth century the division of the equatorial belt into twelve signs or stations of 30°-long sections on a great circle as a reference for tracking solar and planetary motion in relation to the stars. It was apparently devised for computing purposes. The stations, designated by their familiar (Greek) names, ranged from Aries the Ram, residing in the first mansion or month of the New Year, through Pisces the Fish, residing in the twelfth mansion when agriculture resumed. The Greeks later called this belt the *zodiac* from *Zodion* (small animal).

Political turbulence, a slackening in social pressure for new research, and the recasting of its nascent mathematics as a handmaid of religion arrested the advance of mathematics in Mesopotamia for over a thousand years. Never again was the vitality of the Old Babylonian period attained there, although a Seleucid efflorescence included a medial zero in numeration and extensive computation of numerical tables.

A final note on geography. While ancient Mesopotamia's urban civilization, centering on the Tigris-Euphrates River, contributed significantly to a new mathematics, little if anything emerged in nearby areas. Mesopotamia had many other peoples. With the invention of the scratch plow by 3000 B.C., plowing communities sprang up throughout this and adjacent regions near smaller rivers like the Jordan and Karun. The nearby hills had pastoral nomads, who herded larger animals—horses, cattle, and oxen. These nomads were inimical to the peaceful agricultural communities, which they raided occasionally. They were sometimes parasitic. Neither of these rural societies developed more than crude protomathematics. Only at a distance did a development comparable to Mesopotamia's take place—in another great river valley of the Fertile Crescent, the Nile, and ancient Egyptian civilization.

2.3   Ancient Egyptian Civilization: An Introduction

Ancient Egypt owed even more than Mesopotamia to geographical and climatic conditions. As the Greek historian Herodotus (ca. 484–ca. 425 B.C.) states, Egypt was the gift of the Nile. Life in ancient Egypt, with its almost rainless deserts, depended on it. Each year the Nile flooded from July through October, leaving fertile black mud behind when it receded. During the ensuing dry season, an elaborate system of canals and ditches drew the river's waters for irrigation. Black topsoil and irrigation permitted farmers in a good year to raise two or three times the food necessary for survival. This circumstance, combined with carefully organized planting and harvesting, produced in ancient Egypt an agricultural prosperity unmatched elsewhere in antiquity. Along the river, farmers raised grain and flax, maintained vineyards, and grazed cattle. The river, an abundant source of fish and waterfowl, also tempered the intense heat of Egypt and facilitated transportation and communication.

Before 3000 B.C., two kingdoms had emerged in Egypt. The Upper or Southern Kingdom occupied the nearly six-hundred-mile stretch of the Nile from the first cataract at Aswan to the Delta; the Lower or Northern Kingdom, the marshy delta of about 150 miles from Memphis to the Mediterranean Sea. Together these two parts of the Nile Valley are shaped like a long-necked funnel. On both sides limestone cliffs encase the upper river valley, which is never over fifteen miles wide. Beyond the cliffs and flanks of the Delta are vast and nearly impassable deserts of red land. These deserts isolated ancient Egypt, making it relatively immune from foreign invasion, and limiting trade and cross-cultural influences before the domestication of the camel, an animal fit for long-range travel over deserts. The Nile made possible the political unification of the Upper and Lower Kingdoms. The ancient Egyptians spoke of a great king—Menes, the Greeks would later call him—who was perhaps the Egyptian Narmer. Narmer brought both kingdoms under his rule around 3100 B.C. He founded the first ruling family or dynasty.

Thereafter, Egyptian civilization progressed rapidly. It benefitted from limited trade contacts with Sumer. Having an environment similar to that of the lower Tigris and Euphrates, it adapted much from the Sumerian experience, including large-scale irrigation systems, metallurgy, the scratch plow, limited wheel vehicles, writing, an administrative bureaucracy, protomathematics, and monumental sandstone buildings. Egyptian civilization, however, was distinct, and the major distinction lay in its ruler or pharaoh, a term meaning "great house" used after about 1350 B.C. In Sumer, kings were the "tenant farmers of the gods"; in Egypt, pharaoh was a living god, the human form of the falcon god Horus. All land belonged to him, and he was the source of all law. The building of the pyramids indicate the extent of his wealth and power. Ruler worship was the foundation of ancient Egyptian society.

Ancient Egyptian civilization lasted for over three thousand years. While the Egyptians divided their history by dynasties, modern historians divide its course into three great periods: the Old Kingdom, lasting from about 2700 to 2181 B.C., when its culture reached its apex in the building of its most impressive pyramids and its religion achieved classical form; the Middle Kingdom of the years between about 2052 and 1786 B.C., during which the capital was moved from Memphis to Thebes; and the New Kingdom or Empire from 1575 to 1087 B.C., when Egypt's power and foreign contacts increased and a renaissance brought realism to the arts. The intermediate periods between these were marked by internal and external disruptions. Between the Old and the Middle Kingdom provincial governors overthrew pharaonic despotism, while around 1640 B.C. mounted Hyksos ("Rulers of Foreign Lands") invaded Egypt, bringing new bronze weapons and Mediterranean culture.

As Herodotus reports, the ancient Egyptians were "excessively religious." They practiced a simple worship of many gods, some in animal form, while Amon-Re, in time the king of the gods, personified the sun and Osiris the Nile. For a moment at about 1350 B.C., pharaoh Akhnaton's monotheism prevailed, although he was the object of worship for his subjects. Ancient Egyptians were attentive to ethics and social justice and were profoundly concerned with life after death, as their *Book of the Dead* of the New Kingdom indicates. Their concern with the afterlife led them to devise mummification for preserving the body of the pharaoh (with some latter day extensions) and to make the tomb the most characteristic structure of their architecture.

By the time of the Old Kingdom the Egyptians had progressed from simple tombs to modest step pyramids to monumental pyramids. Herodotus maintains that four groups of a hundred thousand men labored three months each over twenty years to build the most famous of the eighty or so pyramids, the Great Pyramid of Khufu (in Greek, Cheops) at Giza. Built around 2644 B.C., it was composed of 2,300,000 stone blocks averaging two and a half tons each. Over 480 feet high and its base covering thirteen acres, it was the largest building of antiquity.

Perhaps the greatest innovation of ancient Egyptian scribes was developing their own complex and subtle system of writing. Before 3000 B.C. they devised the calligraphy the Greeks later called *hieroglyphics* ("sacred carvings"), a complicated pictographic system something like present-day cartoons, which were painted and chiseled on stone monuments. By 2500 B.C. they had developed the first of its two variants, the hieratic (sacred) script—flowing, cursive, and dependent on syllabic forms. In the first millennium B.C. they produced the demotic (popular) script, whose alphabetic characters could be written more rapidly and with greater facility.

Both the hieratic and the demotic characters were written with pen and ink on fine papyrus sheets, a material that Egyptians had devised as early as the First Dynasty; this freed them from dependence on chisel and stone. Our word "paper" perpetuates the memory of this invention as well as much of its technique. The relatively cheap papyrus sheets were the pressed pith of a Nile Delta reed sliced into strips. Although papyrus was a distinct improvement over clay tablets, it dried after a short time and crumbled. Consequently, few papyri have survived. Those that remain were preserved by the warm, rather dry climate of Egypt.

Thousands of hieroglyphics painted or carved on the walls of temples and tombs and some papyri comprise the documentary source of ancient Egyptian civilization and its proto- and early mathematics. In modern times, deciphering these long forgotten languages was a puzzle until the Rosetta Stone, discovered in 1799 by the Napoleonic Expeditionary Force near Alexandria, provided the key. This polished black basalt slab contains a message written in Greek, hieroglyphic, and Ptolemaic demotic. Still, the decipherment was difficult. Knowing Greek and assuming that the Rosetta Stone had the same message written in parallel forms was not enough. After recognizing also that special symbols on the Rosetta Stone referred to the rulers Ptolemy and Cleopatra, the French Egyptologist Jean François Champollion cracked the code of hieroglyphics in 1821. The grammar he compiled and his *Dictionnaire Egyptienne*, published in 1842, opened the way for new studies.

2.4   The Mathematical Record in Ancient Egypt

Pharaonic Documentary Sources

Documentary sources for pharaonic mathematics may be divided into two groups. Records of early counting and measuring remain on pertinent hieroglyphic engravings on temples, tombs, tombstones, and calendars. But the main sources of our present knowledge of later counting, measuring, and reckoning are a very few, perhaps a dozen, mathematical papyri. Chief among these are two hieratic papyri: the *Rhind*, or *A'hmosè* after the scribe who copied it, presently housed in the British Museum in London, and the *Golenischev* in the Museum of Fine Arts in Moscow. Each is named after a former owner. Henry Rhind, a Scottish archaeologist, purchased the Rhind papyrus in the Nile resort town of Luxor in 1858.

The Rhind Papyrus, a scroll almost 18 feet long and 13 inches wide, is the larger of these two. A'hmosè, who copied it about 1650 B.C., claimed it was based on an older document dating from two hundred years earlier.[166] Some of its materials may even date from the Old Kingdom and the building of the Great Pyramid.

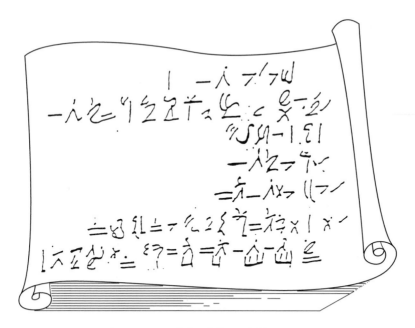

A segment of the A'hmosè Papyrus

A practical handbook containing eighty-seven problems and tables, the A'hmosè Papyrus is arranged not logically, but rather for convenient references by employers of labor, granary overseers, and brewers. The older Golenischev or Moscow Papyrus was copied less carefully by an unknown scribe about 1850 B.C. It contains twenty-five problems. Of these, eleven concern proper portions for making different strengths (*pesus*) of beer and bread, while three deal with the area of a triangle. As will be seen, Moscow problems 10 and 14 are significant in computing what seem to be a curvilinear surface and the volume of a truncated pyramid.

In addition to these two papyri are about a dozen fragments preserved in museums, universities, and other institutions that are germane to protomathematics. These include the Akhmin tablets, which were composed at about 2000 B.C.; a number of papyri approximately dated (the Berlin and the Kahun) at about 1850, the Reisner at 1800, and the Harris around 1164 B.C.; and various hieratic, demotic, Greek, and Coptic ostraca from 2000 B.C. to A.D. 600. Collectively, these papyri and ostraca suggest that there was an early creative period in ancient Egyptian protomathematics prior to roughly 1850 B.C., followed by perhaps one and a half millennia of stagnation.

Numeration, Arithmetic, and Rhetorical Algebra

The ancient Egyptians developed three decimal systems of numeration, which were associated with developments in their language. Each had its shortcomings: two are not positional; the third gives evidence of having only a primitive positional system that includes signs to distinguish tens (st'wt) from units (st't);[167] and none has a zero.

Like the Sumerians, the Egyptians began with a grouping system of numbers. Grouping systems have names for 1 and powers of their base b, for example, in the case of base 10, 10^1, 10^2, 10^3 .... Hieroglyphs for numerals included

1	10	100	1,000	10,000	100,000
Stroke	Loaf	Snake	Lotus Flower	Bent Finger	Burbot Fish (or Pollywog)

The Egyptians formed intermediate values by repeating each symbol the required number of times, thus **IIIIInnn** represents 35. Herodotus reports that the Egyptians customarily wrote numbers and calculated from right to left. Large numbers occurring in records of royal estates and the logisitics and engineering of the pyramids required much space.

	1	2	3	4	5	6	7	8	9
Units									
Tens									
Hundreds									
Thousands									
Tens of thousands									

Egyptian multiples of powers of 10

The later hieratic system, as in the A'hmosè Papyrus, and the demotic decimal system were ciphered. This means that Egyptian scribes gave names not only to 1 and powers of base 10, but also to multiples of those powers 20, 30, ..., and 200, 300, and so on. This represents an important development in numeration, because it eliminates the cumbersome repetition of symbols in forming numbers. It remains an integral part of modern numeration. In the A'hmosè Papyrus the number 1025 would be written (from left to right) as

$$\text{ろ } \hat{\lambda} \text{)}$$

The Harris Papyrus, an inventory of the receipts of the reign of Rameses III, demonstrates that the hieratic form could represent large numbers with receipts for 310,000 sacks of grain, 12,000 pounds of gold, 2,200,000 pounds of silver, and 5,200,000 pounds of copper.

To the ancient Egyptians, fractions were an enigma. Probably to avoid computational difficulties, they worked only with unit fractions, fractions with 1 as a numerator, with the sole exception of 2/3, which had the sign ⲣ in hieratic and ⲣ in hieroglyphic. Initially, ancient Egyptian scribes denoted most unit fractions by placing above the regular numeral the hieroglyph �, which is pronounced *ro* and means "part." Thus, in hieroglyphics

$$\underset{\text{III}}{\overset{\bigcirc}{\underset{\text{II}}{}}} = 1/5, \qquad \underset{\text{IIII}}{\overset{\bigcirc}{\underset{\text{IIII}}{}}} = 1/8 \qquad \overset{\bigcirc}{\underset{\cap}{}} = 1/10, \text{ and} \qquad \overset{\bigcirc}{\underset{\cap\cap}{}} = 1/20$$

In hieratics a dot replaced the oval. A few fractions, probably those occurring most often in daily life in counting and measuring, had special symbols: in hieratics:

$$\textbf{>} = 1/2, \qquad \textbf{✓} = 1/3, \qquad \text{and} \qquad \textbf{X} = 1/4.$$

The ancient Egyptians did not eschew the nonunit fractions that they encountered. Rather, they decomposed them into unit fractions. The $2/n$ table of the A'hmosè Papyrus discloses how they did this.[168] This conversion table reduces the division of 2 by odd numbers, from 3 to 101, to sums of unit fractions. This splitting method begins

n	$2 \times \bar{n}$, where $\bar{n} = 1/n$		
3	$\bar{2} + \bar{6}$	or	$2/3 = 1/2 + 1/6$
5	$\bar{3} + \overline{15}$	or	$2/5 = 1/3 + 1/15$
7	$\bar{4} + \overline{28}$	or	$2/7 = 1/4 + 1/28$
9	$\bar{6} + \overline{18}$	or	$2/9 = 1/5 + 1/18$
..	......		
61	$\overline{40} + \overline{244} + \overline{488} + \overline{610}$	or	$2/61 = 1/40 + 1/244 +$
			$1/488 = 1/610$
65	$\overline{39} + \overline{195}$	or	$2/65 = 1/39 + 1/195$

In A'hmosè's table the plus sign does not appear, but is understood. Nor does the modern fractional representation occur. In order to obtain these decompositions, A'hmosè generally utilizes the verbal equivalent of what is today called the splitting identity:

$$2/n = 1/[(n+1)/2] + 1/[n(n+1)/2].$$

If $n = 7$, for example, then

$$2/7 = 1/8/2 + 1/7(8)/2 = 1/4 + 1/28 = \overline{4} + \overline{28}.$$

Showing a preference for $2/3$, for all fractions of the form $(2/3)p$, A'hmosè employs another general rule: $(2/3)p = (1/2)p + (1/6)p$. The case where $p = 5$ gives the result

$$\frac{2}{15} = \frac{2}{3} \cdot \frac{1}{5} = \frac{1}{2 \times 5} + \frac{1}{6 \times 5} = \frac{1}{10} + \frac{1}{30} = \overline{10} + \overline{30}.$$

This $2/n$ table was quite important in Egyptian reckoning, in part because of the dyadic nature of the multiplication, which is described below. It also reveals skillful scribes seeking to simplify and make a more uniform table. To achieve these goals, scribes adopted additional organizing rules. They preferred even denominators, smaller denominators not greater than 1000, simpler decompositions having no more than four terms, smaller denominators to appear first, and a larger initial denominator only if this reduced the remaining one in a decomposition.

Containing the results of adept work with the four arithmetical operations of addition, subtraction, multiplication, and division, mathematical papyri offer no clues as to how scribes and clerks added and subtracted. Historians assume that ordinary additions merely involved accumulating symbols, while subtraction contracted them by removing from any subset those symbols that exceeded the desired limit. Since few scribal errors exist in addition or subtraction, some scholars believe that tables were used for many of these. No such tables have survived.

Ancient Egyptian arithmetic was basically additive: multiplication and division were reduced to a summing process. Clerks and scribes multiplied by a combination of "duplation," successive doubling of the larger number in a problem, and a parallel "mediation," successive halving of the smaller number. Addition of the multiples on odd number lines gives the desired product, a dyadic technique that digital computers use today to multiply. Consider 22×9 in hieroglyphic notation both by duplation and mediation, given in brackets, or simply by duplation.

❙❙∩∩	[9]*	1	22 taken once
❙❙❙❙∩∩∩∩ ❙❙❙❙∩∩∩∩	[4]	2	$22 \times 2 = 44$
❙❙❙❙∩∩∩∩ ❙❙❙❙∩∩∩∩	[2]	4	$44 \times 2 = 88$
❙❙❙∩∩∩∩∩ ❙❙❙∩∩∩	[1]*	8	$88 \times 2 = 176$
❙❙❙❙∩∩∩∩∩ ❙❙❙❙∩∩∩∩∩	[9]* + [1]*		
		$1 + 8 = 9$	$22 + 176 = 198$

Ancient Egyptian scribes and merchants took duplation for granted and did not question its validity. Today we know why this procedure works: every positive integer can be expressed as a sum of powers of 2, including $2^0 = 1$.

For division the Egyptians again used duplation, which had the pedagogical advantage of not introducing a new mathematical operation. In modern numerals the ancient Egyptians divided 219 by 9 in this way:

$$
\begin{array}{cc}
1 & 9 \\
2 & 18 \\
4 & 36 \\
8 & 72 \\
16 & 144 \\
32 & 288
\end{array}
$$

The scribe then simply selected the doubled numbers in the right-hand column whose sum equaled or closely approximated the dividend. The sum of the parallel numbers in the left-hand column plus any remainder expressed as a unit fraction provides the answer. The asterisks indicate the components.

$$
\begin{array}{ccl}
8 & 72 & \\
16 & 144 & \\
24 & 216 & \text{with a remainder of 3.}
\end{array}
$$

Since $3/9 = 1/3$, the answer is $24 + 1/3 = 24\frac{1}{3}$.

The habit of performing multiplication and division by repeated doubling accompanied an interest in numbers arranged serially, especially 1, 2, 4, 8, 16, ..., and other geometric and arithmetic series. By inductive reasoning, scribes made basic discoveries about the sums of selected terms. For geometric series, where the first term is the common multiplier, the sum of n terms equals the common multiplier times the sum of the previous terms plus 1. Thus, in the series 7, 49, 343, 2,401, ... the sum of the first three terms is $7(7+49+1) = 7(57) = 399$. Problem 79 of the A'hmosè Papyrus sums the first five terms of this series, finding $7 \times (2,801) = 19,607$. This problem is fancifully considered the antecedent to the famous Mother Goose rhyme that begins:

> As I was going to St. Ives,
> I met a man with seven wives,
> Each wife had seven sacks,
> Each sack had seven cats...

Problem 40 reflects results of trial and error work in summing the third through the fifth term of an arithmetic series. For the arithmetic series 1, $6\frac{1}{2}$, 12, $17\frac{1}{2}$, 23 with a common addition of $5\frac{1}{2}$, A'hmosè finds that $12 + 17\frac{1}{2} + 23 = 7\left(1 + 6\frac{1}{2}\right)$, that is, a multiple times the sum of the first two terms.

Although ancient Egyptian duplation and notation precluded sophisticated calculations by hand with any degree of complexity, multiplication and division with them were not so clumsy as some authors maintain. Egyptian computational methods met well existing practical needs, for example, distributing bread, beer and pay in given quotas. Easily comprehended and mastered, they were in use by scribes and clerks alike. For these reasons they had wide appeal among later civilizations. Egyptian methods were passed on to the Greeks, Romans, and even Byzantines, each of whom also lacked place-value notation.

Ancient Egyptian scribes also dealt with problems now classified as algebraic. They solved problems comparable to modern linear equations of the form $x + ax = b$ or $x + ax + bx = c$, where a, b, and c are knowns and x is unknown. Practical needs from daily life prompted such problems as preparing bread and beer of given strength from cereals of differing quality, mixing feed for cattle and fowl, dividing estates left as an inheritance, and distributing salaries for workers and temple personnel. Scribes also considered the next higher type of equations, quadratics with squares of unknowns. They could solve the simplest form, $ax^2 = b$, as well as those quadratic equations in two unknowns, for example, $x^2 + y^2 = c$ (constant), in which one unknown can be eliminated to reduce it to the simplest form. The Berlin Papyrus solves in verbal form these simultaneous equations:

$$x^2 + y^2 = 100 \qquad x^2 + y^2 = 400$$
$$4x - 3y = 0 \qquad 4x - 3y = 0$$

Confined to primitive reckoning techniques and limited numerical notation and symbolism, Egyptian scribes could not handle quadratic equations with two or more irreducible unknowns and higher order equations.

Problems 24 through 27 of the A'hmosè Papyrus give the oldest and most universal Egyptian method for solving linear equations with one unknown. They find by the method of false position or false assumption the unknown quantity h or heap, pronounced as "aha." Problem 26 is typical. It states that "a quantity whose fourth part is added to it becomes 15. What is the quantity?" In modern notation, this means

$$x + x/4 = 15.$$

According to the papyrus editor T. E. Peet, A'hmosè first assumed that $x = 4$ to make the quarter value equal 1. But this gives $x + x/4 = 5$, which is not the desired answer. To obtain 15, multiply $5 \times 1 = 5$, $5 \times 2 = 10$, $5 \times 3 = 15$. This means that the correct answer is $3 \times 4 = 12$. J. Tropfke's *History of Elementary Mathematics*, published in 1980, assumes that A'hmosè first divided the unknown into four equal parts, making five of those parts equal 15. Thus, every part must be 3. That the scribe multiplies 4×3 and not 3×4 supports this interpretation. A'hmosè checked this result by a verbal process that amounts to this modern form:

$$
\begin{array}{rcccl}
x & = & 4 \times 3 & = & 12 \\
x/4 & = & 12/4 & = & 3 \\
\hline
x + x/4 & = & 12 + 3 & = & 15
\end{array}
$$

The rule of false position was adopted and became popular again in the Latin West during the Middle Ages.

Problems 24 through 27 suggest that Egyptian mathematics was neither purely practical nor based entirely on trial-and-error procedures, as some scholars today hold. Each of these problems sought a general numeric solution, not associated with pay or loaves of bread or preparation of beer. The method of false position was possibly the result of trial and error, but the scribe chose the simplest possible integral answer with his first guess, that is 7 in problem 24 and 2, 4, and 5 based on the fractions replacing 1/7 in the next three problems. Finally, the scribe's check of his answer suggests a groping toward the notion of mathematical proof—a logical, step by step argument for accepting one position over another without ambiguity..

Problems 28 and 29 give a glimpse into the play element in Egyptian mathematics. These are perhaps the earliest "think of a number" problems, not unlike those found two millennia later in Diophantus's *Arithmetica*. Scribes challenged one another to find a number, such that after performing a number of operations on it the player obtains the original number by "magic." Here is a rough translation of problem 28: "Think of a number and add two-thirds of it. From the new total subtract one-third. Suppose the result is 10. Then subtract 1/10 of 10 giving 9, and 9 was the first number." These steps in modern notation are $9 + 6 = 15$, $15 - 5 = 10$, and $10 - 1 = 9$. If the result in step 2 is 60 rather than 10, the number 54 is the solution.

Modern symbolism simplifies solving algebraic problems of the ancient Egyptians, which makes it easy to underestimate their capacities and achievements. Like all other ancient peoples, they had a verbal algebra with few symbols. This is a great handicap. A'hmosè advanced algebraic symbolism slightly by representing addition and subtraction by the hieroglyphs $\wedge$ and $\wedge$, the legs of a person coming and going (the Egyptians wrote from right to left), and he denoted square root by $\urcorner$. But even these few symbols were not commonly accepted, and scribes had no standard terms for addition, subtraction, multiplication, division, and so on. A'hmosè himself denoted these by different words, such as might translate into the English as "count" and "reckon." Such variable terminology was confusing. Ancient Egyptian scribes and merchant clerks, nevertheless, proved pertinacious and uncommonly skillful with the awkward computational methods and notation available to them.

Pragmatic Geometry

Both the material culture and the documentary record disclose substantial Egyptian contributions to geometry, the only area of mathematics in which, so far as our evidence indicates, they appear to have exceeded the Mesopotamians. Ancient Greek historians and philosophers judged the Egyptian work highly. Herodotus traces the origins of geometry to the yearly surveys made for tax purposes after the Nile flood obliterated farm boundaries. Indeed, the

Greeks used this account as the basis for the word geometry (*geo* = earth and *metria* = measure). In the *Phaedrus,* Plato relates that the Egyptian god Toth invented geometry; Aristotle maintains that Egyptian priest-scribes did so. The specialization of labor in ancient Egypt, by Aristotle's accounting, gave priest-scribes time for amusement and pure speculation, which in turn fostered intellectual curiosity. From their mental exercises, together with their utilitarian agricultural and building concerns, they invented geometry.

The classical Greeks, then a bold and eager people less tradition bound than the Egyptians, seeking and capable of assimilating older knowledge and techniques, were overly generous in these appraisals. The beginnings of geometry cannot be limited to one place. The Greeks themselves invented demonstrative geometry, or what is known today as geometry. Demonstrative geometry includes an axiomatic method with proofs and theory. The ancient Egyptians had none of these. They rarely enunciated general rules in their papyri. Their contributions, therefore, lie not in demonstrative geometry, but in the preliminary and less abstract forms of descriptive and pragmatic geometry developed by scribes and *harpedonaptai* (chord stretchers or surveyors), geometry as a practical tool, with most of its results based empirically on trial and error rather than theory. To this may be added a refined descriptive geometry. Patterns of triangles and alternate squares or checkers appear in the textiles, mosaics, basketry, pottery, jewelry, and bas-reliefs of the Egyptians. In their largely geometric decorative art, intersecting circles as well as squares and triangles inscribed in circles were popular. Craftsmen, artists, and architects, who produced these designs, fully appreciated the place of form in decoration but only minimally in protomathematics.

As in Mesopotamia, pragmatic geometry was not an independent field of mathematical research, but a branch of mensuration algebra. *Harpedonaptai* depicted in hieroglyphics tied knots or made marks on ropes at equal intervals and used these tape measures to make systematic, official annual surveys for taxation purposes. The twenty-six problems that may be classified as geometric in the A'hmosè and Golenischev Papyri demonstrate this. To solve mensurational problems, most of them involving land areas and volumes of rectangular and cylindrical granaries, scribes relied on arithmetical and algebraic rather than formally geometric techniques. Thus, scribes devised prescriptions to determine areas of triangles, rectangles, and trapezoids, along with volumes of cubes, boxes, cylinders, and other figures. Egyptian scribes dealt equally well with rectilinear and curvilinear figures.

Problem 50 of the A'hmosè Papyrus with its method for computing the area of a circle represents the highest quality in Egyptian two-dimensional geometry. It calculates that area with the equivalent of the formula $A = (8d/9)^2$, where d = diameter. While the Egyptians did not recognize the invariance of the ratio of the circumference of a circle to its radius, or what today is denoted as π, a student can extract that value from Egyptian data. When compared with the modern formula $A = \pi r^2 = \pi d^2/4$, the Egyptian formula gives $\pi = 3.1605$ or approximately 3 1/6, a value more accurate than the Mesopotamian $\pi = 3$, but not so accurate as the Mesopotamian value of 3 1/8 or the 3 1/7 often used today. Here is the computation of the ancient Egyptian value and the variance of it, as well as three close fractional approximations from the modern figure:

	Approximation to π	Error
$A = 64d^2/81 = \pi d^2/4$	$3\frac{1}{8}$	(−)0.01659
or $64(4)/81 = \pi$	$3\frac{1}{7}$	0.00126
$3.1605 = \pi$	$3\frac{1}{6}$	0.0251
	3.1605	0.0189

Without knowing that π is an invariant ratio, how did Egyptian scribes compute the area of a circle so exactly? Kurt Vogel proposes that problem 48 of the A'hmosè Papyrus in its rough diagram of what appears to be an octagon inscribed in a square suggests an answer. The octagon, whose area is calculable, approaches the area of a circle inscribed in the same square. According to the translation of R. A. Parker, problem 32 of the Cairo Papyrus written in the first century B.C. sought to turn a square parcel of land into a circle. The Cairo attempt at circling of the square is the inverse of the later Greek efforts at squaring of the circle. In these two problems, the Egyptians thus approximated the circle with a straightedge figure, an approach not unlike the later Greek method of exhaustion, which presaged the theory of limits.

The inscribed octagon was drawn by constructing a square of 9 units, trisecting the sides, and deleting the four corner isosceles triangles formed by connecting the trisection points nearest the corners. The area of the square is 9^2 and each corner triangle is $3^2/2$. Thus the area of the octagon is:

$$A = 9^2 - 4(9/2) = 81 - 18 = 63.$$

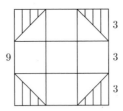

Octagon inscribed in a square.

The ratio 63/81 is nearly 64/81 or $(8/9)^2$, which appears in the ancient Egyptian rule for computing the area of a circle. Certainly, Egyptian scribes knew their method was not exact. Still, this method makes A'hmosè the first known circle-squarer.

Early Egyptian scribes inventively examined triangles, primarily for architectural and surveying reasons. A general misconception holds that they likely knew what mathematicians today call the Pythagorean theorem relationship. No extant documentary evidence from before the end of the New Kingdom confirms that assumption.[169] Still, scribes and surveyors achieved much in their study of triangles. Most modern historians, Dirk Struik and Carl Boyer included, believe problem 51 of the A'hmosè Papyrus calculates the area of an isosceles triangle exactly by a verbal form of the modern rule: half of the base (*teper*) times the altitude (*meret*); in modern symbols $A = bh/2$. Problem 52 calls into question the minority opinion that *meret* simply means side. The rule it presents for computing the area requires considering the isosceles triangle to be a composite of the equivalent of two right triangles, such as $\triangle ABD$ and $\triangle CBD$ in this drawing. Here are the vague beginnings of the notion of congruence, the quality of coinciding.

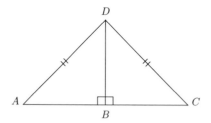

The notion of similar triangles appears in problem 56. A'hmosè seeks a uniform *seqt*, resembling the ratio of "run" to "rise" indicating steepness of downward slope in pyramids. Although the measuring units differ, it corresponds to what architects today call "batter." In a pyramid, first divide half the base by the height. In this illustration, half the base, AH, is 200 cubits and its height, CH, 350 cubits. This gives $200/350 = 4/7 = 1/2 + 1/14$. Next, multiply the result by 7 to obtain the number of hands per cubit. In this example, the *seqt* is $7 \times 4/7 = 4$ hands per cubit. That of the Great Pyramid, whose height is 280 cubits and width 440 cubits, is $220/280 \times 7 = 5\frac{1}{2}$ hands per cubit. A'hmosè always expresses the *seqt* as a length; he thus fails to recognize that the ratios simply as quotients of numerical quantities constitute a valid measure. Concrete geometrical thinking remained dominant.

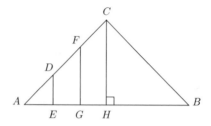

In building the Great Pyramid, Old Kingdom architects and engineers applied this concrete knowledge with amazing exactness. Casting blocks averaging two and a half tons each were fitted with a joint one-fiftieth of an inch; the margin of error in the near perfect square base with sides of mean length 755.8 feet was 0.09 percent on the north and south sides and 0.03 percent on the east and west sides, or roughly $4\frac{1}{2}$ inches per side. This precision, worthy of a jeweler, the ancient Egyptians did not again achieve.[170]

Bizarre mathematical speculations and mysticism are attached to the Great Pyramid that mainstream Egyptological studies do not confirm. One such claim holds that the Great Pyramid was built so that twice the side of the base divided by the height was equal to π or $3.141592\ldots$. Although $2\,(755.78)/481.2 = 3.1412\ldots$ is close to π, this appears accidental. Likewise questionable are beliefs that the Great Pyramid builders knew the regular pentagon and designed the pyramid to give ratios of height, slope, and base that are the important golden ratio, that is, $\frac{1}{2}(\sqrt{5}-1)$ or the ratio of the radius of a circle to the side of an inscribed decagon.

Among the twenty-five problems in the Golenischev Papyrus, problem 14 is ostensibly the masterwork of ancient Egyptian geometry. It contains the equivalent of the correct formula for the volume of a truncated square pyramid or frustum, that is $V = \frac{h}{3}(a^2 + ab + b^2)$, where $h =$ altitude and a and b the sides of the square base and top. Figures for problem 14 are altitude $= 6$, base side $= 4$, and top side $= 2$. The volume is 56. A poor drawing that resembles an isosceles trapezoid accompanies a computational form of the formula, which could not have been derived purely empirically.

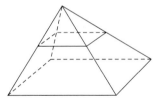

Truncated square pyramids, which are shaped like cornerstones with sloping sides, were granary silos. Stimulated by a ritual significance, continued on the verso side of the U.S. dollar bill, requiring great accuracy in constructing the early pyramids, the Egyptian scribal mathematicians had found at least one specific case of the correct volume formula. It is not known how.

Early hieratic Egyptian prescriptions for simpler rectilinear areas were generally correct or sufficiently accurate to meet the needs of daily life, but prescriptions from later times were less exact. Practical needs dictating these prescriptions may account for the degree of accuracy. Scribal mathematicians devised them to estimate rents and taxes on fields in a system requiring bailiffs and tax collectors to have only a general idea of revenues. Absolute accuracy, therefore, was unnecessary.

Field deeds listed in the dedicatory inscription of the Horus temple in Edfu, which date at about 100 B.C., contain representative later prescriptions. These address numerous four-sided fields. By denoting the sides a, b, c, d, where a and c and b and d are opposite sides, deeds give the field areas as $(a+c)/2 \times (b+d)/2$ or $\frac{1}{4}(a+c) \times (b+d)$. For triangular fields there is no d, and the rule becomes $(a+b)/2 \times c/2$. These later rules are exact for rectangles and squares, but only crude approximations for the more common triangles and trapezoids.

Astronomy and Time

Egyptian like Mesopotamian astronomy extensively utilized protomathematics. Astronomical knowledge was essential in ancient Egypt for calendar keeping that could predict the annual flood and foretell the seasons as a guide to agricultural operations, as well as for making astrological predictions. No astronomical papyri survive, but funerary papyri provide a glimpse into this work. Priests were the pharaonic astronomers. They carefully observed, recorded, and calculated movements of the moon, sun, planets, and stars, including heliacal risings (their first visibility shortly before dawn). Founded in awkward arithmetic and fractions, Egyptian scribal calculations were crude. Far inferior to Mesopotamian computations, these calculations subsequently disappeared. They contributed practically nothing to the development of mathematical astronomy by the Greeks and their successors.

The calendar was the chief achievement of the Egyptians in astronomy. By 2776 B.C. they had invented the 365-day civil calendar for the purpose of official record keeping. It applied little to daily life.[171] Lunar months along with the rise and fall of the Nile regulated agricultural activities and the lives of the masses. Since the Nile year was slightly erratic, sometimes exceeding the solar year and at other times falling short of it, and the lunar year did not match the solar, the state sought a more precise way of keeping its records. Through the centuries scribes recorded the number of days between each high Nile, determined the average, 365 days, and established a canonical duration. They pegged the beginning of the year on the heliacal rising in July of the Dog Star *Sothis* (in Latin, *Sirius*), the brightest star. This shortly preceded the rising of the flood waters of the Nile. Taken from the Greek, the term heliacal rising refers to the phenomenon of a star seen rising over the horizon at the start of dawn and rapidly disappearing with the onset of daylight.

By the Middle Kingdom the Egyptians divided the 365-day civil calendar into twelve months of thirty days each and five year-end (epagomenal) days dedicated to the birthdays of their chief gods. This calendar appears as an attempt to reconcile a variant of older lunar calendars, regulating festivals according to lunar phases, with the solar year. The twelve-month year consisted of three seasons of four months each, with an inundation and sowing, a growing, and a harvesting season. This division suggests its agricultural significance and mainly nonastronomical character. Perhaps because of sacerdotal opposition, the Egyptians failed to intercalate an extra day into their civil calendar every four years, even though they later knew the $365\frac{1}{4}$ day year to be more accurate. The civil calendar thus slipped backward by one day every four years and by thirty days over 120 years, thereby losing its relationship to the seasons. Since the civil calendar was only for official purposes and did not change noticeably in a human lifetime, no significant pressure existed to correct it. The civil calendar set itself right in $365 \times 4 = 1,460$ years, an interval known by Egyptian astronomers as the Sothic cycle. The Egyptian civil calendar served as the cornerstone for the Julian and Gregorian calendars.

Another major Egyptian contribution to astronomy was the division of the day into selected units. Astronomers began with a system of what the Greeks later termed *decans*. When Sirius rose heliacally, there were twelve *decans* at night based on other star risings, but in the wintertime there were more. These *decans* became "hours" or "watches," whose length varied slightly with the seasons. Sundials came to measure day watches, and water clocks (clepsydras) night watches. Not until Hellenistic times were seasonal hours replaced by equinoctial hours of constant length. Our modern division of the day into twenty-four hours of sixty minutes thus began to arise in ancient Egypt and was modified by Hellenistic astronomers.

3

Beginnings of Theoretical Mathematics in Pre-Socratic Greece

Geometria est archtypus pulchritudinis mundi.
(Geometry is the archetype of the beauty of the world.)
— Johannes Kepler

The Greeks were the next people in the Mediterranean West to leave a record of notable advances in mathematics. From about 600 to 400 B.C., they began to transform a collection of intuitive and almost entirely empirical rules inherited from the Near East into an orderly, theoretical science. Geometry, central to this breakthrough, progressed from crudely intuitive insights toward formal valid proofs consisting of a logical step-by-step argument leading to a single unambiguous conclusion. The evolving theoretical geometry with its illustrative diagrams was characteristic of a people who, as their pottery and city designs show, appreciated order and saw form and clarity in nature. Theoretical geometry's insistence on rigor and exactness led pre-Socratic Greeks to intellectual expansion, not constriction. In describing numbers as geometric forms and discovering incommensurable magnitudes, the ancient Greeks also pioneered the formation of a geometric theory of numbers.

3.1 Ancient Greece from 1200 to 600 B.C.: An Introduction

Historians generally agree that during their classical period from about 479 to 338 B.C. the Greeks originated a major share of the ideas, institutions, and values essential to what we call western civilization. To the present, Greco-Roman learning and institutions and Judaeo-Christian religious traditions are its pillars. An appreciation of the nature and extent of the classical Greek achievement before Socrates, whether in culture or mathematics, requires some knowledge of the prevailing Greek ethos, the transmission lines of earlier knowledge, and the social and economic framework of their scholarship.

After establishing monumental cities on Crete and the Balkan mainland, the ancient Greeks had what is known as their dark age, a period lasting from roughly 1200 to 750 B.C. Few artifacts or documents survive from that time: hence the darkness. The dark age began with the disintegration of Mycenaean culture as a result of internal battles among Mycenaean kings, the fragility of an overspecialized agricultural base, and disruption of commerce caused by the near collapse of the Egyptian empire. After the Mycenaen collapse, bands of less civilized Dorians, a Greek-speaking tribe from the north, entered the Aegean region at the time iron was gradually replacing bronze in the eastern Mediterranean after 1200 B.C. Armed with iron weapons, the Dorians settled the Peloponnesus, Crete, and southwest Asia Minor. They founded the city of Sparta.

During the years that followed, warfare was endemic and a subsistence economy replaced trade. Whatever writing there was has disappeared. For self-preservation, many ancient Greeks—or Hellenes—migrated eastward. Some emigrants settled in Attica, but most continued eastward to the islands of the Aegean Sea and to northwest Asia Minor. These dispersed people with a distinctive dialect became known as Ionians. Thus not all the Hellenic people were located in what is today Greece. A "very migratory people," to use Herodotus' words from his *History* (I, 56), they settled on the coast of the Greek peninsula and Asia Minor "like frogs around a pond." By the eleventh century B.C., the pond and cradle of their civilization was the Aegean Sea. The Aegean, generally calm except in winter, served as a bridge rather than a barrier to travel and communication. Amid its dense scattering of islands, seamen were never more than forty miles from some shore. The ancient Greeks thus developed into a seafaring as well as an agricultural people.

The Greeks during their hard times managed important developments. They assimilated numerous gods and modified beliefs to form their pantheon of twelve gods on Mount Olympus. These included the sky-god Zeus, father of the others, and Apollo the sun god. Therein the Greeks exercised one of their primary characteristics, a facility for cultural borrowing and improving on what they had borrowed. The *polis*, a self-governing state, burgeoned as the major form of political organization. It was constituted for mutual defense and as a source of prosperity and honor among its members.[172] The Greek *polis* (whence our word politics) had explicit and formal ideas of law and justice, as well as a remarkable social life distinguished by conversation and open argument. Together these properties of the *polis* have been enduring legacies to western civilization. Around 750 B.C., Athens, which had escaped Dorian incursions, led a cultural revival that ended the dark age.

In the period that historians term its archaic age, from 750 to 479 B.C., Greek civilization quickly revived and spread. As Athens, Sparta, and Corinth rose to prominence among the Greek *poleis*, the Greeks undertook a second colonial expansion, this time across the whole of the Mediterranean and Black Seas. Overpopulation and the search for rich, new agricultural soil may have prompted the emigration. It began with the founding of a Chalcidian colony on the bay of Naples in 750 B.C., probably a better starting date for documented

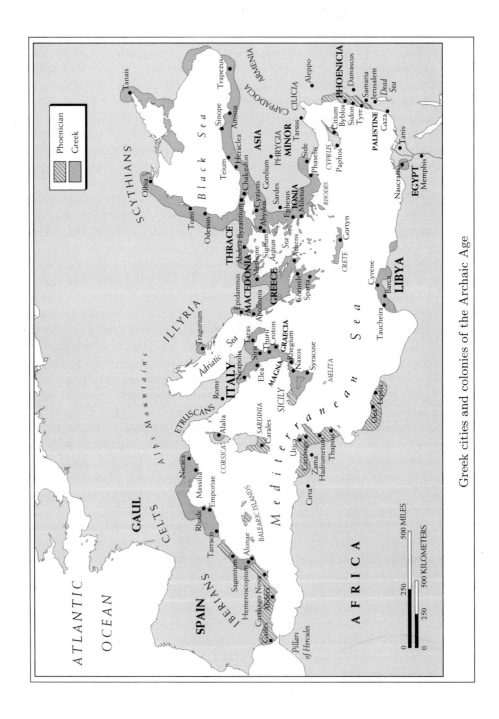

Greek cities and colonies of the Archaic Age

Greek history than the traditional 776 B.C., the date of the first Olympic victor lists. During the next two centuries, the Greeks planted hundreds of colonies from Asia Minor in the east through southern Italy and Sicily (*Magna Graecia* or "Great Greece") to Marseilles and Spain in the west. The second expansion made the Greeks a Mediterranean, not simply an Aegean, people, and the Mediterranean became the center of the ancient western world.

Early in the archaic period, a cultural blossoming occurred among the Greeks. It probably centered in Ionia on the west coast fringes of Asia Minor. Introduction of the Phoenician alphabet and its sharp transformation, which required letters to stand for certain sounds and added vowels, as well as the availability of papyrus scrolls by the mid-eighth century B.C. surely helped to stimulate it. The cultural flowering had its most important origin and expression in the appearance of the *Iliad* and the *Odyssey*, two epic poems, whose examination of motives and expression of daring and honor together with love, suffering, and endurance molded subsequent Greek mentality. Also notable are two surviving poems by the Boeotian author Hesiod, *Work and Days* and *Theogony*.

Little is known about the sources or author of the *Iliad* and the *Odyssey*. Ultimately, the sources of these poems, dependent on memory for transmission, derived from an oral poetic tradition dating from the Mycenaean age. At athletic and musical festivals, *aoidos*, professional minstrels with a lyre, chanted poems, which were artfully reworked compilations of stock phrases and themes. Homer's reputed blindness is less an obstacle when text did not have to be read. Although the epics later came to be written in canonical form and bore the name of only one author, "Homeros" (Homer or Hostage), some scholars believe that two authors created the canonical texts and that the author of the *Odyssey* was a woman. The Hellenes referred to the author simply as "the poet," a sign of his high standing. Most likely Homer resided in Ionia, or perhaps Chios, about 750 B.C.

The *Iliad* and *Odyssey* were crucial in forming the Hellenic view of life. In Hellas, which was not a single country, but a union of languages and religions, these epics helped give the citizens of the different, often warring *poleis* a sense of collective identity. For the ancient Greeks, who had no individual sacred text, they became a Bible. They introduced alphabetic writing to the Greeks. This factor and their aristocratic code of values, together with poetic genius and Greek love of the past, assured them a place as the staple of Greek literary education for centuries.

The *Iliad* describes the siege of the city we know as Troy, apparently in the twelfth century B.C. The shorter, less majestic *Odyssey* recounts Odysseus' wanderings on his way home to Ithaca and his spouse Penelope after the war. Both epics depict an established order in which the gods and men are "of one race," to quote the Greek poet Pindar. The gods, often bickering, lusting, and erring, do not dwarf the heroes, among them Achilles, Ajax, Hector, and Odysseus. But the gods are immortal, as well as capricious. "As flies to wanton boys, so are we to the gods." The mortal characters strive for *arête*, human excellence embracing honor and courage. The Greek has no equivalent for *machismo*, but the term fits their warriors. They employ their own abilities and resourcefulness to strive against privations and learn more about the world. Odysseus, who symbolizes man's questing spirit, pits initiative and ingenuity against lethal dilemmas and the will of the deities. Such was the standard that classical civilization set from the beginning.

Unlike Homer, who recounted heroic deeds of the Mycenaean Age four centuries past, Hesiod roots his writings in his own time. For this reason, Hesiod has been called the earliest Greek poet of the archaic age. Living on the western side of the Aegean (Homer was from the east) in the area of central Greece later known as Boeotia, Hesiod was a small farmer whose poems exalt hard work as the only way to prosperity and distinction and criticize the *basileis*, or royalty, along with the noble *aristoi* in the *poleis*. Hesiod's *Work and Days* describes life on a small farm, gives homely precepts for farming, and contrasts the sensible life of the husbandman with the dangers that merchants face in seagoing travels. In Hesiod's best known work, the *Theogony* or *A History of the Birth of the Gods*, placed at about 700 B.C., the mythic perception of the origin and operation of the world culminates among the ancient Greeks. In its hymn of praise to Zeus, Hesiod labors to organize earlier diverse myths into a coherent whole and give them some relevance to the human condition.

Though exalting the honest toil of subsistence farmers, Hesiod's words do not mask their travails and anxieties. One myth from *Work and Days* assigns human woe to feminine frailty. It is the tale of Pandora, who out of curiosity opens a forbidden jar, letting loose many evils on humanity. The *Theogony* is gloomy, its story set amid overpowering winter storms and the parching heat of summer.

The seventh century B.C. did not produce another Homer or Hesiod. Very few centuries have. It was an era of preparation rather than of departures in politics and culture. Still, in politics the gulf that separated increasingly wealthy aristocratic clans from merchants, artisans, and farmers widened. All social orders though now looked to *eunomia*, good social ordering and respect for law, rather than individual interests. The theory of a democratic state, whose citizens have unquestionable rights, was taking hold. It profited from an expansion in the size of armies, which increased the power of the populace, the writing down of codes for laws previously consigned only to memory, and commercial prosperity based on exports of grain and wine. More personal forms of poetry developed, including elegies along with iambic and lyric poems. The final groundwork was laid for the mythic frame of mind that Xenophanes of Colophon (ca. 545 B.C.) articulates:

> Truly the gods have not from the beginning
> revealed all things to mortals, but by long seeking
> mortals make progress in discovery.

The springtime of Hellas was over, and the summer of new learning was about to begin.

3.2   The Ionian Nascence and Deductive Reasoning

In the sixth century B.C., Ionia gave birth to classical Greek civilization, their use of deductive reason, and a rather continuous tradition leading to the formal organization of theoretical mathematics. That this civilization came into being on the west coast of Asia Minor and its offshore islands, rather than on the Balkan mainland, is attributable to the presence there for centuries of Hellenic cities untroubled by the hardships on the rugged mainland, possessing a healthy agrarian economy nourished by a pleasant climate and fertile soil, and connected with older high cultures.

Among the Ionian *poleis*, Miletus was then ascendant. In the first half of the sixth century its citizens enjoyed a remarkable prosperity,[173] resting partly on plentiful crops and on Miletus' location at the crossing of important trade routes between the Greek West and the Near East. Milesian trade in woolens, textiles, pottery, and, above all, wine and olive oil flourished. The Milesians, who were racially mixed peoples, included refugees from Athens and Mycenae on the Greek mainland. By the sixth century they were known for a sophistication that included a candid individualism, humorous cynicism, sensitivity to feelings of fellow citizens, and tactful speechmaking. Diverse in agricultural, commercial, and artisanal pursuits, they shared nevertheless a unity of belief in the Olympian gods, choosing Apollo as their tutelary deity.

The Egyptian breaking of Assyrian power in the Fertile Crescent with the help of Ionian mercenaries in 663 B.C. provided access to older learning that was a further probable reason for the Ionian cultural nascence. Milesians and other Greeks opened contacts with the major peoples of the Crescent. As one of the southernmost Greek *poleis* in Asia Minor, Miletus had a close caravan route overland to Syria and Mesopotamia. Shipping with Egypt soon became extensive. The timing for transmission of knowledge was propitious. The Chaldean or neo-Babylonian empire was at its peak in Mesopotamia. Although documentary evidence is scant, it is reasonable to surmise that the widely traveled Ionian Greeks now learned much about the mythology, scribal rules, legal systems, medical practice, geographical and astronomical lore, protomathematics, and nascent mathematics of the Mesopotamians, Syrians, and Egyptians, as well as their architecture, pottery, and textile designs.

Early in the sixth century B.C., several bold Ionian thinkers turned with some abruptness from mythical and fetishistic explanations and began a search for a rational accounting of the world. Contradictions among myths of different religions may have helped open the way for this shift. Among them *logos*, meaning word, became rational inquiry evolving toward a connected inquiry into the general causes of things, permitting only answers consonant with the knowable natural universe. *Logos* offered an alternative to myth (*mythos*), a dramatic story that explains the workings of the universe by reference to the actions of gods. In general, *mythos* invokes capricious, supernatural interventions. Even those Ionian sages who were more naturalistic than others did not agree on the roots of things. Their initial speculations, while radical, were also naive. It would be another century before the Aristotelian doctrine of causality and a systematic approach to theoretical studies were to mature. Aristotle in *Metaphysics* I was to reduce to a small number of "causes and first principles" the multiplicity of natural phenomena.

In the Mediterranean region, the Ionian shift from *mythos* to *logos* was sudden, pronounced, and apparently original with the Greeks. The speculative or abstract method of inquiry following rigorous and often self-corrective intellectual processes has no precedent in any history known to modern scholars. Source documents are scant and fragmentary, and the writings of Aristotle and his followers are secondary works. Consequently, neither the initial intensity of the new thinking nor the extent of the break with the past is certain. It appears that the shift occurred principally in natural science (roughly *physis* and *episteme*), in history ($\iota\sigma\tau\rho\iota\alpha$ or *historia*, which means learning), and in mathematics, a field for which the Greeks had a particular bent. History did not yet have its modern meaning. Aristotle later defined history to be a systematic explanation of natural phenomena, as the modern phrase *natural history* suggests. Not until the European Renaissance, when the Latin word *scientia* designated systematic studies of natural phenomena, did history generally come to apply solely to human affairs and chronological ordering.

Deeply curious about the physical world (*macrocosmos*), Ionian thinkers devoted great attention to the study of what are today called natural science and cosmology. In the beginning, science and philosophy were the same. The specialization that characterizes today's higher learning would not transpire for another two millennia. The term natural philosophy is used here to denote the intersection of philosophy and natural science. It should be kept in mind that a respect for the mysterious and irrational persisted as the new Ionian physical studies were undertaken.

The study of *physis* (nature) by the Milesian school of natural philosophy, founded by Thales, swiftly swept away most of mythopoeic thought. While Thales, so Aristotle reports in *De Anima* (A 5), thought "all things in the world are full of gods," his explanations of physical phenomena omit the gods as active agents of change. He thereby began to emancipate physical theories from creation myths and religion. For this revolution in thinking, the ancient Hellenes honored Thales with the title "Father of Philosophy." The Greek word *cosmos* for the universe suggests the core of Milesian thought. It meant regularity or order in the physical world, as well as adornment or beauty.

Milesians posited a primal substance out of which Earth is formed. Thales made water the fundamental constituent of the world and theorized that Earth floats on the waters.[174] That water has three observable states—solid, liquid, and vaporous—and its rarefaction and condensation were essential to the felting of woven materials in Miletus may have influenced his theory. According to Aristotle, the natural moisture of nutriment and semen could have suggested it. Perhaps knowledge gained from travels in Egypt also influenced Thales. The concept of the formation of the world from water could have come analogically from the silting of the Nile, the notion of the Earth's riding upon water from the Egyptian myth of Nun. Undoubtedly, contacts with other cultures enriched Milesian thinking. Whatever the source of Thales' ideas, Milesian natural philosophy spawned a group of ingenious physical theorists in the late sixth and the fifth century B.C. that included Anaximander and his pupil Anixemenes, Pythagoras, Parmenides and his pupil Zeno, Anaxagoras, Heraclitus, Empedocles, and Leucippus and his pupil Democritus.

During the fifth century B.C., the discipline of *historia* began to emerge and methods for achieving its objective, the discovery of historical truth, were improved. History built on older record keeping and anecdotal accounts but was a major departure from them. The inquisitive Ionian traveler Herodotus of Halicarnassus (ca. 480 to ca. 425 B.C.) transcended the level of opinion and traditional tales that logographers (chroniclers) of the time accepted.[175] He raised history to an art by treating sources and the vicissitudes of human fortune in a dispassionate and not overly credulous manner, by seeing his subject within a larger natural process, and by coherently incorporating into the history of wars between Greeks and Persians a broad range of materials. Setting himself outside the dominant poetic tradition, he expressed history in prose. His younger contemporary Thucydides (ca. 460 to ca. 400 B.C.) in his history of the great Peloponnesian War between Sparta and Athens achieved a greater mastery through a more systematic and dispassionate investigation. His book is still a model of political history. For his rigor in judging evidence and his quest for precision, Thucydides has been called the first modern historian.

Sixth and fifth century Greek thinkers were not concerned only with reason and the pursuit of truth, whether in the natural or the social sciences. Perhaps the major discipline then was rhetoric or elegant speech, aiming at persuasion more by eloquence than by truth. The works of Aristotle are a good example of both: quackery is the enemy.

3.3 Seeds of Theoretical Mathematics in Archaic and Classical Greece

Sources

The scanty sources do not reveal exactly how theoretical mathematics was emerging within sixth-century B.C. Ionian intellectual life. Usually, either an attraction to structure or the imperative of logic is assumed to have prompted its development. As Milesian natural philosophy and the evolving theory of a democratic state demonstrate, the ancient Hellenes were fond of theory. Perhaps the taste carried over to geometry. Among a people drawn to elegance in speech, moreover, at least a few would likely have enjoyed the elegance of theoretical mathematics. But Milesian theory was limited, while the early development of theoretical mathematics suggests a higher sophistication. That, along with the study of early Greek problem solving, suggests that a complementary, basically utilitarian motivation is more plausible. The construction, application, and improvement of astronomical devices, such as sundials, sectors, angle-measuring devices, and sliding marked rulers, together with close attention to harmony and proportion in architecture and pottery, may have encouraged the initial growth of demonstrative geometry. In addition, Miletus' expanding trade and commerce must have demanded finer computational techniques and the resolution of cross-cultural differences in measuring and number systems.

The men believed to have created, shaped, and practiced mathematics in ancient Greece are reported to have left voluminous writings. Only two fragments dating from before 400 B.C. have survived, the tract of Hippocrates of Chios on lunes and that by Archytas of Tarentum on duplicating the cube. Hippocrates' tract was reproduced by Eudemus and the sixth-century A.D. writer Simplicius in his commentary on Aristotle's *Physics*. For several reasons, other older writings did not survive. Papyrus documents are fragile and quite perishable in the humid northern Mediterranean climate; a number of great Greek libraries were destroyed in wars; and Euclid's *Elements*, an immensely successful compilation, superseded earlier texts. Euclid's text has become the primary source for pre-Socratic Greek geometry. By comparing differences in style and content in its thirteen books, scholars have been able, in some cases, tentatively to identify the authors of specific contributions.

The most important commentator on pre-Euclidean Greek mathematics is Proclus (d. A.D. 485), who headed the Academy in Athens. His *Commentary on Euclid, Book I* includes a few pages formerly known as "the Eudemian Summary," after the lost *History of Geometry* compiled late in the fourth century B.C. by Eudemus of Rhodes, who was a pupil of Aristotle. This extract, now known as the "Catalogue of Geometers," traces the history of geometry from its introduction into Greece from Egypt to the work of Euclid. The closing date of the extract suggests that Proclus' source was not Eudemus, but a condensation of Eudemus' narrative by a later author.

In painstakingly reconstructing the beginnings of theoretical mathematics in classical Greece, scholars have used segments from Eudemus on Hippocrates of Chios and Archytas, as well as comments on mathematics by major Greek thinkers, such as Herodotus, Plato, and Aristotle. They have also drawn on fragments from the third-century B.C. Alexandrian sage Eratosthenes of Cyrene's *Platonicus*, authentic portions of his letter to Ptolemy Euergetes on the duplication of the cube, and, above all, editions of Euclid's *Elements* and commentaries on it. While their complicated task is far from complete, the outlines of the history seem clear.

Thales, Anaximander, and the Demise of Miletus

Thales of Miletus (ca. 625 to ca. 547 B.C.) begins the history of mathematics in ancient Greece. He was the first of the "seven wise men" there who, while not demigods, continued the mythological tradition of making valuable discoveries *de novo*, as Hephaestus had done in metallurgy.

About the historical Thales very little is known. He is said to have been the son of a distinguished Milesian family whose mother was possibly a Phoenician. A cosmopolitan who spent his early career in Miletus, Thales is believed to have afterward traveled to Egypt and Mesopotamia, where he carried on business and studied their more ancient knowledge. When he returned to Miletus, he became a commanding figure as a statesman, entrepreneur, engineer, astronomer, and mathematician.

While these details are shaky, surviving anecdotes are apposite but probably apocryphal. One tradition holds that Thales was the epitome of practical ingenuity. Herodotus in I, 75 and 170 of the *History* skeptically reports that he diverted the river Halys, the frontier between Lydia and Persia, to enable a Lydian army under Croesus to cross. In A 11 of the *Politics*, Aristotle relates with some disbelief that when criticized for impracticality Thales correctly forecast a plentiful olive crop one year after several bad crops, bought all olive presses around Miletus, and made a huge profit by renting them. Yet in his commercial and personal conduct Thales was reported to be a scrupulous man. When asked how to lead a righteous life, he replied, "by refraining from doing that which we blame in others." Plato gives a different account of Thales from Aristotle's. In 174 A of *Theaetetus*, Plato portrays him as a dreamer, perhaps to show that philosophy involved more than utilitarian concerns. He recounts that Thales, while walking and stargazing one evening fell into a ditch, whereupon a pretty Thracian girl mocked him for trying to learn about the heavens when he could not see what was lying at his feet.

Among his contemporaries, Thales was probably more famous for his work in astronomy than for his mathematics. Herodotus in I, 75, credits him with predicting a solar eclipse for May 585. To perform this feat, he probably would have had to use the Babylonian saros, a cycle of 223 lunations after which solar and lunar eclipses repeat themselves with little change. But he could not have known the Babylonian saros, which, despite its antique name, was not invented until 1691 by the English astronomer Edmond Halley.[176] If Thales did make the prediction (which is doubtful), it may have been the result of intelligent guesswork on his part, rather than formal scientific computation. Because he had been the leading Greek thinker and astronomer when the solar eclipse happened, later authors would assume he was able to foretell it. Such accounts show that *mythos* and *historia* were closely intertwined and illustrate the imaginative appeal and tenacity of mythic tales.

While in Egypt to study, Thales reportedly astonished his hosts by demonstrating how to calculate the height of the Great Pyramid indirectly by means of shadows. Plutarch says in the *Convivium* that "he showed that the height of the pyramids is to the length of his staff held perpendicular to the ground in the same ratio as their shadows," or $h_1/h_2 = s_1/s_2$. He thus assumed that sides of equiangular triangles are proportional. The sides of the base of the Great Pyramid were known to be the equivalent of 756 feet. When Thales' staff of 6 feet casts a 9 foot shadow, the pyramid casts a shadow whose length measured from the center of the pyramid is $\frac{1}{2}(756)$ plus the distance to the tip of the shadow, 342. The length of its shadow is thus $378 + 342$ ft. This gives $h_1/6 = (378 + 342)/9$ or $h_1 = 6(378 + 349)/9 = 480$ ft. According to Eudemus and Proclus, Thales introduced Ionian Greeks to the study of geometry, bringing it from Egypt.

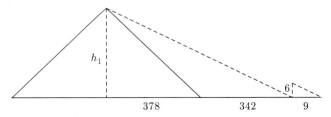

Herodotus and Proclus identify Thales as the first to search for a logical foundation of geometrical theorems. The "Catalogue of Geometers" ascribes to him these five propositions in elementary geometry:

1. A circle is bisected by its diameter (Euclid I.17).
2. The base angles of an isosceles triangle are equal (Euclid I.5).
3. If two straight lines intersect, the vertical angles are equal (Euclid I.15).
4. Two triangles are congruent if they have one side and two adjacent angles equal (Euclid I.26).
5. Every angle inscribed in a semicircle is a right angle (Euclid III.31).

If Eudemus' account is accurate, these five theorems while perhaps known in part to the Egyptians (although they were not formally proved) could indicate the thinking of a pioneer who attacked "some problems with greater generality and others more empirically."

Thales is thought to have proposed a proof of theorem 1, but we have no record of it. He may have provided an empirical rather than deductive demonstration and simply folded a paperlike fibrous circle along a diameter. Possibly, he stated only theorems 2 and 3, which, again, appear intuitively evident on folding.

The conjectured source of proposition 4 is a practical application of geometry to determine the distance of ships from shore. Accurate eyeballing of distances between ships and shore points was a well developed skill among mariners and required correlating vertical heights to horizontal distances. Possibly, Thales observed ships from a tower and applied proportionality of sides of similar right triangles. In the drawing, point E is where the line of sight cuts the line drawn through the foot of the observer parallel to the sea level line, which is drawn perpendicular to the other through its foot. This gives $x/y = (i + h)/i$ or $x = y(i + h)/i$. For this finding, theorem 4 is not needed. If he followed this argument, Thales may not have had congruence theorem 4, as Proclus implies.

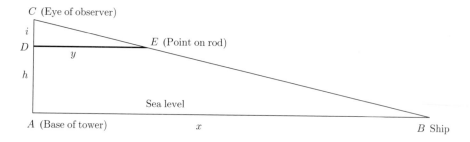

Thales' geometry clearly differs in character from the rule-of-thumb techniques of practical mensuration of Egypt and Mesopotamia.[177] Neither of those traditions explicitly formulated properties of figures as general statements. Nor did they apparently have Thales' primitive idea of proof, which advances by means of demonstration from theorem to theorem. Thales' notion of proof, though incomplete and intuitive, was a seminal idea that called for elaboration and refinement.

Because the historical claims Eudemus makes in the "Catalogue" seem to have been based on inference, views of what Thales actually achieved widely diverge.[178] Relying on Proclus' account of the "Catalogue," B. L. van der Waerden has asserted that Thales transcended Egyptian pragmatic geometry and initiated the development of theoretical geometry. Otto Neugebauer has viewed Thales as a less original thinker.

Between Thales and Pythagoras lived one geometer worthy of note, Anaximander of Miletus (ca. 610 to ca. 546/545 B.C.). A pupil of Thales, Anaximander, as Herodotus reports him in II, 109, made an outline of geometry and introduced into Greece the gnomon, a right-angled instrument, the prototype of the carpenter's square, whose shadow indicates the sun's direction and height. He also drew on a tablet a map of the inhabited world. His map was circular. It had the Greek world at the center surrounded by known parts of Europe and Asia, with the ocean forming the outside boundary. It is the crude, first-known attempt at geodesy. In natural philosophy Anaximander substituted for Thales's water the *apeiron*, "the boundless," the source of all. Worlds emerge from it, and when they perish, they are reabsorbed into it. Anaximander may have been the first to attempt to determine distances between Earth and other planets, which would indicate a secular attitude toward things then commonly assigned to divinity.

Unlike Thales, who may have drawn mostly on Egyptian knowledge as a beginning of his inquiry, Anaximander relied principally on Mesopotamian learning. That both those Near Eastern cultures had at the time essentially an empirical mathematics made probable differing techniques and results. Mesopotamian scribes, for example, approximated the area of a circle by the verbal equivalent of the rule $3r^2$, while one Egyptian method can be expressed by the formula $(8d/9)^2$. Perhaps the Greeks, as might be expected of people with frequent cross-cultural and trade contacts, consciously made comparisons between Egyptian and Mesopotamian findings in order to try to determine the more accurate, a practice that could have stimulated more exact deductions, new techniques, and geometrical demonstrations.

At Anaximander's death in the mid-sixth century, the political, economic, and intellectual center of the Greek world was moving from Miletus and other Ionian cities to the west. The rising power of Persia produced the shift. Ionian cities had prospered for more than two centuries free of a major military threat from the outside. Then, shortly after 540, the Persians, led by Cyrus, who reigned from 557 to 529 B.C., conquered the neighboring Lydian empire and turned against the rich Ionian cities. Miletus was forced to surrender. Although Cyrus and Darius, ruling from 521 to 486, were tolerant of their subjects' beliefs and traditions, under Persian domination the intellectual and commercial life of the Ionians deteriorated. Restive Ionian Greeks in Asia Minor rose in revolt against the Persians in 498 and burned Sardis, seat of the local Persian governor. In reprisal the Persians destroyed Miletus in 494. On Ionian islands not subject to Persia, mystical religious sects flourished as people turned inward in the face of an increasing external threat.

Pythagoras and the Early Pythagorean School

Late in the sixth century B.C., two great Greek schools arose in southern Italy, the Pythagorean and the Eleatic. Set beside the enthusiastic Italian scholars in the train of the powerful mystery cult of Dionysius, so Friedrich Nietzsche has proposed, the rationalistic Milesian sages may be considered the Apollonians.

While the Milesians invented the cosmos, an ordered universe tractable to reason, and defined a primal substance constituting the universe, the Pythagoreans combined philosophical speculation and mysticism in their inquiries into the basic forms and structures of the world we experience. Milesian cosmologists shunned anthropocentric reasoning; their Pythagorean counterparts gave philosophical legitimacy to ethics and politics.

Pythagoras (ca. 560 to ca. 480 B.C.), an exile from the Ionian island of Samos, was a mystic, philosopher, prophet, geometer, and sophist, that is, a teacher of wisdom. Since he followed the ancient Near Eastern practice of making his teachings entirely oral and the Pythagorean community required secrecy among its initiates, no firsthand written records of the master are left. What information is available comes chiefly from secondhand records, nearly contemporary to him, and from writings of later Pythagoreans, the Roman compiler Plutarch, and commentators, such as Proclus. Only a few fragments of records from shortly after Pythagoras' death survive, and these few are contradictory, mainly because of a split among his followers. When these sources discuss the work of Pythagoras, moreover, they generally attribute solely to the founder the collective achievements of early Pythagoreans. Modern scholars, attempting to extract historical detail from the legends, have determined or surmised with some degree of confidence basic facts about Pythagoras' career and part of his personal achievement.

Pythagoras supposedly went to Miletus to study, perhaps under Thales, but more likely under Anaximander. Next he journeyed to Egypt, where he may have been captured by the Persians under King Cambyses in 525. Either as a prisoner or voluntarily, he went to Babylon for a seven years' stay. There, scribes acquainted him with Mesopotamian and Egyptian mystical rites, numbers, music, and protoscience. Afterward, he returned to his birthplace on Samos, only to have to flee it around the age of fifty, perhaps because the tyrant Polycrates banned him. He moved westward to *Magna Graecia* and settled at Croton, a prosperous Dorian seaport in southern Italy.

At Croton, Pythagoras founded his religious and philosophical society, which was to spread to other Greek cities in Italy. One of the many mystery cults of the time, it had a highly selective membership of about three hundred young aristocrats. The Pythagoreans followed a rigid ascetic and monastic discipline. They were vegetarians, did not drink wine, held all goods in common, and believed in the transmigration and reincarnation of the soul. These practices and beliefs differed little from those of other mystery cults. What most distinguished the Pythagoreans from the others was their belief that the elevation of the soul to union with the divine occurs by means of mathematics. In contrast to the materialistic Ionians, they proclaimed that any true understanding of the world comes through numbers, which underlie and determine all things in the universe. God, Who has ordered the universe by means of them, is unity; the world is plurality. Without number, existence is impossible: numbers separate from the void all that exists.

The Pythagoreans invoked a common program of study, centering on four *mathemata*, or subjects: what Archytas would later call arithmetic (in the sense of number theory), geometry, *harmonia* (music), and *astrologia* (astronomy). Plato and Aristotle were to make these subjects fundamental to a liberal arts education. In the Latin Middle Ages, they became known as the *quadrivium*.

Following hierarchical principles, Pythagoras divided his followers into two groups that reflected the two components of Pythagorean doctrine, the mystical-religious and the scientific. The *akousmatikoi* (listeners) listened to the master in attentive silence and accepted his teachings as revelation. The disciple's dictum, "He himself has spoken," expresses ultimate authority. Translated into Latin as *ipse dixit* and retained in English, the phrase declares the self-justifying character of a master's assertion. These were the less advanced followers to whom Pythagoras only revealed cryptic prescriptions and doctrines. After three years' training, advanced followers were allowed into an inner group, the *mathematikoi*. The *mathematikoi* not only received the master's ideas more completely and clearly, but also expressed their own opinions and elaborated the teachings of Pythagoras. Legend holds that the early *mathematikoi* included twenty-eight women and that late in life Pythagoras married one of them. Ancient sources give the name of Theano for his wife or daughter.

The Pythagorean community had two crucial secret symbols. The *tetractys* is a sacred fourfoldness, lodged in arithmetic and illuminating the perfect number 10. Ten, the pivot on which the entire decimal system turns, is (wondrously) generated by summing the sacred first four numbers $1 + 2 + 3 + 4$. Number 1 serves as the generating element for point, 2 for line, 3 for surface, and 4 for body. Later this succession came to be the four elements: earth, air, fire, and water. The equilateral triangle with each side of 4 units, ⋰⋱, represents the *tetractys* and is, of course, the perfect triangle. Pythagoreans considered the *tetractys* quartet sufficiently sacred to invoke it in oaths and prayer.[179] The other symbol, the five-pointed star or regular pentagram, was known in Babylon and perhaps transmitted to their founder. By this sign, Pythagoreans from different communities recognized one another.

Critics made life unpleasant for early Pythagoreans. Stage presentations ridiculed them as superstitious, filthy vegetarians to whom the bean was sacred and forbidden. The early Pythagorean order nevertheless briefly had a major role in the political life of *Magna Graecia*. Local aristocrats at first found the hierarchical Pythagorean order an ally against expanding democracy. Later, these same patricians opposed the Pythagoreans. About 500 B.C. the Crotoniates forced Pythagoras to retire to the neighboring city of Metapontum, where he died. When the democratic revolution swept *Magna Graecia* about 450, the "aristocratic" Pythagoreans were set upon and their meeting houses burned.

Early Pythagorean *Arithmetica* and Application of Areas

Scribal protomathematicians in earlier high cultures, even Milesian sages, had cultivated mathematics more for its practical applications in surveying, commerce, astronomy, architecture, and temple and palace record-keeping than for its inherent interest. Pursuing the mystical over the scientific component and perhaps building upon a transmission of earlier Near Eastern knowledge, Pythagoras and his followers are generally believed to have added to the conceptual framework of early theoretical mathematics. They seem more conscious than their predecessors of the abstract nature of geometrical figures and, to an extent, numbers, though instead of completely detaching numbers from the physical world, they regarded numbers as the essences or primal constituents of material objects, so Aristotle notes in I, 5 of his *Metaphysics*. The Pythagoreans viewed mathematics as a self-contained realm, reached by contemplative ascetic exercises and yielding something of a geometer god's higher-order, eternal, number-based designs underlying the entire physical, metaphysical, and moral universe. Only by reference to the divine building blocks of numbers, Pythagoreans believed, could the universe be interpreted. In pursuit of this mystical vision founded in number, they helped to create pure mathematics. Moreover, by founding geometric theorems and principles on chains of logical reasoning, they contributed fundamental elements to the notion of formal proof.

The Pythagoreans specifically held that beneath physical objects lies geometry and beneath geometry, number. Number meant a positive integer. Contributing to this conviction that "everything is number" and firing their imagination undoubtedly was the notion of figurate numbers, which linked form or the shape of things to number. When arrayed as dots, numbers formed triangles, squares, rectangles or pentagons. The direct or indirect sources of these diagrams can probably be traced to widely known tabulations of arithmetic and geometric progressions performed in ancient Egypt and Mesopotamia. Awe of numbers must have found additional reinforcement in a discovery attributed to Pythagoras that acoustical pleasure in music is reducible to number. Harmonic consonance arises from ratios among integers. As Paul Tannery has conjectured, this musical theory enhanced with a multiplicative treatment the additive manipulation of fractions by the Egyptians. This suggests a fundamental, flexible shift in attitude to fractions, bringing their treatment closer to that of integers.

Pythagorean enthusiasts envisioned a program whereby all aspects of nature, whether physical, metaphysical, or moral, would be exposed as distinctive patterns of integers, as music had been. To this end, they promoted number as the language appropriate to science (including numerology). Their numerical speculations underlay an *arithmetica* comprised of numerological assumptions and the beginnings of a geometric theory of numbers. Among them and other pre-Socratic Greeks, the term *arithmetica* referred only to the abstract mathematical properties of integers. Classical Greeks referred to the computational aspects of mathematics as *logistica* and treated it separately. The two terms were socially as well as theoretically distinct. *Arithmetica* was the province of a leisured, intellectual order; *logistica* was the concern of trades, merchants, bankers, artisans, and slaves. Pythagorean number theory was the basis for the arithmetical books, VII through IX, of Euclid's *Elements*, where the results appear in a strictly deductive fashion devoid of the original number mysticism. This emphasis is even stronger by the time of Nicomachus' *Introductio Arithmetica* (ca. A.D. 100).

Pythagorean number mysticism, which included as staples sacred and lucky numbers, probably derived mostly from Mesopotamia and the secretive cabala of Near Eastern magi. Throughout the fifth century B.C., number mysticism remained an integral part of Pythagorean *arithmetica*. Numerology and number theory would later go their separate ways. As Aristotle informs us in A 5 of the *Metaphysics*, Pythagoreans associated each number with human attributes or abstract concepts. Number 1 stands for reason, since reason produces only one body of truth. Number 1 was considered not the first odd number, but rather the generative source of all numbers. Number 2 stands for man and opinion; number 3, woman and harmony. Even numbers after 2 are feminine, since they are unstable about a center and pertain to the ephemeral earthly; odd numbers after 3, which are stable about a center, are masculine and the enduring spiritual. Number 4 is the symbol of justice, because it is the product of equals. Number 5 represents marriage, because it is the sum of the first feminine and masculine numbers, and so forth.

The idea that numbers have mystical properties had occurred in many cultures. In their *arithmetica*, early Pythagoreans systematically developed the theory of special classes of numbers. This work falls between mystical speculation and theoretical science.

The notion of perfection in numbers flowered in classical Greece, but perhaps had begun earlier. Pythagoreans admitted two kinds of numbers to the class of perfect. The number 10, the sum of the first four numbers or *tetractys*, alone occupies the first class. The other kind of perfect numbers are those integers equal to the sum of their proper divisors. Two perfect numbers by this Pythagorean reckoning are $6 = 1 + 2 + 3$ and $28 = 1 + 2 + 4 + 7 + 14$. Other examples are 496 and 8,128. The last theorem of arithmetic in Euclid's *Elements*, IX, 36, belongs near the apex of ancient number theory. It asserts that if $2^{n+1} - 1$ is prime, then $2^n(2^{n+1} - 1)$ is a perfect number, where $n \geq 1$ is an integer.[180] For $n = 1$, the number 6, the smallest perfect number, is obtained in this way: $2(2^2 - 1) = 2(3) = 6$. For $n = 2$, the number 28 results from $2^2(2^3 - 1) = 4(8 - 1) = 4(7) = 28$.

Similar numerological interests prompted sporadic searches for extensions of perfect numbers in pairs of "amicable" or "friendly" numbers. Two numbers are amicable if each is the sum of the proper divisors of the other. Perhaps because of the name and the importance Pythagoreans attached to friendship, such numbers gained considerable attention, but they are of little importance to latter-day number theory. By the end of antiquity only one pair of amicable numbers was known, 220 and 284.[181]

A like quest was for Pythagorean triples of integers that satisfy the rule $a^2 + b^2 = c^2$. In our notation, an algorithm that Proclus maintains Pythagoras devised for obtaining triples (a, b, c) meets the rule: Let $a = 2n+1$, $b = 2n^2+2n$, and $c = 2n^2 + 2n + 1$, where $n \geq 1$ is an integer. Pythagoras possibly arrived at this rule by recognizing these relations between a square number and its next smaller square: $(2k-1) + (k-1)^2 = k^2$. Assuming that $2k-1$ is a perfect square, as it often is, and having $2k - 1 = p^2$ permit this solution for k: $k = (p^2 + 1)/2$ and $k - 1 = (p^2 - 1)/2$. Substituting these values in the relation between a square and its next smaller square gives $p^2 + [(p^2 - 1)/2]^2 = [(p^2 - 1)/2 + 1]^2$, with $a = p$, $b = (p^2 - 1)/2$, and $c = (p^2 + 1)/2$, satisfying the Pythagorean equation for any odd integer p greater than 1. The value p must be odd, because $p^2 = 2k - 1$ is odd. Letting $p = 2n + 1$ for n greater than 1 or equal to 1 yields results that are supposedly Pythagoras'. Calculations that use those results give right triangles with the hypotenuse exceeding the longer leg by 1. When $p = 5$, for example, $b = 12$ and $c = 13$. While it seems doubtful that Pythagoras derived this rule, he may have learned it in Babylon, where such triples had been computed at least a thousand years earlier.

In arriving at the figurate, or polygonal, numbers that were so significant in the Pythagorean *arithmetica*, Pythagoras and his followers may have depicted numbers pictorially as dots drawn in the sand or as groups of pebbles. Building on diagrammatic depictions and the belief that numbers are the source of basic structural forms, they classified numbers according to the geometric shapes that the dots could form. Their figurate numbers included triangular, square, rectangular, pentagonal, and higher numbers. In our illustration, n_5 means the sides are five equally spaced dots, and n_3, three equally spaced dots.

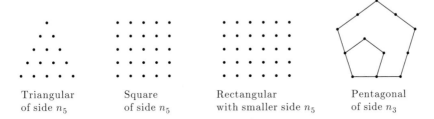

| Triangular | Square | Rectangular | Pentagonal |
| of side n_5 | of side n_5 | with smaller side n_5 | of side n_3 |

Pythagoreans are believed to have obtained figurate numbers from the sum of simple arithmetical progressions. Thus, either a transmission of older Near Eastern learning or their own derivation led them to associate an arithmetical series with each figurate number and to sum the numbers.

Adding successive integers (1, 2, 3, 4, ...) gives triangular numbers (3 ∴,

6 ∴, 10 ∴, ...), because the total number of dots can be arranged as equilateral triangles. Separating dots in this configuration with a slash reveals that the next triangular number emerges from adding to the previous triangle another row containing one more dot than the prior row does. Thus, if a_n represents the nth triangular number, then

$$a_n = a_{n-1} + n = 1 + 2 + 3 + \ldots + (n-1) + n.$$

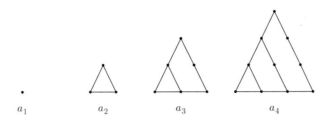

$a_1 \qquad a_2 \qquad a_3 \qquad a_4$

The Pythagoreans also had sums generating square, rectangular, and pentagonal numbers. Summing successive odd integers (1, 3, 5, 7, ...), they found, gives total numbers of dots that can be arranged as squares, for example 4 ::, 9 ⋮⋮⋮, 16 ⋮⋮⋮⋮. Since odd numbers can be represented by $2n-1$, they determined that $1 + 3 + 5 + 7 + \ldots + (2n-1) = n^2$. Rectangular numbers were found by adding one to each term of a square number's series $(1+1, 3+1, 5+1, 7+1, \ldots)$, which gives the even numbers $(2, 4, 6, 8, \ldots)$. Pentagonal numbers are composites of a square number of side n and a triangular number of side $n-1$ or of three triangular numbers and a line. The first four pentagonal numbers are 1, 5, 12, and 22. In modern terms, the sum of the pentagonal numbers from 1 to the nth such number is

$$p_n = 1 + 4 + 7 + \ldots + (3n - 2) = n(3n - 1)/2.$$

The theory of figurate numbers attracted fruitful studies two millennia later. In his *Treatise of Figurative Numbers*, which appeared in 1665, Blaise Pascal conjectures that all positive integers are the sum of three or fewer figurate numbers. In 1801, Carl Gauss proved this conjecture.

The theory of even and odd was the most developed part of Pythagorean *arithmetica*. Its early stages included these three theorems:

1. The sum of two even numbers is even.
2. The product of two odd numbers is odd.
3. When an odd number divides an even number, it also divides its half.

These theorems were perhaps derived from the search for Pythagorean number triples. The theory of even and odd subsequently proved helpful in the discovery of incommensurable magnitudes.

Incommensurable magnitudes were the most important discovery of the early Pythagoreans. This discovery must have disturbed at least some of their leading thinkers. It contradicts their assumption that proportion is limited to ratios of positive integers, which builds on their conviction that "everything is [whole] number." Beginning with any two line segments, Pythagoreans believed that it should be possible to find a third line segment, perhaps exceedingly small, comprising a common measure that, when sufficiently repeated, could mark off the whole of each segment exactly. Ratios that cannot be expressed by whole numbers with a common measure were called incommensurable or *alogos* (unutterable or without ratio) or *arrhetos* (inexpressible). The Greek word *logos* (speech) denoted the ratio of two integers. Incommensurable magnitudes posed a serious anomaly to the Pythagorean theory of proportions and argued for recasting it. The discovery of *alogos*, usually rendered into English as the "irrational," marks the beginning of the study of irrational numbers.

The discovery by the Pythagoreans of incommensurable magnitudes probably stemmed from their investigation of the two sides and diagonal of a unit square or, to put it another way, of an isosceles triangle with legs 1 and hypotenuse $\sqrt{2}$, which is incommensurable with 1. Some scholars believe, however, that it was the study of the pentagon that led them first to discover $\sqrt{5}$. An early Pythagorean leader, Hippasus of Metapontum, is credited with discovering incommensurable magnitudes. Legend relates that the Pythagoreans were at sea at the time and threw Hippasus overboard to drown for making a discovery that denied the doctrine that all existence can be reduced to whole numbers and their ratios.

According to Aristotle in *Prior Analytics* I.23, the Pythagoreans proved by *reductio ad absurdum* that length $\sqrt{2}$ is incommensurable with unit length 1. This method, which assumes the negation of the conclusion sought and derives a contradiction from that assumption, dominated mathematical argument in early classical Greece. The Pythagorean proof is worth repeating for its simplicity and beauty. It begins by assuming that the ratio of the hypotenuse of an isosceles right triangle to its leg can be expressed as the ratio of two whole numbers, a and b, that share no multiplicative factor. Thus, either a or b or possibly both must be odd. In the case of the isosceles right triangle with leg 1, this means that

(1) $\sqrt{2}/1 = a/b$, where a and b are relatively prime and $b \neq 0$.
   Squaring both sides yields

(2) $2 = (a/b)^2$, and by cross multiplication

(3) $a^2 = 2b^2$.
   The number a, then, cannot be odd, since odd times odd is odd. Hence, a is even, or $a = 2c$, and b must be odd. But substituting the new value of a in (3) gives

(4) $4c^2 = 2b^2$ or (5) $2c^2 = b^2$,

which shows that b is even. But a number cannot be both even and odd. So $\sqrt{2}$ cannot be expressed as the ratio of two integers and hence is irrational.

   Modern mathematics retains this oldest known proof that $\sqrt{2}$ is irrational. It is the only proof of the incommensurability of line segments attributed to the early Pythagoreans. It was interpolated into older editions of Euclid's *Elements* in Book X as proposition 117. Since this proof does not appear in what are considered the earliest extant versions of Euclid's text, modern editions omit it.

   Although the discovery of one or two cases of incommensurability probably did not provoke a major crisis for the Pythagorean theory of integral proportions, it did pose a fundamental problem for geometry, because it implied that geometric magnitudes (lengths, areas, and volumes) are *continuous* in character rather than *discrete*. After this discovery, Pythagoreans could no longer manipulate geometric magnitudes in proportions as always consisting of discrete numbers. In order to multiply and divide lengths, areas, or volumes, they may have devised a mathematical procedure known as the application of areas. It became the principal procedure employed in what is considered the geometric algebra in Books II and VI of Euclid's *Elements*. Geometric algebra seeks to solve algebraic equations geometrically. The construction problem shown here, in which line segments represent numbers, is an example of application of areas.

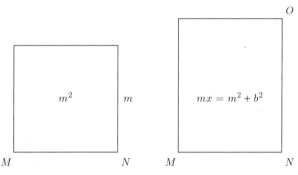

Given a line segment $\overline{MN}$ of length m, construct a rectangle with base $\overline{MN}$ and side $\overline{NO}$, which is larger than the square of m by the square of a given length (b^2). In modern notation, this solves by geometric divisions the equation $m^2 + b^2 = mx$. In this division, the area m^2 is applied to the rectangle $\overline{MN}$ by $\overline{NO}$. The method of applying areas allowed geometers to compare any two geometric magnitudes, since any rectilinear figure can be transformed into a rectangle of the same base and a height, giving the rectangle the same area as the figure. Results building on this Pythagorean method appear in Books II and VI of Euclid's *Elements*.

Early Pythagorean Geometry, Music Theory, and Astronomy

Modern scholars have determined rather accurately the contributions of the Pythagoreans to arithmetic and the origins of geometric algebra. Their achievements in plane geometry remain a question.

The name of Pythagoras is, of course, indelibly linked with the famous theorem, Euclid I. 47, about the hypotenuse of a right triangle. How he came by this theorem is not known. Since the Mesopotamians of Hammurabi's time knew the rule for obtaining what are now called Pythagorean triples, he could have learned it during his Babylonian sojourn. Proclus and the Greek historian Plutarch credit Pythagoras personally with the discovery. Plutarch quotes the poetic distich "When Pythagoras discovered his famous figure, for which he sacrificed a bull." But Plutarch's account seems unreliable, since Pythagoras was a vegetarian who opposed the killing or sacrificing of animals, particularly cattle. Nor does Plutarch have a reputation for critical judgment. Clearly, Pythagoras could not formally prove his theorem for a right triangle, $a^2 + b^2 = c^2$. He did not have the complete mathematical tools required for a formal proof.[182]

Early Pythagoreans, knowing the properties of parallel lines and their transversals, were able to show that the sum of the three angles in a triangle is 180°. The demonstration proceeds as follows: Consider two parallel line segments A' and B' cut by two transversals that form the triangle CDE. Since the opposite interior angles formed by the transversals cutting two parallel lines are equal, that is, $\alpha = \alpha'$ and $\beta' = \beta$, we get $\alpha + \beta + \gamma = 180°$.

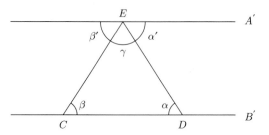

Early Pythagoreans also investigated the five regular polyhedra, the convex solids whose edges form congruent regular polygons. Proclus conjectures that

Pythagoras knew how to put together or construct all five, but the scholium to Book XIII of Euclid's *Elements* seems more accurate when it states that Pythagoreans could construct only three: the tetrahedron or pyramid, having four faces, the hexahedron or cube, of six faces, and the dodecahedron, having twelve faces. One tradition ascribes these three constructions to Pythagoras, another to Hippasus. Plato's contemporary Theaetetus reportedly discovered the other two: the octahedron, of eight faces, and the icosahedron with twenty faces.

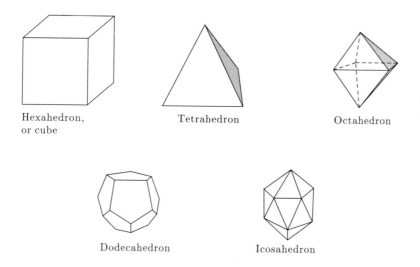

Hexahedron,
or cube

Tetrahedron

Octahedron

Dodecahedron

Icosahedron

Five regular polyhedra

It may have been a relatively common mineral that aroused the Pythagorean curiosity about regular polyhedra. Iron pyrite crystals found in southern Italy have faces composed of regular pentagons. In addition, an Etruscan cult near Padua had made before 500 B.C. a magical object by carving soapstone in the shape of a dodecahedron. This is a possible link to the magical practices of the Italian Pythagoreans, because the twelve faces of a dodecahedron are regular pentagons whose diagonals form a star-pentagon.

Among the early Pythagoreans the star-pentagon, a pentagram, became a symbol of health and a secret mark of identification for fellow Pythagoreans. A mathematically interesting feature of the five pointed star is that the five lines divide each other in a mean and extreme ratio, as the golden section in Euclid IV. 10 and 11 demonstrates. In the diagram here, this implies that $(a - x)/x = x/a$. Since a firm knowledge of similar triangles is required to prove this division, early Pythagoreans are assumed not to have established this geometric relation.

From their study of numerical ratios, the pentagon, and musical intervals, the Pythagoreans created a theory of special means. Early members knew three: the arithmetic, geometric, and harmonic means. From Archytas in the fourth century B.C. to Nicomachus in the second century A.D., they defined seven new types of means.

In his commentary on Nicomachus' *Introductio Arithmetica*, Iamblichus asserts that Pythagoras learned the arithmetic mean in Babylon, but this cannot be proved. The classical form of the arithmetic mean is $a - b = b - c$. Thus, 6 is the arithmetic mean of 4 and 8 because $8 - 6 = 6 - 4$. In modern times, the arithmetic mean of two numbers is defined as half their sum. The geometric mean of numbers a and c is the number b, such that $a : b = b : c$. That is, the ratio of the first term to the second equals the ratio of the second to the third. Put another way, the geometric mean of two numbers is the square root of their product. For example, 8 is the geometric mean of 4 and 16.

The harmonic mean or subcontrary, as it was originally called, of a and b, where $a < b$, is represented by the number h, such that $(h - a)/(b - h) = a/b$. The harmonic mean of two numbers is thus established when the difference between the harmonic mean and the first number divided by the difference between the harmonic mean and the second number equals the ratio of the first number divided by the second. One example the Pythagoreans found was that 8 is the harmonic mean of 6 and 12, because $(8 - 6)/(12 - 8) = 6/12$. Since the cube has six faces, eight vertices, and twelve edges, they called it a harmonic body.

Although the sources and lines of development of early Pythagorean research in geometry are not known exactly, its importance is unquestioned. Johannes Kepler (d. 1630), who, to gauge the ratios of the planets' distances apart ingeniously placed the orbits of the known planets on thick spheres encasing the five regular polyhedra nested within one another, speaks glowingly of their achievements:

> Geometry has two great treasures: one is the theorem of Pythagoras; the other, the division of the line into extreme and mean ratio. The first we may compare to a measure of gold; the second we may name a precious jewel.

Early Pythagoreans invented music theory and contributed to astronomy as well. These "kindred sciences," as Plato calls them, were studied less by manipulating musical and astronomical instruments than by "ascending to generalized problems." Since deduction was so important to this method, early Pythagorean kindred science had strong connections with mathematics, especially the theory of integral proportions. The music theory of the Pythagoreans seemed to verify the doctrine that "everything is number." By listening to the lyre and flute and perhaps by experimenting with partially filled glass vessels, they discovered that the preceding arithmetic, geometric, and harmonic means dictate basic intervals (*diastema*) of music. Two taut, alternatively vibrating strings produce harmonic sounds when one has twice the length of the other; the notes of the shorter string are said to fall at an octave (2 : 1) above corresponding notes of the longer string. When the longer is one and one-half times the length of the shorter, the harmonic notes of the shorter are at intervals of a fifth (3 : 2) above those of the longer. If the difference is a third, the intervals are at a fourth (4 : 3). Octave, fifth, and fourth give the (inclusive) count of how many tones of the diatonic (eight note) scale are involved. These musical ratios reconfirm the centrality of the *tetractys* (1, 2, 3, 4). The length of strings on a lyre had been 12 to 9. Among the many feats attributed to Pythagoras was that of shortening them in the ratio 8 to 6, which yields different pitches of tone. If so, it is the oldest known example of a natural law discovered empirically.

Perhaps drawing on data from Babylonian observations and Anaximander's theories, Pythagorean astronomers sought a harmonious relationship between planetary periods of revolution and interplanetary distances. In their search for proper numerical order in the heavens, they correlated periods of revolution with musical intervals, imagining, at least metaphorically, a musical harmony of the spheres.

Shortly after Pythagoras died, his followers split into two factions, the *mathematikoi* led by Hippasus and the *akousmatikoi*. The *mathematikoi* continued to pursue new *mathesis* or learning; hence their name. Their subsequent speculations and research accelerated the transformation of mathematics in ancient Greece into a deductive science. They were thus the first known group of theoretical mathematicians. Fittingly, we derive our words mathematics and mathematician from the term *mathematikoi*.

Hellenic Numeral Systems: The Attic and Ionic

While Pythagoreans were developing *arithmetica*, important change was occurring in the numerals of *logistica* (computation). Daily transactions in commerce requiring weights, measures, and sums of money and expanding needs in administration, law, and mathematics moved the ancient Greeks toward new numeral systems. By the fifth century B.C. they had written acrophonic systems with regional variations. Acrophonic systems designated numerals by the first letter of Greek number-words. Their alphabetic base distinguished acrophonic numerals from older Near Eastern numerals. Otherwise, they were similar to

Egyptian or later Roman numerals in being primarily an additive string to obtain their total values.

The Attic or Athenian is the best-known acrophonic numeral system. Like languages, written numeral systems change over time. Introduction of the archaic Phoenician alphabet in Homer's day and knowledge of Egyptian ciphered numeration seem to have influenced the formation of Attic numerals. The systems had special signs for 5, 10, 100, 1,000, and 10,000.

Numerical Sign		Value	Initial of the Word
Γ	[letter pi. archaic form of π]	5	Πεντε [pente, "five"]
Δ	[letter *delta*]	10	Δεκα [deka, "ten"]
Η	[letter *eta*]	100	Ηεκατον [heckaron, "hundred"]
Χ	[letter *khi*]	1000	Χηιλιοι [khilioi, "thousand"]
Μ	[letter $\overline{m}u$]	10,000	Μυριοι [murioi, "ten thousand"]

Besides constructing larger numerals by adding symbols, for example, $\Delta\Delta\Delta = 30$, the system had a multiplicative principle. Placing a symbol inside the acrophonic numeral for 5 quintupled its value, and so forth. Some numerals combined both techniques, such as $X\digamma\Delta\Delta\Gamma = 1525$

50	$\ulcorner^{\Delta}$	$=$ $\Gamma . \Delta$	5×10
500	$\ulcorner^{H}$	$=$ $\Gamma . H$	5×100
5000	$\ulcorner^{X}$	$=$ $\Gamma . X$	5×1000
50,000	$\ulcorner^{M}$	$=$ $\Gamma . \mathrm{M}$	$5 \times 10,000$

Epigraphical evidence from monument inscriptions suggests that the Attic system was widely used in the fifth century B.C., contended with the Ionic in the fourth, and was superseded in the third century B.C.

The later and more sophisticated Ionic or alphabetic numeration followed the Egyptian ciphered model more closely than the more advanced Babylonian place value. It used twenty-seven signs consisting of the twenty-four letters of the classic Greek alphabet, which was derived from the Phoenician, adding three obsolete letters: digamma, koppa, and san. These twenty-seven signs were divided into three groups of nine; the first group represented unit digits, with diagamma for 6; the second group, multiples of 10, with koppa for 90; and the third, multiples of 100, with san for 900. Thus, the Greek alphabetic system had a base 10.

UNITS				TENS				HUNDREDS			
A	α	alpha	1	I	ι	iota	10	P	ρ	rho	100
B	β	beta	2	K	κ	kappa	20	Σ	σ	sigma	200
Γ	γ	gamma	3	Λ	λ	lambda	30	T	τ	tau	300
Δ	δ	delta	4	M	μ	mu	40	Y	υ	upsilon	400
E	ε	epsilon	5	N	ν	nu	50	Φ	φ	phi	500
Ϛ	Ϛ	digamma	6	Ξ	ξ	xi	60	X	χ	khi	600
Z	ζ	zeta	7	O	o	omicron	70	Ψ	ψ	psi	700
H	η	eta	8	Π	π	pi	80	Ω	ω	omega	800
Θ	θ	theta	9	Ϙ	ϙ	koppa	90	ϡ	ϡ	san	900

To distinguish numerals from words, the ancient Greeks frequently drew a bar over a numeral. To represent numbers larger than 999, scribes initially placed a hooked or looped line above the numeral to indicate thousands. This notation gave way by the fourth century A.D. to an oblique stroke at the lower left or right of the numeral or directly under it.

$_{\prime}\alpha$	$_{\prime}\beta$	$_{\prime}\gamma$	$_{\prime}\delta$	$_{\prime}\epsilon$	$_{\prime}\varepsilon$	$_{\prime}\zeta$	$_{\prime}\eta$	$_{\prime}\theta$
1000	2000	3000	4000	5000	6000	7000	8000	9000

Writing numerals over a *mu* (μ), for myriads or tens of thousands, originally indicated larger numbers ranging from tens of thousands to tens of millions. This convention was probably the source of systems representing very large numbers set forth by Apollonius, which was based on the myriad, and by Archimedes, based on the myriad myriad (ten thousands times ten thousands). In Daniel 7:10 and Revelation 5:11 of the Christian Bible, the myriad myriad was the multitude that no man can number. The interest in very large numbers was theoretical, rather than practical, and restricted to mathematics and mathematical astronomy.

In their geometrized understanding of numbers, classical Greeks dealt with ratios but did not develop common fractions (m/n, where m and n are integers and $n \neq 0$) and operations with them. They added a variety of symbols— strokes, bars, or dots—as fraction indicators. Thus, a stroke (essentially an accent) added to the basic *arithmoi* converted β or 2 to a half, γ or 3 to a third, δ or 4 to a fourth, ϵ or 5 to a fifth, and so forth. The ancient Greeks made a part or parts (of a given whole) basic, rather than requiring the decomposition of fractions into sums of unit fractions. The sequence of parts beginning with β, the two-parts, expresses two-thirds. More complicated fractions were compound parts, such as $\gamma'\delta$ for $\frac{3}{4}$ and $\epsilon'\zeta$ for $\frac{5}{6}$. In our terminology, the numerator in these appeared first, and terms in the numerator and denominator almost always were in decreasing order.

The Eleatic School: Parmenides, Zeno, and Democritus

Elea was the site of the second school in *Magna Graecia* that examined the foundations of demonstrative geometry. Fleeing before the Persians in the mid-sixth century B.C., Ionians had settled in southern Italy, where Elea was a western Greek colony. Perhaps one of these refugees was Xenophanes of Colophon. Parmenides (ca. 515 to ca. 450 B.C.), the founder of the school, was born in Elea. Through Xenophanes he was introduced to Ionian cosmology. Since Elea is near Croton and Metapontum and communication between these cities was not difficult, he came under the influence of the early Pythagoreans. Plato reports that an elderly Parmenides and his pupil Zeno visited Athens around 450 B.C. and met the young Socrates.

Convinced like the Pythagoreans that sense perception is untrustworthy, Parmenides regarded as derivative and defective the world perceived by the senses. Instead, he required abstract reasoning in long, tightly woven arguments as the basis for conclusive demonstrations. His hexameter poem, perhaps entitled "On Nature" with sections on "opinion" and "truth," brought a new degree of rigor in logical deduction to the Greek pursuit of truth. In an ambitious attempt to deduce all that could be known about the universe of Being, Parmenides effectively uses the law of the excluded middle, which asserts that every meaningful statement is either true or false. Then by adroit *reductio ad absurdum* arguments he shows that denying his proposition led an adversary into contradiction and so proves the adversary is wrong. He urged readers and listeners—he sang his poems—to follow abstract reason and spurn the seductive appeal of sense knowledge.

The central doctrine of Parmenides' rational metaphysics is a rigorous monism that conflicted with the less stringent position of the Pythagoreans. They apparently taught that the interplay of numbers (Being) and the void (non-Being) defines or limits what exists. Parmenides was unwilling to confer the status of existence on that which is nonexistent. He rejected the Pythagorean category of non-Being as sterile, inherently impossible, and repugnant to reason. How, he asks, could Being arise from non-Being? Perhaps Parmenides believed further that the discovery of incommensurable magnitudes had shown the falsity of the Pythagorean numerical-natural philosophy based on a particulate collection of units. Incommensurability and the infinite divisibility of lines may have suggested to him that the universe of Being is fundamentally continuous despite appearances to the contrary.

Through strictly applied logic, Parmenides finds the universe of Being to be a necessarily unique, eternal, changeless, and spherical (and hence bounded) body (the One). The concept of the sphere allows him to deal with crucial philosophical issues of the bounded and the unbounded, later seen as the finite and the infinite. Current scholars disagree over whether he posits this sphere as a reality or as a metaphorical ideal; most accept the latter interpretation. In opposition to Heraclitus, he argues that the plurality, change, motion, and time of the world of mortals are vague, contradictory, and presumably partly unbounded. They are merely illusions of the senses. Here was a radical new metaphysics, which holds that deductive reason when faithfully pursued to the end denies the manifold world of the senses. Eleatic reflections on how demonstrations and proofs relate to reality later bore further fruit in the dialectic of Socrates and Plato.

Theories of motion depend on theories of the nature of space and time. If space and time are infinitely divisible, they frame a smooth and continuous field. The Pythagorean universe is composed instead of a plurality of discrete units: a granular space, in effect, and cinematographic motion. Attempting to identify the smallest unit of length, a few Pythagoreans after the discovery of incommensurables may have resorted to some form of infinitely small, elementary line segments, *lineae indivisibles* or roughly infinitesimals, as though sufficiently diminished units would evade or resolve the impasse. That would not have satisfied Parmenides. His bounded spherical One admits no spatiotemporal fragments within it.

Plato's *Parmenides* (127 ff.) tells us of four paradoxes, in which Zeno of Elea (ca. 490 to ca. 425 B.C.) employs *reductio ad absurdum* proofs to deduce internal contradictions regarding motion. Parmenidean metaphysics, however, prohibits internal contradictions, such as accepting as true both a statement and its denial. Thus, Zeno, a philosopher and logician, does not try to prove directly (as opposed to *reductio*) the existence of the spherical One, real and unchanging, or that motion, time, and plurality are simply illusory. His four paradoxes of motion, known as the Dichotomy, Achilles and the Tortoise, the Flying Arrow, and the Stadium or Moving Rows, instead demonstrate that the possibility of the real existence of motion and plurality seemingly lead to a logical impossibility. Proclus asserts that, assuming their existence, Zeno arrived at forty paradoxes. Either alternative—denial or acceptance of motion and plurality—leads to strange and discomfiting consequences. Such damned-if-you-do, damned-if-you-don't intellectual choices are called antinomies.

The Dichotomy and Achilles challenge kinematic, or spatiotemporal, theories of motion. Can a moving body traverse a physical interval covering each of its subdivisions in a finite time, if the series interval has dense or infinitely many subdivisions: $1, 1/2, 1/4, 1/8, \ldots, 1/2^n, \ldots$ in the Dichotomy and $1, 1/100, \ldots, 1/100^n, \ldots$ in Achilles? Zeno finds it easy to move across the first points in theory, but encounters difficulties as the intervening distances become smaller and the number of points grows ever greater. He implicitly appeals to an eternity for covering these points one by one. The Dichotomy and Achilles need not in themselves militate against the existence of motion, as a more subtle form of the argument holds: they might simply support the thesis that the time-interval division is also dense, that is, possibly infinitely divisible, and the total time finite. They can thereby permit a continuity theory of motion.

Similarly, the Arrow and Stadium paradoxes test the notion that space and time are comprised of finite, discrete units or "nows." If this is true, in each successive instant is the arrow at rest or in motion? Can each atom of arrow be mapped to a specified instant of time at every point of the trajectory? The solution, which awaited modern times, lies again in infinite series. Zeno finds it logically impossible to find a terminal instant of motion in any subinterval in an infinite progression. For the paradox of the Stadium, whose foreground-moving, background-stationary rows present the first awareness in antiquity of the relativity of motion, a logically acceptable solution depends on constructing a logically tight definition in infinite series for betweenness and alignment and rephrasing the original thesis, so as to satisfy the new requirements.

Zeno's idea of infinite processes gave him the gist of a mathematical theory of the continuum. His paradoxes are an early step in the study of such processes. Their resolution would emerge from the nature of the linear continuum, which Georg Cantor two millennia later was to determine contains a nondenumerable (uncountable) infinite number of points. The concepts of convergence and limits, of course, were also still centuries in the future. It should be noted that Zeno clearly understood that the sums of some infinite series are finite. He knew, for example, that the sum of infinite series arising from the continued bisection of a unit cannot exceed the unit. He demonstrates this in summing the series named after him, $1/2 + 1/4 + 1/8 + 1/16 + \ldots + 1/2^n + \ldots = 1$.

Through the ages, Zeno's paradoxes have been extensively and often fruitfully studied. Commentaries on them range over time from Aristotle's dismissal of them, possibly in a revised form, in his *Physics* as fallacies to such twentieth-century refinements as Herman Weyl's conception of an "infinity machine" to cover the Dichotomy, Bertrand Russell's view that motion for the arrow occurs between the instants, and Adolph Grünbaum's proposal of mind-independent time, which is granular, and mind-dependent time, which is a continuum, as well as his observation on the relation of the Zenonian paradoxes to quantum mechanics. Grünbaum's proposal on the nature of time follows William James and A. N. Whitehead's assertion that time is not a continuum but pulsating.

Zeno's place in the history of ancient mathematics is still debated. B. L. van der Waerden has found little influence of his ideas on most geometers of his day. But it seems no small coincidence that Eudoxus later studied in Athens at Plato's Academy and, before he developed his general theory of proportions that accurately handles infinitesimals, learned and participated in debates about Zeno's work as presented in Plato's *Parmenides* and Aristotle's *Physics*. In the *De Generatione et Corruptione* (325a23 ff.), Aristotle reports that Eleatic and Pythagorean doctrines influenced the atomist Democritus (ca. 460 to ca. 379 B.C.). A wealthy citizen of Abdera in Thrace, Democritus did indeed study under Leucippus, a student of Zeno. Leucippus taught him the essentials of an atomism built on the Pythagorean numerical model and Zeno's criticisms.

Zeno apparently had argued that Pythagoreans confused geometric points with physical atoms. Their generation of extended lines and sensible bodies from extensionless geometric points troubled Democritus. Basing the universe on solid corporeal atoms and the infinitely extended void, Democritus apparently attempted to resolve the Zenonian dilemma by granting first that all geometrical magnitudes are infinitely divisible and that a geometric point has no magnitude. But atoms, he argued, are neither points nor solids of geometry. Atoms are instead physical bodies that are impenetrable and physically indivisible, which is the meaning of *atomos*. The atoms, infinite in number and differing in size and shape, are mostly small and invisible, but one may be the size of the cosmos. No atom could ever be reduced to sheer nothingness. Historians disagree over whether Democritus denied that they were mathematically infinitely divisible in an effort to remove the Pythagorean confusion between physical atoms and geometric points.

Democritus also made an important discovery in solid geometry. Pythagorean research on the five regular polyhedra, along with studies of the three famous construction problems that are examined in the next section and attempts to achieve perspective in staging Aeschylus' tragedies in Athens, made it a major field of Greek mathematics at the time. The Stoic Chryssipus had Democritus investigate whether two contiguous surfaces obtained by horizontally slicing a cone are equal or unequal. His finding is not known. Some historians argue that he found the sizes of these surfaces a dilemma; others that he considered them mathematically equal; and others that he worked with a stepped physical cone. In the introduction to *On the Method*, Archimedes states that "Democritus ... was the first to ... [assert] ... that the [volume of] the cone is a third part of [that of] the [circumscribing] cylinder, and the pyramid of the prism, having the same base and equal height." His study of pyramids and prisms may have begun with the Egyptian scribal mathematicians' rule that the volume of a pyramid equals one-third of the base times the altitude. Next he could have divided a triangular prism into three triangular pyramids of equal height and base area and begun to confirm their relationship. But Democritus, Archimedes observes, did not prove this assertion rigorously; that was left for Eudoxus. According to Plutarch, Democritus determined the volumes of pyramids and prisms (or, likely as a corollary, cones and cylinders) by viewing them as solids theoretically sliced into infinitely many, infinitely thin sections that are triangular (or circular) sections parallel to the base. Proof of the validity of infinitesimal methods lay far in the future.

3.4   Mid-Fifth to the Fourth Century B.C.: Principal Subjects of Study

Theoretical mathematics had its greatest growth in ancient Greece during the classical period, which began about 479 B.C., and ended politically in 338 B.C. with Philip's conquest of most of Hellas. For intellectual development it may be extended to Aristotle's departure from Athens in 323 B.C. During these years mathematicians from across the Greek world, but mainly from *Magna Graecia* and Athens, pursued at least four major subjects: the theory of numbers; metrical geometry (later epitomized by the Alexandrian Hero) to compute areas and volumes; nonmetrical geometry, especially three famous construction problems;

and musical theory. Logic followed an autonomous as well as ancillary course, as it had in Elea. Such attempts to improve mathematical demonstration, or proof theory, led to the axiomatic method as found in Euclid's *Elements*. Number and musical theory were the special, but not the exclusive, province of the Pythagoreans.

Three famous construction problems in nonmetrical geometry were long important. They are:

1. *The quadrature (squaring) of the circle*, constructing a square equal in area to a given circle.

2. *The duplication of the cube*, the so-called Delian problem, determining the edge needed to construct a cube that is twice the volume of a given cube.

3. *The trisection of an angle*, dividing an arbitrary angle into three equal parts.

By the time of Euclid, constructions solely with an unmarked straightedge and a compass held a privileged status, as Book I of his *Elements* attests. Aesthetic reasons may have prompted this restriction, since the straight line and the circle were considered the only perfect curves. According to tradition, it is Plato who made the straightedge and compass the sole authoritative instruments. Although classical Greek geometers could find no solution to the three problems satisfying these conditions—they are, in principle, insolvable—their investigation led to numerous mathematical discoveries, including mechanical solutions.

Fruitful study of these problems outlived classical Greece. The squaring of the circle especially captured the attention of major mathematicians into modern times. In his work leading to the invention of a first stage of calculus in late 1675, Gottfried Leibniz examined the arithmetical quadrature of the circle. Solving this problem depends on constructing a line segment $\sqrt{\pi}$ times the given circle's radius. Not until the late nineteenth century did Ferdinand Lindemann, when he proved that π is a transcendental number, show that $\sqrt{\pi}$ is not constructible. This means it cannot be the root of any polynomial equation having rational coefficients. Nineteenth-century mathematicians also showed that, when restricted to the basic instruments of straightedge and compass, the two other famous construction problems are also impossible.

Hippocrates of Chios (ca. 460 to 380 B.C.), the most famous geometer of the mid-fifth century B.C., is not to be confused with his contemporary, the celebrated physician Hippocrates of the nearby island of Cos. Slightly younger than Zeno, Hippocrates the geometer was born on the Ionian island of Chios. Since Samos, the birthplace of Pythagoras, is nearby, he may have come early under the influence of the Pythagoreans, with whom he would have a lifelong association. Like Thales, Hippocrates began his adult career as a merchant. Aristotle claims that he was "lacking in sense" as a merchant and was swindled out of his property by crooked Byzantine customhouse collectors. More likely, he lost his property when Athenian pirates captured his ship in the Samian War of 440 B.C. When he went to Athens to prosecute the pirates, he was obliged to remain for perhaps two decades. The outcome of his complaint is not known. In Athens, Hippocrates probably met some Pythagoreans and became proficient in geometry. He was one of the first in Athens to earn a living by charging fees for teaching mathematics.

The three special construction problems had engaged the attention of Athenian Sophists and geometers. Hippocrates addressed at least the squaring of the circle and duplication of the cube, producing imaginative results. By the mid-fifth century many geometric theorems were being established, and he must have systematically gathered these theorems and tightened their proofs. He represented his results by composing the first textbook on geometry that systematically presented theorems and proofs, the now lost *Elements of Geometry*, a title Euclid would later use.

Circle measurement had a long history, starting partly in the search for empirical rules to approximate more accurately volumes of cylinders and cones in commerce. The attempt to achieve the quadrature of the circle, yielding results on which Hippocrates' reputation primarily rests, was a prominent enough problem to gain reference in Aristophanes' *The Birds* of 414 B.C. as an issue among Greek geometers: "by laying out I shall measure with a straight ruler, so that the circle comes square to you."[183] The Athenian sophist Antiphon had attempted to obtain an exact measurement by successively doubling the sides of regular polygons inscribed in a circle, beginning with the square and adding isosceles triangles to each side to proceed to the next stage. But Antiphon only carried out this procedure for a finite number of doublings, and a substantial treatment of the concepts of continuity of magnitude and limits awaited Eudoxus.

According to Aristotle's *Prior Analytics* and Proclus, Hippocrates did not square the circle, but put this problem into a new form and employed related techniques to solve a related problem, the squaring of the lune. We know this is true, because the sixth-century A.D. Aristotelian commentator Simplicius preserved a fragment of Hippocrates' writings on the subject. This is the only sizable remnant of pre-Socratic Greek mathematics still preserved. The lune, a plane figure bounded by two intersecting circular arcs of unequal radii and concave in the same direction, resembles a crescent moon (*luna* = moon). Thus, Hippocrates proved the area of a rectilinear figure equal to the area of a curvilinear figure.

Aristotle's commentator Alexander of Aphrodisias explains that Hippocrates' attack on the problem begins with the simplest case. He starts with two right isosceles triangles circumscribed by a semicircle. Their hypotenuses are the sides of a square. He next constructs a semicircle having a hypotenuse as its diameter. The lower boundary arc of the lune is also a segment of the arc of the larger semicircle.

In the proof for squaring a lune, $\overline{AB}$ is the side of a square of area $\overline{AB}^2$. $\overline{AC}$ is a diagonal of the square and a diameter of the circle in which the square is inscribed. *AEB* is a semicircle having *AB* as its diameter. By the Pythagorean

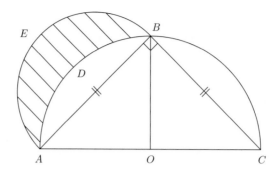

Lune $AEBD$ represented by the shaded area.

theorem, $\overline{AC}^2 = \overline{AB}^2 + \overline{BC}^2$. Since the right triangle is isosceles, $\overline{AC}^2 = 2\overline{AB}^2$, Hippocrates reportedly knew what would become Euclid XII.2: the areas of two circles are proportional to the squares of their diameters, or

$$\frac{\text{Area} \cap ABC}{\text{Area} \cap AEB} = \frac{\overline{AC}^2}{\overline{AB}^2} = \frac{2\overline{AB}^2}{\overline{AB}^2} = \frac{2}{1}$$

where $\cap$ means semicircle. Hence, the area of quadrant $AOBD = $ area $\cap AEB$. Subtracting the part common to both figures, ADB, yields the result lune $AEBD = \triangle ABO$, where $\triangle$ means triangle. The result is a quadrature; a curvilinear figure has been reduced to a rectilinear one.[184]

Having squared the lune, Hippocrates proceeded to attempt to square the circle similarly. For this purpose, he inscribes an isosceles trapezium in a semicircle. The trapezium consists of the diameter of the circle as its base and three consecutive equal sides. The ratio of the altitude to one of its equal sides is $\sqrt{3} : 1$ in this half of the regular hexagon. Next, semicircles are drawn with sides AB, BC, and CD as diameters. What follows is a variation of Hippocrates' findings working with a half regular hexagon $ABCD$ inscribed in a semicircle having center O. The three outer semicircles with equal diameters AB, BC, and CD form lunes at their intersections with the circumference of the original semicircle about O.

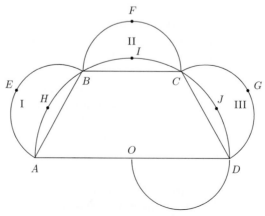

Departing for the sake of brevity from Hippocrates procedure, construct another semicircle with a diameter equal to radius OD of the original semicircle. Since the sides of the regular hexagon equal the radius of the original semicircle, it follows from Euclid XII.2 that $\cap ABE / \cap ABD = (\frac{1}{2}d)^2/d^2 = \frac{1}{4}$. Thus, the sum of the areas of the four smaller semicircles equals the area of the semicircle about O. Subtracting areas ABH, BCI, and CDJ gives the area of the half-hexagon $ABCD$. If any three lunes can be squared in the same fashion as Hippocrates' first lune, we could construct a rectangle equal to the semicircle about O. Since any rectangle can be reduced to a square of the same area, the result would be quadrature of the circle. But not every lune can be squared: only the special case given by Hippocrates with an isosceles triangle is possible. To overcome the limitation of straightedge and compass, in constructing lunes he introduced the concept of *neuousa* (inclining), in which a line moves, verging toward a given point. This *neusis* construction technique was to be important in ancient Greek problem solving, including, as we shall see, the trisection of an angle.

While Hippocrates seemingly put his findings in the most favorable light, he was too skilled a geometer to claim that he had proved the quadrature of the circle. In A 2 of *Physics*, Aristotle correctly observes that the supposed quadrature of the circle by lunes involves the fallacy that every lune can be squared. Nevertheless, several ancient commentators, such as Eudemus, less skilled in logic and geometry, credited Hippocrates with squaring the circle.

Another problem that Hippocrates examined, the duplication of the cube, romantic legend attributes to two sources. We know both through accounts of them by the Alexandrian Eratosthenes and the sixth century A.D. commentator Eutocius.

One poetic report is that King Minos was dissatisfied with a cube-shaped tomb that he had ordered built for his son Glaucus. The cube, equivalent to 100 feet on a side, the king found too small and he ordered its sides doubled while retaining the original proportions. Doubling its sides, of course, increases the volume eight times. According to Eratosthenes, the tragic poet who conceived this story erroneously thought that doubling the sides would merely double the volume. To double the volume requires increasing the sides exactly in the ratio $\sqrt[3]{2}$, a feat beyond the theoretical capabilities of Minos' architects. But cube duplication does not seem the point of the tale. In seeking to discover the origins of the cube duplication problem before Hippocrates, Eratosthenes probably misconstrued the story.

Eratosthenes' other account has the people of the holy island of Delos, the reputed birthplace of Apollo, suffering from a devastating pestilence about 380 B.C. An oracle told them that to rid themselves of it they must double the size of the existing cube-shaped altar of Apollo, while keeping its form. Hence the problem was called the Delian problem. The pestilence worsened when workmen doubled the edge. When a deputation from Delos went to Plato in Athens to ask his advice, he stated that the god speaking through the oracle was interested not so much in doubling the altar as a rostrum for prophesy as in reproaching "the Greeks for neglecting mathematics and making little of geometry."[185] But there were no cubic altars in Delos or anywhere else in ancient Greece.[186]

More likely, the source of the Delian problem for Hippocrates was an effort to take the doubling of the square and carry it to three dimensions by constructing a new square on the original square's diagonal. Study of the star pentagon had led Pythagoreans first to discover mean proportionals and then to recognize that the problem of doubling a square can be reduced to finding one mean proportional between two line segments. Probably by extending such research, Hippocrates reduced the duplication of the cube problem to finding two mean proportionals, x and y, between a given line segment (a) and another twice its length ($2a$), in modern notation, $a/x = x/y = y/2a$. Cross multiplying the first pair of ratios gives $x^2 = ay$, the second pair $y^2 = 2ax$. Squaring the first and substituting for y^2 will give $x^4 = a^2y^2 = 2a^3x$ or $x^4 = 2a^3x$ or $x^3 = 2a^3$, as presented in Euclid VII.11 and 12. The volume of the cube with side x is double the volume of the cube with edge a. Hippocrates' finding was thus not limited to lines in a 2 : 1 ratio.

Although Hippocrates' finding gives the desired answer, it cannot be constructed by straightedge and compass alone. By the later Platonic canons of ancient Greek geometry, his reduction was therefore inadmissible as a rigorous solution. His finding is significant nevertheless, because it reduces a problem of solid geometry to a form to which he applies a new group of plane geometric techniques, those of proportion theory. This was to be the form in which Archytas, Eudoxus, and Menaechmus, among others, attacked the problem of cube duplication.

While no detailed data exist on Hippocrates' lost textbook, *Elements*, his proofs related to the three famous construction problems along with remarks about Hippocrates' work by Simplicius indicate its subject and methodology. Since the three famous problems were fairly advanced in scope and method, Hippocrates' text probably contained some theorems that later appear in the first two books of Euclid's *Elements* and most of the content of the third and fourth books. In plane geometry he likely knew about congruence, areas, the Pythagorean and related theorems, circles, angles in circles, proportions, and numerous constructions. He also had at least an elementary working knowledge of the concepts of ratios of areas and similarity. According to Simplicius, Hippocrates deduced, but did not rigorously prove, that the ratio of the areas of two circles is the same as the ratio of the squares of their diameters or radii, as Euclid XII.2 does. Presumably he reached this result by inscribing regular polygons in the two circles and then increasing the number of sides of the polygons. At each stage he found the ratio of the two inscribed polygons to be the same as the ratio of the squares of the radii of the two circles.

Hippocrates' known proofs, though not flawless, reveal that enormous progress had been made toward demonstrative geometry in the 150 years since its supposed, modest beginnings in Thales' five theorems. Most historians trace to his lost *Elements* the start of systematic ordering of geometric theorems wherein the author distinguishes between more and less important theorems. Although neither Hippocrates nor his immediate successors established adequate starting points for deductive proofs, they apparently carried out substantial foundational studies that made possible the later findings of Plato's circle.

Hippocrates had little success with the third construction problem, the trisection of an angle. The sophist and polymath Hippias of Elis (fl. 400 B.C.) fared better. As a second-generation sophist, Hippias taught for a fee in the town of Elis in the northwest Peloponnesus. He visited Athens twice, and it has been conjectured that he wrote about Thales, providing the source of Aristotle's knowledge of that sage. Hippias was known for his exceptional memory. Plato criticizes his humorless boastfulness and thirst for flattery.

According to Proclus, Hippias invented the transcendental curve (it would be later known as the quadratrix) and used it to trisect "every rectilinear angle." It is the first known curve plotted point by point and not by straightedge and compass. Hippias begins with a square $ABCD$ circumscribing a quadrant of a circle BED. Two lines move with a constant velocity: radius AB rotates clockwise toward AD and BC approaches its parallel AD. These lines intersect at point F, and BFG, the path traced by F, is the transcendental curve.

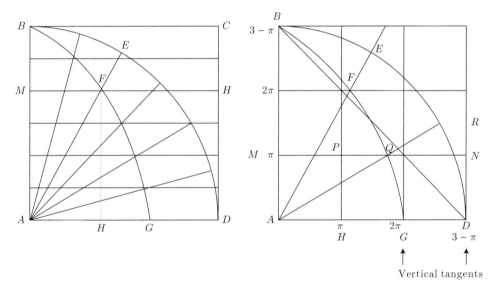

Vertical tangents

To trisect angle EAD, drop a perpendicular from F to AD, which meets AD at point H. Divide FH into three equal parts, with HP the first third. Draw a straight line parallel to AD through P meeting the transcendental curve at Q. Line AQ then trisects angle EAD.

The curve, if fully exploited, can obtain any fractional part of an angle. Though ancient commentators claimed otherwise, Hippias apparently lacked the sophisticated knowledge of it to attempt to square the circle. Probably Dinostratus, a century later, first applied the quadratrix and a double *reductio ad absurdum* argument in an effort to square the circle.

The Pythagorean Archytas of Tarentum (ca. 438 to ca. 347 B.C.), who continued the advance toward theoretical mathematics into the fourth century, was a political and pedagogical leader of the *mathematikoi* in southern Italy. The career of Archytas may have provided Plato with a model of the philosopher king. After the Sicilian Syracusan tyrant Dionysius drove most Pythagoreans out of nearby southern Italy, Archytas maintained the Dorian city of Tarentum (now Taranto) as the chief seat of their school. Although there was a one-term limit, the people of Tarentum elected him seven times to govern them as *strategos*, a member of the board of generals. Under him the Pythagorean program of study flourished in Tarentum. He is credited with first dividing Pythagorean mathematical studies into four parts: arithmetic, geometry, music, and astronomy. Plato and Eudoxus were his two most illustrious students. When Plato was held prisoner by Dionysius, Archytas' letter to the tyrant probably saved his student's life.

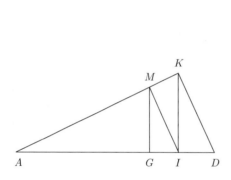

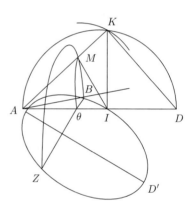

Figure 1   Figure 2

Archytas contributed to solid geometry. Eratosthenes states that his mechanical solution to the Delian problem is by means of an ingenious three dimensional construction. His method, which follows that of Hippocrates, reduces the problem to that of finding two mean proportionals (AI and AK in Figure 1) between two given line segments (AM and AD) located in similar right triangles. Exhibiting his exceeding stereometric insight, it defines the required proportionals by locating the intersection point of three rotating solids: a semicylinder havings as its base in Figure 2 semicircle ABD', a right cone generated by AB as semicircle ABD' rotates on axis AD', and a torus, a doughnut-shaped anchor ring, in which the inner diameter is zero, formed by an equal semicircle perpendicular to ABD' as AD turns on the pivot point. The common pivot point of the three solids is A. Archytas next draws a triangle, AKD, having as vertices their intersection point, the pivot point of the three solids, and a point whose distance from the pivot point equals the length of the larger line segment. As these two figures illustrate, he drops perpendiculars from the intersection point and a point at length $\overline{AM}$, the given line segment, from the pivot point. The result is three similar triangles and four lines, $\overline{AM}$, $\overline{AI}$, $\overline{AK}$, and $\overline{AD}$, in continued proportion. Archytas, rather than Hippocrates, also probably composed the proofs to justify Hippocrates' algebraic solution that are incorporated into Euclid's *Elements* Book VIII, 10–12.

True to the Pythagorean tradition, Archytas explored number theory as well. Most theorems and proofs in Book VIII on continued proportions, geometric progressions, numerical ratios, and the theory of irrationals are likely his. These proofs suffer from faulty logic. They build on a theory of divisibility and numerical ratios found earlier in Book VII of the *Elements*, which van der Waerden ascribes to Pythagoreans prior to Archytas.

Foundations of the ancient exact sciences and their interconnections and applications also engaged Archytas. He maintained that *arithmetica*, not geometry, is the fundamental science and perhaps rediscovered the Babylonian iterative process for finding square roots. He expanded the rudimentary Pythagorean number-theoretical foundation of music theory and applied the arithmetic and harmonic means to explain further the harmonic properties of scales. He viewed mathematics as the basis of astronomy and presented the first mathematical treatment of mechanics. Machines fascinated Archytas. The Roman architect Vitruvius claims that Archytas designed a wooden dove that could fly. Keep in mind in studying Archytas and other ancient Greek geometers that the method of Greek chroniclers was to credit important discoveries to heroic individuals as single step exploits. Modern historiography of mathematics rejects such interpretations.

3.5   Athens, the School of Hellas

In the two centuries separating Thales from Archytas, migrations and birth had placed Hellenic geometers in widely scattered places across the Greek world. Responding to Greek conquests in the west and Persian conquests in the east, the center of Greek mathematics and learning shifted in the late sixth century B.C. from Miletus and other cities in Ionia to the west to *Magna Graecia*, where Pythagoras and the Eleatics located. By the mid-fifth century the centrum moved again, this time to Athens, which was experiencing its later Golden Age from 479 to 404 B.C.

Nurtured by a strong economy and the political and academic freedom of the *polis*, Athenian culture flowered. Through commercial agriculture and trade, especially of wine and olive oil, together with shipbuilding, soldiering, construction, mining, metallurgy, and tribute from colonies in the Delian League, Athens grew rich. The underside of its economy was brutal slavery in the silver mines of Attica by prisoners of war. In a surge of political boldness, the Athenians extended their limited form of democracy to establish the freest government in the world. Its citizenry became intensely involved in public affairs.

The achievement was impressive, even though a majority of Athenian residents were not counted as citizens.

Under Pericles (d. 429 B.C.), a leader of the democratic faction, Athens became the cultural center of the Mediterranean. A patron of the arts, he not only subsidized drama competitions but, rather undemocratically, compelled citizens—as many as 30,000—to attend plays. In return the tragedians Aeschylus, Sophocles, and Euripides educated Athenians in politics, religion, human emotions, ethics, and morality. In acerbic satires such as *The Clouds*, the comedic genius Aristophanes ridiculed political demagogues and intellectual pretensions. The Periclean building and sculptural program beautified the city. On the *Acropolis* or upper city, the most brilliant of Greek sanctuaries assumed its final form. To the right of center along its Sacred Way is the Parthenon, the now ruined temple of gently curving columns and impressive statuary, including the gold and silver statue of Athena Promachos by Phidias.

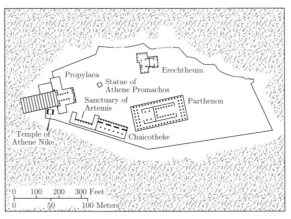

Diagram of the Acropolis

Parthenon and model of the Acropolis

By about 450 B.C., poets, artists, and such geometers as Hippocrates of Chios were drawn to Athens from all over the Greek world to participate in its stimulating intellectual life. Pericles boasted that Athens was now the School of Hellas, an image immortalized by Thucydides and reinforced by the sixteenth-century artist Raphael in his painting *The School of Athens*. This painting centers on the idealist Plato and down-to-earth Aristotle, each of whom advocated mathematical education.

Major wars, the Persian and Great Peloponnesian, framed Athens' Golden Age. During the first, the Persians were defeated on land at Plataea and on the sea near Micale in 479 B.C. This repulse of the Persian threat gave new confidence to the Greeks and unleashed a torrent of creative activity. Afterward, the Greeks divided into two major blocs. Athens, the leading naval power, and its Delian League enjoyed hegemony in the Greek world. Athens stood for freedom, and its opponent Sparta, for military discipline and order. This was a time of tension among the *poleis* and within each. Imperialistic expansion by Athens precipitated the ruinous hostilities of the Great Peloponnesian War from 431 to 404 B.C. During that war, Athens' trade was destroyed, her democracy overturned, her civic pride and virtue shaken, and her population ravaged by pestilence. In 404 B.C. she was forced to capitulate to Sparta.

4

Theoretical Mathematics Established in Fourth-Century Greece

Let no man unversed in geometry enter.
— Inscription above the gate of Plato's Academy

After two centuries of germination, theoretical mathematics took definitive shape in classical Greece during the fourth century B.C. The primary accomplishments were in arithmetic and geometry. In arithmetic, a few more precise and reflective mathematicians sought a general theory of proportionality that, unlike Pythagorean proportionality based on integers, could accommodate the new theory of incommensurable magnitudes or irrationals. But classical Greek literature does not support the conjecture that the inapplicability of Pythagorean proportionality to the new incommensurables provoked a "foundations crisis" in mathematics.[187] Instead, the incompatibility of incommensurables with integral proportions spurred technical achievements in geometry. Success in new methods of construction apparently influenced Eudoxus of Cnidus, who introduced a general theory of proportions by using continuously varying magnitudes. Eudoxus thereby set the geometric preferences and methods of theoretical mathematics for the rest of antiquity.

Geometry was the centerpiece of classical Greek mathematics. Probably, philosophy and geometry together intensified at Plato's Academy the quest for rigor in proofs. There Plato, young Aristotle, and Eudoxus provided heretofore missing starting points for formal geometric proofs in definitions, postulates, and common notions or axioms. Combining technical results with logical, step-by-step, and unambiguous formal proofs, the classical Greeks consolidated demonstrative geometry. They also extended geometrical models and relations into astronomy and mechanically solved the Delian problem. By sectioning solids, they generated a few special curves as well as theorems in higher geometry for curves more complex than the circle and straight line. The geometer Menaechmus most likely devised conic sections. Still another accomplishment was the fashioning of the complementary but reverse geometric methods of analysis and synthesis. Analysis, beginning with the general, derives conse-

96

quent constituent properties while synthesis begins with constituent properties and proceeds to establish the general.

Mathematics continued to be driven, in part, by practical needs, as in ancient Near Eastern protomathematics, but its scope now went far beyond the purely utilitarian. Mathematics fared well in a culture that encouraged critical inquiry, and the shift from oral to written tradition made it a staple in higher education. While thinkers in outlying Cyrene, Elis, and Cnidus produced some theoretical mathematics, Athens was the hub. Plato, who could recall when Athens relied mainly on oral instruction, incorporated mathematical studies into the curricula of his Academy, as did Aristotle for the Lyceum. The Academy and later the Alexandrian Museum made these studies fundamental in Greek higher education and transmitted them to generations of students.

4.1   The Glory That Was Greece: Fourth-Century Athens

Although her days of imperial sway had ended, the brilliant cultural and intellectual life of Athens continued in the period from her loss to Sparta through her conquest by Philip of Macedon in 338 B.C. and up to the death of his son Alexander the Great in 323. Indeed, when people repeat "the glory that was Greece," they generally include the cultural achievements of Athenians in the fourth century as well as those of Periclean times. With a population estimated at less than 150,000, the probable number at the start of the Great Peloponnesian War, Athens remained the most populous *polis*. In the midst of their troubled age, Athenians accomplished much. Their Middle and New Comedy explores domestic tragicomedy. Their sculpture, exemplified in *Apoxyomenos* ("the scraper"), turned from ideal abstraction to the depiction of the ordinary. Their temples are triumphs of balance, order, and proportion. Above all, Athens was the home of many distinguished philosophers, most notably Socrates, Plato, and Aristotle.

Philosophy was the first love of Socrates (469 to 399 B.C.), a stonemason by profession. Troubled by the loss of ethics in Athens during the Peloponnesian War, he set out to transform philosophy from a theoretical study primarily of the natural world into a study of human nature and meaning. Holding that rational principles, as opposed to traditional beliefs and local customs, are necessary for living in a world of disorder and corruption, Socrates claimed that real knowledge is within the mind of each person and needs only critical examination to bring it forth. That examination, as he taught and employed it, consists of a persistent technique of question and answer.

A Hellenistic statuette of Socrates

One of the two principal interpretations of mathematics today is as an introspective discipline, and so Socrates, deepening the introspective component of Greek thought, indirectly contributed to the expansion of mathematics. Though his technique had antecedents and Socrates derived it from Zeno, it is now known as the Socratic method. To the best of our knowledge Socrates, like Jesus, never wrote a line. His pupils, including Xenophon and Plato, transmitted to the far future the Platonic version of his thought. But his contemporaries judged him harshly as a corruptor of conventional ways and beliefs, citing as an example the conduct of his pupil, the opportunistic traitor Alcibiades.

4.2   The Theory of Irrationals: Theodorus and Theaetetus

Early Pythagoreans, limited to an *arithmetica* consisting of the theory of odd and even, as later expressed in Book VII of Euclid's *Elements*, had been able to prove only that the line segments $\sqrt{2}$ and perhaps $\sqrt{5}$ are incommensurable with unit length 1. They did not have the mathematical tools to prove the generalization that "Line segments, which produce a square whose area is an integer, but not a square number, are incommensurable with the unit length." In modern number theory this means that "The square root of an integer that is itself not the square of an integer is irrational." By the early fourth century B.C., two geometers were to generalize the theory of incommensurable line segments by recasting the foundations of mathematics as Pythagorean proportionality based on integers had set them. Jean Itard has put forth a provocative thesis that these two geometers, Theodorus of Cyrene and his student Theaetetus, were merely fictitious characters born in the fertile literary imagination of Plato.[188] Itard's claim, however, does not have the assent of most scholars, and at any rate narrative convenience argues for treating them as real historical figures.

Armed with the findings of Hippocrates on lunes and Archytas on number theory, Theodorus (ca. 465 to ca. 399 B.C.) and Theaetetus (ca. 417 to 369 B.C.) demonstrated that square roots from 3 through 17, excluding 4, 9, 16, are incommensurable with 1. Their insistence on commensurability with unit length 1 perhaps kept them from recognizing that incommensurables such as $\sqrt{18}$ and $\sqrt{8}$ are commensurable with each other because they have a common factor $\sqrt{2}$, a point Euclid was later to prove in the *Elements* X. 9. Their findings supported the division of all numbers into two figurated groups—"square" numbers and "oblong" numbers—and implied that, as Plato claims in 147C through 148B of *Theaetetus*, the square roots of all oblong numbers are incommensurable with 1; that is, these square roots are irrational and "evidently infinite in number." Contemporaries did not realize that the two were not merely extending Pythagorean research and thought but doing something quite new. Plato's employment of the term "incommensurable" in the dialogue *Theaetetus* for ratios previously described as *alogos* (inexpressible) or *arratos* (terms without a ratio) suggests the recognition that mathematics was on a new path.

Theodorus of Cyrene in North Africa, the mathematical tutor of Theaetetus and Plato, may have been a Pythagorean and contemporary of Hippocrates. The most reliable information about him comes from the *Theaetetus*. Here, in 164 and 165, Plato reports that Theodorus had been a disciple of the Sophist Protagoras, but turned from abstract humanistic speculations to theoretical geometry. Plato also informs Socrates that he learned geometry, arithmetic, harmony, and astronomy from Theodorus.

According to Plato, Theodorus was the first to demonstrate that the square roots of nonsquare integers from 3 to 17 are incommensurable with 1. Why begin with $\sqrt{3}$? Why end with $\sqrt{17}$? Plato's remark that "for some reason [Theodorus] stopped" at the square root of 17 would have been unacceptable to Socrates. The answers to both questions seem clear. Theodorus must have known the Pythagorean proof of the incommensurability of $\sqrt{2}$ with 1. A reason for stopping at $\sqrt{17}$ is that $\sqrt{18} = 3\sqrt{2}$ simply repeats, while the computation of $\sqrt{19}$ is very complex.

Theodorus may have used the spiral given here. This snaillike figure, proposed by the mathematician J. H. Anderhub only seven decades ago, employs diagonals of successive right triangles to provide base sides of new triangles. Each of these diagonals is the square root of the next integer. The construction provides each triangle with an outer leg that is 1, as well as the second leg that is the hypotenuse of the previous triangle. Mathematicians have called this figure the Ur-spiral, meaning the grandfathe of all spirals.[189] It is at $\sqrt{18}$ that the spiral first cuts across its initial axis. But since the spiral construction does not prove the irrationality of any hypotenuse, Theodorus' demonstration almost certainly proceeded in another manner. No account of his proof has survived, but since he was dealing with line segments he probably followed a geometrical demonstration.

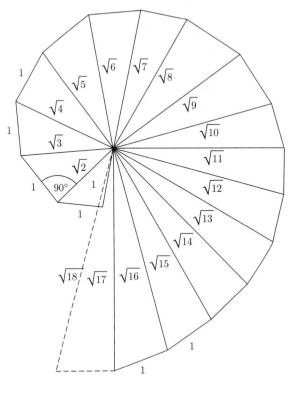

Ur-spiral

Theodorus' proof most likely dealt separately with each quantity $\sqrt{3}$, $\sqrt{5}$, $\sqrt{7}, \ldots, \sqrt{17}$, as Plato's dialogue suggests, by a repeated subtraction procedure based on what became Euclid X. 2: "If when the lesser of two unequal magnitudes is continually subtracted from the greater, the remainder never equals the one before it, the magnitude will be incommensurable." Pythagoreans had known that commensurable line segments are integral multiples of a common measure. Their procedure for finding the common measure was the same repeated subtraction technique, which ends after a finite number of steps. But for incommensurable magnitudes, the procedure suggested here as one that Theodorus probably followed would lead to the areas of a pair of lunes, whose ratio is the same as that of the starting pair of line segments. This suggests that the process does not end, an arithmetic truth that Theodorus did not prove. Aristotle calls this repeated subtraction the operation of *antihypophairesis*, or reciprocal subtraction. Application of the anthyphairetic procedure would explain Theodorus' stopping at $\sqrt{17}$. Computing what is essentially the continued fraction for $\sqrt{19}$ is formidable even with modern notation and techniques. The period recurs only after six stages of computation. The resulting anthyphairetic ratio for $\sqrt{19}$, moreover, cannot be drawn, especially in sand.[190]

The anthyphairetic procedure is similar to the repetitive procedures now associated with the Euclidean algorithm for obtaining a greatest common divisor and the determination of continued fractions, but is distinct from them. The Euclidean algorithm refers to divisions, not subtractions, and may have hastened the demise of the anthyphairetic approach.[191] Determining continued fractions involves a sophisticated generalization, which the ancient Greeks lacked.

Theaetetus of Athens, the leading geometer of the early fourth century B.C., was the son of a patrician. He met Theodorus either in Athens or in Cyrene. His luminous mind, like that of Socrates, resided in a lumpy body highlighted by a snub nose and slightly protruding eyes. His contemporaries widely respected and admired him. Plato, who esteemed him as second only to Socrates, made him a principal speaker in two dialogues, the eponymous *Theaetetus* and the *Sophist.* Our knowledge of Theaetetus' life chiefly derives from the dialogue that goes by his name. It is set on the day that Socrates faced his accusers in the year 399 B.C. The characters in the dialogue speak of the amiability, modesty, and logical acumen of Theaetetus, who was then a teen-ager. They pay him a particular compliment, rare for the Greeks, calling him a gentleman. Speaking (through Plato) of his penetrating mind, his teacher Theodorus states, "... this boy advances toward learning and investigation smoothly and surely and successfully, with perfect gentleness, like a stream of oil that flows without sound."

Theaetetus did not have long to complete his life's work. He taught geometry and astronomy at Heraclea on the Black Sea before returning to Athens about 375 B.C. Possibly Plato introduced him into the Academy. A war between Athens and Corinth in the summer of 369 cut short his brilliant career. Upon conscription into the Athenian army, he fought gallantly but was wounded in action near Corinth and contracted dysentery in camp. Carried back to Piraeus in serious condition, he died probably more from dysentery than from his wound.

4.3 Plato's Circle

Plato: Ideal Forms, Logical Realism, and the Academy

It was Plato (429–347 B.C.) who dominated Greek intellectual life in the early fourth century B.C. Drawn to the depth and subtlety of abstract thought in a time of turmoil, he compounded into an orderly whole many earlier, diverse philosophical and mathematical ideas from the Pythagoreans, the Eleatics, Socrates, and others. Except for a few letters, Plato transmitted his thought in the form of dialogues, making Socrates the leading philosophical debater in all of them but the *Laws,* his last and longest work. His preeminence derived not only from the power of his intellect and writings, but also from the long-lasting influence of the Academy, the school that he founded in Athens.

Plato was the son of two aristocrats, Ariston and Perictione. From his youth he was conscious of his social position and probably aspired to participate in public affairs. As a privileged lad in Athens, he received a fine education. When he was about twenty, he met Socrates, becoming one of his few regular students for the next eight years. After Socrates was condemned to death in 399 B.C., Plato and other disciples retired briefly to the town of Megara, located between Athens and Corinth, to pursue truth and master the "art of argument" (*logon techne*). Plato probably next traveled about Greece. According to Cicero, he visited Egypt before journeying to southern Italy and Sicily about 390. In Syracuse, the tyrant Dionysius the Elder listened to his discourses on morality and a sober life, and he may have briefly imprisoned Plato. The falling out with Dionysius resulted in cordial relations with Archytas in the nearby mainland city of Tarentum. That contact facilitated the improvement of Plato's knowledge of Pythagorean education, probably his chief reason for the trip to the West. Study under Archytas and conversation with Pythagoreans in the Peloponnesus deepened his knowledge of mathematics. Afterward, his work showed a strong Pythagorean influence.

Permanently impressed by the condemnation of his revered master as a public enemy, Plato returned to Athens in 388 B.C. his formative period over. Holding teaching and training to be the highest service man can render to his fellows, he sought to school his fellow Athenians in a knowledge of the Good: essentially virtue and right conduct. His work directed philosophy away from the cosmological interests of the Ionians.

Like Socrates, Plato maintains that the Good is real, an enduring, absolute standard, not relative and transitory, a matter of opinion as Sophists held; that it can be apprehended only by a trained mind cleared of prejudices, rigorously examining cherished opinions, and defining terms exactly; and that genuinely to know the Good necessarily compels the knower to be virtuous. Justice is equivalent to the application of the Good in civil life.

Correcting, in his view, the doctrine of the sixth-century Greek sage Heraclitus of Ephesus that "all things are in flux and nothing is stable," Plato, speaking through his reconstruction of Socrates, agrees in *Cratylus* 402A that we cannot gain true knowledge from "material things," for they are constantly in the process of becoming or perishing. No stable substance inhabits the sensible world. Plato will not, however, settle for a Heraclitean flux. To introduce stability, he adroitly reconciles elements of different philosophies into a new epistemology.

Platonic thought holds that independent of the material world of sensible things exists a spiritual realm of unchanging Ideas (ιδεεατ) or Forms (ειδη).[192] Forms constitute the ultimate reality: for each notion, kind of being, or material object, there corresponds a patterned Form, abstract and eternal. Like number in the Pythagorean universe, the Forms impose intelligible structure on the universe. The ideal Form of the horse, for example, is the archetype of all possible horses and makes it possible to recognize Trigger as a horse. The supreme ideal Form is the Good. Other exalted Forms are those of beauty, proportion, and justice. The doctrine of Forms, of course, which according to A.1.987 of Aristotle's *Metaphysics* continued the Socratic search for precise definitions begun in the sphere of morals, is purely concerned with understanding the higher Goods, not horses and fingernails.

Like Heraclitus and Parmenides, Plato insists that material objects and the senses are sources of opinion and error, offering merely evanescent appearances, on phenomena. True knowledge is of the archetypal Forms, accessible alone to the trained intellect. Final knowledge of triangles in geometry, for example, comes not from the imperfect imitations we draw with straightedge, but from mental apprehension of the perfect, immutable Form that is the archetype of all possible triangles. Only through strenuous, properly trained reasoning can the human mind discover Forms behind their ephemeral and shadowy images presented by the senses. Human knowledge is remembering more than discovery, a recall of prenatal contact with the ideal world of Forms awakened by glimpses of their flawed replicas in the sensible world. Plato's Realism—modern Idealism—is an uncompromising pursuit of basic, constant patterns of order in the universe.

Plato's *Politeia*, the *Republic*, offers strict and careful theoretical prescriptions for the moral, physical, and intellectual education of the elite after initial grounding in mathematics and the Socratic dialectic. Only through mathematics and the dialectic can aristocratic youth come to know the Good and avoid deceptions from what appears to the senses. Ignorance is the sole cause of vice. The guardians (soldiers), Plato announces in Book VI, also need knowledge of mathematics for effective military planning.

Shortly after he returned to the city in 388 B.C., perhaps as early as the next year, Plato set up his Academy. He probably lived in the neighborhood and taught there until his death. The Academy, with the motto "Let no man ignorant of geometry enter" inscribed over its gate, was one of the earliest European institutions of higher learning. Its students attended lectures either inside or outdoors. Lodgings were available. Like the Pythgoreans, students took meals in common. During Plato's tenure, Theaetetus, Aristotle, and Eudoxus studied at the Academy, and Theaetetus and Aristotle also taught there.

After preparatory studies and gymnastics, at age twenty highly promising students were selected to continue formal studies. Their interrelated, five-part curriculum was restricted to mathematical sciences: the quadrivium and solid geometry, as Book VII of the *Republic* describes them. Formal instruction in each of the five subjects could illuminate the theory or application of concepts, such as ratios. The quadrivium, previously restricted to Pythagoreans and other geometers, now became an integral part of the curriculum for Greek higher education. Plato disdained applied mathematics and protested the use of moving mechanical instruments in geometry. His proof theory was limited to drawings made with compass and straightedge. For those capable of coming to know the Good and the doctrine of Forms, the study of theoretical mathematics was a necessary beginning to transcending the knowledge of the experiential world.

The five-part theoretical mathematical studies consigned to a privileged few were preparatory to the dialectic (logic) stage. At age thirty after a second selection process, the brightest students advanced to a five-year course of sharpening their minds through dialectic exercises. The dialectic is a prolonged process of persistent, critical questioning that peels away falsehood or error until the practitioner arrives at a truth. One purpose of the dialectic was to compare and elucidate common elements in different disciplines and guide students to generalizations. As the selection process shows, not all citizens had the acumen or discipline needed for study of the dialectic. Plato recommends for the general citizens a less technical program of studies.

The Academy remained influential in Greek education long after Plato's death. While the Museum in Alexandria shortly after 300 B.C. supplanted it in mathematical leadership, the Academy was active in teaching philosophy until A.D. 529, when an edict of the Christian emperor Justinian disbanded it for being pagan and dangerous. In his *Decline and Fall of the Roman Empire,* historian and cynic Edward Gibbon, who portrays freedom as "the first blessing of our nature"[193] and extols Roman toleration, gives his perception of the edict of Justinian as fatal to classical antiquity and its exercise of deductive reason, making way for dependence on written revelation and the persecution of infidels and skeptics.

After sifting through Plato's dialogues, including fragments of *Platonicus* on the duplication of the cube preserved by Theon of Smyrna, Plutarch, and Eutocius, as well as authenticated portions of a letter by Eratosthenes on that topic,[194] scholars have long disputed Plato's technical skills in theoretical mathematics. Some consider him little more than a mathematical dilettante; others argue that he mechanically solved the Delian problem or substantially contributed to number theory. The appraisal of Harold Cherniss as modified by D. H. Fowler is broadly accepted today. Cherniss portrays Plato as an intelligent critic of method and formulator of broad problems but not a technical mathematician.[195] Fowler asserts that Plato possessed a "detailed knowledge of important characteristics and problems of technical mathematics."[196] He dismisses Jean Itard's suggestion that Theodorus of Cyrene and Theaetetus were merely Plato's inventions, which, if correct, would make Plato-Theodorus-Theaetetus nearly the equal of Eudoxus as a mathematician.

From Proclus to Fowler, classicists and other scholars have held that Plato identified for the literate sector of society major mathematical problems that required further study. His *Parmenides*, for example, refers to the Zenonian paradoxes, *Meno* to possible cube duplication, and *Theaetetus* to incommensurability. A passage in 819 and 820 of the *Laws,* dismissing as a "disgrace" the notion that all magnitudes are commensurable, adds gratuitously that people who have never heard of the theory of incommensurables "hardly deserve to be called human." Plato admits, however, having encountered the theory himself only late in life. Since he had discussed incommensurability in the dialogue *Theaetetus*, this admission in the *Laws* must refer to the time when he first learned of Eudoxus' geometric theory of proportions.

Fowler, like Cherniss, finds Plato a keen critic of method. His search for an axiomatic foundation for mathematical proofs is of fundamental importance. To emphasize the certitude of mathematical proofs against mere probability and opinion, he pursued missing starting points for formal proofs. The *Meno* and the mathematical excursus in *Theaetetus* argue the necessity of framing careful definitions and establishing these as an essential component in proofs. In 86e through 87b of *Meno*, 92d of *Phaedo*, and the *Republic* starting at 510c, Plato's treatment of reasoning from hypothesis, or what Hippocrates referred to as reduction, suggests that requisite starting points other than axiomatics remained vague in his thought.

Fowler departs from Cherniss in crediting Plato with serious collaborative research on mathematics with Academicians Amyclas and the brothers Menaechmus and Dinostratus, as well as a few original contributions. In number theory Plato reportedly derived the verbal equivalent of our formula $(2n)^2 + (n^2 - 1)^2 = (n^2 + 1)^2$ for obtaining Pythagorean triples, and he elaborated upon Theaetetus' work by showing that a rational number can be the sum of two irrationals. The *Timaeus*, his major scientific writing, constructs the five regular solids and associates them with the four Empedoclean elements—earth with the cube, air with the octahedron, fire with the tetrahedron, and water with the icosahedron—and the cosmos associated with the dodecahedron. Since he holds that the regular solids are the basic constituents of the cosmos, they are sometimes called "cosmic" solids. They are better known as Platonic solids.

Aristotle: The Theory of Statements and Potential Infinity

Plato's most famous student and rival was the more empirical Aristotle (384–322 B.C.). He was from Stagira, an Ionian colony near the Balkan kingdom of Macedon, where his father was personal physician to the kings. At about his seventeenth birthday, he came to Athens, where, among several Ionians and Macedonians, he enrolled in the Academy. He was to spend twenty years there. When Plato died in 347 B.C., Aristotle, who did not succeed to headship of the Academy, left Athens, staying away for twelve years. In 342 B.C. he went to Pella in Macedonia to tutor the thirteen-year-old prince, Alexander. After Alexander completed his tutorial in 340, he retained Aristotle as a trusted adviser. When Alexander put Athens under Macedonian control in 335 B.C., his teacher returned and set up a school called the Lyceum. For twelve years, Aristotle lectured at the Lyceum, where he introduced written examinations into Western education. The teacher and students often walked about in the open air discussing issues: hence their nickname, the Peripatetics. When Alexander died in 323 B.C., Aristotle's Macedonian sympathies drew threats against him in Athens. Saying, in light of Socrates' fate, that he did not want Athens to have the added disgrace of murdering another philosopher, he withdrew in voluntary exile to Chalcis, where he died the next year.

Aristotle classifies pure mathematics, by which he means arithmetic and geometry, as part of theoretical systematic knowledge in philosophy, together with

A Roman copy of a Greek bust of Aristotle

with metaphysics and physics, whose subdivisions include astronomy, cosmology, harmonics, mechanics, and optics. Mathematics is the most exact of the three, he claims in K. 7 of the *Metaphysics*, and offers a model for well-organized science. But Aristotelian science seeks efficient, material, formal, and final causes for physical effects and properties. To the extent that mathematics deals only with logical relations and abstract objects, it is arguably an inferior level of knowledge. Unlike his Pythagorean and Platonic predecessors, Aristotle, who attempts to segregate the sciences, does not present physics as a mathematical science. But mathematics does apply to physics and the subordinate disciplines of the quadrivium as a middle or mixed science. Although Aristotle allows mathematical abstraction in physical studies, he employs it not as a means of inquiry but mainly as an aid for subsequent exposition. He advocates the austere practice, followed in most surviving fragments of Greek mathematics, of suppressing the workshop apparatus, the differing techniques attempted and mistakes committed in the course of achieving a polished proof. Archimedes later supplemented the presentation of mathematics only in its finished show window form, or what Aristotle calls its didactic form. As transmitted by the notes of his disciples, Aristotle's scientific inquiries never reached the refined systematic stage: he left behind a record of preliminary steps in mathematical research.

Aristotle mostly accepts the Pythagorean and Platonic view of the contemplative, theoretical nature of pure mathematics and the beauty of its orderly arrangement. But he finds unsatisfactory some of Plato's other views on mathematics and its realm of application. Plato had ranked the Socratic dialectic higher than mathematics as a method of reaching sound conclusions. By the time he left the Academy in 347 B.C., Aristotle disagreed, in *Posterior Analytics* I and 1 and 2 of *Topics* I, insisting on the priority of mathematics. Proof theory in Greek logic and mathematics impressed him as more effective than the dialectic in activating the memory of prenatal forms. Although he accepts the existence of a world of Plato's ideal Forms, such as the Good, Aristotle does not concede to universal Forms a self-standing independence of the things of the sensible world. In a rupture with Plato, he argues that knowledge originates through the intellect's interaction with concrete, sensibly perceived particulars. Only through the concrete can the student attain to abstraction and a recognition of Forms. Learning mathematics entails actively abstracting concepts from observations of regularities in the behavior and nature of physical phenomena, not on a summoning up of prenatal memories. These regularities, Aristotle explains in ii, iv, and xxx of *Posterior Analytics* I, are described in the study, again abstract, of motion and change as these relate to space and time. The chief aims of science are to record, explain, and generalize. Synthetic mathematical proofs, which proceed from the particular to the general, allow the inquirer to achieve all three. In introducing syllogistic logic (*Topics* is a manual for it) and the controlled use of analogies, Aristotle extended the means of developing satisfactory proofs.

Aristotle, who concentrates on the method proper to various disciplines, articulates in *Posterior Analytics* an important theory of statements. A major step in the effort to axiomatize mathematics, it was to support a drive for disciplinary autonomy for the subject. This theory was the first to delineate the starting points for logical proofs in classical Greece. Recognizing that formal proofs must begin with basic statements that are indemonstrable, for otherwise proofs would involve an infinite regress, Aristotle distinguishes between two classes of such statements. "Common notions" or *axioms* underlie logical thinking as fundamental assumptions. They are universally true for any deductive science dealing with quantities. The statement "when equals are added to equals, the sums are equal" is an Aristotelian axiom. "Special notions" or hypotheses, later called *postulates*, apply only to a particular science and need not otherwise be self-evident. The statement "through every two points a straight line may be drawn" is an Aristotelian postulate taken to be true for geometry, but not to have the same application elsewhere. The collection of axioms and postulates should be the minimum number required for solving all problems in a given body of knowledge, and precise definitions will set the boundaries to that body. Definitions have to involve operational terms whose meanings are known prior to that of the word to be described. If that procedure is impossible, some operational terms can be defined.

Aristotle's theory of statements advanced the Greek formalization of mathematics, the more so when Euclid followed Aristotle's prescriptions in the axiomatic system of the *Elements*. An acceptable theory of relations, the second basic component of deductive reasoning, lay far in the future.

While method was his principal interest, Aristotle also carefully examined three related "objects" of mathematics: incommensurables, infinity, and continuity. These objects describe the behavior or condition of physical things, but are not themselves composed of matter.

Aristotle rejects the Pythagorean claim that numbers are primal atoms.

Study of Pythagorean number theory, however, fostered his fascination with incommensurable magnitudes. Although he discusses no surds in detail except for $\sqrt{2}$, whose irrationality he proves by *reductio ad absurdum*, he must have known about the work of Theodorus and Theaetetus, because he names and describes the reciprocal subtraction method of *anthyphairesis*. Though his mathematical background was apparently limited, Aristotle also demonstrates in M 2 1077 of *Metaphysics* a recognition that Eudoxus' general theory of proportions resolved the anomaly in Greek geometry provoked by the discovery of several incommensurable magnitudes.

Aristotle's study of incommensurability may have contributed to his speculation about infinity. Since Theodorus had based on anthyphairetic ratios his proof of incommensurable magnitudes, it seemed that analytical proofs can require an infinite number of steps. Added to this challenge to finite analysis was Plato's discussion in *Parmenides* of Zeno's paradoxes that involve infinite processes. In his review of the status of infinity in *Physics* III, Aristotle distinguishes between potential infinity pursued by an ongoing mathematical process, an exercise within unfettered imagination of increasing or subdividing without end, and actual infinity, the infinity reached by physically achievable steps. Rejecting the existence of actual infinity, he observes:

But my argument does not anyhow rob mathematicians of their study, although it denies the existence of the infinite in the sense of actual existence as something increased to such an extent that it cannot be gone through; for, as it is, they do not even need the infinite or use it, but only require that the finite [straight line] shall be as long as they please... . Hence it will make no difference to them for the purpose of demonstration. (Chapter 7, 207 b 27)

The concept of potential infinity allowed Aristotle to define continuity and thereby, at least to his satisfaction, resolve Zeno's paradoxes. *Physics* V places the concept of the continuous under the more general category of the contiguous, that which is successive and touches. Space, time, and motion, being continuous, are infinitely divisible. If space and time are infinitely divisible, as Aristotle stipulates in discussing Zeno's paradoxes, then motion can occur. Aristotle's discussion represents early efforts to unravel the two "objects" that they are rooted in: infinity and continuity.[197] It is also a part of his program to refute Parmenides' monism by constructing an extensive logical bridgework to legitimate passage from the One to the Many. In *Physics* VI, Aristotle dismisses Zeno's paradoxes as fallacies.

Eudoxus of Cnidus: General Theory of Proportions, Method of Exhaustion, and an Astronomical Model

Eudoxus of Cnidus (ca. 391–338 B.C.) was the foremost mathematician associated with Plato's circle. In antiquity he ranks second only to Archimedes. Probably a generation younger than Plato, he was born in the prosperous island-city of Cnidus on the Black Sea, perhaps as early as 408 B.C., but more likely in 391. A remark in his treatise on geography referring to Plato's death gives credibility to that date. Eudoxus died in his native town apparently at the age of fifty-three, highly honored as a lawgiver and teacher. For contemporaries, his most spectacular discoveries came in astronomy. He also won acclaim as an orator, philosopher, geographer, and physician; his friends called him Eudoxus the renowned.

Eudoxus began his distinguished, widely traveled career in education by studying the quadrivium under Archytas and medicine with Philiston in Sicily and at the medical school at Cnidus, which rivaled the school built by several generations of the family of Hippocrates of Cos. At the age of twenty-three he accompanied the Cnidian physician Theomedon to Athens. Theomedon found him lodging in the nearby seaport of Piraeus, where living was cheap. Eudoxus was determined to study philosophy and rhetoric under the great Athenian philosophers. To attend Plato's discussions, which he found stimulating, he had to walk daily from Piraeus to Athens and back, a two-hour trip each way. After two months of studies, he returned to Cnidus and probably was a physician.

About 365 B.C. Eudoxus visited Egypt for fourteen months, in part on a diplomatic mission. According to the Roman philosopher Seneca, writing in the first century A.D., Eudoxus acquired from the priests of Heliopolis a knowledge of planetary motions and Chaldean astrology. While in Egypt he made observations, prepared a map of the sky, and composed his eight year calendric cycle, *Oktaetris*. The inclusion in Eudoxus' sky map of the star Canopus, which he calls "the star visible in Egypt," and some constellations in the southern celestial hemisphere visible from Egypt give some support to Seneca's account. From Egypt, Eudoxus returned to Ionian Greece, specifically to Cyzicus on the Sea of Marmara, where he opened a school. Here he completed his book *Phaenomena* on risings and settings of constellations, his edited improvement of that work, *The Mirror*, and perhaps his greatest but lost writing, *On Speeds*, on motions within our solar system.

Several years before 350 B.C. Eudoxus turned over the direction of his school to a pupil and paid a second visit to Athens, this time as a master teacher accompanied by several disciples. An attraction to the Athenian masters and perhaps a failure of the Cyzicus currency prompted the visit. Plato gave a banquet in his honor. The two men respected each other. But on important points they differed. Plato opposed the Cnidian's doctrine that Forms or Ideas are "blended with observable things," along with Eudoxus' insistence that pleasure is the highest good, pleasure being not merely sensual gratification, but the satisfaction educed by a rational adherence to justice, honor, and moderation in all things. Out of his debates with Eudoxus concerning mitigations of the Good, Plato wrote *Philebos,* criticizing Eudoxean hedonism. In X.2 of the *Ethics*, Aristotle praises the character of that doctrine's Cnidian founder.

Shortly after Eudoxus arrived in Athens, the people of Cnidus overthrew their ruling oligarchy and established a democracy. The Cnidians moved the side of their city closer to the coast and appealed to Eudoxus to draw up a constitution for them. He returned to his native city and prepared the necessary legislation. He taught within the school of medicine and built an observatory, where he worked to improve his astral calendar. He also labored to keep in agreement with the basic calendar additions of feast days to various luni-solar calendars of different towns. His observatory was still seen two centuries later.

In mathematics, Eudoxus, concentrating on problem solving, improved methodology. His work is associated with the most creative period at Plato's Academy. Undoubtedly, Eudoxus witnessed and engaged in Academic debates between his own student Menaechmus (fl. 350 B.C.) and Speussippus, Plato's nephew and successor as director, over the primary concern and nature of mathematics. A strict Platonist, Speussippus held method to be paramount and preferred "theorems" over, for example, "problems," since theorems are closer to perfect contemplation. Mathematics is certain, being based on axioms inherent in the ontological structure of the universe. By contrast, Menaechmus believed that problems and constructions are primary. Viewing theoretical mathematics as hypotheses of the human mind to be tested rather than certitudes, he based geometry on postulates. Eudoxus agreed with Menaechmus. In methodology Eudoxus systematically applied to mathematics Aristotle's theory of statements and his own reflections on the debates between Speussipus and Menaechmus. Eudoxus carefully refined Aristotle's definitions, special notions, and common notions into the definitions, postulates, and axioms that are the foundation of Euclidean axiomatics.

Yet, the well-deserved mathematical reputation of Eudoxus rests chiefly not on his contributions to method, but on his general theory of proportions, his invention of the process later known as the method of exhaustion (the inexact term "to exhaust" was introduced only in the seventeenth century), and his application of spherical geometry to astronomy. The results of his work on proportions make up much of Books V and VI of Euclid's *Elements*; his studies of the method of exhaustion constitute a substantial portion of Book XII.

Eudoxus' general theory of proportions, which, from our vantage, amounts to a theory of real numbers, resolved the anomaly that the discovery of several incommensurables had introduced into Greek mathematics. Pythagorean proportionality was based on establishing, in our terms, the equality of two ratios between integers $a/b = c/d$. This had become inadequate because incommensurable line segments and irrational numbers are not ratios of integers. Within the educated public the incompatibility was not generally known, nor was it prominently noted among geometers. Still, members of Plato's Academy, including Aristotle, pondered whether geometers could validly extend to incommensurable quantities proofs regarding commensurable lengths, areas, and volumes. By an original departure from Theaetetus, Eudoxus salvaged and generalized the theory of proportions.

Proceeding as an algebraist, Theaetetus had considered every line segment, whether commensurable or incommensurable, as a separate entity with definite mathematical properties. Eudoxus, in introducing the general notion of magnitude, proceeds instead as an analyst. He ingeniously defines not only line segments but also angles, areas, and volumes by referring to magnitudes that may vary continuously. Continuously varying magnitudes can approach limits and be approximated with arbitrary closeness. Whereas Theaetetus had determined a ratio by means of a sequence of integers in arithmetic, Eudoxus determines a ratio from its place among rational ratios enclosing it on both sides. His new concept of ratio reinforced the earlier view that magnitudes to be compared must be of the same kind, for example, in constructing a ratio an area cannot be properly compared with a line.

Eudoxus' definition of geometric ratios by these magnitudes survived as Definition 5 of Book V of Euclid's *Elements*. It states:

> Magnitudes are said to be in the same ratio, the first to the second and the third to the fourth, when, if any equimultiples whatever be taken of the first and the third, and any equimultiples whatever of the second and the fourth, the former equimultiples alike exceed, are alike equal to, or are alike less than, the latter equimultiples taken in corresponding order.[198]

This passage means that, given arbitrary whole numbers α and β and the proportionality $a : b = c : d$, then

$$\alpha a > \beta b \quad \text{implies} \quad \alpha c > \beta d,$$
$$\alpha a = \beta b \quad \text{implies} \quad \alpha c = \beta d, \text{ and}$$
$$\alpha a < \beta b \quad \text{implies} \quad \alpha c < \beta d.$$

This definition, along with the concept of continuously varying magnitudes, constituted a theory both more general than *anthyphairesis* and more elegant. *Anthyphairesis* continually subtracts two unequal magnitudes in finding the greatest common denominator. That method appears in Euclid VII.1 and X.2.

Eudoxus' chief sources for this theory of proportions and its generalization, his method of approximation, or what is now called "exhaustion," were the description of infinite processes in Zeno's paradoxes, the squaring techniques of Hippocrates and Archytas, and Democritus' procedures to determine the volume of the cone and pyramid. Hippocrates' results were most important. To reach them, classical Greek geometers must have intuitively assumed that areas of simple curvilinear figures are geometric magnitudes of the same type as areas of polygonal figures and recognized two natural properties in areas of both. One is *monotonicity*: if polygon S is contained in curvilinear figure T, then, in modern symbols, area $(S) <$ area (T). The other is *additivity*: if R represents the union of figures S and T that do not overlap, then areas $(R) = \text{area}(S) + \text{area}(T)$. These properties are essential beginnings for the method of exhaustion. Eudoxus was the first to discover proofs for the geometrical conjectures of Hippocrates and Democritus about curvilinear areas, and therein he provided a rigorous, precise alternative to Hippocrates' vague notion of summing curvilinear areas S by filling them with a large sequence of polygons.

The Eudoxean theory of proportions and method of exhaustion contain the germ of the fundamental process of higher mathematics known today as the theory of limits. This process seeks to determine unknown quantities by an approximation procedure, based on known quantities, that never arrives at a final limit, but allows the inquirer to approach arbitrarily near to it. The theory of limits was a great discovery with few parallels in the history of mathematics. By applying Eudoxus' infinite approximation process within the sequence of rational numbers, later mathematicians were able, for example, to approach sufficiently close to the irrationals to prove theorems involving them.

For the most part, the Eudoxean theory of proportions facilitated rapid progress in Greek geometry, especially by Euclid, Archimedes, and Apollonius. But it had drawbacks. Number, the province of arithmetic, was now sharply separated from the spatial magnitudes of geometry. Magnitudes, which alone accommodate both commensurable and incommensurable quantities, are not to be considered numbers. Previously, Pythagoreans had stressed number theory. Henceforward, classical and Hellenistic Greeks paid less attention to *arithmetica* and number, preferring to convert algebraic problems into geometric problems. After Eudoxus, demonstrative geometry gained the deepest attention of Greek mathematicians—and their Latin medieval successors—and became the basis of rigorous mathematics.

According to Archimedes, Eudoxus also employed his technique of progressive, successive approximation to find areas and volumes of curvilinear figures. Theoretically, the area of a circle can never be completely exhausted by inscribed rectilinear figures. But Eudoxus' proofs do not require going beyond a finite number of steps. His proofs, which appear in Book XII of Euclid's *Elements*, are based on his general theory of proportions, approximations, and inequalities and double *reductio ad absurdum* arguments.

Consider Eudoxus' proof of Hippocrates' theorem as presented in Book XII.2 of Euclid's *Elements* that the areas of two circles (A_1 and A_2) are to each other as the squares of their diameters (d_1 and d_2); symbolically, $a(A_1)/a(A_2) = d_1^2/d_2^2$.

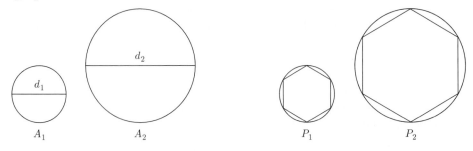

Eudoxus first supposes that $\frac{a(A_1)}{a(A_2)} < \frac{d_1^2}{d_2^2}$ or $\frac{a(A_1)}{a(A_2)} > \frac{d_1^2}{d_2^2}$ or $\frac{a(A_1)}{a(A_2)} = \frac{d_1^2}{d_2^2}$. Given an arbitrarily small area (ϵ), he chooses a similar regular polygon of a sufficient number of sides, P_1, to inscribe in A_1, where the area of P_1 differs by less than (ϵ) from $a(A_1)$. This suggests the same can apply for the area P_2 from $a(A_2)$. By sufficiently increasing the number of sides of the regular polygons, Eudoxus proves that either case of the inequality leads to a contradiction. If $\frac{a(A_1)}{a(A_2)} < \frac{d_1^2}{d_2^2}$ and $\frac{a(P_1)}{a(P_2)} = \frac{d_1^2}{d_2^2}$, applying cross multiplication followed by division by d_1^2 gives

$a(A_2) > a(A_1) \times \frac{d_2^2}{d_1^2} = S$ and $a(P_2) = a(P_1)\frac{d_2^2}{d_1^2}$. Let $a(A_2) - S = \epsilon > 0$ and $a(A_1) - a(P_1) < \epsilon$ and $a(A_2) - a(P_2) < \epsilon$. But since $a(A_2) - S = \epsilon$, this suggests that $a(P_2) > S$, which is a contradiction. Proceeding from the first inequality of less than and a lemma of equality thus gives

$$1 > S/a(P_2) = a(A_1)\frac{d_1^2}{d_2^2}/a(P_1)\frac{d_1^2}{d_2^2} = a(A_1)/a(P_1) > 1.$$

The second inequality of more than also produces a contradiction. Therefore, $a(A_1)/a(A_2) = d_1^2/d_2^2$. Using upper and lower bounding with similar regular polygons, Eudoxus proves a disproportionality in a finite number of steps.

Rigorous proofs of Euclid XII.2 and the entire method of exhaustion hinge on being able to make arbitrarily small the difference between the area of a curvilinear figure S and that of an inscribed regular polygon by sufficiently increasing the number of sides of the polygon. The author of XII.2 found the property of arbitrary smallness to be a consequence of repeated application of Euclid X.1:

> Two unequal magnitudes being set out, if from the greater there be subtracted a magnitude greater than its half, and from that which is left a magnitude greater than its half, and if this process be repeated continually, there will be left some magnitude which will be less than the lesser magnitude set out.

If the proof of XII. 2 originated with Eudoxus, as Archimedes claims in the introduction to *On The Method*, then there is good reason to refer to X.1 as Eudoxus' principle. As the leading geometer in Plato's circle, he at the least must have known it, for Aristotle nearly duplicates X.1 in *Physics* VII.10, 266. Some scholars have pointed out that Eudoxus also probably knew the fourth definition at the beginning of Euclid V: "Magnitudes are said to have a ratio to one another which are capable, when multiplied, of exceeding one another." But this statement requires a proof case by case to determine whether two magnitudes have a ratio. It lacks the generality of X.1, without which many of Eudoxus' proofs cannot be rigorous.

Archimedes subsequently refined Eudoxus' first stage of the method of exhaustion. Consider the status of Eudoxus' principle as a case in point. The principle is proved in Euclid X.1. Archimedes first realized that the assumption about arbitrary smallness cannot be proved and must be given as a postulate. In the treatise *On the Sphere and Cylinder*, he used the so-called Archimedean postulate, repeating arbitrarily small distances to measure off the distance between two points on a line segment and eventually to exceed it.[199]

With the embryo of a theory of limits and an intuitive grasp of the Archimedean postulate in the method of approximating to any desired degree, Eudoxus had broken the impasse created for Zeno in traversing the infinite subdivisions of a line. The ultimate arrival at geometrical points or zeros in the infinite subdivision process had occupied Zeno, because it raised the question of how an extended world could arise from extensionless points in the void. Eudoxus responded by basing the constructions and proofs of mathematics not on extensionless, geometrical points, but on distances between points that could be made as small as the geometer wishes. These are early indivisibles. For heuristic but not yet formal demonstration, Archimedes' *On the Method* next subdivided geometric figures into their indivisibles by parallel sectioning, therein essentially having an early form of infinitesimals.

At a time of intense study of celestial mechanics, provoked partly by Plato's *Timaeus*, Eudoxus became the first mathematician seriously to attempt to describe the intricate motions of celestial bodies, both daily and annually, as seen from the central Earth by using a mathematical model based on spherical geometry. *The Phenomena*, more exactly *The Mirror*, and perhaps with most refinement *On Speeds*, present an ingenious geocentric system with twenty-seven rotating homocentric spheres within a hollow sphere. The outer sphere carries the fixed stars, and the others account for motions of the sun, moon, and five known planets, including irregularities in planetary motion as viewed from the

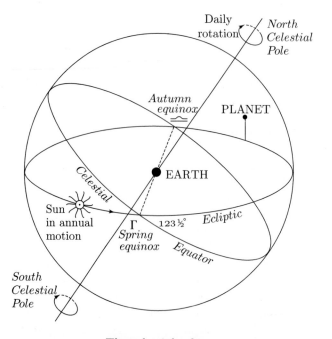

The celestial sphere

Earth. A suitable combining of rolling and unrolling spheres crudely approximates the observed motion of planets. Most notably, Eudoxus represents as a result of spherical rotations the apparent reversals of direction on the part of the planets observed from Earth, the phenomenon termed retrograde motion occurring against the background of the fixed stars. Superimposed rotations of inclined concentric spheres bear planets on trajectories that from Earth resemble what he terms a *hippopede* ("horse fetter," an elongated figure-eight curve), which contains periods of seemingly reverse motion. Eudoxus' successful application of spherical geometry opened the way for geometrical representations of the planetary system.

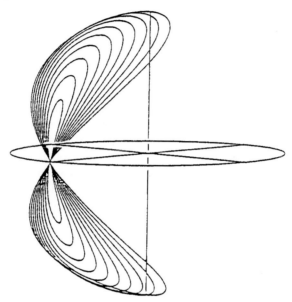

The hippopede

While the Eudoxean astronomical model testifies to its author's geometrical skills and attention to observation, it had its faults. It lacked exactness. Eudoxus and other Greek astronomers regarded luminosity as a measure of the distance of stars and planets from Earth. But each planet varies in brightness at different points in its orbit and therefore does not appear always equidistant from the Earth, as it should in a circular orbit. Nor was each planetary retrogradation identical with the previous one, as the hippopede suggests. Mathematical tools to describe motions of celestial bodies numerically did not yet exist. Astronomy, moreover, was not yet a science having a collection of increasingly accurate and properly ordered observable numerical data sufficient to establish the greater precision of any proposed theory.

Eudoxus' astronomical model might have disappeared had not Aristotle incorporated Callipus' correction and extension of it into his cosmology. Callipus added second spheres for Mercury, Mars, and Venus. In his schema, the planets rotate on a sphere turning on the poles inside another sphere. What Eudoxus may have proposed simply as an abstract geometrical representation,

Aristotle turned into a cumbrous physical mechanism to propel the heavenly bodies. By the dictates of his physics, celestial spheres are mechanically connected, rotating physical substances to cause motion, not merely mathematical constructs that describe it. Added retrograde concentric spheres and differing rates of movement fine tuned the apparatus to improve explanations, especially of irregularities in planetary motions. Through Aristotle, the Eudoxean model persisted and influenced astronomical thought through antiquity, the Middle Ages, and the European Renaissance.

4.4   Menaechmus: Conic Sections and the Method of Analysis and Synthesis

Before geometers could provide an adequate geometrical model for astronomy, they had to develop new mathematical knowledge, particularly of higher curves. Perhaps the most crucial discovery for astronomy was that of conic sections. Merely to recall Kepler's discovery of the elliptical orbit of Mars in the early seventeenth century is to appreciate their importance. Although Eudoxus did not discover conic sections, his attempt to solve the Delian problem of doubling the cube prepared the way. His student Menaechmus, continuing the search, as an interesting by-product apparently discovered conic sections.

Menaechmus, who along with his brother Dinostratus was a student at the mathematical school in Cyzicus, probably later became head of the school. Most likely through an introduction from Aristotle, who had close ties with the geometers there, he met and became a tutor to Alexander the Great. According to Stobaeus, when Alexander asked Menaechmus to teach geometry by an easy method he replied, "O king, for traveling through the country there are private roads and royal roads, but in geometry there is one road for all."[200]

Menaechmus knew that Hippocrates and Archytas had solved the Delian problem algebraically by finding two mean proportionals x and y between a and $2a$. In modern symbols, $a : x = x : y = y : 2a$. It follows that $x^2 = ay$, $y^2 = 2ax$, and $xy = 2a^2$. Modern coordinate geometry and notation show that x and y are the intercept points of two parabolas or a parabola and a hyperbola. Menaechmus certainly did not know that equations in two unknowns determine a curve. Possessing only a poor algebraic notation and lacking graphing techniques, the ancient Greeks had no simple way of finding a geometrical interpretation of the algebraic solution of the Delian problem. Instead, Menaechmus, so the commentator Eutocius notes, like Archytas found the mean proportionals by appropriate intersections of two conic curves traced by a mechanical device. The curves probably included the three types of cones: right angled, acute angled, and obtuse angled. The angle at the vertex defines these cones.

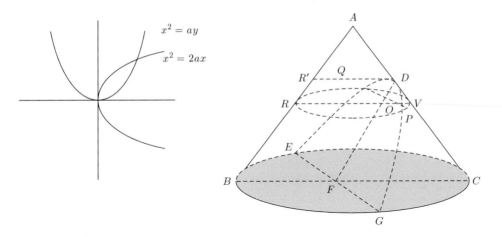

Plato criticized this mechanistic approach and called instead for a purely theoretical solution. But Eratosthenes suggests and Proclus states that Menaechmus discovered *ab initio* from this study that cutting the cone with planes produces three "conic lines" or curves. Noting that these curves were not yet generated as triplets, Wilbur Knorr disagrees. He contends that Menaechmus constructed these curves pointwise, possibly working with a means-finding apparatus.[201] Assuming that Menaechmus did treat these curves as sections of a cone, he supposedly deduced the asymptotic property of a rectangular hyperbola. Otto Neugebauer has speculated that the construction of sundials may have led to the investigation of conic sections.

That Aristaeus the Elder (fl. 330 B.C.) probably introduced the terms "section of a right-angled cone" for what came to be called a parabola, "section of an obtuse-angled cone" for a hyperbola, and "section of an acute-angled cone" for an ellipse has raised doubts that Menaechmus earlier knew these curves as plane sections of a cone. We do not know definitely even whether Menaechmus introduced the three types of "conic lines"and, if so, whether he considered them not as plane loci but as conic sections. But Eratosthenes and Geminus attribute both accomplishments to him, and this is the most plausible interpretation today. All three of Aristaeus' terms refer to the vertex angles formed by the sides with the axes. Cutting any right circular cone by a plane perpendicular to a generator produces a conic section, a procedure that early Greek geometers utilized. Apollonius subsequently coined the Greek terms parabola, hyperbola, and ellipse. In crediting Menaechmus with the terms "orthotome" (straight cut), "amblytome" (blunt or obtuse cut), and "oxytome" (sharp or acute cut), most ancient authors acknowledge that he constructed the three types of curves. Signs of knowledge of conic sections before Menaechmus are lacking, while after the appearance of his work their study blossomed.

By the late fourth century B.C., Greek geometers were creating the geometric method of analysis and synthesis that was to become a potent method of discovery and exposition for solving construction problems, particularly cube duplication. As Eutocius comments, Menaechmus adopted this method, whose origins lie in Hippocrates' reductions. In *Collection* VII, Pappus later described the method of analysis as admitting a figure as given and deriving consequences that follow from its constituent elements, while its reverse order of exposition, the method of synthesis, starts from constituent properties that are arranged so as to arrive at the desired construction.

5

Ancient Mathematical Zenith in the Hellenistic Third Century B.C., I: The Alexandrian Museum and Euclid

> Euclid alone
> Has looked on Beauty bare. Fortunate they
> Who, though once only and then but far away,
> Have heard her massive sandal set on stone.
> — Edna St. Vincent Millay

> When I think of Euclid even now
> I have to wipe my sweaty brow.
> — Swedish poet C. M. Bellman echoing
> the despair of past generations of schoolboys

During Hellenistic times from 338 to 133 B.C., advances continued in theoretical mathematics in the central and eastern Mediterranean, but unevenly. Geometers, most of Greek descent, began strongly. Indeed, theoretical mathematics may have been the highest point in the general autumn flowering of ancient Greek thought and culture. Like literature and the visual arts, it was innovative, but Hellenistic scholars also labored to assemble definitive compilations of classical sources. Highly innovative mathematicians attempted to fit together in harmonious patterns historically disparate mathematical ideas. Hellenistic geometers, like poets and visual artists, sought beautiful patterns. Most notably, they developed canons of beauty that expressed a refined understanding of geometrical shapes and acoustical laws of music. Their standards are very much alive. "Beauty," the British mathematician G. H. Hardy was to write two millennia later, "is the first test: there is no permanent place in the world for ugly mathematics."

Whether or not it dominated the intellectual flowering of the third century B.C., theoretical mathematics did reach its zenith in antiquity during that century. In his classic textbook, the *Elements*, Euclid collected and codified the results of Plato's circle and its Hellenic sources. This text was a vital link in the transmission of mathematical ideas from classical Greek thinkers to geometers of the third century and beyond. Three brilliant geometers came shortly after Euclid. Archimedes extended and skillfully exploited the Eudoxean method of approximation; the polymath Eratosthenes founded mathematical geography and invented a "sieve" for finding prime numbers in arithmetic; and Apollonius exhaustively developed the theory of conic sections. This level of mathematical achievement was not to be equaled again until the seventeenth-century scientific revolution.

Although the central and eastern Mediterranean remained the geographical setting for mathematical activity during Hellenistic times, the intellectual center was no longer Athens or any other city on the conquered and depopulated Balkan mainland of Greece. Athens had been quickly relegated to the periphery of the intellectual and artistic world. Three other great cities supported by thriving commerce and enlightened, or simply ambitious, rulers had supplanted it. Syracuse in Sicily, the Athens of the West; and Pergamum, the capital of the Attalid dynasty in western Asia Minor, only briefly nurtured theoretical mathematics. Alexandria, the capital and chief seaport of Ptolemaic Egypt, quickly emerged as the principal center of mathematical activity and would retain that status throughout the entire Hellenistic and Roman eras. Almost every noteworthy mathematician of both ages studied, taught, or lived there. Some gaps of more than a century between leading names suggest that the mathematical tradition was not uniformly continuous.

5.1   The Hellenistic Age

The term "Hellenistic" was first coined in the nineteenth century to demarcate the post-Hellenic period of more than two centuries, when Greek culture spread to Egypt and well into Asia. The new age was largely the outcome of the sudden rise to eminence of the kingdom of Macedon in the north of the Balkans.

By the fourth century B.C., continuing suicidal warfare and the loss of western markets for their wine, olive oil, and manufactured goods had undermined the Greek *poleis*, now unstable and vulnerable to external attack. The Greeks did not have long to wait. After military tutelage in Thebes acquainted him with Greek vulnerability, Philip II reorganized and enlarged his army and unified Macedon. Under Philip, who reigned from 359 to 336 B.C., Macedon conquered all coalitions on mainland Hellas except for Sparta. Beginning with lavish gold and silver bribes and an army of forty thousand men, Philip steadily extended his power southward. Meanwhile, in fiery attacks known eponymously as "Phillipics," the great Athenian orator Demosthenes warned that Philip was a dangerous enemy. Demosthenes rallied the Greeks too late. In 338 B.C., Philip defeated a coalition of Athenian and Theban forces at Chaeronea. His eighteen-year-old son Alexander led the decisive blow, a massed cavalry charge. Chaeronea marks the end of classical Greek freedom and autonomy and the beginning of the Hellenistic age.

The hard-driving and hard-drinking Philip II did not live to enjoy the fruits of his conquests. In 336 B.C., while preparing for a Persian campaign, he was assassinated, a not uncommon practice of ancient statecraft. After the Macedonian fashion, the army elected his son Alexander king.

Aristotle teaching Alexander

The twenty-year-old Alexander, who was to reign from 336 to 323 B.C., embarked on a brilliant but ephemeral career. Probably lured by the enormous resources of the vast but weak Persian Empire, he proceeded with his father's plan to invade it. In 334 B.C. he crossed the Hellespont (Dardanelles) into northwestern Asia Minor with only about 30,000 infantry, 5,000 cavalry, and few supplies. Rejecting a proposed scorched earth retreat that might have defeated him, the Persians stood and fought.

Alexander won battle after battle. Entering the city of Gordium, according to a famous anecdote, he came upon an intricate, sacred knot tied around a pole. When told that the man who could untie the sacred knot would rule all of Asia, he reportedly simply slashed it with his sword. The poet-classicist Robert Graves says this sacrilege symbolizes the decline of Greek wisdom. In 333 B.C., Alexander defeated the main Persian army under Darius III at the river Issus in Syria, sending Darius fleeing eastward. After Alexander destroyed the previously impregnable Phoenician city of Tyre, the Egyptians welcomed him as a liberator-god. At the Canopic mouth of the Nile, he founded the first and most splendid of seventeen cities that would be called Alexandria. Leaving Egypt in pursuit of Darius, Alexander continued his conquests in the east. In 331 B.C. he defeated the Persians at Gaugemala, near ancient Nineveh. The following year he entered Babylon and the Persian capital, Persepolis. At Persepolis, he seized the royal treasury, ending his fiscal difficulties. After burning Persepolis, probably in revenge for Xerxes' invasion of Greece and Macedon, Alexander continued to pursue Darius, who was murdered by his own nobles. He proceeded across the steppes of southern Russia to Bactria (Turkestan), where in 327 B.C. he married the princess Roxane.

Alexander's troops crossed the Indus River into present-day Pakistan, but weary from fighting and confronted by military elephants, they refused to go farther. The enraged monarch was forced to turn again to the West. In 325 B.C.

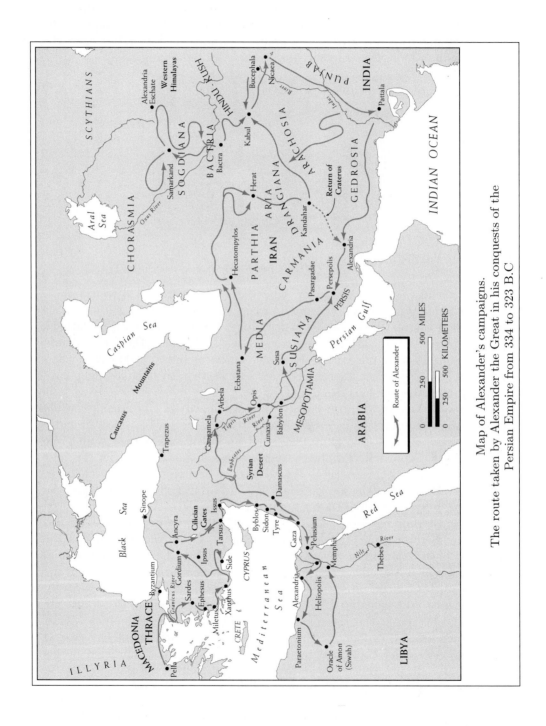

Map of Alexander's campaigns.
The route taken by Alexander the Great in his conquests of the
Persian Empire from 334 to 323 B.C

he was back in the Persian city of Susa, arranging the marriage of over nine thousand Macedonian officers and troops to Persian and Median princesses and commoners to symbolize the new amalgamation of Persians and Greeks as the ruling groups of his empire. In 323 B.C., while in Babylon, Alexander caught a fever that grew virulent. He died suddenly at the age of thirty-three.

Political unity was never achieved. When Alexander lay dying, an aide asked to whom his crown should be left. The answer was a single word "Kratisto" (to the strongest).[202] Shortly after Alexander's death, the generals on whom the conqueror had depended began to fight among themselves. During the next half-century, three successor empires emerged: the Ptolemies ruled Egypt; the Seleucids, Asia Minor and Persia; the Antagonids, Macedon. The Attalids also soon carved a small kingdom around Pergamum in Asia Minor.

Alexander's career deserves those terms like "turning point" or "watershed" that historians so relish. Hellenistic social and political life differed greatly from that of classical Greece. Vast monarchies replaced Greek *poleis* as the central unit of political life. Eclectic cosmopolitanism that appreciated the unity of the Mediterranean world replaced Greek ethnocentrism, which distinguished between Greek and barbarian. As democracies faded, ruler worship appeared, perhaps traceable to the cult of the pharaohs and to Greek adulation of heroes, but it did not take root in Greece and Macedon. Alexander's conquests, moreover, opened vast territories in the Middle East stretching into India to a two-way transmission of culture.

Hellenistic high culture spread thinly and unevenly across the Middle East. It made its greatest impact in large cities, the foci of the Hellenistic world. The new culture drew heavily on classical Hellenic language, ideas, institutions, arts, sports, and manners. Widespread use of Greek made inevitable the diffusion of Hellenic thought. Cultural borrowing increased as *koine*, a form of Greek common speech, became the international language of government and commerce, displacing Aramaic. Alexander's "Successors," or *Diadochi*, enlisted Greeks and Macedonians for their bureaucracies, armies, athletics, free professions of commerce, medicine, and architecture, and high cultural activities. Cosmopolitanism encouraged intermarriage in the urban upper classes, whose members easily acquired a Greek education and learned Greek manners. The commercial prosperity of early Hellenistic times expanded an urban middle order, many of its members acquiring a degree of Greek learning.

Greeks meanwhile confronted a new array of Eastern mystery religions, emotional in conviction and centered in personal as opposed to civic associations. The chief mystery religion of the Hellenistic Age was an amalgamated cult derived from the Egyptian goddess Isis, which may suggest an improved status for women among the upper strata of society. Now Isis became a Greek goddess bestowing civilization, law, fertility, and immortality. Another presence was monotheistic Judaism, reproving the antics of Olympic gods, of whom the Greeks themselves had become skeptical. Jews, at least in corners of the Hellenistic world such as Egyptian Alexandria, practiced a moralism incompatible with the Greek habit of nudity in gymnasia and athletics.

Philosophy shifted from the extroverted politics and confident humanism of the fifth century B.C. to an introspective attentiveness and a resignation that Platonism had to a degree anticipated. Athens remained a center of philosophic studies. While Aristotle's Lyceum engaged in historical studies, Plato's Academy turned toward skepticism. These schools were joined by two influential groups, the Epicureans and the Stoics. Epicurus (d. 271 B.C.) pursued not knowledge as its own reward, but *ataraxia*, a state of pleasant well-being attained by rational disciplined moderation; he withdrew from politics and concentrated on friendship, rather than competition. Believing the universe to be composed of Democritan atoms, Epicurus took sense perception to be the basis of human knowledge.

Stoics also concerned themselves with individual conduct and a proper understanding of nature. Their founder, the Phoenician Zeno (d. ca. 263 B.C.), proclaimed as a model "the Wise Man," who accepts freely his role in a divinely ordered universe and recognizes all men as brothers. Wise men are to pursue a virtuous life, which includes temperance, courage, and fairness, in living in reasoned conformity to nature. Zeno's doctrines carried well beyond the *Stoa Polike* (painted Portico) in the Athenian marketplace where he taught. His legal and ethical ideas strongly influenced Romans, normally suspicious of Greek philosophy, along with medieval philosophers and jurists, the humane savants of the eighteenth-century Enlightenment, and western adherents to the ideal of the brotherhood of humanity.

In the new temples and government buildings and gridiron plans for cities like Alexandria, Rhodes, and Pergamum, architecture in the Hellenistic monarchies retained the Hellenic sensitivity to line and composition. Sculpture, on the other hand, gradually moved away from the refined idealized representation of the fifth century B.C. Sculptors now mastered expression and physiognomy in an emotional and realistic mode of representation. A statue expressing the new realism well is "the Dying Gaul," which portrays every detail of the warrior's agony.

The few surviving fragments of Hellenistic historical writings suggest that most historians of the period emphasized sensational and biographical detail and that none equaled Thucydides in rigorous analysis. The greatest Hellenistic historian, the Greek Polybius of Megapolis (d. ca. 120 B.C.), conscientiously recorded the rise of Rome, while nearly two millennia later English historian Edward Gibbon was to chronicle its fall. Polybius attributes the growth of Rome to Fortune or Chance no less than to design, the military discipline of the Romans, and their administrative efficiency. Although less charming than Herodotus and less incisive than Thucydides, Polybius has a broader perspective than both.

5.2   The Alexandrian Bridge

Alexandria, then, was the intellectual and commercial hub of the Hellenistic world.[203] It linked classical Hellenic learning and culture with that of later times. From shortly after Alexander the Great founded it in 331 B.C. until its conquest by the Arabs in A.D. 642, Alexandria was the capital and chief seaport of Egypt. Astutely placed at the westernmost distributary of the Nile, it was well suited for trade along the widest channels where the Nile entered the Mediterranean. Its location connected the rich Nile Valley with the entire Mediterranean world. For centuries it was to dominate trade and communication between Mediterranean Europe and east Africa, Arabia, India, and China. Alexandria had scant competition at the time of its founding, for Alexander had destroyed Tyre, and Egypt had no other important coastal port.

During the Hellenistic age, the Ptolemaic dynasty ruled Alexandria and made it the chief city of the richest and strongest successor empire. Alexandria's political, commercial, and intellectual importance would not lessen appreciably when the Ptolemaic dynasty ended with the death of Cleopatra in 30 B.C. Egypt then became a Roman province of which Alexandria remained the capital. Under the Romans, it was the leading provincial capital, ranking second in political importance only to imperial Rome. It was also the granary of the Empire. For all its opulence and cultural grandeur, Alexandria was not immune from recurring times of trouble. Like other ancient cities, it experienced cycles of depression along with continuing corruption. Its demise in antiquity, however, did not set in until the late third century A.D.

Under the Ptolemaic dynasty, whose first members promoted trade and industry, Alexandria grew into a flourishing commercial center. By ruthless predation and taxation combined with the perennial fertility of the Nile Valley, the Ptolemaic government accumulated immense surpluses of grain and foodstuffs from Middle Egypt. An effective royal monopoly over grain sales accounts for much of its wealth. As Aristotle observes in 1330a of the *Politics*, the great cities of antiquity often lived at the expense of the industrious peasantry. Most Egyptian peasants were no more than hereditary serfs bound to large estates. From every part of the Greek world and later from the Roman, buyers flocked to Alexandria to purchase grain. The city's artisans, most of them of Greek descent, manufactured papyrus, glassware, linen, woolen goods, herbal medicines, spices, cosmetics, and perfume. The city was also a port of entry for wood from the Near East, spices from Arabia, copper from Cyprus, gold from Abyssinia and India, tin from Britain, elephants and ivory from Nubia, silver from the northern Aegean and Spain, carpets from Asia Minor, and silk from China. Many of the imported materials were reworked into luxury articles. Ancient Alexandrian traders fostered across the Mediterranean a taste for luxuries in the Oriental and the Aegean style.

Alexandria could not have achieved its success in international commerce and industry without the multiracial migrants who came from across the eastern Mediterranean and swelled its early population to hundreds of thousands. From its beginning the city's population was divided into three nations. Foremost were the ruling elite of Macedonians and Greeks. Ancient Alexandria was a Greek and not an Egyptian city. From the start the Macedonians and Greeks dominated commercial life with the assistance of two other free nations, thinly Hellenized Jewish merchants and ethnic Egyptians. Polybius notes that the

Greeks for centuries denied citizenship to the Egyptians in Alexandria. The city also had large numbers of slaves. Within the polyglot city, knowledge was transmitted and occasionally fused through intellectual exchanges enriched by trade with Phoenicians, especially Seleucid Babylonians, or Chaldeans. Although no demographic records exist, the population of Alexandria likely exceeded two hundred thousand within a century of its founding. By the time of Caesar Augustus, it was surely over 600,000 and perhaps as large as 1 million.

Alexandria attained a beauty and magnificence that probably surpassed Rome's. In 295 B.C., Ptolemy Soter (Savior), the first Ptolemy, had set about the task of building Alexandria into a beautiful city. He and his successors succeeded admirably. On a low hill in the Greek section, or royal quarter (*Basileia*), overlooking the eastern harbor were impressive royal palaces and temples with walls of alabaster and shimmering white marble. Courts of justice were connected with groves and flower gardens, and the Museum complex, probably located near the juncture of Nabi Daniel Street and Al-Harraya Avenue, had many arcades. The city possessed canals, broad colonnaded boulevards, and spacious parks. Its large northeastern harbor offered a fine approach to ships. At this harbor's entrance stood the famous twenty-four story cylindrical Pharos lighthouse. Built by the engineer Sostratus of Cnidus about 270 B.C., it became one of the seven wonders of the ancient world.

During the Hellenistic Age, the immense financial resources that the Ptolemies gave to scholarship made Alexandria the foremost center of learning. No other Hellenistic court extended comparable financial support. The early Ptolemies considered their patronage of scholarly studies a royal virtue. They invested their money wisely. Their central project was proposed by Demetrius of Phalerum, a pupil of Aristotle's Lyceum who later served as governor of Athens and was a member of Ptolemy Soter's court. It was for the building of a research institute modeled on Aristotle's Lyceum and similarly dedicated to the nine female divinities, patronesses of the arts known as the Muses. Hence the name of the temple complex was the *Mouseion* or, in Latin, Museum. Its first task was to carry out a systematic inventory of knowledge. Ptolemy Soter, who reigned from 305 to 285 B.C., like Alexander a student of Aristotle, founded the Museum; his two successors Ptolemy Philadelphus (d. 246 B.C.) and Ptolemy Euergetes (d. 221 B.C.) enlarged it. The Museum was to have a companion *bibliotheke*, or library, to house manuscripts.[204] The idea of a symbiotic library probably derived from Assyria, Babylon, and most of all pharaonic Egypt, as well as from Aristotle, who amassed the first sizable private library. Before Alexander's time, few Greek tyrants had collected books. The Ptolemies attracted leading scholars from across the Greek world to work in the expanding Museum. The cities of Athens and Cyrene were major sources of talent. In this international recruitment, the Ptolemies encouraged the unrestricted movement of men of learning.

The members of the Museum established the highest scholarly standards. Librarians collected and edited Hellenic classics and compiled catalogues of them. Grammarians, lexicographers, and critics seriously examined different versions of a classic to establish its correct text. These scholars then composed a biography of each author and wrote commentaries. They paid particular

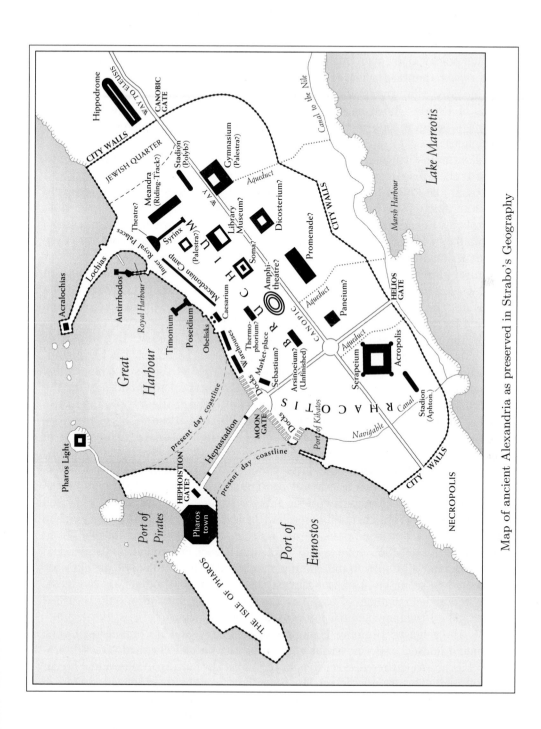

Map of ancient Alexandria as preserved in Strabo's Geography

attention to preparing correct texts of Homer's *Iliad* and *Odyssey*, a basis of Greek education despite the decline of the Olympian gods. Their work gave rise to the science of philology. Another Alexandrian group developed a distinctive poetry that, even with the strength of mystery cults, appealed to the mind rather than the heart. The new verse ranged from charming to pedantic and included the *Argonautica* about Jason and the Golden Fleece by Apollonius of Rhodes (born ca. 295 B.C.). Also at the Museum were historians, botanists, geographers, medical men, astronomers, natural philosophers, and mathematicians.

Like the Institute for Advanced Study at Princeton or All Souls College at Oxford, the Museum was a philanthropic foundation. The scholars lived and worked at the king's expense. The fortunes of the Museum and its research, therefore, fluctuated from one generation of monarchy to another. By the time of Ptolemy Euergetes, the Museum consisted of residence halls, a refectory, colonnaded walks or peristyles, rooms for physicians and surgeons to carry out experiments in anatomy and physiology, an astronomical observatory, a botanical garden, and a zoo. Its learned men did not have to wander outside the Museum grounds. One ancient satirist, Theon of Philus, sourly called the Museum "the birdcoop of the Muses." Its critics charged that like chickens pecking for corn and grain, they pecked for royal subvention.

At the north end of the Museum complex was the library. Before part of it was destroyed in a fire in 47 B.C., the library and its annex, the Serapeum, apparently held between 400,000 and 700,000 papyrus scrolls. When Julius Caesar ordered his occupation forces to set fire to the Ptolemaic fleet in the eastern harbor, and the city's Etesian winds spread the conflagration, the flames reached the library. The ancient Alexandrian library had no modern inventory system, and the size of its collection is only a rough estimate. The extent of the loss in 47 B.C. remains in dispute. Seneca suggests that 40,000 scrolls perished, while Plutarch would later claim that the complete collection was lost. That neither of two important contemporaries, the gossip Cicero or the possible eyewitness Strabo, mentions this development suggests that Plutarch is exaggerating.[205] In their bibliomania, Ptolemaic officials had instructed sea captains to seek out new books, so one section of the library was labeled "from the ships." An anecdote relates that Ptolemy III once ordered all ships searched for books, and all books found there confiscated. Other Ptolemaic agents were instructed to scour book markets in Athens, Ephesus, Rhodes, and Syracuse. They demanded originals of major works for copying, but often only returned copies. The Ptolemaic acquisition of books was rapacious.

By the reign of Ptolemy Euergetes, the library and its expansion to the Serapeum housed papyrus scrolls containing the best of classical Greek literature. Librarians gathered extensive materials on the culture of pharaonic Egypt as well, including its mysticism. In time they brought together the corpus of Hermes Trismegistus, or "Thrice Greatest," supposedly a god who had given the ancient Egyptians their arts and sciences; but modern scholars now believe that Hermes was an Egyptian sage of the second century A.D. The *Corpus Her-*

meticum, a product of Hellenistic enterprise dealing with astronomy, astrology, medicine, alchemy, and Greek philosophy, includes some elements of Egyptian magical and occult lore.   Librarians also assembled and produced works on Babylonian astronomy and religion and, by the late Alexandrian period, the Jewish cabala.   Cross-cultural influences were strong.   Ptolemy Philadelphus invited Jewish scholars to translate into Greek the first five books of the Old Testament.   This began the effort to have a Greek version of the entire Hebrew Scriptures that was later to be called the Septuagint, since about seventy scholars started the project.   Later Christian translations of the Old Testament were based on the Septuagint.   Beyond preserving original texts outside the Greek tradition, the library commissioned translations into Hellenistic Greek of Chaldean, Elamitic, Ethiopian, Persian, Phoenician, and in time Latin works.

Alexandria's thriving commerce gave resident mathematicians part of the practical basis and financial support for research that, in some instances, achieved spectacular results.   Mercantile enterprises used an elementary applied mathematics, probably initially indebted to Athenian practices.   In prosperous years as many as 20 million bushels of grain were shipped from the city.[206]   Scribes recorded costs and computed profits.   The Alexandrians needed well-built ships as well as sae roads for the unhindered passage of caravans. The Ptolemies, who controlled sea transport, financed substantial programs for both.   They retained naval architects and technical advisers to design and build roads.   Some of these nonartisanal builders must have had mathematical training.

The decision in Greek Alexandria to emulate the earlier Athenian model of basic education provided a strong theoretical foundation to mathematical study. Book VII and the rest of Plato's *Republic*, which covers the quadrivium—arithmetic, geometry, harmony, and astronomy—and the dialectic, together with 1337b of Aristotle's *Politics* and Isocrates' essays on oratory, constituted for the early Ptolemies and Museum scholars a base of education.   Both Plato and Aristotle had accorded mathematics a fundamental role in higher education.   By the early third century B.C. the Alexandrians were retaining more of this component of the Athenian educational tradition than were the Athenians themselves.   Both the Platonic Academy and the Aristotelian Lyceum were turning away from studies in the mathematical sciences and engaging primarily in philosophical work.

Royal patronage and a fruitful interaction of mathematicians with other scholars provided a supportive environment for research in theoretical mathematics that occasionally went far beyond practical needs.   This was probably most true during the early Hellenistic age.   Persuaded that geometric studies added luster to their court, the early Ptolemies and a few successors likely gave geometers considerable freedom.   Royal support perhaps continued on a lesser level or intermittently after Ptolemy Euergetes II, who ruled from 145 to 116 B.C., appointed military officers to direct the Museum, and the Romans later made it a sinecure for political favorites.   In early Alexandria, dogmatic religion imposed no fetters on scientific research.   Neither Greek nor Egyptian priests had powerful political influence at court.

From the beginning, Alexandrian geometers were in close touch with two other disciplines—geography and astronomy. Geographers had to search for improved trade routes for the Ptolemies. This took them to areas bordering the Red Sea, East Africa, and the Arabian peninsula. They looked to geometers, astronomers, and cartographers for assistance in map making. Like the classical Greeks, Alexandrian scholars recognized the importance of astronomy for time telling and for agricultural, navigational, and religious purposes. They recorded positions of celestial bodies, studied the flourishing astronomy of the Babylonians, or Chaldeans, developed new geometrical models of the universe, and worked to create a quantitative astronomy with which to compute and predict positions and paths of celestial bodies. Not until Roman times would the evolving synergism between astronomy and geometry give rise to spherical trigonometry, including the basics of plane trigonometry.

5.3   Euclid: A Conjectural Life

Euclid (ca. 330 to ca. 270 B.C.) began the great mathematical tradition in ancient Alexandria. Some classical scholars, such as Englishman Ivor Bulmer-Thomas, regard him as the founder at the Museum of a school of mathematics that was without peer in antiquity.[207] The assertion about a formal school is plausible, but basic source materials have not survived and only a few Hellenistic geometers are definitely known to have worked at Alexandria. Euclid must have worked on mathematical problems with associates at the Museum and gathered a group of students around him. He was broadly interested in the mathematical sciences, writing on astronomy, optics, music, and probably mechanics. Above all, he compiled and systematically arranged the unrivaled mathematical text *Elements* in thirteen parts, or what he called books. The English translation of the *Elements* without commentaries fills a single large volume.

Neither Euclid's parentage, birthplace, and birthdate nor whether he was a freeman or a slave is recorded. Only two biographical details have survived, and they come from Pappus (or perhaps one of his interpolators) and Proclus almost seven hundred years after Euclid died. Pappus declares in VII.35 of his *Collection* that Apollonius spent a long time in Alexandria with the disciples of Euclid. The inference placing Euclid's life span in the period between the mature Plato and Archimedes is drawn from Proclus, who in the "Catalogue of Geometers" declares that Euclid was "younger than Plato's circle but older than Eratosthenes and Archimedes."

A search of the extant mathematical literature by Euclid's immediate predecessors and immediate successors has been made to support or revise the conclusion of Pappus and Proclus. But a reference to the *Elements* in I.2 of Archimedes' tract *On the Sphere and Cylinder*, long believed to confirm the later date, has proved to be a marginal gloss added by a later interpolator. The presumed sequential relationship between the works of Aristotle and those

Euclid, a sixteenth-century depiction by André Thevet

of Euclid seemingly makes him younger or later than Plato. While Euclid's *Elements* skillfully employs Aristotle's theory of statements with its axioms and postulates, Aristotle's discussion of the equality of the base angles of an isosceles triangle ignores the neat Euclidean formulation, which strongly suggests that Euclid I.5 was not at hand. Was Euclid then an Aristotelian? Proclus maintains, to the contrary, that his encouragement to the study of mathematics and his ending of the *Elements* with the construction of the five Platonic solids were the works of a Platonist. That may be so. But the testimony of so zealous a Neoplatonist as Proclus hardly establishes the affiliation.

If Euclid was the man the ancients believed him to be, there are materials for a conjectural biography. These include the early history of Alexandria, the writings of Pappus, Proclus, and Stobaeus, and internal evidence from the *Elements* itself.

Apparently, Euclid did not come to Alexandria in 332 B.C. when Alexander founded it. Nor did he arrive a decade later when the city began to grow. Demetrius of Phalerum perhaps invited Euclid to Alexandria shortly after 300 B.C., when Demetrius was recruiting scholars for the Museum and library. While governor in Athens, Demetrius may have learned of the reputation of the young geometer, who also lived there. Apart from the testimony of Proclus, it seems reasonable to believe that Euclid studied in Athens at the Platonic Academy, which was the chief center of mathematical learning in the late fourth century B.C. At the Academy he would have studied not only Aristotle's theory of statements but also the original geometrical ideas of Theaetetus and Eudoxus. He may also have come across early compilations of geometry that were useful as points of departure for the *Elements*, possibly the works of Hippocrates of Chios and Leon, a member of Plato's circle. Like Aristotle, he probably studied the mathematical compilation of Theudius of Magnesia, which from what we know was the immediate precursor of his *Elements*.

If Euclid arrived in Alexandria shortly after 300 B.C., he must have quickly advanced in mathematical studies and gained a group of disciples. Presumably, his teaching and writing had already established his reputation. We do not know what his first books were. The only internal or external evidence on their order comes from the citation of his *Optics* in the preface to his *Phaenomena*. Perhaps Euclid completed the *Elements* early in his stay in Alexandria, since it is the fundamental work.

All that we know about the character of Euclid as a teacher and scholar comes from anecdotes of dubious authenticity. According to Proclus, Ptolemy Soter, after visiting a class, asked Euclid whether there was "a shorter way to the study of geometry than the *Elements.*" Fearlessly, Euclid told the king that "there is no royal road to geometry!" (Stobaeus, remember, recorded a nearly identical colloquy between Menaechmus and Alexander.) A more persuasive anecdote has Pappus (or an interpolator) praise Euclid as "most fair and well disposed toward all who were able in any measure to advance mathematics, and although an exact scholar not vaunting himself." Though this statement is part of a passage intended to denigrate Apollonius, there is no reason to question its assessment of Euclid's character. In *Eclogues* II.31 Stobaeus relates that, after having understood only the first theorem, a beginning student asked what was to be got out of learning such things. Calling a slave, Euclid said, "Give [the student] three obols, since he must need make gain out of what he learns." That Euclid was aware of the precarious position of intelligence in a world of power and money is believable.

Euclid's name became a synonym for geometry. His fame rests almost exclusively on the *Elements*, which became a classic soon after its publication. Archimedes refers to it as the standard textbook of basic mathematics, and older students and schoolchildren were to study it for the next two thousand years. Ancient and modern authorities have singled out Euclid as "the Writer of the *Elements*" or simply "The Geometer." Only in the early twentieth century was his supremacy ended.

5.4   The *Elements (Stoichia)*

Content and Axiomatics

The *Elements* has enormously influenced the western mind. Except for the Bible, no other book has had as many translations, editions, and commentaries. Since its first printing in 1482, over a thousand editions of the *Elements* have appeared. Its geometrical conception of mathematics greatly influenced the natural sciences—in medieval Arabic and Latin natural philosophy as well as in Isaac Newton's *Principia* of 1687, which follows the format of the *Elements*.

Its conceptions were basic to Kant's *Critique of Pure Reason,* published in 1781. Forty-year-old Abraham Lincoln mastered its first six books to make him a more exact reasoner, and Albert Einstein at twelve experienced Euclidean geometry as "a second wonder" in his life. Einstein later said about the incident, "it is marvelous that man is capable at all to reach such a degree of certainty and purity in pure thinking, as the Greeks showed us for the first time to be possible in geometry."[208]

At present, scholars disagree about the purpose of the *Elements.* Was it a treatise for learned mathematicians or a text for students? Its omission of three fields of mathematics well known at its time of composition suggests an answer. It treats neither logistics nor computation and hence was not a practical aid for business. And the absence of conics and spherical geometry strongly suggests that the author did not have in mind a treatise simply for learned mathematicians. Rather, he sought to prepare a school text, a view reinforced by the author's skillful progression of materials from the simple to the complex, which perhaps shows a pedagogical as well as a logical preoccupation.

Popular misconceptions about the content of the *Elements* are more serious. It is not a treatise containing "elementary" theorems, meaning those that are simple and uncomplicated. Instead, its "elements" are fundamental theorems. Euclid follows Aristotle, who says in 998 a 25 of *Metaphysics* that "We give the name 'elements' to those geometrical propositions the proofs of which are implied in the proofs of all or most of the others." Proclus correctly likens the relationship of Euclid's theorems to geometry to that which the letters of the alphabet have to language. Nor does the *Elements* deal only with plane geometry. Plane geometry is the subject of only seven of its books, and these include geometrical algebra and proportions. Three of the books treat arithmetic or number theory and another three solid geometry. Book XII on solid geometry deals with the method of exhaustion: the beginnings of the basic process of higher mathematics now known as integration.

In brief, the subjects of the thirteen books of the *Elements* are:

Plane Geometry

I. Preliminary principles (definitions, postulates, and axioms) along with congruence theorems and geometry of straight lines and rectilinear figures

II. Transformation of areas

III. Major propositions about circles

IV. Construction of regular polygons of three, four, five, six, and fifteen sides

V. Eudoxus' theory of proportions applied to commensurable and incommensurable magnitudes

VI. Application of this general theory of proportions to similar figures

Theory of Numbers

VII. Pythagorean theory of numbers

VIII. Series of numbers in continued proportion

IX. Miscellany on the theory of numbers, including products and primes

Plane Geometry

X. The classification of certain incommensurable magnitudes (irrationals)

Solid Geometry

XI. The geometry of three dimensions, particularly parallelepipeds

XII. Areas and volumes found by the method of exhaustion

XIII. Inscription of the five regular solids in a sphere

In later antiquity it was a frequent practice to attribute to famous authors books that they had not written. Older editions of the *Elements* added two books with further propositions on regular solids. Both postdate Euclid. Book XIV was written by Hypsicles (fl. ca. 150 B.C.). Its chief finding is that the ratio of the surface of a dodecahedron to that of an icosahedron inscribed in the same sphere is the same as the ratio of the volumes. The third section of Book XV, a lesser work on angles of inclination, may have been written out by a pupil from the lectures of Isidorus of Miletus. Isidorus, who lived in the sixth century A.D., was one of Justinian's two architects of the temple of Hagia Sophia (Holy Wiom), a museum in present-day Istanbul.

Essentially the *Elements* culminated the classical Greek tradition in theoretical mathematics. Most of all, Euclid appears to be a talented editor who recognized, compiled, and skillfully arranged a vast body of independent materials from earlier Hellenic thinkers. It is believed that parts of Books I, II, VII, VIII, and IX are based on the work of Pythagoreans and parts of Books III and IV on that of Hippocrates of Chios. Euclid also drew on materials from Plato's circle. Books X and XIII likely stem from Theaetetus and Books V, VI, and XII from Eudoxus, while the concept of exact definitions is from Plato and the basic outline of the axiomatic method from Aristotle and Eudoxus. The *Elements* appeared at a time that allowed Euclid to systematize mathematical results of Plato's circle. With the exception of advanced geometry, it put into final form classical Greek geometry.

The quality of Euclid's exposition is uneven, which seems mainly related to his sources and slightly to the variety of scribal hands that the *Elements* passed through before printing. Books in which he excels draw on excellent sources. One of the two most admirable books of the *Elements*, the ingenious Book V, is apparently indebted to Eudoxus, and the other, the richly subtle Book X, to

Theaetetus. Three nineteenth-century responses to Book V typify its esteem among mathematicians. Arthur Cayley observed: "There is hardly anything in mathematics more beautiful than the wondrous fifth book."[209] Among Germans, Karl Weierstrass employed definition 5 to define equal numbers, and Richard Dedekind to define irrational numbers.[210] By contrast, Book VIII, based on the Pythagoreans, contains cumbrous enunciations, some repetition, and even logical fallacies. For some modern critics this unevenness among books raises questions about Euclid's ability. Already in antiquity, Pappus proposed another plausible interpretation. Euclid had historical sensibilities and thus respected the traditional teaching of arithmetic, choosing to include it in Books VII through IX with few revisions.

Euclid was not merely a compiler and editor of the *Elements*; he made original contributions to demonstrative geometry. Most important is his refinement of a comprehensive axiomatic method with its notion of formal proof and a strictly deductive ordering of theorems.

As noted in Chapter 4, the roots of the axiomatic method may be traced in classical Greece at least to the fifth century B.C., when the Eleatics used reason as a criterion to reach philosophic truths, and Hippocrates of Chios made the first known attempt systematically to arrange geometric theorems. As late as Plato, the axiomatic foundations of mathematics remained vague. Material beginning at 86 c of *Meno* and 510 c of *Republic* joins with 92 d of *Phaedo* and 162 e of *Theaetetus* in demanding mathematical proofs that contrast to the mere probability of opinion. Plato stimulated an interest in exploring the role of primary principles in the foundations of mathematics. Aristotle's theory of statements accepts Plato's insistence on careful definitions and adds the concept of common notions or axioms and that of special notions or postulates. Eudoxus includes all three, emphasizing postulates in probably the first mathematical axiomatics.

Seeking to avoid circular arguments and provide sound starting points for mathematical reasoning, Euclid begins the *Elements* with a set of three types of first principles that correspond to those of Aristotle and are concordant with Eudoxus' approach. His first principles consist of twenty-three explicitly stated (although sometimes vague) definitions together with five postulates and five common notions, or what Proclus would later call axioms. These definitions, postulates, and axioms are the foundational rules of the game. Building on them, Euclid derives almost exclusively by deductive reasoning an orderly progression of 465 propositions. His demonstrations of these propositions employ high standards of consistency, rejecting any deduction of contradictory theorems from the axioms and postulates. Proofs of each proposition conclude with the Greek formula for "which was to be proved," known to students by its Latin abbreviation Q.E.D. (*quod erat demonstrandum*); proofs for problems conclude with "which was to be done," the familiar Q.E.F. (*quod erat faciendum*).

Book I of the *Elements* begins abruptly with a list of twenty-three definitions taken from accepted classical Greek ones. These definitions may suggest the state of geometrical knowledge before the *Elements*. Here are definitions 1 through 5, 17, and 23:

1. A point is that which has no part.

2. A line is length without breadth. [The Greek word that English renders "line" means "curve."]

3. The extremities of lines are points. [The *Elements* has no line extending to infinity.]

4. A straight line is a line that lies evenly with the points on itself. [In English, this definition is tautological. A mason's level or looking along a line with one eye may have suggested it.]

5. A surface is that which has length and breadth only.

   . . .

17. A diameter of the circle is any straight line drawn through the center and terminated in both directions by the circumference of the circle, and such a straight line also bisects the circle.

   . . .

23. Parallel straight lines are straight lines that, being in the same plane and being produced indefinitely in both directions, do not meet one another in either direction.

Although no substitute for extensive experience in working with mathematical concepts, these initial definitions do successfully condense these concepts into short sentences, conveying to each reader of the *Elements* the same pictorial sense.

From the viewpoint of twentieth-century mathematics, these definitions have faults. Definitions 1 and 3 are vague, describing only what the terms are not. Definition 2 introduces the concept "length without breadth"; definition 4 "lies evenly"; and definition 17, the "circumference of the circle." These are more complex than the original terms, require further definition, and lead possibly to circular definition. According to twentieth-century axiomatics, it is a mistake to attempt to define all terms. Euclid understandably failed to appreciate the necessity of leaving a few primitive terms undefined. Introducing in definition 22 terms like "oblong," "rhombus," and "rhomboid" that he never uses thereafter in the text departs from his usual economy. Neither does he formally define parallelogram after his definition of parallel lines.

Euclid next gives his ten grand assumptions for constructing the edifice of geometry. These assumptions consist of five postulates and five common notions or axioms. Euclid accepts the distinction that Aristotle had made in *Posterior Analytics* and Eudoxus refined. Roughly speaking, postulates are fundamental truths of geometry and axioms are self-evident truths dictated by the nature of our logical thinking and applying to all of mathematics and science. In this view, to deny axioms is to deny logical thinking. Over time the distinction between the two has blurred. Most modern mathematicians list all ten of Euclid's assumptions as axioms or postulates, but, more in line with Eudoxus, skeptically accept them as arbitrary statements to be tested, rather than as self-evident truths.

Thomas Heath's translation of J. L. Heiberg's text of the *Elements* is the most reliable English rendition. Its presentation of Euclid's postulates and axioms renders them:

POSTULATES

1. [It is possible] to draw a straight line from any point to any point.
2. [It is possible] to extend a finite straight line continuously in a straight line.
3. [It is possible] to describe a circle with any center and radius.
4. All right angles are equal to one another.
5. If a straight line falling on two straight lines makes the interior angles on the same side less than two right angles, the two straight lines if produced indefinitely will meet on that side on which the angles are less than two right angles. [The Parallel Postulate]

COMMON NOTIONS

1. Things equal to the same thing are also equal to one another.
2. If equals be added to equals, the wholes are equal.
3. If equals be subtracted from equals, the remainders are equal.
4. Things that coincide with one another are equal to one another.
5. The whole is greater than the part.

To be meaningful, a set of axioms must by modern standards be complete, containing eveything to be used in a theory, so that no tacit assumptions are made. They must also be consistent, allowing no internal contradictions. Each axiom and postulate must be independent, incapable of being proved from the others. From Euclid's time until the mid-nineteenth century the chief axiomatic question was whether the fifth (parallel) postulate is independent. Subsequent geometers would seek unsuccessfully to demonstrate it from the other postulates. Euclid's recognition of its necessity reflects his genius. Strictly from the standpoint of axiomatics and although it is a legitimate interest, this has no bearing on the logical validity of Euclidean geometry.

Modern axiomatics instead criticizes Euclid's axioms and postulates most for their lack of completeness. Beginnings of such criticism can be traced at least to the time of Pappus. Consider common notion 4 and postulate 4. Postulate 4, which states that all right angles are equal to one another, establishes the right angle as a determinate magnitude normative for measuring other angles. At the same time it tacitly assumes congruence, and both it and common notion 4 assume that, after displacement or superposition, a geometric figure will remain the same or be invariable. Bertrand Russell called "a tissue of nonsense" the assumption that without further conditions a triangle can be moved without any alteration in its structure. Proposition 1 of Book I of the *Elements* on the construction of equilateral triangles tacitly assumes that intersecting circular arcs have one point in common. Other tacit assumptions occur in Euclid. All are usually ascribed to oversights resulting from extensive familiarity with his subject. Or they may stem from a perceptive intuition based not on a methodological blind spot, but on a vision of geometry shared with leading members of Plato's circle. Modern mathematics, after two thousand years of studies of Euclid's axiomatic method, has discovered problems also with its definitions and arrangement of axioms and postulates. "The value of . . . [Euclid's] work as a masterpiece of logic," Russell announced in 1902, "has been . . . exaggerated." Common notion 4 has a geometric nature and seems better located with the postulates. In 1899, German mathematician David Hilbert's *Grundlagen der Geometrie (Foundations of Geometry)*, presenting the first complete set of axioms for Euclidean geometry, revealed major flaws along with subjects requiring further development. He found that Euclidean geometry rests on fifteen postulates and six primitive or undefined terms.

Hilbert's book has not detracted from Euclid's singular influence on mathematics. To the contrary, it continued and refined the Euclidean axiomatic method and deepened our understanding of the nature of mathematics, while establishing greater uniformity in it.

Originality, Selections, and a Pristine Search

Euclid is credited with two major original contributions in the *Elements* beyond the refined axiomatic method. They are a sound theory of parallel lines and a famous proof of Pythagoras' theorem.

A passage beginning at 65 a 4 of Aristotle's *Prior Analytics* observes that his contemporaries think that "they describe parallels," while in fact their theory contains circular reasoning. Later logicians would call such reasoning a *petitio principii*. Recognizing that the concept of equidistance from definition 23 of Book I does not provide the essential criterion for parallelism, Euclid wisely opts for *nonmeeting* or *nonsecancy*. That he lists as a postulate his basic statement about parallelism reveals that he knows it to be indemonstrable and fundamental to his entire geometry.

Today the fifth postulate is most widely known in the form Scottish mathematician John Playfair gave in 1795 in Section 11.1 of his *Elements of Geometry*.

> Given a line l and a point P not on l, there exists one and only one line m containing P and no points of l. (Playfair's Postulate)

As the drawing indicates, a consequence of Playfair's postulate is that l and m' will always intersect if $\angle\alpha + \angle\beta < 180°$, as the fifth postulate states, or if $\angle\alpha + \angle\beta > 2 \times 90°$. In Book I, Proposition 28, Euclid proves that m is parallel to l when $\angle\alpha + \angle\beta = 180°$.

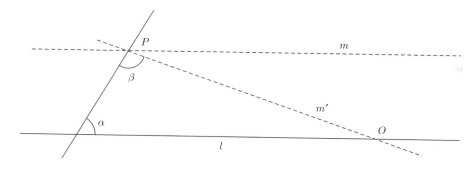

Propositions 27 through 31 of Euclid's Book I discuss the theory of parallels. Proposition 27, which is the converse of the parallel postulate, states: "If a straight line falling on two straight lines makes alternate angles equal to each other, the straight lines will be parallel to each other." Since Euclid is able to prove this proposition, some geometers believed the parallel postulate could also be proved. In proving Propositions 1 through 28 of Book I, he does not use the parallel postulate. Its first necessary employment is for proving Proposition 29: "If two parallel lines are cut by a third line, then alternate interior angles are equal, exterior angles [are] equal to the interior and opposite angles, and interior angles on the same side of the transversal equal two right angles." Combining the second part of Proposition 30 with Proposition 31 essentially gives Playfair's postulate.

The proof of Pythagoras' theorem appears in Book I, Proposition 47 of the *Elements*. This fundamental proposition states that "In right-angled triangles the square of the side subtending the right angle is equal to the [sum of the] squares on the sides containing the right angle." To prove it, Euclid uses only the application of areas and the contents of Book I. Considering its reasoning unduly intricate, German philosopher Arthur Schopenhauer (d. 1860) dismissed it as not an argument but a "mousetrap." The proof nudges the reader forward conclusion by conclusion until abruptly slamming shut its door, as in an ambush. It is not subsequently possible to gnaw through the solid bars of its logic. Today this elegant demonstration is commonly known as the "mousetrap proof."

PROPOSITION 47.

In right-angled triangles the square on the side subtending the right angle is equal to the squares on the sides containing the right angle.

Let ABC be a right-angled triangle having the angle
5 BAC right;

I say that the square on BC is equal to the squares on BA, AC.

For let there be described on BC the square $BDEC$,
10 and on BA, AC the squares GB, HC;   [I. 46]
through A let AL be drawn parallel to either BD or CE, and let AD, FC be joined.

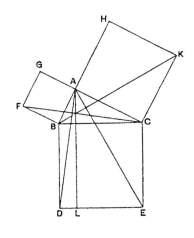

15   Then, since each of the angles BAC, BAG is right, it follows that with a straight line BA, and at the point A on it, the two straight lines
20 AC, AG not lying on the same side make the adjacent angles equal to two right angles;

therefore CA is in a straight line with AG.   [I. 14]

25   For the same reason

BA is also in a straight line with AH.

And, since the angle DBC is equal to the angle FBA: for each is right:
let the angle ABC be added to each;
30   therefore the whole angle DBA is equal to the whole angle FBC.   [C. N. 2]

And, since DB is equal to BC, and FB to BA,
the two sides AB, BD are equal to the two sides FB, BC respectively,
35   and the angle ABD is equal to the angle FBC;
therefore the base AD is equal to the base FC,
and the triangle ABD is equal to the triangle FBC.   [I. 4]
Now the parallelogram BL is double of the triangle ABD, for they have the same base BD and are in the same parallels
40 BD, AL.   [I. 41]

And the square *GB* is double of the triangle *FBC*,
for they again have the same base *FB* and are in the same
parallels *FB*, *GC*.   [1. 41]
    [But the doubles of equals are equal to one another.]
45   Therefore the parallelogram *BL* is also equal to the
        square *GB*.
    Similarly, if *AE*, *BK* be joined,
the parallelogram *CL* can also be proved equal to the square
HC;
50   therefore the whole square *BDEC* is equal to the two
        squares *GB*, *HC*.   [C. N. 2]
    And the square *BDEC* is described on *BC*,
    and the squares *GB*, *HC* on *BA*, *AC*.
    Therefore the square on the side *BC* is equal to the
55 squares on the sides *BA*, *AC*.
    Therefore etc.
                                                    Q. E. D.

The mousetrap proof. The figure is sometimes called
the Franciscan cowl or the bride's chair.

Using the technique of application of areas, Euclid speaks of squares constructed on the sides of a right-angled triangle, not as we do of square numbers. The proof thus follows the Pythagorean tradition and adheres to the distinction made by Eudoxus: geometric proportions between numbers and geometric magnitudes. The proof, moreover, makes no appeal to similarity of figures, for similitude depends on finding a proportional, and Euclid does not present until Book V Eudoxus' general theory of proportions covering incommensurables. Pythagoras' theorem had established a metric that is used for Euclidean space, in modern notation $ds^2 = dx^2 + dy^2$.

The belief that this proof of Pythagoras' theorem was original with Euclid is extrapolated from a statement by Proclus in his commentary on Euclid: " ... the writer of the *Elements* ... gave a very clear proof of this proposition ... ." We lack corroboration. No comment at the time that would attribute originality to Euclid's proof, as Aristotle's theory of parallels does for his work on that subject, is to be found. Neither does any record exist of a remark by Euclid on the inadequacy of prior attempts by Pythagoreans. Proclus' statement that Euclid in Book VI, Proposition 31, provides an "irrefutable argument" for his generalization of Pythagoras' theorem puts into slight question the claim that the proof of I.47 originated with Euclid. Book VI, Proposition 31 holds, "In right-angled triangles any figure [described] on the side subtending the right angle is equal to the [sum of the] similar and similarly described figures on the side containing the right angle." As Netherlands historian E. J. Dijksterhuis has shown, Euclid's proof is "irrefutable" only for rectilinear similar figures. It does not apply to curvilinear figures nor is it complete even for rectilinear ones.

Euclid devotes a major portion of the *Elements* to the application of areas. The theory in Books II and VI of line segments and areas synthesizes the ancient Greek attempt to treat the irrational by means of geometry. Although their propositions and lemmas appear geometric, their content is algebraic. According to van der Waerden, Euclid creates geometric algebra, a modern term coined by the Danish mathematician H. G. Zeuthen, which puts algebra into geometric form, with numbers visualized as line segments and geometric constructions replacing algebraic operations.[211] Historian Sabetai Unguru strongly opposes interpreting this work as geometric algebra, arguing that it begins instead with al-Khwārizmī.[212] Van der Waerden insists, however, that from its origins algebra was closely linked to geometry. Euclid's intricate transposition of the Pythagorean application of areas and inherited Babylonian mixed algebraic calculations designed to find rational numbers, he contends, amount to geometric algebra. In their problem texts, the Old Babylonians treated numbers as lengths, widths, and areas. The medieval central Asian scholar Umar al-Khayyām, van der Waerden holds, recognized that Euclid, using geometrical constructions, had solved quadratic equations. Euclid reduces quadratics to geometric equivalents of these components:

$$x(x + a) = b^2, \quad x(x - a) = b^2, \quad x(a - x) = b^2.$$

Proposition 4 of Book II is an example of Euclid applying areas or, put more strictly, constructing a figure to solve an equation. It states:

> If a straight line be cut at random [into two parts a and b], the square on the whole is equal to the square on the segments and twice the rectangle contained by the segments. [In modern notation this gives the identity $(a + b)^2 = a^2 + 2ab + b^2$.]

To solve this equation, Euclid first constructs a square $ABCD$ with sides $AB = BC = b$ and then lays off $AE = a$ in extending AB. A line drawn from BD to meet the extension of EK forms the larger rectangle $EBFH$, in this case a square. It turns out that $KH = CF$, which must be the desired quantity a, because the bordering rectangles are ab and their sides b are perpendicular.

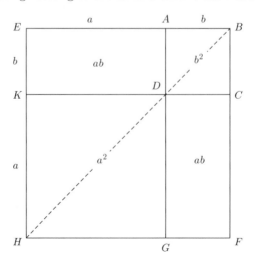

The Greek application of areas and Euclid's early geometric algebra, together with the isolation of scholars and practitioners working on the same subject, restricted the growth of algebra in the West until the appearance of printing and its sixteenth-century translation into early modern symbols, which combined separated algebra from geometry.

Euclid's theory of divisibility, the oldest and chief part of number theory and an important contribution to it, centers on prime numbers. He recognizes that the number 1 divides each number and that each number divides itself. Natural numbers greater than 1 with only these two trivial divisors are called primes. Books VII and IX of the *Elements* lay the groundwork for a theory of divisibility.

As an initial step to the theory of divisibility, Euclid presents VII.31: "Every composite number is measured by a prime" (where the word "measured" pertains to the measurement of geometric magnitudes and not division). Its more general statement is "Every natural number greater than 1 is a product of primes." This assertion is an almost immediate consequence of the definition of primes. In IX.14, Euclid proves the fundamental theorem of arithmetic. In modern form it reads "Every natural number can be written as a product of primes in exactly one way" (though the order of the prime factors may be rearranged.)

Shortly afterward, in a display of intellectual audacity and ingenuity, Euclid proves in IX.20 a mathematical proposition of the highest caliber. The original form of IX.20 holds "The primes are more [in number] than any assigned quantity of primes." This proposition can be stated in modern terms as "The number of primes is infinite." Its proof, while invoking the process of the potential infinite, can thus be read as concluding with an infinite collection, which runs counter to the supposed Greek horror of the actual infinite.

Although the proof of IX.20 means that no catalogue listing of all primes is possible, human curiosity generated many catalogues, beginning with a procedure known as the sieve of Eratosthenes. When Eratosthenes' sieve showed that primes become more and more rare among large natural numbers, the question of "how rare" arose. It took another two millennia before the prime number theorem of Carl Gauss and its proof by Jacques Hadamard answered this question. Put roughly, this theorem holds that there are about $N/\log_e N$ primes that come before a large number N.

Besides the three major propositions on primes, the *Elements'* theory of divisibility has as a basic component the Euclidean algorithm. This algorithm, appearing in VII, 1–3, finds the greatest common divisor of two natural numbers. VII.2 gives a procedure that algebra textbooks relate in this way:

To find the greatest common divisor of 16 and 76, for example, subtract 16 from 76 enough times so that a remainder results that is less than 16. This requires four subtractions. Next, repeat the procedure by subtracting the new remainder of 12 from 16. Repeat the procedure until a zero answer results. The greatest common divisor (GCD) is the last remainder (or divisor) before zero is reached.

$$
16\overline{)76}
$$

$$
\begin{array}{r}
4 \\
64 \\
\hline
12
\end{array}
$$

$$
\begin{array}{r}
1 \\
16 \\
12 \\
\hline
4
\end{array}
$$

$$
\begin{array}{r}
3 \\
12 \\
12 \\
\hline
0
\end{array}
$$

$76 = 4 \times 16 + 12$
$16 = 1 \times 12 + 4$
$12 = 3 \times 4 + 0$

The GCD $(76, 16)$ is 4.

VII.2 declares that every pair of nonprime numbers a and b and VII.3 that every three nonprime numbers a, b, and c have a unique greatest common divisor d, while VII.1 holds that every pair of numbers with a common factor of 1 only

PROPOSITION 20

Prime numbers are more than any assigned multitude of prime numbers.

Let A, B, C be the assigned prime numbers; I say that there are more prime numbers than A, B, C.

For let the least number measured by A, B, C be taken, and let it be DE; let the unit DF be added to DE.

Then EF is either prime or not.

First, let it be prime; then the prime numbers A, B, C, EF have been found which are more than A, B, C.

Next, let EF not be prime; therefore it is measured by some prime number [VII. 31].

Let it be measured by the prime number G.

I say that G is not the same with any of the numbers A, B, C.

For, if possible, let it be so.

Now, A, B, C measure DE; therefore G also will measure DE.

But it also measures EF.

Therefore G, being a number, will measure the remainder, the unit DF; which is absurd.

Therefore G is not the same with any one of the numbers A, B, C.

And by hypothesis it is prime.

Therefore the prime numbers A, B, C, G have been found which are more than the assigned multitude of A, B, C.　　Q. E. D.

are relatively prime. The Euclidean algorithm contains everything that modern number theory considers essential about the divisibility properties of integers. Keep in mind that Euclid's thinking was geometric, and his proofs here again deal with line segments and common measures.

Book XII on relations between areas and volumes freely employs the method of exhaustion. Eudoxus had invented the early stage of this procedure for computing areas of plane figures bounded by curved lines and volumes of bodies bounded by curved surfaces. The base of his procedure is expressed in Euclid X.1, a theorem implying the continuous divisibility of a continuum. In constructing the formal proofs of Book XII, Eudoxus or Euclid adroitly circumvents conceptual hurdles posed by Eudoxus' separation of continuous magnitude from discrete number. Since the Greeks easily summed finite geometric proportions, they presumably could have handled infinite series without difficulty. But instead of attempting the formal summing of infinite series, the author uses elaborate double *reductio ad absurdum* arguments. Since the section on Eudoxus discusses XII.2 on areas of circles, I shall comment here only on principal propositions concerning volumes of pyramids and cones.

Democritus had discovered and Eudoxus first proved that the volume of a pyramid with a triangular base as well as the volume of a circular cone equals one-third the area of its base times its height. Modern notation renders this as $v(P) = \frac{1}{3}Ah$. Propositions 5 and 7 of Book XII are crucial for pyramids.

Proposition 5 states:

> Pyramids that are of the same height and have a triangular base are to one another as their bases. That is, if a pyramid with the triangular base area A_1 has a volume V_1 and another with a base area of A_2 has a volume V_2, then $V_1/V_2 = A_1/A_2$.

Euclid's proof divides an arbitrary pyramid of height h and base area A into two equal pyramids similar to the whole with height $h/2$ and base area $A/4$ and two equal prisms. This means that the two prisms are greater than half the volume of the whole pyramid. After dividing the second pyramid the same sufficiently large number of times, Euclid demonstrates by a double *reductio* argument almost identical to that in XII.2 that assuming either that the first ratio is less than the second ($V_1/V_2 < A_1/A_2$) or that it is greater ($V_1/V_2 > A_1/A_2$) leads to a contradiction.

Consider $V_1/V_2 < A_1/A_2$. This implies that $V_2 > V_1 \times A_2/A_1 = S$, and by X.1 we may let V_2 differ from S by less than any given quantity (ϵ), that is, $\epsilon = V_2 - S$. Label as $v(P_2)$ the union of dissections of the second prism. Then $V_2 - v(P_2) < \epsilon = V_2 - S$, or $v(P_2) > S$. But this suggests that the union of prisms contained in V_2 is greater than V_2 itself, which is a contradiction. The assumption that $V_1/V_2 > A_1/A_2$ also leads to a contradiction.

From this proof that the two adversary assumptions lead to contradictions and from constructions in Proposition 7 showing that any prism with a triangular base divides into three pyramids with equal volumes, Euclid has demonstrated that $v(P) = \frac{1}{3}Ah$.

Let there be a prism in which the triangle ABC is the base and DEF its opposite: I say that the prism $ABCDEF$ is divided into three pyramids equal to one another, which have triangular bases.

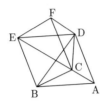

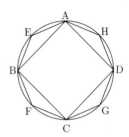

In XII.10, Euclid establishes the same volume relationship for circular cones: the circular cone is one-third the cylinder with the same base and height. Here is Euclid's proof in outline:

Given a circular cone T, having vertex O, base circle C, and height h, inscribe a sufficiently large number of pyramids to exhaust the circular cone. One form of the pyramid is a regular polyhedra or Platonic solid. The number of inscribed regular polygons is increased by bisecting the base sides for new constructions. The resulting sequence of regular polygons filling the base circle C may be denoted as $(P_1, \ldots, P_n)$ and the pyramids with vertex O and base P_n as $(T_1, \ldots, T_n)$. It follows that $v(T_n) = \frac{1}{3}a(P_n)h$. From Eudoxus' principle and given $\epsilon > 0$, Euclid knew that $M_n = v(T) - v(T_n) < \epsilon$. The same holds in the case of cylinder Q with base C and height h filled with a sufficiently large number of inscribed prisms Q_n with base P_n and height h, $v(Q) - v(Q_n) < \epsilon$. Euclid then proves by *reductio* arguments that $v(T)$ can be neither less nor greater than $\frac{1}{3}v(Q)$. It follows that $v(T) = \frac{1}{3}v(Q) = \frac{1}{3}Ah$.[213]

To cite again the English mathematician and classicist Thomas Heath: the *Elements* was to become, despite its slight "imperfections," the "greatest mathematical textbook" in history. After its widespread adoption and the destruction of ancient texts, all previous Greek compilations on theoretical mathematics, including the *Elements'* immediate predecessor by Theudius of Magnesia, disappeared. This selective preference combined with accidental and destructive losses of sources through the effects of climate, neglect, and strife turned the *Elements* into the major source of mathematical knowledge in the western world. It decisively established geometry as the dominant field of mathematics in the West. Its influence was so strong that until the nineteenth century mathematicians believed that no other consistent geometry that applies to the physical world could exist.[214]

To the problem of what Euclid borrowed must be added the problem of what Euclid himself actually wrote. Since no manuscript written directly by Euclid is extant, reconstructing the original *Elements* has required comparing the numerous emendations, recensions, commentaries, and scholia.

By late antiquity the Alexandrian book copiers, who were poorly versed in mathematics, had introduced or copied errors into numerous editions of the *Elements*. In addition, the earliest commentary by the Alexandrian geometer and mechanist Heron and a more extensive and important one by Pappus were lost, except for fragments. The same is true of lesser commentaries by Porphyry and Proclus. In the fourth century A.D., Theon of Alexandria made a slight emendation of the *Elements*, altering its language and providing alternative proofs of theorems. His redaction was a written version of his lectures. Until the nineteenth century, all surviving editions of the *Elements*, except for one, were based on Theon. His recension was the basis of the first Arabic translation of the *Elements* by al-Hajjāj in the ninth century, the first English translation by Sir Henry Billingsley, published in 1570, and the Latin translation issued in 1572 by Frederigo Commandino, the most astute of the European Renaissance editors of Euclid.

In modern times the search for a pristine or pre-Theonine Euclid has apparently met with some success. In 1808 François Peyrard, whom Napoleon sent to Italy to gather valuable manuscripts, discovered in the Vatican Library a Greek manuscript of the *Elements* now known as version P, which is thought to be based on a text more ancient than Theon's. Late in the century, Danish philologist Johan Ludvig Heiberg, by a painstaking comparative study of almost all known and pertinent secondary sources on the *Elements,* edited out alterations that Theon's text had apparently incorporated. In general, he found that later Arabic translations, based on lost Greek texts, were inferior to surviving Greek manuscripts. Issued in installments between 1883 and 1888, Heiberg's definitive, critical edition of the *Elements*, following the Peyrard text, is the product of meticulous, learned research. Two decades later Thomas Heath published *The Thirteen Books of Euclid's Elements*, which incorporates Heiberg's findings and gives a masterly treatment of Book X.

But the debate over which texts of the *Elements* are closest to Euclid's original has not ended. Scholars dispute the Greek wording that Euclid used in proofs. Resolving that question will suggest which texts most reflect the original.

5.5   Remainder of the Euclidean Corpus

Although little of his writings besides the *Elements* has survived, we know that Euclid wrote about a dozen other works ranging over most, if not all, branches of mathematics of his time. They certainly covered elementary and advanced geometry and applied mathematics. These works include a Greek version of his *Data*, which is closely related to the first six books of the *Elements*. Students who had mastered those books used the *Data*. There are also an Arabic translation of the treatise *On the Division of Figures,* four books on conics that were perhaps the basis for the first four books of Apollonius' *Conics*, two recensions of the *Phaenomena*, which was a text on spherical geometry for use in astronomy; and two recensions of *Optica*, an elementary treatise on perspective. Proclus believed Euclid to have also written the lost *Surface Loci* in two books, *Porisms* in three books, *Pseudaria* on fallacies, *Catoptrics* on mirrors, and *Sec-*

tio Canonis or *Theory of Musical Intervals.* Current scholarship suggests that Euclid wrote *Catoptrics.* Arabic sources also credit Euclid with a lost book, *On the Balance,* and *A Book on the Heavy and the Light,* a text on mechanics. His authorship of these is doubtful, because only one treatise apparently from antiquity, *De Canonio,* attributes to him any writings on mechanics.

At least two lost works expanded on the proper application of the axiomatic, deductive method. The advanced *Porisms* pursued new geometrical discoveries, while *Pseudaria* instructed beginners in avoiding by disciplined method any pitfalls in geometric reasoning and in discovering paralogisms.

In ancient Greek geometry, the word that English renders "porism" stemmed from the verb for "procure" and usually meant "corollary." According to Pappus, Euclid referred instead to a consideration based on construction and falling between a theorem and a problem. Recognizing that inaccurate drawings give misleading results, Euclid likely presented porisms as auxiliary theoretic proofs showing that a specified point or line must exist and satisfy a certain condition, for example, that a point or points must lie on a straight line or that given lines must intersect at only one point. Euclid's porisms, then, doubtless examined topological rather than mensurational aspects of a problem. He required a porism to prove that the three bisectors of the angles of a triangle intersect in one point, something that an accurate drawing cannot exactly demonstrate. Euclid's *Porisms,* perhaps the earliest text on projective geometry, may have been based on his study of conics.

Pseudaria in Proclus' explanation is a "purgative and disciplinary" book that sets out some means for detecting and refuting certain geometrical fallacies. The fallacies seem to follow by strict reasoning from established scientific principles, but actually conceal, sometimes cleverly, an erroneous assumption. *Pseudaria* presents side by side for comparison numerous true and false methods of proof. It appeals for a clear understanding of the axiomatic method and gives practical illustrations side by side. It also contains an expository appeal for clear-sighted understanding of the axiomatic method and practical illustrations of it. Without these helpful references, a beginning geometer might accept wrong conclusions and be deprived of the power of the axiomatic, deductive method.

Here is an example of a geometric fallacy. It purports to be a proof of the proposition that "every triangle is isosceles." First, draw an arbitrary triangle ABC. Next obtain point P at the intersection of the bisector of angle $A(AP)$ and the orthogonal bisector (DP) of the opposite side BC. By applying to triangles BPD and CPD the side-angle-side congruence theorem, we can prove that $BP = PC$. Next, from point P drop perpendiculars to sides AB and AC. By angle-side-angle, $AF = AE$ and similarly $FP = EP$. Then, by the Pythagorean theorem, $FB = EC$. Axiom 1 asserts that equals added to equals give equals. Hence $AF + FB = AE + EC$ or $AB = AC$ in every triangle. All the congruences in this demonstration are valid in the abstract, but not necessarily the last addition. Why? The figure may suggest that point P falls inside every triangle. Actually, in a nonisosceles triangle, the intersection of the

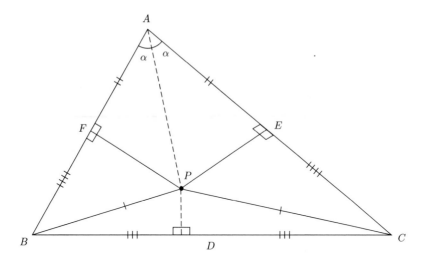

bisector of angle A and the orthogonal bisector of the opposite side BC must fall outside the triangle.

The mathematical achievements of Euclid so impressed Arab mathematicians that they came to believe his name, which they variously pronounced as *Uclides* or *Icludes*, stemmed from *ucli* (a key) and *dis* (a measure). They thus portrayed him as the "key of geometry." Arabic lore also held that every Greek school had posted over its door the words "Let no one come to our school who has not learned the *Elements* of Euclid."

6

Ancient Mathematical Zenith in the Hellenistic Third Century B.C., II: Archimedes to Diocles

Give me a place to stand on,
and I will move the Earth.
— A boast of Archimedes from
Pappus' *Collection*, Bk. VIII, Prop. 11

6.1 Archimedes: Life and Work in Outline

Shortly after Euclid, compiler of the definitive textbook, came Archimedes of Syracuse (ca. 287–212 B.C.), the most original and profound mathematician of antiquity. Archimedes, who was probably also the greatest physicist of the age besides being a talented mechanical engineer, wrote a remarkable collection of mathematical treatises that E. J. Dijksterhuis, Thomas Heath, Ivo Schneider, and most recently Wilbur Knorr have basically arranged in three main groupings: (1) pure geometry with measurements of curvilinear plane and solid figures; (2) systematic inquiries into geometrical mechanics and hydrostatics; and (3) miscellaneous works on such topics as method and arithmetic. Thomas

Archimedes, a sixteenth-century depiction taken from a Roman coin by André Thevet

Heath has declared his treatises to be "without exception, models of classical mathematical exposition." They surpass all others in their depth, precision, elegance of demonstration, and, in his major areas of study, completeness of treatment of subject matter. In Chapter XVIII of *Parallel Lives*, the mathematically unversed Plutarch writes of Archimedes' treatises that

> it is not possible to find in all geometry more profound and difficult questions treated in simpler and more lucid terms. Some attribute this success to his genius; others think it due to incredible labor that everything he did seemed to all appearances to have been performed without labor and with ease. (Eulogy for Archimedes in *Parallel Lives: Marcellus*, Chapter XVIII)

Neither time nor critical history has diminished his reputation. Galileo Galilei calls him "divine Archimedes, superhuman Archimedes," and Carl Gauss restricts to the ancient Syracusan along with Isaac Newton the term *summus*.

Although more biographical information exists about Archimedes than about any other ancient mathematician, it is scanty and mixed with legend. A biography by his friend Heracleides is lost. We have to rely on personal information from his dedicatory prefaces to his treatises and scattered references in classical literature by such authors as Polybius, who in the second century B.C. drew directly on eyewitness accounts of the Second Punic War, and derivative though seemingly reliable comments by Livy in the first century B.C. and Plutarch three centuries later. With their information we can reconstruct an outline of his life.

In I.9 of *Sand-Reckoner*, Archimedes states that his father was the astronomer Phidias. Archimedes himself was born in Syracuse, the largest Greek settlement in Sicily, and grew up there. His year of birth, usually given as 287 B.C., is not known for certain. Knowing that he died in 212, historians project backward from the assertion in *Chiliad* 2, history 35 of the twelfth-century Byzantine verse chronologist John Tzetzes that Archimedes "worked at geometry until old age, surviving seventy-five years." In antiquity, Polybius claimed only that Archimedes had been elderly when he died, and Proclus believed him to have been the same age as Eratosthenes, which suggests that his birth came as late as 276 B.C. Archimedes' father probably taught him Eudoxean geometry, which had the greatest influence on his mathematical work, as Wilbur Knorr has convincingly shown.[215] As a young man, Archimedes must have visited Egypt and studied mathematics under disciples of Euclid at the Alexandrian Museum. His later correspondence with Conon of Samos, Dositheus, and Eratosthenes, all of whom resided in Alexandria, supports this belief. While in Egypt he apparently invented the now famous Archimedean screw or water snail (*cochlias*), a device to raise canal water from the Nile over levees for irrigation. It is a long cylinder open at both ends with a continuous spiral piece running its length. When the tube is tilted down into water and the spiral inset rotated, water is carried to the top and flows out. After completing his studies, Archimedes returned to Syracuse, where he devoted himself to significant research in pure geometry and its application. Polybius and Plutarch assert that he was related to King Hiero II of Syracuse. Whether this is correct or not, Hiero was his patron and he dedicated *Sand-Reckoner* to Hiero's son Gelon.

Departing from pure geometry, Archimedes pioneered geometrical mechanics and constructed some ingenious mechanical devices. Several anecdotes exist about these ventures.

Of dubious authenticity is the well-known story from chapters III and IX of Roman architect Vitruvius' *De Architectura* that King Hiero, suspicious of a goldsmith, asked Archimedes to assay the gold in a crown or wreath that he had commissioned without defacing the workmanship. After puzzling over how to do this, Archimedes discovered a method while bathing in a public bathhouse that so excited him that

> he did not delay, but in his joy leapt out of the tub, and rushing naked towards his home, he cried out in a loud voice that he had found what he sought. As he ran he repeatedly shouted in Greek, "Eureka! eureka!" ("I have found it! I have found it!").

While stepping into a nearly full bathing pool, so the tale goes, Archimedes had observed that water overflowed the sides. He reasoned that the bulk of his submerged body causing the overflow could be correlated to the volume and weight of the spilled water. Procuring equal weights of gold and silver, he noted the volume of water each displaced when immersed in a vessel full of water and compared these volumes with that of the water spilled by the submerged crown. Since this amount of water differed from the volume displaced by an equivalent weight of gold, an alloy was suggested. Further comparisons of water displacement apparently gave the relative contents of gold and silver in the crown.

Vitruvius' story attempts to reconstruct and capture the moment of discovery when a major problem is solved. Therein it contrasts in tone and technique to Archimedes' treatise *On Floating Bodies*, which proceeds in Euclidean fashion. After stating fundamental assumptions concerning the properties of fluids, Book I demonstrates propositions that may, in part, be consequences of the bathing pool experience. Proposition 5 of Book I states that

> Any solid lighter than a fluid will, if placed in the fluid, be so far immersed that the weight of the solid will be equal to the weight of the fluid displaced. (Heath's paraphrase)

Proposition 7 on solids denser than a fluid expresses the principle of buoyancy, now called the "principle of Archimedes":

> Solids heavier than the fluid, when thrown into the fluid, will be driven downward as far as they can sink, and they will be lighter

[when weighed] in the fluid [than their weight in air] by the weight of the portion of fluid having the same volume as the solid. (Clagett's paraphrase)

Put in more succinct modern terms, "a body that is wholly or partly immersed in fluid is buoyed up with a force equal to the weight of fluid displaced by the body."

Another anecdote illustrates Archimedes' reputation in engineering. After recognizing the power of a lever to move a great weight with a small force, he boasted to King Hiero: "Give me a place to stand on [and a fulcrum]! And I will move the entire Earth." When challenged by the king to prove his boast, the story goes, he moved along the beach by a system of levers and compound pulleys a three masted ship laden with passengers and freight, drawing "her towards him smoothly and safely, as if she were gliding through water." According to Proclus, the astounded king used a series of compound pulleys to launch the *Syracosia*, the largest ship in antiquity and Hiero's gift to Ptolemy Euergetes. The *Syracosia* was an estimated 4,200 tons.[216]

Despite legend and fame, Archimedes' attitude toward mechanical inventions is in question. His credits include the water snail, a hydraulic organ, and, according to the Roman orator Cicero in I, XIV, 21 and 22 of *The Republic*, the construction of a model planetarium to demonstrate the Eudoxean system of astronomy, including the apparent motions of the Sun, Moon, and planets about the fixed Earth. Ancient Mediterranean writers treated his planetarium as a wonder.[217] Archimedes seems to have prized it above his other technical achievements. His only book on mechanical devices, the lost *On Sphere Making*, must have dealt with the subject. In his final years, Archimedes also contributed devices to defend Syracuse, especially against the Roman siege. Nevertheless, Plutarch's eulogy claims that Archimedes disdained the practical, "regarding the work of an engineer and every art that ministers to the needs of life as sordid and ignoble." This disdain appears to reflect Plutarch's own prejudice, but Archimedes apparently left no written record about his mechanical inventions except *On Sphere Making*. (Plutarch's contempt for an engineer's work and ignorance of it typifies the attitude of the educated elite in antiquity. This attitude found weighty support in the writings of Plato and Aristotle.) Archimedes died in his native Syracuse in 212 B.C. during the Second Punic War.

The events that brought into this conflict between Carthage and Rome the prosperous commercial city of Syracuse,[218] its population numbering in the hundreds of thousands, go back at least to 265 B.C., when the Syracusan general Hiero defeated Mamertime (Italian) mercenary brigands occupying nearby Messana. For this, Syracuse elected him military *strategos* or king. The Mamertimes called for assistance from Carthage, the strongest naval power in the Mediterranean. When Carthage intervened in Sicily, Rome responded, possibly fearing Carthaginian control of all the island. Two years after the start of the First Punic War, lasting from 264 to 241 B.C., the Romans concluded a fifteen-year alliance with Hiero that became the key to maintaining their position in southern Italy. After Carthage signed a treaty surrendering claims to Sicily in 241, a fragile peace held until the army elected Hannibal as leader of Carthage in 221. A territorial dispute in Spain, where Carthage had vital trading colonies, provoked in 218 the sixteen-year Second Punic War. At Cannae in 216 B.C., Hannibal dealt the Roman army its worst defeat in history—almost eighty thousand men killed or captured. Despite the continuing loyalty of allied Italian tribes, some Greek cities in southern Italy defected from Rome. Two years later Claudius Marcellus, the Roman commander in Sicily, was ordered to take Syracuse. At this time Syracusan supporters of Rome murdered King Hieronymous, who had succeeded Hiero in 215 B.C.; they also killed his family. This treachery led the city to break completely with Rome.

For nearly three years Marcellus besieged Syracuse, but defenses bolstered by devices invented by Archimedes and built under his supervision repulsed attacks on its outer walls by the Roman fleet. According to literary embellishments, not archeological artifacts, these devices probably included battering rams, cranes, grapnels and compound pulleys to move ships easily on shore and to lift them from the sea and drop them from a height, and catapults and other ballistic instruments. Roman soldiers came to fear and flee from Archimedes' "deviled contraptions." Ancient authors also refer to paraboloidal glass mirrors that reflected the sun's heat and set afire Roman ships within a bowshot of the city's walls.[219] Perhaps the mirrors ignited the ships only in the imagination of biographers. Temporarily withdrawing to lessen the guard, the Romans on a feast day in 212 B.C. breached and captured the outer walls, exploiting a weakness in the defense that they had discovered with the aid of a regicide. A few months later the rest of the city fell to allied mercenaries. Once Marcellus secured royal treasures, he turned the city over to pillage by Roman soldiers. During the looting, a Roman soldier slew Archimedes.

Death of Archimedes

Accounts of Archimedes' death by Livy, Plutarch, Valerius Maximus, and Tzetzes are picturesque and dramatic. Most conjecture that he was engaged in a geometrical problem when a soldier surprised him. He either asked the soldier for more time to complete a proof or, in later legend, ordered the soldier not to disturb the geometrical diagram that he had just drawn in a sand board. (The modern version of the last statement is "Do not spoil my circles.") Insulted, the soldier drew his sword and killed Archimedes. This episode, which has become a subject for artists, may be an exaggeration intended to contrast the pragmatic Romans and theoretical Greeks.[220] As Alfred North Whitehead has written, "no Roman ever died in contemplation over a geometrical diagram."

The death of Archimedes upset Marcellus, who erected a small column to mark his grave. Following the wish Archimedes had expressed to friends, it was carved with the figure of a right circular cylinder whose height equaled the diameter of the sphere it circumscribed. An inscription giving the ratio between the two volumes (his favorite theorem and discovery, see below) accompanied the figure. In Book V, xxiii, 64 through 66 of *Tusculan Disputations*, Cicero, who was governor of Sicily in 75 B.C., tells of recognizing this emblem in searching for the grave in an overgrown area of a cemetery near the Agrigentine gate in Syracuse. He found the stone marking the site with drawing intact and about half the ratio epigram legible. Cicero had the neglected area cleared and the face of the column marker restored. It soon disappeared again, to be rediscovered in 1965 during an excavation for a Syracuse hotel.

Archimedes' prefaces and remarks by commentators suggest something of his personality. His high order of genius gave him the ability that Newton would also possess to concentrate exclusively on a scientific problem over an extended period of time. Plutarch says that, when absorbed in scientific thought, Archimedes

> forgot to take food and he neglected the care of his body; and when, as was often the case, he was forcibly driven to the bath and to chrism, he would trace geometrical figures in the ashes of the fire, and diagrams in the oil on his body.

Sometimes popular legend has interpreted such behavior as "absent mindedness," an unfortunate choice of terms that suggests ridiculous vacuity, rather than profound and sustained absorption in the fundamentals of mathematics and natural philosophy. A wry sense of humor, openness in sharing his method, and an optimism about scientific discovery tinged with personal humility are other conjectural personality traits.

Archimedes' humor, or perhaps a research split with Alexandrians, can be detected in his occasional practice of sending mathematical theorems to colleagues in Alexandria without demonstration so that they might have the pleasure of discovering these. Annoyed by colleagues who accepted his theorems without proof, perhaps as their own, he states in the preface to *On Spirals* that he includes false theorems in his correspondence so that "those who claim to discover everything, but produce no proofs of the same, may be confuted as having actually pretended to discover the impossible."

Others do not see the subtle humor of this correspondence. Renaissance authors interpreted his conduct as self-seeking. For centuries mathematicians, such as John Wallis, believed that Archimedes had shared Aristotle's shop window approach to mathematics and wished to keep his methods secret. Although the Renaissance view persists and cannot be disproved, Heiberg's discovery in 1906 of the Greek text of *On the Method*, previously considered lost, makes it doubtful. In an adroit piece of detective work, Heiberg discovered a parchment in a monastic library in Constantinople whose Archimedean writing by a tenth-century hand had been washed and scraped off, perhaps in the thirteenth century, for reuse of the erased sheets for a book of prayers and ritual. Such a manuscript is called a *palimpsest* (from the Greek for rescraping). With the aid of a magnifying glass, Heiberg made out most of its expunged tenth-century material. Introductory remarks to *On the Method* may suggest Archimedes' openness and optimism about scientific discovery.

> I deem it necessary to expound the method partly because I have already spoken of it and I do not want to be thought to have uttered vain words but equally because I am persuaded that it will be of no little service to mathematics, ... either of my contemporaries or of my successors, ... to discover other theorems ..., which have not yet occurred to me. (Heath's translation)

This treatise shows Archimedes' method of discovering theorems by a mental weighing balance, an ingenious juggling and juxtapositioning of geometric figures along the arm of an imaginary balance until a state of equality occurs. This balance, although mental, is also mechanical. By separating heuristics from airtight proofs or apodictics, Archimedes both respects and refutes the Platonic disallowal of mechanical solutions.

Archimedes had no worthy successor until the seventeenth century.

6.2   Archimedes' Writings: Method of Exhaustion, Geometrical Mechanics, and Extension of Numerical Calculation

The range, originality, and importance of Archimedes' scientific achievements are illustrated in his surviving writings. He took as his province all the mathematical sciences known in his time, investigating not only geometry and arithmetic, but also statics, optics, astronomy, and engineering. He may even have been the first to base hydrostatics in mathematics, for no record has been discovered of any precursor. Unlike Euclid and Apollonius, Archimedes wrote no massive textbooks, but prepared smaller tracts, often about problems raised in his correspondence with Alexandrian mathematicians. None of these tracts was a compilation; each was a highly original contribution to mathematical science.

Of course, Archimedes drew on works of predecessors—especially Euclid's *Elements* and *Conics*—and scrupulously acknowledged debts to Aristeus, Aristarchus, Democritus, Euclid, Eudoxus, and others. At the same time, he cited his own proofs of many theorems previously demonstrated inadequately. The simple and elegant Proposition One in *Dimension of the Circle*, otherwise known as the circle theorem, seems such a case. It states that

> Every circle is equal to a right-angled triangle of which the (line) from the center [that is, the radius of the circle] is equal to one of the sides about the right angle, and the other to the perimeter [circumference of the circle] [$A = \frac{1}{2}rC$].

This proposition has two constants, say j associated with the ratio of the area of the circle to its radius and J associated with the circumference of a circle to its diameter. Archimedes shows that they are equal when $j = A/r^2$ and $J = C/2r$. That j and J are related is not apparent. Brilliantly connecting the two, Archimedes proceeds to determine that if circumference of a circle can be found by working with close approximations of π, which is proven elsewhere, its area can be equated with that of a right-angled triangle.[221] This falls in the scope of traditional Greek geometric research. Clearly, Archimedes also distinguished his discoveries from past knowledge. The introduction to *On the Sphere and Cylinder* asserts, for example, that this tract presents theorems never before demonstrated.

Ten complete or nearly complete collections of mathematical writings of Archimedes deriving from later Greek copies that include editorial revisions and scribal confusion have survived, together with fragments of Latin and Arabic translations of others. Philologists continue to sift carefully through these works and attempt to separate out his actual work from later revisions and corruptions. Again, Archimedes' writings may be loosely classified according to these three groupings: [A] theorems and computations relating to curvilinear plane and solid figures; [B] statics, hydrostatics, and application of statics to geometry, and conversely; and [C] miscellaneous studies. The Indo-Arabic numeral after a title suggests its place in the chronological order of Archimedes' extant corpus.

Works in pure geometry, [A], seemingly begin with a fragment of *Dimension of a Circle* (1a or 9). The second of its three propositions is seemingly a later insert, since its proof requires the result of the third. This fragment seems only a small portion of what must have been a comprehensive treatise on the circle. Uncertain about its chronological placement, Archimedes' modern editor J. L. Heiberg has placed it last or ninth among studies in pure geometry. But since it serves to introduce Eudoxus' method of limits or bounds, it probably belongs ahead of his refinement of that method given in [A]5, *On the Sphere and Cylinder*. Next comes *On the Quadrature of the Parabola* (2) consisting of twenty-four propositions, the first seventeen being addressed to statics rather than pure geometry, and *On the Sphere and Cylinder* (5) in two books with fifty-three propositions. For its great rigor in using two-sided convergence of bounding figures and its inventive proofs, modern mathematicians judge *On the Sphere and Cylinder* Archimedes' masterpiece. The tract *On Spirals* (6), containing twenty-eight propositions, boldly presents a new, nonstatic curve known as the Archimedean spiral and gives a method of drawing tangents to it that presaged the later development of differential calculus. *On Conoids and Spheroids* (7), which consists of forty propositions, has a general slicing-construction procedure to sum different volume problems. A conoid generated by revolution of a conic section includes a paraboloid or hyperboloid, and Archimedes' spheroid is now called an ellipsoid of revolution. This summing work was essential in forming integral calculus two millennia later.

Among Archimedes' writings in statics, [B], is *Plane Equilibria* (1), Book I, which consists of fifteen propositions. Although previous thinkers, such as Archytas, had studied the lever and wedge, Archimedes was apparently the first to demonstrate systematically the elementary theorems of statics by applying the principles of the center of gravity and equilibrium determined (as disclosed in *On the Method*) by a conceptual weighing with an ideal balance. *Plane Equilibria* (3), Book II, contains eight propositions on centers of gravity, while the incomplete *On the Method of Mechanical Theorems* (4), of which some portions are in pure geometry, summarizes procedures for discovering proofs of theorems by using an ideal balance. Archimedes' proofs here follow the indirect method of limits that is characteristic of Eudoxean geometry. The two-book *On Floating Bodies* (8) founded the science of hydrostatics. In these two books Archimedes emphasizes symmetry obtained through geometrical analysis.

Works that cannot be assigned to pure geometry or statics include *Psammites* (*Sand-Reckoner*, 10). It gives his system for representing large numbers essentially beyond the range of ancient Greek alphabetic numeration. It is also one of the few extant sources of early Greek astronomy. *The Cattle Problem* concerns a problem famous in antiquity, which leads to solving in integers what is now called Pell's equation, $x^2 - Ay^2 = 1$. The fragmentary *Stomachion* deals with a mathematical puzzle.

Ancient Mediterranean authors also refer to seven tracts now lost. These include *Archai* or *The Naming of Numbers*, probably a preliminary work to *Sand-Reckoner*; *On Polyhedra*, an investigation of the thirteen semi-regular solids; *Elements of Mechanics*, perhaps Archimedes' first account of centers of gravity; and *Optics* or *Catoptrics*, the theory of mirrors. Medieval Arabic authors attribute to Archimedes several additional planimetrical writings. The most notable is *Lemmata*, or *Liber Assumptorum* ("Book of Lemmas"), which exists only in a Latin translation taken from the Arabic version of Thābit ibn Qurra. Thābit translated all known works of Archimedes from Greek. A few of its fifteen propositions closely resemble others in *On Spirals*. In its present form *Lemmata* is probably not by Archimedes, however, for several proofs refer to his name.

Even genuine extant works pose problems for the historian attempting to determine Archimedes' exact contributions to mathematical science. None exists in the Sicilian-Doric dialect in which he wrote. While most have some linguistic transformation, *On the Sphere and Cylinder* and *Dimension of the Circle* are almost entirely purged of the original language. In general, partly because of such linguistic difficulties his writings may not be understood without commentary.

That Archimedes' primary interest was theoretical geometry is illustrated in the prefatory letter to *On the Sphere and Cylinder,* which announces proofs of two propositions. Again, since he had figures and ratios for these propositions inscribed on his grave marker, he must have considered them his greatest discoveries:

> ○ The surface area of any sphere is equal to four times that of its great circle [in modern terms, $S = 4\pi r^2$]. (Proposition 33)
> ○ The volume of a circumscribed cylinder with base equal to the great circle of a sphere and with height equal to the diameter of the sphere is one and a half times the volume of the sphere. [By modern symbolism, $V = 4\pi r^3/3$ for the sphere and $V = 2\pi r^3$ for the circumscribed cylinder.] (Porism to Proposition 34 in Dijksterhuis' translation)

The porism, which in ancient Greek geometry means corollary, exceeds a proof from Eudoxus or Euclid given in the *Elements.* Eudoxus-Euclid had found that the volume of a sphere is proportional to the cube of its radius, or $V = kr^3$, where k is a proportionality constant. Archimedes discovered that $k = 4\pi/3$. Working with a related problem in *On the Sphere and Cylinder,* Archimedes demonstrates that the surface of the circumscribed cylinder, including its two ends, is $3/2$ the surface (or S) of the sphere and that the volume of the circumscribed cylinder is $3/2$ the volume (or V) of the sphere. Hence, the ratio of the surface area of the sphere to that of the circumscribing cylinder is the same as the ratio of their volumes. Again, Archimedes' request that results from Proposition 33 and the porism to Proposition 34 be carved on his grave marker, and his presentation of a second proof of them based on a statical procedure in Proposition 2 of *On the Method* suggest their importance to him.

On the Method states that Archimedes first proved the porism to Proposition 34 and from it conceived the proof of Proposition 33. Here is the essence of his statical argument. Composite elements comprising the volume of a sphere of radius R, and of a cone of base radius and height $2R$ suspended through their conjoined center of gravity on a moment arm of R length, are found to balance a companion cylinder of radius R and height $2R$ located on an opposing moment arm of $2R$ length. The volume of a cylinder is known from Euclid. In modern symbols,

$$[V_{sphere} + V_{cone}] = [V_{sphere} + \tfrac{4}{3}V_{cylinder}] = 2V_{cylinder}$$

and

$$V_{sphere} = \tfrac{2}{3}V_{cylinder} = (\tfrac{4}{3})\pi R^3.$$

Within theoretical geometry, Archimedes succeeded in making the method of exhaustion a powerful mathematical procedure having wide application. He was able to apply it to curvilinear surfaces other than circles, spheres, and hyperbolas, as well as to pyramids from Eudoxus-Euclid. Obtaining areas is inherently more complicated for curviplanar than for rectiplanar surfaces, and Archimedes' Greek predecessors had reached only an elementary stage of the method of exhaustion. In *On the Quadrature of the Parabola*, Archimedes first applied his enhanced version of the method. The preface acknowledges that earlier geometers had found areas of segments of a circle and hyperbola, but adds that no one had yet attempted the quadrature of a parabolic segment for which the conventional method was adequate. Perhaps this omission indicates neglect by geometers to study the parabola; the parabola did not even have a technical name. Earlier Greek geometers had classified it as a conic section and named it the right-angled cone. Archimedes analyzes this curve in greater detail than his predecessors. His proof of Proposition 24, "Every segment bounded by a parabola and a straight line (chord Qq) is equal to four-thirds of the triangle which has the same base as the segment and equal height," is the fruit of his deeper understanding of the method of exhaustion.

Extending the application of the method of exhaustion, Archimedes significantly improved its foundation. Euclid's *Elements* is a model for the axiomatic-deductive method: Book I of Archimedes' *On the Sphere and Cylinder* gives the model its highest degree of rigor in antiquity.

On the Sphere and Cylinder I begins by limiting the class of surfaces to which the method of exhaustion applies. With acute geometric insight Archimedes introduces convexity axioms governing simple, closed curves or surfaces that define the concept of surface area. These novel axioms assert that if a convex or bent curve (or surface) with end points in a plane contains within it another convex curve (or surface), the included is the lesser. The first convexity axiom assumes that the straight line is the shortest curve joining two end points.

Of greater importance for the new foundation was the postulate that Archimedes created. Eudoxus-Euclid had based the embryonic method of exhaustion on Proposition X.1 of the *Elements*, which implies infinite divisibility of a continuum. Archimedes formulated a more fundamental statement and perceptively assigned it to the postulates, not to the theorems. The new postulate is Assumption 5 of *On the Sphere and Cylinder*. (See also the preface to *On the Quadrature of the Parabola* and *On Spirals*.) It assumes

> that of two unequal lines, unequal surfaces, and unequal solids the greater exceeds the lesser by an amount such that, when added to itself, it may exceed any assigned magnitude of the type of magnitudes compared with one another. (Clagett's paraphrase)

Here is the Archimedean postulate for a line: Given two points A and B on a line and a third point C, no matter how far distant on that same line, repeatedly measuring off AB will eventually go beyond C. Only with such a postulate can Proposition X.1 of the *Elements* be justified, for it guarantees that magnitudes from successive divisions always have a ratio in the Euclidean sense.

$$A \quad\quad B \quad\quad\quad\quad\quad\quad\quad\quad C$$

Although the Archimedean postulate resembles and has been occasionally identified with Definition 4 of Book 5 of the *Elements*, it has greater importance. To be sure, it embraces that definition, which states that

> Magnitudes are said to have a ratio to one another that are capable, when multiplied, of exceeding one another.

But note that the postulate, unlike the definition, can be understood as excluding indivisibles, an early infinitesimal, from Euclid X.1, while permitting the continuous divisibility of the continuum.

Building on his postulate, Archimedes develops two types of the method of exhaustion that go beyond the embryonic state of Eudoxus. Dijksterhuis has labeled these types the "compression" and the "approximation" method. Archimedes almost exclusively employs the compression method, which he subdivides into two forms—decreasing differences and decreasing ratios. Eudoxus had exhausted curvilinear areas simply by inscribing them with regular polygons of successively larger numbers of sides. Archimedes makes a two-pronged attack. He compresses the length, area, or volume of a curvilinear figure between successively larger *inscribed* and successively smaller *circumscribed* regular polygons to a point at which these polygons nearly merge with the curvilinear figure to be measured. Examples of the decreasing difference form of the compression method appear in Propositions 1 and 3 of *Dimension of the Circle* and Propositions 24 and 25 of *On Spirals*, while the ratio form (*circumscribed polygon/inscribed polygon*) appears in Book I: Propositions 13, 14, 33, 34, 42, and 44 of *On the Sphere and Cylinder*. Though both methods suggest appreciation of our notion of approaching limits, neither Archimedes nor any other ancient Greek seems to have envisaged a process with an infinite number of steps but were content with finite steps that meet required accuracy.

Proposition 3 is the most important in *Dimension of the Circle*. It states that

> the circumference of any circle is three times the diameter and exceeds it by less than 1/7 of the diameter and by more than 10/71 of the diameter. (Dijksterhuis's paraphrase)

It was the first statement of which evidence remains to establish explicitly an accurate inequality for the ratio of the circumference of a circle to its diameter, the ratio we know as π a designation made standard by Euler. Here in modern symbols is Archimedes' approximation:

$$3 \ 10/71 < \pi < 3 \ 1/7.$$

To determine the length of the circumference of a circle with increasing accuracy, he simultaneously inscribes and circumscribes regular polygons of 6, 12,

24, 48, and 96 sides. It is natural to begin with the inscribed hexagon (p_6), since the length of its chords equals the radius, and tangents drawn to its points on the circle form the circumscribing hexagon (P_6). Repeated construction of bisectors of the regular polygons' sides will reach 96 sides and beyond. The circumference here lies between the sequence of the perimeters of the two, being greater than the inscribed polygons' perimeters yet less than that of the circumscribed polygons, or $p_6 < p_{12} < \ldots < p_{96} < C < P_{96} < P_{48} < \ldots < P_6$. At each step, therefore, this Archimedean process provides an upper bound and a lower bound, the two bounds approaching each other ever more closely. C is the greatest lower bound of the P_is and the least upper bound of the p_is. But there is no glb and lub at each step. This procedure has an advantage over Cauchy's later single sequence, where $c_m - c_n < \epsilon$ for m, $n > N$: the investigator knows at every step that the desired quantity lies between two magnitudes.

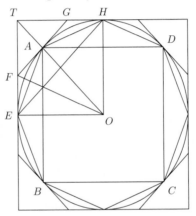

Knowing the differences between the two sequences alone did not suffice to compute the approximation of the final inequality, however. In the process of reaching those bounds, which differ by very little, Archimedes observes that the harmonic and geometric mean of P_n is P_{2n} and that of p_n, p_{2n}. Thus, $P_{2n} = 2p_n P_n/(p_n + P_n)$ and $p_{2n} = \sqrt{p_n P_{2n}}$. Beginning with diameter d, the perimeter $p_6 = 2\sqrt{3}d$ and $P_6 = 3d$. As an upper bound approximation for the irrational $\sqrt{3}$, Archimedes assumes that $1 : \sqrt{3} \lessapprox 153 : 265$, but does not explain how he reached these figures. He proceeds to find $265/153 < \sqrt{3} < 1351/780$, possibly by computing by recursion formulas through four stages that $(3 + 10/71)d < 96 \times 66/2,017\frac{1}{4} < p_{96}$, obtaining the lower bound of π, and $P_{96} < 96 \times 153/4,673\frac{1}{2} < (3 + 10/70)$, for the upper bound. He simplifies the results from the larger multiplications to obtain the simpler fractions given here: $3\frac{10}{71}$ and $3\frac{1}{7}$. To reach these figures he at each stage of his computation rounded upward for the upper bound and downward for the lower, which produces a cumulative error. Still, his final results are only 0.002 apart, that is, $3.1408 < \pi < 3.1428$. This inequality closely approximates the numerical value of π, which is $3.14159265\ldots$.

The inequality of Proposition 3 suggests that Archimedes computed the final bounds while working with 96-gons. There is strong reason to doubt this.

In I.26 of *Metrica*, Hero gives another estimate of bounds supposedly from Archimedes that turns out to be poorer. But basing his claim on the precision of the computation of irrationals and the large quantities that Archimedes rounds off in ratios to reach simple fractions, historian Wilbur Knorr has argued that Archimedes made even more accurate approximations to π by following the compression method, beginning with regular decagons inscribed in a circle and circumscribed about it and ending with 640-gons and continued fractions.[222] This could yield $333/106 < \pi < 377/120$.

Archimedes calculates π via a spreadsheet program

No. of sides	inside area	outside area
	A	B	C
1	s	a	A
2	2	2	4
3	8	2.82842712475	3.31370849898
4	16	3.06146745892	3.18259787807
5	32	3.12144515226	3.15172490743
6	64	3.13654849055	3.14411838525
7	128	3.14033115695	3.14222362994
8	256	3.14127725093	3.14175036917
9	512	3.14151380114	3.14163208070
10	1024	3.14157294037	3.14160251026
11	2048	3.14158772528	3.14159511775

Archimedes calculates π using perimeters of polygons

No. of sides	outside P	inside p
s	P	p
3	5.19615242271	2.59807621135
6	3.46410161514	3
12	3.21539030917	3.10582854123
24	3.15965994210	3.13262861328
48	3.14608621513	3.13935020305
96	3.14271459965	3.14411838525
192	3.14187304998	3.14145247229
384	3.14166274706	3.14155760791
768	3.14161017660	3.14158389215
1536	3.14159703432	3.14159046323

Archimedes chose to recognize the value of small-scale precision, with an amount that met practical purposes. The value $3\frac{1}{7}$ or 22/7 was adopted uniformly throughout the tradition of subsequent ancient metrical geometry. Today it is called the Archimedean value of π.

Archimedes' approximation form of the method of exhaustion appears in only one of his extant writings. Propositions 18 through 24 of *On the Quadrature of the Parabola* make up its second part. Proposition 23 is a lemma on the sum of a finite number (k) of terms within a series of areas of triangles inscribed in a parabolic segment for measuring the segment's area. Beginning with a triangle that has the same base and height as the parabolic segment and shares the vertex point, Archimedes inscribes a series of triangles, starting with the sides of the original triangle and, drawing lines from their endpoints to the new vertex to form triangles in each new parabolic segment.

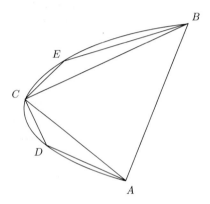

Such triangles can be drawn similarly as far as desired. From properties of the parabola, Archimedes had found in Propositions 21 and 22 that the new triangles at each stage equal one-eighth of the original triangle and hence together are a fourth of its area. In modern notation, he is summing in Proposition 23 a series of areas that amount to the infinite geometric series of ratio 1/4 or

$$A_1\{1 + 1/4 + (1/4)^2 + (1/4)^3 + \dots\} = \tfrac{4}{3}A_1.$$

Only in modern times were techniques of series summations developed that allow a direct computation, but it is tempting to believe that Archimedes had grasped them intuitively. To prove Propositions 23 and 24 rigorously, he employs a corollary to Proposition 20, which states that

> it is possible to inscribe in the parabolic segment a polygon such that the segments left over are together less than any assigned area [which depends upon Euclid X.1],

and a double *reductio ad absurdum* argument applied to the preceding polygons. In this fashion he demonstrates that the area of the parabolic segment could be neither greater nor less than $\tfrac{4}{3}A_1$, because both lead to a contradiction.

More original than the compression and approximation procedures was Archimedes' method of discovery. To determine areas and volumes, he ingeniously brought to bear the mechanical construct of an imaginary balance and that of reducing lines, areas, and volumes to an ideal center of gravity, compounding these with the concept of plane laminae. From the results he derived theorems later proved rigorously by the compression or approximation procedure. The plane laminae, akin to successive slices through a cone, treat plane figures as the sum of parallel line segments of unit density located unlimitedly close together. Presumably, these line segments are infinite in number. Using conceptual manipulations, Archimedes imagines the statical consequences when these abstract line segments comprising an unknown area are located at a known distance from a fulcrum on a given side of a weightless lever. The law of the lever determines conditions of equilibrium with corresponding line segments of the same magnitude from a plane figure whose area is already known. Then the unknown area can be found. Analogously, Archimedes imagines a solid to consist of plane elements (indivisible slices) of unit density. In *On the Equilibrium of Planes* (wherein he generalizes the law of the lever) and Propositions 6 through 16 of *On the Quadrature of the Parabola*, he applies this mechanical method to solve geometrical problems. *On the Method* describes and applies this heuristic laminar technique.

The mechanical method of Archimedes is remarkable for several reasons. Ancient Greek geometers normally followed Aristotle's shop window approach, removing most traces of presystematic analysis involved in the process of discovery and communicating only the final synthesis of theorems in a deductive system. Archimedes' *On the Method* was, to a degree, one exception to the rule.[223] He clearly shared with his contemporaries, however, the view that the discovery of a theorem does not compare in importance to its proof. Even by modern standards, his insistence that his mechanical method was simply one of discovery and not of proof is unusual. A modern mathematician might consider a mechanical argument adequate for proving a theorem. Archimedes' refusal to do so bespeaks a high standard of rigor. In addition, he must have recognized that his mechanical arguments for deriving theorems generally constituted incomplete proofs. This does not detract from their importance to his work. The mechanical method, moreover, demonstrates the fruitfulness of the interplay of the two sides of Archimedes' genius: the theoretical and mechanical. Perhaps only Newton possessed these two types of genius on a comparable scale and combined them as skillfully as Archimedes had.

Although his method of exhaustion is impressive, Archimedes clearly did not invent infinitesimal calculus. Nor should it be expected of him. What elements of the future calculus had yet to be developed? The method of compression gives the least upper and the greatest lower bounds, but nowhere does Archimedes introduce the limit concept explicitly; for the most part he avoids the use of infinite series. Nor does he recognize the directly inverse relationship between quadrature or obtaining areas under curves and tangent problems, having examined the latter in Propositions 18 through 20 of the tract *On Spirals*. The discovery of this relationship by Newton and Leibniz was crucial in the invention of calculus. And in Archimedes' time no general computational algorithm existed for finding curvilinear areas and volumes, the center of gravity or centroid of a triangle, and the areas enclosed by his spiral. He did not establish a connection among these problems. In modern notation they all depend on the integral $\int f(x)\,dx$. Dependence on geometric algebra, the lack of an adequate theory of irrational numbers, the power of the deductive method, and the paucity of symbolic notation in later antiquity all worked against the development of general procedures. An expanding and deeper exploitation of the analogies of the ideal balance and center of gravity was missing, along with new analogies crucial to founding infinitesimal calculus two millennia later.

Another of Archimedes' mathematical achievements, the analysis of the spiral, took him into investigation of the role of motion in mathematics, not the most comfortable of topics for ancient Greek geometers. They studied in detail only a few curves. Most of these curves were static and were defined by simple locus conditions. The Greek mind gave a privileged position to being in a set place over the role of motion, or becoming, in mathematics. Among generated curves only those depending on uniform motion, either linear or circular, gained close attention. It was probably his study of a parallelogram of velocities applied to tangents that led Archimedes to discover his mathematical spiral, a transcendental curve composed of both types of uniform motion. He wrote:

> If in a plane a straight line (half ray) of which one extremity remains fixed, is made to revolve at a uniform rate until it returns to the position from which it set out, and if at the same time that the line revolves, a point moves at a uniform rate along the line, starting from the fixed extremity, the point will describe a spiral in the plane. (J. F. Scott's translation of Definition 1 of *On Spirals*)

If this curve is defined by modern polar coordinates, its characteristic property requires that the rotary motion of the radius vector be at constant angular velocity and the radial motion of the point at constant linear velocity. Building on Proposition 14 of *On Spirals*, the equation for the Archimedean spiral in modern polar coordinates is $r = a\theta$, where $a > 0$.

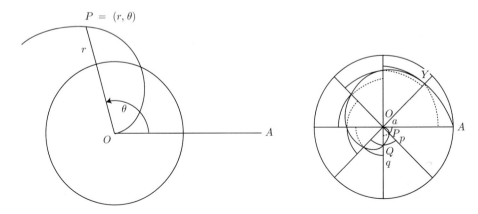

On Spirals is a rich source of ancient mathematics. As Archimedes' approximation of π and the concomitant construction of a right triangle equal in area to a given circle illumine the historic problem of the quadrature of a circle, *On Spirals* in part addresses the quadrature of its new curve. In what modern mathematicians consider his most fascinating achievement, Archimedes calculates the area of the segment enclosed by the first loop of the spiral using the method of compression with numerous circumscribed and inscribed circular sectors. He reaches the equivalent of $A = \frac{1}{3}\pi(2\pi a)^2$, a problem easily solved by integral calculus, but of prohibitive difficulty before. Archimedes also constructs a tangent to the spiral, perhaps the first tangent drawn to a curve other than a circle.

Greek geometers before Archimedes had moved from the bisecting of angles to attempting to trisect them. Propositions 5 through 9 of *On Spirals* employ *neusis* (verging by using an idealized sliding ruler) constructions. Greek *neusis* is mechanical; a line segment is inserted and made to verge toward a point. An example is given below. The Archimedean spiral permits trisection of every angle. Begin with a spiral and draw OM from the origin to intersect the terminal side. Next trisect OM, giving points N and P. If circular arcs are drawn with radius OP and ON, they will intersect the spiral at N' and P'. The lines drawn from the origin to N' and P' trisect $\angle AOM$.

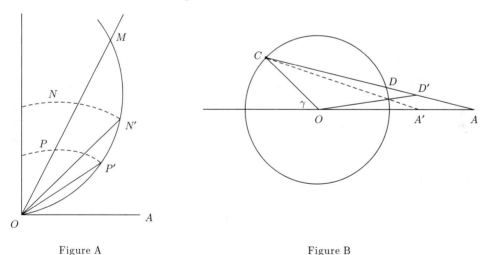

Figure A   Figure B

Although the *Lemmata* does not come from Archimedes in its present form, Proposition 8 has such a strong Archimedean character that he appears to be its substantial originator. It also applies a *neusis* construction to trisect an arbitrary angle mechanically (see Figure B): An angle O is given having radius OC and angle γ formed between OC and the diameter. Extend the diameter to a point A well beyond the circle and extend chord CD to form line segment CA. Next, let line segment CA sweep down until $D'A' = OD' = OC$ (the radius). Then $\angle\alpha$ formed by radius OD' and line OA equals $\frac{1}{3}\angle\gamma$. The proof is

$$\angle \gamma = \angle OCD' + \angle CA'O = \angle CD'O + \angle \alpha = 2\angle \alpha + \angle \alpha = 3\angle \alpha.$$

The introduction to *Sand-Reckoner*, a work that addresses not geometry but astronomy and arithmetic, summarizes the heliocentric theory propounded by Aristarchus. Archimedes rejects it for not conforming to available observation. A moving Earth requires an annual stellar parallax, a shift in the positions of foreground bodies relative to the backdrop of distant ones from opposite points in the Earth's orbit. The great astronomical distances defeat the power of the naked eye, however, so ancient astronomers could observe no annual parallax of fixed stars.[224] To account for the absence of parallax, Aristarchus proposed a universe of large dimensions: his value for the radius of the Earth was minute compared with that of the sun's distance from the Earth, and the sun's distance was a point compared with the distance of the fixed stars. Archimedes finds this proportion unsound mathematically: no proportion can exist where there is no magnitude (but only a point).

In rejecting Aristarchus' hypothesis, Archimedes agreed with the majority. Stoics were scandalized by the displacement of the sacred Earth from the center of the universe. Natural philosophers objected to an orbiting Earth that contradicted Aristotelian physics, in which the center of the Earth is the necessary center of gravity for all heavy objects, and required a spatial displacement or lag from the point of origin to the point of return for objects detached from the terrestrial moving surface, when observations showed that bodies fall vertically. To have the Earth orbit in a circular path, moreover, fails to explain the differing lengths of seasons. Ancient astronomers, moreover, soon found Apollonius' epicycles and a geocentric orientation more advantageous, particularly for explaining retrograde motions. A stationary Earth seemed better to fit experience. Only the Chaldean astronomer Seleucus of Seleucia (ca. 150 B.C.) reportedly agreed with Aristarchus.

While *Sand-Reckoner's* astronomy has become obsolete; its arithmetic was far in advance of the times. Archimedes expounded a new notational system for expressing very large numbers that went beyond Greek alphabetic, nonpositional numerals. He counts in units of ten thousand myriads of the myriad-myriadth numbers: in modern notation, $P = 10^8$ (octads). He extends this system to the second octad to 10^{16}, the third octad to 10^{24}, and so forth.

To demonstrate his system, Archimedes undertakes to count the number of grains of sand needed to fill "a mass equal in magnitude to the cosmos," that is, a finite sphere bounded by the fixed stars. If the diameters of the sun, moon, and Earth be represented, in our notation, by D with subscripts, Archimedes' assertion becomes

$$D_{sun} = 30 D_{moon} < 30 D_{Earth}.$$

Most astronomers of the time had a factor of 20 rather than 30 in the last two positions. Archimedes allows for a larger cosmos. Working cleverly with the perimeters of a regular 812-sided polygon and a chiliagon, a regular thousand-sided polygon, inscribed in a greatest circle of the cosmos, he finds that

$$3D_{cosmos} < 1000D_{sun} < 30,000D_{Earth}.$$

Astronomers believed that the distance from the Earth to the sun was the diameter of the cosmos. If the Earth's circumference is assumed to be the accepted 300,000 stades, the diameter had to be less than 100,000 stades. Multiplying 30,000 times 100,000 gives 3×10^8. Multiplication by a safety factor of 10, so as not to be too small, gives

$$D_{cosmos} < 10^{10} \text{ stades}.$$

Assuming that the volume of a poppy seed contains no more than 10,000 grains of sand and that from twenty-five to forty poppy seeds exceed a finger breadth, Archimedes takes the conservative figure 40 to find that he needs 64,000 poppy seeds or 640 million grains of sand to fill a sphere with the diameter of a finger breadth. The latter figure is less than one billion grains. If the length of a stade is taken to be less than 10^4 finger breadths, filling the sphere will require fewer than $10^9(10^4)^3 = 10^{21}$ grains of sand. To fill the sphere of the cosmos with a diameter of 10^{10} stades, Archimedes needs a mere $10^{21} \times (10^{10})^3 = 10^{51}$ grains of sand.

The pursuit of precise numerical computation so striking in *Sand-Reckoner* pervades all the mathematics of Archimedes. Propositions 6 and 7 of *On the Equilibrium of Planes I* strengthened the foundations of number theory. Appealing to geometric symmetry and lever equilibrium using the ideal balance, Archimedes in Proposition 6 put commensurable magnitudes in equilibrium and in Proposition 7 offered an incomplete proof of the same for incommensurable magnitudes.[225] Proposition 3 of *Dimension of the Circle* further shows his adeptness at manual calculation. The Arab scholar al-Bīrūnī credited Archimedes with discovering the Heronian formula for computing the area of a triangle (at least a right triangle) by means of its sides. That formula is

$$A = \sqrt{s(s-a)(s-b)(s-c)},$$

where s is the semiperimeter and a, b, and c are the sides. Known titles of Archimedes' lost work provide some basis for this conjecture. If al-Bīrūnī is correct, here is another evidence of Archimedes' mathematical boldness. By multiplying four lengths, Archimedes cast aside restrictions seemingly imposed by three-dimensional Euclidean space.

The writings of Archimedes, unlike Euclid's *Elements*, were little known in later antiquity and medieval Europe. While Hero, Pappus, and Theon of Alexandria quoted from them, a textual history of collected Archimedean tracts began only with Eutocius (b. ca. 480), who was not an original mathematician. Eutocius wrote commentaries elucidating three tracts: *On the Equilibrium of Plane Figures, On the Sphere and Cylinder I* and *II*, and *Dimension of the Circle*. From the early fifth to the thirteenth century the Archimedean corpus was preserved primarily either in Byzantium or in the Islamic world. Isidore of Miletus and Anthemius of Tralles (d. ca. 534) began to study and teach the commentaries of Eutocius in Constantinople. In the next four centuries, Byzantine authors recovered several treatises by Archimedes. Leon of Thessalonica's collection in the ninth century included all extant works except *On Floating Bodies, On the Method, Stomachion,* and *The Cattle Problem.* In the ninth-century Islamic world, the Banū Mūsā in *Verba filiorum* incorporated proofs from Archimedes, especially on the measurement of the circle and *neusis* construction; Thābit ibn Qurra translated all known Archimedean texts into Arabic.

During the twelfth and thirteenth centuries, the Latin West received knowledge of Archimedes' work from Arabic and, to a lesser extent, Byzantine sources. Gerard of Cremona (ca. 1114–1187) transmitted to western Europe part of the Archimedean corpus in Arabic garb. His translations were from the Arabic and relied on commentaries by Arabic scholars. The Dominican William of Moerbeke (ca. 1220–1286) later translated Archimedes from the Greek of Byzantine sources. Even so, the line of transmission from Arabic sources prevailed. Appreciation of the exact and extensive recovered Archimedean corpus from the original Greek together with sophisticated new mathematical studies based on it awaited the late European Renaissance.

6.3   Eratosthenes, Apollonius, and Diocles

Among the other eminent Hellenistic mathematicians were Eratosthenes, Apollonius, and Diocles. All were younger than Archimedes.

Eratosthenes (ca. 276 to ca. 195 B.C.), a correspondent of Archimedes, was born in Cyrene (now Shabhat, Libya). About 260 he traveled to Athens to study philosophy. He found the Platonic, Aristotelian, Epicurean, and Stoic schools in constant dispute. The Cynics, who minimized material possessions, were influential as social and moral critics. Eratosthenes listened to leading proponents of every school and came to admire especially two teachers: the Peripatetic Ariston of Chios, a critic of the Stoics' moral casuistry who held to uncompromising ethics, and Arcesilaus, who reorganized and revived studies at the Academy. Arcesilaus chose not to introduce students directly to the views of his school. Rather, he followed Plato's original program, requiring each student first to learn mathematics thoroughly and then to study the dialectical method before addressing more abstract problems or making particular philosophical commitments. Eratosthenes received his higher education at the Academy, where the geometer and astronomer Autolycus of Pitane was probably still active.

Around 246 B.C. Ptolemy Euergetes invited Eratosthenes to Alexandria to tutor the crown prince, Philopator. He remained there for the rest of his life. When Apollonius of Rhodes, the chief librarian, died about 235 B.C., Ptolemy appointed Eratosthenes to head the famous library. In a city where polymaths were greatly admired, he exhibited a breadth of learning that almost rivaled what Aristotle had shown in the previous century.

Located at the intellectual center of the thriving Ptolemaic Egyptian kingdom, Eratosthenes collected and systematically organized extensive geographical data and founded a mathematical as opposed to exclusively descriptive geography. Drawing on this research, which set a precedent for the related studies of Strabo and Claudius Ptolemy, Eratosthenes wrote his three-book *Geography,* which divides the terrestrial globe into zones, estimates many distances along roughly defined parallels that are antecedents of latitude and longitude, and carefully maps the southeastern portion of the known inhabited world, superceding traditional Ionian maps that depicted the southeastern quadrant as round and surrounded by a circular ocean.

Eratosthenes' book also describes peoples and places. He computed rigorously the perimeter of the spherical Earth, and he suggested the possibility of a circumnavigation of the globe. His *Geography* long remained an authority. In VI, 24 of *Gallic Wars,* Julius Caesar cited its passage on the Hercynian forest (variously sited near German mountains). Eratosthenes also made chronology more critical and exact by developing a system of dating based on olympiads and placing events in each olympiad solely by use of authentic documents and reliable estimates. He scrupulously eliminated unverifiable reports. Much of the dating in his books *Chronology* and the more popularized *Olympic Victors,* such as that of the fall of Troy in 1184–1183 B.C., the start of the Dorian migration in 1104–1103, and the first olympiad in 777–776, have so far not been improved on.

Eratosthenes also wrote on grammar, literary criticism, philology, philosophy, astronomy, and theoretical mathematics. Constant citations in scholia to Aristophanes' plays establish him as an authority on Old Comedy. As head of the largest library in antiquity, he must have kept closely in touch with the rich source materials in his custody. Eratosthenes also composed poetry and was a distinguished athlete whom the students at the Alexandrian Museum called *Pentathlus*, the five-sports champion. His associates knew him as an erudite, industrious, and conscientious man of artistic taste.

According to the encyclopedic *Suda Lexicon,* Eratosthenes was portrayed as an "All-Rounder," "Another Plato," and *Beta* (the Greek numeral two). The nickname *Beta* may suggest that he was second only to Archimedes or that he attained superior rank in many fields rather than the highest rank in one. But perhaps it meant only that a poll of scholars would accord him the "vote of Themistocles" in every branch of knowledge.[226] The Greek defeat of the Persians in the sea battle of Salamis depended greatly upon the stratagems, valor, and fortitude of the Athenian leader Themistocles. After the victory when the generals voted to decide who was the bravest in battle "each voted for himself as the most valorous [*alpha*] and Themistocles as the second [*beta*]." Eratosthenes' subsequent reputation faltered. In the second century B.C., Hipparchus declared Eratosthenes' geography not mathematical enough, failing to use fully astronomical data in preparing maps. In the early Christian era, Strabo's *Geography* (94, cf. 15) called him a mathematician among geographers

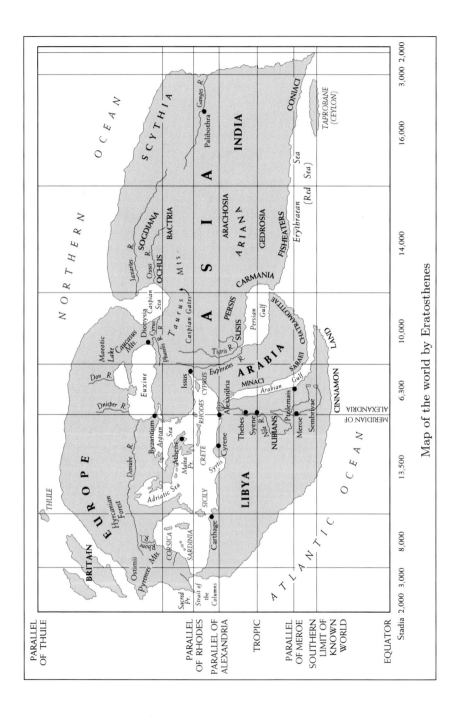

Map of the world by Eratosthenes

and a geographer among mathematicians. Strabo spurned mathematical geography and decried Eratosthenes' emphasis on it, reserving praise for his skillful drawing together of the extensive data available at the Museum to make a reliable map of the southeastern portion of the known world. If these unfavorable assessments of his application of mathematics to geography are correct, it is a puzzle why so able a mathematician as Archimedes saw fit to dedicate to him the tract *On the Method*, to comment favorably on his potential in research in its preface, and to send him the *Cattle Problem* to circulate among Alexandrian mathematicians.

In old age Eratosthenes became almost blind from ophthalmia and reportedly committed suicide by voluntary starvation, a practice not uncommon in antiquity and known as the "philosopher's death."

Eratosthenes' *Geography* contains his most celebrated achievement, a simple procedure for calculating the circumference of the Earth. Not the first attempt at such a measurement, it appears the most accurate. Eratosthenes' calculation is based on estimates of the arc of the great circle passing through Alexandria and Syene (modern Aswan) in Upper Egypt. He observed that, while the sun cast no shadow from an upright gnomon in Syene at noon on the summer solstice, the shadow cast at the same time at Alexandria indicated a light ray's angular deflection from the zenith. The gnomon used for obtaining a precise measure was probably a thin stylus or pointer placed at the center of a hemispheric bowl. Expressed in the Alexandrian system of angle measurement, the deflection or inclination of the sun's rays with the vertical would be 1/25 of the hemisphere or 1/50 of the full circle (four right angles), that is, 7°12′. (Eratosthenes did not know the familiar angular division into 360°, which Hipparchus later introduced into Hellenistic Greek science.)

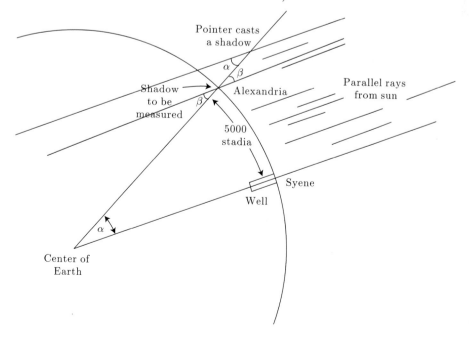

According to II, 183 of the Roman author Pliny's *Natural History*, Eratosthenes believed that Syene (23°44′ N) and Alexandria lie on the same meridian, and he accepted a rounded off value of 5,000 stades as the distance from Syene to Alexandria; it was the distance determined by trained counters-of-steps known as *bematistes* or surveyors. Employing Euclid I.15 on the equality of vertical angles, that is $\beta = \beta'$, and I. 29 on the equality of alternative interior angles, that is, $\alpha = \alpha'$ and $\alpha = \beta$, he calculated the Earth's circumference by a means that modern notation expresses as

$$\frac{C}{360°} = \frac{5000 \text{ stades}}{(1/50)(360°)} = \frac{5000 \text{ stades}}{7°12'},$$

$$C = 250,000 \text{ stades}.$$

Hipparchus maintains that Eratosthenes later modified the value of C to 252,000 stades in order to obtain a number easily divisible by 60, because he had divided the circle into sixtieths. But fresh observations could have brought him to the change. Eratosthenes' computation of Earth's circumference places him within the tradition of scholars from at least Aristarchus a century earlier to Archimedes, who sought measures of the size of Earth and the cosmos.

While Eratosthenes' method is sound in theory, the accuracy of his final result depends on the precision of the basic data and which of several possible stades he employed as a unit of measurement. The length of the stade varied from $7\frac{1}{2}$ to 10 stades to the Roman mile. If Eratosthenes used the Egyptian value just above 10, the 252,000 stades become 24,662 Roman miles. Since the Earth's actual circumference is 24,888 miles, this gives an error of only 226 miles and hence a remarkable accuracy within 1%. More likely, Eratosthenes used $7\frac{1}{2}$ stades, which means that his 252,000 stades equaled 29,000 miles, which is more than 16% too high.

Another estimate of Earth's circumference made at the time based on the sea distance between Rhodes and Alexandria may suggest Eratosthenes' goal. Its computation gave 180,000 stades, or 48 ×3750. Since 252,000 stades reveal 700 stades as the estimate of the distance along 1° of arc on the great circle of Earth, while 180,000 gives 500 stades, Eratosthenes seems to be seeking not precision so much as a handy order of magnitude to employ in geographical calculations.

Eratosthenes' principal work in mathematics is a dialogue *Platonicus*, of which only a few extracts survive. It discusses proportions and progressions and provides a mechanical solution to the Delian problem of the duplication of the cube. Eratosthenes' history of the problem mentions solutions by Archytas, Eudoxus, Menaechmus, and pseudo-Plato. The method that Archytas devised involving a sequence of triangles likely set the precedent for Eratosthenes, who advances a mechanical solution depending on a bronze instrument called a mesolabium, from *mesalabos* or "taker of means." It had three rectangular plates, marked with their diagonals, that slid between three grooves between a fixed and a movable ruler. Eratosthenes employs it to construct two mean proportionals, here BZ and GH, between two given lines, AE and $D\Theta$, and thus to solve the cube-duplication problem. His orientation of the mesolabium diagonals reverses that of Archytas, perhaps to enhance the mechanical operation.

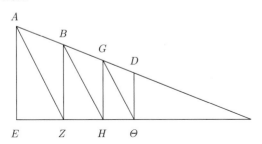
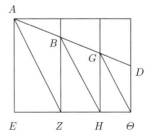

Eratosthenes was not satisfied with providing only the geometrical rationale for the mesolabium. He also gives specifications for its construction. Perhaps he exaggerates its practical application to commercial computations involving weights and measurements of volumes, as well as the design and production in scale of ships and military engines. He, like Archimedes, appreciated both the theory and the practical application of geometry.

I, 13 of the *Introductio Arithmeticae* of Nicomachus of Gerasa reveals Eratosthenes' chief accomplishment in theoretical mathematics: the invention of a process for finding all primes less than a given number N. That process is now called the "sieve (*koskinon*) of Eratosthenes." Book IX of Euclid's *Elements* had pointed up the importance of primes. Proposition 14, the fundamental theorem of arithmetic, holds prime numbers to be the multiplicative building blocks of all other integers. Even though Euclid IX.20 asserts that the number of primes is infinite and thus that no complete catalogue of them is possible, mathematicians beginning with Eratosthenes sought a catalogue. His sieve involves a simple procedure: write out the positive integers and then strike out all multiples of 2, 3, 5, 7, ... in succession until all composites are eliminated and only primes remain. Consider 2 to 30.

$$
\begin{array}{cccccccccc}
 & 2 & 3 & \not{4} & 5 & \not{6} & 7 & \not{8} & \not{9} & \not{10} \\
11 & \not{12} & 13 & \not{14} & \not{15} & \not{16} & 17 & \not{18} & 19 & \not{20} \\
\not{21} & \not{22} & 23 & \not{24} & \not{25} & \not{26} & 27 & \not{28} & 29 & \not{30}
\end{array}
$$

The sieve of Eratosthenes identifies the primes but does not provide a simple rule for determining how often they occur. Only if the number of primes below a large number N is known can the sieve yield a cumbersome formula for determining the number of primes below $\sqrt{N}$. The concise prime number theorem was not discovered until the early nineteenth century by Adrien-Marie Legendre and Karl Gauss.

Perhaps a generation younger than Eratosthenes was Apollonius (ca. 240 to ca. 174 B.C.). A contributor to all ancient branches of pure and applied mathematics, especially astronomy, he ranks with Eudoxus, Euclid, and Archimedes

as a leading mathematician of antiquity. Yet, surviving references about his life
are meager and sometimes untrustworthy. They report that he was born in the
small Greek city of Perga in southwestern Asia Minor when Ptolemy Euergetes
ruled Egypt and that he became a noted mathematical astronomer in Alexandria during the reign of Ptolemy Philopator, who ruled from 221 to 203 B.C.
Pappus in VII of the *Collection* claims that Apollonius studied under pupils of
Euclid at the Museum. But this seems unlikely; there is no confirming evidence.
The best source of biographical information on Apollonius comes from prefaces
to the separate books of his *Conics*. These indicate that as a young man he
resided in Pergamum in western Asia Minor and visited Ephesus and that he
lived most of his adult life in Alexandria.

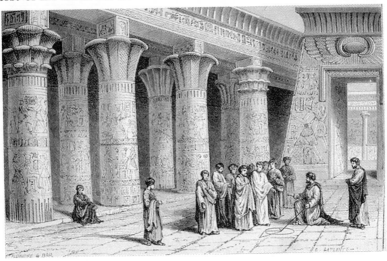

Apollonius at the Alexandrian museum

Apollonius' credentials and residence in Pergamum and Alexandria gave
him accessibility to the two chief Hellenistic libraries and to important intellectual circles. Attempting to turn his kingdom into a successor of fourth-century
Athens, King Attalus I of Pergamum during his reign from 214 to 197 B.C. had
commissioned a large library and attracted Athenian scholars to his court. It
became a new center of learning modeled on the Museum and library in Alexandria. Close contact with a major library such as Pergamum's was important
to leading Hellenistic mathematicians. Perhaps in preparation for writing the
Conics at the court of Attalus, Apollonius held discussions with other scholars:
he later dedicated the first three books to the geometer Eudemus, whom he
had met there.

Prefaces to the *Conics* reveal Apollonius' association with other major
mathematicians of the day and give strong evidence of when the text was written. After Eudemus died, he dedicated Book IV and the remaining books to a
certain Attalus. It is unlikely that this was the Pergamese king, for the omission of the title "King" would be a gross breach of judgment. The Attalus of

the introduction was probably a Macedonian mathematician. Apollonius also mentions Conon, a correspondent of Archimedes. In the preface to Book I, Apollonius remarks that he hastily composed an early version of *Conics* for the geometer Naucrates, who was visiting him in Alexandria. Afterward, most likely during the first years of the second century B.C., he completed the revised version that we know today. Mentions of his adult son Apollonius in the preface to Book II and the death of Eudemus in the preface to Book IV suggest that he wrote *Conics* when he was older. These prefaces are the basis for the chronology of the completion of the *Conics*. It agrees with the assumption that Apollonius had access to Archimedes' writings on the parabola and conoids.

The first-known nearly exhaustive monograph on a specific mathematical subject and the chief reason for Apollonius' modern reputation is the master work *Conics,* consisting of eight books containing 487 propositions. In the preface to Book I, Apollonius points out the limitations and inadequacies of his predecessors, apparently to justify the appearance of his new textbook, although some readers, including Pappus, pronounced him boastful and envious. Books I through IV have survived in Byzantine Greek manuscripts dating from the twelfth and thirteenth centuries; Books V through VII and a fragment of Book VIII, in an Arabic translation made in 1290. Most of Book VIII is lost, but Edmund Halley's seventeenth-century restoration based on lemmas of Pappus in *Collection* VII indicate its probable content on conjugate diameters.

By the mid-fourth century B.C., Eudoxus' student Menaechmus, as noted in Chapter 4, had discovered conic lines and begun investigating their mathematical properties. Eutocius' commentaries tell us that through the appropriate intersection of two conic lines (they are now called a hyperbola and parabola or possibly two parabolas) Apollonius determined the required two mean proportionals to solve the Delian problem. Current studies, however, suggest that Diocles, not Apollonius, was the subject of Eutocius' reference. Afterward Aristaeus introduced the names "section of a right-angled cone," "section of an obtuse-angled cone," and "section of an acute-angled cone." These names suggest orthogonal conjugation, the producing of a conic section by cutting though any right circular cone a plane perpendicular to a generator. Archimedes employed this traditional method to obtain new results on the subject. The preface to Book I of *Conics* mentions that Euclid, in *On the Division of Figures*, had attempted unsuccessfully to provide a more general formulation for this branch of higher geometry. In *Conics*, Apollonius, who mastered earlier sources, succeeds admirably with the synthetic Euclidean program.

As the preface to Book I states, the first four books rework the corpus of elementary theorems on conics, modeled primarily on the content and style of Euclid.[227] *Conics* also elaborates and extends Archimedes' researches, which Apollonius rejects, perhaps to prevent his originality from being overshadowed by his great predecessor. The subjects of its first four books are:

Book I: Methods of Generating Conic Sections, Basic Properties, and Tangents

Book II: Asymptotic Properties, Diameters and Axes of
            Conic Sections, and Methods of Drawing Tan-
            gents to Conics
Book III: Figures Generated by Diameters and Tangents,
            Harmonic Properties of Pole and Polar, and Fo-
            cal Properties of Conic Sections (45–52)
Book IV: More Properties of Pole and Polar (including a
            proof that two conic sections can intersect in no
            more than four points)

To these Apollonius added four new, more specialized books. They present particular extensions and appear largely original with him. Their subjects are:

Book V: Normals (Maxima and Minima)
Book VI: Equal and Similar Conic Sections
Book VII: Determinations of Conjugate Diameters of a
            Central Conic
Book VIII: Determinate Conic Problems, possibly more on
            Conjugate Diameters

The novelty, originality, and thoroughness of Book V wins it among the books the highest rating by modern mathematicians. Whereas today we consider normals to be lines drawn perpendicular to the tangent at a point of tangency, Book V deals with them as (relative) maximum and/or minimum line segments drawn to the curve from any point located within or without the curve. Thus, the minimum straight line segment that can be drawn to a conic at point P from an outside point G is perpendicular to the tangent at P. Book V shows that the extension of a normal to an ellipse or parabola will remeet opposite sides of these curves. It also investigates the locus of what are now called evolutes, the envelope of normals that provides their limiting positions.

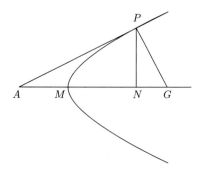

The method of generation of conic sections in *Conics* was radically new. Apollonius abandoned the traditional method of orthogonally cutting either a right-angled, acute-angled, or obtuse-angled right circular cone. In Propositions 11 through 14 of Book I he proposes a more general approach to generating all three curves from one circular cone, either oblique or right. This makes an ellipse the locus of a point P moving in such a way that the sum of the distances from two given points, the *foci*, remains constant. For a hyperbola the difference between the distances remains constant. This formulation allows the three conic sections to be treated by the application of areas in Book VI of the *Elements*. Thus, conic sections entered a mathematical domain now called geometrical algebra, from which it later passed into algebra in the form of second-degree equations. Since Apollonius and geometrical algebraists present no algebraic symbolism, many of their linguistic circumscriptions are difficult for the modern reader to translate into algebraic expressions. Another advantage of the Apollonian method is that it produces oblique representations for curves instead of simply orthogonal lines. Apollonius' predecessors had apparently used perpendicular sections, because they wanted to treat the converse problem of ratio of areas. Although his coordinates of curves are based on the diameter and its tangent rather than on central axes x and y, as in the later Cartesian system, his use of oblique representations is close to a general system of coordinates.

Equally innovative were the definitions in *Conics*. Adapting early Pythagorean terminology of the application of areas, Apollonius introduced the terms "parabola" ("exact application"), "hyperbola" ("exceeds"), and "ellipse" ("falls short"). For a parabola this meant that the area of a rectangle with sides equal to the abscissa and a constant p (later called a focal diameter) equals that of the square on the ordinate y.[228] In addition, Apollonius was the first geometer known today to recognize both branches of the hyperbola and systematically develop a theory for them. Previous writers had considered only one branch. Apollonius' definitions lead to these characteristic Cartesian equations, when the focal diameter of a curve is designated as p and the corresponding diameter is d:

$$y^2 = px \qquad \text{(parabola)}$$
$$y^2 = px + px^2/d \qquad \text{(hyperbola)}$$
$$y^2 = px - px^2/d \qquad \text{(ellipse)}$$

The *Conics* made the new nomenclature standard.

In *Conics*, Apollonius puts into rationally ordered sets theorems previously existing in disconnected and haphazard ways and adds theorems to form a comprehensive system. (In *Géométrie*, Descartes states otherwise. See Chapter 14.) True to his claim, Apollonius' novel method of generation is also a more general formulation than any that his predecessors had presented. The result is a systematic textbook. Until recently, it was believed that *Conics* was not designed to include all known theorems. The omission from it of the focus-directrix property of the parabola was so interpreted. Historians of ancient mathematics had generally attributed to Euclid the recognition of a conic as a locus of points, wherein the ratio of the distance from a fixed point (focus) and fixed line (directrix) remains constant. In his edition of Diocles' *On Burning Mirrors*, published in 1976, G. J. Toomer argues that the attribution is untenable, claiming that Diocles (see below), whose work was apparently unfamiliar to Apollonius, discovered it.

Conics quickly became an authoritative, even canonical text, making preceding writings on the subject superfluous. Since by the end of the Byzantine era its predecessors had disappeared, its superiority in mathematical consistency, generality, and thoroughness cannot be judged by direct comparison. Yet the superiority must have been conspicuous. According to Geminus of Rhodes in the first century B.C., contemporaries in their admiration for *Conics* gave Apollonius the title "The Great Geometer." Later mathematicians and astronomers concur. Kepler drew on *Conics* to describe orbits of planets; Descartes and Roberval admired its thorough treatment of second-degree curves; Newton provided new proofs of selected theorems; and Halley prepared a new edition that, on the basis of Pappus, restored Book VIII.

In the higher geometry of conic sections and other areas of geometry, Apollonius' work was said to be extensive. But his only surviving writing besides *Conics* is the *Cutting off of a Ratio* in two books, and this exists only in an Arabic translation. Pappus claims in Book VIII of *Collection* that Apollonius also wrote *Cutting of an Area, Determinate Section, On the Burning Mirror, Plane Loci* in two books, *Tangencies*, and two books on *neusis* constructions, entitled *Inclinations*. Pappus' summaries are the only form in which these lost treatises exist.

In addition to pursuing geometry, Apollonius refined arithmetic and numerical calculation. Chiefly from Pappus' commentary on Book X of Euclid's *Elements*, we know that Apollonius extended its theory of irrationals. Eutocius credits him with calculating in *Easy Delivery* the value of π to a more precise inequality than the $3\frac{10}{71}$ and $3\frac{1}{7}$ bounds given by Archimedes.[229] He probably did this by extending the Archimedean compression method. Pappus and Eutocius also relate that Apollonius expressed large numbers by a place-value system with a base of 10,000 (a myriad). Perhaps this was a refinement of Archimedes' ideas from *Sand-Reckoner*. Thereby, he too overcame limitations of the Greek alphabetic numeral system. Pappus in Book II of the *Collection* gives detailed rules for applying the Apollonian numerical system.

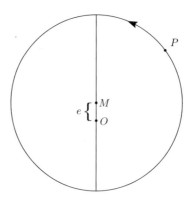

To account for celestial motions, Apollonius employs the twin mathematical models of eccentric circles and epicycles later used in Ptolemy's *Almagest*, or *Syntaxis Mathematikos*. In a geocentric universe, a body P moves about an eccenter whose actual center M is distant from the central East O by an amount e, which is the eccentricity. Apollonius combines the two models by ingeniously using the same circle as eccenter and as epicycle: a reading of *Conics* III, 37, and Ptolemy's *Almagest* III, 3, would indicate that he transformed the epicyclic model into the eccentric model by using an inversion on a circle. His combined models were one of two striking astronomical ideas introduced in the third century B.C. The other was Aristarchus' heliocentric theory. Since no writing of Apollonius on astronomy has survived, we do not know whether theoretical interests or physical arguments most influenced his work. But its significance is not in doubt. Apollonius' innovations made possible a rational astronomy, as opposed to the speculative cosmogony of his predecessors. He probably closely studied lunar theory. In the second century A.D., commentator Ptolemaus Chennus was to remark, seriously or not, that Apollonius was called epsilon because the shape of the Greek letter ϵ resembles that of the moon.

In pure mathematics, Apollonius had no ancient successors who were his equal. Work on a level with his would appear only in the seventeenth century with Descartes, Fermat, Desargues, and Pascal. In astronomy it was different. Particularly distinguished among the ancient astronomers who succeeded Apollonius was Claudius Ptolemy, who elaborated his models in the *Almagest*. Consequently, in later antiquity he was more famous in astronomy than in pure mathematics.

Almost an exact contemporary of Apollonius was Diocles (fl. 210–190 B.C.). Although Diocles was of lesser stature, his contributions to the theory of conics made him more than an epigone of others. Most inferences made before the 1970s about his contributions are inadequate or erroneous, because his only surviving writing, *On Burning Mirrors*, had been lost. Eutocius reformulated three excerpts in antiquity. The recent discovery of an Arabic translation of the entire text and its scholarly editing by G. J. Toomer in 1976 has filled a gap in the historical record.

The only reliable details of Diocles' life are a few references in his text *On Burning Mirrors*. Since he not only cites Archimedes and his Alexandrian correspondents, but also refers to a visit by the Athenian Zenodorus, who wrote on isoperimetric figures, we can place him at least at the beginning of the second century B.C. His and Apollonius' failure to refer to each other strengthens this dating, indicating that neither was at work long after the other became famous. Diocles resided at least for a time in Arcadia: he states that Zenodorus visited him there.

On Burning Mirrors is an important Hellenistic source on higher geometry containing a basic discovery about conics and distinctive examples of major problems that Hellenistic geometers addressed. A heterogeneous collection of sixteen propositions, it treats conics in the manner of Archimedes and his followers and not Apollonius. Most notable is its method in Propositions 4 and 5 for constructing a parabolic mirror of given focal length. It amounts to constructing a parabola from focus and directrix. The proofs of these two propositions have led Toomer to infer that Diocles discovered the focus-directrix property of the parabola. In Propositions 7 and 8, Diocles divides a sphere into segments of a given ratio, a problem set out in Archimedes' *Sphere and Cylinder* II, 4, which amounts to a cubic equation. He, and likely Archimedes, solved it by the intersection of a hyperbola and an ellipse. Propositions 9 and 10 offer two methods for cube duplication by finding two mean proportionals. Diocles' solution using the intersection of two parabolas, to which Eutocius refers, is usually wrongly attributed to Menaechmus. The second method, found in Propositions 11 through 16, uses a special curve known today as the cissoid, because seventeenth-century geometers identified it with the ivy-shaped class of curves. It comes to points, as in ivy leaves, and shares characteristics with a class of curves from ancient sources by establishing an angle with itself, as in the hippopede and conchoid, and being a mixed curve, like spirals.

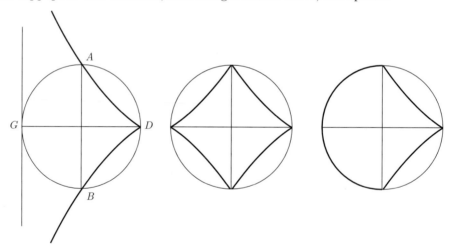

As Knorr and Toomer argue, however, the identification of the curve of Diocles with the ivy-shaped class of curves is almost certainly incorrect. On the basis of his reading of Proclus, who gives its characteristics, and Pappus, Knorr thinks that Diocles' curve was probably a circle-based conchoid, which originated with Nicomedes using *neusis* constructions.[230] That neither Theophrastus (ca. 300 B.C.) in his study of plants nor ancient Greek artists, who drew ivy leaves, had a single ivy shape is presented in support of this interpretation.

Except for three excerpts in Eutocius, Diocles' writing was lost to later antiquity. One excerpt taken from Proposition 10 mentions the curve of Diocles. Studies by Descartes, Fermat, Roberval, Huygens, and Newton generalized the cissoid curve and discovered many of its properties.

By about 175 B.C. an intense period of mathematical achievement apparently was ending. It had lasted for more than a century in the work of a

Ivy forms
Hedera helix on the left is common to western and central Europe,
Hedera canariensis on the right is found in Northwest Africa, and
Hedera colchica, also on the right, is found in the Caucasus.

few brilliant men: Euclid, Archimedes, Apollonius, and, to a lesser extent, Eratosthenes, Diocles, and the epigones, who were guided largely by Archimedes' methods and problem-solving techniques. The works of Zenodorus on isoperimetric figures, Perseus on the anchor ring, and Dionysodorus on the measurement of curvilinear figures, for example, are indebted to the ancient Syracusan. Together, these third-century geometers attained a level of mathematics beyond which advances could hardly be accomplished without new methods and improvements in notation. Their collective work constitutes perhaps the greatest—it was certainly the most permanent—accomplishment of ancient Greek science.

Their work brought mathematics to a certain maturity as an independent discipline. The early Pythagorean *mathematikoi* had examined a broad range of subjects, and Plato's *Republic* employed the term *ta mathemata* ("learning") to include both what we call mathematics and his studies of the Good, before Aristotle systematically distinguished between *mathematikē* and *physikē*. Aristotle then organized mathematics around a primary abstract branch including arithmetic and plane and solid geometry and a lesser physical branch: astronomy, mechanics, music, and optics. Hellenistic mathematical practitioners gave a sharper focus to Aristotle's definition by confining themselves to more specialized studies than had their predecessors. By pursuing subjects different from those of ethics and natural philosophy, they established the independence of their field. In essence, they added the title of mathematician to that of philosopher.

7

Mathematics in Roman and Later Antiquity, Centering in Alexandria

> We [Romans] have established as the limit of this art [of mathematics] its usefulness in measuring and counting.
> — Cicero

> [Mathematics] arouses our innate knowledge, brings to light the concepts that belong essentially to us, takes away the forgetfulness and ignorance that we have from birth, sets us free from the bonds of unreason ...
> — Proclus

In late antiquity, theoretical mathematics developed only sporadically. The level of discoveries generally fell, in demonstrative geometry precipitously. No hint of significant new methods or of improved notation was forthcoming in the Mediterranean world. The dominant Romans, affecting a rugged simplicity adverse to theoretical science, did not support abstract or abstruse mathematics. As they limited mathematics more and more to architecture, civil engineering, geography, optics, and surveying, the Romans produced only practical manuals, handbooks, encyclopedias, and literary asides to the subject. Under Roman rule mathematics, receiving lessened and more infrequent patronage, continued to center in Alexandria. Prior to A.D. 200, Alexandrian mathematical astronomers, still intimately connecting astronomy and geometry, created trigonometry, while Alexandrian geometers reedited older works and furthered mensuration and arithmetic. A revival during the period from the mid-third through the fourth century, with the work of Diophantus in algebra and number theory and Pappus in geometry, is often described as the Silver Age of Hellenistic mathematics. But it is not for pure geometry, at any rate, that the age is honored.

7.1   Early Roman Civilization and Mathematics

By the second century B.C. a major shift in the political, intellectual, economic, and social life of the Mediterranean world influenced the course of learning, which included slowing the pursuit of theoretical mathematics. Governmental corruption and cycles of famine in the Hellenistic era eased the way for Rome to gain hegemony, a position it was to hold for the rest of antiquity.

After sorting various discordant legends, the historian Varro (d. 27 B.C.) set the city's foundation at 753 B.C., which became the base year of the official calendar. Located fifteen miles from the mouth of the Tiber River at a fordable spot, it was a natural center for trade and communication north to south and east to west on the Italian peninsula. Rome's rise to importance in Latium (central Italy) began during the sixth century B.C. with its capture by the Etruscans. Skilled in the use of chariots, the Etruscan aristocracy conquered most of Latium. The victors operated within a political structure that gave extraordinary power to an elected king and Senate. It was from Etruscans, perhaps, that Romans learned to personalize their own polytheistic deities, presenting them in sculpture and assigning roles similar to those of classical Greek deities. For Romans were prone to amalgamate cults and rites onto more primitive layers of animism and polytheism. This amalgam was inseparably bound to patriotism.

Roman subjects were so enraged by the outrageous behavior of Etruscan monarchs that in 509 B.C. they expelled them and established a republic. For a millennium no Roman magistrate, however dominating, dared aspire to the title of king. By means of an excellent army, a network of military roads, and a diplomacy based on force and bargaining, the republican Romans by 265 B.C. had captured the entire Italian mainland south of the Po. Their society was composed of patricians by birthright, who initially held all power as *senators*, plebeians (from the Latin *plebs* for commoners); and slaves. The plebeians included small farmers, merchants, professionals, artisans, clients of the political welfare system, and urban poor. The cement of this social order was a mentality represented on coins by the letters SPQR, *Senatus Populusque Romanorum*, the official republican emblem of state. "Senate and People of the Romans" signified two houses, separate but respectful. Social reality overtook political theory, and plebeian pressure led in 451 B.C. to the codification of Roman law and custom in the Twelve Tables, which restricted the exercise of arbitrary power by the patricians, but not their role of governing elite.

From the third century B.C. the republican citizenry, triumphant and stable at home, built a fleet and entered or was unwittingly drawn into overseas expansion. Carpenters fitted ships with a *corvus* (crow's beak), which resembled a gangplank with a spike on the end and allowed Roman marines to board enemy vessels. Between 265 and 146 B.C., Rome conquered its archrival Carthage in three bloody Punic Wars. During the second, which lasted from 216 to 202 B.C., the Carthaginian general Hannibal invaded Italy and annihilated the Roman army at Cannae, but in 202 B.C. Scipio Africanus decisively defeated him at

Zama. Scipio imposed a humiliating treaty that left Rome in command of the central and western Mediterranean. In 146 B.C., Carthage was destroyed and the last major Greek polis, Corinth, fell to Rome. Rome then rapidly extended its power through the Hellenistic eastern Mediterranean. By 129 B.C., the republic had incorporated as provinces turbulent Macedon, Greece, and Pergamum. After Julius Caesar conquered Gaul in the west, Seleucid Asia Minor and Ptolemaic Egypt fell to Octavian Caesar in 30 B.C. The Mediterranean was, except for pirates, a Roman lake.

Imperial expansion, institutionalized spoils of warfare, and a new harshness toward both allies and vanquished changed Rome's political structure and society. Romans survived parasitically on plunder, indemnities, taxes, and slavery. Officials who matured by passage through a strict career path produced and exported organizational efficiency, but the organizational skill could not heal the city's lack of a sound economic base. Consequently, late republican Rome lagged while the Mediterranean fringes prospered. Within the city, ambitious military commanders came to the fore, a corrupting arrangement of patronage undercut the old system of political orders, and a few gained fortunes by overseas contact, while an impoverished mob grew in size. At the beginning, a hardy peasantry with small landholdings had been Rome's bulwark. When its wars became international with military expeditions lasting for several years, many farmer-soldiers, who could not return to work their fields, fell into debt and upon returning from prolonged military service drifted into the city. Aristocrats purchased their land and forced out other dirt farmers to amass huge agricultural estates (*latifundia*), on which cattle grazed and grapes were grown. Prisoners of war were imported to work as slaves on the *latifundia*. Shopkeepers, artisans, and households in the city also widely used slaves. Manual labor came to be disdained everywhere. As the gulf between rich and poor grew, rulers adopted a policy that Juvenal called "bread and circuses" to keep the city's masses in check.

Hellenistic civilization reached rough-hewn Romans through ambassadors, merchants, artists, writings, and methods of education and increasingly influenced them during the republic's last two centuries. Translations (often transliterations) from Greek rhetoric and poetry enriched Rome's language. Latin, like the Roman legions, dominated the central and western Mediterranean, but never displaced Greek in the east. Except for a brief flirtation with philhellinism, the conservative Romans mostly absorbed and adopted those segments of classical Greek and Hellenistic thought fitting to their ethos of duty, prudence, and *mos maiorum*—patriotic respect for tradition. They adopted teachings of the Platonic Academy together with Epicurus and most of all Stoic philosophy. During the high point of Roman culture in the first century B.C., Lucretius and Cicero were the two leading exponents of Greek thought. Lucretius' poem *De Rerum Natura* (freely translated, "On the Nature of the World") is Epicurean in its atomic theory and its teaching of liberation from fear. Cicero's Stoic ethical philosophy founded stability and liberty on natural law and natural rights reflective of a rational God. This he coupled with the Roman tradition of active political life in state service. Classical Greeks had prized *arête,* excellence in virtue, courage, sports, and art; and Hellenistic Greeks, exuberance, variety, and introspective self-possession. At the center of the Roman system was the absolute power of the father over his extended family, *pater potestas.* The Roman preference was for *ordo,* law and order, based in precedent and supplemented outside the Italian peninsula by *ius gentium,* the indigenous law of subject peoples. Often the Romans saw in ancient Greek writings, especially in ethics and politics, what their own values told them to find there, while they ignored the theoretical sciences and portions of Hellenistic realistic art.

As Hellenization proceeded, civil war was to change government from republic to imperial monarchy, without that unacceptable name and with the senate maintained as a facade. Roman culture briefly flourished from 27 B.C. through the Principate, ending in A.D. 180. It was the time of the *Pax Romana.* The first half of the period is generally designated the Golden Age of Roman literature, the second the Silver Age. Notwithstanding strong institutions of state, Rome's high culture was thin and fragile, depending on a handful of creative people and royal patronage. It centered in respect for law, positive and Stoic, that was acclaimed in the *De Legibus* of Cicero, and in an imaginative literature graced by poetry.

The Golden Age of Roman poetry under Augustus, as Caesar Octavius renamed himself, reflected new social conditions. Republican patronage had come from individual aristocrats; under Augustus, who reigned from 27 B.C. to A.D. 14, all patronage flowed from the *princeps* and his chief cultural adviser. Patronage supported writing, especially that praising the leader. The most important Augustan poets, Vergil and Horace, both lauded Augustus for ending civil war and portrayed him as a leader destined to bring about great things. But deference to the ruler did not damage the integrity of their work. Vergil wrote the pastoral idylls *Eclogues, Georgics,* a paean on the greatness of Italy, and his master work, the *Aeneid,* the Latin equivalent of the Greek *Iliad* and *Odyssey.* The *Aeneid's* human hero, Trojan warrior Aeneas, embodies the Roman ideals of duty and patriotism. Horace, who relished life on a small farm, wrote *Satires (Sermones),* lyric *Odes (Carnuna),* and *Romanita,* which celebrate the virtue of moderation. A near equal of Vergil in talent, but opposite in his fortunes under the emperor, was Ovid. Offended by the unblinking look at social mores in *Metamorphoses* and the explicit and casual sexuality of the *Ars Armatoria,* the puritanical Augustus exiled its author.

From the last decade of the first century through the first quarter of the second, Plutarch and Suetonius created the basic antique canons of biography, while Tacitus wrote his trenchant *Histories* and *Annals* about the successors of Augustus. Written with ironic wit, the short tract *Germania* of Tacitus tells of the burning, pillaging, and killing that accompanied the extension of the *Pax Romana.* He records the observation of a barbarian chieftain, "They create a wilderness and call it peace."

The Romans excelled in massive architectural and civil engineering projects and public services. These gained prominence under Augustus, who boasted, "I found a city of brick and left one of marble." Two notable projects are the *Ara Pacis* (Altar of Peace), dating from 9 B.C., and the Colosseum, from A.D. 80, Rome's largest ampitheater, seating 50,000 spectators. Sadistic crowds in the four-tiered Colosseum took delight in circuses, animal hunts, gladiatorial combat, sea battles, and other bloody spectacles. To construct the Colosseum, the Romans added the principle of the semicircular arch borrowed from the Etruscans and began massively using a new building material, concrete. The Romans built perhaps twelve aqueducts that brought into the city of one million, 300 million gallons of water daily. They designed an effective public sewage system, the *Cloaca Maxima*, and in the empire a network of 50,000 miles of roads. The Romans rank with the Old Kingdom Egyptians as the most skilled engineers in antiquity. That achievement was attended by so slight an interest in sophisticated mathematical technique that they frequently employed Greeks to construct complicated aqueducts or siege engines.

Absorbed in law, government, military conquest, and monumental engineering projects, the urban Roman elite believed that the pursuit of theoretical natural knowledge was not innocent frivolity but a culpable diversion from obligatory moral concerns. Even after Greek teachers were imported with considerable effort in the second century B.C., many Romans remained ambivalent to classical Greek and Hellenistic learning, especially antipathetic to theory in the natural sciences. Education of their upper class males for public service centered in rhetoric and literature. Roman education emphasized utilitarian *prudentia*, as opposed to philosophic *sapientia*. Among the Romans, Greek *logos* became *ratio*.

The Romans embraced in manuals, handbooks, and encyclopedias the Hellenistic tradition of summarized nontechnical knowledge. Marcus Terentius Varro (116–27 B.C.) introduced this tradition into Roman literature. In encyclopedic seventy-four works, he popularized versions of the liberal arts. Varro became the most authoritative figure known among the spate of pedestrian Latin authors. He had numerous imitators, like Pliny the Elder (A.D. 23/4–79). Pliny's comprehensive *Natural History* of thirty-seven books, which refers to over 473 authors, is the compilation of Roman science. An important but uncritical source of ancient lore, it is concerned more with curiosities and wonders than with systematic observations.

Vitruvius Pollio (d. ca. 25 B.C.) wrote the best known Roman technical manual, drawing upon mathematics. His ten-book *On Architecture* is dedicated to the author's patron, Caesar Augustus. Although Vitruvius attempts to encompass architectural theory and practice, he presents the theory of only a few Greek architects, whose ideas he had not satisfactorily assimilated. Roman architects worked within an agrimensorial, that is, practical surveying, tradition without recourse to theoretical treatises. Vitruvius argues that architects should have substantial education in theoretical geometry, but this seems a conventional gesture of respect, not a call to serious study. Vitruvius' comments on mathematics and theoretical astronomy are rudimentary. He unsuccessfully tries to reduce temple building to strict numerical rules and gives the mathematical proportions underlying harmony and aesthetic design. His two approximations for π, 3 and $3\frac{1}{8}$, are crude. Perhaps following a Greek manual, he opens book IX with the three chief prior mathematical discoveries: the 3, 4, 5 right triangle, the incommensurability of the diagonal with the sides of a unit square, and Archimedes' hydrostatic solution of the golden crown problem. Vitruvius' section on pre-Ptolemaic astronomy uses primitive notions to explain lunar phases and planetary retrogradations.

Cicero's grasp of mathematics and disposition toward it set the standard attitude within the Roman intelligentsia. This urbane man of letters took his mathematics from Plato's *Timaeus*, which he translated, and the *Republic* as well as Eratosthenes' *Geography*. In Book I, ii, of his *Tusculan Disputations*, Cicero succinctly contrasts Greek and Roman views of mathematics:

> The Greeks held the geometer in the highest honor; accordingly nothing made more brilliant progress among them than mathematics. But we have established as the limit of this art its usefulness in measuring and counting. (Morris Kline's translation)

Sections I.3–6 and 11–16 of his master work, *On the Orator*, lack a sense of the rigor and critical methods of mathematical science. There Cicero contrasts the abstruseness and obscurity of mathematics with the eloquence of oratory as Aristotle and Isocrates define it. On the ground that many Greek thinkers had contributed to perfecting geometry, Cicero suggests that any serious student can achieve some success as a geometer.

Possessing a rudimentary arithmetic and, for engineering and surveying,[231] approximate geometric rules added to formulas borrowed from the Alexandrian Greeks, the Romans counted with numerals lacking place-value notation, and they computed with different forms of the abacus, with fingers, and with arithmetical tables. Among the familiar Roman numerals are:

I	V	X	L	C	D	M
1	5	10	50	100	500	1000

Scholars disagree about their origins. Traditional theory held that these Roman alphabetic numerals are of Greek origin. Lucien Gerschel and anthropologist Franjo Skarpe in his study of Dalmatian shepherds persuasively argue, however, that the three oldest Roman numerals, I, V or Λ, and X, derive from widely used markings on notched tally sticks transmitted through the Etruscans.[232] The letters L, C, D, and M, which date from after 89 B.C., Georges Ifrah reconstructs as graphic transformations of archaic Etruscan signs, such as Λ to L for 50 and $\mathbf{\bar{X}}$ to $\supset$ or C (also the initial letter of *centum*) for 100.

Rules for combining numerals were simple. When a smaller numeral precedes a larger, the operation is subtractive; thus IV = 5 − 1 = 4. When a smaller follows a larger, the operation is additive; thus CX = 100 + 10 = 110. Roman fractions were in base-12 with words and abbreviations for 1/12, ..., 11/12, 1/24, 1/48, ... Empiricism and limited theoretical residue from Hellenistic times sufficed to solve problems the Romans faced.

Archaeological evidence and Roman authors, such as Cicero in *Epistulae ad Atticum* 5, Juvenal in *Satires* 10, Ovid in *Fasti* 3, Firmicus Maternus in *Mathesis* 1, and Macrobius in *Saturnalia* 1, indicate that the Greco-Roman world extensively employed finger numbers and finger reckoning.[233] Hilton Turner and Charles W. Jones have shown that Chapter 1 of Venerable Bede's *De Temporum Ratione*, dating from the eighth century, accurately describes a common system of finger numbers existing from the late Roman Republic.[234] Those for one to ninety-nine are:

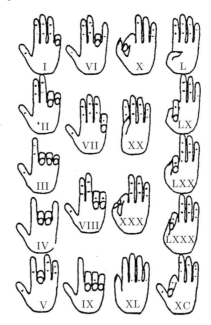

Roman finger numerals.

(From J. Hilton Turner, "Roman Elementary Mathematics: The Operations," *Classical Journal*, 47 (1951): 66.)

Historians have studied the reasons for their continuing use. Burma and Richard Williams have found untenable the view of James Gow and David E. Smith that merchants needed them to overcome language barriers in trade. Surviving multilanguage trade accounts do not support that hypothesis. Rather, the depiction of finger numbers in surviving texts, such as Pliny the Elder's *Naturalis Historia* 2 and 34 and Suetonius in *Claudius* 21, suggest their being integral to everyday arithmetic and particularly for allowing buyers and sellers to overcome the noise in the marketplace, much like hand signals on the current stock exchange, and impediments to clear verbal communication among surveyors caused by distance.

Fairly accurate measuring sufficed for land surveys and large-scale architectural and engineering projects. This was the principal use for geometry in Rome. Land surveys adjusted city boundaries and fixed boundary lines of small farms, estates, and temple areas. New cities and military camps were built on the grid plan. Vast engineering projects constructed public buildings and impressive roads, bridges, and aqueducts across Europe. Roman surveyors and civil engineers preferred to measure with congruent triangles and simple instruments.

Aesthetics produced a rare instance of a Roman work whose subsequent serious study was important for theoretical mathematics. A Roman mosaic of the third century contains a coiled ribbon that may be described as a Möbius band. In its simplest form a Möbius band is a rectangular strip of paper given a half twist, 180°, and connected at the ends. Possessed of only one side and one edge, this band was closely studied in the nineteenth century and contributed to modern topology.[235]

One notable result of practical Roman mathematics, an early version of the Julian calendar, depended on cumulative, careful reckoning based possibly on Babylonian and Egyptian astronomy, distributed over at least seven hundred years, as well as that of classical Greece and Alexandria.[236] Establishing an absolute calendar having astronomically fixed dates, rather than a relative chronology, requires a fairly accurate theory of eclipses. Such a theory and observations could be found in Mesopotamian archives dating from the Assyrian to the Greek period. Before Julius Caesar (100–44 B.C.), the basic Roman calendar consisted of twelve months with a total of 355 days. Every other year an intercalary month of twenty-two or twenty-three days was added between the twenty-third and the twenty-fourth of February. Thus, the average Roman year in the Republic was $366\frac{1}{4}$ days. As time passed the civic and religious calendar slipped out of agreement with the seasons and therefore ceased to provide an adequate basis for setting celebrations. Nor did planting and harvesting occur at the same time yearly. Plutarch in 59 of *Caesar* claims that Julius Caesar asked leading mathematical astronomers and philosophers to improve the calendar. According to Pliny in 18.211 and 212 of *Natural History* and the nineteenth-century classicist Theodor Mommsen, the mathematician Sosigenes probably introduced one or two calendar reforms.

Sosigenes, who may have been Alexandrian, most likely helped bring the solar year back into agreement with the passage of the sun by making the 365-day Egyptian calendar, whose astronomical basis Eudoxus and Callipus had refined, more accurate by adding an additional day every fourth year. The day could be intercalated at different places in the calendar. The uninterrupted use of the Egyptian calendar for over a millennium together with parameters and observations from Mesopotamian lunar theory made for greater precision in the expression of intervals between eclipses in years, months, and days. It also supported Sosigenes' addition of a rational system of intercalation to the calendar.[237] In 45 B.C. the early Julian calendar was adopted for religious and official use in Rome.

It would be a mistake to equate Sosigenes' calendar with the modern Julian version. The development of historical chronology was more complex. A very accurate eclipse theory awaited Ptolemy's *Almagest*.[238] During later antiquity, the Middle Ages, and the early modern era in the Latin West, Egyptian years were to be a favorite unit. In *De Emendatione Temporum*, published in 1583, Joseph Justus Scaliger was the first to construct precise Julian days set in undisturbed planetary weeks.[239]

7.2   Crucial First-Century Changes Affecting Mathematics

The Roman conquest of Ptolemaic Egypt was the first of three events affecting the course and nature of mathematics that were concentrated near the intersection, or (nonexistent) zero point, between the times designated "before Christ" and the era whose years are designated *anno Domini*. During a skirmish between troops of Julius Caesar and local forces in Alexandria in 48 B.C., so it seems, a fire destroyed much of the Library but not the entire Museum complex. While Cleopatra received the Pergamum library from Mark Antony as a replacement, it was housed in the daughter library at the temple of Serapis, rather than at the Museum. The relocation suggests that the power of religious officials had increased in Alexandria and indicates a shift of interest toward religion away from theoretical mathematics.

Two decades later the embattled Roman republic ended. The formation of the empire was largely the work of the shrewd Julius Caesar and his heir Octavian, later called Augustus. Once in power, Augustus confiscated the Alexandrian treasury and used it to support building projects and literature. Since the early Ptolemies, theoretical mathematics had probably not received substantial royal patronage, but the full loss became evident in Roman times. By the reign of Augustus, there was no small cadre of mathematical experts such as that assembled in early Alexandria. Without them the oral tradition was lost; and in an age lacking concise formulas and symbols, a virtually continuous oral tradition was indispensable to the development of theoretical mathematics. Able oral expositions could highlight essentials of proofs or suggest clues on how to find new proofs, provide sound instruction in geometric methods and prepare students to go beyond them, and foster the formulation of new problems. With the vanishing of this tradition, even in Alexandria, with its vast library resources available to scholars, theoretical mathematics waned.

Then, in about 6 B.C. according to scholarly calculations, came the third crucial event, the birth of Jesus. The rise of Christianity was to transform western civilization and affect the course of mathematics.

The life of Jesus came nearly at the midpoint of the religious revolution stretching from Confucius and the Buddha in the fifth century B.C. to Mohammed in the sixth century A. D. Jesus, who preached in Roman Palestine, left no writings, but by means of the excellent Roman communications system his first followers, above all Paul, spread his message of brotherhood, humility, and a kingdom of peace. Breaking with the exclusivist Jewish tradition that might have defined Jesus as the expected Messiah of the Hebrew people alone, Paul spoke of Jesus as the Christ (from the Greek *christos*, or "Anointed One"), promised salvation to all peoples through God's grace, and carried Jesus' message of love notably in the passages that today appear as Corinthians 13. The first members of this obscure sect expected the imminent return of Jesus. They cultivated community, centering it in the celebration of the Lord's Supper. They preached monotheism. Paul's epistles and the four Gospels, written in the first and second centuries, provided a primary source of Christian beliefs that quickly acquired the status of a divinely inspired revelation. By the third century, Christians, who were already prominent in Damascus and other cities of the eastern Mediterranean, had a center in imperial Rome. Rome's position of primacy in the early Church dates from the time when St. Peter served as its first bishop (*episkopos*). Verse 16:18 of the Gospel of Matthew reports that Jesus said, "Thou are Peter and upon this rock, I will build my church."

The pagan Romans, who esteemed piety, distrusted as atheists these people who could worship neither pagan gods nor the emperor. That many Christians were also pacifists and refused to serve in the army branded them as traitors. The historian Tacitus refers to Christianity as "disastrous superstition." From the time of Nero, there were sporadic, local persecutions of Christians in Rome and provincial cities.

Still, Christianity spread, at first primarily among the common people and the downtrodden in the slums and slave quarters of Rome and provincial cities. Unlike the old, elite pagan religions, Christianity accorded dignity to slave as well as freeman, to women and men alike. Christianity also preached an ethos of hope and striving that contrasted to the passivity and fatalism common among pagan sects. From communal resources they administered a vigorous program of social services for widows, orphans, and indigents. Persecution strengthened their resolve. Diocletian, whose rule ran from 284 to 305, was particularly harsh. Restoring order and stability to an empire threatened by inflation, plagues, and shifting military factions, Diocletian ordered priests imprisoned, churches leveled, and properties confiscated. Brutal repression shortly gave way: Constantine, reigning from 306 to 337, became the first Roman emperor to embrace Christianity. After reportedly seeing a miraculous sign in the sky (or in a dream)—a cross-shaped insignia with the Greek inscription, "In this sign conquer"—Constantine triumphed over his rival Maxentius at the battle of Milvian Bridge in 312 just north of Rome. A year later he issued the Edict of Milan, one of a series of laws making Christianity a tolerated religion throughout the empire.

By the late empire, the meaning of the term "mathematics" was affected by the environment of anxiety and fatalism that was growing in imperial Rome, where crime, plagues, accidents, filth, and political conspiracies abounded. Adding to these internal problems was the eruption of Vesuvius in A.D. 79

that left thousands dead. At the same time that Christianity emerged amid
the uncertainties and the dread, astrology became almost universal. While the
term "geometer" now meant an applied mathematician, as astrology spread the
Latin term *mathematici* came to mean astrologers. The new connotation fits
in with the role of omens and soothsayers in Roman history. The intelligentsia
had long viewed much of the work of Archimedes as that of another magician, a
wonderworker. Christian preachers had effectively presented Jesus and St. Peter
as wonderworkers. Ptolemy was known for the *Tetrabiblos*, or Four Books, presenting rules for astrological predictions, as much as for the *Almagest*. Ptolemy
considered the two texts to be complementary. The *Almagest* gave methods for
predicting positions of celestial bodies, while the *Tetrabiblos* related the physical effects of these bodies on human beings. Connecting mathematics with
astrology created dangers for mathematics. Sensing that self-fulfilling prognos-
tications undermine the *imperium*, Diocletian, in a law that would also apply
in medieval Europe, forbade as damnable and evil the "art of mathematics"
or "astrology." The Roman distinction between "mathematician" and "geometer" continued in some fashion past the Renaissance to as late as the eighteenth
century.

7.3   Alexandria to 200: Trigonometry, Mensuration, and Arithmetic

During Roman and later antiquity the development of mathematics continued mainly in Alexandria. Revenues from a flourishing grain trade until the
fourth century A.D. and maintenance of a large library in the Serapeum and a
remnant of the Museum enabled theoretical mathematics to be advanced occasionally under the enfeebled educational tradition set by the early Ptolemies.
Alexandria's mathematical astronomers largely created trigonometry, and her
geometers commented on older works, reedited them, and further developed
mensuration and arithmetic.

Trigonometry, the systematic study of relations between angles and arcs of
circles and subtending chords, was the chief mathematical innovation of late
antiquity. Its origins lie in the geometric models that Eudoxus, Aristarchus,
Archimedes, and Apollonius had applied to astronomy. The Greeks called such
studies spherics: its future became entwined in a debate about the role of astronomy. One Aristotelian school argued that the proper role was "to save the
appearances": to preserve the fit between observed celestial movements and the
physical machinery producing them, geometry being a secondary resource. Another school chose to ignore "appearances" and work within a geometry tied to
no particular celestial-mechanical model. Trigonometry initially belonged more
to astronomy than to theoretical mathematics. Its method and proofs, while
far more stringent than those of the Epicurean and Stoic philosophies of the
time, did not have the rigor of demonstrative geometry. The three mathemat-

ical astronomers most responsible for founding trigonometry are Hipparchus (fl. ca. 175–125 B.C.), Menelaus (fl. ca. A.D. 100), and Ptolemy (ca. 100–178), who brought it to a mature state in the *Almagest*. Menelaus and Ptolemy resided in Alexandria. We are ill-informed about the steps in the development of trigonometry prior to Ptolemy, because most of the relevant writings are lost.

Of the numerous writings of Hipparchus, one of the foremost astronomers of antiquity, only the comparatively minor "Commentary" on the *Phaenomena* of Aratus and Eudoxus survives. It deals mainly with the simultaneous risings and settings of stars. This text, brief references by Greek and Latin authors, Nicaean coins, and, above all, passages from the *Almagest* citing his astronomical observations are the sources of our knowledge of the life and achievements of Hipparchus.[240] These give five reliable biographical data: he was born in Nicaea, Bythnia (now Iznik, Turkey), conducted early astronomical observations in Bythnia (presumably in Nicaea), and after 141 B.C. moved to Rhodes, where he spent the rest of his career. Ptolemy also mentions his last two recorded astronomical observations made in 126 on lunar motion. Later these observations were related to conditions when the moon is near apogee, its farthest point, and perigee, its nearest point, on its epicycle. Scholars agree on Hipparchus' profound importance to astronomy.

The Greek predecessors of Hipparchus had developed geometrical models to describe the motion of the five known planets, the sun, and the moon, including anomalies posed by the last two, and to compute and predict the positions of these heavenly bodies applied to their model's numerical parameters in the Babylonian fashion. Hipparchus determined more precise numerical parameters than his predecessors. To describe celestial motions, he skillfully utilized models with epicycles and eccentric circles from Apollonius, and he demonstrated their equivalence. Recognizing the insufficiency of a simple epicyclic model, Hipparchus borrowed lunar eclipse data, arithmetical methods, such as what are today called linear zigzag functions, and numerical tables of Babylonians from the Seleucid and earlier periods for his lunar theory. Linear zigzag functions give points that lie on the same straight lines of differing slopes. Franz Kugler was the first to show this borrowing around 1900.[241] Expanding on both his Greek and Mesopotamian sources, Hipparchus made systematic and remarkably accurate, naked-eye observations. In this way he made a very good approximation of the mean synodic month and computed the distance of the moon and the sun from Earth in the right order of magnitude. His mean distance for the moon was $67\frac{1}{3}r$ (where $r=$ Earth's radius), while that for the sun was $490r$.

Probably inspired by a pioneering work on stellar constellations by Eudoxus and a poem based on it by Aratus, Hipparchus undertook making a detailed catalogue of about 850 fixed stars. Here, as in his "Commentary," his accuracy depended on his invention and skillful use of a diopter, a sighting device with a six-foot wooden rectangular cross section mounted on a bronze disc, and a plane astrolabe. Although Hipparchus may have marked the positions of stars on a celestial globe and had a ring indicating local horizons to give degrees of the ecliptic for risings and settings, it is unlikely that he assigned latitude and longitude coordinates to all or any stars. If he had, Ptolemy probably would not have lacked such a procedure in Book VIII, 3, of the *Almagest*. To quote Pliny in Book II, 95, of *Natural History*, the star catalogue "left the heavens as a legacy to all mankind." It was valuable in preparing Ptolemy's star catalogue in Books VII and VIII of the *Almagest*. But Hipparchus' work was distinctive and not totally assimilated into or identical with that of Ptolemy. It cannot be reconstructed completely from the *Almagest* alone.

Late in his career, Hipparchus made his greatest discovery, the precession of the equinoxes. Perhaps by comparing his observations with those of Aristyllus and Timocharis made just over 160 years earlier, he found that those points where the Earth's ecliptic intersects the celestial equator (these are the equinoxes) are not constant, but move slowly in respect to the fixed stars. Thus, the sidereal year, the time in which the sun returns annually to the same equinox or solstice in the zodiac, is shorter than the tropical or solar year, the time Earth takes to complete its annual orbit. The attraction of the sun and moon on the equatorial bulge of Earth causes this wobbly discrepancy. Hipparchus computed the sidereal year to be $365\frac{1}{4} + \frac{1}{144}$ days and the tropical year $365\frac{1}{4} - \frac{1}{300}$ days. A single precessional revolution occurs in 27,600 years. This estimate is quite accurate, differing from the correct value of 25,000 by 7%. The period of the single revolution is called the great or Platonic year.

To determine numerical parameters as an aid to prediction, Hipparchus developed an embryonic stage of plane trigonometry. Again, he went beyond earlier Greek astronomers. Essential to their computation of parameters was the solution of plane triangles. Theon of Alexandria reports that Hipparchus prepared the first table of chords or half chords subtended by arcs of a circle. His table is lost, but we can reconstruct it. He began with a circle whose circumference was divided in normal Babylonian manner into 360 degrees of 60 minutes each. The radius (R) of the circle became $R = 360 \times \frac{60}{2\pi} \approx 3438'$. The chords of Hipparchus are related to our sine function: $(\text{Crd} \, 2\alpha)/2 = 3438 \sin \alpha$. Using right triangles and linear interpolation, he computed the values of chords at intervals of $1/48$ of a circle (or $7\frac{1}{2}°$ intervals). Trigonometry, which literally means measurement of triangles, was under way. Hipparchus' table of chords, together with his computations and observations, substantially transformed astronomy from a descriptive qualitative to a predictive quantitative science.

The next major contributor to the emerging science of trigonometry was the astronomer Menelaus of Alexandria, who lived two centuries after Hipparchus. A contemporary of Plutarch, he made astronomical observations at Alexandria and at Rome during the reign of Trajan. Pappus, in the *Collection* VI.110, says that Menelaus used trigonometry to treat unequal times of risings and settings of signs of the zodiac. His table of chords in six books has not survived. Presumably, his chord AB was related to the sine function in the same manner as Ptolemy's: the subtended chord AB of an angle equals twice the radius (R) times $\sin \alpha/2$.

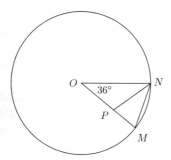

Ptolemy's calculation of crd 36°.

Menelaus' chief contributions to trigonometry are in his *Sphaerica*. The text, which survives only through an Arabic translation probably taken from a Syriac version, consists of three books. In it Menelaus severs trigonometry from spherics and astronomy, making it a separate science. Book I provides a basis for spherical triangles analogous to Euclid's plane triangles. The intersections of three arcs of great circles on a sphere form spherical triangles. Some theorems of Book I have no analogue in plane geometry: spherical triangles, for example, are congruent if the angles of one equal respectively the angles of the other. After dealing with astronomy in Book II, Menelaus essentially founded spherical trigonometry in Book III. Proposition One of Book III is "Menelaus' theorem," which is fundamental to the spherical trigonometry and astronomy of Ptolemy's *Syntaxis*. This theorem in plane geometry asserts that if a transversal intersects three sides of triangle ADC or their extensions in the points L, M, and N, the points of intersection must lie in a straight line and the ratio of the lengths is $CN/NA \cdot AL/LD \cdot DM/MC = -1$. Menelaus' theorem led to many interesting propositions, including Proposition 5 of Book VI, which assumes the anharmonic property for transversals of great circles: in the projection of a great circle on to another great circle, the cross ratio is not altered.

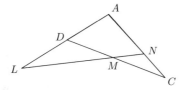

Figure for Menelaus' theorem.

The development of classical Greek and Hellenistic astronomy and trigonometry culminated in the writings of Claudius Ptolemaus (Ptolemy). Although he ranks with the Pergamese physician Galen as one of the two giants of second-century science,[242] we know little of his life. It is said that he was born Ptolemais Hermiou in upper Egypt near Akhmin and died at the age of seventy-eight.[243] These statements can neither be confirmed nor disproved.

The surname "Ptolemaus" suggests that he lived in Egypt and was descended from Greek or Hellenized forebears. The forename Claudius indicates that he held Roman citizenship, a status possibly conferred on an ancestor by Claudius or Nero. Examination of the series of his astronomical observations listed in the *Almagest* suggests when and where he lived. The observations date from 124 to 141 in Alexandria. To 132 he collaborated with a certain Theon, who may have been his teacher. Otherwise, we are ignorant of his education or who his colleagues were. Undoubtedly, he lived most of his life in Alexandria. Ptolemy worked at the Museum, where he continued the highest tradition of early Hellenistic scholarship in applied mathematics. He served the Alexandrian royal court but was not a member. So great was his influence in the West in later antiquity and the Middle Ages that he was referred to as "godlike" and mistaken for having been an Egyptian king.[244]

Claudius Ptolemy

The reputation of Ptolemy rests chiefly on his *Syntaxis Mathematikos* ("Mathematical," that is, astronomgical, "Compilation") in thirteen books. Neugebauer considers this his first writing and dates it before A.D. 146/147, when his *Canobic Inscription* appeared.[245] The *Syntaxis* completed the pioneering work of Hipparchus and Menelaus in plane and spherical trigonometry. It essentially did for mathematical astronomy what Euclid's *Elements* had done for demonstrative geometry and Apollonius' *Conics* for conic sections. Probably to separate the *Syntaxis* from earlier Greek works on elementary astronomy, later ancients referred to it as the great compilation (*Syntaxis Megale*). The Arabs combined the article *al* with the superlative *majastī* to call it "al-majastī," which in medieval Latin became "almagesti."[246] From this distortion of the Arabic comes the title by which it is best known today, *Almagest*, meaning the very great or greatest.

The *Almagest* of Ptolemy covers the whole of mathematical astronomy as conceived by the ancient West. This large and technological manual is superior in clarity and method. Ptolemy skillfully synthesizes work from predecessors, drawing on Babylonian arithmetical techniques, classical Greek concepts, and the kinematical model of Hipparchus in the Hellenistic world. Part of the components of the fascinating hybrid had, of course, originated in Roman Alexandria.[247] Ptolemy also adds many original contributions, including new observations, a refined lunar theory, and improved mathematical tools. The result was that as Euclid's *Elements* had supplanted prior writings in demonstrative geometry, the *Almagest* replaced previous treatises in mathematical astronomy.

The *Almagest* accepts the traditional Greek view of the universe with a fixed, immovable Earth at the center. The Pythagorean and Platonic perception of the circle as perfect form, along with Aristotelian physics and Hipparchus' arguments, convinced Ptolemy that observable motions in the heavens are reducible to a kinematic model of uniform circular motion. His representation of celestial motions with epicycles, revolving circles whose centers revolve on the circumference of another circle, is an elegant geometric model of the heavens. To avoid a double epicycle, Ptolemy uses an eccentric circle to account for anomaly in planetary motions. He introduced the point E, in modern notation, which medievals called an equant. It is removed from the center (C) of the circle by length e. The equant is Ptolemy's most original addition to the geometric model applied to astronomy. Observational materials given by Ptolemy have led one modern to hypothesize that he fudged his data, but most, notably historian of astronomy Owen Gingerich, think that the most plausible explanation is that he had a much larger observational data base that was not very precise at his disposal.[248]

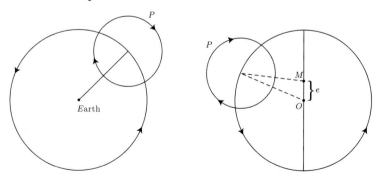

Epicycle and equant

Recognizing that the value of a chord (crd) on a circle bears a fixed relation to the sides and angles of both plane right and spherical triangles, Ptolemy constructed a table of chords (or half-chords) of an angle. His table appears in chapters IX through XI of Book I and in Book II.

Arcs	Chords	Sixtieths	Arcs	Chords	Sixtieths
$\frac{1}{2}$	0;31,25	0;1,2,50	6	6;16,49	0;1,2,44
1	1;2,50	1;1,2,50	47	47;51,0	0;0,57,34
$1\frac{1}{2}$	1;34,15	0;1,2,50	49	49;45,48	0;0,57,7
2	2;5,40	0;1,2,50	72	70;32,3	0;0,50,45
$2\frac{1}{2}$	2;37,4	0;1,2,48	80	77;8,5	0;0,48,3
3	3;8,28	0;1,2,48	108	97;4,56	0;0,36,50
4	4;11,16	0;1,2,47	120	103;55,23	0:0,31,18
$4\frac{1}{2}$	4;42,40	0;1,2,47	133	110;2,50	0:0,24,56

Ptolemy gave the first clear procedures for computing a table of chords, requiring only elementary geometry, in particular the Pythagorean theorem, and his own theorem on a quadrilateral inscribed in a circle (see below). The method begins with values of readily obtainable chords, such as crd 60°, which equals the radius, and combines these with formulas derived by use of Ptolemy's theorem for $\mathrm{crd}(\alpha + \beta)$, $\mathrm{crd}(\alpha - \beta)$, and crd $\frac{1}{2}\alpha$, when crds α and β are known. The first two of these formulas are equivalent to $\sin(A + B) = \sin A \cos B + \cos A \sin B$. Ptolemy then approximates crd 1° and cleverly uses inequalities to compute a table of chords from $\frac{1}{2}^\circ$ to 180° in steps of $\frac{1}{2}^\circ$ and to three sexagesimal places.

Ptolemy's chord function relates in this way to our sine function, which derives from the Indians: crd AB of angle $\theta = 2R\sin\frac{\theta}{2}$ or $\sin\frac{\theta}{2} = 1/2R \times \mathrm{crd}\,AB$ of angle θ. In Ptolemy's case, $R = 60$ equal parts. With an embryonic sine function, Ptolemy could solve all problems concerning plane right triangles.[249] But the absence of a tangent function made some solutions cumbersome.

Like most Hellenistic astronomers, Ptolemy bases his calculations on sexagesimal angular measure and the sexagesimal, place-value numerals of Babylonians, although Greek alphabetic numerals and Egyptian fractions appear on occasion. His sine tables are therefore founded on a circle whose circumference is divided into 360 equal parts with subdivision of parts or degrees into minutes (each sixtieth of a degree) and seconds (each sixtieth of a minute). Triangles inscribed within the circle depend on the central angle given and a diameter divided into 120 equal parts.

In addition to offering the first clear procedures for the computation of chords, Ptolemy's table includes some of the most accurate extractions of irrationals in antiquity. Computing the length of the chord of 90°, for example, produces a close approximation of $\sqrt{2}$. Where P refers to parts or units, it proceeds:

$$\mathrm{crd}\,90^\circ = \sqrt{(60^P) + (60^P)} = 60\sqrt{2^P}.$$

The chord is found to equal $84^P 51'10'' = 84.8526$. Dividing 84.8526 by 60 gives 1.414.

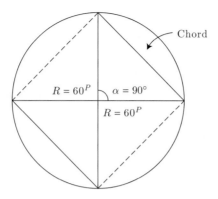

(The length of chord 90° is the length of the side of a square inscribed on a circle with a diameter of 120^P.)

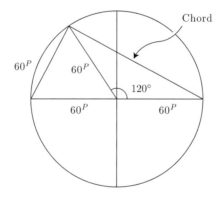

The chord of 120° similarly gives an approximation of $\sqrt{3}$.

$$\text{crd}\,120° = \sqrt{(120^P)^2 - (60^P)^2} = 60\sqrt{3}^P$$
$$= 103^P 55'23'' = 103.9224$$

Dividing 103.9224 by 60 gives 1.732.

With the aid of Euclid XIII.9–10, Ptolemy obtained crd 36°, the side of a regular inscribed decagon, as $30^P(\sqrt{5} - 1) = 37^P 4'55'' = 37.083$ and crd 72°, the side of a regular pentagon, as $30^P\sqrt{(10 - 2\sqrt{5})} = 70^P 32'3'' = 70.536$.

To extend his table of chords, Ptolemy developed several trigonometric identities. One is equivalent to $\sin^2\alpha = \frac{1}{2}(1 - \cos 2\alpha)$. Perhaps the best known identity is for any arc less than 180°: $(\text{crd}\,S)^2 + \text{crd}(180° - S)^2 = (120^P)^2$. This is equivalent to the Pythagorean relationship expressed in a modern unit circle as $\sin^2\alpha + \cos^2\alpha = 1$, where α is any angle.

The *Almagest* also contains a lemma, now usually called Ptolemy's theorem, although it was discovered earlier. The theorem states that if *ABCD* is any quadrilateral inscribed in a circle, then the sum of the products of the opposite sides equals the product of the diagonals. In the accompanying figure, $AB \times CD + BC \times AD = AC \times BD$.

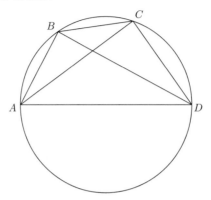

When two arcs and their chords are known, this theorem makes it possible to compute the chords of the difference or sums of the arcs. In modern notation, Ptolemy's finding resembles the formula $\sin(\alpha - \beta) = \sin \alpha \cos \beta - \cos \alpha \sin \beta$. Ptolemy notes this in stating that if he were given crd $72°$ and crd $60°$ he could find crd $12°$.

Ptolemy's *Almagest* was long the authority in mathematical astronomy in the West, as were the physiological treatises of his contemporary Galen in medicine. The *Almagest's* geometric models and first trigonometric tables strengthened the belief in the reliability of geometric models and mathematical predictions applied to celestial phenomena. As a result, the Ptolemaic epicyclic geocentric model of the universe dominated astronomy in Europe and the Near East until they were displaced in the early seventeenth century by a reform of astronomy in which Kepler's elliptical orbits supported the Copernican helio-centric model. The *Almagest* also gave definitive form to plane and spherical trigonometry for over a millennium. Indian sages developed trigonometry from the fifth century and Arab scholars from the late ninth.

Ptolemy wrote widely on mathematical science, astrology, and geography. He systematically continued the *Almagest* in two works translated as *Canobic Inscription* and *Handy Tables*, while *Planetary Hypotheses* is a popular sum-mary of the Ptolemaic system. They refine parameters given in the *Almagest* and offer physical models of celestial motions in place of his previous purely geometric models. Ptolemy also wrote *Phases of the Fixed Stars* on heliacal risings and settings and weather predictions, consisting of two books, and two smaller treatises, the *Analemma*, with an ingenious graphical technique now called nomographic, and the *Planisphaerium*, which maps circles of the ce-lestial sphere onto a plane. This mapping is the mathematical basis for the plane astrolabe. In related fields Ptolemy composed *Harmonics* and an *Optics*, which examines light, color, and the theory of vision, reflection, and refrac-tion. Later in his career Ptolemy completed *Geographike Syntaxis (Geography)* in eight books. Possibly he had planned it at the same time as the *Almagest* and wrote book VIII before completing the list of "Important Cities" in his earlier work, *Handy Tables.* The *Geography,* a summary of knowledge of the known world, which for Mediterranean peoples consisted of Europe, Asia, and Africa, became the authority in geography. It contains a general map of the world and twenty-six detailed regional maps.

Ptolemy divides the globe into 360 parts (later called degrees), and across its surface places a conic projection network of meridians and parallels, both of which are curved lines. The curved meridians converge at the poles. He then systematically lists some eight thousand places by their supposed latitude and longitude, and places his arbitrary prime meridian at the westernmost part of the known world in the Fortunate (Canary) Islands. Ptolemy was off by seven percent in their distance from the Continent. Had he not rejected Eratosthenes' estimate of Earth's circumference (subject to qualifications noted on page 333), his globe would have been less distorted. The distance from Europe westward to Asia was too small and geodetic measures to correct it did not yet exist.

As late as its translation into Latin in 1409 and the addition of its maps in 1478, the *Geography* was treated with great deference. A reading of the *Geography* probably led Christopher Columbus, on his discovery voyage of 1492, to believe that he had reached southern India. Thus he called the people Indians. Although Ptolemy's conic section technique was an unsuccessful attempt to map spherical surfaces onto a plane, the *Geography* stimulated the search for a more successful mapping in sixteenth century cartography.

Shortly before Menelaus, another Alexandrian advanced Hellenistic mathematics by improving mensuration and working imaginatively in mechanics. Hero's own or a contemporary's observations on a lunar eclipse reported in chapter 35 of *Dioptra* show that he flourished in A.D. 62. No other lunar eclipse occurred in that region during the five hundred years centered on that date. The firm dating of that event by Otto Neugebauer has settled a long debate variously placing Hero's activity between 150 B.C. and A.D. 250.[250] At the Alexandrian Museum he studied the work of Ctesibius (fl. 270 B.C.) and wrote treatises on mensurational geometry, surveying, and numerical calculation as well as mechanics, pneumatics (the theory and application of air pressure), and catoptrics (on mirrors). Thirteen books under his name are extant.[251] These writings show a concern with pedagogy and center on practical problems. In particular, they attempt to provide a scientific foundation for land surveying and engineering, rather than for astronomy.

Hero's contribution to mathematics and thus his place in its history are still debated. Does his work represent a decline of Hellenistic Greek theoretical mathematics or was he a technician who ignored fine points of theory because they were not critical to his work? The sources do not easily resolve these issues. Extensive editing has altered the original form of Hero's writings. The *Mechanica* survives only in an Arabic translation, the *Optica* in Latin. Hero's surviving texts divide into technical and geometrical categories. From the *Pneumatica,* based partly on Ctesibius and containing technical descriptions in a disordered array of materials, they range in style to the *Mechanica*

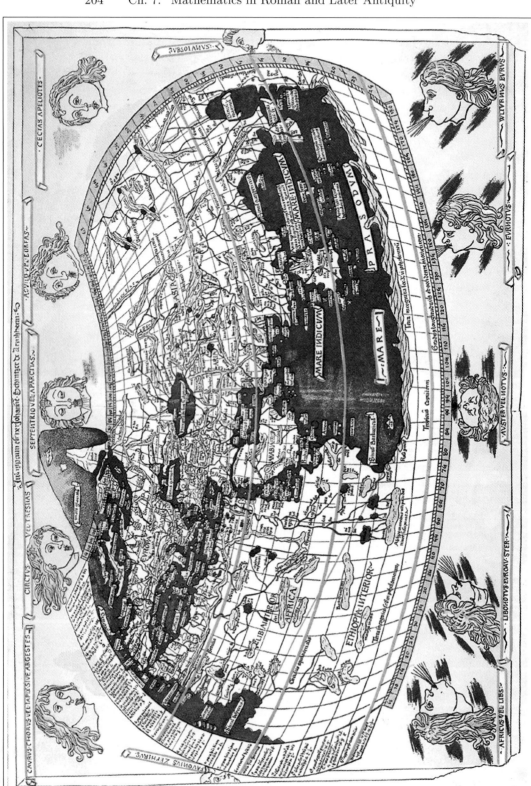

The habitable world according to Ptolemy

and *Dioptra* (on geodesy), which have a limited theoretical component, to the geometrical treatises *On Definitions* and *Metrica*. All seek greater efficiency in operations with minimal attention to underlying physical theories. Even though Hero utilizes Archimedes and modifies Euclid in a commentary that reduces the axioms to three and gives simpler, alternative proofs to *Elements* I.2 through 4 and II.5 through 10, his grasp of theory does not appear firm. Viewed exclusively within the gallery of Hellenistic theoretical mathematicians, Hero makes a small figure. Perhaps he can, however, be seen more appropriately as mixing early Hellenistic geometry with an older mensurational tradition whose primary aim was to calculate and improve practical applications. Collectively, the Heronic writings reveal a well-educated applied mathematician, a link in the tradition of practical mathematics, with procedures and rules for approximate calculation, originating in Babylon and Egypt and continuing through medieval Islam to Renaissance Europe.[252]

Hero's chief geometrical work, *Metrica*, shows both a wide knowledge of mathematics and skill in its application. This popular mensurational treatise in three books was rediscovered in 1896 in a manuscript copy in Constantinople. Euclid had dealt with line segments, circles, polygons, and solids without speaking of areas and volumes; Hero's rules and theorems concentrate on numerical values for plane areas and surfaces and volumes of common solids.

Book I builds on Democritus, Eudoxus, Euclid, and most notably Archimedes, whose results Hero freely utilizes. The book sums up in three computational formulas the state of the art of geometry.

The first of these formulas is an iterative technique for making an ever closer approximation of the value of the square root of n, where n is nonsquare. Hero, however, usually stops at the first approximation. His verbal form of these problems without symbols was known in old Babylon and to Archytas and is frequently used in computers today (see Chapter 2, page 31).

Hero's two other computational formulas may derive from Archimedes. The so-called Heronic formula finds the area of a triangle from its sides. Expressed in modern symbols, it is $A = \sqrt{s(s-a)(s-b)(s-c)}$, where s is the semiperimeter or half the sum of the sides of length a, b, and c. Hero gives the example of a triangle having sides 7, 8, and 9. This gives $s = 24/2 = 12$ and $A = \sqrt{12(5)(4)(3)} = \sqrt{720}$. The Heronic formula, which may seem odd in its use of the square root and semiperimeter, gives precise and unique values. From Euclid I.8 on side-side-side congruence, Hero knew that triangles with equal sides must be equal in area. Three sides of a triangle thus give one and only one value for the area. And when the height of a triangle is not known and thus the simple, standard formula for computing its area [$= \frac{1}{2}$(base) $\times$ height] cannot be used, the Heronic formula is invaluable. In *Journey through Genius*, published in 1990, William Dunham describes Hero's ingenious, although circuitous, proof of this formula. Hero's third formula correctly says that the area of every circle segment is more than four-thirds the area of the inscribed triangle with the same base and height. He states this after citing the Archimedean ratio of the circumference of a circle to its diameter as "greater than 211,875 to 67,441 and less than 197,888 to 62,351, but since these numbers are not easily handled, they are reduced to 22 to 7 (*Metrica* I.25)." Paul Tannery and B. L. van der Waerden have conjectured that Hero knew Apollonius' closer estimate of $\pi : 3.1416$.[253] Hero also gives a proof quite similar to that of Archimedes for the segment of a parabola, which equals four-thirds the inscribed triangle.

Book II of the *Metrica* considers volumes of solid figures. It begins with cones and cylinders and proceeds to prisms, parallelepipeds, pyramids, regular and irregular frusta, and the sphere. To find the volume of a prism or cylinder, Hero multiplies the base area by the height. The volume of a cone or pyramid equals one-third the volume of a prism or cylinder with the same base and height. With the aid of a result from Archimedes, Hero determines the volume of a sphere to be two-thirds of the volume of a circumscribed cylinder. Book II concludes with calculations of the volumes of the five Platonic solids.

Book III of *Metrica* includes Hero's probable method of calculating approximations to cube roots. His division of cones, conical frustums, and pyramids led to cube roots of noncube n's. His exposition gives only one numerical example, finding the cube root of 100, by considering how much it exceeds 64 and falls short of 125. While leaving his method ambiguous, Hero arrives at the final result of $4\frac{9}{14}$. Some scholars interpret this figure as having been derived empirically, while others discern a valid general rule. In attempting to reconstruct his method from a single numerical result, one substantial conjecture is that he begins with two integers, such that $a^3 < N < (a+1)^3$. If Hero then computes $N - a^3 = d_1$ and $(a+1)^3 - N = d_2$, linear interpolation should produce the approximation $a + d_1/(d_1 + d_2)$ between a and $a + 1$. Instead, Hero apparently arrives at a more complex rule. If he assumes $a^3 = N$, he would have found that $(a^3 - N)/a = a^2 - N/a$, a more nearly linear function. The value for a then becomes $-d_1/a$, and for $a + 1$, it is $+d_2/(a + 1)$. According to Thomas Heath, linear interpolation between the original two values gives the equivalent to Hero's value as

$$\sqrt[3]{N} \approx a + [(a + 1)d_1]/[/(a + 1)d_1 + ad_2].$$[254]

Recognition of his excellent mathematical sense and the substantial corruption from many editors and frequent copied editions of his text underly crediting him with a general rule here.

Throughout the *Metrica*, Hero presumes an elementary knowledge of geometry, apparently as needed for effective mensuration, and gives few formal proofs. Here, just as in the *Dioptra*, he seems to accept a distinction between theoretical geometry and geodesy, which is taught outside of liberal education to carpenters, masons, and surveyors.

Hero's *Pneumatica* illuminates the state of Greek mechanics in late antiquity and his ingenuity in applying simple mechanical and pneumatic principles. Its two books containing eighty chapters are his longest writing. With no discernible plan of chapter arrangement, it appears to represent a preparatory gathering of materials leading to the writing of a textbook. Its prologue distinguishes between the two aims of applied mechanics: to "supply the most necessary wants of human life" and to "produce astonishment and wonder." Utility and amusement seem to be conventional justifications; Pappus and Proclus designated as a separate branch of mechanics the construction of wonderful devices. Drawing on writings now lost of Strato of Lampascus, Philo of Byzantium, and Ctesibius, as well as the statics of Archimedes, Hero describes such practical implements as a water organ and a "siphon used against conflagrations," a double-force fire pump with a system of valves, cylinders, and pistons. A comparison of these with instruments that other writers discuss will reveal Hero's capacity for refinement and invention. III. 19 of *Mechanica*, for example, presents the double screw press for cutting screws, and the *Dioptra* an improvement on the surveyor's transit.

Some of the more than dozen "wondrous" gadgets described in *Pneumatica* have religious applications. Temple doors controlled by steam pressure in II.18 appear in the company of a drinking horn with hidden interconnected pipes and siphons for dispensing water and wine separately in a given proportion, a theater with moving puppets, and in I.12 a hollow ball filled with water made to rotate on pivots by steam escaping from attached bent tubes. The last gadget has some elements that reappear in the modern steam engine, but it was not a prototypical steam turbine as has been claimed.

While Hero exemplified the geodetic and applied mechanical lineage, Nicomachus of Gerasa (fl. ca. A.D. 100) reasserted the Pythagorean arithmetical tradition.[255] Nicomachus compiled his sizable *Introductio Arithmeticae* in two books. The *Introductio* does not simply separate arithmetic or the theory of numbers from geometry, but maintains that arithmetic is more fundamental, which suggests further that it is the source for the other three subjects of Plato's quadrivium.

The comprehensive, orderly *Introductio* treats the arithmetic of whole numbers and their ratios. Nicomachus drew heavily on the work of early Pythagoreans, dividing numbers into even and odd (prime, composite, and relatively prime), plane (square, rectangular, and polygonal), and 3-dimensional solids. Equality and inequality are the most basic relations of numbers, and inequality is divided into less and greater. While Nicomachus includes little that is original, he rediscovers progressions in summing the odd numbers: 1, 3, 5, 7, ... The first number in this series equals the cube of 1, the sum of the next two numbers the cube of 2, the sum of the next three the cube of 3, and so forth, which early Pythagoreans had understood. Nicomachus gives the four perfect numbers, 6, 28, 496, and 8,128, with Euclid's formula for finding them (but could not be expected to know that they are the only ones), and he describes ten types of proportions, including the musical $a : (a+b)/2 = (2ab)/(a+b) : b$, and discusses the sieve of Eratosthenes for quickly finding prime numbers. His work demonstrates a persisting Pythagorean influence in late antiquity.

The *Introductio* makes serious mistakes in certain problems and has shortcomings in its method of exposition. One error appears in the characteristics it gives in II.28 for subcontrary proportions and another in the classification of composite numbers as a species of odd rather than a class of all numbers. In treating numbers as quantities rather than line segments, Nicomachus breaks away from the arithmetic of Eudoxus-Euclid but does not exploit his innovation. Thus, while Euclid represents numbers by letters, such as A or MN standing for line segments, Nicomachus writes numbers as words. This makes his work clumsier. He also abandons abstract proofs, such as those Euclid presents in Books VII through IX of *Elements*, and illustrates principles only by specific numerical examples. Nor does he pursue practical applications for computations, as had Hero in mensuration.

Notwithstanding these drawbacks, the *Introductio* was the standard text in arithmetic until the late sixteenth century. Scholars in Latin Christendom favored its preference for arithmetic and widely copied it. Nicomachus even gained a reputation for being a leading mathematician. That indicates the limitations of arithmetical studies in the Latin West until the late Renaissance.

7.4 Revival and Demise of Ancient Alexandrian Mathematics

With the exception of a few instances of innovation by extraordinary individuals, after the year 200 conservation rather than development principally set the agenda in theoretical natural science and medicine.[256] The history of science in antiquity suggests that scientific advances depended crucially upon some subsidy for intellectually gifted individuals. Why did the numbers decline so significantly? Why were no portions of the works of Ptolemy and Galen extended that later connected to modern science? The change in the Roman empire after 200 when comparative calm and prosperity gave way to turbulence offers a clue.[257] By 300 trade had weakened in Alexandria. As Alexandrian fortunes worsened, interest in divination, such as expounded in Ptolemy's manual, apparently grew. The rise of Christianity also brought a preference for otherworldly studies. The privileged status of geometry, moreover, had constricted the western channels of mathematical research: recognition of the range and power of algebra was needed.

The shift in mathematics to commentaries and reediting of older works, begun with Hero, Geminus, and Menelaus in the first century A.D. as part of the Museum's compiling tradition, continued with Theon of Smyrna's *Expositio* ... ("Mathematical Knowledge Useful for the Learning of Plato") in the early second century and became dominant after 350. About midway in this shift the studies of Diophantus in algebra and number theory and Pappus in theoretical geometry managed partially to revive Alexandrian mathematics.

Despite urban disturbances, the city's economy was relatively stable: turmoil between pagans and Christians had not yet become especially destructive and Alexandria's intellectual resources were still ample. In addition to holding prior Greek texts, the Serapeum with its over 300,000 manuscript scrolls must have housed ancient Egyptian papyri such as those of A'hmosè, old Babylonian tablets, and Seleucid (new Babylonian) materials. Probably the techniques of computational arithmetic contained in the old and new Babylonian scrolls influenced mathematical thinking in the city. Neopythagorean arithmetical studies must also have been known in Alexandria, as were embryonic speculations that later took shape in the Cabala.

Of the life of Diophantus (fl. 250), who produced the highest level of algebra in antiquity, almost nothing is known aside from his residence for some time in Alexandria. No biographical details have survived from personal records or comments by contemporaries. Presumably, Diophantus was a Greek or perhaps a Hellenized Babylonian. A letter and an epigram written centuries later provide our only source of biographical information.

The belief that Diophantus flourished in the mid-third century A.D. rests entirely on a letter written by the Byzantine Neoplatonist and rhetorician Michael Psellus (1018 to ca. 1090), who was preserving writings of Iamblichus and Proclus. Psellus writes that Anatolus, who became bishop of Laodicea (in what is now Turkey) in 270, dedicated a tract on Egyptian computation to his friend Diophantus. This dating supports the assumption that the Dionysius to whom Diophantus dedicates the *Arithmetica* is the person who became the Christian bishop in Alexandria in 247.

An arithmetical epigram in the *Greek Anthology* offers the only other information on Diophantus' life. The *Greek Anthology*, a collection of some 3,700 epigrams or short poems, was compiled in the tenth century.[258] Its fourteenth section contains epigrams, a popular form of puzzle, generally with problems of little depth, that permit us to construct an equation. This one states that the boyhood of Diophantus comprised one-sixth of his life, that his beard grew after one-twelfth more, that he married after one-seventh more, and that the couple had a son five years later. The son lived only half as long as the father and the father died four years after his son. Assuming that x is the age when Diophantus died gives the equation:

$$(1/6)x + (1/12)x + (1/7)x + 5 + (1/2)x + 4 = x.$$

If the epigrammatic data are correct, then his boyhood lasted fourteen years, he grew a beard at twenty-one, and he married at thirty-three. He also had a son who died at forty-two when the father was eighty. The father died at the age of eighty-four.

Five treatises are attributed to Diophantus. *Moriastica* and *Porismata* are lost, *Teaching of the Elements of Arithmetic* is conjectural, and *On Polygonal Numbers* and *Arithmetica* are extant but incomplete.

Moriastica and *Porismata* are known only from remarks of later commentators who studied Diophantus' work. *Moriastica* perhaps dealt with computation with fractions, while *Porismata* may have proved propositions in the *Arithmetica*. Although *Porismata* was most likely an independent work, Paul Tannery has speculated that its lemmas originally formed a part of the *Arithmetica* lost by later commentators. Recently, the Greek mathematician Jean

Christianidis has strongly argued for the existence of *Teaching of the Elements of Arithmetic* and its serving as a theoretical base for the diverse problems of the *Arithmetica*.[259] Christianidis cites the style and scholia along with problem 59 of the *Arithmetica* and comments by the tenth-century Persian scholar Abū al Wafā and the Byzantine Maximus Plaanudes.

Small fragments of *On Polygonal Numbers* show that, unlike the *Arithmetica*, it included deductive geometric proofs. The fragments display little originality; the author reviews a topic long known to earlier Greeks. Most of the *Arithmetica*, the *magnum opus* of Diophantus in thirteen books, has been preserved. Ten books are known. Six of them exist in the Greek original from a thirteenth-century copy. These were assumed to be the first six books. The discovery in the 1970s of four new books from an Arabic translation and commentary forced a revision of that ordering. The original order now appears to be I through III in Greek, IV through VII preserved in the Arabic rendering, and three more in Greek.

The reputation of Diophantus rests chiefly on the *Arithmetica*. Despite its title, this text does not principally treat elementary number theory (*arithmetica*) in the customary Greek geometric mode. It appears as the first treatise on an independent algebra largely free of geometry. Its author apparently has two goals: to teach diverse computational methods of solving problems from logistics and to find positive rational solutions satisfying given equations. In the *Arithmetica*, Diophantus is a tireless calculator who makes few mistakes. His calculations usually proceed from simple to difficult according to the degree of an equation and its number of unknowns. Throughout he is clever and imaginative in devising a separate method for attacking each of the 189 heterogeneous problems in the six Greek books. Some interdependent groups of problems and well-ordering appear in the 101 problems of the four books surviving in Arabic translation. Diophantus does not attain generalization on a deeper level, but relies on ad hoc illustrations. These and an absence of geometric proofs put the *Arithmetica* closer to procedural texts of ancient Egypt and Babylon, such as the *Rhind Papyrus*, than to the systematized deductive geometry of Euclid, Archimedes, and Apollonius.

In seeking positive integers and rational fractions satisfying equations and algebraic inequalities, Diophantus contributed to the beginnings of algebraic number theory. Equations associated with his problems differ in their type and degree. The two distinctive types of problems that he covers are determinate equations, for example in modern notation $4x^2 = 9$ having as its positive solution $x = 3/2$, and indeterminate equations, such as $xy = 18$ having an unlimited number of solutions $(1/2, 36)$, $(1, 18)$, $(2, 9)$, and so forth.

Most of the problems in Book I lead to linear and determinate equations in two or more unknowns. Problems in the next five Greek books are chiefly quadratic and indeterminate, of the type reducible in modern notation to $Ax^2 + Bx + C = y^2$. To solve these, Diophantus chose the form of the problem in which one of its given terms immediately meets a required condition, and the other number is to be found. In the preceding quadratic, if we assume a projection of its symmbolism into Diophantus' mathematical language, he assumes that A is a square number and lets $y = ax + m$. He also treats several particular cases of what is now known as the Pell equation, such as $y^2 = 1 + 30x^2$.

Only one problem in the Greek books requires a special cubic equation. The Arabic books contain most of the higher-degree equations, which reach as high as the ninth power. By skillfully choosing numerical parameters, the author can reduce these to a lower degree. Diophantus' algebraic approach to these problems suggests Babylonian influence or origins, but he also differs from them in seeking only exact and not approximate solutions.

While much of the *Arithmetica* is an attempt to consolidate and extend earlier computational knowledge, the treatment of indeterminate equations of the second degree in the last five Greek books is innovative. Indeterminate equations did not originate with Diophantus, as Archimedes' cattle problem that derives from Book 12 of Homer's *Odyssey* testifies. That problem seeks the number of bulls and cows with hides in four colors, which produce eight unknowns, and under a given condition. Diophantus, however, was the first to give an extensive and masterly account of the study of rational solutions of indeterminate equations. Consequently, he is considered the founder of indeterminate analysis. This branch of algebra is now called Diophantine analysis and the equations are known as Diophantine equations. Today, solutions of these are usually restricted to integers.

Here are six of the problems that Diophantus solves:

I:17   Find four numbers such that the sum of all sets of three of them is one of four given numbers. Consider the given numbers 20, 22, 24, and 27. If x is the sum of all four numbers, the numbers are $x - 20$, $x - 22$, $x - 24$, and $x - 27$. Setting x equal to the sum of these four numbers, gives $4x - 93 = x$ and $x = 31$. Diophantus arrives at the answer 11, 9, 7, and 4.

II:28   Find two square numbers such that their product added to either gives a square. The answer is 9/16 and 49/576.

III:6   Find three numbers such that their sum is a square number and the sum of any pair is a square number. Diophantus' answer is 80, 320, and 41. A later interpolated solution is 840/36, 385/36, and 456/36.

IV:21   Find three numbers in geometric progression and such that the difference between any two is a square number. Diophantus' answer is 81/7, 144/7, and 256/7.

VI:1   Find a (rational) right-angled triangle such that the hypotenuse minus each of the sides gives a cube. Diophantus forms two sides of the required triangle from the basic elements x and 3. Working with compounds of them gives hypotenuse $x^2 + 9$, perpendicular $6x$, and base $x^2 - 9$. This should mean that $x^2 + 9 - (x^2 - 9) = 18$ is a cube, but it is not. Since $18 = 2 \bullet 3^2$, he replaces 3 by m and seeks to know when $2 \bullet m^2$ is a cube. The immediate solution is $m = 2$. Diophantus then forms the sides of the required right triangle from compounds of x and 2, obtaining $x^2 + 4$, $4x$, and $x^2 - 4$ as one condition and $x^2 - 4x + 4 =$ a cube as the other. The second condition means that $(x - 2)^2$ is a cube, let us say $8^2 = 4^3$. Then $x - 2 = 8$ and $x = 10$. The sides of the triangle from condition one are 40, 96, and 104.

VI:16   Find a rational right triangle such that the number representing the intercepted portion of the line bisecting an acute angle within the triangle is rational. This holds for a right triangle with perpendicular 28, base 96, and hypotenuse 100.

Diophantus is satisfied with finding the simplest solution and accepts only positive rational roots. In cases having two such roots, he gives only the larger. Since he does not recognize roots that are irrational, negative, or zero, he eschews equations that produce such roots. Thus in V:2 he rejects the equation $4 = 4x + 20$. Yet on occasion he treats negative quantities, stating for example that "minus times minus makes plus." Mathematicians in the Latin West did not accept negative numbers until the Renaissance.

In a major innovation, Diophantus took algebra beyond the existing rhetorical stage, in which equations were written in words. In their place his work employs for the most part a shorthand or stenographic abbreviations. His mixture of abbreviations, a few symbols, and words in equations historians term a second or syncopated stage of algebra. It preceded the stage of fully symbolic algebra, which did not emerge until the late sixteenth century. To denote the unknown x of our notation, Diophantus introduced the symbol ς, perhaps a distortion of the final sigma in the word *arithmos* or a joining of the first and last symbols. That he uses only one symbol for the unknown limits the flexibility of his work. Powers of the unknown from 2 to 6 are represented with capitalized abbreviations:

$$\Delta^{\Upsilon} \ (= \textit{dunamis}, \text{ power or square}) \text{ for } x^2$$

$$K^{\Upsilon} \ (= \textit{kubos} \text{ or cube}) \text{ for } x^3$$

$$\Delta^{\Upsilon} K \ (= \text{square-square}) \text{ for } x^4$$

$$\Delta K^{\Upsilon} \ (= \text{square-cube}) \text{ for } x^5 \quad \text{and}$$

$$K^{\Upsilon} K \ (= \text{cube-cube}) \text{ for } x^6.$$

Diophantus has no sign for addition, but indicates it by putting alongside one another at the beginning of a problem all terms to be added. Terms to be subtracted follow the sign Λ, described as an inverted *psi* (ψ). This subtraction sign may be a stenographic abbreviation of the Greek verb *leipsis* ("to want"). The first letter (τ) of the Greek word for equals does the work of our $=$ sign. Where we write a constant, Diophantus writes the abbreviation $\overset{\circ}{M}$ for *Monades* followed by the numeral. This indicates that the following numeral is purely a number and not a coefficient for the unknown or any of its powers. Using this notation and Greek alphabetic numerals, Diophantus writes the polynomial

$$x^6 - 2x^5 + x^3 - 2x - 3 \quad \text{as} \quad K^{\Upsilon}K\alpha K^{\Upsilon}\alpha \Lambda \Delta K^{\Upsilon}\beta\varsigma\beta\overset{\circ}{M}\gamma$$

Such abbreviations for the unknown and its powers through the sixth and for subtraction, equality, and reciprocals gave theoretical mathematics a formal simplicity missing in Greek algebra in its rhetorical stage. While Euclid had used letters in problems, Diophantus pioneered the recognition of notation as being not merely a shorthand but an operational device for solving algebraic problems more efficiently.

Diophantus had no worthy ancient successor. His breakthrough could not arrest the waning of Hellenistic culture in the late third century. The *Arithmetica* did not greatly influence theoretical mathematics until a much later time. The commentary of the Alexandrian Hypatia on six of its books was soon lost. Arab scholars, who had at least portions of this text, and their Latin medieval counterparts, who did not, still practiced the rhetorical form of algebra. In the late fifteenth century, Regiomontanus, the first to translate the *Arithmetica* into Latin, recognized its importance, asserting that it encompassed the entire "flower of arithmetic, the *ars rei et census*, called algebra by the Arabs." Its penetrating propositions on prime numbers and representations of numbers as the sum of two squares in II:8 and V:9, of three squares in V:11, and four squares in IV:29 and 30 and V:14 captured the interest of Pierre de Fermat. Fermat's study of V:9,

> To divide unity into two parts, such that, if the same given number be added to either part, the result will be a square,[260]

especially led to his famous last conjecture or theorem that it is not possible to obtain positive integral solutions for $x^n + y^n = z^n$, where n is an integer larger than 2 and x, y, and z are not zero.[261] In their enterprise to found algebraic number theory, mathematicians from Fermat through Euler to Gauss examined, proved, and generalized propositions from the *Arithmetica*.[262]

Among the mathematicians who lived during the five hundred year interval following the death of Apollonius, Pappus of Alexandria (fl. 300–330) reached the highest level of achievement in theoretical geometry. The last major geometer of antiquity, he sought to advance the demonstrative branch of the mathematical tradition begun in Alexandria by Euclid.

The principal treatise of Pappus is the *Synagoge* or *Mathematical Collection*, including eight known books. A large handbook intended to accompany original works, the *Collection* contains in part a synopsis of the whole of ancient Greek geometry, including materials on proportions, higher plane curves, solid geometry, numeration, and mechanics. Of its eight known books, the first and part of the second are missing. In Book VII Pappus states that his treatise contains twelve books. Each book was probably written at a separate time and the whole later collected by Pappus or one of his students, as the title suggests.

Discussing more than thirty previous classical and Hellenistic Greek geometers, the *Collection* attempts to consolidate earlier geometrical knowlege. It remains our richest and in some respects our only source of much of ancient Greek geometry, especially the lost works of Euclid and Apollonius. Its preface to Book III also cites wrong or unworkable methods of a teacher named Pandrosian, whom Alexander Jones argues was a woman geometer.[263]

Our information on the three famous problems of antiquity comes from Book IV, for example, where Pappus claims to square a circle by using Archimedes' spiral or Nicomedes' conchoid and trisects an angle by Apollonius' *neusis* constructions. Pappus asserts that the three problems as stated cannot be solved by straightedge and compass alone, which would first be formally proved in the nineteenth century.

In preparing the two-part Book V, which primarily addresses isoperimetry—figures with equal perimeters or surfaces—Pappus seems to have consulted a version of Archimedes' *Dimension of a Circle*, but was apparently most influenced by the lost tract *On Isoperimetric Figures* by the Alexandrian geometer Zenodorus.[264] Its first part compares figures with equal bounding perimeters, the second part, volumes of solids with equal bounding surfaces. Book V's proof that the sphere has a greater volume than any regular solid with an equal surface follows Zenodorus. The preface of Book V praises the wisdom of bees for making their honeycombs hexagonal, because for fitting together of components the hexagon is the figure containing the greatest area. Such treatment of problems of maximum and minimum is unusual in antiquity. Book V also covers the thirteen semiregular solids, that is, solids contained by equilateral and equiangular polygons, but not all similar ones. The discovery of these solids is attributed to Archimedes but does not appear in his extant writings. Book VI, which is an introduction to Ptolemy's *Almagest*, corrects misrepresentations of Euclid.

In addition to preserving older geometrical knowledge, the *Collection* contains commentaries on earlier works, alternative proofs, novelties that generalize several earlier theorems, and some discoveries. Whatever its limitations, it is also a work of originality.

Pappus' manner in Book III of inscribing the five regular solids in a sphere differs markedly from that given in Book XIII of Euclid's *Elements*. His proof for a theory of means (arithmetic, geometric, and harmonic, based on a construction in a single circle) he claims as being his own, but this is not clear. Proposition One of Book IV generalizes the Pythagorean theorem more than does Euclid VI.31: Given any triangle ABC, draw parallelograms on sides AB and BC. Next extend sides DE and FG to meet in H. Draw HB and extend it to meet AC at J, and then draw AL and CK parallel to HJ. It is then possible to show that the sum of parallelograms drawn on two sides of a triangle, $ABDE$ and $CBGF$, equals the parallelogram drawn on the third side, $ACKL$.

Book IV also gives Pappus' discovery that a cone of revolution intersects a right cylinder with Archimedes' spiral as its base in the quadratrix of Hippasus. The second section of the book contains a series of novel propositions on

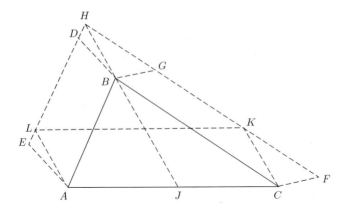

inscribing circles with the "shoemaker's knife" or arbelos that Archimedes had introduced in his *Book of Lemmas.*

Book VII is the most notable of the *Collection.* Its deserved reputation derives from its preservation of the *Domain of Analysis,* its original insights, and its subsequent influence on mathematics. Prior to Pappus, Greek geometers had employed both the method of synthesis, highlighted by Euclid, and the method of analysis. Pappus provided the first extant systematic study of the method of analysis. Next to Euclid's *Elements,* the *Domain* (also known as the *Treasury*) was the work most essential to professional mathematicians. Of the nine treatises it reviews, eight are by Apollonius. It also discusses works by Euclid and Aristaeus the Elder in describing the method of analysis and synthesis. Again, most of the treatises that it covers, such as Euclid's *Porisms* and Apollonius' *Determinate Section, Inclinations, Plane Loci,* and *Tangencies,* are otherwise lost.

Book VII's digression on Apollonius' *Conics* contains the second mention in antiquity of the focus-directrix property of conics. Diocles' *On Burning Mirrors* provided the first. Pappus' reference occurs in a problem first called the locus with respect to three or four lines and later the problem of Pappus. If the angles made by lines from M to the four lines are oblique and kept constant and the product of the distances from point M to two lines is proportional to the product of the distances to the other two, $m_1 m_3 = k m_2 m_4$, where k is a constant, then the locus of the moving point M is a conic section. Employing compound ratios, Pappus generalizes this problem to six lines, but finds it difficult. For the case of six lines he obtains a higher degree curve that in modern notation involves a cubic equation in two variables. A later attempt to generalize this famous locus problem led Descartes to modern coordinate or analytic geometry. By substituting new algebraic symbols for geometrical entities, Descartes avoided the troubles that had beset Pappus. Descartes was then able to demonstrate that the locus with respect to three or four lines is a second-degree equation, that is, a conic section, while that of five, six, seven, or eight is the next higher curve, a cubic, that of nine, ten, eleven, or twelve, a quartic, and so forth.

Among algebraic geometrical nuggets in Book VII is Pappus' description of Euclid's *Porisms*. Three of the thirty-eight lemmas he provides to improve an understanding of *Porisms* are remarkable. They concern any system of straight lines that intersect each other two by two. Lemma 3, proposition 129 proves the equality of what was called a century ago anharmonic ratios or today is referred to as the cross ratios from any two transversals cutting a pencil of four lines. In the accompanying drawing, two transversals emanating from the same point, *HBCD* and *HEFG*, cut line segments *AG*, *AF*, *AL*, and *AH*. Pappus demonstrates that

$$\frac{HE \cdot GF}{HG \cdot FE} = \frac{HB \cdot DC}{HD \cdot BC},$$

which means that the cross ratios of these collinear points are invariant under projection. This later became the fundamental theorem of projective geometry.

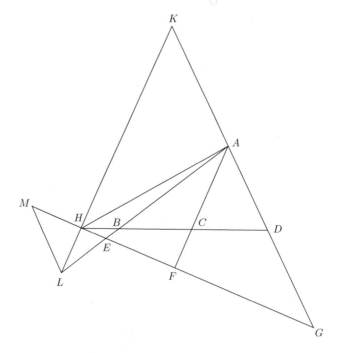

Collinear points lie on the same line. Lemma 4, proposition 130 firmly extends the relationship to any transversal intersecting pairs of opposite sides of a quadrilateral and its diagonals in three pairs of conjugate points, or what Desargues later named points of involution. Lemma 13, Proposition 139 gives another important result known today as Pappus' theorem. Simply stated, if *A*, *B*, *C*, and *D*, *E*, *F* are two sets of collinear points, then the points of intersection *G*, *H*, *I* of the lines connecting them are collinear.

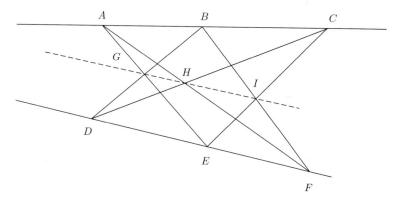

Although the two lemmas to *Porisms* on cross ratios and involution suggest that geometers just prior to Euclid had some understanding of these concepts, an appreciation of the significance of these lemmas and of Pappus' theorem awaited Girard Desargues (1591–1661), who did pioneering work in projective geometry. Book VII states an important theorem on axes of rotation about centers of gravity. This theorem was later named for Paul Guldin, who independently discovered it in the seventeenth century. In Book VII, where he claims that mechanics is a mathematical subject and not limited to utilitarian concerns, Pappus gives the only explicit definition surviving from antiquity for center of gravity.

Only fragments of Pappus' other writings survive. These treat geography, hydrostatics, possibly alchemy, and all fields of the quadrivium. The fragments include commentaries on Euclid's *Elements* and on Ptolemy's *Almagest* and *Harmonics*. From the commentary on the *Elements*, only a two-part section on the tenth book survives in a tenth-century Arabic translation. It carefully delineates how earlier Greek thinkers from the initial Pythagoreans and Theaetetus contributed to the study of irrationals. The elementary commentary on the *Almagest* may be Pappus' first work. Extant portions of Books V and VI methodically detail for beginners Ptolemy's astronomical theories. Pappus improves on Ptolemy's lunar theory by correcting the anomalous location of the moon at quadratures when it is not near either apogee or perigee. This irregularity is now called evection. Pappus also explains conditions for predicting eclipses and writes that he had observed an eclipse in 320, the observation that is our chief basis for dating his life. Pappus wrote another influential treatise, titled in English *Description of the World*. The *Description* uses Ptolemy's world map as a starting point for fourth-century geographical knowledge.

Arabic sources claim that Pappus also studied hydrostatics and made an instrument for measuring liquids, which an ancient oath attributes to him (see G. Grumer, *Isidis*, 1807). This instrument suggests that Pappus had an interest in alchemy and that his religious views may have blended Christian and late pagan beliefs.

Like Diophantus, Pappus had no worthy successor in antiquity. After his time, original lines of mathematical inquiry nearly disappeared in Alexandria and the Mediterranean world. For this, along with the rigor with which the *Collection* covers the whole of ancient Greek mathematics, it has been called the requiem for ancient Greek geometry. The first significant successors of Pappus appeared during the seventeenth-century scientific revolution, above all Descartes in his invention of analytic geometry and Isaac Newton, who in Lemma XIX of the *Principia* discusses Guldin's theorem.

By the late fourth century the fragile supports for the mathematical tradition in Alexandria had broken: the social and economic framework for subsidized scholarly activity appears to have precipitously deteriorated and Greek intellectuals increasingly immersed themselves in otherworldly concerns. To the warfare, plague, famine, and corruption of the third and early fourth centuries A.D. were now added the collapse of the Alexandrian economy as trade declined and an increase in religious hostilities as paganism retreated before Christianity. Particularly damaging for the transmission of pagan higher learning was the sacking and burning of the library of over 300,000 volumes at the shrine of Serapis in 391 or 392. Surely it was after the Roman emperor Theodosius I proscribed cult practices, which opened the way for Christians to besiege pagan cult places, including the Serapeum. A group of Alexandrian pagans who defended this shrine made forays out against the Christians, capturing and killing a few of them. Some Alexandrian intellectuals opposed to destroying the magnificent statues and cult objects of the Serapeum joined the unsuccessful pagan defenders. The few individuals still carrying on the tradition of mathematics were teachers and commentators.

At this time Alexandria was home to Theon (ca. 335–ca. 405) and his daughter Hypatia (ca. 355–415), who was Theon's most learned pupil and most esteemed collaborator. These two teachers of the Greek and Egyptian heritage were committed to the multilingual tradition of higher learning in Alexandria. They were drawn to rational Hellenism, notably in the quadrivium, and steeped in astrology, divination, and related writings of Hermes Trismegistus and Orphic poetry. Interest in astrology, which was taught in the schools, was especially widespread in a city notorious for its many fortune tellers. Theon and Hypatia were perhaps the most distinguished teachers at what remained of the *Mouseion*; Theon refers in his writings to companions (*hetairai*) who were fellow instructors and students. Theon's extant works, which are based on lectures and intended for student use, show the continuing influence of Ptolemy, Pappus, and Euclid. Working with close associates under his direction, Theon edited and wrote commentaries on their classics or prepared new editions, primarily in mathematics and astronomy, a project that Hypatia continued as a mature scholar on her own.

Internal references taken from Theon's writings reveal that his first and longest commentary was on the thirteen books of Ptolemy's *Almagest*. Sections on Book II and most of Book V are lost, and lacunae probably exist in the remaining ones. For the most part, the commentary is trivial. Interpretive rather than critical, it does not add substantially to our understanding of Ptolemy. Its section on Book VI shows extensive borrowing, if not lifting, from Theon's earlier compatriot Pappus. Theon's commentary on the *Almagest* and two later ones, the *Great Commentary* in five books and the *Little Commentary* in one on Ptolemy's *Handy Tables*, all of them dedicated to Epiphanius, perhaps Theon's son, offer convenient means for sexagesimal computations and are replete with them. The preface to the large commentary on the *Handy Tables* contains an important observation on the condition of theoretical learning. Theon states that he prepared this work because most students could not follow geometrical proofs in the commentary on the *Almagest*, a testimony to the caliber then of mathematics students. The Theonic version of the *Handy Tables*, the extant version of which Hypatia probably prepared, is significant for transmitting the theory of the astrolabe in the seventh century to Islamic astronomers, while his treatise *On the Small Astrolabe* explains its construction. The most common pretelescopic sighting device, the astrolabe has movable arms set atop a coordinate grid that allows the observer to determine the altitude of a celestial body and locate its correct position in a chart of the heavens. The medieval Arabs in turn passed knowledge of the astrolabe to late medieval European astronomers. Theon is also our chief source of the *Handy Tables*.

Theon's edition of Euclid's *Elements* is his most famous writing. This rework of the original makes mostly slight additions and corrects a few mistakes. Its alterations are minor and hardly improve on the original. Theon's edition of the *Elements* nevertheless became extremely influential. It superseded earlier editions and consigned them to near oblivion. It was the source of all editions of the *Elements* in the Latin West until the late nineteenth century. The portrait of Theon that emerges from this edition along with his commentaries is that of a competent but unoriginal mathematician.

The tenth-century encyclopedic *Suda* suggests that, more than the rational inquiry of the mathematical sciences, it was the secrets or *arcanae* of our physical world that drew Theon's interest. These he investigated through astrology and the interpretation of omens, both important to the Neoplatonists, and the revelations of Hermes and the poet Orpheus.[265] *Suda* tentatively lists Theon as the author of at least two poems on these themes that now appear in the *Greek Anthology* and *Corpus Hermeticum*. After ascribing intelligence and power to stars, one poem maintains that they determine human temperaments and psychic states from birth. Another poem expresses Theon's devotion to the perfect starry heavens beyond the sphere of the moon, and it praises Ptolemy as an elect of the gods who was able to tear himself away from mundane earthly existence and pierce with the splinters of divine knowledge within the human mind the immutable laws of the heavens.

The 156 surviving letters of Synesius of Cyrene, who became a Christian bishop, offer the most substantive information on Hypatia's teaching and circle, as well as on political and social life in Alexandria.[266] Theon's student Philosorgius and her sixth-century biographer Hesychius of Miletus support the claim that Hypatia exceeded her father as a scholar, and Theon and the lexicographer Suidas referred to her as "the philosopher," while Michael Psellus calls her "the Egyptian wise woman."[267] The scant information of this second woman mentioned by name in the history of mathematics has credited her, among other things, with collaborating with Theon on his commentary on Books III and IV of the *Almagest* and his edition of the *Elements*. But an analysis of writing styles, including a use of the present tense of verbs, and the exactness and orderliness of its tabular computations, especially its divisions at the close of Book III to several sexagesimal places perhaps by her or another using an abacus, have recently led classicists such as Alan Cameron to assert that Hypatia corrected not her father's commentary but the text of the *Almagest* itself for a new edition.[268] Books I and IX, for example, are loose and improvised; Books III and IV tight and orderly. These last two books may have been prepared when Hypatia alternated with her father in teaching the astronomy course. As a young scholar and a woman, she was under even greater pressure to be precise and thorough. A notice in Suidas also attributes to her a commentary on Apollonius' *Conic Sections* and Paul Tannery another on Diophantus' *Arithmetica*. Tannery and recent classicists, such as T. Perl, credit to her commentary the survival of most of the *Arithmetica* and contend that it served as the basis for many interpolations, scholia, and adaptations of Diophantus. If the last view is correct, her Diophantine commentary must have been interpretive.

The principal Neoplatonic philosopher of her time in Alexandria, Hypatia probably taught a traditional middle Platonism that began with an examination of the thought of Plato and Aristotle. Likely it drew more on the metaphysics of Porphyry, which tended to mysticism and the thought of Iamblichus. Alexandrian scholars were not yet receptive to the firm schema of the metaphysics of Porphyry's teacher Plotinus. Among the required readings in Hypatia's courses were biographies, especially Porphyry's *Plotinus* and Iamblichus on Pythagoras. Apparently donning the black and white philosopher's cloak, she taught courses in her home to a small group of regular students, often in dialogue form, and gave public lectures that drew civic officials to her home and to public lecture halls around Alexandria. Her classes included philosophy, astronomy, mathematics, and religious literatures. According to J. M. Rist, Hypatia adopted Cynic tactics, reportedly using blunt and occasionally shocking gestures to quiet "frivolous male detractors" who questioned a woman's place in the traditionally male profession of higher education.[269] Not a devoted pagan, she avoided the cultic elements that marked Hellenism in her time. Her eclectic lectures on religious literature included Christian texts, and in striving to reach higher levels of cognition, some of her classes sang hymns. Hypatia did not alienate but instead attracted as students such Christian leaders as Synesius, who praised her learning.

Still, Hypatia's Neoplatonism along with her becoming enmeshed in a complex political struggle would seem to have been the reason for her death. In 412, Cyril was elected bishop of a divided Christian community, and three days of street fighting preceded his installation. Hypatia apparently supported the new Roman prefect of the city, Orestes, who was a Christian. In extending his authority, Cyril attempted to expel the Jews from Alexandria and incited monks against Orestes. When Hypatia remained a staunch supporter of Orestes and local officials, Cyril's followers criticized her for elitism and arrogance, and they charged her with practicing witchcraft and sorcery. On her way to a public lecture hall during Lent of 415, when she was probably nearly sixty years of age, a fanatical Christian mob led by a lector named Peter dragged her from her carriage, stripped her, dragged her to a church, killed her by stabs with pottery shards, scraped her flesh from the bones, and burned the remains of her body.

Belletristic literature and historical conjectures, rather than the critical study of ancient sources, have nourished a legend about the ghastly death of Hypatia that has captured the ancient and modern popular imagination. This legend is well expressed in the novel, *Hypatia, or New Foes with an Old Face.* This anti-Catholic Victorian romance by Charles Kingsley, published in 1853, builds on the reductive accounts of Voltaire in his *Dictionnaire Philosophique* of 1764 and of Edward Gibbon in Volume 5 of *Decline and Fall of the Roman Empire.*[270] Kingsley's Hypatia, young and beautiful as well as learned and obdurately anti-Christian, is the lover of the Roman prefect Orestes, who is the enemy of the power-hungry Christian patriarch Cyril. As hostility grows between the two men and more generally between Christians and pagans, Cyril inflames a mob of fanatic Christian monks, who throw Hypatia from her carriage and tear her to pieces.

The murder of Hypatia by thugs, according to a ninetenth-century romanticized French engraving

The savage death of Hypatia has been perceived incorrectly as demarcating, at least in Alexandria, the passing of the age of ancient Greek mathematics[271] or, more generally, as testifying to an irreversible historical drama: the decline of the teachings of Plato, Aristotle, and the Neoplatonists and the victory of a new Christian order. This broader interpretation traces to two eighteenth-century authors, John Toland, who wrote a treatise on Hypatia, and Voltaire, both of them influenced by Neohellenism.[272] The death of Hypatia did add to the assault on higher learning that had contributed to the burning of the Serapeum twenty-five years earlier. But pagan religion did not yet end, and an eclectic Neoplatonism based on Plotinus flourished in Alexandria shortly before the Arab conquest in the seventh century. A limited pursuit of mathematics and the physical sciences continued at the remnants of the Museum through the late sixth-century work of John Philoponus, who wrote a critical commentary on Aristotle's *Physica*. Philoponus rejects Aristotle's potential infinity, believing that it leads necessarily to an actual infinity.

The West after Hypatia no longer achieved even the derivative level of mathematics that the fourth-century geometer Pappus had attained. A possible exception is the work of Eutocius of Ascalon (b. ca. 480). Eutocius' commentaries on Apollonius' *Conics, Books I* through *IV*, on Archimedes' *Dimension of a Circle* and *On the Sphere and Cylinder* were important to their survival. He also wrote on the three classical problems. Apparently, in editing *Conics* I through IV, Eutocius corrected errors in variant copies. Further critical study might determine the extent of his contributions. From the mid-fifth century, however, the flame of mathematics otherwise became barely a flicker in the West in the commentaries of Proclus and Simplicius, who was an adversary of Philoponus, and the handbook of Boethius. Of these scholars, Proclus (d. 485), who headed Plato's Academy in Athens and had access to many works since lost, produced an extensive volume of writings. His *Catalogue of Geometers* remains an invaluable source for Greek geometry from Thales to Plato's circle, and he also gave what would become Playfair's version of Euclid's parallel postulate.[273] Before mathematics was again fruitfully pursued for its own sake in the West during the Renaissance, it survived at a low level in medieval Christendom, but experienced in medieval Islam a period of high achievement.

8

Mathematics in Traditional China from the Late Shang Dynasty to the Mid-Seventeenth Century

> The method of calculation is very simple to explain but has wide application. This is because "man has a wisdom of analogy," that is, after understanding a particular line of argument he can infer various kinds of similar reasoning .... Whoever can draw inferences about other cases from one instance and can generalize ... really knows how to calculate .... To be able to deduce and then generalize ... is the mark of an intelligent man.
> — *Arithmetical Classic of the Gnomon*

> [I]n this world there is something that exists beyond existence. Beyond existence are the natural numbers ....
> [N]atural reasoning [can be employed to]...understand [them] ....
> — *Sea Mirror of Circle Measurement*

In ancient China and India, mathematics long developed independently of Old Babylon, pharaonic Egypt, Seleucid Babylon, and classical Greece and Alexandria. India did not experience until the second century A.D. a major intrusion of mathematical and astronomical knowledge from Babylon and Greece,[274] and Chinese mathematics remained principally indigenous until the seventeenth century. Concentrating on algebra and giving little attention to geometric proofs, sages in China and India stood closer to Babylon than to Greece and Alexandria.

Mathematics in traditional China began in written form in the late Shang dynasty at least 1300 years before the Christian era. Natural barriers to outside contacts kept Chinese protomathematics practically independent of any foreign influence to the fourth century B.C. Responding chiefly to daily needs in commerce, calendar making, government record keeping, architecture, and surveying, ancient Chinese scholars centered their effort in computational arithmetic and mensurational geometry. Potsherds, oracle bones, and bronze script, which convey the few fragments on mathematics that survive from China before the Han dynasty that ruled from 206 B.C. to A.D. 220, confirm that Chinese masters had on their counting rods a decimal notation that had developed by Han times into a decimal place-value system without a zero. Two Han classics, *Zhoubi Suanjing* ("The Arithmetical Classic of the Gnomon and the Circular Paths of Heaven") and *Jiuzhang Suanshu* ("The Nine Chapters on the Mathematical Art"), are our major sources of ancient Chinese mathematics. The technical meaning of gnomon is the pointer on a sundial.

From A.D. 220 to the 1650s the main thrust in Chinese mathematics continued to be in computational arithmetic and algebra. The development was intermittent and often connected with astronomy. After Liu Hui revised and commented on *The Nine Chapters* and a system of Chinese mathematics appeared in the *Ten Books of Mathematical Classics* by the tenth century, arithmetical algebra blossomed from the eleventh through the middle fourteenth century. Using a triangular array of numbers now known as the Pascal triangle and iterated multiplication, Chinese algebraists extended root extraction methods to solve higher degree equations beyond cubics. Their algebra reached its pinnacle in the thirteenth-century work of Qin Jiushao, Li Zhi, Yang Hui, and Zhu Shijie. By the fifteenth century counting rods had evolved into the abacus. Theoretical geometry remained virtually unknown in China until the Jesuit Matteo Ricci and Xu Gaunqi translated into Chinese the first six books of Euclid's *Elements* in 1607. This was part of a broader transmission of western mathematics into China that included trigonometry and logarithms.

8.1   Ancient Chinese Civilization to 220: An Introduction

The origins of Chinese civilization date from the fifth millennium B.C. Its two major centers were the Yellow River and its tributaries in the north and the 3,200-mile Yangtze River to the south. Archaeological artifacts and scientific analysis of pollen grains preserved in peat have established the southern highlands of the Yellow River as the cradle of Chinese civilization. A temperate climate with warm moist summers and cold dry winters together with a fertile soil, loess, that retains scarce rainfall favored agriculture and encouraged permanent settlements around 4000 B.C. Early permanent Chinese farmers raised millet, buckwheat, and, to a lesser extent, rice, and they kept pigs and other small animals.

For millennia Chinese civilization was practically independent of foreign influences. China, like ancient Egypt, was isolated by formidable natural physical barriers—an ocean to the east; the Yunnan plateau, jungles, and the Tibetan massif to the south and west; and the Gobi desert to the north. Its nuclear area, the highland tributaries rather than the floodplain of the lower Yellow River, lacked the irrigation so important in the Fertile Crescent and contributed to its autochthony. Sea travel did not overcome its separation. Contacts with India and western Asia began in the thirteenth century B.C., but trading was limited until the Bactrian camel appeared in the fourth century B.C.

Chinese civilization is distinguished by its striking continuity of language, culture, and geography. A succession of conquering groups was absorbed into its traditional culture and learning, altering themselves in the process.

Chinese civilization begins with the Xia dynasty, lasting from approximately the twenty-first to the sixteenth century B.C., and the bronze-age Shang, from about the sixteenth to the eleventh century B.C. Shang China was a small area around its capital of Anyang. A reconstruction of its history is based on in-scriptions on oracle bones (steer shoulder bones or tortoise undershells) used for divination and on bronze script. Shang scribes invented a writing brush and carved on bamboo strips joined by two strips of cords to form scrolls. These were stored in boxes. Rapid disintegration of bamboo strips and a scarcity of durable writing materials may account for the absence of samples of Chinese writing before 1300 B.C. In the stratified Shang slave society, a warlike aristocracy controlled bronze weapons and some chieftains were independent of central control. Not far removed from primitive nature worship, Shang society practiced human sacrifice.

China's classical age covers the western and eastern Zhou dynasties, reigning between about 1050 and 256 B.C. The Zhou quasi-feudal society engaged in agriculture and war. Commerce with road networks rose, towns multiplied, and population grew. A new cavalry with crossbows replaced chariots, and Zhou forces extended their conquests to the Yangtze Valley so that Zhou lands were called the Middle Kingdom, a name still applied to China. Following recurrent wars, a line of great ethical thinkers appeared, including Confucius (551–479 B.C.), whose philosophy dominated later Chinese thought, and the shadowy Lao-Tzu (ca. fourth century B.C.), the founder of Taoism. Known through the collected sayings in his *Analects*, Confucius advocated universal education and taught loyalty, moderation, respect for elders, superiors, and the past, and the exercise of responsibility in government. The rivals of Confucius and his disciples, the Taoists, incorporated much of earlier magic, mystery, and protoscience. The Tao, or Way, is how the universe works, and Taoist speculations formed the basis of ancient Chinese science. By the late Zhou period, the twin pillars of Chinese society up to 1911, scholar officials in a bureaucratic government and peasant farmers, were essentially in place. During the eighth century B.C., work on the first Great Wall began and iron was introduced to China.

After Zhou authority disintegrated during the Warring States period from 475 B.C., strong Qin rulers in the northeast triumphed in 256 B.C. The Qin dynasty ruthlessly curtailed the power of the old nobility and developed large-scale public works, such as irrigation projects. With iron weapons and severe military codes, they had unified China briefly by 221 B.C. The Qin tyrant Shih Huang Ti standardized the characters in Chinese writing and established uniform weights and measures for trade. He also attacked the education base of the old bureaucracy and thought scholars dangerous to the regime. Confucian teaching was banned, and all books, except technical manuals, which include those on agriculture, were ordered burned. Only one copy of Confucian books was preserved in the Imperial Archives. Few books escaped the pyre. Overextending itself in reshaping China, the Qin dynasty collapsed a generation after unification.

The Han dynasty, extending from 206 B.C. to A.D. 220, reunified China, extending its authority over much of Central Asia, and developed a centralized government with a large civil service to deal with a population that by A.D. 2 had reached 60 million people. The early Han carefully avoided the excesses of the Qin. Books were reconstructed from fragments and schools reopened. Education, not birth, became the chief qualification for entrance into the imperial bureaucracy, and Confucianism was gradually accepted as the official ideology. Taxes in the form of money, a portion of crops, or corvée labor supported the government. In corvée labor, each farmer spent about a month each year working for the government free on canals, roads, or palaces and occasionally providing military service.

Han China experienced commercial prosperity and extensive cultural growth. By the first century A.D., the silk trade and cultural contacts were stretching to Iran, Alexandria, and Rome in the West. Characters for writing have barely changed from those of Han scribes. The two classics of ancient Chinese mathematics, *The Arithmetical Classic of the Gnomon* and the *Nine Chapters on the Mathematical Art*, were now completed.

8.2   Numeration and Major Literary Sources

The beginnings of decimal notation in China are extremely ancient. The University of Chicago historian Ping-ti Ho cites isolated symbols from Yangshao pottery of about 5000 B.C., which are perhaps the earliest known numerical symbols. He detects numerals for four of the first nine digits:

Yangshao potsherds	I	II			X		+			
Indo-Arabic	1	2	3	4	5	6	7	8	9	10

The first written record of decimal numerals from China comes much later, however, from late Shang oracle bone script. Oracle bones were tortoise undershells or came from large animals. On them Chinese scribes carved such information as numbers of prisoners captured or killed in battles or birds and animals acquired on hunts. The following table gives the decimal numerals included among the roughly five thousand characters used in late Shang Chinese writing.

Late Shang numbers read from left to right or bottom to top. In oracle bone script, the number 2656 is ⪕⬡⍍⋂, or $2000 + 600 + 50 + 6$. The values of larger numerals not powers or multiples of 10 are thus determined by the additive, but not yet the multiplicative, principle. Both principles underlay place value. From late Shang to Han times, Chinese decimal numerals evolved slowly and a place-value system without a 0 appeared. As this table of digits from each of these eras indicates, numerals changed slightly and stabilized in form:

Shang Oracle Bone (1300–1000 B.C.)
Bronze and late Zhou (500–250 B.C.)
Han and Jin (150 B.C.–A.D. 316)

The Han period also had two series of counting-rod numerals:

The number of separate numerals had substantially lessened. Oracle-bone numbers, which range from 1 to 30,000, include ideograms not only for units and 10. They also have compound words consisting of three characters for powers of 10 through 10,000 and for multiples of 10 and its powers to 30,000.

In the Han counting-rod numerals, note that the first series begins with vertical strokes and the second with horizontal strokes. In the first series the horizontal bar represents 5 in numerals 6 through 9, while in the second series the vertical bar does. As will be seen, these contrasting features were required for working on counting boards. If the strokes were not properly spaced, there could be confusion. With the vertical Han series, certain numbers, such as 12— I II —could be confused with 3 or 21, 32 with 5 or 41, and so forth.

The chief method for calculation in ancient China was by manipulating counting rods, called Chou, small bamboo, wood, or ivory rods configured to represent numerals—Chou suan. No later than the Warring States period, the Chinese were familiar with computational techniques of counting rods. The unearthing of counting rods in 1978 by archaeological excavations on the Warring States and Han periods and references to them in the sacrificial and archery sections of *Zhou Rites* confirm this. In the history of computation, Chou are distinctive to Chinese civilization. By contrast, ancient Mesopotamians computed directly with numerals on clay tablets, and Egyptians on papyrus, while both Indian and Islamic calculators utilized sand or dust boards.

Counting rods were placed in a top, middle, and bottom row on a board that stabilized the system. That the rows not be confused, units in the first row were vertical, the second row horizontal, and the third row vertical strokes. Columns represented powers of ten from left to right: 1 10, 10^2, 10^3, ... Techniques for manipulating rods in performing arithmetical operations column by column were developed that put carry overs into the next column. To subtract 789 from 1245, first put out 1245 and then subtract 7 from the hundreds' place, 8 from the tens', and 9 from the units' place:

$$\left\{ \begin{array}{cccc} 1 & 2 & 4 & 5 \\ & 7 & 8 & 9 \end{array} \right. - \left\{ \begin{array}{cccc} 1 & 2 & 4 & 5 \\ & 7 & & \end{array} \right. - \left\{ \begin{array}{ccc} 5 & 4 & 5 \\ 8 & & \end{array} \right. - \left\{ \begin{array}{ccc} 4 & 6 & 5 \\ & & 9 \end{array} \right.$$

─ II ⩴ ⫪	⪉⫫ ☰ IOI	⪉⫫ ⊥ ◨◫	⪉⫫ ☰ T
	5 4 5	4 6 5	4 5 6

II ≐ ⫫⫫

For China before the Han, the *Book of Crafts* and other chapters from the *Record and Rites of the Zhou Dynasty* and the *Book of Master Mo* seem to represent well the sources and nature of mathematical knowledge. Both texts were probably written during the Warring States period.

While late Shang and Zhou art, clothing, jewelry, and pottery would demonstrate a secondary aesthetic influence, a fondness for patterns in illustrative geometry, the *Book of Crafts* treats the practical tasks of making walled cities along with canals, carriages, boats, bows, and arrows as the major reasons for mensurational geometry. It presents information on fractions, angle measurement, and uniform measures. Its equivalent of inches is divided into tenths (*mei*

or later *fen*), depicted as the commonest fractions. Angle measures include 90°
and increments of a fourth of that value. Surveyors had probably worked with
right angles and right triangles from at least the sixth century B.C. Arcs of circles are measured by the number of bows (three, five, seven, and nine) needed
to make the circumference of a circle. Among the standard measures are a
square and cube of their foot.

Education as the *Zhou Rites* discusses it has similarities to Castiglione's
schooling for the Italian Renaissance courtier. Children of the Zhou nobility
must study the "six gentlemanly arts": ritual, music, archery, horsemanship,
calligraphy, and mathematics. At the age of six, children learn numbers, counting, and directions. At nine, they learn days and dates from a lunar calendar
scheme based on ten heavenly stems and twelve earthly branches that repeats
after sixty pairs. Sixty is the least common multiple of ten and twelve. This lunar calendar work thus provided a sexagesimal recording system to complement
decimal numeration.

Book II of *Zhou Rites* reveals the development of two groups of government
officials, the *Sihuai*, who were trained in arithmetical calculation for compiling
statistical records, and *Chouren*, who studied astronomy and made star charts
and calendric computations; Book II also suggests problems for mathematical
innovation. The *Sihuai* had to keep records of all state properties, expendi-
tures, and taxes, compile the census, make maps, and maintain all records and
documents in archives. The army had parallel officials with a rank called *Fa-
suan* to keep military statistics. By the Han Dynasty, the *Sihuai* and *Chouren*
occasionally fell from favor and were dispersed by the central government. Since
the *Chouren* were hereditary, mathematical work suffered in generations having
little mathematical aptitude or interest.

The *Book of Master Mo* planted seeds for theoretical geometry in ancient
China. It begins with definitions of basic terms and statements about different
concepts linked by logic, mathematics, or physics. Among the definitions are:
flat: same height, same length; match up exactly, center; point of same length;
and circle: one center with same length. Master Mo further defines a point,
line, surface and solid, and parts of solids. Paradoxes range from self-negating
verbalisms, such as "a white horse is not a horse," to provocative parallels of
Zeno. Debater Gongung Long states: "a ... stick, though half of it is taken
away each day, cannot be exhausted in ten thousand generations." This is
resonant with the series

$$1/2 + 1/2^2 + 1/2^3 + \ldots + 1/2^n + \ldots = 1.$$

Modern Chinese mathematicians often use this "taking half daily" to help explain the notion of limit.

The disciples of Master Mo, the Mohists, who preached universal love and
drew abstract conclusions from physical particulars, were overwhelmed during
the Warring States period, and Taoist naturalists accepted only segments of
their thought. In Chinese literary culture, Confucians applied rational methods chiefly to social equations, while Taoists preferred mysticism to theoretical

exploration. Neither group passed the threshold to theoretical geometry.

The two Han landmark works, *Zhoubi suanjing* and *Jiuzhang suanshu* (hereafter *Nine Chapters*), are extremely valuable sources that provide much of our knowledge of ancient Chinese mathematics. The older Han classic *Zhoubi* has sections on mathematics, while *Nine Chapters*, a specialized writing, is the more important and most representative classic in ancient Chinese mathematics. Largely because of the Qin burning of books in 213 B.C., the two were until recently the oldest known Chinese writings on mathematics. Recent excavations of the Zhangjiakou Han tombs have unearthed two earlier Chinese writings on the subject, *A Book on Arithmetic*, placeable at some time between 200 and 150 B.C., and the *Law of Fair Taxes*, dating from before 100 B.C. Both are written on bamboo strips. The *Nine Chapters* incorporates materials similar to those in *A Book on Arithmetic*, which computes with fractions, calculates field areas and fair taxes, and apportions rice. Neither *Zhoubi* nor the *Nine Chapters* has as yet been conclusively dated.

Completed shortly before A.D. 100 during a period of rapid improvement in agriculture, science, and technology, the extant version of *Zhoubi* must include materials from the Zhou, Warring States, and Qin eras and refers to them. The title of the text and a dialogue with the Duke of Zhou near its beginning have led to the mistaken notion that *Zhoubi* is a Zhou text. The absence of any mention of it in the astronomy manual *Huai Nan Zi*, completed near the end of the second century B.C., and the popularity of theories from the Qin period in both suggest that *Zhoubi* and perhaps a preceding text with much of its content were put into book form between 100 B.C. and A.D. 100. The oldest surviving edition was published shortly after 1213 and is housed in the Shanghai Library.

According to Joseph Needham, the title *Zhoubi* means Zhou shadow gauge. It can also mean perimeter. If Needham's translation is correct, which the book's content supports, the title comes from Chinese observers of the heavens using shadow gauges, which were upright gnomons or verticals of sundials. From their capital city of Zhou, these observers precisely noted the passage of the sun by shadow lengths and their directions. Shadow lengths on these gauges supplied basic data for time and calendric calculations, and upright gauges were also used for surveying.

Primarily an astronomical text, *Zhoubi* covers shadow reckoning, calculations of the annual movement of the sun, and observations of the moon, pole star, and other stars. It surveys these stars and the land using the Gougu or what is called in the West the Pythagorean theorem. *Zhoubi* is the representative work on Gai Tian theory, the first of three schools of astronomy in the Han dynasty. The second school is Xuan Ye and the third Hun Tian. Gai Tian theory considers Earth to be shaped "like an upturned basin" bounded by four seas, with the sky like an umbrella hat or concentric hemisphere, while the more progressive Hun Tian theory posits a rotating sphere centered on Earth, as in Eudoxean astronomy.[275] From Gai Tian theory, the phrase "within the four seas" became a synonym for China. The astronomy and calendrics in *Zhoubi* drew on substantial mathematics and probably furthered it.

The beginning section of *Zhoubi* discusses the art of numbering and a special case of the Gougu theorem. The Gougu theorem arises from shadow gauges, in which the vertical gauge is Gu, the shadow cast by the sun Gou, and the hypotenuse Xian, so that $Gu^2 + Gou^2 = Xian^2$. The Gougu theorem is stated and "proved" by the "piling up of rectangles." The text begins with a 3, 4 rectangle and draws a square on the diagonal 5. The square on the diagonal is circumscribed by 3, 4, 5 triangles to form a square of side 7. Subtracting the four outer 3, 4, 5 triangles gives $49 - 24 = 25$, the area of the original square. The proof is thus a computational check. No Euclidean proof based on axioms, postulates, and deduction is given.

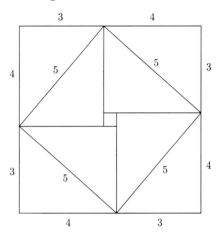

In the later sections of *Zhoubi*, the Gougu theorem is applied in general, and not just for special cases. The text states that the legendary Emperor Yu "quells floods, he deepens rivers and streams, surveys high and low places ... [by using] the Gougu theorem." In Gai Tian theory, Chinese masters calculated the distance of the sun from the Earth. Following an approach similar to that of the Alexandrian Eratosthenes in computing the circumference of Earth, they found (if their measures be rendered into our nearest equivalents) that at summer solstice an eight-foot gnomon at noon casts no shadow at site A, while an equal but imaginary gnomon 60,000 miles south casts a 6 foot shadow. From similar right triangles, they concluded that the height of the sun or leg of the larger right triangle is 80,000 miles, and the hypotenuse or distance of the sun from the observer at site B must be $\sqrt{60{,}000^2 + 80{,}000^2} = 100{,}000$. Mathematically, the computation is correct, but the assumptions that the shadow disappears over 60,000 miles and that Earth has a flat surface were wrong. Apparently, the height of the gnomon and the location of the imaginary site B were so selected as to make the solar distance equal to 100,000 units. This work and accurate land measurements using two similar right triangles show that Chinese sages early mastered the Gougu theorem.

The *Zhoubi* also has relatively complicated calendric computations, which involve fractions. Using counting rods, Chinese masters determined that the

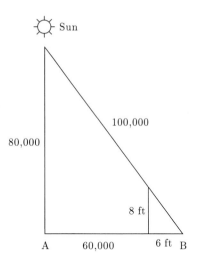

length of the year is $365\frac{1}{4}$ days and that each nineteen year cycle has to add seven lunar months. Each year, then, averages 12 and $\frac{7}{19}$ lunar months, and each lunar month $365\frac{1}{4} - 12\frac{7}{19} = 29\frac{499}{940}$ days. Starting with this result, the masters computed the positions of the moon.

The shadow length section of the *Zhoubi* gives directions for multiplying and dividing with decimal fractions and approximates square roots, but does not give the process. In the second part of *Zhoubi*, Master Chen lectures on the nature, methods, and teaching of mathematics. Master Chen asserts that its wide application comes from the ability through analogy and reasoning to infer from special cases and generalize. He is thus the first known sage in China to refer to the abstract character of mathematics and emphasize training in deductive and inductive thought for learning it.

After the Qin tyrant burned all books in 213 B.C., Zhang Cang (d. 152 B.C.) and the Han Minister of Agriculture Geng Shouchang (fl. 73–49 B.C.) recovered the remnants of prior works on "calculation of nines" and filled gaps. Since *Nine Chapters* is not listed in the arts and crafts section of the *History of the Han Dynasty*, these materials were probably not put into book form before the end of the first century B.C. The work of Zhang Cang and Geng Shoucang was apparently reedited and augmented before the scholar Zheng Zhong (fl. 150) put the book into essentially its present form. The preface by Liu Hui (fl. 250) as well as recent archaeological excavations indicate that this Han classic, with its sections on computational arithmetic, mensurational geometry, and algebra, was the culmination of the gradual development of mathematics from the Zhou to the Qin dynasty.

Like the work of A'hmosè in ancient Egypt, *Nine Chapters* is a practical handbook that is not arranged in a systematic, logical fashion. It was meant to assist in the daily tasks of architects, astronomers, civil and military engineers, imperial and local officials, surveyors, and tradesmen. The *Nine Chapters* contains 246 problems in question and answer form. That form, perhaps earlier and certainly later, was standard in ancient Chinese mathematical writings. Each chapter opens with a rule for solving one or more problems. The titles and subject of the chapters are:

1. *Fang tian*, "Field Measurement." Rules for calculating how many square units fields with the shape of rectangles, triangles, trapezoids, circles, and annuli contain; also arithmetical operations with fractions.

2. *Sumi,* "Cereals." Refers to millet and rice; proportions, especially in exchanging cereals.

3. *Cui fen,* "Distribution by Proportion." Proportional taxation and distribution of properties.

4. *Shao gong,* "What Width?" Given an area or volume, to find the length of a side or diameter; also a trial-and-error method for extracting square and cube roots.

5. *Shang gong,* "Construction Consultations." Rules for computing exact and approximate volumes of prisms, pyramids, cylinders, circular cones, and tetrahedrons.

6. *Jun shu,* "Fair Taxes." How to allocate the amounts of grain and corvée labor that taxpayers owe on the basis of the size of populations in principalities and provinces and distances from the capital over which grain must be delivered.

7. *Ying buzu,* "Excess or Deficiency." How properly to distribute payments by linear equations using the rule of false position and how to determine phases of the moon.

8. *Fang chen,* "Rectangular Arrays." How to solve simultaneous linear equations and calculate with positive and negative numbers.

9. *Gougu,* "Right Angles." The Pythagorean theorem and introduction of general methods for solving quadratic equations.

A principal contribution of *Nine Chapters* is its sophisticated treatment of fractions. Though the *Zhoubi* discusses complicated computations with fractions, the initial eighteen problems of *Nine Chapters* were the first among surviving eastern texts systematically to apply to fractions the four arithmetical operations. Using counting rods, scribes found the largest common factor of a numerator and denominator by a method of successive subtraction of the smaller number from the larger, not unlike Euclid's algorithm. Once the lowest common denominator is known, it is possible to add or subtract fractions. For example, $1/2 + 1/3 + 1/4 + 1/5 = 30/60 + 20/60 + 15/60 + 12/60 = 77/60 = 1 + 17/60$. To multiply fractions, *Nine Chapters* multiplies numerators and denominators as we do today. But it divides fractions, such as $b/a \div d/c$, by finding a common denominator and dividing, so that $b/a \div d/c = bc/ac \div da/ac = bc/ad$, which amounts to cross multiplying.

"Cereals," the first of four chapters that pursue a variety of problems on proportions, begins with a problem that assumes an exchange rate of fifty measures

of millet for thirty measures of unpolished rice and for twenty-seven measures
of polished rice, and it then determines how much unpolished rice to get for
one dou of millet. The author multiplies the given quantity of one dou by the
exchange rate thirty for rice and divides the result by the exchange rate of fifty
for millet. This amounts to cross multiplying, where $50/30 = 1$ dou/x or 30
$dou = 50x$ or $x = (30$ $dou)/50$. "Distribution by Proportion" divides game
killed in the ratio of $5 : 4 : 3 : 2 : 1$. Proportion was also crucial for calculating taxes and allocating corvée labor. The chapter "Fair Taxes" introduces
compound proportions. For Emperor Wu, who reigned from 140 to 87 B.C.,
Sang Hongyang had developed a tax system for collecting grains or an equivalent payment that was directly proportional to the size of the population and
inversely proportional to the distance from the capital.

To cover its two hypotheses, "Excess or Deficiency" introduces a feature
later known in the Latin West as the method of double false position. Ancient
Egyptian scribal mathematicians had invented this problem-solving technique
by guessing a value for an unknown quantity in a linear equation and adjusting
it upward or downward according to how the answer it gives compares with the
original value sought. The Chinese apparently independently invented this rule
and employed it to solve both determinate and indeterminate linear equations.
In the late Middle Ages, Arab mathematicians were to transmit to Europe
what Latin terms the *"regula falsae positionae."* The thirteenth-century Italian
mathematician Fibonacci is credited with introducing the West to this method,
which he called the method of *elchataym*. This name probably derives from the
Arabic name for China or Cathay, *al-khata'ain*.

The *Nine Chapters* is also generally impressive in mensurational geometry
but juxtaposes accurate and inaccurate results. Most geometric calculations
appear in the first chapter, on field measurement. In modern notation, the
scribe calculates a square field area as $S = a^2$, where S is the area and a the
side, a rectangular field as $S = ab$, a triangular field as $S = \frac{1}{2}ab$, and a planting
field or trapezoid as $S = \frac{1}{2}(a+b)h$. The area of a circular field is $S = P/2 \times D/2$,
where P is the perimeter and D the diameter. The volume of a pyramid (for
a roof or square cone) is $V = a^2h/3$, where a is the side of the base and h the
height, and that of a frustum or square pavilion is $V = h(a^2 + b^2 + ab)/3$, where
a is the side of the upper square and b the side of the lower square. In the
planning of a circular fort, the volume of a cylinder is $V = P^2h/12$, where P is
the circumference of the circle. If π is 3, as Chapter Five assumes, this formula
equals the modern $V = \pi r^2 h$. The volume of a circular cone is $V = P^2h/36$,
with P the circumference of the base circle.

The principal innovation in algebra in the *Nine Chapters* is its orderly
method for solving a system of linear equations. This method seems to reflect
among Chinese sages a profound interest in patterns. It appears in eighteen
problems in Chapter Eight, "Rectangular Arrays." Eight of these problems have
two unknowns, six have three unknowns, two have four unknowns, one has five
unknowns, and one indeterminate system of five equations has six unknowns.

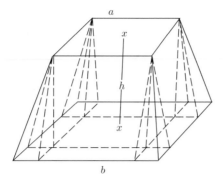

Patchwork method for volumes.

The term rectangular array refers to how counting rods are arranged.

The first problem in Chapter Eight deals with mixing three grades of rice by volume in what amounts to these equations:

$$3x + 2y + \;\; z = 39,$$
$$2x + 3y + \;\; z = 34,$$
$$x + 2y + 3z = 26.$$

While counting rods list each equation in a vertical column, the modern horizontal equivalent listing the coefficients is

$$\begin{matrix} 3 & 2 & 1 & 39, \\ 2 & 3 & 1 & 34, \\ 1 & 2 & 3 & 26, \end{matrix}$$

The goal is to reduce several coefficients in this matrix to zero. Chinese scribes took the first coefficient of 3 and multiplied it times the second row to obtain

$$\begin{matrix} 3 & 2 & 1 & 39, \\ 6 & 9 & 3 & 102, \\ 1 & 2 & 3 & 26, \end{matrix}$$

and then subtracted twice the first row from the second to get

$$\begin{matrix} 3 & 2 & 1 & 39, \\ 0 & 5 & 1 & 24, \\ 1 & 2 & 3 & 26. \end{matrix}$$

Next, the third row is multiplied by 3 and the first row subtracted from it, which gives

$$\begin{matrix} 3 & 2 & 1 & 39, \\ 0 & 5 & 1 & 24, \\ 0 & 4 & 8 & 39. \end{matrix}$$

The coefficient 5 from the middle row is then multiplied by the third row and four times the second row is subtracted from it, with the result

$$\begin{array}{cccc} 3 & 2 & 1 & 39, \\ 0 & 5 & 1 & 24, \\ 0 & 0 & 36 & 99. \end{array}$$

The solutions are now at hand. The third row yields $36z = 99$ or $z = 2\frac{3}{4}$; the middle row, $5y + z = 24$ or $5y + 2\frac{3}{4} = 24$ or $5y = 21\frac{1}{4}$ and $y = 4\frac{1}{4}$. Similarly $x = 9\frac{1}{4}$. This method, which is used in modern linear algebra, was not known in the Latin West until the sixteenth century.

The *Nine Chapters* introduces the method of positive and negative to add and subtract positive and negative numbers. In treating finances, Chapter Eight associates positive numbers with balances and receiving and attaches the character *zheng*, while negative numbers are associated with deficits or paying and have the character *fu* attached. In astronomy the ancient Chinese spoke of strong and weak numbers. Problem three in Chapter Eight gives these correct rules:

> for subtracting—same signs take away, different signs add together, positives from nothing makes negative, negative from nothing makes positive; for adding—different signs take away, same signs add together, positive and nothing is positive, or nothing and negative makes negative.[276]

The Chinese had red or triangular-section counting rods to represent positive numbers and black for negative numbers. Almost a millennium before the Indians and Arabs, they possessed negative numbers and correct rules for adding and subtracting with them. Correct rules for multiplying and dividing with them awaited the thirteenth century in China.

Using counting rods, Chinese scribes extracted square roots essentially as moderns do. Consider the number 55,225. The Chinese assumed that $\sqrt{N} = a + b + c$ and that the last two digits on the right give the units' place, the next two digits on the left the tens' place, and the next two digits on the left the hundreds' place. Begin with the number divided into components as $5'52'25$, and find a square just smaller than 5. The number $2 = a$, or the square root of 4, goes in the hundred's place. Double a to get 4 and subtract this from 5. The number 4 is the tens' digit to divide into 152. The divisor has to be 43, since 43 times 3 equals 129, which is close to 152. So 3 is the tens' digit needed in the answer The divisor for 2325 is $46x$. Here x must equal 5, because 465 times 5 equals 2325. The square root is 235. In shifting counting rods, scribes essentially followed these steps.

$$\begin{array}{l} \quad\quad\quad 2\ 3\ 5 \\ \quad\quad\quad \overline{5'\ 52'\ 25'} \\ \quad\quad\quad 4 \\ 43\quad\quad \overline{|1\ 5\ 2} \\ \quad\quad\quad 1\ 2\ 9 \\ 46x\quad\quad \overline{|2\ 3\ 2\ 5} \\ x = 5\quad\ \ \underline{2\ 3\ 2\ 5} \\ \quad\quad\quad\quad\ \ 0 \end{array}$$

As a corollary to their method of extracting square roots, the Chinese devised a general method for solving quadratic equations. Each arrangement of their counting rods in extracting square roots from the lowest to the uppermost row may be treated as coefficients in a quadratic equation.

8.3 Civilization in Imperial China, 220–1644: An Introduction

After the Han empire collapsed in 220, China experienced another three and a half centuries of internal disorders, barbarian invasions, and political fragmentation. Buddhism, which central Asian missionaries had brought to China in the first century A.D., spread rapidly, as did Christianity in the West. Buddhism looked to nirvana, a state of release from suffering to be attained by a disciplined mind and ethics, and its monasteries practiced devotion to ancestry. More capable than Taoism of spreading its teachings, Buddhism incurred occasional persecution by Taoist emperors, but by the sixth century was established across China. Massive migrations of refugees from the north greatly increased the population to the south and forced the swift development of rice agriculture in the Yangtze plain. Rice now replaced wheat and millet as China's main foodstuff.

The times of disorder ended when the Sui dynasty, ruling from 598 to 618, restored the Chinese empire. Sui rulers revived the Han centralized bureaucratic government, basing appointments on examinations, and built a grand canal linking the rice-growing Yangtze basin to north China. The resentment of over 5 million peasants forced to build this canal, however, and near bankruptcy after failures to conquer northern Korea and parts of central Asia led to the overthrow of the Sui. They were succeeded by the cosmopolitan T'ang dynasty, which lasted until 907.

The empire of the early T'ang rulers expanded and contracted much like an accordion. A bureaucracy of merit built on a public examination system favoring the aristocracy was the basis for T'ang political effectiveness. An Imperial Academy, founded in 754, prepared scholars in Confucian classics for public service. By the eighth century, the T'ang capital of Ch'ang-an had a population of almost 2 million, making it the largest city on the planet. From this center of trade, caravans with Bactrian camels went west to India and the Near East. Under T'ang rulers, who were regarded as "sons of Heaven," Chinese society flourished.

The T'ang economy rested on a system of land tenure and monopolies of basic industries. The land tenure system was designed to curb the growth of large estates and provide peasants with land. New crops, such as tea and "wet rice," and improvement in transportation of foodstuffs stimulated agricultural productivity. The 650-mile Grand Canal was completed with post houses, granaries, and taverns. The canal provided an important means of transportation between north and south. The carefully regulated T'ang economy was based chiefly on monopolies in the new papermaking industry and in ironmaking, pottery, and silk manufacture, which grew to meet the demands of foreign trade. Chinese culture and scholarship meanwhile flowered anew.

T'ang culture adopted music and musical instruments encountered in trade with central Asia. For the first time, sculpture became a major medium in China. Buddhists made statues of gods, and secular sculptors made lifelike statues of dancing girls, singers, horses, ducks, deer, camels, and other animals. Chinese ceramicists made the first true porcelains.

The invention of block printing by Chinese artisans about 600, fostered probably by the demand for more widespread circulation of Buddhist scriptures, or *sutras*, Taoist prayers and secular calendars, made for a proliferation of writing. Carved wooden block printing satisfactorily reproduced the calligraphy and ideographs of the Chinese language, which requires thousands of characters. In the first century A.D., the Chinese had discovered how to make paper, so they had a good medium on which to print.

T'ang literature centered in historical research and poetry and gave attention to Taoist studies in alchemy and cartography. Within the reestablished scholarly and bureaucratic complex, historical knowledge and training had the practical advantage of preparing men for civil service examinations. Early T'ang emperors had court historians prepare comprehensive institutional histories. "By using antiquity as a mirror," one claimed, "you may learn to foresee the rise and fall of empires." Such works as *The Understanding of History*, produced in the eighth century, emphasized the critical analysis of materials as well as narration of events. T'ang poets appreciated life's impermanence and ironic humor. Of the more than two thousand, perhaps the greatest was Li Po (701–762), who responded to the instability of his world with poetry that is powerful, passionate, and sensitive to beauty:

> Far and wide the Tartar troops were spreading,
> and flowing blood mired the wild grasses
> Where wolves and jackals all wore officials' caps.[277]

After the mid-eighth century, corruption spread in the T'ang government and wealthy landowners evaded taxes, which placed increasing tax burdens on the peasants. To gain lost revenues, the government confiscated Buddhist monasteries, which numbered about 4,600. After halting T'ang expansion, attacks by Turks and Mongols now posed serious external dangers. In 907, military commanders deposed the last T'ang puppet emperor.

Following another half-century of disunity, the Song dynasty began in 960 a rule that was to last until 1279. It established an autocracy, which marks a fundamental transition to late imperial China. Under these rules, agriculture, the economy, and technology advanced. Early-ripening rice doubled agricultural production and, for the first time, tea and cotton were widely cultivated. Manufacturing activity matched that of agriculture: pottery and porcelain making expanded, the coal and iron smelting industry provided improved tools and weapons, and foreign trade surged, especially in silks. A commercial revolution attended the development of banking and paper and metal currency. Cities of more than 100,000 quadrupled. Song inventions included the magnetic compass, the sternpost rudder, and a form of gunpowder for use in celebrations, rockets and grenades.

Under Genghis Khan (1167-1227), whose original name was Temujin, nomadic Mongols located to the north of China invaded and overthrew the Song there. The Mongols wanted China's riches. Genghis' grandson, Kublai Khan, moved his capital to Beijing in 1264 and adopted in 1271 the Chinese name Yuan for his new dynasty that was to rule until 1368. He proceeded to defeat the Song in southern China. Kublai Khan hosted the Venetian merchant Marco Polo from 1275 to 1292 along with other Westerners and was tolerant in religion, inviting Nestorian Christians and Tibetan Buddhists to China. Despite the rapacious, minority rule of Yuan warlords, Chinese science advanced and ancient Chinese mathematics now reached its zenith.

In their turn, the Yuan fell to the Ming dynasty, which reigned between 1368 and 1644. A reduction in autocratic controls and a new prominence of the scholarly gentry liberated cultural energy. Expanded irrigation, fertilizers, higher yielding strains of rice, and new crops such as corn, sweet potatoes, and peanuts from America enriched agriculture. An increase in food supply contributed to pushing the Chinese population from roughly 60 million in 1368 to about 125 million in 1644. Late in the Ming period trade with Europe grew and Western missionaries, especially the Jesuits, gained admittance to China. The Jesuits studied Chinese and Confucian classics. This preparation, along with the receptivity of prominent Chinese officials to western mathematical astronomy, calendrics, geography, and firearms, earned the Jesuits entry to the Imperial Court and appointment to the State Observatory in Beijing.

8.4   Mathematics in China, 220–1653

Although well integrated into Chinese learning from the time of the *Nine Chapters*, mathematics from the fall of the Han to the tenth century developed only occasionally. The two foremost mathematicians before 500 were Liu Hui (fl. 260), author of *Commentary on the Nine Chapters* and *Sea Island Mathematical Manual*, and the engineer and astronomer Zu Chongzhi (429–500). To improve astronomical and calendric computations, Chinese sages from the sixth to the tenth century developed an algebraic method of interpolating second differences, and the *Ten Books of Mathematical Classics* used at the Imperial Academy shaped a complete system of Chinese mathematics.

We know essentially nothing about the life of Liu Hui, whose *Commentary on the Nine Chapters* reworks that text into a comprehensive form that endures to the present and shows him to be an original mathematician. The *Nine Chapters* had little explanation of general methods for calculation. Liu Hui provides principles, creates diagrams and geometric models, and systematically applies them. All this comprises a brief form of proof of the accuracy of calculations. The *Commentary* solves linear equations with the rule of double false position, Diophantine equations with the matrix method, and higher-degree equations following the *Nine Chapters'* square root method. Liu Hui also gives in problem six of the "What Width?" chapter a decimal place-value notation and provides an early notion of limit in computing the value of π. The most original work of Liu Hui in solving problems was probably in computing the value of π and volumes of solid objects.

Approximations of π had a long history in China. Chinese masters had no single term to indicate π, but spoke of the ratio of the circumference of a circle to its diameter. They early assumed that the area of the circle is $\frac{3}{4}d^2$ or $\frac{1}{12}c^2$. These rules give their ancient value of π as 3, but from the first Christian century Chinese sages persistently searched for a more accurate value. In constructing a bronze vessel, Liu Xin used $92/29$ or $\pi = 3.1547$, and in finding the volume of a sphere, Zhang Hang (78–139) had $\pi = \sqrt{10} = 3.16$, and Wang Fan found $\pi = 142/45 = 3.1556$.

In discussing problem 32 of the "Field Measurement" chapter, Liu Hui introduces his "method of circle division" to calculate π. He first shows that the traditional value of 3 for the ratio gives a regular twelve-sided figure (dodecagon), not a circle. He then inscribes a hexagon in a circle and proceeds to double the number of inscribed sides up to a regular 96-gon and a 192-gon. He does not circumscribe regular figures about the circle to obtain upper as well as lower limits, as had Archimedes. Apparently, then, he was not influenced by that geometer. By repeatedly applying the Gougu theorem, he computes the sides of regular $2n$-gons from 24 to 96. He arrives at the verbal equivalent of the general formula

$$l_{4n} = \sqrt{\{r - \sqrt{\{r^2 - (l_{2n}/2)^2\}^2} + (l_{2n}/2)^2}.$$

Letting the circle radius equal one Chinese foot renders Liu Hui's calculation as $a(s_{192}) = 3.14$ and $64/625$ square feet. This gives $\pi = 3.141024$. For practical purposes, he lets π equal 3.14 or $157/50$.

Increasing the sides of inscribed regular polygons by $2n$, Liu Hui recognized, means a sequence of ever finer cuts or "short arrows," $1/2$, $1/4$, $1/8$, ... In problem 16 of the "What Width?" chapter he makes extremely fine cuts, so that the amount left over in the figure "is not worth mentioning." Here and in computing areas of irregular figures, Liu Hui analyzes early infinitesimals and a notion of limit that derives from them, two components essential to modern higher mathematics.

Using the Gougu theorem and a patchwork rearrangement of composite solids, Liu Hui inventively computed the volumes of the frustum of a pyramid and a tetrahedron, and he searched for an accurate computation of the volume of a sphere. His rearrangement of composite solids is similar to his rotating and translating of figures in computing irregular areas. Problem 10 of "Construction Consultations" divides the frustum of a pyramid into a middle cube, four prisms at the sides, and four pyramids at the corners to prove that $V = \frac{1}{3}(ab + a^2 + b^2)$, which is correct. Liu Hui also finds that a special tetrahedron with two opposite perpendicular edges is one-sixth of a cube with those edges. Problem 11 of "Construction Consultations" calculates the ratio of the volume of a circular pavilion to that of a circumscribing square pavilion as $3\pi : 4$, and it attempts to find the ratio of the volume of a sphere to that of its circumscribing cylinder. Liu's approach is correct, but he at first mistakenly believed that the ratio was that of the area of the circular cross section to the area of its circumscribing square. He soon recognized his error and awaited another "capable man to solve" this problem.

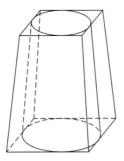

Finding the volume of a circular pavilion.

The capable man appeared two centuries later in the brilliant engineer and mathematical astronomer Zu Chongzhi, who in later life worked with his son Zu Geng. Apparently, the Zu family had studied astronomy and calendrics for generations before Chongzhi, who studied works of Zhang Heng and Liu Hui in his youth and corrected mistakes in them. In 462 he constructed an improved calendar that court officials opposed. They insisted that the calendar was the product of thousands of generations of sages and not to be amended by one man. They accused Chongzhi of blaspheming against the Confucian classics. It was a moment in the conflict between dogma and the corrosive restlessness of science. Progress in astronomy and calendrics, Chongzhi declares in his "Riposte," comes "not from spirits or from ghosts," but from careful observations and accurate numerical computations. People must be "willing to hear and look at proofs in order to examine truth and facts." Still Chongzhi's opponents blocked the acceptance of his calendar until after his death. His engineering, in which he improved the gear mechanism for chariots, did not face the same trouble; nor did his mathematics. He also wrote ten books in a genre that in the eighteenth-century West was to become the novel. Chongzhi's computation of π accurate to seven decimal places and his formula for the volume of a sphere made him the worthy successor of Liu Hui. His work is the most notable exception to the largely static Chinese mathematics in the first centuries after the Han.

Proceeding from Liu Hui's method of circle division for computing π and subsequent ratios, Chongzhi was dissatisfied that no accurate upper and lower bounds existed in China for the value of π. In accordance to custom, he first used two fractions for π, the approximate ratio of 22/7, which is accurate to two decimal places, and found a closer ratio of 355/113 or 3.1415929, accurate to six. Possibly he obtained the last figure by realizing that $3 < \pi < 22/7$ and setting the equality $\pi = (22x + 3y)(7x + y) = 3.1415929$, where $x = 16$. Working with a circle with a 100,000,000-part diameter, laborious nine-digit computations, and hundreds of square root extractions leading to an inscribed regular 24,576-gon, Chongzhi computed in his *Memoir on the Calendar* that $3.14159127 < \pi < 3.1415926$. His calculation of these deficient and excessive values established an inequality not surpassed for a millennium. The ratio 355/113 was not to be discovered in Europe until the late sixteenth century and deserves the name of Zu's ratio.

Surpassing Liu Hui's *Commentary on the Nine Chapters*, Zu Chongzhi and his son correctly computed the volume of a sphere. Their calculation begins with constructing within the sphere a cube with sides equal to its radius r. This equals one-eighth the volume of another cube circumscribing the sphere. Next construct two cylinders with radius r cutting the front and side faces of the small cube. This cuts the small cube into four parts. One part within the cylinder equals one-eighth of a "part umbrella." The intersection of the sphere and the two cylinders makes for three surfaces. The Zus compare with an inverted pyramid the three areas outside the part umbrella, finding, though, that the shapes are not the same, that matching membranes or cross-section surfaces at the same height are equal. The total part umbrella equals two-thirds the volume of the large cube, whose sides equal diameter D, and the ratio of the square cross section of the part umbrella to the inscribed sphere at identical heights is the ratio of the square to the circle. From this, Chongzhi and his son correctly calculate

$$V = (\pi/4)(2D^3/3) = (\pi/4)(2 \times 8r^3/3) = 4\pi r^3/3.$$

The theorem that equal cross sections at the same height imply equal volumes is usually attributed to the seventeenth-century Italian mathematician Bonaventura Cavalieri.

As pressure grew by the sixth century for increased accuracy in calendar making and astronomy, especially in forecasting lunar eclipses and determining positions of the sun, moon, and planets, the astronomer Liu Zhuo (544–610) introduced another major mathematical innovation, the method of interpolating second differences. Early Han astronomers had believed that each celestial body moves at a uniform speed, but their late Han successors knew that the speed of the moon changes. By 600, Chinese astronomers realized that the same is true of the sun. In preparing his New Imperial Calendar for 600, Liu Zhuo worked with vast amounts of observational data collected over time and applied his interpolation formula over equal time intervals to calculate the position of the sun and the moon. In modern symbols, given constant intervals $m, 2m, 3m, \ldots$ and $m + s$, where $0 < s < m$, he found $f(m + s)$ by the equivalent of $f(m+s) = f(m)+s\triangle+s(s+1)\triangle^2/2!$. To obtain the values of $\triangle$ and $\triangle^2$,

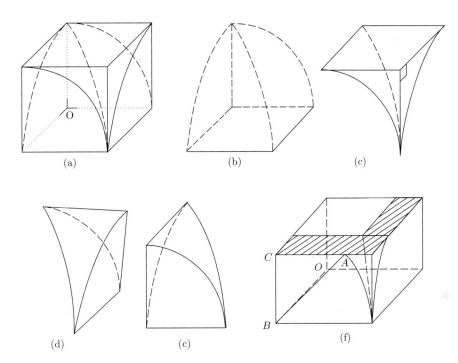

The cross-section of the "part umbrella"

let $\triangle_1^1 = f(2m) - f(m)$ and $\triangle_2^1 = f(3m) - f(2m)$ and $\triangle_1^2 = \triangle_2^1 - \triangle_1^1$ and $\triangle_2^2 = \triangle_3^1 - \triangle_2^1$. In this array, $\triangle$ is the first difference and $\triangle^2$ the second.

T'ang astronomers refined Liu's interpolation formula. In the Da Yan Calendar of 727, the monk Yi Xing employed an unequal interval in his formula for interpolating second differences. He assumed that I_1 and I_2 are unequal time intervals. Times m, $m + I_1$, $m + (I_1 + I_2)$ give resulting observations $f(m)$, $f(m + I_1)$, and $f(m + [I_1 + I_2])$. Letting $d_1 = f(m + I_1) - f(m)$ and $d_2 = f(m + [I_1 + I_2]) - f(m + I_1)$, Yi arrived at the equivalent of $f(m + s) = f(m) + s(d_1 + d_2)/(I_1 + I_2) + s(\triangle_1/I_1 - \triangle_2/I_2) - s^2/(I_1 + I_2) \times (\triangle_1/I_1 - \triangle_2/I_2)$. In compiling the Xuan Ming Calendar of 822, Yi Xing simplified the formula for unequal differences. To compute the position of the moon, he essentially gave this version of the formula for second differences over equal intervals:

$$f(m + s) = f(m) + sd_1 + s(d_1 + d_2)/2 - s^2(d_1 - d_2)/2.$$

This formula is equivalent to the first three terms of what is now known as Isaac Newton's interpolation formula:

$$f(m + s) = f(m) + s\triangle + s(s - 1)\triangle^2/2! + s(s - 1)(s - 2)\triangle^3/3! \ldots.$$

During the Sui and T'ang dynasties mathematical education occasionally evolved, occasionally was suppressed, and was located at either the Imperial Academy, the Institute of Astronomy, or the Institute of Records. The Sui *Book of Officials* reports that mathematics was taught at the Imperial Academy and once had two scholars, two teaching assistants, and eighty students. The Imperial Academy was like a state university today. Adopting Sui rules, T'ang officials established mathematics at the Imperial Academy shortly after 628, but abolished it in 658 because "mathematics leads to trivial matters,"[278] and transferred it for a time to the Institute of Records. The T'ang encyclopedia lists great changes in mathematics at the Imperial Academy and two institutes, but mathematical educators had only a small number of students and taught lower ranks of officials along with commoners.

According to the T'ang encyclopedia, an education in mathematics took seven years. Thirty students were divided into two groups of fifteen. One group learned basic, practical mathematics; the other, more complicated methods of interpolation as well. Graduates could sit for T'ang civil examinations in a subject called "understanding mathematics." The student group confining itself to basic problems had to solve problems from the *Nine Chapters* and single questions on the *Sea Island* and *Master Sun* and seven others; the advanced group had seven questions on the *Method of Interpolation* and *Continuation of Ancient Mathematics*. Six correct answers out of ten was a pass. The latter group also had orally to complete sentences picked at random from the *Ten Mathematical Manuals*. These examinations stressed memorization, not originality.

These examinations confirm that in the *Ten Books of Mathematical Classics*, also known as the *Ten Mathematical Manuals*, late T'ang mathematics had evolved into a complete system. Culled from more than thirty-five mathematical titles that had existed during the millennium before the T'ang dynasty, these chiefly practical works formed the core of mathematics education at the Imperial College. They include the *Zhoubi, Nine Chapters, Sea Island Manual,* and the three-volume *Master Sun's Mathematical Manual* of about the third century, which supplements the *Nine Chapters* with arithmetic problems that include finding square roots. The most famous problem of Master Sun is problem 26 of Volume 3 on linear congruences. Solving its equivalent of three simultaneous congruences means obtaining the smallest positive whole numbers that satisfy them.[279] This problem provided a way to compute the position of the sun, moon, and planets starting from midnight of the winter solstice. Another of the ten classics is the sixth-century *Mathematical Manual of the Five Government Departments* in three volumes by the Buddhist Zhen Luan. This applied arithmetic text calculates rectangular, triangular, trapezoidal, and circular field areas more systematically than does the *Nine Chapters* . It also has chapters on customs, silk trade, warehouse problems, and such military matters as army transport and supply problems.

From the tenth through the fourteenth century, Chinese mathematics underwent a splendid period of development, primarily in arithmetical algebra. Contributing to its flowering were intermittent support at the Imperial Academy and even more a significant shift of mathematical emphasis to solving higher-degree equations.

From Han times, square and cube root extraction and the solution of algebraic equations had been associated with geometric slicing techniques. In the *Nine Chapters* the "corollary to extracting square roots" solves the equivalent of quadratics, $x^2 + ax = b$, and in *Continuation of Ancient Mathematics* the "corollary to extracting roots" solves cubics equivalent to $x^3 + ax^2 + bx = c$. In inventing "the method of extracting roots by iterated multiplication," Jia Xian (fl. ca. 1050) improved on the methods for extracting square and cube roots. Discovering the intimate relation between the pattern of procedure for extracting roots and the array of numbers that are binomial coefficients, he began the process of turning his method into a general way of extracting higher-degree roots for equations with positive coefficients. If in modern notation $x^3 = b$ and $x = h + z$, the root of x is found in this fashion:

R_3	$-b$	$-b + h^3$	$-b + h^3$	$-b + h^3$
R_2	0	h^2	$3h^2$	$3h^2$
R_1	0	h	$2h$	$3h$
R_0	1	1	1	1

The final column contains the third degree binomial coefficients.

Geometric slicing techniques did not solve equations higher than the third degree. Jia Xian's method did. In order to find the root where $x^5 = a$, he set $x = h + z$ and transformed the equation to $z^5 + 5hz^4 + 10h^2z^3 + 10h^3z^2 + 5h^4z + h^5 + C = 0$, where C is a constant. The coefficients 1, 5, 10, 5, 1 are from the so-called Pascal triangle. The lost manual *Si shu suo suan* ("The Key to Mathematics") gave the equivalent of this:[280]

A_5	1	1	1	1	1	1
A_4	5	4	3	2	1	
A_3	10	6	3	1		
A_2	10	4	1			
A_1	5	1				

Each column gives the coefficients of the expansion of $(a + b)^n$. Summing from left to right the numbers in the next column that are parallel to and above it gives the result. Counting from left to right, where the first column is 0, gives the fifth column coefficients, or $(a + b)^5 = a^5 + 5a^4b + 10a^3b^2 + 10a^2b^3 + 5ab^4 + b^5$. This chart puts into vertical columns with the absolute term at the top the counting rod configuration of the binomial coefficients. This is not unlike Thomas Harriot's practice of writing equations in seventeenth-century England. Reading the numbers in the vertical rows beginning at the right yields the familiar Pascal's triangle. Jia's step-by-step method of solution by iterated multiplication is the same as that introduced in Europe by William Horner and Paolo Ruffini at the start of the nineteenth century.

Removing the restriction that the leading coefficient be positive was the next step to generalizing the method for extracting roots of higher degree equations. It is reported that in the lost *Discussion of the Old Sciences*, Liu Yi (fl. 1080–1120) introduced a method that applies to both positive and negative coefficients. The counting rod configuration of its problem 18 amounts to the equation $-5x^4 + 52x^3 + 128x^2 = 4{,}096$. Here is Liu Yi's solution: Multiply the first coefficient, -5, by the highest power, 4. Add the result, -20, to the coefficient of x^3 to obtain 32. Next multiply 1 by the coefficient of the x^2 term and add it to that number. The result is 256. Now multiply 256 times the highest power, 4, to obtain 1,024. Multiplying this by 4 gives 4,096. Addition of this to $-4{,}096$ yields 0, so the root must be 4.

$$
\begin{array}{rrrrr}
-5 & +52 & +128 & 0 & -4{,}096 \\
   & -20 & +128 & +1{,}024 & +4{,}096 \\
\hline
-5 & +32 & +256 & +1{,}024 & 0
\end{array}
$$

Studies by Jia Xian, Liu Yi, and other twelfth-century Chinese mathematicians culminated in the next century in the works of four masters: Qin Jiushao (ca. 1202–ca. 1261), Li Zhi (1192–1279), who changed his name to Li Ye, Yang Hui (fl. ca. 1261–1275), and Zhu Shijie (fl. 1280–1303).[281] George Sarton has placed Qin and Zhu among the greatest of all mathematicians,[282] and the half-century from Qin Jiushao's *Mathematical Treatise in Nine Sections* of 1247 to Zhu Shijie's *Precious Mirror of the Four Elements* in 1303 is considered the peak of the golden age of Chinese mathematics.

These four men worked in places remote from one another. Qin and Yang were from the south, and Li and Zhu from the north. They had no known contacts with one another, though Marco Polo visited Li. Their work, therefore, was independent. Qin, Yang, and Li were scholarly civil servants and Zhu a celebrated itinerant scholar and teacher.

The polymath Qin Jiushao is an intriguing character who studied mathematics, astronomy, music, and architecture at the Imperial Astronomical Bureau. Athletically gifted, he was known for his love affairs, which rivaled in number those of ibn Sīnā. Qin, who rose to be district commander in Sichuan, was not reluctant to poison opponents. His major mathematical writing is the *Mathematical Treatise in Nine Sections,* which should not be confused with the *Nine Chapters.* Written during a time of turbulence and personal hardship, for several centuries his *Mathematical Treatise* existed only in manuscript form. It has nine problems in each of its nine sections. The section subjects are indeterminate analysis, heaven and season (calendrics), field boundaries, surveying, corvée labor, money and foodstuffs, architectural design, and military matters. The problems solved are far more complex than those contained in bureaucratic manuals of that time.

Most significantly, Qin's *Mathematical Treatise* proposes in its first section a method for solving systems of simultaneous linear congruences that improves on prior work in China on the subject by permitting solutions where moduli are not necessarily pairwise prime. With this method, indeterminate analysis reaches its highest level in premodern China. Qin solves only specific indeterminate equations and offers no general solutions. Section two on calendric cycles investigates common factors of numbers and how to find least common multiples. In modern form, Qin's findings amount to $N \equiv R_i \pmod{A_i}$, where A_i has common factors with A_j, A_k, $\ldots$. This is an intermediate stage in solving indeterminate equations.

More than twenty problems in the *Mathematical Treatise* require solving equations. Twenty of these are two-term quadratic equations. Qin also solves one cubic, four quartics, and one tenth-degree equation, which is problem four in section five. The cubic in problem six of section six is $4{,}608x^3 - 3{,}000{,}000{,}000 \times 30 \times 800 = 0$. The answer is 25,000. Qin lays out these equations on counting boards and gives diagrams detailing computations for each step leading to the solutions. Coefficients could be positive or negative and integral or fractional. Qin extends Jia Xian's extraction of roots by iterated multiplication to solve higher-degree numerical equations. Problem three of section four, which finds the area of a pointed field, solves the equivalent of this quartic, which begins with a negative coefficient:

$$-x^4 + 1{,}534{,}464x^2 - 526{,}727{,}577{,}600 = 0 \quad (x = 720 \text{ steps}).$$

Qin solves simultaneous equations by eliminating a term and subtracting a newly found root in the original equation.

The *Mathematical Treatise* is instructive on arithmetic, geometry, and the nature of numbers in late Song China. As solutions for equations, Qin accepts positive and negative numbers. Positive numbers were represented by a red counting rod and negative numbers by a black counting rod or an oblique line over a numeral. The *Mathematical Treatise* introduced a round zero symbol. Arithmetic progressions describe matters ranging from finances to the collection of rainwater, a subject important to Song agriculture. The *Mathematical Treatise* is not so strong in geometry as in algebra and computational arithmetic. It discusses right-angled triangles for surveying and gives three values for π: 3, $22/7$, and $\sqrt{10}$. On the character of numbers themselves, Qin gives a hint of that mystical cosmology that in the West was expressed by the Pythaogreans. Numbers, he says, have two primary applications. In quantifying, they can "manage worldly affairs" and "analogiz[e] all created objects." In numerology, numbers also allow human beings to reach the deities, and it is therefore, Qin suggests, the more important application. Numerology, to be sure, is closer to divination that to a natural philosophy. How close Qin may have veered toward a perception of a universe ordered by number is not clear from this text.

In 1248, one year after the *Mathematical Treatise* appeared, Li Zhi in northern China completed the *Sea Mirror of Circle Measurements*. This twelve-chapter text with 170 problems was not published until thirty years later. Its problems deal with relationships between a right triangle and the circle inscribed in or circumscribing it. When Li Zhi discovered that his name was that of the third T'ang emperor, he changed it to Li Ye. Until the Mongols captured Junzhou in Henan province in 1232, he had been a magistrate there. Before the capture, he fled and became a famous recluse scholar, living with two friends near the foot of Mount Fenglong. Known as the "three friends of Fenglong Mountain," they frequently lived in poverty.[283] Kublai Khan granted Li Ye several audiences and named him to several positions, obliging him, for example, to join the Nahlin Academy in 1264, but pleading old age Li resigned each within a few months and returned to teaching. Li Ye lectured to people in barns and stables. At the age of eighty-eight he died in Hebei near Mount Fenglong.

To set up equations from given conditions, Li Ye in the *Sea Mirror* copies from the Shansi mathematician P'eng Che the second great invention of Chinese algebra of the era: the technique of the celestial element. "Celestial element" refers to the unknown in a polynomial equation. This method systematically configures algebraic equations in vertical columns on counting boards, with increasing powers above the absolute term, generally called *shih*, and reciprocal powers below it. The modern equation $-x^3 + 74x + 786 + 123x^{-1}$ becomes

$$-1$$
$$0$$
$$74$$
$$786$$
$$123$$

Li expanded on the use of standard notation, partly by denoting a zero by a circle and negative numbers by an oblique line over the final digit or the words *i* or *hsu* before the term. Both of these notations had evolved during the twelfth century in China. Li has the word *lien* refer to the coefficient of x^2, first *lien* to that of x^3, and so forth. His listing of solutions of linear, quadratic, cubic, and quartic equations as well as those of the fifth and sixth degree, omits the means of solution by the iterated multiplication method, the other great algebraic invention, which suggests that its generalization was then well known in China.

The 170 problems in the *Sea Mirror* find diameters for circles mainly by means of quadratic and cubic equations. Again, all circles are inscribed in a right triangle or circumscribe them. For example, problem two of chapter two states

> Two people, *A* and *B*, start from the western gate [of a circular city wall]. *B* first walks a distance of 256 *pu* eastward. Then *A* walks a distance of 480 *pu* south before he can see *B*. Find [the diameter of the circular wall] . . . .

A cubic equation solves a similar problem, number four of chapter 3:

> *A* leaves the western gate [of a circular city wall] and walks south for 480 *pu*. *B* leaves the eastern gate and walks straight ahead a distance of 16 *pu*, when he just begins to see *A*. Find [the diameter of the city wall] . . . .[284]

The cubic equation $x^3 + cx^2 - 4cb^2 = 0$, where $c = 16$ *pu* and $b = 480$ *pu*, gives the diameter of the city wall. Li Ye rejected answers in which the diameter x is a negative number.

Li Ye later wrote a text for beginners on the celestial element method, the three-chapter *New Steps in Computation*. Its sixty-four problems combine a circle and a square or a circle and a rectangle. It portrays natural numbers as reflecting natural laws, another intimation of a similarity to Pythagoreanism, and denies that mathematics is a "dying" enterprise within what was called "the Nine-nines."

The scant information about the life of Yang Hui comes from prefaces to his writings. A native of what is today Hangzhou, Yang Hui studied under the mathematician Liu I. Most likely, he was a civil servant who collaborated with a small circle of friends in his studies. Basing himself on these studies, Yang Hui completed in 1261 *A Detailed Analysis of the Mathematical Methods in the "Nine Chapters"* and in 1274 and 1275 a seven-volume collection known under the title *The Method of Computation of Yang Hui*, which was first printed in 1378. These books reflect methods used in trade and commerce in China. The writings of Yang Hui, who collected older methods, including techniques for extracting roots, became the chief source of their transmission even into the twentieth century. His work is thus crucial to understanding the development of Chinese mathematics. His acceptance of negative coefficients also strengthened the solution of equations, and in his computations he freely employed decimal fractions. Yet, until the recent translations and commentary of Lam Lay Yong, among modern scholars Yang Hui was the least studied of the four great thirteenth-century Chinese algebraists.[285]

A Detailed Analysis of the Mathematical Methods in the "Nine Chapters" is the best known but not the most original writing of Yang Hui. Its twelve chapters revise the classical nine chapters of Liu Hui and add his results. Only five classical chapters and two others survive. To assist mathematics students, Yang Hui reclassifies Liu Hui's 246 problems according to progressive difficulty and examines in detail 80 selected problems. The careful reclassification of problems, the step-by-step diagrams, and the explanations of components of geometric arguments proceed toward proof theory, which was unusual in China. Yang Hui's detailed examination of selected problems built more closely on traditions. He sums algebraic series, such as $1 + 3 + 6 + \ldots + n(n + 1)/2 = (n/6)(n + 1)(n + 2)$, and offers the first extant illustration in the form of a diagram of Jia Xian's triangle for binomial coefficients of $(x+1)^n$ to $n = 6$. Yang Hui solves numerical equations higher than the second degree by continuously approximating the general method of extracting roots that employs iterated multiplication.

Among the equations solved is the cubic $x^3 - 1{,}860{,}867 = 0$. Here is the approximation method for finding its roots: Put the number 1,860,867 in the second row of a five-row counting board and insert the coefficient of x^3, that is 1, in the fifth row. The root x will ultimately appear in the first row.

Since x lies between 100 and 150, the numeral 1 goes in the hundreds' place in the first row and is multiplied by the value in the fifth row. The result, 1, appears in the fourth row and multiplied by 1 gives 1 for the third row. The third row value is subtracted from the same column of the second row, as shown

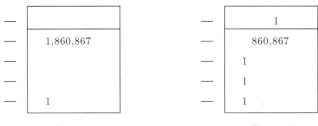

Figure 1   Figure 2

in Figure 2.

Next, the number in the fifth row is multiplied by the number in the first row, and the result added to the fourth row is 2. This 2 is multiplied by 1 from the first row, and the result added to the first row yields 3. For a third time, the value in the fifth row is multiplied by that in the first row, and the result is added to the fourth row, the total being 3. The value in row 3 is shifted one place, that in row 4, two places, and that in row 5, three places to the right, and the outcome is Figure 3.

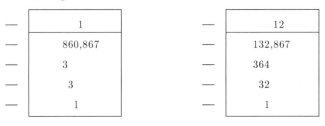

Figure 3   Figure 4

In the next approximation, x must be between 120 and 130, because $2^3 = 8$. Thus 2 is listed in the ten's place of the root and the multiplication process is repeated. That is, $2 \times 1 = 2$, which added to 30 makes 32, and $2 \times 32 = 64$, which added to 300 makes 364. Then $364 \times 2 = 728$, which is subtracted from the same column of the second row. The result is 132,867, as shown in Figure 4.

To obtain the third or units' digit of the root, first multiply the fifth row by 2 and add to the fourth row, getting 34, which multiplied by 2 and added to the second row results in 432. Multiply the fifth row by 2 a third time, and add the product to 34 to get 36. Shift the results to the right in the sequence of 1, 2, and 3 places shown in Figure 5.

Figure 5   Figure 6

The units' digit must be 3. Place 3 in the units' column of the top row. Three times 1 gives 3, which added to 360 gives 363 and $3 \times 363 = 1,089$, which added to 43,200 gives 44,289. In turn, $44,289 \times 3 = 132,867$, which leaves no remainder when subtracted from the second row. The root is therefore confirmed to be 123, as shown in Figure 6.

For historians of mathematics, the seven-volume *Method of Computation of Yang Hui* is a rich source. The first volume gives a study program for beginners in mathematics. The second discusses multiplication and division, including the avoidance of division through multiplying by reciprocals. The theory of equations is prominent in remaining volumes. To find positive roots of quadratics, Yang Hui gives the equivalent of the formula $x = (-b \pm \sqrt{b^2 - 4ac})/2a$. His method for solving quadratics appears more flexible than those in the West. He solves simultaneous linear equations, indeterminate equations, and higher-order equations by iterated multiplication.

Besides advancing the theory of equations, Yang Hui sums geometric progressions, gives formulas for finding areas and volumes, and in *Continuation of the Ancient Mathematical Methods for Elucidating the Strange [Properties of Numbers]*, dated 1275, presents magic squares through order 10. Fond of patterns, the Chinese constructed row and column diagrams or magic squares, which occupy a relatively exotic part of number theory. The magic squares consist of n^2 successive natural numbers put in $n \times n$ squares, such that sums of any column, row, or main diagonal are the same, except in the 10×10 square, in which the diagonal does not have a matching sum. Magic squares were known in China at least since Hsu Yo's *Shu Shu Chi I* of 190 and were developed in the early northern Song era. These were part of the early Song *luo-shu* or Luo river writing. Supposedly, a Luo river turtle in the time of the legendary Emperor Yi brought the original magic square to man. Yang Hui denies any originality for these squares. His preface to the *Continuation* claims that they are taken from older manuscripts. The 3×3 square presented below was associated with mysticism. This text is the oldest surviving Chinese work to give magic squares of an order higher than the third. The second magic square of Yang Hui given here is of order 5.

4	9	2
3	5	7
8	1	6

12	27	33	23	10
28	18	13	26	20
11	25	21	17	31
22	16	29	24	14
32	19	9	15	30

Zhu Shijie, the other thirteenth-century Chinese algebraic master, was probably born near modern Beijing. For more than two decades after the Mongol conquest of 1279, he traveled widely, the first known Chinese professional mathematician to do so. After returning home to present-day Yangzhou, he flourished as a mathematics teacher. Joseph Needham observes that people are reported to have come "like clouds from the four quarters to ... learn from him."[286]

Zhu Shijie wrote two surviving texts. Intended for beginners, *Suan-xue Gimeng* ("Introduction to Mathematical Studies") of 1299 consists of three chapters with 259 problems. It covers all fields of mathematics known to the Chinese but emphasizes algebra. The *Introduction* begins with multiplication and division and proceeds to root extraction and the celestial element technique. Problems are arranged according to their level of difficulty. Zhu Shijie's other text, the three-chapter *Suyan Yujian* ("Jade Mirror of the Four Elements") of 1303, contains 288 problems. This general introduction to algebra demonstrates the extraordinary mathematical skill and imagination of the author. His new four-elements method obtains higher-degree polynomial equations with up to four unknowns—*t'ien* (celestial), *ti* (earth), *jen* (human), and *wu* (material thing)—with the constant term or *t'ai* and solves these equations. The four unknowns and their powers are represented on a counting board arrayed above the constant *t'ai*. The method of four elements could solve $w + x + y + z = 0$ or their square. Essentially, it is a method of elimination. Consider the cubic equation $x^3 - 574 = 0$. Zhu Shijie's *fan fa* method obtains the transformation $y^3 + 24y^2 + 192y - 62 = 0$. To find the coefficients for the transformed element, the calculator first guesses that $x = 8$ and performs the computations given on the right here, where the top horizontal row has the equation's coefficients and constant while the second row contains 8, its square, and its cube. The figures in the vertical columns are added. The root then is 8 plus 62 divided by the sum of the amounts at the bottom of the first three columns or $x = 8 + 62/(1 + 24 + 192) = 8\frac{2}{7}$.

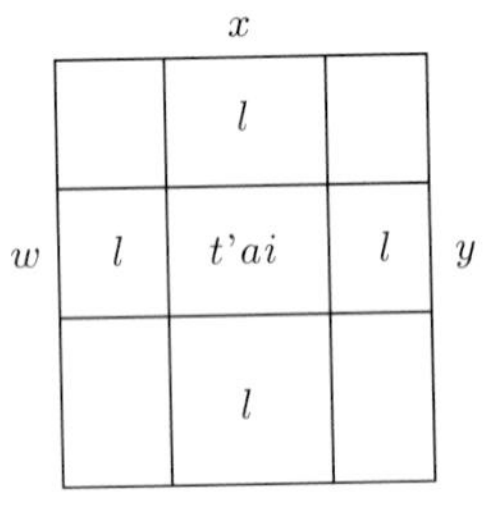

Figure 5

The *Jade Mirror of the Four Elements* carries algebra to its apogee in medieval China. The *Nine Chapters* had solved a system of linear equations having several unknowns, and the method of four unknowns appears a natural extension of the celestial element method. Zhu Shijie's generalization thus exceeds the work of Qin and Li.

The *Jade Mirror of the Four Elements* opens with a diagram of Jian Xian's triangle for binomial coefficients of $(x + 1)^n$ from $n = 1$ to $n = 8$. This extends the diagram of Yang Hui in *Detailed Analysis*, which goes to $n = 6$. In his diagram, which he properly refers to as the *ku fa* (old method) associated with root extraction, Zhu Shijie uses rod numerals and a round zero. His command of this triangle surpassed his predecessors'. Zhu observes, if his comments are put into modern form, that values along the triangle's right side are the coefficients of the highest power of x, while those along the left are the absolute term, and he indicates that the horizontal rows give the coefficients of the polynomial. Here is his diagram to $n = 6$, rendered into Indo-Arabic numerals:

$$
\begin{array}{ccccccccccccc}
 & & & & & & 1 & & & & & & \\
 & & & & & 1 & & 1 & & & & & \\
 & & & & 1 & & 2 & & 1 & & & & \\
 & & & 1 & & 3 & & 3 & & 1 & & & \\
 & & 1 & & 4 & & 6 & & 4 & & 1 & & \\
 & 1 & & 5 & & 10 & & 10 & & 5 & & 1 & \\
1 & & 6 & & 15 & & 20 & & 15 & & 6 & & 1
\end{array}
$$

Zhu Shijie's *Jade Mirror of the Four Elements* also advances the study of arithmetic and geometric series. From the eleventh century, Chinese masters had examined the problem of equal differences, that is, arithmetic series, and the "technique of small increments," known as the problem of piling up stacks. In his *Dream Pool Essays* Shen Kuo (1032–1095), who invented the technique of small increments, had discussed both. Shen Kuo opened to interpolations in astronomy the study of higher-order differences and their connections, and Yang Hui extended this work. Although no proofs appear in the *Jade Mirror*, Zhu Shijie in summing each of these series, which he called triangular piles, was the first to devise rules for obtaining the whole in a single step. For arithmetic series, his rule in modern form is

$$1 + 2 + 3 + 4 + \ldots + n = n(n+1)/2!.$$

Zhu names it "pile of reeds." The technique of small increments, piling up stacks, he reduces essentially to this formula:

$$1 + 4 + 10 + 20 + \ldots + n(n+1)(n+2)/3!$$
$$= n(n+1)(n+2)(n+3)(n+4)/5!,$$

a triangular pile of scattered stars. Zhu's summing of squares and cubes includes $1^2 + 2^2 + 3^2 + 4^2 + \ldots + n^2 = n(n+1)(2n+1)/3!.$

Zhu improved the application to astronomy of the piling-up-stacks method. To find arbitrary constants in formulas describing celestial motions, astronomers since Tang times had employed it. In modern terms, it works this way: $S = U_1 + U_2 + U_3 + \ldots + U_n$, where

$$\begin{aligned}
\triangle &= U_1 & \text{(upper difference)} \\
\triangle^2 &= U_2 - U_1 & \text{(second difference)} \\
\triangle^3 &= U_3 - (2\triangle^2 + \triangle) & \text{(third difference)} \\
\triangle^4 &= U - [3(\triangle^3 + \triangle^2) + \triangle] & \text{(lower difference)}
\end{aligned}$$

In *Work and Days Calendar* of 1280, Wang Xun (1235–1281) and Guo Shoujing (1231–1316), in calculating lunar and solar motion, developed the third-order interpolation method. Another of their innovations was to give differences not on a daily basis but on average over several weeks. They also adopted Zhu's methods to sum power series.

Chinese masters might have continued to develop into the fourteenth and fifteenth centuries the celestial element and the four unknowns techniques, the summing of higher-order difference series, and interpolations. They did not. These techniques rarely applied to daily life in China. Social, economic, and agrarian conditions did not require the solution of higher than quadratic equations. Prevention of river flooding, for example, involved only quadratic equations. After Wang Xun and Guo Shoujing, moreover, even State Observatory astronomers who used their tables seemingly did not know how to derive them. Consequently, by the fifteenth century understanding of the thirteenth-century mathematical innovations was largely lost in China.

During the Ming Dynasty, which ruled China between 1368 and 1644, the major advance in Chinese mathematics was the invention of the abacus. Buoyed by a population doubling from roughly 60 million to 125 million, trade and commerce generally fluctuated upward, and crafts and efficient governmental finances expanded. Growth generated expanding data and complexity in computation and pressures for more convenient computations. Addition and subtraction requires continual adjustment of counting rod configurations, which slows calculations. Multiplication and division originally demanded three rows of counting rods, top, middle, and bottom, which is awkward. The abacus admirably met the new computational needs.

From eighth-century T'ang society, the demand had grown for faster and simpler multiplication and division. T'ang masters devised ways, based on halvings and doublings, to reduce counting rod multiplication and division from three rows to one. In thirteenth- and fourteenth-century China, versed rhymes primarily simplified daily mathematics and prompted the evolution of the abacus. The *Dream Pool Essays* of Shen Kuo, *Precious Reckoner* . . . of Yang Hui, *Jade Mirror* . . . of Zhu Shijie, and other texts contain verses. Here are two verses of Yang Hui. The first is a decimal conversion rhyme.

> Finding 1, omit a place 625; finding 2, go back to 125; finding 3, write 1,875; finding 4, change it to 25; finding 5 is 3,125; finding 6 liang the price is 375; finding 7, put 4,375; finding 8, change it to 5.[287]

This verse means that $1/16 = 0.0625$, $2/16 = 0.125$, and $3/16 = 0.1875$, ... The counting rods required the student to check the quotient. The verses give a direct form of division: simply step back one place in the process to find the quotient. A division verse is

> 9 as divisor, seeing 45 quotient is 5; 8 as divisor, seeing 4[0] quotient is 5; 7 as divisor, seeing 35 quotient is 5; 6 as divisor, seeing 3[0] quotient is 5; 5 as divisor, seeing 25 quotient is 5; 4 as divisor, seeing 2[0] quotient is 5, ...[288]

Yang Hui employed only single-digit divisors, but spoke of computations with two-digit divisors as "flying division." Two-digit division was not generally in use at that time. Zhu Shijie worked in steps from each leading digit in a divisor. If the divisor is 436, first use 4. Find the digit in the quotient; then multiply the remaining digits by the quotient and subtract from the dividend. The modern Chinese essentially call this "division in nine parts."

By the late fourteenth century, counting rods no longer sufficed. The hand movements required for reconfigurations made them slower than reciting rhymes or mental reasoning. To avoid these reconfigurations, the abacus was developed. There is no record of a single inventor. The oldest extant book on abacus calculation is Xu Xinlu's collation of 1573, the *Method of Calculating on the Abacus*. It claims sources over a century old, and the discussion in Tao Zongyi's *Talks While the Plough Is Resting* in 1366 of "beads on the calculating board" suggests the abacus. Pertinent Ming literature suggests that it was widely used in China by 1400. The abacus consisted of a series of strings with five beads in parallel columns. The early abacus had an upper compartment, where the bead above a transverse represented 5, like the upper counting rods, and a lower compartment, where each remaining bead represented 1. For computations, counting rods had to be put in or taken out; in the abacus simple movements of beads were sufficient.

During the early seventeenth century, western mathematics entered China, primarily to emend the inaccurate Da Tong and Islamic calendars. In predicting such events as solar and lunar eclipses, these two calendar schools often made glaring errors. A wrong prediction of a solar eclipse by the Chinese Calendar Official in late 1610 led officials to propose calendar emendation as a way to curry favor with the emperor. Some officials supported the adoption of western calendric methods. In consequence, Chinese mathematics turned to Euclid's *Elements*, trigonometry, logarithms, and calculation by pen and pencil.

In the early seventeenth century, contacts with Europe were increasing. Western Europeans had earlier embarked on transoceanic expansion. Since Muslims had controlled the overland spice trade with the Orient for centuries, the Portuguese Vasco da Gama rounded Africa's Cape of Good Hope and sailed to India from 1497 to 1499, seeking a safer sea route for trade with the Orient. Western Europeans expanded trade with China and soon attained commercial supremacy. Following the Reformation, Europeans also sent abroad missions of evangelization.

Among others, the Jesuits reached China in the late sixteenth century. To gain entrance into educated circles, the Jesuits dressed first as Buddhist monks and, when this proved a poor choice, donned robes of Confucian scholars. With the Jesuits came clocks, sundials, maps, celestial spheres, and books. Prominent Chinese officials sought both calendar emendation and training with Portuguese cannon from Macao to defend their northern frontier. The Jesuits taught gunnery along with western mathematics, calendar computation, and astronomy.

Matteo Ricci (1552–1610), the first Jesuit to arrive in China, landed in

Macao in 1582, was in Canton the next year, and moved overland to Beijing. In Ricci, the Chinese had an early exposure to western mathematics. Ricci translated into Chinese the first six books of his teacher Christoph Clavius's edition of *Euclidis elementorum libris XV* and Clavius's *Epitome of Practical Arithmetic*. In 1607 Ricci dictated the plane geometry books of the *Elements*. His Chinese co-worker was the scholar Xu Guanqi (1562–1633) from what is today Shanghai, who was educated in astronomy and calendar computation and stood next to the emperor in importance. Ricci and Xu referred to Clavius as Mr. Ding. This arose from the translation into Chinese of the Latin *clavius*, or nail.

The translation of the first six books of the *Elements* was not a simple task. No technical dictionaries existed and such terms as point, curve, parallel line, obtuse angle, and acute angle did not yet exist in Chinese. The *Elements* introduced the axiomatic method, deductive proofs following from a few axioms, postulates, and definitions. The method contrasted to the inductive form of mathematics in the traditional Chinese *Nine Chapters*. The *Elements* brought theoretical geometry to China. In his "Discourse," Xu Guanqi lauded the axiomatic method for its simplicity, clarity, and ingenuity. He predicted that in time "everyone will study" the *Elements*.

Xu wanted Ricci to complete the translation of the *Elements*, but Ricci left that to others. That translation did not come soon. Chinese scholars wrestled with the new terms and unfamiliar deductive proofs of the *Elements*. Not until 1857 did Li Shanlan and Alexander Wylie translate the last nine books from a French edition of Euclid. By that time western schools were operating in China and theoretical geometry was a compulsory subject, fulfilling Xu's prediction.

In 1607 or shortly after, Ricci dictated to the Chinese scholar Li Zhizao (1565–1630), who transcribed it, Clavius's eleven-chapter *Epitome of Practical Arithmetic*. Li Zhizao added problems and Chinese methods of calculation. The result was his larger *Treatise on European Arithmetic* in three volumes, completed in 1631, which transmitted western arithmetic to China. The *Treatise* introduced addition, subtraction, multiplication, and division with pen and paper. Li calculated with pen and paper, using Indo-Arabic numerals essentially as we do today. One difference is that the numerator and denominator in fractions are inverted. The *Treatise* also includes the summing of various arithmetic and geometric series, methods of root extraction, and a feature generally missing in traditional Chinese mathematics: the casting out of nines and other methods of checking computations. The *Treatise* reveals that Chinese arithmetic had stagnated, while Chinese algebra in solving higher-degree equations remained on a level superior to that in the West.

Pressure for calendar emendation increased in 1629, when three schools differed in their predictions of the time for a solar eclipse. Added to the older Da Tong and Islamic Calendar schools was the New Method School, in which Xu Guanqi applied Western techniques. When only Xu's forecast agreed with the actual observation the emperor appointed him to direct the calendar emendation and Li Zhizao to assist. Four Jesuits, Niccolo Langobardi, Giacomo

Rho, Johann Terreng, and Johann Adam Schall von Bell, helped Xu gradually revise the calendar by improving on the western method. They pursued a thorough understanding of the technique, beginning with the work of making translations.[289] After Xu Guanqi died in 1633, Li Tianjing directed the emendation in stages.

The Imperial Court tested the forecasting accuracy of four schools, especially on solar and lunar eclipses. Two of the schools were an eastern and a western branch of the New Method school. In 1643 the Ming Court adjudged the Western method far more accurate. Its predictions were based on two massive texts, *Chong Zhen Reign Treatise on [Astronomy and] Calendric Science*, and the Jesuit Johann Bell's revision of the thirteen-volume *Treatise on [Astronomy and] Calendrical Science according to the New Western Methods*. The Court appointed Jesuits to the State Observatory. But frequent disputes continued over old and new calendric methods: opposition to the western system often reflected hostility to foreigners. In 1664 the first Qing ruler denounced westerners, confined western missionaries to the Court, and sought other calendric methods. After frequent errors in astronomical predictions during his reign, the Jesuits were reinstated to the State Observatory, which they controlled for roughly the next two centuries.

The *Chong Zhen Reign Treatise* and the *Treatise on ... New Western Methods* collected all books on western astronomy, calendrics, and mathematics taken to China during the early seventeenth century. These texts are thus crucial sources in the study of the transmission of western mathematics to China. They brought plane and spherical trigonometry to China as well as associated tables and logarithms, all essential to calendric computations. Indian spherical trigonometry, transmitted to China during the T'ang dynasty, had evoked little interest. Western missionaries and traders now also introduced western calculating devices, such as Napier's bones, which the Chinese called western counting rods.

Plane and spherical trigonometry appeared chiefly in three books: in 1631 the *Complete Theory of Surveying*, a part of the *Chong Zhen Reign Treatise*, and in 1653 two texts by Nicolas Smoguleski, who had arrived in China in 1646, and Xue Fengzuo, *True Course of Celestial Motions* and *Method of Trigonometric Calculations*. These books define the eight trigonometric lines according to central angles and arcs of a circle: sine, cosine, tangent, cotangent, secant, cosecant, versine, and coversine. In modern notation their formulas include

$$\sin a \times \csc a = 1, \qquad \cos a \times \sec a = 1,$$
$$\tan a \times \cot a = 1, \qquad \tan a = \sin a / \cos a,$$
$$\cot a = \cos a / \sin a, \qquad \sin^2 a + \cos^2 a = 1,$$
$$a / \sin A = b / \sin B = c / \sin C,$$
$$\sin a = \sin(60° + a) + \sin(60° - a), \text{ and}$$
$$\sec a = \tan a + \tan(90° - a)/2.$$

The *Complete Theory of Surveying* gives sine, tangent, and secant tables at 15-minute intervals. These tables are accurate to four decimal places. Using linear interpolations, Chapter 6 of *Treatise on Calendric Science* gives tables at 1-minute intervals to five decimal places. The three works also address methods and formulas of spherical trigonometry.

In 1653, Smoguleski and Xue Fengzuo's *True Course of Celestial Motions* and *Logarithm Tables with Explanations* introduced logarithms. The Scot John Napier and the German Joost Bürgi had recently independently invented these in Europe. Smoguleski and Xue do not adopt the word logarithms, which Napier coined, but call them "corresponding numbers" or "power numbers." Logarithms, they knew, reduce multiplication to addition and division to subtraction, save "six or seven tenths of the work compared with the earlier procedures,"[290] and reduce errors. Chapter 6 of *Logarithm Tables* gives base 10 logarithm tables to six decimal places for numbers from 1 to 10,000.

Finally, western missionaries introduced western calculating devices in to China. In *Napier's Bones* (1628), another part of the *Chong Zhen Reign Treatise*, the Jesuit Giacomo Rho discusses that calculating device, which converts multiplication of large numbers to a grid of single-digit multiplications and additions in diagonals across the grid. This resembles the *gelosia* method of multiplication (see page 281). Division proceeds similarly with simple divisions and subtractions. Missionaries brought other calculating devices, similar to Edmund Gunter's calculating ruler, which uses logarithms to speed computations. The Gunter ruler lacked a cursor, but was a forerunner of the slide rule. The digital calculator, invented in 1642, was also carried to China.

9

Indian Mathematics: From Harappan to Keralan Times

> Like the crest of a peacock, like the gem on the head of
> a snake, so is mathematics at the head of all knowledge.
> — *Vedanga Jyotisa* (ca. 500 B.C.)

Although the surviving literary records dealing with mathematics in India through medieval times are few and fragmentary, they offer an outline of the development of mathematics there. Most of these extant writings address mathematical astronomy and include few sections on mathematics. Apparently, rudimentary mensurational geometry and computational arithmetic existed in India during its Harappan and Vedic periods, together running from approximately 2300 to about 400 B.C. Amid a transmission of Hellenistic astronomy, Indian mathematical astronomy emerged during the late Jaina period, which lasted to A.D. 320. Afterward, Indian sages intermittently developed astronomy and mathematics, often in an interlocking fashion. In trigonometry, classical Indian mathematician-astronomers introduced the sine, cosine, and versed sine (in modern notation $1 - \text{cosine}$) half-chords. They utilized algebraic identities with these, such as $\sin^2 \theta + \cos^2 \theta = 1$, to compute astronomical tables and, during the late medieval period, devised infinite series expansions to compute half chords. In algebra, Indian sages provided general solutions for linear and quadratic indeterminate equations, and in computational arithmetic they had developed by A.D. 900 a decimal place-value number system with a zero. Florian Cajori was to call that system "their grandest achievement."[291]

9.1 The Nature and Sources of Ancient and Medieval Indian Mathematics

Indian mathematics differs markedly from that in ancient Greece and Hellenistic Alexandria. Indian sages normally eschewed the deductive, axiomatic proof theory inherited from the Greeks and its pictorial components. Perhaps the stress on phonetics and grammar of the Indian language in their verbal algebra during the Vedic period was a reason they emphasized not demonstrative

geometry but algebraic arithmetic. That preference may trace as well to the
demand for accuracy in pronouncing every syllable in Vedic sacrificial prayers
and to Babylonian influence during the late Jaina period. After the astronomer
Āryabhata (b. A.D. 476), algebra became a separate branch of mathematics.
Indian sages skillfully solved problems that led to quadratic equations, for which
they recognized two roots, as well as simple cubic and quartic equations. They
found roots arithmetically rather than by intersection points of conic sections.
It was not until the sixteenth century that European algebraists using an arithmetical approach were first to achieve general solutions for cubic and quartic
equations.

Classical and medieval Indian algebra was most accomplished in solving
linear and quadratic indeterminate or Diophantine equations—those with two
or more variables. In India as in China, Diophantine equations arose partly
in response to needs of commerce and inheritance. They were also especially
employed for determining planetary orbits and when constellations become visible in the heavens. Diophantus had accepted only one rational solution for
the linear case; Indian algebraists after Brahmagupta (b. 598) pursued all integral solutions. Brahmagupta named Diophantine analysis *kuttaka* and Pruth-
dakaswarmi (ca. 850) renamed it *bijagnita*, the science of calculating with unknown elements.

Indian algebraists also enlarged the domain of numbers. Free from the
demands of proof theory, they did not have to discriminate rigidly between
rational and irrational roots of equations. After accountants introduced negative numbers to represent debts and positive ones for assets, Indian algebraists
slowly put both types of numbers on an equal level. Negative coefficients help
in algebra, because the three types of quadratics classified by Diophantus,
$ax^2 + bx = c$, $ax^2 = bx + c$, and $ax^2 + c = bx$ for a, b, and c positive, can
be reduced to one form $px^2 + qx + r = 0$. Still, negative numbers met continuing rejection elsewhere, since a negative quantity seemed to run counter
to reality. Another Indian mathematical innovation, a daring and significant
intuition into the nature of number, was to treat zero as a number rather than
a place holder and gradually to incorporate proper operations with zero into
their computations.

Indian algebraists, lacking interest in Euclidean axiomatic geometry with
its strict definitions and systematic logical order, and therefore unhampered
by conceptual difficulties that beset the ancient Greeks in the treatment of
magnitude and number, extended the domain of number. Yet the absence of
formal proof theory kept their geometry at an elementary mensurational stage
employed in surveying and architecture, particularly in constructing Vedic sacrificial altars, temples, and granaries. For this work, they, like the ancient
Egyptians, had cord stretchers. As will be seen, Indian geometry rested on a
few unproved recipes and ritualistic prescriptions that often were inaccurate approximations. For this neglect of demonstrative geometry, George Rusby Kaye
called the Indian attitude toward learning "decidedly unmathematical."[292] But
continuing accusations that Kaye had a bias against Indian mathematics make

this assessment controversial.

The fragmentary literary evidence on mathematics from Jaina, classical, and medieval India illuminates the cultural pressures and personal motivations for innovation, as well as the strengths and weaknesses of Indian mathematics of that time. Among the extant writings from mathematical astronomy are the five surviving *Siddhāntas,* which have a few sections on mathematics. Their chief impulse for developing mathematics was to devise and simplify—though sometimes they needlessly complicated—computational techniques for a more accurate astronomy. The *Siddhāntas* largely improved the calendar. To a lesser extent, business accounting, as well as inheritance and construction, prompted mathematical activity.

Within this mostly utilitarian setting, Indian authors seem to have taken delight in pursuing complicated, masterful numerical manipulations. They did not usually seek theoretical innovations. Instead, they devoted themselves almost exclusively to computations and reducing these to rules of thumb that elaborated basic formulas or prescriptions. For the Indians, as for the Babylonians and Chinese, such work comprised the heart of mathematics, although they included traces of theoretical knowledge from Seleucid Mesopotamians and Hellenistic Greeks. The word *ganaka,* a calculator, came to denote "astronomer" as well as "mathematician," while the word *ganita,* the science of calculation, stood for mathematics.

Except for computational originality, Indian astronomy was repetitive and failed to recognize the possible usefulness of its methods for solving some problems in physics. Astronomers preserved older knowledge in a way that subordinated internal consistency to a reluctance to part with any segment of religious convention. Undoubtedly, this situation prevented the evolution of systematic theory in astronomy and probably in mathematics. Nor was there a tradition of observational astronomy to provide a framework or impetus for basic theory. Without occasional transmissions of theories from the West, which were absorbed and adapted, Indian mathematical astronomy might have developed far more slowly.

Vedic texts containing materials on mathematics and astronomy have an obscurity traceable to their poetic style of presentation and preservation dating from the *Brahmanas* of about 1000 B.C. and subsequently adopted by scientific and philosophical schools. To facilitate memorization, to conserve scarce writing materials, and probably to confine knowledge to a caste elite, Vedic scribes composed texts in verses, as did the pre-Socratic Greeks in natural philosophy. The usual Vedic poetic form, the *sutra,* sought the utmost brevity by avoiding verbs as much as possible and by extensively compounding nouns, reducing rules and results to aphorisms. *Sutras* are easily memorized, but without extensive commentaries make little sense to the uninitiated. India's caste system limited this knowledge to certain privileged families, which impeded mathematical progress. Vedic word numbers were variously rendered into metrical syllables, and attempts to follow meter often meant omitting important mathematical prescriptions. Such expression clearly lacked the precision of technical terminology. *Sutras* repeated and texts copied by members of a later generation, who had little aptitude for mathematics or interest in it, lost unstated procedures required for solving problems. For many problems that Indian literature presents, we can today only guess at the given conditions and invent a plausible solution with procedures to fit the context of the entire work.

Located between China and the lands of the Middle East and western Mediterranean, India suffered many military invasions through the centuries, had long-range trade, and between 500 B.C. and A.D. 1200 its intellectual contacts included three major transmissions of astronomical thought from the West. Knowledge of foreign influences and how Indian sages modified them is scanty. This further limits our ability to analyze Indian mathematical texts. This chapter briefly summarizes the course of Indian civilization from Harappan through medieval times and the mathematics within it.

9.2   The Harappan, Vedic, and Jaina Periods

From about 2300 to about 1750 B.C. on the fertile floodplain of the Indus River, the earliest identifiable urban civilization on the Indian subcontinent evolved and disappeared. Each of its two principal sites, Mohenjo-Daro and Harappa, probably had a population of over 35,000. The area of the Indus or, more precisely, Harappan culture far exceeded that of ancient Mesopotamia or Egypt. Although scholars are not yet able to decipher the Harappan script and thus have limited knowledge of its protomathematics, archaeological findings indicate that the stable, traditional culture of its states, which were perhaps theocratic, had standard weights and measures with scales and instruments for measuring lengths. This suggests a numerate culture. A ruler etched on a shell has accurate gradations between its lines of 1.32 inches, which is called the "Indus inch." The Harappan kiln-fired brick technology suggests a high-craft tradition, which possibly developed mensurational geometry. The standard kiln-fired brick proportion was 4 : 2 : 1, which is optimal for effective bonding.

During at least two centuries after 1800 B.C., waves of Aryans, a people speaking an Indo-European language, migrated from central Asian plateaus into the Punjab and upper Ganges areas of northern India. Armed with bronze axes, skilled in the use of the horse and chariot, which they introduced into the region, and toughened by migrations, the Aryans overpowered fortified cities. They took their name *Arya*, born free, to separate themselves from the conquered peoples. For lack of evidence, the hypothesis that they destroyed the Harappan civilization is no longer accepted. Perhaps abnormal floods contributed to the disintegration of the earlier culture. As Aryan tribes shifted from pastoral seminomadism to a settled agriculture in interdependent villages, they refounded Indian civilization.

Our knowledge of the rural Aryan civilization rests not on archaeological artifacts, but almost exclusively on four compilations of literature, known collectively as the *Vedas*. The word *Veda* means knowledge, and these ritual, priestly, and speculative texts ostensibly convey enduring wisdom gleaned by ancient seers. The four stages of *Vedas* are the *Samhitas*, especially the hymns and prayers of the *Rig Veda* composed between 1700 and 1000 B.C.; the *Brahmanas*, commentaries described as a practical handbook for conducting sacrifices, composed between 1000 and 800 or 600 B.C.; the *Aranyakas*, dated about 700 B.C.; and the *Upanishads*, elaborations of Vedic literatures, completed between 800 and 500 B.C. The *Upanishads* presented a new world view, which placed meditation and knowledge above sacrificial ritual as the way to attain access to *Brahman*, the sacred.

Responding to the needs of religion, building crafts, surveying, commerce, and perhaps music and gambling, Vedic scribes set out with a few abbreviations their mensurational geometry, computational arithmetic, and verbal algebra. The chief sources of Vedic mathematics are not the *Vedas* themselves but appendices called *Vedangas* on astronomy and rules for ritual and ceremonies together with ritual literature known as *Kalpasutras*. Both are in *sutra* form. Within the *Kalpasutras* are *Strautasutras* on constructing sacrificial fires at different times of the year, and these include the important *Sulbasutras*, influential between 800 and 200 B.C., the chief source of Vedic mathematics. Originally *sulbasutra* meant a rule governing sacrificial rites, but *sulba* came to refer to a chord of rope used for measuring altars. Like ancient Egypt, then, Vedic India had rope stretchers who transmitted primitive geometrical lore.

We know little about the authors of *Sulbasutras*, the *sulbakaras*, except that they were scribes and probably priest-craftsmen. The three most important mathematically are Baudhayana, who wrote sometime between 800 and 600 B.C., Apastamba, and Katyayana, who wrote at least two centuries later. Katyayana seemingly added little to the work of his predecessors.

The *Sulbasutras* contain simple, disconnected verbal prescriptions for finding areas and volumes. Some of these prescriptions are accurate, others approximations, and most of them are based on corrupted foreign rules. Construction of Vedic altars for sacrifices demanded precision, and thus there was accuracy in working with the geometric shapes most used in the construction of temple, home, and cemetery altars: squares, rectangles, trapeziums, circles, and semicircles. Elsewhere, the volume of a prism or cylinder is given as base times height and a pyramid as half the product of base times height. *Sulbakaras* failed to distinguish between accurate and approximate solutions.

Emphasizing the precise construction of altars with a minimum number of tools, *sulba* geometry gives versions of the Pythagorean theorem, whose relations were known from Vedic surveying or from Mesopotamian or Egyptian sources. "The rope ... stretched across the diagonal of a square," Baudhayan states, "produces an area double the size of the original square." Katyayana's version is more generalized: "The rope [stretched along the length] of the diagonal of a rectangle makes an [area] which the vertical and horizontal sides make together."[293] Vedic Pythagorean triads for altar construction include triples of integers: $\{5, 12, 13\}$, $\{12, 16, 20\}$, $\{8, 15, 17\}$, $\{15, 20, 25\}$, $\{12, 25, 37\}$, and $\{15, 36, 30\}$; triples with fractions $\{2\frac{1}{2}, 6, 6\frac{1}{2}\}$; $\{7\frac{1}{2}, 10, 12\frac{1}{2}\}$; and triples with irrationals: $\{1, 1, \sqrt{2}\}$, $\{5\sqrt{3}, 12\sqrt{3}, 13\sqrt{3}\}$, and $\{15\sqrt{2}, 36\sqrt{2}, 39\sqrt{2}\}$. Limited to word numbers until the fourth century B.C. and afterward to Kharosthi and Brahmi numerals that are somewhat like Roman numerals,[294] ancient Indian sages did not compute large Pythagorean triples, as had ancient Babylonian scribes, possibly out of playfulness.

Concentrating on inscribing a circle in a square and its companion effort to square the circle in Vedic altar geometry, Apastamba obtained an accurate value for $\sqrt{2}$. Circumscribing a circle about a square with side $a = 1$, Apastamba found $\sqrt{2}$ to equal the original measure of 1 plus its third and its third times one-fourth less than a thirty-fourth of the last term. In unit fractions this equals $1 + 1/3 + 1/(3)(4) - 1/(3)(4)(34)$, or $1.4142156\ldots$, which is accurate to five decimal places. According to B. Datta, Apastamba reached this approximation by a geometric procedure for doubling the square. It requires dividing a second identical square into three equal parts and adding the first two parts to the original square, almost doubling it, and then adding parts from the third section of the second square as indicated in the drawing here:

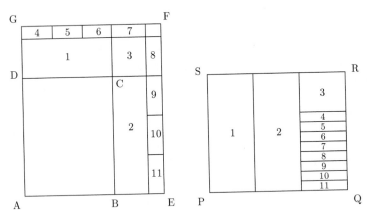

How doubling the square may have prompted the Indian
approximation of $\sqrt{2}$

In attempting to square the circle, Apastamba and other Vedic geometers also made several approximations for the ratio of the circumference of a circle to its diameter, or π. Working with a circle circumscribed about a square with side $2a$ and a smaller circle that nearly equals the square in the area, Apastamba found that the radius of the smaller circle equals $a + (a/3)(\sqrt{2} - 1)$ and, for a equal to 1, $\pi = 4a^2/[a + (a/3)(\sqrt{2} - 1)] = 3.0883$. Other results for π were 3.004, 3.0885, 3.16, and perhaps $\sqrt{10}$, called the Jaina or Hindu value.

In none of Vedic geometry does proof appear; all rules are empirical. Vedic sacrifices disappeared in the late first millennium B.C., removing the interest in work on altar geometry.

The rise of new religion in the sixth century B.C., partly in reaction to excesses of Vedic religion and social organization, is important to the course of Indian mathematics. So are the emergence of the Mauryan empire and two massive transmissions of foreign astronomy.

Both Jainism stemming from Hahavira and Buddhism from Gautama challenged the authority of the *Vedas*, were congregational, preached nonviolence, and held compassion to be the great virtue. Both gained support from wealthy landowners and merchants, Jains becoming the leading financiers. The Jains required elementary mathematics in educating their leaders and prepared canonical texts, not codified until the sixth century A.D., containing mathematical and astronomical information.

Formation of monarchical states and commercial towns in northern and western India from the seventh century B.C., together with the power vacuum left in the northwest after Alexander's campaigns in 327 B.C., encouraged the emergence of the Mauryan empire in the Indus region and central India by 321 B.C. Perhaps the greatest Mauryan emperor was Ashoka, who reigned from 272 to 232 B.C. His rock edicts are important documentary sources for early Indian history. These edicts suggest that Ashoka pursued Buddhist *dharma*, the attainment of heaven through the merit of good actions. He became the ideal of Buddhist and Hindu kingship, the *chakravartin* or universal monarch who rules with justice and wisdom. Under the Mauryan empire, contacts between the Indian subcontinent and the eastern Mediterranean, which were growing closer since Achaemenid rule in Persia in the sixth century B.C., became more varied. These contacts contributed substantially to the development of Indian mathematics.

The first significant mathematical achievement in ancient India is the emergence there of mathematical astronomy. It had roots in late Vedic times, but did not reach completion until the *Siddhāntas* of the late Jaina and early classical periods.

The earliest of three transmissions of astronomical knowledge from the West strongly influencing this development was from Mesopotamia during the late sixth and the fifth century B.C., when Achaemenids ruled northwestern India. A crude, calendric Indian astronomy arose in the *Gargasamhita* and *Vedanga Jyotisa* about 500 B.C. To determine times for performing seasonal sacrifices, Vedic literature simply treats *yugas* (periods), variations in *rtus* (seasons), equinoxes, solstices, *paksas* (half-lunations), *naksatras* (groups of stars), and *adhimasas* (intercalary months), though no systematic method for intercalation existed in them. While this literature does not achieve a formally mathematical astronomy, it establishes traditions into which later Indian astronomers attempted to mold foreign systems.

The paucity of mathematical sources at the beginning stage of Jaina study limits the observations that can be made on continuities, shifts, and innovations in Jaina mathematics. But it can be determined that a decline of sacrifices and Vedic altar geometry greatly altered the animus for applied mathematics. Scribes of the Jaina period chiefly addressed practical secular daily problems of wages, profits, loss, impurities in precious metals, and speed required for covering distances in given times. The Greco-Babylonian astronomical transmission, the second major transmission of western knowledge, was to elevate problems related to astronomy and to foster the emergence of theory. Once contacts with the Hellenistic West broadened and grew frequent after Alexander's invasion, Indian astronomy attained a mathematical stage. Jaina literature examined planetary conjunctions, computed planetary longitudes for use in casting horoscopes, and computed solar and lunar eclipses, which were possibly associated initially with omens. While in cosmology Vedic *Puranas*, appearing after 550 B.C., described the earth as a flat-bottomed circular disc, Jaina canonical texts accepted the Greek spherical earth.

Like Vedic sages, Jaina authors were fascinated with the contemplation of large numbers. They used these to measure time and great distances traversed by the gods. One measure of time was a *shirsa prahelika*, which is equivalent to $756 \times 10^{11} \times (840,000,000)^{28}$ days. The Jains divided numbers into three groups, enumerable, innumerable, and infinite, with each of these divided into three subgroups. Perhaps for reasons of theological prudence or aesthetic symmetry, they rejected the concept that all infinities are equal. Their third groups of numbers had two infinities, the smaller *asmkhyata* or rigidly bounded infinite and the larger *ananta* or loosely bounded, infinitely infinite. Not until the nineteenth century was mathematics to achieve, in the work of Georg Cantor, the first sound distinction among transfinite numbers.

The Bakshali manuscript, a handbook of arithmetic and algebra, is perhaps the best source of late Jaina mathematics. Written on birch bark in old Sanskrit, this manuscript is in sutra form. Discovered near the northwest Indian border in 1881, it is now housed in the Bodleian Library at Oxford. Its date of origin is still disputed, George R. Kaye having considered it a twelfth-century work and Rudolph Hoernle, a later copy of a manuscript from the first Christian centuries. If Hoernle's dating is correct, as seems more likely, this manuscript helps fill the gap between the *Sulbasutras* and India's classical period, which historians locate roughly during the Gupta era extending from A.D. 320 to about 657. If so, a more complete study of it is needed to identify all later additions to the original manuscript.

The Bakshali manuscript, a treasury of information on Jaina arithmetic, is the first surviving Indian text systematically to treat the rule of three and compute square roots. The rule of three manipulates proportions. Here is an example: if four apples cost 60 cents, how much will fifteen cost? The answer is $(0.60) \times (15)/4 = \$2.25$. In the Latin West, this rule was later known as the golden rule. The manuscript finds square roots of nonsquare numbers by first breaking a nonsquare number A into $a^2 + r$. In modern notation, the square root is $\sqrt{A} = \sqrt{a^2 + r} = a + r/2a - (r/2a)^2/2(a + r/2a)$. Thus, for $A = 3$, $\sqrt{3} = \sqrt{1 + 2} = 1 + 2/2 - 1/2(2) = 1\frac{3}{4}$. This method is not so accurate as the iterative technique of the ancient Babylonians and Hero. The Bakshali manuscript also has a limited decimal place-value system with an improved variant of Brahmi numerals and a dot for zero. Its notation, a + sign or a dot over a numeral for minus, a large dot for the unknown x, and the abbreviation *bha* for *bhadanga* for division, appears to be a later addition to the original work.

The algebra of the Bakshali manuscript solves the equivalent of quadratic equations, introduces the equivalent of indeterminate equations in three unknowns, and discusses arithmetical and geometric progressions. The *Sulbasutras* had introduced problems leading to quadratic equations in India but gave no solutions. In modern notation, the Bakshali manuscript solves the equation $ax^2 + bx + c = 0$ in this way: $x = (\sqrt{b^2 - 4ac} - b) \div 2a$, a rule first explicitly stated in India a half-millennium later by Sridhara (ca. A.D. 900). Indeterminate equations were soon to become central to Indian algebra. One solved in the Bakshali manuscript is:

> One person possesses seven asavas, another nine hayas, and another ten camels. [Asavas and hayas are breeds of horses.] Each gives two animals, one to each of the others. They are then equally well off. Find the price of each animal and the total value of the livestock possessed by each person.

In symbolic terms, it lets $5x_1 + x_2 + x_3 = x_1 + 7x_2 + x_3 = x_1 + x_2 + 8x_3 = k$. Solving this requires finding the least common multiple of 4, 6, and 7, which is 84. Hence, the price of the asava is 42, the haya 28, and each camel 24, and the value of the livestock of each person is 262. The Bakshali manuscript gives not only rules for solving specific problems, but also demonstrations or proofs of the rules, a feature missing in Vedic literature.

9.3   India's Classical Period

A class of reformed astronomical compendia called the *Siddhāntas* probably dates from shortly before the Christian era through the fifth century A.D. Two important transmissions of astronomical knowledge from the West, the Greco-Babylonian between 200 and 400, during which Babylonian influence was possibly greater, and the beginning of the Greek from 400 to 1200, were crucial to this development. As many as eighteen *Siddhāntas* were written, but only portions of five survive.

During India's classical age, Gupta rulers unified northern and parts of western and central India through conquests, astute marriages, a relatively small and effective administration, and a cultural revival that produced several finished astronomical *Siddhāntas*. Extending from Kashmir on the west to Bengal on the east, the Gupta realm became the largest on the Indian subcontinent since Ashoka's six centuries earlier. During his reign from ca. 375 to 415, Chandragupta II turned his kingdom into an empire. At its height, Gupta rulers controlled ports on the Arabian Sea and possibly had the monarchs of Sri Lanka as vassals. Their economic strength lay in the control of rich veins of iron and the Ganges as an avenue for north Indian commerce. Peasants were taxed a fourth of their harvests during an age of bountiful agriculture. The peasantry raised

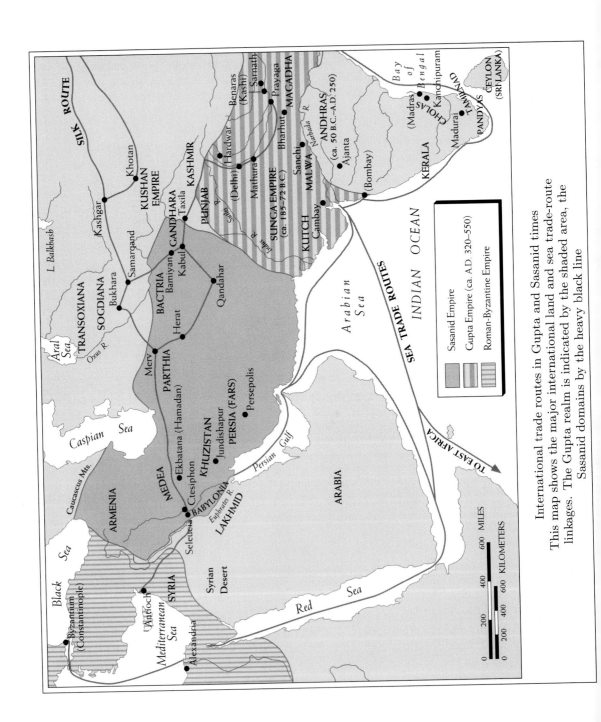

International trade routes in Gupta and Sasanid times
This map shows the major international land and sea trade-route
linkages. The Gupta realm is indicated by the shaded area, the
Sasanid domains by the heavy black line

rice, wheat, sugar cane, fruits such as pears, plums, peaches, and grapes, and spices, among them ginger and mustard. Gupta rulers and traders benefited from trade contacts with southeast Asia and China by sea route and camel caravan. Buddhist missionary zeal and new navigational knowledge helped stimulate this trade and Indian commercial settlements in southeast Asia. The Gupta age brought material prosperity for an urban elite and was characterized by internal peace and stability along with personal freedom. As Rome and China waned, the Gupta empire was the most peaceful and perhaps most civilized one on the planet.

The Guptas chiefly supported Brahmanic traditions in religion and the classical Hindu form of religious, social, and cultural life assumed shape, eclipsing Buddhism and Jainism. They encouraged all three religions, but preferred Hinduism and its four-caste social hierarchy. Brahmans were the top caste, followed by nobles, tradesmen, and servants, with outcasts or untouchables at the bottom. This hierarchical caste system was deeply embedded in Indian civilization. Brahmans, descendants of the Aryan priestly class, had accepted the Buddhist principle of *ahimsa*, noninjury to all living things, and were vegetarians. Sacredness of the cow illustrates their belief in *ahimsa*. The resurgence of Hindu polytheistic religion increased the use of Sanskrit, the sacred language of the Brahmans, and a related cursive Gupta script with a thirty-seven letter alphabet was developed.

The Gupta era represents a cultural peak of Indian civilization, when classical norms were established in literature, poetry, drama, music, painting, architecture, plastic arts, and scholarship. It is compared to Periclean Athens and Augustan Rome. The drama and poetry of Kālidāsa (fl. 400–415), known as the Shakespeare of Sanskrit letters, and an array of folk tales and fables that would influence Chaucer, Bocaccio, and the Grimm brothers, enriched Sanskrit literature. Kālidāsa's most famous drama, *Sakuntala Recognized*, a love idyll about a king, with characters who represent aristocratic ideas, elegantly depicts the beauties of nature. Painting and sculpture proliferated as central elements of temple decorations. The iconography of Hindu myth was set out, and temple frescoes included erotic scenes to reveal the beauty of the human form. Cave art reached its apogee in the Ajanta caves. To support scholarship, the Guptas founded a major center of learning at Nalands.

Two of the foremost intellectual achievements in the Gupta and medieval eras are the evolution of mathematical astronomy and associated contributions to trigonometry, numeration, and Diophantine equations. These accomplishments depended greatly on the Gupta establishment of regular channels of communication across the vast area of India and from abroad, which facilitated the Greco-Babylonian and Greek transmissions to India. These suggest that Indian trigonometry was built on Greek spherical astronomy, probably from the later Alexandrian period. Apparently, Indian astronomers drew mostly on Aristotelian sources, with a few additions from a corrupted form of Ptolemy. They seem unaware of most of Ptolemaic astronomy.

During Gupta and medieval times, Indian mathematical astronomers and mathematics had a limited institutional base concentrated in paksas, or schools of astronomy. The two most famous schools were Brahmapaksa, founded about 400, which centered in Ujjain, and Āryapaksa, founded around 500 in Kusuma-pura. Both cities were centers of learning in the north. The deity Brahma Paitāmaha reportedly revealed the idea of Brahmapaksa. Its basic text, *Paitā-maha Siddhānta*, which was completed in the early fifth century, continued the Indian penchant for large numbers. It has Mahayugas that consist of 4.32 million years and Kalpas that are 1,000 Mahayugas. The Brahmapaksa was influential in the northwest and west of India for over 1500 years, and the Āryapaksa in the northwest and south of India for even longer. Our chief concern is with the work of a quartet of Indian scholars: Āryabhata I, who founded Āryapaksa with his *Āryabhatiya*, Brahmagupta and Bhāskara II, who belonged to Brahmapaksa, and Madhava from Kerala in the south.

The mathematical ideas of the astronomer Āryabhata I (b. 476) appear in a slender volume on astronomy, entitled *Āryabhatiya*. This volume was completed in the thirty-six hundredth and first year of Kaliyuga when, the author mentions, he was twenty-three. Since scholars begin the Kaliyuga era in 3102 B.C. and this work was completed in 3601, a reasonable dating approximates A.D. 499. Aryabhata's statement places his birth in 476. He apparently wrote *Āryabhatiya* in Kusumapura (Pātaliputra) near the modern city of Patna in Bihar. Kusumapura was then the Gupta imperial capital and had been a major center of learning since the Jaina period.

Partly by refining sine tables, Āryabhata I codified and recast results that *Paitāmaha* and other earlier *Siddhāntas* had arrived at in trigonometry. The *Romaka* and *Pauliśa* have various prescriptions for determining the sine values, such as $\sin 45° = \sqrt{R^2/2}$, while the *Paitāmaha* and Āryabhata use a sine function where $R = 3,438$ units. This value of R derives from $c/d = [21,600 (\text{or } 360 \times 6)]/6,876$. As the *Pañcha Siddhānta* and subsequent work of Varāhamihira illustrate, Indian astronomers generally based sine tables on a radius of 120 parts, or double the value Ptolemy uses to compute chords. Āryabhata I was perhaps the first to associate the half-chord (*jiya andha*) with half the arc of a full chord, recognizing the equivalent of $\mathrm{Sin}_{120}\,\theta = \mathrm{Crd}_{60}\,2\theta$. With that recognition and his development of the equivalent of the equation $\sin(n+1)\theta - \sin n\theta = \sin n\theta - \sin(n-1)\theta - (1/225)\sin n\theta$, Āryabhata accurately computed half-chords for the quadrant of a circle divided into twenty four equal parts. The smallest arc is thus $3°45'$. These are sine tables of Varāhamihira, Āryabhata, and modern values:[295] Āryabhata also introduced early cosine and versine half-chords.

The three-part *Āryabhatiya* also includes a cookbook form of algebra with an emphasis on computation. Its mathematics, making up only thirty-three of its one hundred eighteen verses, gives sixty-six elaborate verbal instructions that build on alphabetic numerals. These rules are a potpourri of correct materials for arithmetical operations and algebra together with inaccurate approximations in geometry. The work lacks the symbols needed to make its algebraic procedures easier to follow, as well as a general logical method to provide coher-

Angle θ	Varāhamihira		Āryabhata		Modern value of $\sin \theta$
	$r \sin \theta$ $(r = 120')$	Computed $\sin \theta$	$r \sin \theta$ $(r = 3438')$	Computed $\sin \theta$	
3°45′	7′51″	0.06542	225′	0.06545	0.06540
7°30′	15′45″	0.13056	449′	0.13060	0.13053
11°15′	23′25″	0.19514	671′	0.19517	0.19509
15°	31′ 4″	0.25889	890′	0.25962	0.25882
18°45′	38′34″	0.32139	1105′	0.32141	0.32143
22°30′	45′56″	0.38278	1315′	0.38249	0.38268
26°15′	53′ 5″	0.44236	1520′	0.44212	0.44229
30°	60′	0.50000	1719′	0.50000	0.50000
33°45′	66′40″	0.55556	1910′	0.55556	0.55556
37°30′	73′ 3″	0.60875	2093′	0.60878	0.60876
41°15′	79′ 7″	0.65931	2267′	0.65910	0.65935
45°	84′51″	0.70708	2431′	0.70710	0.70711
48°45′	90′13″	0.75181	2585′	0.75189	0.75184
52°30′	95′13″	0.79347	2728′	0.79348	0.79335
56°15′	99′46″	0.83139	2859′	0.83159	0.83147
60°	103′56″	0.86611	2978′	0.86620	0.86602
63°45′	107′38″	0.89694	3084′	0.89703	0.89687
67°30′	110′53″	0.92402	3177′	0.92408	0.92388
71°15′	113′38″	0.94694	3256′	0.94706	0.94693
75°	115′56″	0.96611	3321′	0.96597	0.96593
78°45′	117′43″	0.98097	3372′	0.98080	0.98079
82°30′	119′	0.99167	3409′	0.99156	0.99144
86°15′	119′45″	0.99792	3431′	0.99796	0.99786
90°	120′	1.00000	3438′	1.00000	1.00000

ence. Āryabhata, nevertheless, provides successive verbal steps for solving linear and quadratic equations, and he is the first among Indians to originate explicit first-degree indeterminate equations ($ax + c = by$, where a, b, and c are positive integers and x and y are unknowns) to determine planetary orbits. These he solves by continued fractions. Euclid's algorithm leads in the direction of continued fractions. Where $m > n$, for example $m = m_1 n + r$ and $m/n = m_1 + r/n$. Continuation of the process gives $m/n = m_1 + 1/[m_2 + 1/m_3 + \ldots]$. In a similar fashion, this process applies to the case $m < n$. Āryabhata's greatest interest was in computation, and he gauged the correctness of method by the utility of answers. He gives correct rules for summing integers, squares, and cubes. The persistence of this interest in computation is shown by his fifteenth-century commentator Nilakantha Somayaji, who, by piling rectangular strips in a series of sizes in inverse rows, reconstructs the rectangle and demonstrates in modern notation, that $\sum_{i=1}^{n} i(i+1)/2 = n(n+1)(n+2)/6$.

Āryabhata's work reflects the poverty of geometric thought in Indian mathematics. He was dependent on rules that inaccurately approximate the volumes of pyramids: $v_p = \frac{1}{2}(A_{\text{base}}h)$; and spheres: $v_s = (A_{\text{greatcircle}})^{3/2}$. Three of his four values for the ratio of the circumference of a circle to its diameter are $3\frac{3}{4}$, $3\frac{27}{191}$, and $\sqrt{10}$. His other approximation, meticulously applying the method of exhaustion, achieves greater accuracy:

> Add four to one hundred, multiply by eight and then add sixty-two thousand. The result is approximately the circumference of a circle of diameter twenty thousand. By this rule the relation of the circumference to diameter is given.

This gives 62,832 divided by 20,000 or $\pi = 3.1418$, a value close to Ptolemy's. The circumference is nearly triple its 21,600 minutes of arc. Yet Āryabhata in practice used the neat value of $\sqrt{10}$, a practice followed by Indian sages until the twelfth century.

Current research on Jaina mathematical astronomy indicates that Āryabhata may not be, as was once thought, the pioneer of Indian mathematics, but instead was mainly an influential elaborator. As the central text of the Āryapaksa, the *Āryabhatiya* was devotedly studied and had commentaries written on it for centuries. The talented mathematician and astronomer Varāhamihira (505–587) of Ujjain improved on it. Among the three Indian trigonometric half-chords, he discovered relations that in modern symbolism are $\cos\theta = \sin(1/2\pi - \theta)$, $\sin^2\theta + \cos^2\theta = 1$, and $\sin^2\theta = \frac{1}{4}(\sin^2\theta + ver\sin^2 2\theta) = \frac{1}{2}(1 - \cos 2\theta)$, and he computed a more accurate sine table using a radius of 120 parts, as the preceding chart shows. Perhaps the foremost commentator on the *Āryabhatiya* was Bhāskara I, who prepared the *Bhasya*, that is, large commentary, in 629. Later commentators include Somesvara in 1040, Suryadeva Yajvan in 1191, and Milakantha Somayaji, who died in 1545. India's first artificial satellite, launched by the USSR in 1975, was named Āryabhata.

Brahmagupta (598–ca. 665) was the next eminent Indian mathematician of the Ujjain school after Varāhamihira. His reputation rests mainly on his comprehensive astronomical text *Brāhmasphuta Siddhānta*, the second representative text of the Brahmapaksa. He completed this volume in 628, when he was thirty and resided in a capital city called Bhillamala (modern Bhimal) in southern Rajasthan. Among its twenty-four chapters are four and a half on trigonometry, algebra, arithmetic, and mensurational geometry. Brahmagupta addresses these subjects with an ability ranging from brilliant in algebra through proficient in trigonometry to pedestrian in geometry. From the *Maha Siddhānta*, a text based on the *Brāhmasputa Siddhānta*, medieval Arab authors first learned Indian astronomy and had their interest in trigonometry aroused. At the age of sixty-seven, Brahmagupta wrote a second treatise in astronomy, the *Khanda Khadyaka*. The Muslim Abbāsid *Zij al-Arkand* and the *Pauliśa Siddhānta* are based on its astronomical parameters.

Brahmagupta obtains sines of intermediate angles in a sine table by a method equivalent to the Newton-Stirling interpolation formula. Here in modern symbols is his calculation of the sine of $67°$: let $h =$ the interval or $15°$ in this case, the residual angle $\Delta\theta = (67 - 60) = 7°$ or $420'$, and the known tabulated difference between the sines of $45°$ and $60°$ is D_p or 24, while the difference between $60°$ and $75°$ is D_{p+1} or 15. Brahmagupta's interpolation formula is

$$\Delta\theta/h[(D_{p+1} + D_p)/2 + \Delta\theta/h(D_{p+1} - D_p)/2].$$

These figures give $7/15[(15 + 24)/2 + 7/15(15 - 24)/2] = 8.12$. Brahmagupta's Indian sine (*jya*) or $67° = 130 + 8.12 = 138.12$, which is near to $150\sin 67° = 138.08$.

In spherical trigonometry, Brahmagupta expanded on materials from both the *Paitāmaha* and the *Āryabhatiya*. His rule for determining the solar altitude is quite exceptional, though not the simplest. In modern symbols, it is equivalent to

$$\sin\theta = \Big[(R^2/2 - \sin^2\alpha) \times 12^2/(72 + s_0^2)$$

$$+ 12s_0 \times \sin\alpha^2/(72 + s_0^2) + 12 \times s_0\sin\alpha/(72 + s_0^2)\Big]^{1/2},$$

where R is the radius and s_0 the base of a gnomon shadow triangle that depends on the height of the sun in the sky. This rule is only valid when the sun is situated in the northern hemisphere and makes an angle of $45°$ between the shadow and east-west line. The equal sides of the shadow right isosceles triangle are s_0.

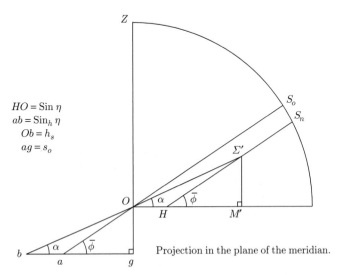

$HO = \text{Sin }\eta$
$ab = \text{Sin}_h\,\eta$
$Ob = h_s$
$ag = s_o$

Projection in the plane of the meridian.

Brahmagupta's greatest inventiveness, however, is in the theory of equations in chapter 18 of *Brāhmasphuta Siddhānta*. The title of the chapter, "Kuttaka," literally means "pulverizer" but is generally taken to mean algebra. Working with a nearly syncopated algebra consisting of words and abbreviations but no symbols, Brahmagupta solves quadratic equations of the type $px^2 + qx = r$. He first takes the number of the opposite side of the equation from the square (r) and multiplies it by four times the coefficient of the square $(4pr)$. He then adds the square of the coefficient of the x term (q^2), takes the square roots of this value minus the coefficient of the x term, $[\sqrt{(4pr + q^2)} - q]$, and divides that value by twice the coefficient of the square term $(2p)$ to obtain two roots or unknown x's.

Brahmagupta excelled in solving linear and quadratic Diophantine equations. He provided the first general solution of the linear form $(ax + by = c)$. He grasped that this equation is solvable if and only if the greatest common divisor of a and b also divides c. Diophantus had given only one rational solution; Brahmagupta offers a method for obtaining all possible integral solutions. His lengthy examination of quadratic Diophantine equations correctly identifies as essential to a general solution those of type $y^2 = ax^2 + 1$ (now called the Pell equation). He begins his solution of the equation $y^2 = 8x^2 + 1$ by assuming a smallest and largest root. If $x = 1$, then $y = 3$. He next arranges this material crosswise

Crosswise multiplication and addition of products gives $3 + 3 = 6 = x$, and $y = 17$. This yields the new crosswise set

Repeating this procedure results in $17 + 18 = 35 = x$ and the new set

Brahmagupta's solution sets are $(1, 3)$, $(6, 17)$, $(35, 99)$, $(204, 577)$, ... Once he had the first solution, infinitely many integral solutions are possible. This crosswise procedure was later attributed to Leonhard Euler in the West. Brahmagupta applies this method to solve problems of finding planetary longitudes. His treatment of Diophantine equations was not surpassed until five centuries later by Bhāskara II. By itself, it is enough to ensure him a prominent place in the history of mathematics.

As a preface to the theory of equations in the *Brahma Sputa Siddhanta*, Brahmagupta introduces rules for operating with negative numbers and zero that are mostly correct. Positive numbers divided or multiplied by positive or, similarly, negative by negative give affirmative results; and negatives subtracted from zero give positive results. Uninfluenced by Greek formalism, Brahmagupta was not afraid to introduce arithmetical operations with an early form of zero. Unaware of subtleties in operating with zero, he assumed erroneously that a cipher divided by itself is nought $(0/0 = 0)$ and that a positive or negative number divided by zero is a fraction $(\pm a/0)$. At the same time, he recognized that dividing by zero does not yield a unique result, as is the case in multiplying any two other numbers. Any finite number multiplied by zero gives the same answer of zero. Today we say that division by zero leads to an undefined result.

Brahmagupta's numerical calculations with elaborate rules produced substantive results. He poses and solves mainly practical problems, such as this:

> 500 drammas were loaned at a rate of interest not known. The interest on the money for four months was loaned to another at the same rate and amounted in ten months to 78. Give the rate of interest of the principal.

A rule for solving the problem precedes each solution. Since no general explanation of rules or sources is given, presumably many were derived either by trial-and-error procedures or imported. We thus do not know the origins of Brahmagupta's rules for summing series of square and cube integers:

$$\sum n^2 = [n(n+1)(2n+1)]/6 \quad \text{and} \quad \sum n^3 = [\tfrac{1}{2}n(n+1)]^2.$$

Brahmagupta's geometry is superficial and generally formalistic. He gives Hero's formula for computing the area of a triangle by use of its sides, $A = \sqrt{s(s-a)(s-b)(s-c)}$, where $s =$ the semiperimeter and a, b, and c the sides, as well as a rule for quadrilaterals, $A = \sqrt{(s-a)(s-b)(s-c)(s-d)}$, but fails to note that the latter is correct only for cyclic quadrilaterals. Hero's formula is a corollary of it. In that case, $d = 0$. Brahmagupta approximates the gross areas of triangles and quadrilaterals. To find the area of an isosceles triangle, for example, he multiplies half the base times one of the equal sides, which differs considerably from the correct value. For ordinary calculations involving the ratio of the circumference of a circle to its diameter, he lets the value $\sqrt{10}$ or 3 suffice.

9.4   Medieval Indian Mathematics

Theoretical mathematics continued to evolve in the writings of Mahavira (fl. ca. 850), Sridhara (fl. ca. 900), Āryabhata II (fl. ca. 900), and commentators at paksas from about 650 to 1100, culminating in part by 900 in a mature decimal position system of numbers built on nine Gwalior symbols for the unit digits and a zero. Stone and temple wall inscriptions, epigraphs, the *Āryabhatiya*. and the *Brāhmasphuta Siddhānta* reveal, however, that word and alphabetic numerals long continued in use.

Two parts of the decimal positional system, the decimal base and place value (without the zero), are very ancient. In India the decimal base had first evolved in Harappa and was the only base in Sanskrit literature. The source of the place-value arrangement among Indian scribes is disputed. George Joseph and the Netherlands historian Hans Freudenthal assert that it was an indigenous development. Joseph maintains that the Indian fascination with large numbers, up to 10^{62} in the *Vedas* and *Ramayana* and even beyond among the Jains, was crucial to the emergence of place value. With word numbers, a numerical value had come to be associated with the order from left to right or right to left of the separate words for tens, hundreds, thousands, and so forth. Other scholars conjecture that place value came from abroad during the Greco-Babylonian transmission from about 200 to 400, which conveyed the sexagesimal positional system with a small omicron for a medial zero, or from Chinese pseudo-positional rod numerals.[296]

Symbols for numbers, up to the Jaina period the missing element for decimal positional numeration, had early predecessors in Kharosthi numerals, which appear in inscriptions dating from the fourth century B.C. to the second century A.D. Kharosthi numerals, which resemble Herodianic and Roman numerals, use a repetitive or grouping principle and read from left to right. There are symbols for 1, 4, 10, and 20, with numbers up to 100 built additively and special symbols for higher powers of 10. In this efflorescence of learning during Ashoka's reign (d. 232 B.C.), the more developed Brahmi symbols appeared. Brahmi numeration has symbols for digits 1 and 4 through 9, multiples of 10, and multiples of 100 up to 900. The Brahmi system thus resembles the ciphered Greek Ionian system. After Brahmi writing became the mother Indian alphabet, Brahmi numerals superseded the Kharosthi.

	1	2	3	4	5	6	7	8	9	10	...	20	...
Kharosthi	I	II	III	X	IX	IIX	IIIX	XX		⌐	...	⌇	..
Brahmi	—	=	≡	Ⴟ	Ⴌ	⅙	7	Ⴙ	Ⴢ	Ⴀ	...	℗	..
Gwalior	٦	₹	₹	Ⴔ	५	⊂	⊃	⎰	९	٦o	...		..

The Kharosthi numeral for nine is not known for certain.

Three types of Indian numerals given in their chronological order.

The late *Sulbasutras* and the Bakshali manuscript had limited decimal position systems using Brahmi symbols, but it was Āryabhata's work and that of Varāhamihira in 587 that made it standard. Āryabhata constructed his Brahmi numeral system so that "from place to place each [numeral] is ten times the preceding." In his system, the number "eight hundred fifty six" was written no longer as 800′50′6, as in the old Brahmi manner, but as the Brahmi equivalent of 856. Only after the detailed work of Varāhamihira on the subject did this decimal positional system with Brahmi numerals gain wide acceptance within Indian learned circles.

The place-value principle was thus early added to the Brahmi numerals, but it took nearly three-quarters of a millennium to develop a widely accepted system with refined Brahmi numerals. Encipherment with Brahmi numerals for 1 and 4 through 10 was a step toward place-value notation, whose general extension was impeded until numerals other than the unit digits were discarded. To the seventh century, Indian mathematician-astronomers also lacked symbols for 0 and all digits 1 through 9 with which to express any number no matter how large.

Not until the late ninth century did Indian numeration mature into a full-fledged decimal positional system with an operational 0. Whether a single stimulus or a combination of factors produced an operational zero is not known. The West had transmitted the Neobabylonian medial zero and Ptolemy's o for *ouden* (nothing) indicating a missing place. Joseph Needham cites Indochinese studies of Taoist emptiness as another possible source. The zero concept was congenial to the void of Hindu philosophy, and the word *sunya* for zero appears in the *Siddhāntas*. During extensive cultural interpenetrations between India and southeast Asia, small zeroes appeared in a Malay inscription in Sumatra and a Khmer inscription in Cambodia in the seventh century and another in eighth-century Vietnam. An inscription made in 876 (Samvat 933) shows that the Brahmi symbols had evolved into decimal place-value Gwalior numerals for all nine digits and a 0 similar to our modern symbol. Gwalior numerals, in turn, were ancestors to west Arabic or gobar numerals, of which the symbols for 1, 2, 3, 7, and 9 had a marked likeness to the Gwalior equivalents. By the late fifteenth century, the gobar numerals were prominently used in Europe, largely popularized by Simon Stevin. In the early sixteenth century, Albrecht Dürer gave them their present design.

The last major Indian mathematician and astronomer of the Ujjain school during the medieval period was Bhāskara II (1115–ca. 1185), or Bhāsharacha-raya, as he is better known in India. Acharaya means master or teacher. Bhāskara II was perhaps the most original and influential Indian scholar of the medieval period. Closing verses of his *Siddhānta Śiromani* of 1150 and an inscription at Patna describe him as the son of Mahesvara of the Sandilya *gotra*; that is, he was descended from Brahmans. His home city apparently was Vijjadavidda, now Bijapur in Mysore state. He became the director of the astronomical observatory in Ujjain and led the Brahmapaksa. Like his father, he was also a prominent astrologer. His younger contemporaries and immediate successors judged his work so highly that a *matha* (educational institution) was endowed in 1207 for the study of his writings, beginning with the two part *Siddhānta Śiromani* ("Head Jewel of an Astronomical System"). The first is a detailed, systematic study of astronomy, and the second part examines the sphere covering its nature and employment in geography, astronomical models, spherical trigonometry, and astronomical computations.

The *Siddhānta Śiromani*, which is based on Brahmagupta's *Brāhmasphuta*

Siddhānta and the subsequent Brahmapaksa tradition, is the third and last text of the Brahmapaksa.

This text, which assumes that the prime meridian passes through Lanka and Ujjain, gives circumferences and radii of planetary epicycles from the *Paitāmaha*. It makes more accurate the representation of the daily motions of the moon and the mean apparent diameters of the planets given by Brahmagupta and the computation from the *Paitāmaha* of times of heliacal risings and settings of planets. Bhāskara corrects but still offers crude values of mean daily motions of planets. His text also covers cosmography, geography, and astronomical equipment.

The *Siddhānta Śiromani* presents an advanced trigonometry and is more theoretical than its Indian astronomical predecessors. Its Goladhyaya section especially explains the theory behind astronomical computations as well as algebraic, numerical, and trigonometric procedures. It begins with the *Paitāmaha's* sine table, except for an improved value of $\sin 60°$, but expands on it to offer this more accurate sine table with $R = 120$ at $10°$ intervals:

	$\operatorname{Sin}\theta$	$R\operatorname{Sin}\theta$
10	21	20.8
20	41	41.0
30	60	60
40	77	77.1
50	92	91.9
60	104	103.9
70	113	112.7
80	118	118.1
90	120	120

Bhāskara gives trigonometric rules superior to those in previous Indian work. For example, he states the equivalent of $\sin(\alpha \pm \beta) = \sin\alpha\cos\beta/R \pm \cos\alpha\sin\beta/R$, where $R = 3438$. The Persian Abū al-Wafā had earlier known this rule. Bhāskara's advanced trigonometric knowledge suggests an influence from medieval Islamic astronomy, but direct evidence for this is lacking.

An interesting mathematical feature of the *Siddhānta Śiromani* is its differing values for π. In the section, entitled "Ganitadhaya," Bhāskara computes the radii of the orbits of the sun and moon with a ratio equal to $3{,}927/1{,}250$ or 3.1416, a ratio he initially found by inscribing regular polygons with as many as 384 sides in a circle. At another point, he computes the ratio of the circumference of the Earth to its diameter as $4{,}967/1{,}581 - 1/24$ or 3.1413. Both are less accurate than Brahmagupta's best value.

Two earlier works, *Lilavati* and *Bijagnita*, had established Bhāskara's reputation. *Lilavati* ("The Beautiful") has thirteen chapters on algebra, arithmetic, and geometry, while the more systematic *Bijagnita* ("Seed Counting" or "Root Extraction") consists of twelve chapters on numbers and algebra. The name Lilavati refers to a lady, possibly Bhāskara's daughter. The fifteenth-century

Persian poet Fyzi, who translated this text, writes that Bhāskara's meddling through the casting of a horoscope ended his daughter's only chance for marriage. From her horoscope, Bhāskara discovered the best time for her marriage. He then placed a cup with a small hole at the bottom in a vessel of water, so that it would sink at the propitious hour. But when Lilavati out of curiosity bent over the vessel, a pearl dropped and blocked the hole. The propitious hour passed without the cup's sinking. To console her, Bhāskara promised to name his first book after her, "which will last to the latest times." *Lilavati* and *Bijagnita* are sometimes thought to have been part of the *Siddhānta Śiromani*, but that is questionable. Together with that text, they comprise the most complete presentation of Indian mathematics before modern times.

Among Bhāskara's improvements on Brahmagupta and the Brahmapaksa's efforts in algebra is his introduction of some symbols into what had been basically rhetorical problems. To represent subtraction or negative numbers, he places a dot above numerals, "pricking" them in his phrase. He indicates the second power by an initial for the word "square" and the third power by that for "solid," and he represents square roots by abbreviations, such as *ka* for *karana* (square root). He does not limit unknowns to a single symbol, but designates them by initial syllables of words for colors, such as red, blue, and yellow. These do the work of our x, y, and z. This notation is closer to modern symbolism than is Diophantus'.

With a place-value number system of positive and negative numbers and his limited notation that allows a better treatment of equations, Bhāskara reduces quadratic equations to a single type and solves them by completing the square or by radicals. His greatest strength lies in solving first and second degree indeterminate equations. His problems, some of them fanciful and meant for pleasure, are more varied than Brahmagupta's, and his work is filled with ingenious procedures for solving them. In the *Bijagnita,* he found the general solution of the linear diophantine equations that in modern symbols is $ax + by = c$, when a, b, and c are integers. For a and b relatively prime, the solutions are $x = p + mb$ and $y = q - ma$, where m is any integer. In the *Bijagnita* he also solves a more general problem of the type $ax + by + cz = d$, where a, b, c, and d are integrals. The problem is:

> The horses belonging to four people are 5, 3, 6, and 8. The camels belonging to the same are 2, 7, 4, and 1. The mules belonging to them are 8, 2, 1, and 3 and the oxen 7, 1, 2, and 1. All four people being equally rich, tell me the price of each horse and the rest.

Knowing that various answers are possible, Bhāskara gives one result as 85 for horses, 76 for camels, 81 for mules, and 4 for oxen. Intrigued, like previous mathematicians, by the relationship we know as Pell's equation $(ax^2 + 1 = y^2)$, he makes it a focus of his discussion of quadratic indeterminate equations. He solves five cases of it, where

$$x = 6, \qquad\qquad a = \;\; 8, \quad \text{and } y = 17;$$
$$x = 3, \qquad\qquad a = 11, \quad \text{and } y = 10;$$
$$x = 3, \qquad\qquad a = 32, \quad \text{and } y = 17;$$
$$x = 5{,}967, \qquad\quad a = 67, \quad \text{and } y = 48{,}842;$$
$$x = 226{,}153{,}980, \quad a = 61, \quad \text{and } y = 1{,}776{,}319{,}049.$$

The last two calculations are quite difficult. Along with its provision of a limited symbolism, Bhāskara's advanced treatment of indeterminate equations makes his work the capstone of medieval Indian algebra.

In computation, Bhāskara skillfully explains operations with zero and negative numbers, giving rules known since Brahmagupta and detailed by Sridhara (ca. 900) in the standard work on the subject, *Pataganita*. The *Lilavati* presents the correct rules for addition, subtraction, and multiplication with zero and retains the problematic status of division by zero. Division by a cipher, he says, is questionable because, in the reverse process of multiplication, "the product of a cipher is nought," rather than a unique value. Bhāskara was later to exhibit deeper insight into the consequences of permitting division by zero. His *Bijagnita* holds that a quantity divided by zero becomes a fraction, "which is termed an infinite quantity." Thus, 5 divided by 0 equals infinity ($5/0 = \infty$). Bhāskara holds these unusual fractions to be singular entities, whose form could not be changed if inserted into computations or extracted from them. But his further comment that $p/0 \cdot 0 = p$ shows that he still lacks a firm understanding of this situation. Bhāskara's inclusion of negative numbers among solutions of equations and his acceptance of both positive and negative square roots encouraged the acceptance of negative numbers. Recognition of these remained troublesome, as his solution of this problem, which is expressed in a utilitarian form, shows:

> The fifth part of a troop less three, squared, had gone to a cave and one was in sight, having climbed on a branch. How many were these?

In modern notation, this becomes $(x/5 - 3)^2 + 1 = x$ or $x = 50$ and 5. The second solution of 5, however, that gives $(x/5 - 3) = -2$ is not allowed, since a negative number of troops is meaningless. But Bhāskara's working with negative numbers and his correct grasp of the convention of signs—minus times minus equals plus, and plus times minus equals minus — make him a perceptive mathematician.

Among the many computational rules in the *Lilavati* and *Bijagnita*, some are geometric. Pythagorean triples are computed by what amounts to $2mn$, $m^2 - n^2$, and $m^2 + n^2$, a general rule for finding them, known to Brahmagupta and Mahavira and effectively having appeared in Euclid's *Elements* X. From Vedic times, right-angled triangles had interested Indian mathematicians. A problem involving them is this:

A snake's hole is at the foot of a pillar and a peacock is perched upon its summit. Seeing a snake at a distance thrice the pillar's height and gliding to its hole, he pounces obliquely upon him. Say quickly, at how many cubits from the snake's hole do they meet, both proceeding an equal distance?

Inspired by the Chinese *Nine Chapters*, Bhāskara provides a paper folding proof of the Pythagorean theorem. He divides the square of the original hypotenuse (c) into four right triangles with the sides of the square as their hypotenuses, which leaves a central square with side $(a - b)$. Reassembling these figures, he forms the areas of two squares, as the drawings indicate.

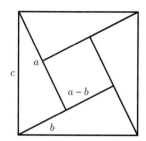 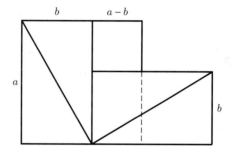

In another refreshing turn, Bhāskara criticizes mathematicians who use diagonals to compute areas of quadrilaterals without first checking to see whether the diagonals are determinate. People who blindly follow computational rules without adequate conceptualization he calls "blundering devils." His geometric rules include prescriptions for computing the surface and volume of a sphere, which Archimedes had discovered and highly prized.

Lilavati gives five methods of multiplication, including the *gelosia* (lattice or grating) method, as it was later named. The *gelosia* method generally avoids having to carry over amounts mentally until the final addition. To multiply 375 by 294, first draw a three-by-three grid with diagonals beginning in the square on the bottom right. Next, multiply the numbers at the head of each column and write the results in the appropriate squares.

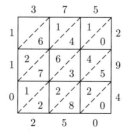

Thus, in the first row, $3 \times 2 = 6$, and $7 \times 2 = 14$, and $5 \times 2 = 10$. Begin the second row with $9 \times 3 = 27$. The diagonals separate two digit numbers.

Next, add the digits in the diagonal rows beginning in the lower right corner and, when necessary, carrying over amounts to the next row. This gives 0, and $5 + 2 + 8 = 15$, so write down 5 and carry the 1. A continuation of this process gives the numbers 1, 1, and 0 on the left side. The answer is 110,250. This *gelosia* method passed to the Arabs and from them to Renaissance Italy. It was employed until the 1600s, when it mainly disappeared, perhaps because it was more difficult than other forms to put into the new print.

The death of Bhāskara ends most of the original contributions to mathematics in Ujjain and northern India for the rest of medieval times, but not on the southwest coast of India in Kerala. Sheltered by high mountains to the east and the Arabian Sea to the west, Kerala enjoyed a measure of tranquility, while the rest of India underwent political upheaval. After 1200, new studies occurred there in astronomy and trigonometry with series expansions for computing values of half chords. We do not know how the Kerala astronomer-mathematicians came to work on these series.[297] No obvious external source exists, but perhaps some influence stemmed from China.

Kerala was a major center of the maritime spice trade that exchanged goods with Babylon and the Arabian peninsula to the west and with southeast Asia and China to the east. Vasco da Gama visited Kerala on his pioneering voyage to India. Donald Lach has uncovered a later technology transfer from Kerala to Europe, and it remains to be discovered whether there was a similar transmission of mathematical knowledge.[298]

Apparently, the central figure in Kerala mathematics is the astronomer and mathematician Madhava of Sangamagramma (ca. 1340–1425). Although his mathematical writings are lost, his surviving astronomical writings and comments upon his trigonometric series expansions by such later Kerala scholars as Nilkantha (1445–1545), Jyesthadeva (ca. 1550), Narayana (ca. 1500–1575), Sankara Variar (ca. 1500–1560), and Putumana Somayaji (ca. 1660–1740) reveal his mathematical originality.

The study of inverse trigonometric functions was critical in the late seventeenth-century creation of infinitesimal calculus in Europe. A significant finding was James Gregory's assertion that $\tan^{-1} x = x - x^3/3 + x^5/5 - \ldots$ for $-1 \leq x \leq 1$. The *Yuktibhasa* of Jyesthadeva, dated at about 1550, indicates that Madhava had already discovered the Gregory power series for arctan x:

· The first term is the product of the given sine and radius of the desired arc divided by the cosine of the arc. The succeeding terms are obtained by a process of iteration when the first term is repeatedly multiplied by the square of the sine and divided by the odd numbers 1, 3, 5, ... The arc is obtained by adding and subtracting (respectively) the terms of odd rank and those of even rank. It is laid down that the (sine of the) arc or that of its complement whichever is smaller should be taken here (as the given sine). Otherwise, the terms obtained by this above iteration will not tend to the vanishing magnitude.[299]

Letting the Indian sine and cosine equal $r \sin \theta$ and $r \cos \theta$ gives

$$r\theta = r(r \sin \theta)/r \cos \theta - r(r \sin \theta)^3/3(r \cos \theta)^3 + 4(r \sin \theta)^5/5(r \cos \theta)^5 - \ldots$$

or $\theta = \tan \theta - \tan^3 \theta/3 + \tan^5 \theta/5 - \ldots$, which is equivalent to Gregory's series. According to *Tantra Sangraha* by Nilakantha and *Karana Paddhati* by Putumuna Somayazi, Madhava derived an important result from that series. If in radian measure $\theta = 45° = \pi/4$, the series gives $\pi/4 = 1 - 1/3 + 1/5 - 1/7 + \ldots$, or the Leibniz series. Leibniz's precise computation of π that had this connection to the arctangent series was a major breakthrough leading to the early calculus. The incommensurability of π had likely fostered these series expansions in late medieval India. In *Kriyakramakari*, Sankara Variar states that Madhava, working with a circle with a diameter $d = 9 \times 10^{11}$, computed π to be 3.14159265359. This is correct to eleven decimal places.

The *Yuktibhasa* suggests that, two and a half centuries before Newton, Madhava derived the power series for the sine and cosine. He worked with arc a, radius r, and angle θ, as shown here.

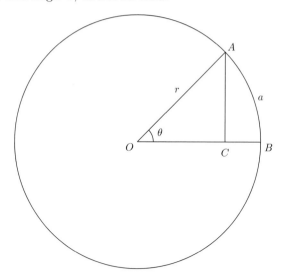

In modern notation, Madhava derived the result

$$r \sin \theta = a - a^3/3!r^2 + a^5/5!r^4 - a^7/7!r^6 + \ldots.$$
$$\text{For } r = 1, 0 \to \theta, \text{ and } 0 \to a$$
$$\sin 0 = 0 - 0^3/3! + 0^5/5! - 0^7/7! + \ldots.$$

Similarly for the Indian cosine,

$$r \cos \theta = r - a^2/2!r + a^4/4!r^3 - a^6/6!r^5 + \ldots,$$

and letting $r = 1, 0 \to a$, and $0 \to \theta$ gives

$$\cos 0 = 1 - 0^2/2! + 0^4/4! - 0^6/6! + \ldots.$$

Possibly, Madhava used these series to compute a sine table for twenty-four equal intervals in the quadrant of a circle. Almost all his table values are correct to eight or nine decimal places, an accuracy not achieved in Europe for two more centuries.

The long neglected Kerala mathematics is only in an early stage of study, largely begun in the mid-twentieth century by C. T. Rajagopal. Many basic points still need clarification. Like the Pythagoreans, Indian tradition attributes to a master the discoveries of his school. Madhava's exact contributions must be determined. Kerala mathematicians lacked a facile notation, a concept of function in trigonometry, and advanced work on conic sections. Did they nonetheless recognize the importance of inverse trigonometric half chords beyond computing astronomical tables and detect connections that Newton and Leibniz saw in creating two early versions of calculus? Apparently not.

Part III

The Islamic World, Latin West, and Maya America from the Middle Ages to ca. 1500

10

Mathematics in the Service of Religion

Tolle numerum omnibus rebus et omnia pereunt.
[Take away the number from all things and all things
shall perish.]
— Isidore of Seville

Only a trickle from the voluminous reservoir of Greek mathematics passed to
the Romans and their Latin Christian successors. Concurrently, Greek-speaking
Byzantines, who served as the legacy's curators, facilitated its transfer to receptive Islamic scholars. In the early medieval Latin West, long neglect nearly
obliterated the theoretical foundations of mathematics, disuse made operations
arcane, and content shrunk to an elementary level. Practical mathematics continued mostly in the hands of artisans. Despite this severe contraction, the
prestige of mathematics preserved vestiges of content and method from absorption into magic and myth.

This chapter will sketch the decline of mathematics in the early medieval
Latin West as revealed in attitudes of Church Fathers; in texts—the handbooks, most of them Boethian, and surveying manuals for monastic and palace
school students; and in leading figures: Bede, Alcuin, and Gerbert. A brief appendix to this chapter addresses Maya mathematics during essentially the same
period, roughly between 275 and 900. Geographically and culturally isolated
from Eurasia, talented Maya scribes established within an even closer bond between religion and mathematics, a high culture employing place-value notation,
including zero symbols as well as sophisticated calendars and chronologies.

10.1   Early Medieval Europe

In antiquity the development of mathematics depended largely on specialized
research within a small community of savants with financial support. Impover-
ishment of mathematics, already considerable in late Hellenistic times, increased
with the Roman annexation of Egypt in 30 B.C. While Alexandria, the jewel of
Egypt, had been a magnet for eminent geometers—recall Euclid, Eratosthenes,

Archimedes, and Apollonius—it was even more a nursery for cults of mystical illumination. Both strains appear in the second Christian century in Claudius Ptolemy, a native citizen, eminent in astrology and mathematical astronomy. The practical yet deeply superstitious Romans accepted the goddess of the Alexandrian temple, Isis, but neglected the legacy of theoretical geometry.

Twenty-five years before the Roman annexation of Egypt, Cicero, who was executed in 43 B.C., had made in *On the Orator* the first known Latin reference to Euclid. There he is mildly condescending toward mathematics, depicting it as an esoteric and constricting pursuit. Elsewhere, he cautions that mathematics can be "vicious" if it diverts students from moral philosophy. Yet he found and restored the lost tomb of Archimedes, seeking public recognition not for a neglected mathematical genius, but for a great patriot and engineer, who in foiling an entire Roman attack force was a model for Roman citizens. Plutarch's *Parallel Lives* presents Archimedes as akin to a wonder-working magician. Cicero represented and advocated a late antique education that for its promotion of civic virtue placed the study of rhetoric above such branches of learning as mathematics. In this setting, translators lacked the incentive to prepare Latin versions or paraphrases of Greek mathematical works. Roman translators remained untutored in higher mathematics and no translations were made. Whoever heeded the rhetorician Quintillian (fl. 90) to study Greek before Latin could still read the texts in the Alexandrian library, though very few actually did.

By the end of the fourth century, ethnic tensions and economic stress had undermined the appeal of bilingual education in the West and the resources for it. Byzantine culture was older, richer, and deeper than that of the Latin West, particularly that of northern Europe. The East was superior in wealth and trade. Unlike its Roman counterpart, the Byzantine aristocracy had not abandoned the cities for country estates. The East had more defensible borders as well as superior armed forces. Disparity between a wealthy East and an impoverished West bred resentments. The initial, experimental military and administrative division of the empire in 285 had deepened in 330 when Constantine founded Constantinople as a new imperial center to replace Rome. Formal partition came in 395 upon the death of Theodosius. Except during Justinian's sixth-century reign, this partition was permanent. Within a few generations in the West, literacy declined, vulgar Latin replaced the classical form, Germanic languages became common, and command of Greek—necessary for access to Greek higher mathematics—disappeared. At Constantinople, Latin was considered a barbarian tongue and fell into disuse after the reign of Justinian (d. 565).[300] In the six turbulent centuries after 395, Christianity successively displaced the old civic religion of the empire and cults of barbarian migrants. These migrants, arriving in a double wave dating from the fifth and the ninth century, transformed ancient Mediterranean culture, though without destroying it.

Under Constantine, Christianity, having gained tolerance in 311 and his support in legislation two years later, became a wedge between East and West.

The Church, liberated and triumphant in the fourth century, immediately broke into rival sects that branded one another heretic and pronounced mutual excommunications. A series of Ecumenical Councils was able to impose doctrinal unity, most notably against the Arians who denied the divinity of Christ. Yet differing traditions and styles continued to feed rivalries between Latin and Byzantine Christianity, presaging the cleavage into Greek Orthodox and Latin Catholic churches.

After A.D. 180, imperial institutions, especially in Rome, had become increasingly dysfunctional. With their crumbling, the ordered peace that could have made for new developments in mathematics and other humane endeavors weakened.

Rome was long a tattered symbol when in 395, the year of partition, a figurehead Roman emperor Honorius in a relocated capital at Ravenna took orders from a barbarian Vandal military commander. During the same reign in 410, Visigoths under Alaric sacked Rome itself. Finally Odovacer, another Germanic chieftain, deposed Emperor Romulus in 476. Although it terminated the line of Western emperors, this coup was not viewed as a catastrophe outside Italy. Romulus had never been recognized in Constantinople.

One of the two Germanic groups who, migrating from Baltic homelands in the first Christian centuries, made a major barbarian impact on the Latin West was the Goths. Visigoths (meaning "wise" or perhaps "western") had penetrated the Balkans before their sack of Rome and went on to establish a Germanic state in southern Gaul (France) about 466. From the Danube basin, Ostrogoths ("brilliant" or "eastern") also invaded the Italian peninsula. They had strong ties to Constantinople, where Theodoric their king had been educated. By 493 he had overthrown Odovacer, the first Germanic ruler in Italy. Theodoric patronized the arts and engaged as distinguished courtiers Boethius and Cassiodorus. Despite his Byzantine training and able leadership, his *civilitas* (humane politics) failed. The heretical Arian faith of his Goths blocked their amalgamation with Latin Catholics.

The other Germanic people were the Franks. Their conversion to Catholic orthodoxy made them acceptable, and they were able to blend German and Latin cultures. Their effective warrior society was knit by loyalties to military commanders and kinfolk rather than institutions of state. For instance, they avenged rather than adjudicated uncompensated blood crimes. Germanic customs became a foundational element of medieval feudal society. After Frankish warriors under Merovech defeated the Huns at Chalons in central France, his grandson Clovis converted to Christianity in 496 and, allied to the Gallo-Roman bishops, made Paris his capital. Within several generations the Franks had shifted the West's political and military center far to the north of Rome and produced the visionary leader Charlemagne.

The shift from paganism to Christianity further affected the intellectual setting within which mathematics had to find a place. The redirection was especially telling among Greek Christian scholars, whose predecessors had nourished abstract mathematics. Seekers for entrance into the kingdom of heaven had theological interests at odds with the concerns of pagans. By Theodosius' reign the Christian community, which in 303 as 10 percent of the imperial population had suffered vicious persecution, was able to turn persecutor. In 391 a Christian mob destroyed the Alexandrian Serapeum and parts of its library. Theodosius, who had outlawed pagan sacrifices, congratulated the hoodlums. In 415 another religious riot at Alexandria took the life of Hypatia. In 529 Justinian, for religious reasons, closed the Platonic Academy in Athens, ending its near millennium of association with mathematics. Yet Christianity was prepared to make its own accommodation to classical studies.

Roughly two centuries after Jesus, when Christians no longer lived in expectation of imminent doomsday, the brilliant Alexandrian theologian Origen (early 200s) had made such an accommodation, insisting that his pupils study some mathematics. Successors cautiously surveyed classical learning for help in propagating their religion. Drawing selectively on antique learning, Christian scholars known as Fathers of the Church cumulatively produced a comprehensive doctrinal corpus to supplement the Bible. Language and rhetoric were preeminent interests, since these subjects aided the understanding and promulgation of religious doctrines. There were misgivings: to St. Jerome, whose fourth-century Latin Vulgate is the approved Catholic Bible, fondness for Cicero brought nightmares.

The most influential of the Latin Fathers, St. Augustine (354–430), bishop of Hippo in north Africa, had taught rhetoric. By the metaphor of an eternal heavenly Jerusalem and an ephemeral earthly Babylon, his classic *The City of God* relates the sacred to the secular. Peerless among contemporaries in the rhetorical educational tradition, he advanced a master plan of education. His *On Christian Doctrine* supports the liberal arts curriculum with its mathematical component as a means of sharpening youthful minds in preparing for sacred studies. Augustine develops the Neopythagorean doctrine that number is the root of all material things. For that claim he uncovers in Wisdom 11: 20 a minor biblical warrant: [God] "arranged all things by measure and numbers and weight." Numerology, which Alexandrians had used as a tool for interpreting poetry, treats biblical numbers, for example, the ten commandments and the twelve apostles, as encoding God's deeper designs. Scriptural and Augustinian backing provided a place more secure but marginal for the mathematical curriculum.

In his *Marriage of Philology and Mercury*, Augustine's contemporary Martianus Capella (fl. 365–440), allegorizes the mathematical arts of arithmetic, geometry, astronomy, and music and the arts of communication: grammar, dialectical logic, and rhetoric. Much studied throughout the medieval period, this text set the canon of the seven liberal arts. Because these cultivate the mind, they appropriately educate free men with leisure to study. By contrast, mechanical arts require the manual skills of a laboring class. The introduction of *Marriage* sets the tenor of the work: Philologia, a bride with the seven liberal arts as bridesmaids, ascends to heaven to marry the god of eloquence.

Martianus' *Marriage* has eulogistic chapters on geometry and arithmetic.

But its closing ten-page chapter on Euclidean geometry is flawed in mistranslating the definition of a point, with which the *Elements* opens, and equating a diameter with a half-circle. These lapses, whether original or copied, document a prevailing mathematical subliteracy. The geometry chapter deals mostly with surveying and has few fragments from the *Elements*. The arithmetic chapter, the longest in the text, presents elementary number theory: definitions, ratios, classifications that include primes and figurates, and thirty-six of Euclid's arithmetical propositions. Martianus gives numerical illustrations but no proofs for these. Not Euclid but primers on the *Elements* and a digest of Nicomachus' *Introductio* were probably Martianus' sources. His coverage illustrates the shift introduced by Nicomachus, who moved the status of arithmetic from subordinate to equality with geometry. The loss of the mathematical *Disciplinae* of Varro in the early Middle Ages leaves the *Marriage* as the best source for reconstructing ancient Roman mathematics.

Allegory and personification in this popular handbook replace method and calculation. Earlier Romans had adopted handbooks, a Greek invention to popularize technical subjects. While some Greek authors had understood the texts that they condensed, Roman compilers like Martianus juxtaposed already gutted mathematical texts with lists of authorities known only from secondary works. Listing of authorities satisfied their requirements for proofs.

The Roman patrician and compiler of handbooks Anicius Manlius Severinus Boethius (ca. 480–524) was the major figure in transmitting mathematics. Probably in recognition of his abilities, as well as in an attempt to strengthen ties with Roman nobility, Theodoric appointed him minister at court. But Boethius defended and possibly connived with the Roman aristocracy whom the king charged with treason. His espousal of Roman ideas of freedom was also suspect, and Theodoric imprisoned and executed him in 524.

Aware of the desperate need for basic texts and convinced by the age of twenty that mastery of some mathematics was a prerequisite for philosophy, Boethius had projected a handbook digest of each branch of the liberal arts. He Latinized as *quadruvium*, four-part juncture, the four methods of Nicomachus. The term, corrupted to *quadrivium*, became generic for the mathematical curriculum subjoined to a three-path communication arts *trivium*, a later designation.

Boethius did not make original contributions to mathematics. His chief work, *De Institutione Arithmetice*, paraphrases or approximately translates the *Introductio Arithmetica* of Nicomachus, adding occasional materials and condensing portions of the *Introductio*.

Boethius supplies terminology for multiples, factors, fractions, ratios, and proportions, whose very convolutedness opened a way for later medieval scholars to redirect Eudoxean-Euclidean proportion theory. According to this (inherited) terminology, a larger number in a ratio is "superparticular" if the ratio contains a smaller number as a factor, such as 9 : 6. "Boethian arithmetic" introduced the quadrivium, which lasted for more than a thousand years: his labored language of subsuperpartient ratios, now long replaced, still appears in

Isaac Newton.[301]

Boethius' geometry, which has not survived intact, followed the encyclope-dists' practice of deleting sections of the *Elements*. He translated only portions of Books I through IV, giving definitions, axioms, postulates, and enunciations of theorems without proofs. By following Nicomachus, Boethius lost Euclid's generality and rigor. No longer represented as lettered line segments, numbers surrender their general character. Nicomachean linear numbers are strings of specified points in one direction constituting data for computations. These com-putations substitute for general proofs. Euclid's geometry was further reduced when some of Boethius' translated sections disappeared almost immediately. By 800, only fragments of his books I, III, and IV remained, yet these were long the authority of theoretical geometry in the Latin West. Boethius' hand-book, stemming from Nicomachus and Ptolemy, for a music based mainly on a catalogue of numerical proportion fared better and has largely survived, but conclusive evidence that he wrote a handbook on astronomy that derived from Ptolemy's *Almagest* is lacking.

Had Boethius lived longer, Latin resources for philosophy and mathematics would have been richer. He sought to translate the complete works of Plato and Aristotle accompanied by commentaries. He had the requisite command of Greek and contacts with Constantinople, but the task was too ambitious. He translated only Aristotle's main logical works with his own commentaries. Those translations and commentaries, however, established a Latin logical vo-cabulary and became the basis for study of the subject in Europe until the twelfth century. While imprisoned, Boethius composed his classic, *The Con-solation of Philosophy*. It blends pagan detachment and Christian resignation without direct appeal to revealed truths.

Boethius' friend Flavius Magnus Aurelius Cassiodorus (ca. 480 to ca. 575) found his consolation in the monastery that he founded on his retirement from court at Vivarum in southern Italy. Disturbed that so many libraries and manuscripts had perished, he collected manuscripts and set up a *scriptorium* for copying them. His *Introduction to Divine and Human Letters*, a handbook with a critical bibliography for educating monks, became definitive. Like Augustine, he urges the importance of selected pagan humane learning. His comments on arithmetic, subordinated as usual to rhetoric and dialectical logic, derives from Boethius. As further incentive for its study by monks, he points to Jesus' remark that the hairs in the head are numbered.

Geometry was not confined to textbooks. Surveying (*ars gromica*, from *groma*, a surveyor's measuring rod) existed as an independent oral tradition. About 450 technical manuals containing some theoretical geometry and other relevant materials appeared. By the sixth century the *ars gromica* had acquired some associations with education in the liberal arts. Perhaps young students turned to gromatic materials as a source of simple theoretical geometry; or perhaps surveying was taught as a liberal art (as medicine and architecture had been before Martianus Capella's classification).

After Boethius, encyclopedic geometry continued along the decline that was evident in Martianus. Forced to rely on defective secondary sources, Isidore of Seville (d. 636) a century later nevertheless attempted a compilation of universal knowledge, the *Etymologies*. It defines any four-sided plane figure as a square and a cylinder as a square figure with a semicircle above it. Notwithstanding its mathematical weakness, one cautious modern historian has called the *Etymologies* "one of the outstanding feats of scholarship of all times."[302]

Despite their shortcomings, these handbooks are ancestors to the survey textbook, legal brief, and scientific resumé. After Roman neglect, they helped restore teaching of the quadrivium. Boethian arithmetic preserved the ideal, as distinct from the model, of a theoretical science when the urban culture of the Latin West had so crumbled that advanced mathematics was incomprehensible there.

Urban populations exposed, among other things, to new diseases carried from Asia dropped precipitously: in the sixth century not one of the once populous Western cities had 20,000 residents and many cultivated fields were abandoned. Early medieval Europe had become rural and dependent on Benedictine monasticism for cultural cohesion. In 529 St. Benedict, a former aristocrat turned hermit, founded a monastery at Monte Casino between Naples and Rome. Monks ("solitary") had practiced ascetic Christianity in Egypt and Syria. Benedict's contribution was to channel their diffused energies into a regulated life of common prayer, work, and meals under a paternal abbot. His rule was strict but moderate, enjoining propertyless celibacy and silence, permitting a siesta and ration of wine, and encouraging hospitality. Successors to cities, economically self-sufficient abbeys spread rapidly. Monastic labor kept alive building and craft skills and improved agricultural techniques and empirical medicine.

Monastic prayer was tied to the calendar. Its pivot was Easter, a moveable date determined by lunar movements. Around it sacred cycles of the ecclesiastical or vulgar year revolved. Forecasting the correct date of Easter was inexact and spawned controversies during the seventh century. Benedictines looking to Rome observed one date: Irish monks drawing on an Alexandrian (that is, Dionysian) Greek tradition sometimes observed another. In Northumbria, where Celtic and Roman influence mingled, King Oswy (d. 671) was banqueting for Easter, while his Kentish-born queen was fasting and observing Palm Sunday.[303] This calendar crisis had been brewing underground for some time and its surfacing prompted the king to ask church authorities to convene the Synod of Whitby in 664.[304] The calendar difference arises since dating depends upon solar and lunar movements that are incommensurable. With a strong sense of order and concord but without benefit of astronomical and mathematical understanding, the fathers imposed the Roman reckoning based on the Dionysian cycle of 19 solar years and 235 lunar months.[305]

In 725 Bede (673–735) called "the Venerable," an Anglo-Saxon monk at Jarrow near Newcastle in England and the foremost western intellectual of his time, wrote the first reliable text of computing procedures to fix the date of Easter. Bede was so bookish that he described the ancient Roman wall, an easy walk from Jarrow, by means of classical authors without archaeological inspection. Jarrow's library was large for those times, and Bede enjoyed a rare fluency in Greek, as well as mastering Celtic sources. Theodore of Tarsus, a refugee Greek monk who was archbishop of Canterbury, had introduced him to Greek and brought valuable manuscripts and "church arithmetic" (probably the Byzantine *computus*) to Britain. Bede's *De Temporum Ratione (On the Reckoning of Time)* addresses a branch of monastic study called the *computus*. The book combines mathematical rules and tables with descriptive astronomy, records, and chronological materials. Before the *Reckoning*, about six hundred Irish monks had used these multiple sources, probably assembled in Spain, to align the lunar and ecclesiastical or vulgar with the civil and solar calendar.

To determine the number of lunar months in a solar year, we now divide 365.2422 (days in a solar year) by 29.5306 (days in a lunar month) to obtain 12.3683. The decimal fraction 0.3683, representing the part of a lunar month that a calculator must add at the end of any solar year to align it with the lunar, can be approximated by the common fractions $3/8$, $4/11$, and $31/84$, and most accurately by $7/19$ ($= 0.3684$). Thus 3 lunar months intercalated into an interval of 8 years, 4 into an interval of 11 years, 31 into 84, and 7 into 19 will align the two calendars essentially as accurately as was possible. Although Meton had defined the 19-year cycle in Athens five centuries before the Christian era, monks confusedly used all the cycles in combination. Computists also disagreed about the time at which a lunar month began and the date of the spring equinox.

Working as a textual scholar, Bede compared about three hundred documents, untangled existing mathematical inconsistencies, and recovered essentials of the Dionysian cycle intercalating procedure, which requires subtracting the *saltus lunae* or one day near the end of the lunar cycle. Dionysius had extended the 19-year cycle to 95 years, and others had made it a 532 cycle (19×28), which Bede calls the great paschal cycle. To give added weight to his findings, he reported that the angel Pachomius had dictated the method for calculating Easter. Still, he knew that Dionysian reckoning produces discrepancies, for example having the the solar eclipse of 664 occur on May 3 instead of May 1, but further calendric improvement was not achieved in the Latin West until the eleventh and thirteenth centuries.[306]

Interested in chronology as well as *computus*, Bede sought to unify astronomical and historical time and successfully campaigned for a second calendar reform to replace localized regnal dating by the now universal dating of events from the year of Christ's birth. Other scholarly interests of Bede included metrology (whose units doubled as fractions) and finger counting, for which he gave the first written description.[307] This orally transmitted method permitted the calculator to record a subtotal temporarily while doing mental arithmetic. Monks calculating the Easter date needed a way to record intermediate results, and this method also supplemented reckoning on the *abacus* counting board. Finger counting, although unwieldy, implicitly incorporates a place-value system with the capability of computing up to values of 9999. Beginning from the little finger, the computer formed units and powers of 10 by prescribed joint movements and extensions of the fingers on the left hand. To form 10, he touched his index finger nail to the middle of the thumb. Multiples of 100s and 1,000s were read off the right hand—at the observer's left—in an ascending series from right to left. The technique requires supple joints—there were complaints that dancing fingers were necessary above 9,000—and interpreting the many combinations required much practice. By additional body language, Bede expanded finger counting to an upper limit of 1,000,000. (To form 10,000, rotate the left palm outward and place it at mid-chest with the fingers extending upward.)[308]

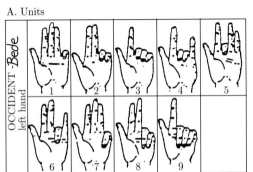

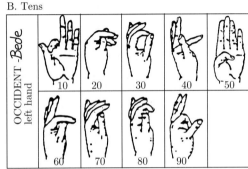

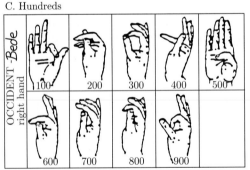

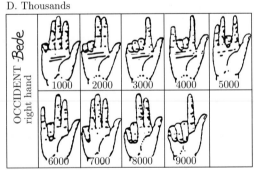

While some monks learned the *computus*, others, plotting their monastic enclosure as holy ground, sacralized the practical geometry of ancient surveyors (*agrimensores*). Religious symbolism and agrimensorial utility blend in a blueprint of a monastery from the ninth century preserved at St. Gall in Switzerland. A 40-foot square, the governing module throughout the St. Gall grid, was a continuing reminder of the duration of Lent, the days of the Biblical flood, and the resurrected Lord.

The many multiples of 40 also made it an excellent number for architects, whose measuring depended on pace, its doublings and halvings. In sixteen single-step paces (*gradus* = 2.5 feet) an architect readily stepped off the 40-foot side of his module. Through doublings of the 40-foot modular side, he laid out an overall length of sixteen module sides, 640 feet. By switching to a three-step pace (7.5 feet), the architect produced a width of 480 feet that kept the integrity of his 40-foot modules and added new units of measurement through various doublings and halvings. The 4 : 3 ratio of 16 module lengths to 12 modular widths permitted him to square his rectangle by fixing the diagonal at 20 modular lengths. The Christian cross, recurring through the design, depended on the Pythagorean triple 3 : 4 : 5 for various constructions.

Easier intercalations than those for Easter aligned units of architecture and field surveying. Surveyors, like architects, worked by doublings and halvings. Progressive halvings of the Roman mile, 1,000 paces, containing 5,000 feet (stepped off by two-step marching paces) were not in agreement with progressive doublings of the architect's unit of measurement, the *gradus*. Thus the length of St. Gall's at 640 feet corresponded to the field surveyor's eighth-mile *stadium* at 625 feet. Ninth-century field surveyors had learned to correct these anomalies by intercalations of the foot or pace at remembered intervals. Their pace measurement rested on the smaller anatomical unit, the foot (*pes*), which linked measurements of bench and field.

Monastic culture partially filled the void created in the eighth century, when Islam broke the last connection between Italy and Byzantium. The Latin West rallied about the papacy as a unifying center. To the north, Carolingian dynasts became a new political and military force. Near Tours in west central France the cavalry commander Charles the Hammer in 732/3 crushed overextended Spanish Saracens. His son, Pepin the Short, deposed a puppet Merovingian king and Pope Stephen II consecrated Pepin as king of the Franks in 754. Succeeding Pepin was his 6 foot $3\frac{1}{2}$ inch son, Charles the Great or Charlemagne, whom Pope Leo III crowned Emperor of the Romans on Christmas day of the year 800.

Charlemagne sought nothing less than to restore the order of the vanished Latin empire. He envisioned an educational revival creating a corps of record-keeping secretaries and a literate clergy. This barely literate warrior, who supervised the royal egg count and paddled a schoolboy for a minor grammatical blunder, a solecism, seems an unlikely choice as an educational reformer. But tremendous energy more than compensated for Charlemagne's personal limitations: a better informed king might never have undertaken a task of such magnitude.

Eighteen years before his coronation, Charlemagne had recruited the learned monk Alcuin (ca. 735–804), headmaster of York's cathedral school where regular clerics learned to read.[309] Alcuin instructed the Frankish king, members of his family, and prominent young nobles in rudimentary reckoning and astronomy, set up a palace school at Aachen, dreamed of a new Athens in France, and labored to make clerics literate: able, for example, to distinguish *ara*, altar, from *hara*, pigsty. A problem text of fifty-three computational puzzles in Roman numerals, *Propositions for Sharpening Youthful Minds*, is sometimes erroneously attributed to Alcuin. Some of its problems demand mathematical ingenuity. In the most famous, three men and their three sisters must cross a river in a boat holding only two people, and to cross safely each girl must have her brother in the boat.

At his palace in Aachen and the monastery at Tours, Charlemagne under Alcuin's guidance established centers for studying, collecting, and systematizing scattered and fragmented Latin texts. By about 789, the program at these and satellite monastic centers required capable copyists, whose work was hampered because long neglect had debased both handwriting and grammar. National hands, such as Visigothic and Irish, had proliferated: monks transferred from gardens to *scriptoria* were more at home with obedience than with literacy. The schools trained copyists who developed the Carolingian minuscule, a standardized lowercase lettering from which the type on this page descends. Most surviving Latin classics trace to a Carolingian copy.

Although Alcuin ranked poetry above science and this bootstrap revival, sometimes called the Carolingian Renaissance, centered on grammar and rhetoric, Charles also mandated a school of computation within every monastery and diocese. Corbie in northeastern France from 661 became an important center for the study and diffusion of geometry in the eighth and ninth centuries. Monks with access to French, Irish, and Italian sources began to anthologize items from gromatics or field measuring and theoretical geometry. Liberal arts students studied these compilations. This may be the earliest effort in the Latin West to rediscover the theoretical basis of geometry.

With the retirement of Alcuin from the Frankish court in 796 to be abbot of St. Martin of Tours, a major leader for educational reform was lost. But Charlemagne, who subjugated the Saxons and conducted diplomacy with Baghdad, had come a far distance from Frankish tribalism. A visionary program to restore Latin peace and order could enlist the energies of restless warriors and aggressive missionaries.

The bold efforts of Charlemagne's revival could not bind for long the culturally, politically, and economically disparate Latin and German peoples. Nor could succeeding emperors with less ability dominate lower lords in an ill-defined command structure. Salic law dividing the succession among royal sons contributed to the fission, and communication was inadequate. The Carolingian dynasty petered out in Germany, in Italy, and then in France. In addition, a second, more brutal wave of migratory raiders spanning a century from about 850 undid the Carolingian reforms. Saracen freebooters raided the coasts of France and Italy, while Magyars from the East occupied Hungary and attacked south Germany, eastern France, and northern Italy. The worst assault though came along the Atlantic rivers by Northmen in shallow-draught boats. These raiders destroyed Celtic monastic culture, penetrated to the outskirts of Constantinople and into central France, and sacked the tomb of Charlemagne. Against raiders, localism was an asset: monasteries recovered more quickly than cities and local strongmen learned to fight from castle and ambush.

Agriculture withstood the anarchy and helped restore stability. Since the sixth century the *carruca*, a heavy plough with moldboard, had broken the rich soil of north Europe. In 911 a Carolingian monarch's bequest of lands flanking the Seine to a leader of the roving Northmen tokened incipient stabilization. Soon these resettled pagan Northerners became Christian Norman-French and achieved a prominence in medieval affairs. The coronation at Rome in 962 of Otto I, head of a new line of German-Saxon emperors, marked a resumption of order and a new round of Christianizing the barbarians.

Otto quickly revived education, for he relied on ecclesiastical administrators to curb the power of semi-independent dukes. Unlike Charlemagne, Otto did not direct the renovation personally or concentrate on liberal arts. His brother Bruno set up a chancery school in Cologne, where as archbishop he revitalized other cathedral and monastic schools, such as those at Liège and Hildesheim. The revival drew on the thin tradition of learning that had passed from Jarrow and York to Aachen and on to the satellite centers of Fulda and Reichenau in German lands and Tours and Ferriéres in France. Otto also sent the monk John of Gorze from Lorraine on a three year diplomatic mission to the court of 'Abd al-Rahman III at Cordova. The Cordovan emir patronized learning, collected books, and welcomed scholars, including non-Muslims. During his three-year stay John formed ties with an unknown multilingual Jewish scholar and a Mozarab or Arab Christian courtier, Recemundus, a mathematical astronomer. Recemundus then spent almost a year at Ottonian Frankfurt. His contribution there is unknown, but by the next century Lorraine was a center for mathematical activities derived from the Arabs. Otto's grandson, Otto III, did not fulfill his dream to reunite East and West, but did secure the election of his Frankish tutor to the papacy. Their joint policy led to the religious conversion of the Magyars and most Poles. But it was in mathematics rather than missionary endeavor that the tutor, Gerbert of Auriliac, made his greatest reputation.

Gerbert (ca. 945–1003) had been schooled at Saint Géraud, the Benedictine convent in French Aquitaine. In 967 he traveled to Catalonia to study mathematics and music under the bishop of Vich and probably to investigate Moorish scholarly texts. In 970 he went to Rome and within two years had caught the attention of Pope John XIII and Otto I, who was residing in the city. Assigned to assist the archbishop of Rheims in Lorraine, Gerbert reorganized the cathedral school there. He expanded its library, sometimes requesting manuscripts from Spain, and attracted students by his teaching. Working in the Boethian tradition in mathematics, Gerbert apparently covered some Nicomachean and Pythagorean number theory, some geometric enunciations from Euclid without proof, and agrimensorial rules. For lessons in astronomy, he constructed armillary spheres. The treatise entitled *De Astrolabia*, which reflects Arabic influence, is sometimes attributed to Gerbert. His most important achievement was in restoring and refining techniques for calculating on the abacus counting board that had been used in Roman times.

For operating with Roman numerals, the board's implicit place-value display was easier than finger reckoning to master. Elementary errors, such as one in the computational text of pseudo-Alcuin, occurred easily when positional notation was lacking. On the abacus, calculators positioned pebbles (*calculi*) as unit counters in adjacent columns. These columns from left to right displayed successive powers of 10 up to 10,000,000. Some boards had *calculi* with a five-unit value in a prescribed sector.

Despite this considerable streamlining, calculation remained labored. Routines had to be memorized, zero registered as an empty column intuited, and finger counting retained as a supplement or check. "Iron division" was especially difficult. To reach a final quotient, accountants added a succession of subquotients. Calculators obtained the first subquotient by adding a supplemental number to the original divisor to get the next higher multiple of 10. This permitted easy division. Successive divisions of the sum of any remainders added to products of the supplemental number times the subquotients give the answer. Thus, to find the quotient of 1081 divided by 23, take these steps: $23 +$ the supplemental number $7 = 30$ and $1081/30 = 36 + 1$; $(36 \times 7) + 1 = 253$; $253/30 = 8 + 13$; $(8 \times 7) + 13 = 69$; $69/30 = 2 + 9$; $(2 \times 7) + 9 = 23$; $23/23 = 1$; and the quotient is $36 + 8 + 2 + 1 = 47$.

Gerbert wrote down the method of abacus calculating, heretofore dependent on oral instruction. But his innovation in abacus counting, perhaps borrowed from Spanish Arabs, was to substitute new counters (*apices*) that carried western Arabic symbols nearly equivalent to $1, 2, \ldots, 9$. The first appearance of these numerals in an eleventh-century *Second Geometry* attributed falsely to Boethius and Gerbert would lay the groundwork for acceptance of Indo-Arabic numerals in the Latin West. Besides reducing the number of counters, *apices* forced the accountant to do mental arithmetic as well as count. This contrasts with the monastic reading practice, which required the reader to make lip movements syllable by syllable coordinated with visual identification of the words. Gerbert's computations included the use of $\pi = 22/7$ and $\sqrt{2} = 17/12$, which were probably considered exact values.

During the eleventh century, calculators called *gerbertista* memorialized the proper name of Pope Sylvester II, who reigned from 999 to 1003,[310] and the reputation of the school of Lorraine. Compilers at Corbie rediscovered the *Corpus Agrimensorum*, a collection of elementary surveying geometry, which was adopted into the curriculum at Liège and Cologne. By that time a student of Euclidean materials from Boethius knew most definitions, postulates, and axioms as well as statements of theorems in the first four books of the *Elements* and the three proofs at the beginning of Book I.

Lacking insight and rigor, new geometers worked with determination and enthusiasm. A Liège student named Franco attempted to solve the classic quadrature problem by making a circle from measured pieces of parchment. A written debate between two students at Chartres about reconstructing the proof of the sum of the interior angles of a triangle shows a return to scholarly dialogue without comprehension of basic terminology. The new interest in mathematics spread beyond Lorraine. At St. Gall, Helperic did a new *computus*, and a new board game, *rithomachia*, based on Nicomachean number theory achieved popularity.

A demographic and economic recovery was behind this leisure for mathematical speculating and play. After the Eurasian plague of the mid-eighth century, Europe's population grew from the tenth through the thirteenth century. Climate, agricultural technology, and means of transporting food improved. More remarkable was the increase, from the tenth century, in sources of power beyond human and animal muscle. Throughout the six centuries of this chapter, slaves had been an important source of power and a major export of Europe ("slave" takes its name from Slavic captives). Water and wind were the new power sources. In waterwheels, which increased as much as tenfold in these regions, circular was newly converted to reciprocal motion. Windmills were introduced in areas having steady winds and poor streams. Most windmills had four vertical sails on structures that could rotate and were applied to such tasks as grinding grain.

When Gerbert was elected Pope in 999, Latin Christians could put some hope in the ability of thought and knowledge to aid the human condition. In the tenth century, the papacy had been a shambles. This learned pope helped revive its prestige. More confident about reason, successors in learning turned into a torrent Gerbert's manuscript trickle of Greco-Arabic manuscripts. From Otto's time, chroniclers spoke more of politics and less of divine interventions. As Europe gained knowledge of numbers and shed some intimidations that they had prompted, Gerbert wished the third Otto as many years as his abacus could count.

10.2   A Note on Maya Mathematics and Calendrics

Before western and central European civilization turned to an Atlantic crossing, a mature sister civilization had arisen in Mesoamerica. The Maya differed from Europeans in producing no wheel, except on a few toys, and no metal tools or weapons. Yet from the late third to the early ninth century by Bede's calendar, when medieval Europe was coming to birth, Maya polytheistic agriculturalists enjoyed their classical age.

The Maya lived under kings ruling by supposedly divine mandate. A primitive kingdom was comprised of incipient city states in what are now the Acetin peninsula lowlands of south Mexico, along with Guatemala, Belize, El Salvador, and northern Honduras. Below royalty and a priestly order apparently came hereditary warrior nobles, then merchants and urban artisans—masons, carpenters, potters, sculptors, spinners, and weavers—followed by peasant farmers, who constituted a majority. Productive labor by the urban workers provided stone, wood, textile, and ceramic wares and by farmers an economy built on crops of beans, squash, peppers, hemp, cotton, and cacao. Farmers cultivated these crops without the benefit of large work animals; they had no horses or oxen. The scale of public monuments suggests that their construction depended on subject populations and slave laborers, who completed the social pyramid.

During their classical period, Mayan traders reached markets as distant as Cuba. At home the Maya built imposing palaces and temples of stucco and stone that gracefully dominated cities such as Teotihuacān in Acetin. Most temples were truncated pyramids. The plaza of seven temples in Tikal, one of the largest classical Mayan cities in what is now Guatemala, has notable geometric features and astronomical orientation. The east to west base line stretches from Temple I, which was completed in 700, to Temple III, completed in 810. It forms with the south base line from Temple III to Structure 5D-90 a right isosceles triangle, having the right angle at Temple III. The Tikal temple designs and arrangement suggest an interest in elementary geometry.[311] By the arrival of the Spanish conquistadores, the Maya had undergone a sharp decline, perhaps for reasons of soil exhaustion, warfare, or even a plague. By then, the Maya lived mainly in straw huts in villages. High culture had disappeared among them, and classical monuments were falling into ruins. To the present, Mayan relics continue to draw archaeologists and curious travelers alike.

Archaeology is the principal resource in efforts to gain a greater understanding of the past of these Amerindians, whose oldest village dates from about 1000 B.C. Scholars have probed more than 120 of their sites in Central America and are searching the jungle wilderness for fresh sites. They constantly reevaluate the extensive cache of retrieved artifacts. As the decipherment of five hundred of eight hundred known Maya hieroglyphs that began 170 years ago has proceeded, knowledge and interpretations have accumulated at an impressive rate. Under intensive study are thousands of chiseled stone monuments and thousands more of painted ceramic vessels. Maya literary sources are few, for the Spaniards and Christian missionaries systematically destroyed them as pagan. The major surviving sources for the language are four screen-fold pre-conquest books, called codices, housed in Dresden, Madrid, and Paris.

Like European religion, that of the Maya pervaded culture. Attached to royal courts, the priests were charged with divination and sometimes protection against the anger of the gods, who were believed to control health and the natural environment, and with assisting good gods, like the Sun and Moon, against evil gods who sought to block them in the dark of night. In the Mayan cosmogony, the deified planet Venus had special importance, being recognized as a morning and an evening star. The Venus deity appears iconographically as the menacing spearer. The Dresden codex lists names of primary victims about to be pierced. Ritualistic human torture and sacrifice, which were prescribed as ways of placating the gods, were an honor and guarantee of immortality to the victim.

In his social history of 1566 *Relacíon de las Cosas de Yucatán*, which he may have written partly out of remorse for having ordered in the 1540s the torching of all Mayan writings, Franciscan Diego de Landa (1524–1579) provides extensive materials on astronomy, calendrics, and mathematics.[312] De Landa, the third bishop of Acetin, reports that within the privileged bureaucracy of priests was a subgroup of star watchers, calendar makers, and chronologists that perhaps included calculators who assigned dates to official records and scribal teachers of mathematics. Some scholars believe the Maya worshiped time as a god. The cultures on both sides of the Atlantic put great value in keeping accurate religious calendars, as well as calculating chronologies and tribute records. While Mayan scribal mathematicians were able to construct and manipulate an absolute chronology having 3114 B.C. as its zero point of creation, Christian monks struggled to secure tenuous mastery of a rude computus, arithmetic astronomy.

Since no materials containing formal rules, procedures, algorithms, or sample problems exist, reconstruction of Maya mathematics is limited. Nor is it likely that archaeologists or historians will recover such guides in future excavations. The glyphs in their language are not well suited as a vehicle of mathematics. Mayans did not possess an economical mathematical vocabulary or a set of symbols that would facilitate mathematical expression: the evidence so far indicates little curiosity about mathematics beyond computation, while priestly bureaucrats possibly blocked as magic and perhaps as a challenge to their position the pursuit of a body of theoretical principles. Development of remarkable computational skills—a copious body of which survives—proceeded with little known theory. Having come to understand the Mayan number system well enough to assign values reliably to sums and products, modern scholars have undertaken to reconstruct the steps and algorithms involved in reaching these results: reliance, for instance, on a non-fractional arithmetic and decomposition of numbers into prime factors and their multiples.

The Maya differed from their European counterparts by an indifference to biographical detail. In a culture where scribal mathematicians appear to have been fairly numerous, a distinctive scroll of numbers, usually placed under the arms, identifies them in artifacts. Only two, a male and a female, have been detected so far. The lady has been given the name or title of Ah Ts'ib, which means "scribe." In 1978 Persis Clarkson found them in the same palace scene.[313] For the Maya the font of mathematical knowledge was the god

A female scribal mathematician. Detail from Clarkson, 1978

Pauahtun. On the vase illustrated here, he is pictured together with the Maya god of writing. In such company, names of mortal scribal mathematicians perhaps became supernumerary.

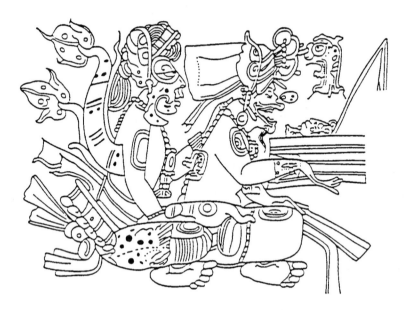

The Maya gods of mathematics and writing

To formulate and communicate their mathematical results, Mayan scribes relied on a clumsy writing system, blending logograhic and phonetic elements. Without benefit of an alphabet and using brush and chisel, scribes fashioned glyphs in meaningful arrays. Glyphs are modified pictograms to which they added complex phonetic and orthographic signs to reflect the spoken language. The products, particularly cartoons, have artistic as well as utilitarian interest. The Maya also had a vigesimal (base 20) place-value number system. That the Mayan equivalent of the word "man" is used for 20 is perhaps a residue of an earlier stage of counting on fingers and toes. Five, the number of fingers on each hand and toes on each foot, was also a special number. Supposedly, the senior god, Hunab Ku, created it.

Classical Maya scribal mathematicians, adept in arithmetic astronomy, astrology, and calendrics, probably devised the place-value system and its numerical notation, in which dots represent units and a bar stroke the number 5. Four bars thus stand for 20.

Here are the number 20 and intermediate numerals formed by combinations of appropriate dots and bar strokes. The austerity of the device, not unlike Morse code, made it easily transcribed and read.

Name	Numeral	Symbol	Name	Numeral	Symbol
huu	1	•	tuluc	11	
ca	2	••	la-ca	12	
ox	3	•••	ox-lahun	13	
can	4	••••	can-lahun	14	
ho	5	———	ho-lahun	15	
uac	6	•	uac-lahun	16	
uuc	7	••	uuc-lahun	17	
uaxac	8	•••	uaxac-lahun	18	
bolon	9	••••	bolon-lahun	19	
lahun	10	═══	hun-cal	20	

Maya numerals from 1 to 20

In the accompanying illustration, which appears on a vase dating from 750, Pauahtun recites to squatting apprentices the numbers from bottom to top 11, 13, 12, 9, 8, and 7.

A classroom scene: The mathematics lecture

This picture and other evidence have led historians to conclude that mathematical instruction was conveyed through schools in a classroom setting. By or during the early classical period, Mayan scribal mathematicians had introduced a zero symbol. Its first known appearance in a place-value context is in a table in the Dresden codex carrying a date equivalent to A.D. 665. Later, zero gained representations besides its shell symbol.

Mayans were fond of synonyms and multiple representations. For forming larger aggregates in base 20, bar-and-dot notation join and partly overlap numbers used in calendars. Eighteen bars constitute one *tun* or 360 days, while 20 days is a *winal*, and each day a *k'in*. Place valuation proceeds vertically from bottom to top with bar-and-dot notation, giving the number of *k'ins*, *winals*, and *tuns*.

No one unified approach to numbering prevailed even during a single historical period. Use of the zero in counting had antedated the place-value system, and scribes who did chronological referencing on civic monuments followed the older tradition without place value. Still other scribes, artists more than routine calculators, illustrated numbers by glyphs that are graceful portrait heads or full figure sketches instead of the prosaic bar-and-dot symbols. Mathematics had something of the luxuriance of its jungle setting.

The power of the Mayan system shows in the sureness with which it handles large numbers. At a time when Europeans rarely quantified dozens, Mayans named powers of 20 as far as 20^6, and their calendar cycles had numerals as large as 306×20^{12}. Largely through management of complex calendrics, Mayan scribal mathematicians developed and extended their proficiency with numbers.

A sacred round calendar of 260 days in 13 vigesimal months coexisted with a civil or chronological calendar of 360 days consisting of 18 vigesimal months. The civil year with 5 added days formed the vague calendar, approximating the sidereal year. No account was taken of a leap year. Correlation of the dates of the sacred and the vague calender involved passing through a complete cycle of 18,980 paired dates. This number arises from the indexing of the two calendars into a calendar round in which the passage of 52 vague years coincides with 73 sacred years. In mathematical terms, 18,980 is the lowest common multiple of 365 and 260, which have 5 as a common factor, and $18,980 = 73 \times 260 = 52 \times 365$. Scribal mathematicians had to be able to compute the matching dates forward or backward in time. The approach of a calendar cycle's end point brought anxiety, and magical interventions were invoked against calamities. Floyd Lounsbury and Michael Closs have reconstructed an algorithm that must have been used for computing dates passing through intervals of less than thirteen *baktuns*. It suggests that Maya mathematics had something in common with a modern mentality. One *baktun* is 144,000 days, and one *katun* 7,200 days. The Maya calendar round that characterized every day for a 52-year time span employed three component cycles. The lesser cycle or 13 months the Spanish later called *trecena*. The names of their 20 days were *veintana*. The calendar round compounds these by adding the third component of the vague or calendar year of 365 days. These can be translated into numerical coordinates (t_i, v_i, y_i). In a chronology in which a terminal date (t, v, y) in the calendar round is reached upon moving forward from the initial date (t_0, v_0, y_0) of that round through n_5 *baktuns*, n_4 *katuns*, n_3 *tuns*, n_2 *winals* or "scores," and n_1 *k'ins* or "units," then t, v, y are given by the formulas

$$t \equiv t_0 - n_5 - 2n_4 - 4n_3 + 7n_2 + n_1 \quad (\text{mod } 13);$$
$$v \equiv v_0 + n_1 \quad (\text{mod } 20); \quad \text{and}$$
$$y \equiv y_0 + 190n_5 - 100n_4 - 5n_3 + 20n_2 + n_1 \quad (\text{mod } 365).[314]$$

To count backward, the scribe simply had to take t, v, y as given. The standard practice in major inscriptions was to list the initial date in the calendar year and lunar calendar as well as its placement in the count. Another ritual cycle of 819 days, involving the rain god, further complicated calendrics. Since the stations of that cycle rotated in the four cardinal directions, scribal mathematicians had to find, for example, the nearest 4×819 or 3,276 day station that precedes a given calendar round date. Inscription dates except for that of the calendar year had to be inserted in a fixed order.

Besides the calendar round, the scribes constructed an absolute calendar. Its zero point of 3114 B.C. by the Gregorian calendar provided a benchmark by which mythological events in negative time could be referenced by a set of distinctively notated "ring numbers" and companion correlate numbers as a nexus to historical time. Scribal mathematicians, then, had no philosophical taboos against negative numbers, but such numbers did not share the status of positive numbers.

While astronomical, magical, and religious elements are blended, no attempt to codify an astrological science appears. Computation is directed to record keeping, in part for listings of predictions of possible times of solar and lunar eclipses visible from the tropics. These tables forewarning of dangers include time slots of less than eleven days for possible eclipses. Maya calendrics and chronologies allowed scribes to prepare an ordering of mythological, historical, and current political and civic events. Bureaucrats, merchants, and managers applied the same techniques to their work.

The Dresden codex tables, oriented to the future rather than the past or present and addressing astronomy, display a higher level of mathematical accomplishment. A few major errors in computations or copying, such as giving the sum of a series as 7.17 that should be 8.17 and another that omits a bar in a number to give 1.15.14 instead of 1.15.19, seems to suggest that this text is a revision of a previous source. A table tracking and predicting the morning and evening appearance of Venus harmonizes planetary and calendrical periodicity. Its forecasts were reliable for a period of over 481 years. To reduce an accumulation of error in the Venus tables, Maya scribes had a mix of a single corrective device of 4 days after every 215 days or double when the first is skipped. Reckoning from the Maya equality of 301 Venus years to more than 481 Earth gives the mean Venus year to be 583.92 days. The Venus table creators preferred negative over positive errors, that is, early over late predictions. The Dresden codex gives a companion for predicting lunar and solar eclipses visible in the tropics for 3,741 years or 1,366,560 days. The Maya eclipse table begins with an eclipse or a 5 month half-year after 23 lunations. To find repetitions of eclipses, it then examines groupings of 135 moons in 23 groups and 405 moons in 69 groups. This scheme is similar to the Seleucid saros cycle. Maya scribal mathematicians decomposed the number 1,366,560 into its prime factors, which in our notation are $2^5 \times 3^3 \times 5 \times 13 \times 31$. The Maya, like the ancient Greeks, seem to have caught the importance of prime numbers. The results in the Dresden codex suggest that Mesoamerican astronomers, or schools if work was done collectively, had in some ways surpassed the Old Babylonians.

It is not surprising that such exacting computation, more laborious without fractions, was beyond the reach of some scribes. Working with twenty *tun* multiples was considered taxing. Commonplace mathematical labor was needed for computing less demanding almanacs and commercial, census, tax, and tribute records. Anonymous scribal mathematicians occupied a significant niche in the advanced Mayan civilization and an honorable place in the history of mathematics.

11

The Era of Arabic Primacy and a Persian Flourish

> Truly, in the creation of the heavens and of the earth, and in the succession of the night and of the day, are signs for men of understanding,
> who standing and sitting and reclining, bear God in mind and reflect on the creation of the heavens and of the earth.   — *Qur'an*, 3: 190–191

> We have tried to express roots by algebra but have failed, It may be, however, that men who come after us will succeed.
>   — Umar al-Khayyām

Classical Greek and Hellenistic mathematics, along with neo-Babylonian and Indian mathematics and astronomy, were transmitted to the dynamic new culture of medieval Islam. Here they encountered a far more receptive response than the Hellenistic tradition had in Rome, Byzantium and medieval Latin Christendom. From about 750 to the mid-1400s, apparently, the revival was limited chiefly to large cosmopolitan cities of Islam. Until the tenth century, the main one was Baghdad. In the eleventh century, the center of mathematical research briefly shifted to Cairo, and afterward major mathematician-astronomers served in royal courts on the peripheries of Islam. These urban capitals were located as far east as Tajikistān and as far west as Moorish Spain, where the palace library at Cordoba contained an impressive 40,000 volumes.

The sources and extent of medieval Arabic contributions to theoretical mathematics are only partially studied and still debated,[315] but it is clear that Arabic thinkers became custodians of two older scholarly traditions, one from the Mediterranean West and the other from neo-Babylonian, Syriac, pre-Sassinid Persian, and Indian knowledge in the East. They painstakingly preserved substantial portions of both traditions by collecting, translating into Arabic, and writing commentaries and amplifications on these classic works, and incorporating them selectively into a new Muslim synthesis. In the later Middle Ages, Arabic translations and commentaries were indispensable to the western Latin rediscovery of large segments of ancient Greek mathematics. Islamic thinkers pressed beyond all this, making distinctive, original contributions to mathematics.

That medieval Arabic scholars did not center mathematics in abstract proofs has often led to contradictory impressions about the originality and depth of their work, for which the historical record is fragmentary. Yet the absence of rigorous standards in proofs coupled with the aim of extending difficult computations may have facilitated creativity. The foremost contribution of medieval Islamic mathematicians stems from a dialectic between algebra and arithmetic. They established algebra as an autonomous discipline, reorganized its foundations, and extended Diophantine analysis to more complex equations. In numeration they introduced the zero symbol (cipher), refined nine Arabic numerals borrowed from Indian sources, put the new numerals within a decimal positional system, further developed operational rules for computing with them, and popularized them. Medieval Islamic scholars also made significant discoveries in plane and spherical geometry and generalized parts of them. In the interplay between algebra and geometry, they further produced elegant geometric solutions of cubic equations and were the first systematically to develop trigonometry.

The initial locus of many of these achievements and other leading work in astronomy, astrology, geography, optics, alchemy, and medicine was Baghdad. From its founding in 762 through the ninth century, this capital city was the creative heir to Alexandria's higher tradition in theoretical natural knowledge. Syrian scholars who had absorbed and translated substantial parts of Hellenistic astronomy and neo-Babylonian mathematics fell under Islamic hegemony. They became the crucial conduit for the passage of both to Baghdad. Again, the new Arabic mathematical tradition received, integrated, and, at some points, advanced this legacy of Greek, Babylonian, Indian, and Persian materials.

11.1   Muhammad, the *Qur'an*, and Early Medieval Islam

Islamic civilization rose to prominence between the seventh century and the ninth. It spread from the Arabian peninsula to the Indus valley in the east and across north Africa into Spain in the west. It thus encompassed a diversity of peoples and most of the Mediterranean world. Within that region, only the Latin West and Byzantine East (the Balkans and part of Asia Minor) stood outside it. The rise of Islam began a series of intense religious, political, and cultural struggles between followers of the Muslim Crescent and champions of the Christian Cross that have yet to end.

Islam, Arabic for "submission" (to the will of God), was born in Arabia with the Prophet Muhammad (ca. 570–632), who preached an uncompromising monotheism. "Muslim," the name for practitioners of Islam, means "true believer." The message of Allah, that is, God, had been conveyed through the Hebrew and Christian traditions from Abraham and Moses to Jesus. Islam holds Muhammad to be the last and greatest of the line of prophets and his message the final articulation of religious truth. With him the list of seminal figures for the monotheistic religious revolution is completed. His followers united around his holy book, the *Qur'an* or *Koran*; the title means "recitation" or "lectionary." The *Qur'an* is considered a literal rendering of a series of visionary revelations to Muhammad from the archangel Gabriel. That Muhammad was illiterate, as Muslims hold, makes miraculous in itself his transcription of these. The *Qur'an*, which consists of 6,200 verses, is slightly smaller than the New Testament of the Christian Bible. It resembles and completes earlier, revealed Scripture as a guide to personal holiness and goes beyond both as an ethical manual and law code.

Before Islam, most Arabs who were neither Jews nor Christians were polytheistic. In Muhammad's time fierce, isolated tribes of nomads, or Bedouins, peopled the Arabian interior, where rainfall is scarce. From the sixth century several trading centers, such as Mecca, developed on Arabia's coastal fringe. Mecca was favorably positioned for trade, with merchants sending large camel caravans as far south as Abyssinia and north to the Byzantine empire. Conjointly, the city was becoming a major pilgrimage center. Each year thousands of pilgrims came to see the Ka'ba, the sanctuary of the sacred black meteorite, which according to legend Gabriel had given to Abraham and his son Ismael, founder of the Arabs. The counterpart of Delphi's *opthalmos*, the shrine held images of about 360 local deities.

Muhammad, who was orphaned at the age of six, grew up in Mecca. Legend relates that he arrived at a monotheistic perspective before the age of twenty, when he joined the caravan service and in Syria spoke with Christians and Jews. During times of meditation by his late twenties, he began to have visions. At first he had little success in attracting followers in heathen Mecca. The material and psychological support of his wealthy wife Khadija made his efforts possible. Besides his family, his initial converts came from among slaves and the poorer orders. City leaders opposed him, in part because they feared that his monotheism would harm the pilgrimage trade at the Ka'ba. A planned assassination forced him to flee in 622 to the northern city of Medina, where he was greeted as an influential chieftain. That flight, the *Hegira*, which means a breaking of old ties, became the beginning date of the Muslim calendar. Once Medina was won over and Mecca subdued in A.H. 8 (after the *Hegira*), Islam spread rapidly. Both city dwellers and the scattered and previously contentious tribes of the Arabian peninsula quickly joined the new faith. Fierce tribalism gave way to a common religious brotherhood. A powerful Muslim force, united despite inevitable sectarian disputes, embarked on expansion.

Within little more than a century after the Prophet's death the Muslims controlled not only Arabia but a far-flung empire. By 642 Muslim armies had wrested from Byzantine control neighboring Syria and Egypt with its naval base at Alexandria. They moved eastward, conquering Mesopotamia and the declined Persian empire, and reached the Indus Valley. To the west, Muslim forces swept across the North African coastline. After the mountain Berbers accepted Islam, they joined a Muslim army that reached the Pillars of Hercules (the Strait of Gibraltar),[316] crossed it in 711, and conquered most of Spain from the Visigoths. The new Moorish capital of Cordoba soon became second to Constantinople among Europe's cities.

Two defeats together with overextension have been cited as a check to the Muslim penetration of Europe. In 717 a massive Muslim fleet and army besieged Constantinople, the Byzantine capital. The next spring, after a cold winter inflicted heavy losses on the Muslim army, Byzantine mariners using Greek fire, akin to napalm with a petroleum base, thrown by catapults surprised and destroyed the Muslim fleet. The besiegers had to retreat. J. B. Bury's two-volume *A History of the Later Roman Empire*, published in 1957, contends that the fall of Constantinople would have exposed the vulnerable West to inevitable domination. In 732 Charles Martel defeated an Arab and Berber force from Spain near Poitiers in southern France. In *The Arabs: A Short History* of 1964, Philip Hitti argues that the Battle of Tours is overrated and that past historians had gotten carried away with Gibbons's poetic imagery of spreading Arabic conquest. The outcome, at any rate, was a significant loss to the invaders. Yet even after these two defeats, Muslims held the Mediterranean, including Sicily, in such a grip until the eleventh century that their naval commanders boasted that it was a Muslim lake.

Under the Ummayad caliphs, who ruled from 661 to 750 from their relocated capital of Damascus, an empire dedicated to increasing trade and commerce gradually developed a new urban constellation. Not only were Aleppo, Antioch, and Damascus revitalized, but military camps and towns were turned into new cities of over 100,000 inhabitants. This occurred while western Europe was a region of petty villages.

Within the new urban centers, a rich and eclectic Islamic culture evolved. Enough Muslims became sufficiently confident in their religion and language to tolerate, if not encourage, diverse ideas. The Ummayad clan had long-standing contacts with wealthy aristocratic Christians. Without a priesthood and formal dogmatic creeds, the new and questing culture was open to knowledge from the past—classical Greek and Hellenistic from the Mediterranean West and neo-Babylonian, Syriac, and Indo-Persian in the East, as well as exchanges of ideas with Zoroastrians, Christians, Jews, and Hindus. Persia, Syria, Egypt, and Byzantium each offered a complement of followers of creeds other than the Islamic. From the start Muslims tolerated two other religions, Judaism and Christianity, having a sacred book by the prophets and worshipping God. Zoroastrianism, another religion recognizing a supreme deity and demanding moral behavior, also enjoyed a measure of Muslim respect.[317] In a nascent high culture for which the Arabic letters in the *Qur'an* were sacred and eternal, the study of language and literature dominated, while the sister sciences of lexicography, philology, and jurisprudence flourished.

Under a new line of caliphs, the Abbāsids, who ruled from 750 to 1258, the social, intellectual, and institutional framework within which mathematics flourished among the Arabs was completed. Their founder al-Mansūr (d. 775) built a new capital city, Baghdad, on the Tigris near the ruined capitals of Old Persian Ctesiphon and Mesopotamian Babylon. It was astrologically correct, conveying positive predictions, and was a junction for caravan routes extending as far west as Spain, north to Armenia, and east to China. The city had a temperate climate and relative safety from mosquitoes and lay on a fertile plain, where vast irrigation projects could increase cultivable land for growing rice, barley, wheat, and millet. As trade and agriculture flourished, Baghdad quickly became a thriving city.

The Abbāsid empire, facing east and filled with converts to Islam, took on a Persian cultural cast. Along with religious infighting, the Ummayad exclusion of all but Arabs from governmental positions pushed from Damascus far west to Spain the center of gravity of the Ummayad empire.[318] Persian army support was decisive in the initial Abbāsid triumph over the Ummayads, and the bureaucracy included many Persians nurtured in traditions of opulence. The Persian luxuriance of the early Abbāsids contrasted to the simplicity of the Prophet and desert Bedouins. The exotic tales of *A Thousand and One Nights* are set in Baghdad during the time of Harūn al-Rashīd, who during his reign from 786 to 809 conducted diplomacy with his Frankish contemporary Charlemagne as a check to Ummayad Spain. Two of its main characters are an enlightened vizier, who calls to mind al-Rashīd, and his plucky daughter Shaharazad. She marries a vile king and tells him fascinating stories each night that end his practice of murdering his brides.

The reign of Harūn al-Rashīd and that of his elder son al-Ma'mūm, who ruled from 813 to 833, attended a splendid age in medieval Arabic learning. After twice defeating his Byzantine enemy in 804 and earning the sobriquet "al-Rashīd" (follower of the right path), he and his family extended royal munificence to attract more artists, poets, musicians, scholars, and philosophers to his court. They came from every part of the empire. Al-Rashīd's caliphal court contrasted with older centers of learning in Islam in not limiting its patronage to Arabic belles-lettres, law, and works of Islamic piety, but embracing natural science and metaphysics. Many of the early students in these last two studies had religious allegiances other than Islam. Natural science, moreover, was pursued cautiously for its practical uses or for satisfying speculative curiosity, rather than for expressing a religious and cultural ideology. Al-Rashīd encouraged literary and learned guilds. Following his example, his retinue of rich courtiers, along with libraries at mosques, promoted learning. Under the influence of Persian calligraphy, a more cursive Arabic script than the old Kufic was developed. The introduction of paper and papermaking techniques from China via the Oxus basin, soon replacing papyrus leaves and parchment, improved the medium for recording learning.

The sources of the transmission of knowledge came of earlier migrations from two directions. When Alexandria declined, scholars had emigrated eastward to more congenial religious environments. Nestorian Christians were welcome in

Antioch, and monasteries elsewhere in Syria, especially Edessa, received them eagerly. Subsequent persecution in Syria at about 489 and in Greece following the closing of the Platonic Academy triggered another exodus of Greek scholars further eastward, this time to Harran and Jundishapur in Persia, where Sassinid emperors offered them refuge and supported their studies.

Right after the second emigration, the rendering of Greek texts into Syriac began. Greco-Syrian translators, such as the Nestorian bishop Severus Sebokht from Kenshra on the upper Euphrates, created a technical terminology whose convertibility helped later Syrian-Arabic translators. Fragments of a book by Sebokht in 662 are the earliest extant text to document the migration of Indian numerals westward, where they encountered arrogant linguistic chauvinists in Byzantium not open to higher learning elsewhere.. He writes:

> I will not say anything now of the science of the Hindus, ... of their subtle discoveries in the science of astronomy which are more ingenious than those of the Greeks and Babylonians and of the fluent method of their calculation which surpasses words. I want to say only that it is done with nine signs. If those who believe they speak Greek had known these things, they would perhaps be convinced, even if a bit late, that here are others who know something, not only Greeks, but also men of different languages.[319]

During the sixth and seventh centuries, Jundishapur was the principal center in western Asia of medical learning and of translations of science classics. Its medical school continued the Hippocratic and Galenic traditions, but also included an older Persian and Indian medical stratum. Efforts were begun to synthesize these different traditions.

The invalid al-Mansūr had invited to his city Nestorian and Monophysite Christian physicians for Jundishapur, and at the start of the ninth century, responsibility for recovery and preservation of ancient Greek and Indian learning was shifting from Jundishapur to cosmopolitan Baghdad. Physicians from Jundishapur, especially the Bukhtishu family, were prominent in the migration of knowledge. Al-Rashīd's tutor and *vizier* (chief minister), Yahya ibn Barmak, a Persian whose family had Buddhist origins, lavishly promoted new studies. (The Arabic patronymic word *ibn* means "son of." It may be supplemented by *Abu*, "father of," and sometimes a tribal or honorific name, such as al-Rashīd). Under the joint patronage of the two, whose relation suggests that between Henry VIII and Thomas More, all forms of high culture poured into Baghdad until the *vizier's* Barmakid family fell in 803. Yet the architect of the golden age, al-Rashīd, was an orthodox Muslim whose court was dominated by severely orthodox theologians. Traditionalists recoiled from portions of Greek literature that commemorated many gods. They complained that the brilliant proponents of Persian culture disdained the *Qur'an* as a bedside book. Admission of Greek elements they saw as a concession to Persian secularism, a camel's nose in the tent of orthodoxy.

During the reign of al-Ma'mūn, Islamic society in the Middle East was established more firmly. Following a brief fratricidal war, al-Ma'mūn and his court restored prosperity and exuberance to Baghdad. In a surge of economic expansion, long-distance trade in luxury items and a regular banking business increased, paper and sugar mills sprang up, cotton was introduced from India, and crafts, such as robe making, prospered. Baghdad was now a dynamic intellectual center with a population approaching a half million. It attained a splendor rare among cities up to that time: refreshing fountains, sumptuous bazaars, magnificent new buildings with their pointed arches. Only Constantinople rivaled it in wealth and culture in the Mediterranean and Middle East. Energies turned inward toward the nurture of a scholarship that included mathematics.

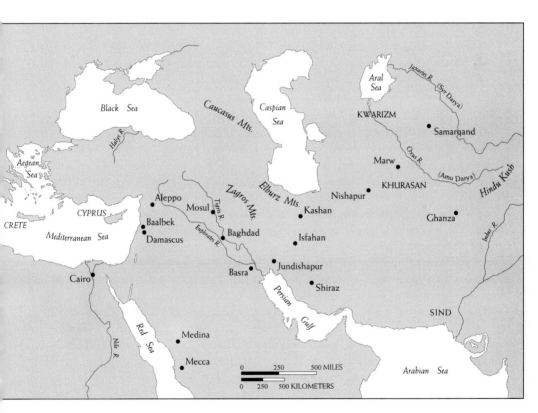

Map showing major cities in Eastern Islam

11.2   Bayt al-Hikmah: Reception of Foreign Knowledge and Translations

Within Baghdad, al-Ma'mūn founded institutions to advance intellectual culture, a crucial step in the support of talented individual and collaborative studies of mathematics. Eager to have thinkers from different schools debate their ideas before him, this keen-minded caliph endowed in 830 a House of Wisdom (*Bayt al-Hikmah*).[320]   He was probably inspired by the Persian academy at Jundishapur. The House of Wisdom was a combined research academy (*Dār al-Hikmah*) and translation bureau where a staff of copyists and binders labored along with translators. A library built by the caliph's father and an astronomical observatory formed part of the complex. The House of Wisdom gave new impetus to the study of theoretical and natural knowledge in Islam.   Leading scholars were brought together in this complex and financially supported by the caliphal treasury. Besides agriculture, alchemy, botany, and medicine, mathematics was a field of study along with subjects closely related to it: astronomy, astrology, cosmology, geography, natural philosophy, and optics. The House of Wisdom quickly became the foremost center in Islam dedicated to the encouragement of such studies.

Within medieval Muslim culture, problems posed by Islamic inheritance laws and astronomy provided a strong impetus for the development of computational and theoretical mathematics. The *Qur'an* requires proper inheritance or legacy rules, and algebra and unit fractions were applied to inheritance problems. Astronomy was highly valued because it provided a reliable means not only for investigating the wonders of God's creation, but also for determining precisely the times for prayers five times a day and tables to locate the direction of the *Ka'ba*, toward which worshipers face. Muslim astronomers had several other civil tasks. They improved on the Muslim lunar calendar and often assisted in preparing horoscopes. Nearly all medieval Muslim rulers consulted horoscopes as al-Mansur had done in picking the site of Baghdad. During and after the Abbāsid period, most mathematicians were also astronomers.

Under al-Ma'mūn interest in ancient Greek thought, especially its logic and science, sharply increased in Islam. This shift to an intense study of *awa'il* sciences, those from the pre-Islamic period, can partly be explained by his massive project of putting into Arabic the entire corpus of classical Greek and Hellenistic manuscripts, especially those of Aristotle. To Syrian Edessa, where Nestorian Christians had sequestered remnants of the Alexandrian library, to Byzantine Constantinople, and to Sicilian Syracuse, Baghdad sent cultural missions in search of these writings. These enterprises were most likely headed by the first director of translations at the House of Wisdom, the Syrian Nestorian physician Hunayn ibn Ishāq (808–873), who possessed a thorough knowledge of Greek. Hunayn's search for a medical treatise of Galen alone took him to Mesopotamia, Syria, Palestine, and Egypt, where he found nothing, and then to Damascus. There he discovered several chapters of the treatise.[321]

In Islam, another pressing reason existed for the sharp increase in interest in ancient Greek thought. Earlier interactions with Indo-Persian cultures had required no fundamental intellectual readjustment by Muslim theologians and jurists. Christians and Jews, two minorities in Baghdad and Damascus, however, handily won debates with Muslim theologians, who were unable to defend their principles in the rigorous Greek manner. This academic failure threatened to undermine religious law in Islamic society, the very basis of authority for the caliphate. There were no independent civil or criminal codes, no rules of primogeniture, and outside religion no definition of the rights of the caliphal office. Greek logic and natural philosophy offered Muslim intellectuals a timely way to safeguard caliphal interests, and their own.

Late in the reign of al'Ma'mūn, the consequences of investigating Greek logic carried caliphal policies beyond abstract jurisprudence. After years of listening to debates, al-Ma'mūn became an adherent to the Mu'tazlites, a group of thinkers who subjected religious beliefs to logical examination. This unorthodox Muslim and son of a Persian concubine, surrounded by Persian advisers, thereby departed from the position of his predecessor and earlier caliphs. Apparently, al-Ma'mūn viewed Greek logic with its mathematical proof theory as another path to truth complementing the path of faith, prophetic sayings and traditions (*hadith*), and the *Qur'an*. Correct use of reason allowed the thinker to distinguish truth from error in religion, to discover hidden religious verities, and conclusively to demonstrate the truth. Court officials and scholars had to adhere to Mu'tazilite doctrine, including the incendiary belief that the *Qur'an* was a created product. Al-Ma'mūn had a way of dealing with people who were insufficiently open minded: he established an inquisition. Some leading jurists and theologians were tortured or killed, and many were imprisoned.

The translation into Arabic initially of Persian and Syriac and next of Greek and Sanskrit texts was, of course, crucial for a higher-level cultivation of mathematics during the Abassid period. It was an instance of the resourcefulness of Abbāsid culture in seeking and rapidly assimilating alien materials. Persian texts were based on neo-Babylonian or Seleucid sources and written in Pahlavi. Their translation introduced technical vocabulary and concepts that facilitated the absorption into Islam of many ancient Greek and Indian scientific ideas. By the reign of al-Ma'mūn, Arabic scholars had not only translated but also commented extensively, and they relied on citation instead of memorization of texts, as had their predecessors. Most translations were completed by the late tenth century. By then original Arabic writings on the sciences had joined the translations to make Arabic the *lingua franca* of the sciences, a position that it held for six centuries. This was most fortuitous. Semitic Arabic, writes J. L. Berggren, proved particularly suitable for "expressing subtle variations" of concepts in the exact sciences and for developing mathematics.[322]

Two generations before al-Ma'mūn, al-Mansūr and his courtiers had inaugurated the tradition of scientific translation among the Arabs. Writing five centuries later, Arab historian al-Qiftī in his *Dictionary of Scholars* would put the beginning at 771 or 773, when a diplomatic mission from Sind in the lower Indus basin of present northern India and southern Pakistan presented al-Mansur with Indian astronomical texts, including the *Mahasiddhānta*, based on the work of Brahmagupta, and a *Sūrya Siddhānta*. Al-Mansūr instructed al-Fazārī and a Hindu pandit to render these texts written in verse form into Arabic and to make this into a handbook for Arabic astronomers. The result was the *Zīj al-Sindhind*. *Zīj*, which apparently distorts the Pahlavi word *zeh* for thread or chord as in a warp of fabric, indicates works on calendar reform or tables of the movement of the sun, moon, and planets, tables with parallel columns and rows that give the appearance of a woven fabric. *Sindhind* refers to the Sind region or possibly to a corrupted form of the Sanskrit *Siddhānta*, which means an astronomical text. Later astronomers called it the *Great Sindhind* and it was the main basis of al-Khwārizmī's *Zīj*. Al-Fazārī also translated sections of the Hindu mathematical treatise *Āryabhatīya*.

The work of al-Fazārī established an early if not completely satisfactory model for Arabic translation, and it introduced Arabic astronomers and natural philosophers to Indian numerals, astronomical parameters, and computational procedures. Derivative and not very systematic, it aroused interest in mathematics and provided a foundation for computational mathematics, but did not by itself stimulate original contributions. Those transpired only after major classical Greek and Hellenistic mathematical manuscripts were translated, selectively absorbed, and revised. Among the first translated were Euclid's *Elements* in 800 and Ptolemy's influential *Mathematical Syntaxis* in 827.

A school of translators who established exacting standards for rendering Greek writings into Arabic first appeared in the House of Wisdom. Among its most active members were the Banū Mūsā sons of the astrologer Moses ibn Shakir: Muhammad, Ahmad, and al-Hasan. They helped to organize the school. Its leading members were the Christian Hunayn ibn Ishāq (Johannitius in the West) and the circle of his disciples who gathered about him, including his son Ishāq ibn Hunayn, his younger Syrian pagan colleague Thābit ibn Qurra, and Thābit's son.

A physician and philosopher who competed with the Bukhtishu family for favor of the caliph, Hunayn was imprisoned for six months before displacing the Bukhtishus as head physician to the caliph al-Mutawakkil (d. 861). Hunayn's standards for translation are not unlike those of modern schools of philology. After collecting as many manuscript versions of a particular Greek text as possible, he collated them to produce a correct manuscript source, made a translation, compared it with an earlier Syriac translation if possible and made his final corrections. This method, combined with his mastery of Greek, led to very literal translations. Hunayn concentrated his attention on medical writings. His principal translations are of writings of three of the founders of Greek and Roman medicine: Hippocrates, Galen and Dioscorides. In cosmology and logic, Hunayn rendered into Arabic or Syriac Plato's *Timaeus*, Aristotle's *Analytica Posteriora*, and preliminary questions to Porphyry's *Isagoge*.

Early in his career, Thābit ibn Qurra had been a distinguished leader in Harran of a religious sect descended from Babylonian star worshippers, who produced leading astronomers and mathematicians. This sect took the name Sabeans after a Chaldean group that had sacred books sanctioned by Book 2, verse 63 of the *Qur'an*. This choice of name gained it tolerance and averted persecution for polytheism. During a visit to Harran, Muhammad ibn Mūsā, oldest of the sons of Moses, met Thābit, who was a money lender there. Impressed with Thābit's scientific learning and mastery of Greek, Arabic, and his native Syriac, Muhammad invited him to Baghdad and became his patron.

Thābit and his younger contemporary Qustā ibn Lūqā from the Lebanese town of Baalbek centered on translating mathematical writings from Greek antiquity. At the House of Wisdom, Thābit translated or revised translations of all available works of Archimedes known in medieval Islam: *On the Sphere and Cylinder, Dimension of the Circle, The Heptagon in the Circle*, and a title translated as *The Lemmas* or the *Book of Assumptions*, as well as Books 5 through 7 of the *Conics* of Apollonius and the *Introduction to Arithmetic* of Nicomachus. He also wrote scholarly commentaries on Euclid's *Elements* and Ptolemy's *Almagest*. Between the *Elements* and the *Almagest*, Thābit began to establish for instruction a collection of intermediate books. These principally addressed spherical trigonometry applied to astronomy and mixed mathematics.[323] The *Book of Assumptions*, containing nineteen propositions that include angle trisection and parallel lines, is usually placed in this category. It is in the same spirit as Euclid's *Data*, another intermediate work, and often confused with it. Thābit's translations and commentaries became standard texts in Muslim lands. Qustā ibn Lūqā subsequently made available in Arabic the writings of Hero and the *Arithmetic* of Diophantus. Notably, Qustā, who translated the title *Arithmetic* as *The Art of Algebra*, made Diophantus known to Muslim scholars in the tenth century after the work of al-Khwārizmī.[324] Many of the most important Hellenistic Greek mathematical works have not survived in the original language and are known today only from these translations.

Here is a chart of Arabic translations from Greek mathematical texts:

Author	Title	Translator	Date/Comments
Euclid	*The Elements*	Al-hajjāj b. Matar	Time of Harun al-Rashid and al-Mamūm.
		Ishāq b. Hunayn Thābit b. Qurra	Late ninth century Died in 901.
	The Data	Ishāq b. Hunayn	
	The Optics		
Archimedes	*Sphere and Cylinder*	Ishāq b. Hunayn Thābit b. Qurra	Revised a poor early ninth-century translation.
	Measure of the Circle	Thābit b. Qurra	Used commentary of Eutocius.
	Heptagon in the Circle	Thābit b. Qurra	Unknown in Greek
	The Lemmas	Thābit b. Qurra	
Apollonius	*The Conics*	Hilāl al-Himsi Ahmad b. Mūsā Thābit b. Qurra	
Diophantus	*Arithmetic*	Qustā b. Lūqā	Died 912
Menelaus	*Spherica*	Hunayn b. Ishāq	Born 809

In his magnum opus *Compendium al-Shifā* ("The Cure" [of Ignorance]), Abū Ali al-Hussain ibn Sīnā or Avicenna (980–1037) portrays among other things the domain of the mathematical sciences in early medieval Islam. This immense four-part work treats logic, physics, mathematics, and metaphysics. *Al-Shifā* compares to Aristotle's *Organon*, which it includes completely, and to al-Fārābī's *Ohsa al-ulum* (*Catalogue of Sciences*), the first comprehensive and detailed enumeration of the sacred and profane sciences in medieval Islam, which along with the *Organon* the *Compendium* expands. In ibn Sīnā's classification, the rational sciences consist of speculative sciences, which seek knowledge that is certain, and practical sciences, which aim for utility and well being. Mathematics, together with physics, astrology, medicine, interpretation of dreams, and alchemy, belongs to the speculative sciences and contains four major and four subordinate sciences. Arithmetic, or the science of numbers, has as its subordinate Indian computation and algebra. Geometry, based chiefly on Euclid's *Elements,* but including materials from Apollonius, Archimedes, and Hero, extends to mechanics, weights and balances, optics, and hydraulics. Geography and astronomy are based on the *Almagest* of Ptolemy. The preparation of maps, astronomical tables, and calendars is subordinate. The science of music and harmonics includes the use of instruments, among them the organ.

The translations had provided one essential element underpinning the establishment of these mathematical sciences. Two other needed components were a critical understanding of ancient Greek ideas and methods along with an occasional departure from them and, subsequently, a continuing fruitful fusion of knowledge from different cultures. Scholars at the early House of Wisdom had all three of these constituents. The result was a period of intense mathematical activity.

11.3   Mathematics at the House of Wisdom and Its Medieval Influence

The development of mathematics at the ninth-century House of Wisdom centered on geometry, numeration, and algebra. Greatly impressed with Hellenistic Greek achievements, its members built Islamic geometry upon the three pillars of Euclid's *Elements,* Archimedes' *On the Sphere and Cylinder,* and Apollonius' *Conics.* The *Elements,* which had been translated into Arabic for Harūn al-Rashīd in 800, underwent two more translations at the House of Wisdom. Any student of mathematics or astronomy had to read this basic text. The classic sources for numeration and algebra initially lay to the east from the neo-Babylonians and Indians. Leading the research in the mathematical sciences at the House of Wisdom were the three Banū Mūsā and Thābit ibn Qurra in plane and solid geometry and al-Khwārizmī in algebra and numeration. Among the Banū Mūsā, Muhammad and al-Hasan concentrated on plane and solid geometry, it appears, and Ahmad emphasized mechanics and translations.

The chief writing of the Banū Mūsā is *On the Measurement of Plane and Spherical Figures.* It is based on Euclid and even more on Archimedes. To find areas of circles and surfaces and volumes of spheres, *On the Measurement* applies Archimedes' compression method of exhaustion but with a departure. Rather than successively inscribing and circumscribing a circle with a sequence of regular polygons whose sides are doubled with each new step, the Banū Mūsā employ the rule of contraries proved in Euclid XII.16 to determine whether an area is greater or less than an area compared to it and whether to omit the transition to a limit value. Extending beyond ninety-six sides the right polygons inscribed in and circumscribed about a circle, they assert that closer upper and lower boundaries can be found for π than Archimedes' $3\frac{10}{71}$ and $3\frac{1}{7}$. The Banū Mūsā also give new but less rigorous proofs of theorems investigated by the ancient Syracusan. Using Euclid XII.17, they demonstrate that the surface of a sphere equals four times its great circle ($S = 4C$). Proposition 7 proved Hero's theorem on the area of a triangle. One important departure from the Greeks bears on the number system. In defining areas and volumes of new curvilinear figures, the Banū Mūsā do not follow the Greek techniques of comparing these with areas and volumes of known figures. Instead, they defined these geometric areas and volumes as products of numerical values. To determine the volume of a sphere, they multiply its radius by one-third its surface. These computations necessitate working more freely with irrationals, treating them equally with integers and rational fractions. The concept of number was later broadened to include positive real numbers. Barely a century after Arab conquerors had passed up spoils, thinking that there was no number larger than 10,000 and at a time when Charlemagne's nobles were learning to write their names, mathematics of this maturity was achieved in Baghdad.[325]

The Banū Mūsā also examine the three classical problems of Greek antiquity. Following Greek procedures, they give mechanical proofs for two. Drawing on Menelaus' *Sphaerica* to solve the Delian problem of doubling a cube, they find two mean proportionals by the intersection of two curved surfaces, and they trisect an angle by means of a kinematic verging construction from Archimedes.

Thābit ibn Qurra (ca. 836–901) was a gifted translator and student of Archimedes. The Banū Mūsā probably worked closely with him: the eldest son was his patron, they too studied Archimedes in detail, and their results overlap in ways that suggest collaborative efforts.

When he translated the preface to *On the Sphere and Cylinder*, Thābit observed that Archimedes had mentioned his discovery of a way precisely to measure segments of a parabola. This reference and his interest in conic sections for spherical astronomy, architecture, and surfaces of sundials prompted Thābit to reconstruct Archimedes' method. Although the treatise by Archimedes on parabolas was unknown in the Muslim world, Thābit devised proofs of Archimedes' results and added a new element to the quadrature of parabolas. In *On the Measurement of the Conic Sections Called Parabolic* and *On the Measurement of Parabolic Bodies*, he computes by summing numerical sequences the area of parabolas and the volume of curvilinear figures produced by rotating parabolas, such as parabolic cupolas. He divides the parabola diameter so as to make its segments proportional to odd numbers and inscribes polygons into these division-point segments, a process that, for us, amounts to valuing the polygons by upper and lower integral sums. Thābit was the first mathematician known today to divide the integration segment into unequal parts. He proves his results by the Archimedean compression method of exhaustion. When instrument makers subsequently complained that his method for constructing a parabola was too verbose, his grandson Ibrāhīm ibn Sīnan (d. 946) enhanced the family's scientific reputation by devising what was the simplest calculation of a segment of a parabola known to the Renaissance.

Possibly derived from a now lost *Elements of Geometry* by Menelaus that Thābit had edited, Ibrāhīm also proposed and proved a neat method for constructing the parabola. He begins with line AG, measures off a segment AB, and perpendicular to AB constructs BE. On BG he then selects as many points, H, D, Z, $\ldots$ as he wishes, and draws semicircles beginning with diameter AH. Through the point T, where the first semicircle intersects BE, he draws a line parallel to BG. Next he constructs a perpendicular to BG through H. These two lines will intersect in K. He there proves that the resulting parallelogram meets the property of a parabola, $KH^2 = AB \times BH$. In the same way, other points L, M, and so on can be obtained. These points describe a parabola having vertex B and axis BG.

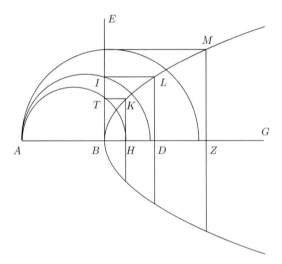

Thābit belonged to the tradition that questioned the independence of Euclid's parallel postulate. Al-Gauhari had first done so at the House of Wisdom by attempting to prove that postulate. His flawed proof amounts to a restatement of Euclid I.33 on lines joining corresponding parts of two parallel and equal lines. Thābit unsuccessfully attempted twice to prove the parallel postulate, the second attempt being more interesting.

The third of five propositions in Thābit's first attempted proof introduced the "Thābit quadrilateral." This quadrilateral has two equal and opposite sides and two base right angles. The equal and opposite sides either approach

or recede from each other equally at both ends. Assuming that its angles are
equal in pairs, Thābit proves that the third and fourth angles are also right angles. Throughout medieval Islam, investigations of the parallel postulate made
prominent use of quadrilaterals with three right angles. They were independently rediscovered in the eighteenth century in the West, where they were
called Lambert quadrilaterals. They were important in creating non-Euclidean
geometries a century later.

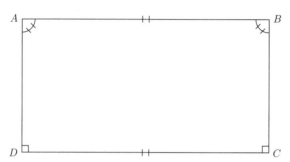

Thābit's second attempted proof of the parallel postulate with seven propositions is based on kinematics. Thābit criticizes Euclid for minimizing the use
of motion in geometry and his successors for perpetuating that neglect. Producing a parallel translation with equidistant straight lines requires, according
to Thābit, a simple linear motion with uniform movement. This appeal to
motion probably originated in Thābit's use of Archimedes' verging technique
for trisecting an angle and constructing two mean proportionals. This work
of Thābit introduced "moving geometry," which became a continuing source
of debate to the Arabs, as it had been to the Greeks. Using Thābit's quadrilateral and motion, al-Haytham later sought to deduce the parallel postulate.
But in rejecting the use of motion in proving theorems, al-Khayyām expressed
the dominant view. In constructing two mean proportionals, Thābit arrived at
what amounts to cubic equations, which he solved by finding the intersection of
a hyperbola and the circumference of a circle. Subsequently, al-Khayyām and
Sharaf al-Tūsī solved by analogous means all cubic equations not reducible to
linear or quadratic forms.

Thābit, who contributed to most known areas of mathematics, disagreed
with Aristotle and Euclid on the concept of number, and he advanced number theory. His popular *Book of Data* criticized Euclid's theory of composite
ratios in books V and VI of the *Elements*. Unlike Euclid, Thābit multiplied
geometrical quantities in composite ratios like common fractions and systematically applied arithmetical terminology to them. Like the Banū Mūsā, Thābit
prepared the way for extending the concept of number to positive real numbers, which al-Bīrunī, al-Karajī, and al-Khayyām were to effect in the eleventh
century. Citing Euclid, Pythagoras and his school, and Nicomachus of Gerasa,
his *Book of Data* discusses the role of prime numbers and the means of generating them, as well as the proof of their infinitude given in *Elements* IX.36.

After presenting a systematic theory of primes that introduces many necessary lemmas on aliquot parts and prime factors, Thābit states his theorem:

> Thābit's Theorem: For $n > 1$, let $p_n = 3 \cdot 2^n - 1$, $q_n = 9 \cdot 2^{2n-1} - 1$; if p_{n-1}, p_n, q_n are prime, then $a = 2^n p_{n-1} p_n$ and $b = 2^n q_n$ are amicable, a is abundant and b is deficient.[326]

He clearly recognizes the central importance of decomposing numbers into their prime factors, and he observes that his predecessors had shown little interest in amicable numbers, such as 220 and 284. These were to have mystical as well as scientific connotations. A treatise of ibn Qurra on the sundial stresses the abstract nature of number and accepts the actual infinite in contradistinction to Aristotle, who had only acknowledged potential infinity.

Islamic Numerals and Computational Arithmetic

Many accounts of the history of mathematics in medieval Islam begin with Muhammad ibn Musā al-Khwārizmī (ca. 780–847), who helped found the Arabic school of algebra. The historian al-Tabarī was to suggest that he had Zoroastrian forebears, in part because his ancestors came from Khwarizm, south of the Aral Sea, in a Persian sector of central Asia. Al-Khwārizmī probably grew up in Baghdad, however, and was an orthodox Muslim. He became a member of the House of Wisdom under al-Ma'mūn and, if the dating of his writings by G. J. Toomer is correct, wrote his major work for that caliph. Little else is known of his life, except al-Tabarī's assertion that al-Khwārizmī was among the astronomers summoned to cast a horoscope for the dying caliph al-Wāthiq, who reigned from 842 to 847. The caliph wanted the horoscope to tell whether he would live or die. Al-Khwārizmī prudently projected another fifty years of life, but the caliph succumbed in ten days.

Al-Khwārizmī's prominence in the history of mathematics stems chiefly from his effective transmission of Indian numerals and computational methods of arithmetic and his contributions to algebra, including his role in initiating its study among the Arabs. To these may be added an achievement long after his death. As will be seen, he was a principal teacher and catalyst for medieval and Renaissance European mathematicians. Origins of his numeration lie in the Indian tradition, while parts of his new algebra seem a skillful synthesis of materials from the neo-Babylonian, Indian, Syriac, and classical and Hellenistic Greek traditions. But neither how far these traditions affected his algebra nor his exact points of departure are yet clearly known. His motive for pursuing the study of numbers, like other mathematicians of early Baghdad influenced by the Greek legacy, was probably philosophical and religious, especially the Neopythagorean and Neoplatonic search for unity in multiplicity. Orthodox Muslims appreciated the doctrine that numbers and figures are generated from the One, to Whom they are ontologically connected. All multiplicity comes from the Creator, Who is One. In the tenth century, scholars in Basra known as the

Brethren of Purity were to expound in epistles Neopythagorean-Hermetic ideas concerning the centrality of number to learning. One of them wrote, "The science of number is the root of the sciences, the foundation of wisdom, the source of knowledge and the pillar of meaning."[327] They also believed that a proper appreciation of numbers requires immediate mystical illumination.

To represent numbers, al-Khwārizmī adopted the nine decimal Indian numerals for 1 to 9, a zero sign called *sifr* (meaning empty) by the Arabs, and place-value arrangement. The word *sifr* is the source of our word cipher. This was a sharp improvement over the Greek alphabetic numeration without place value that the Arabs had inherited and adapted to the twenty-eight letters of their alphabet. In astronomical tables the Arabs had the Babylonian positional system, so place value was not new to their astronomers. Al-Khwārizmī clearly did not introduce Indian numerals to the Arab world: Bishop Severus Sebokht had spread to Mesopotamia in 662 these numerals and associated methods of computation, while al-Fazārī's translations of Sanskrit astronomical texts in the 770s had introduced the Brahmi form of these numerals into Baghdad. (The Gwalior form would not appear until the late ninth century.) Al-Khwārizmī probably learned of Indian numerals from these translations. Indian methods in astronomy strongly influenced him. In an elementary treatise entitled *The Book of   Addition and Subtraction according to the Hindu Calculation*, appearing about 800, Al-Khwārizmī introduced to the Arab world in a systematic fashion the nine Indian numerals and decimal positional system.

Brahmi inscriptions (Nasik)

—	=	≡	Ϟ	Γ	𝟨	7	𝟦	𝟤	⊂
1	2	3	4	5	6	7	8	9	10

Gwalior inscriptions (Bhojaveda, 870)

7	𝟤	𝟥	𝟪	𝟦	⟨	⟩	𝟩	𝟫	7°
1	2	3	4	5	6	7	8	9	10

Hindu numerals

This treatise, now known as the *Arithmetic*, is lost in the Arabic original, undoubtedly because it was superseded by superior texts. The first Arabic arithmetic translated into Latin, it is known today through the partial Latin translation made by Adelard of Bath, Gerard of Cremona, John of Seville, and Robert of Chester in the twelfth century. Al-Khwārizmī presented the new numerals, explained in detail the four basic arithmetical operations with them, employed them to extract square roots, and gave sundry applications. He took pains to demonstrate the usefulness of the decimal place-value system with zero, especially for representing large numbers and in computing.

Within Muslim intellectual circles, Indian numerals and arithmetic were quickly accepted and steadily applied and refined. Among the first to support

them was al-Khwārizmī's younger contemporary at the House of Wisdom, al-
Kindī (ca. 801–866), who wrote four lost treatises on Indian numerals, entitled
Hisabu'l hindi, and another seven on arithmetic. Al-Kindī, wide-ranging in
interests, is sometimes considered the first Arab philosopher. He is also known
for introducing into scholarship the footnote, a convention not employed by
ancient Greek authors.

The first two extant Muslim works on Indian arithmetic date from the mid-
tenth century. They are Kushyar ibn Labban's *Principles of Hindu Reckoning*
and Abu'l-Hassan al-Uqlīdisī's *Treatise on Arithmetic (al-Fusul)*, dated at 952
or 953.

Kushyar, a prominent astronomer born south of the Caspian Sea, presents
the east Arabic form of the nine numerals given here and explains how to add,
subtract, multiply, divide, and extract roots with them and how to check divisions by casting out nines. His explanations are addressed to people computing
with a stick or finger in a tray covered with fine sand, a dust board, rather than
with pen and paper, which calculators had yet to accept. With a dust board in
mind, he avoids having long rows of figures. To add 5625 and 839, for example,
he proceeds from left to right, the opposite of our modern practice, and ends
with the answer at the top right:

$$5625 \ (56 + 8 = 64) \quad 6425 \ (2 + 3 = 5) \quad 6455 \ (55 + 9 = 64) \quad 6464$$
$$\underline{8}39 \qquad\qquad\qquad 8\underline{3}9 \qquad\qquad\qquad 83\underline{9}$$

Fast mental calculators use this technique to the present.

Kushyar's *Principles* also explains computation with the sexagesimal positional system, which he prefers in astronomy because of its unified treatment of
whole numbers and fractions. To this point Muslim astronomers had followed
Ptolemy in adopting the Babylonian sexagesimal system for complex calculations. Hellenistic Greeks, however, had grafted the Babylonian system onto the
twenty-seven letter Greek alphabetic system with notation for the integral part
of a number differing from that for the fractional. By contrast, Kushyar and
other Muslims consistently transplanted the Babylonian sexagesimal system by
not using all twenty eight letters of the Arabic alphabet to represent numbers
but only the first nine to represent one to nine and the next five for ten to
fifty. The sexagesimal positional system represents powers of 60, so no further
numerals are necessary. By going beyond the rigid Babylonian two symbols
but employing fewer than the Greek twenty-seven letters, Muslim astronomer-
mathematicians had thus moved closer to proper cipherization within their own
mode of writing. This widely used Muslim system was simply called "the astronomer's arithmetic."

Al-Uqlīdisī, who resided in Damascus, likely earned a living by making for
sale copies of Euclid's *Elements*. In Arabic, the epithet Uqlīdisī generally applied to such a person. His *Treatise of Arithmetic* seems the most important
of the more than one hundred extant arithmetic texts from medieval Islam. Its
first part presents Hindu numerals and explains place-value notation. By intuitively grasping decimal fractions in his *Treatise*, al-Uqlīdisī began to provide

the component missing in Kushyar. Since these fractions are absent in Indian sources, they seem the invention of Muslim mathematicians, perhaps in a trend going from al-Uqlīdisī to al-Samaw'al in the twelfth century to al-Kāshī in the fifteenth. Al-Uqlīdisī's *Treatise* appeals to decimal fractions to solve only specific problems involving a successive increasing or decreasing of a number by tenths or halving of odd integers. To extract square and cube roots and closely approximate irrational numbers, he employs the rules of zeros, $\sqrt{a} = \sqrt{ak^2/k}$ and $\sqrt[3]{a} = \sqrt{ak^3/k}$, where k is a multiple of 10. This may be stated more generally as $a^{1/n} = (10^{nk}a)^{1/n}/10^k$, where $k = 1, 2, 3, \ldots$. When Kushyar halves numbers, by comparison, he gives answers in terms akin to astronomical degrees for the integral part and minutes for fractions. Al-Uqlīdisī adopted a purely numerical approach. He introduced what seems a forerunner of the decimal point, a vertical mark above a number indicating the unit's place. This allowed him to treat integers and fractions the same in computations. In his system going from right to left, writing before the vertical mark represents tenths, the next hundredths, and so forth. Hence 26 | 587 = 785.62. A comparable notation with decimal fractions did not appear in the Latin West until the early seventeenth century, especially in the work of the Scot John Napier.

Another basic contribution to arithmetic was al-Uqlīdisī's modification of Indian algorithms for computing with Hindu numerals. This appeared in the fourth part of the *Treatise on Arithmetic*. These algorithms required the dust abacus's shifting figures and erasures at each step in computations. In part, al-Uqlīdisī suggested introducing calculating dice instead and in computations superimposing Hindu numerals with dots. His effort seems a first attempt to discard the dust abacus and replace it with paper, pen and ink. A social reason as well as needs of permanent record keeping prompted the abandonment of the dust board. In his *Treatise* al-Uqlīdisī writes:

> Many a man hates to show the dust board in his hands when he needs to use this art of calculations [Hindu arithmetic] for fear of misunderstanding from those present who see it in his hands. It is unbecoming to him since it is seen in the hands of good-for-nothings earning their living by astrology in the streets.[328]

Street astrologers, then, must have carried and used the dust board. By using pen and paper, mathematician-astrologers could avoid being mistaken for wandering horoscopic amateurs, who were sometimes set upon by angry citizens.

Although al-Khwārizmī's *Arithmetic* and Kushyar's and al-Uqlīdisī's books were influential in spreading Indo-Arabic numerals throughout the Muslim world, these numerals were only slowly accepted outside Muslim scientific circles. The Arabs long retained alternative expressions for numerals, including Arabic alphabetic numerals based on the Greek model and their sexagesimal system in astronomy. By the end of the ninth century, moreover, two sets of Arabic numerals with slightly differing shapes had evolved from the Indian inscriptions. Kushyar employed the western caliphate Arabic characters, while the distinct *huruf al-ghubar* (letters of dust) developed in Spain for use on the dust abacus. The chart here shows that our modern numerals are similar to the *ghubar* numerals. The Persian Abū al-Wafā's textbook, *The Science of Arithmetic for Scribes and Businessmen* of about 976, which completely avoids Indo-Arabic decimal numerals, indicates that the eastern caliphate audience outside the sciences still rejected both sets of numerals. Russian historian M. I. Medovoy, who made a critical study of Abū al-Wafā's text, asserts that in general the Arab east still preferred to write numbers as customary verbal expressions.[329]

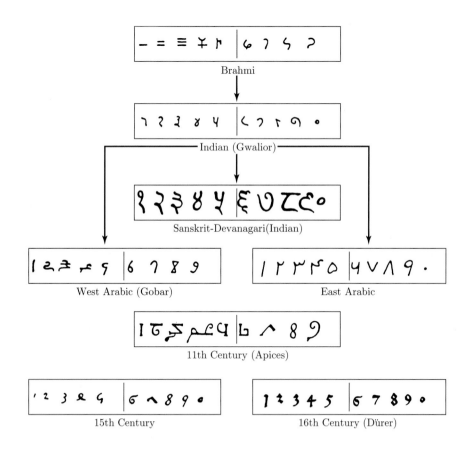

Brahmi

Indian (Gwalior)

Sanskrit-Devanagari(Indian)

West Arabic (Gobar)   East Arabic

11th Century (Apices)

15th Century   16th Century (Dürer)

Evolution of current numerals (Open University)

The transmission to the Latin West of the Indo-Arabic decimal numeration by position did not occur until the high Middle Ages. In the closing years of the tenth century, Gerbert, later Pope Sylvester II, and the monk Vigila of Albelda at the Monastery of Tipoll probably introduced Indian numerals to the West. In the twelfth century four translators into Latin of al-Khwārizmī's *Arithmetic* and *Zij al-Sindhind* repeated the attempt with only a modest impact. In the thirteenth century, Leonardo Fibonacci met with more success in spreading Indian numerals. But a new, more efficient set of numerals posed problems for men of commerce, government, and learning habituated to other numeral systems. Cultural resistance arose from both inertia and the tables of equivalents needed for reducing the possibility of confusion and error. This may partly explain why the Senate of Florence in 1229 forbade the usage of Hindu numerals and required that all numbers be expressed in letters.

Islamic Algebra

Al-Khwārizmī's most important writing was on algebra, not numeration. His *Kitāb al-jabr waal-muqābala* (*The Condensed Book on Calculation by Restoring and Balancing*), appearing at about 830 and dedicated to al-Ma'mūn, is one of the first texts in Arabic on algebra. The title of this manual gave the name algebra to the field. In its three sections, al-Khwārizmī principally discusses what we call linear and quadratic equations. He conceives of algebra as both a theoretical or apodictic science and an applied science, and it begins to emerge as an independent discipline. The first section, which is theoretical, covers how to solve linear and quadratic equations, and it demonstrates the validity of solutions by algebraic rather than geometric proofs. These proofs set al-Khwārizmī apart from the Babylonians and the work of Diophantus, which was apparently unknown to him. The second section is practical, treating land measurements; and the third applies arithmetic and algebra to problems posed by commercial transactions, geometric measurements, and, most of all, testaments or Muslim inheritance laws. In Islam it was typical to orchestrate science and religious law, such as inheritance shares, and al-Khwārizmī's manual became a model for later Muslim reckoners working on this subject.

The two operations this book presents for solving equations are *al-jabr* and *al-muqābala*. In one sense *al-jabr* means restoring an equation by eliminating negative quantities. The process generally means restoring a quantity that is subtracted from one side of an equation to the other, or similarly added, in order to make the equation positive. Thus, it transforms the equation $x^2 = 40x - 4x^2$ to $5x^2 = 40x$. Al-Khwārizmī was not always consistent in using the term *al-jabr*. *Al-muqābala* refers to balancing an equation by subtracting positive quantities of the same power from both its sides, beginning on the side of the larger power. An example of balancing in modern symbols is

$$
\begin{aligned}
50 + x^2 =\ & 29 + 10x \\
-29\quad\ & -29 \\
\hline
21 + x^2 =\ & 10x
\end{aligned}
$$

Until the fourteenth century, when the appellation algebra alone began to be preferred, Christian Europe called the discipline by the two technical terms for these operations. From the sixteenth century, algebra dominated. In English, al-Khwārizmī's manual is today known simply as the *Algebra* from its contracted Arabic title.[330]

In the *Algebra* al-Khwārizmī demonstrates by Indian and Babylonian methods that quadratic equations can have two roots, and he reduces linear and quadratic equations to six canonical or general forms, reflecting the Greek attention to classification and perhaps borrowing from ancient Babylon. This allows him to move toward meeting his goal of understanding three quantities in a problem: simple numbers, roots or the unknown, and the root squared or, in Arabic, *mal*. The demonstration of two roots for quadratics is perhaps his chief contribution to the theory of equations. It is superior to that of Diophantus, who recognized but one root. Al-Khwārizmī divides the six canonical types into two groups of three binomial and three trinomial equations. In modern notation, his classification schema is:

Group I	Group II
$ax^2 = bx$	$ax^2 + bx = c$
mal equals root	number equals root and *mal*
$ax^2 = b$	$ax^2 + c = bx$
mal equals number	number and *mal* equal root
$ax = b$	$ax^2 = bx + c$
root equals number	*mal* equals numbers and roots

All linear and quadratic equations having positive roots can be reduced to one of these six canonical forms. Al-Khwārizmī's classification schema recognizes neither negative numbers nor zero standing alone as coefficients. His schema and rules for solution yield only positive real roots, including some irrationals. Later Muslim algebraists accepted only such roots, whose domain they extended fully to irrationals. But usually they too ignored equations without positive real roots.

In recent decades, historians have debated the sources that al-Khwārizmī possibly employed in preparing the *Algebra*. They have posited contradictory theories regarding the extent of his reliance on three older learned traditions, classical and Hellenistic Greek, Indian, and, most recently, neo-Babylonian. But, as Roshdi Rashed points out, these theories remain in doubt, for no historian has yet established effective links between al-Khwārizmī and the authors of these traditions or his exact points of departure.[331] In general, Greek proof theory was so rigorous that it could be carried to a paralyzing extreme, against which the Indian tradition offered an antidote through its appeal to the "indefinite."

The materials in the *Algebra* argue for some familiarity with Euclid's *Elements*. Geometrical figures in the introduction illustrate equations and some equations are solved by the Hellenic technique of completing the square. The *Algebra* also relies strongly on geometrical justifications of equations, establishing the conditions under which roots exist, much like Book II of Euclid's *Elements*. To give but one example, the solution of an equation that in modern notation is $x^2 + 10x = 39$ al-Khwārizmī justifies by constructing a square composed in part of three areas: x^2, $5x$, and $5x$, as illustrated here. To complete the square, he adds a fourth area of 25, which is based on half the coefficient of x. The area of the large square is $39 + 25 = 64$ and thus its side is 8. The value of x is $8 - 5 = 3$.

$5x$	x^2
25	$5x$

Geometrical algebra was to remain one of two central elements of Arabic algebra during the medieval period. Most Muslim mathematicians greatly admired the authority of their Greek predecessors, as their translations and commentaries indicate.

The hypothesis that Hindu mathematics had a major presence in al-Khwārizmī's *Algebra* can point to its effective numerical processes for solving linear, quadratic, and indeterminate equations, and to the principal aim of the book: an understanding of number, root, and *mal*. The book's use of the rule of three for handling ratios and proportions is consonant with Hindu mathematics. Equally so are many of its mensuration methods, especially that for finding the circumference of a circle from its diameter. Here as in his astronomical studies, al-Khwārizmī appears to draw heavily upon Brahmagupta and through him Aryabhata, at least partly by way of al-Fazārī's translations of Sanskrit astronomical works. The complete exposition of the *Algebra* in words with no symbols (even for numerals) may suggest the influence more of neo-Babylonian and Heronian than of Sanskrit sources, which had abbreviations and first syllables of words in problems. The exposition entirely in words further indicates that Diophantus' *Arithmetica* with its abbreviations and stenographic notation was not yet available in Baghdad.

The place of al-Khwārizmī in the history of algebra is sometimes misunderstood. He did not create the subject nor, in the opinion of Paul Tannery and the Bourbaki, did he carry it to a higher level than Diophantus, an assessment erroneously extended to Arabic algebra. Al-Khwārizmī's *Algebra*, while a solid piece of scholarship, was not brilliant. Probably its thoroughness, its practicality, and its appearance during a golden age of Muslim learning were reasons that it survived when similar works were lost. Through the *Algebra*, al-Khwārizmī proved unusually influential in the development of algebra in medieval Islam, where his immediate successors accorded him precedence,[332] and in the later medieval Latin West.

During the late ninth century, Thābit ibn Qurra, another of the earliest in a line of distinguished Muslim algebraists, solved quadratic equations, giving general proofs, rather than al-Khwārizmī's solutions to particular problems after reduction. His geometrical solutions to what amount to selected cubic equations also make him a leading algebraist. At the same time, the astronomer Abu Abdallah al-Māhānī (825 to ca. 884) proved Archimedes' lemma (*On the Sphere and Cylinder*, II, 4), wherein a plane cuts a sphere into two segments whose volumes are in a given ratio. Archimedes had divided a line segment into parts b and c, so that c is to a given segment as a given area is to b^2. If $b = x$ and $c = a - x$, the proportion requires solving a cubic equation of the type $x^3 + m = nx^2$, which al-Māhānī, like Thābit, failed to solve. Developments in the theory of quadratic equations and important problems in astronomy that translate into cubic equations, for example, in the construction of sine tables, apparently influenced al-Māhānī to seek solutions of cubic equations by trial and error, while al-Bīrūnī extended this work, especially in sine tables. Al-Khāyyam's *Risala* holds that al-Māhānī instituted among the Arabs the tradition of seeking algebraic solutions in the theory of conic sections. Translations of Eutocius' commentaries on Archimedes and Apollonius' *Conics* became important to Muslim algebraists. Abu Ja'far al-Khāzin (d. ca. 971), with the help of conic sections, was the first to solve al-Māhānī's equation. A classification of cubic equations and an attempt at a general arithmetical solution awaited al-Khayyām.

The work of the Egyptian reckoner Abū Kāmil (ca. 850–ca. 930), who belongs to the second generation of Arabic algebraists, is an instance of the arithmetization of algebra that tenth-century Muslim algebraists pursued, building on the work of al-Khwārizmī. Reckoners stressed practical algorithms and practical geometry. Abū Kāmil wrote a commentary on al-Khwārizmī, entitled *Algebra*, that was quite popular. He integrates the numerical procedures of ancient Mesopotamian algebraic geometry with selected works of Hero and Euclid, as well as those of al-Khwārizmī. The result is an extension beyond al-Khwārizmī, providing general statements of rules, rather than confining itself to particular problems. Within the sixty-nine diverse problems of his commentary, Abū Kāmil incorporates irrational numbers as coefficients and roots and devises operations to solve indeterminate equations in several unknowns. He uses part of the rule of signs and the distributive law and begins the theory of algebraic identities, including that for multiplying the square roots of whole nonnegative numbers: $\sqrt{a} \times \sqrt{b} = \sqrt{a \times b}$.

Abū Kāmil's *Algebra* is similar to al-Khwārizmī's in being entirely verbal, having the same classification of six canonical types of equations, and giving geometrical proofs for specific problems. But he departs from al-Khwārizmī's purely verbal algebra in sometimes using Indian numerals in problems and special words for exponents of x up to 8. Since Diophantus' *Arithmetica* was not known in the Arab world before the early tenth century, Abū Kāmil's words for exponents appear to be an original contribution. Beyond this notation, his algebra was generally more abstract than al-Khwārizmī's.

Abū Kāmil's work influenced Abū Bakr al-Karajī, a resident of a suburb of Baghdad in the late tenth century, who joined him as a pioneer in the arithmetization of algebra and its systematization. In modeling the algebra of polynomials on decimal positional arithmetic, al-Karajī's *The Marvelous* furthered the tendency Abū Kāmil promoted to link algebra to arithmetic expressions and operations rather than to those of geometry. Drawing on seven books of Diophantus' *Arithmetica* and his own reinterpretation of Book X of Euclid's *Elements,* al-Karajī refined the theory of algebraic computation. His reinterpreting of geometric magnitudes as numbers basically assigns Book X to the theory of numbers. He then freely applies arithmetical operations to irrational numbers, although not always with justification. To justify procedures and solutions of cubic equations, he provides geometric and numerical proofs, insisting that geometric proofs are no substitute for numerical ones.

Using a form of mathematical induction, al-Karajī's book develops a calculus of radicals with such results as $a^m a^n = a^{m+n}$ and $(ab)^n = a^n b^n$ for m, n positive integers. His algebra studies polynomials (composites) in detail. Though a table of binomial coefficients up to $(a + b)^4$ and $(a - b)^3$ is now lost, his admirer al-Samaw'al describes him as understanding in verbal terms the fundamentals of the formation law of binomial coefficients for integer powers. The rule given in modern notation is $C_m^n = C_{m-1}^{n-1} + C_m^{n-1}$. This seems the first time that the verbal rules for the binomial formation law occurred.[333] Binomial coefficients comprise by rows for each power the so-called Pascal's arithmetical triangle, which is described in Sections 15.3 and 15.6.

In *Al-Bahir fi'l Hisab* (*The Shining Treatise on Calculation*), Al-Samaw'al (d. ca. 1180) in about 1172 corrected defects in and extended al-Karajī's arithmetical algebra. A physician, Jewish by birth, specializing in sexual diseases, he practiced in Baghdad and late in his career around Maragha in the eastern caliphate. *Shining Treatise* is one of the few of his reported eighty-five writings to survive. In his late teens, al-Samaw'al had begun to learn mathematics from Indian computational methods, astronomical tables, and practical techniques, such as surveying. When he found no one in Baghdad who could teach him beyond the first books of Euclid's *Elements*, he completed that work independently and proceeded to study Abū Kāmil's *Algebra*, al-Karajī's *al-Badī*, and al-Wasītī's *Arithmetic*, all on his own. His four-section *Shining Treatise* was first written when he was nineteen and probably reworked several times. Each section is roughly a chapter.

The first two sections of *Shining Treatise* on algebra were not particularly new: they address operations on polynomials (including techniques for extracting square roots), al-Khwārizmī's six canonical equations, indeterminate analysis, and selected applications of algebraic principles. This work suggests the importance that Arab authors attached to obtaining techniques for extracting roots of higher-degree equations through combinatorial analysis. *Shining*

Treatise is clear, insightful, and thorough in evolving the vexing rules of signs as it boldly subtracts from zero, including an excess (positive) number from zero, and takes away deficient (negative) numbers from others; for example, $-5 - (-2) = -5 + 2 = -(5 - 2) = -3$. His rules state:

> If we subtract as excess number from an empty power, the same deficient number remains; if we subtract the deficient number from an empty power, the same excess number remains .... If we subtract a deficient number from a greater deficient number, the result is their subtractive difference; if the number from which one subtracts is smaller than the number subtracted, the result is their additive difference.[334]

Part of the first section of *Shining Treatise* extends division and multiplication from numbers to polynomials and another gives the law of exponents. Through iterations, al-Samaw'al divides what given in modern notation as the polynomial $20x^6 + 2x^5 + 58x^4 + 75x^3 + 125x^2 + 196x + 94 + 40x^{-1} + 50x^{-2} + 90x^{-3} + 20x^{-4}$ by the polynomial $2x^3 + 5x + 5 + 10x^{-1}$ to obtain $10x^3 + x^2 + 4x + 10 + 8x^{-2} + 2x^{-3}$. Apparently, he did this on a dust board. The rows and columns in his accompanying charts show that he had discovered the method for the long division of polynomials and had discerned the law on the formation of coefficients for each term. In this work he recognizes the need for symbols in handling increasingly complex computations in algebra, and his listing of polynomials by sequences of their coefficients is a step toward new symbolism. Another chart gives the operational equivalent of the law of exponents, including $x^n x^m = x^{(m+n)}$ and $x^m x^{-n} = x^{(m-n)}$, the latter for m greater than n. The second section of the *Shining Treatise* presents binomial coefficients, likely from al-Karajī, in a triangular table, now known as Pascal's triangle.

Quite importantly, this book addresses number theory. One of the forty propositions on number theory in the second section establishes that $1^2 + 2^2 + 3^2 + \ldots + n^2 = n(n + 1)(2n + 1)/6$. This result had eluded al-Karajī. The third section mainly classifies irrational numbers following Book X of Euclid's *Elements*.

In pursuing algebraic numerical proofs as well as geometric demonstrations, Al-Samaw'al was at the frontier of mathematics. Although he continued to have the unknown and its powers written out in words, for our x the Arabic term for "root," *mal* for x^2, the Arabic for "cube" for x^3, *malmal* for x^4, and so forth, and the Arabic "part" or p for the inverse, such as *pmal* for x^{-2}, he also assigned each power of the unknown a place in a table and gave in Indian numerals, not words, the sequence of coefficients of a polynomial. These tables with Indian numerals represent a significant step in developing algebraic symbolism. Although in *Shining Treatise* al-Samaw'al failed to meet his goal of systematizing algebra comparably to what the *Elements* of Euclid had done for geometry, in an indirect and restricted way his ideas influenced al-Kāshī in Islam and Luca Pacioli and Girolamo Cardano in the Latin West.

Geometrical algebra continued and its appeal to the theory of conic sections proved fruitful from the late tenth century through the twelfth. Following al-Khāzin's breakthrough in solving the cubic equation deriving from Archimedes' lemma, Sahl al-Din al-Kuhī (fl. 988) solved another cubic equation and thoroughly studied three-term equations. About the same time, Abu al-Wafā (d. ca. 998) solved a quartic equation (in modern terms $x^4 + ax^3 = b$) by obtaining the intersection points of a parabola and hyperbola.

Ibn al-Haytham (965 to after 1040), known in the Latin West as Alhazen, who was to solve selected cubic and a quartic equation by finding points of reflection on conical surfaces, before 1021 had been summoned from Iraq to Egypt, now under the Fatimid dynasty. There he worked at the Cairo Library, the *Dār al-Ilm*, for the ruler al-Hākim. After his initial engineering project to regulate the flow of the Nile failed, he reportedly feigned mental illness to escape the wrath of al-Hākim. After that ruler died, al-Haytham annually copied and sold Euclid's *Elements* and Ptolemy's *Almagest* to supplement his income from teaching. Parts of Apollonius' now lost eighth book of the *Conics* reportedly existed as late as the tenth century, and al-Haytham attempted to restore it entirely. Most famous for his geometrical and experimental theory of light and vision in his major book, *Optics*, in its Book V al-Haytham solves the equivalent of a quartic equation. Since the seventeenth century that equation has been known as "Alhazen's problem." Though lacking a perpendicular coordinate system, he speaks of a circle that cuts one branch of a hyperbola in two points, or is tangential at one point, or falls short of that branch.

Geometrical algebra in medieval Islam culminates in the work of al-Khayyām and Sharaf al-Dīn al-Tūsī. They classified equations up through the third degree and provided geometric solutions for selected cubics.

The transmission of al-Khwārizmī's *Algebra* to the Christian West began in the early twelfth century. Robert of Chester partially translated it into Latin as *Liber Algebre et Almucabala*, which appeared in 1145. Shortly afterward, Gerard of Cremona produced another, less popular version, entitled *De Iebra et Almacubala*. These works were important. Robert of Chester's translation, observe George Sarton and Carl Boyer, initiated European algebra.[335] Indeed, the translation of the *Algebra*, that of the *Arithmetic* as *Liber Alghoarismi* by John of Sacrobosco at about 1240, and the growing use of Indo-Arabic numerals made al-Khwārizmī's name synonymous with the new arithmetic in Christian Europe. Any treatise on that topic was given the corrupted Latin form of his name, *algorismus*, from which comes our word algorism and its corrupted form algorithm.

Islamic Geography and Trigonometry

The new Abbāsid empire straddled continents. Long-distance trade and administration necessitated an accurate knowledge of different lands. The study of geography for the purpose of improving cartography flourished.

Arab intellectuals, like ancient Greek thinkers, believed that Earth is a sphere. After a reading of Eratosthenes, who gave the circumference of Earth as 252,000 stades, and the *Geography* of Ptolemy, which gave a smaller figure of 180,000 stades, and not knowing their exact measure for a stade, al-Ma'mūn, with state needs in mind, commissioned a geodetic survey to measure the length of a degree of meridian on a level plain near Mosul. Multiplying this degree by 360 gives the circumference of Earth. The House of Wisdom sent two survey parties, one proceeding due north and the other due south of a common location. Al-Khwārizmī must have participated in the survey, whose first result is the equivalent of fifty-six miles to a degree. Later, more accurately, another two-thirds of a mile per degree was added. The improved figure puts the circumference of Earth at 20,400 miles: the actual figure is 24,888 miles.

Al-Khwārizmī also assisted a team of scholars constructing a map of the known world for al-Ma'mūn and wrote a geographical book, *The Image of the Earth*. Listing coordinates (latitudes and longitudes) of main cities and regions, it is now known simply as his *Geography*. To prepare it, al-Khwārizmī had to master the method of Ptolemy for projecting a portion of the spherical Earth onto a two-dimensional plane and to apply astronomical observations and computations for determining latitude and longitude. He needed as well to incorporate reliable accounts of travelers. His *Geography* corrects Ptolemy's excessive value for the length of the Mediterranean, improves the descriptive geography of Asia and Africa, and provides accurate coordinates for many Islamic cities. With its maps, the caliph could quickly survey the extent of his empire. Geography and cartography as adjuncts to astronomy helped, moreover, give the direction of Mecca and distances to it for daily prayer and lifetime pilgrimage.

One of the most active members of the House of Wisdom, al-Khwārizmī revised the *Zīj al-Sindhind*, astronomical tables first translated into Latin by Adelard of Bath (d. 1142). Al-Khwārizmī wrote a treatise in 823/24 on the Jewish calendar, two pieces on the astrolabe, and two lost works, *Chronicle* and *On the Sundial*. The *Chronicle* was a survey of the early Islamic period that possibly employed a combination of astrology and astronomy to explain historical occurrences. None of al-Khwārizmī's astronomical writings were prominent.

In medieval Islam the development of trigonometry was closely linked to astronomy. Everywhere in the *Dār al-Islam* worshippers, itinerate or settled, faced Mecca five times daily, and the beginning of each month could be predicted only from the variable first visibility of the crescent moon. These and a need for improved astronomical tables now largely motivated the study of trigonometry. Preparing astronomical tables involved problems ranging from constructing numerical sequences to refining methods in descriptive geometry. Trigonometry in the medieval Islamic world benefited greatly, as had algebra, from integrating Greek models with Indian ideas and computational methods.

Before the ninth century, Islamic astronomy had relied on astronomical tables from Sassanid Persia and India, particularly those translated for Caliph al-Mansur and al-Fazārī. These eastern influences receded shortly after al-Khwārizmī's improvement of al-Fazārī's *Zij*, when Ptolemy's *Syntaxis Math-*

ematikos, Tetrabiblos, and astronomical tables known as *Canones Procheroi* were rendered into Arabic. The *Syntaxis Mathematikos* was translated several times. From the late ninth century, Muslim astronomers were the major elaborators of Ptolemaic astronomy. Among the Hellenistic Greeks the *Syntaxis Mathematikos* or *Mathematical Arrangement* (whence the English "syntax") had become known as the *Great Arrangement.* The Arabs added *al* in front of the Greek word *megiste,* meaning "the greatest." This title was Latinized by twelfth-century western authors as *Almagest.* After reading Ptolemy, Muslim astronomers used a model of a system of epicycles for celestial motions. Until the eleventh century, most of them accepted a geocentric universe.

Two types of trigonometric reckoning were available to the Muslim world: the Hindu arithmetical method directly computes the sine and cosine of an arc not as modern ratios but as lengths, while Ptolemy's geometric method yields chords that convert to sines by a halving of the subtended angles and their chords. Muslim astronomers learned from the verses of *Surya Siddhānta* the Hindu sine values at every $3\frac{3}{4}^{\circ}$ interval of arc to 90° and from the *Almagest* Ptolemy's chord lengths for $\frac{1}{2}^{\circ}$ to 180° at $\frac{1}{2}^{\circ}$ intervals. The Hindu method was based on a half-chord (the Arabic *jiba*)[336] in a quadrant of a circle. For arc AB of a circle, the medieval rule relating the Ptolemaic chord and Hindu sine is $\mathrm{Sin}_R AB = \frac{1}{2}\mathrm{Crd}_R 2AB$. The half-chord, in modern notation $R\sin\theta$, was thus a length depending on the length of radius (or R), which could be 60, 120, 150, or the Hindu 3,438 units. Once the sine was found, the cosine was computed by a trigonometric identity in algebraic form, $\sin^2\theta + \cos^2\theta = 1$ or $\cos\theta = \sqrt{1-\sin^2\theta}$, an identity Ptolemy had established geometrically. Medieval Muslim astronomers refined the Hindu arithmetical calculations of sines and Ptolemaic chords of arcs lying between 0° and 90°. Computing these at intervals as small as $15''$, they obtained far greater precision than their predecessors. Their calculations were accurate from at least three to as many as five sexagesimal places. They also extended the number of trigonometric functions from two to six. With their trigonometric precision they produced superior planetary and lunar tables, which were also used for astrological purposes. It was Thābit ibn Qurra and his younger contemporary al-Battanī (ca. 858–929) who effectively introduced to the Arabs the Hindu sine and other elements of trigonometry. Although al-Battanī was a Muslim, both were descended from Harranian Sabeans, who had developed astral theology.

Thābit's treatise on the sundial has two interesting trigonometric rules. By the equivalent of $\sin h = \cos(\phi - \delta) - versin\, t \times \cos\delta \times \cos\phi$, it defines the height h of the sun by reference to its declination δ, the latitude of the city ϕ, and hour angle t. The versine, which is also called "inverted sine" as opposed to "plain sine," is $1 - cosine$. And by $\sin A = (\sin t \times \cos\delta)/\cos h$, the treatise locates the sun's azimuth A. These two rules are equivalent to theorems for the cosine and sine in spherical trigonometry. Thābit, however, used them only to solve concrete problems. General theorems for both awaited the work of Regiomontanus in the fifteenth century.

The third chapter of al-Battanī's principal work, *Kitab al-Zīj*, appearing about 930 and sometimes called the Sabean *Zīj*, develops his theory of trigonometric functions. As background for his practical text, he made astronomical observations and proficiently built instruments for them at the city of al-Raqqa from 877 until at least 918. At al-Raqqa, which was located on the left bank of the Euphrates, resided several families from Harran. In preparing his *Zīj*, which often cites the *Almagest*, al-Battanī followed the admonition of Ptolemy to read his work critically and remove its errors and discrepancies on the basis of new observations. Among other things, al-Battanī redetermines with greater accuracy the obliquity of the ecliptic, the precession of equinoxes, and lunar mean motion in longitude. Chapter three does not follow the procedure from *Almagest* I.ii of using the chord of twice the angle. Instead, in employing the Hindu sine, cosine, and versine, it adheres to previous Arabic practice. Al-Battanī calculates his sine table at half-degree intervals with an accuracy to four sexagesimal places. Cotangents and tangents are not among his formulas. They appear only in clumsy gnomonics (studies of sundials), for example, in Ptolemy and the Hindu *Siddhāntas*, conceived as lengths of shadows. The derivative Latin term for cotangent, "umbra extensa," means the shadow length of a vertical gnomon and "umbra erecta," tangent, the shadow of a horizontal gnomon. Al-Battanī provides elegant solutions to problems in spherical trigonometry by applying the principle of orthographic projection, a principle Regiomontanus later developed in the Christian West.

The Persian Abu al-Wafā al Būjānī (940–998), who had moved to the Buyid court in Baghdad in 959 and spent the rest of his life there, was the city's last major mathematician-astronomer during medieval times. His handbook *Zīj al-Majisti*, which is obviously based on the *Almagest*, treats trigonometry more systematically than had his Muslim predecessors. Abū al-Wafā includes among his six trigonometric functions the secant and cosecant, known previously to Habash al-Kasib. All six quantities in his and subsequent Arabic works are constant multiples of modern ones. Abū al-Wafā takes an approach that moderns follow by letting the radius $R = 1$ and then, by using the identity $\cos^2 \alpha + \sin^2 \alpha = 1$, calculates the cosine values from sines. For his use of the unit circle and all six trigonometric quantities, historians credit him with being the first calculator of modern trigonometric functions.

By repeated divisions and interpolations, Abū al-Wafā computed a new sine table at intervals of one-half degree (30′) that is accurate to the fourth sexagesimal place. Ptolemy's interpolations, by comparison, were only correct to the third. Abū al-Wafā's $\sin 30' = 0; 0, 31, 24, 55, 54, 55$ when converted to decimal fractions is correct to 10^{-8}.[337] To attain such accuracy Abū al-Wafā employs an inequality based on a theorem of Theon of Alexandria. The inequality stated in modern notation is

$$\sin(\tfrac{15°}{32}) + \tfrac{1}{3}\{\sin(\tfrac{18°}{32}) - \sin(\tfrac{15°}{32})\} < \sin 30' < \sin(\tfrac{15°}{32}) + \tfrac{1}{3}\{\sin(\tfrac{15°}{32}) - \sin(\tfrac{12°}{32})\}.$$

To find the $\sin(15°/32)$, he begins with the known value of $\sin 60°$ and for $\sin(18°/32)$, with the known value of $\sin 72°$. To calculate the sine of half a given angle, he still employs rational operations and a square root extraction. The $\sin(12°/32)$ is $\sin(72°/32 - 60°/32)$. By assuming that $\sin 30'$ is half the bounding quantities above and below in a circle of radius 60, he arrives at his accurate value for it.

In addition to making precise computations, Abū al-Wafā proves two major theorems from which a self-standing spherical geometry has emerged. In place of Menelaus' theorem with its cumbersome "rule of six quantities," he proposes and proves what is known as the "rule of four quantities." Given two spherical right angles, ABG with right angle B and ADG with right angle D, and a shared acute angle at A, then $\sin BG : \sin GA = \sin DE : \sin EA$.

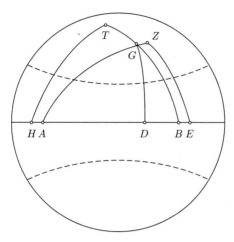

In applying this rule, Abū al-Wafā derives the law of sines for spherical triangles. If ABG is a spherical triangle with sides a, b, g opposite angles A, B, G, then

$$\frac{\sin a}{\sin A} = \frac{\sin b}{\sin B} = \frac{\sin g}{\sin G}.$$

This sine law, which simplifies many problems about arcs on a sphere, is the first theorem on spherical angles. According to al-Bīrūnī, several astronomers proposed this law at about the same time, and a controversy arose over who deserved priority.

The *Hākimī Zīj* of the Egyptian astronomer ibn Yūnus (d. 1009) was named after his second Fatimid patron in Cairo, al-Hākim, who was deeply interested in astrology and valued his accurate predictions. After conquering Egypt, the Fatimids had founded Cairo in 969. Ibn Yūnus, who was from a respected family of historians and scholars of *hadīth*, Muhammad's sayings, had witnessed both. He recorded astronomical observations from 977 to 996, when the new eleven-year-old Caliph al-Hākim ordered them renewed. These observations, which continued until 1003, were completed before that caliph's failed attempts to build an observatory. In computing the many trigonometric tables for his eighty-one chapter *Hākimī Zīj*, an astronomical handbook twice the size of al-Battānī's, ibn Yūnus made two impressive contributions to the field.

One of these, derived probably from ibn Yūnus's practical experience with tables, is a second-order interpolation procedure for calculating at $30'$ intervals in a sine table. He determined that the linear interpolation used for finding $\sin(\phi+k)$ and $\sin(\phi+30')$ always gives too small a value and that the difference between the interpolated and the actual value is greatest or very nearly so at the midpoint when $k = 30'$. Ibn Yūnus seeks a function f that applies from $0' < k' < 60'$ to allow him to compile a table for each minute of arc, such that $f(0') = f(60') = 0$ and has $f(30')$ as the greatest value. Representing $L\sin$ as the Sine value derived by linear interpolation, he then arrives at $L\sin(\phi + k') + f(k')[\sin(\phi+30') - L\sin(\phi+30')]/f(30')$ as a sound rule for interpolation to obtain a more accurate value of $\sin(\phi + 30')$ than the linear interpolation value, $L\sin$, but still not the exact one. From Euclid ibn Yūnus should have known that what in modern notation is $f(x) = x \cdot (60 - x)$ gives the maximum product when $x = 60 - x = 30$. In this case, the maximum value of f is $900'$. As was typical for Islamic trigonometry, ibn Yūnus does not attempt to prove his interpolation discovery from axioms.

Further advancing trigonometry, ibn Yūnus computed auxiliary function tables over a given number of degrees that lessened the tedious computing of astronomical tables. He chose functions that often occur in astronomical tables, such as $\tan\theta \sin\theta/R$, that otherwise had to be calculated step by step with different values for each. With the aid of these auxiliary functions, he computed tables for extensive spherical charts of oblique ascensions at the horizon from $1°$ latitude to $48°$, which cover Cairo at $30°$, giving the times of sunrise and the evening risings of stars. He assumes that twilight begins when the sun sets below the horizon by a certain angle of depression varying from $16°$ to $20°$. These tables were important for civil timekeeping and correctly determining the muezzins' proclamation of prescribed daily times of prayer. The fourteenth-century astronomer Muhammad al-Khalīl generalized and extended them to latitude $50°$. They belonged to the body of astronomical tables employed in timekeeping in Cairo up until the nineteenth century.

The central Asian scholar Abū al-Rayhan al-Bīrūnī (973–1048) consolidated and refined past discoveries in trigonometry. He found the subject an important segment of classical knowledge and that of his times, all of which he set out to master.

After growing up in Khwarizm, al-Bīrūnī corresponded briefly with ibn Sīnā in a dispute over the nature of light. In 997 he exchanged letters with Abū al-Wafā in Baghdad about making simultaneous observations of a lunar eclipse for determining the longitude difference between Kath on the Oxus River and Baghdad. By then al-Bīrūnī was well versed in astronomy, astrology, mathematics, natural science, meteorology, geography, history, chronology, cultural anthropology, and linguistics. His book *Elements of Astrology* became for centuries the standard text for teaching the quadrivium in Islamic countries.[338]

To have financial support for extended intellectual studies and some security at the turn of the eleventh century in the eastern Islamic world, a scholar had to be attached to a royal court. During al-Bīrūnī's twenties, four powers battled each other around Khwarizm. He spent the decade seeking royal patrons and later fleeing from them. At the age of thirty, al-Bīrūnī returned home to serve at the court of Shah Abū l'Abbu Ma'mun. When the army killed Shah Ma'mun, Sultan Mahmud of Ghazna in present day Afghanistan invaded the city and took him prisoner. Although he privately rejected judicial astrology as a way to predict events, al-Bīrūnī was sometimes at these courts an official astrologer. Usually he was a skillful diplomat, known for his "tongue of silver and of gold," who mediated disputes. "I was compelled," he writes, "to participate in worldly affairs, which excited the envy of fools, but made the wise pity me."[339]

Al-Bīrūnī accompanied Sultan Mahmūd on his conquest of northwest India, a section known to Arabs as al-Hind, and was probably exiled there for a time. While in India he developed an interest in Sanskrit and gained a moderate command of it. With this knowledge and the help of Hindu *pandits*, he corrected corruptions in earlier translations of Sanskrit texts. He also gathered materials for *India,* his masterwork of 1030, the leading account of Hindu customs, religion, society, and science written in medieval times.

At the fort of Nandana in India, al-Bīrūnī in *On the Determination of the Coordinates of Localities* elegantly applied trigonometry to cartography. This fort commanded the route taken by conquerors from Alexander the Great through the Moghuls in entering the Indus Valley. Al-Bīrūnī's study begins with a review of Ptolemy's generalizing of Eratosthenes' method, which Ptolemy had accomplished by measuring one degree along a meridian. Muslim astronomers had measured a degree of meridian for Caliph al-Ma'mūm and knew how to determine the height of a mountain. Using a square instrument with a calibrated side, as indicated in drawing (a), he determined the height of a mountain near Nandana. The angle BEZ is called the dip angle d. Al-Bīrūnī drew a tangent (LO) to point L on the Earth's surface and dropped a perpendicular to L, where EL equals the height of the mountain, as indicated in drawing (b). From the law of sines, al-Bīrūnī knew that $EL : LO = \sin(O) : \sin(E) = \sin d : \sin(90° - d)$.

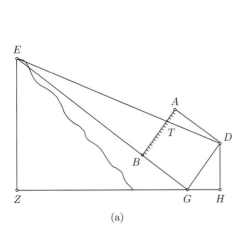

(a)

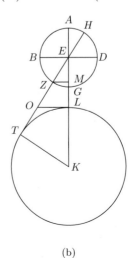

(b)

The Pythagorean theorem gives $EO = \sqrt{EL^2 + LO^2}$, and he knew that $ET = EO + OT$. Applying the law of sines again, he found KT, the radius of Earth. Assuming π to be $3\frac{1}{7}$, he obtained the slightly small value for a meridian length of 55.53 miles to a degree. With reference to al-Ma'mun's survey, he writes: "Here is another method for determining the circumference of Earth. It does not require walking in deserts."

In *Keys to the Science of Astronomy, Shadows, Chords,* and *Mas'udic Canon,* his most comprehensive extant work in astronomy, al-Bīrūnī deals with plane and spherical trigonometry, including numerical tables of trigonometric chords. His *Keys* offers a spirited account of controversies surrounding the near simultaneous discovery of the law of sines in spherical trigonometry and presents a low estimate of the character of Abū al-Wafā. *Shadows* defines tangent and cotangent, discusses conversions of related gnomonic divisions from three different cultures—the Hellenistic 60, Hindu 12, and Muslim 7 or $6\frac{1}{2}$—and calculates tangent and cotangent tables for these lengths. The cotangent table is based on the relation $\cot \phi = \cos \phi / \sin \phi$. *Chords* begins with statements and proofs of geometric theorems and related metric relations between chords helpful for computing a table of chords (or sines). Al-Bīrūnī notes that Archimedes knew Heron's formula and that Brahmagupta's formula holds only for a cyclic quadrilateral. In one problem, he inscribes a nonagon in a circle and then through the relations for the chord x of a 20° angle reduces that problem to solving the cubic equation $x^3 - 3x + 1 = 0$.[340] His approximate sexagesimal solution, 1;52,15,17,13, is accurate. Treatises 3 and 4 of the *Mas'udic Canon* present more precisely than any other source at the time the six standard trigonometric functions from 0° to 90°. Al-Bīrūnī's dedication of the *Canon* to Mas'ud, successor to Sultan Mahmud, won him release from exile and possibly new privileges on his return home. The 146 writings attributed to al-Bīrūnī reveal a polymath who, like Abū al-Wafā, sought greater accuracy and to achieve it appealed more to solid new observations than computational techniques. Of these works, sixty-two are on astronomy, astrology, and chronology, the chief exact sciences of the time; these are followed by nineteen on geography and geodesy and fifteen on mathematics. In their sense of historical sequence, even al-Bīrūnī's histories assume a mathematical perspective. His scientific and historical writings are the product of critical reasoning, meticulous research, and acute insights. These virtues led him to reject Ptolemy's geocentric model of the universe and to propose instead that Earth moves around the sun. His new planetary schema together with the subsequent departure from Ptolemaic astronomy by al-Tūsī and his disciples influenced medieval natural philosophers and probably Nicholas Copernicus.

11.4   Mathematics in Medieval Islam after 1055

Despite kaleidoscopic military change, some hostility to ancient pagan learning, and reduction in financial support, mathematics continued to advance in the eastern Islamic world during the Seljuk period after 1055. The disturbances made development more episodic, but seeds of learning planted in ninth-century Baghdad and Cairo under its Shiite Fatimid dynasty remained fruitful in algebra, algebraic geometry, arithmetic, and trigonometry. Indeed, medieval Islam now produced in Umar al-Khayyām, his disciple Nasīr al-Dīn al-Tūsī, and Jamshid al-Kashī probably its three foremost universal mathematicians. That all three were situated on the eastern periphery of the Islamic world suggests that the chief locus of mathematical development had shifted there.

The disintegration of the eastern caliphate had begun when the Buyids, recent converts from the southwest Caspian region, captured Baghdad in 945, after which Iraq became merely a province in the eastern empire. In 1055 Turkish nomads, the Seljuks, migrated from central Asia into the empire, accepted Islam, and in their turn captured Baghdad. Within two decades the Seljuk sultan was supreme there and the Abbāsid caliph simply a figurehead. In 1096 came the first of nine arbitrarily numbered Crusades from western and central Europe. The last official Crusade would be in 1291. Still another group of invaders, the Mongols from the east Asian steppes, wreaked the most havoc, destroying many centers of high culture in eastern Islam. Ghengis Khan ("Ruler of All within the Seas," ca. 1162–1227) swept into Persia and Iraq around 1206 and, in two decades, captured most of Persia. In 1256 Ghengis' grandson, Hulegu, sacked Baghdad and executed the last Abbāsid caliph by having him trampled to death fully wrapped in carpets so that none of his blood would touch the ground and bring a curse.

Long before Hulegu's capture of Baghdad, the center of learning in Islam had shifted from that city. By the death of Thābit, the House of Wisdom had essentially ceased to exist as an institution, and as the empire crumbled, older Persian cities and capitals of new rulers competed as centers of high culture. The Fatimid ruler al-Hākim built in 1005 the *Dār al-Ilm* (Library) in Cairo and drew al-Haytham and ibn Yūnus to it. Three major cities of Muslim Spain, Cordoba, Seville, and Toledo, were active places of learning and by the twelfth century were transmitting ancient Greek and medieval Arabic knowledge to the Latin West.[341] Scholars are now uncovering a mathematical tradition in the Maghreb in north Africa from the twelfth to the sixteenth century. The twelfth-century savant Ayyash al-Hassar, author of *al-Kamil* on the decimal system, number theory, and extracting of roots of cubic equations, was prominent here.[342] During the twelfth and thirteenth centuries in eastern Islam, Isfahan and Bukhara supported a blossoming of Persian literary culture, especially in poetry. Persian, which Arabic had never completely replaced for high learning, now became the language of polite society.

The intellectual framework for mathematics in Islam after 1055 differed from that in ninth-century Baghdad. Islamic culture had matured, and *falsafa*, drawing mainly on Hellenic and Hellenistic studies but also on neo-Babylonian and Indo-Persian sciences, was accepted. Muslim thinkers had selectively assimilated and expanded on these traditions and added original components. Muslim mathematician-astronomers, then, had for a time a secure niche at courts and in institutions of higher learning (*madrasas*). When not confused with street astrologers, they held in Islamic society a position that, however perilous at times, was better than that of poets, who were often treated harshly, and historians, who were sometimes barely tolerated. Both royal courts and higher schools harbored mathematicians and natural philosophers along with religious scholars (*ulama*). Until the late twelfth century, some representatives of the two groups engaged in generally fruitful dialogues. Both accepted the Iranian heroic tradition, which apparently fostered imaginative literary and scientific studies.

Tensions between sacred and profane sciences intensified throughout the late medieval Islamic period, acerbated by ethnic and sectarian strains. Neither science nor literature was or could be completely separated from religion. The mid-eleventh century triumph of Sunni sectarianism ended the previous Shi'i century of a limited liberty of opinion. *Falsafa*, which had improved the polemical tone of the controversy between Shi'i and Sunni, was subtly attacked. *Falsafa* had triumphed in mathematics and astronomy, tolerated as less important disciplines in which theological implications were minimal, but in higher disciplines of metaphysics and theology it implied—or taught outright—that the material world was not created. Twelfth-century critics charged that *falsafa* was antithetical to the *Qur'an* and public disapproval grew. In response the Andalusian physician ibn Rushd (known in the Latin West as Averroës, d. 1198), the greatest medieval Muslim commentator on Aristotle, maintained that while *falsafa* could not interpret results and images of revelation, it offered a sophisticated alternative to vulgar and simplistic views about the workings of the natural world. His approach accorded with the dialectical theological notion of *Kalam* that religious faith and deductive reason offer distinct but valid grounds for investigation, each addressing its separate domain. But Averroës' influence in the late medieval Muslim world was less than that of the religious critics of *falsafa*.

Ibn Khaldūn (d. 1406), the eminent medieval Arab historian and *Faylasuf*, continued the Muslim Aristotelian classification of the sciences and mathematics. A tutor and later a judge, who moved from Granada to Tunis, his home region, and thence to Cairo, ibn Khaldūn wrote commentaries on ibn Rushd and examined the writings of Nasīr al-Dīn al-Tūsī. His classification of the sciences, which stems from Aristotle's *Posterior Analytics*, appears in the introduction, the *Muqaddimah*, to his history. He divides the sciences into the religious and the philosophical. Like Aristotle, he has philosophy consist of logic, physics, metaphysics, and mathematics. Mathematics is comprised of the numerical sciences of arithmetic, calculation, algebra, commercial transactions, and partition of inheritance; the geometrical sciences of spherical and conical geometry, surveying, and optics; astronomy, which includes astronomical tables and judicial astrology; and music.

During the Seljuk period, official financial support for such mathematics in eastern Islam was more scattered and uncertain than in early Baghdad. Support from *madrasas, waqf* (endowments for libraries and hospitals), and patronage of large-scale building projects dwindled. *Madrasas*, for example, were now generally uninterested in mathematics beyond an elementary level. Al-Samaw'al's failure to find instruction beyond the first book of Euclid in twelfth-century Baghdad is a case in point. Shortly after 1055, funds from the small caliphal retinue and commanders in Syria and Persia for construction of large observatories, schools, bridges, and even mosques decreased sharply. Recurrent economic difficulties and numerous military disruptions dried up these funds. Mongol sultans, Seljuk viziers, and their court members invited savants in astronomy, astrology, calendrics, and theoretical mathematics to their capitals and provided accommodations conducive to study. Amid increasing tumult, however, support was sporadic.

The Persian mathematician-astronomer and poet Umar al-Khayyām (whose anglicized name is Omar Khayyam, 1048–1131) was born in Nishapur, a principal city in Khurasan, to the east of Persia. The epithet Khayyām suggests that he or his father, Ibrahim, was a tentmaker. Shortly after Umar's birth, the Seljuk Turks conquered Persia and Azerbaijan, where they established a vast but unstable military empire. Some historical sources state that he received his early education in Nishapur; others maintain that he studied in Balkh, now in Afghanistan. He expressed an early interest in algebra, arithmetic, and music theory. By the age of seventeen, al-Khayyām was well versed in the many fields of *falsafa*. He became a tutor, whom one student considered narrow-minded and ill-tempered.

To have sufficient funds and time for extended study, al-Khayyām had to attach himself to the court of a magnate or sovereign. But a court did not guarantee a scholar's pursuit of substantial learning. Indifference of the patron, opposition of the monarch, court intrigues that included attacks by theologians, or disruptions of war were possible impediments. About 1070, al-Khayyām traveled in search of financial support to Samarqand, where the chief judge, Abū Tahir, became his patron. While in the service of Abū Tahir, he wrote his great algebraic treatise on cubic equations, *Risala fi'l-barahim 'ala masa'il al-jabr wa'l muqabala* (*Treatise on the Demonstration of Problems of Algebra and Almuqabala*). He had long planned this work. The *Risala* complains of difficulties faced by mathematicians even at a receptive court:

> I was unable to devote myself to the learning of this *al-jabr* and continued concentration upon it, because of obstacles in the vagaries of time which hindered me; for we have been deprived of all the people of knowledge save for a group, small in number, with many troubles, whose concern in life is to snatch the opportunity, when time is asleep, to devote themselves meanwhile to the investigation and perfection of a science; ....[343]

Shortly after 1070 the Seljuk sultan Malik Shah and his vizier invited al-Khayyām to Isfahan to supervise its astronomical observatory. He remained there for nearly eighteen years, participating in its high literary culture and guiding the observatory compilation of tables of ecliptic coordinates and magnitudes of fixed stars. He was also one of eight astronomers commissioned by Malik Shah to reform the Muslim calendar. About 1079 he proposed a new (*Jalali*) calendar that was more accurate than the later Gregorian. It had eight of every thirty-three years as leap years. (The Gregorian calendar has an error of one day roughly every four thousand years, the *Jalali* calendar every five thousand years.) Though he personally rejected judicial astrology, in Isfahan al-Khayyām was court astrologer.

After Malik Shah died and his vizier was murdered by an assassin, al-Khayyām fell from favor in 1092. The new regent, who had opposed the vizier, withdrew her financial support from al-Khayyām and the observatory. Anecdotes suggest that his treatment of other intellectuals made enemies among the *Faylasufa*. When asked by a prince about the Aristotelian commentator Abū'l Barakat's criticism of Avicenna, he replied: "Abū'l Barakat does not even understand the sense of the words of Avicenna. How can he oppose what he does not understand?"[344] A greater danger than *Faylasuf* enemies arose in 1092, when orthodox Muslims critical of freethinking became more influential at the Seljuk court. Al-Khayyām had written a collection of piquant quatrains in Persian, the *Rubaiyat*. The skillful metamorphosis of seventy-five of these into English by the Victorian poet Edward Fitzgerald has led to al-Khayyām's being known in the English-speaking world primarily as a poet. Among over a thousand quatrains that have been attributed to him, many of them long preserved only orally, recent scholarship suggests that he wrote only 121. Authentic verses depict a skeptical man of high moral convictions, torn by doubts about the nature of the physical universe, the passage of time, and man's relation to God. Quatrain 129 of the *Rubaiyat* in the Furughi edition declares that he wants merely "to know who I am." Orthodox Muslims praised al-Khayyām's intelligence and talent but censured him as being among "those unfortunate philosophers and materialists who detached from divine blessings, wander in stupefaction and error."[345] Ash'ari theologians who followed the tradition of al-Ghazali attacked Aristotelian and Neoplatonic philosophy, making Avicenna their particular target, and a wary al-Khayyām apparently refused the appellation *Faylasuf*, recognizing the danger of being an Aristotelian. Perhaps in order to clear himself of the charge of atheism—explicit or implicit—he reportedly made a pilgrimage to Mecca. It may simply be that as a good Muslim he wanted to make such a pilgrimage at least once.

Despite his theologically suspect position, al-Khayyām remained at the Seljuk court. In an attempt to regain support of the observatory from Malik Shah's successors, he wrote a propagandistic work (the *Naruz Nama*), which portrayed ancient Iranian rulers as magnanimous and impartial in their support of education, building, and scholars. It did not have the desired effect. After 1092, al-Khayyām only occasionally served and taught at the Isfahan court. After 1118, he left Isfahan and lived for some time in Marw (now Mary), the new Seljuk capital. He was buried in a still extant tomb in Isfahan, such that "the wind will blow the scent of roses over my grave."

Concentrating on finding short, elegant solutions for cubic equations and their theory, al-Khayyām laid foundations for geometric algebra, and to a lesser extent he contributed to arithmetic and geometry. His work draws upon three traditional sources: ancient Oriental, Hellenistic, and prior Arabic. His calculations employ Indo-Arabic numerals. To extract square and cube roots, he uses Indian methods, which he may have learned through the writings of al-Jili (d. 1029) and al-Nasāwī (fl. 1030). In the *Algebra,* al-Khayyām writes:

> The Indians possess methods for finding the sides of squares and cubes based on such knowledge of the squares of nine figures, that is the square of 1, 2, 3, etc., and also the products found by multiplying them by each other. ...I have moreover increased the species, that is I have shown how to find the sides of the square-square, quadrato-cube, cubo-cube, etc. to any length which has not been made before. The proofs I gave on this occasion are only arithmetic proofs based on the arithmetical parts of Euclid's *Elements.*[346]

That the Chinese *Mathematics in Nine Chapters* influenced al-Khayyām seems doubtful.[347] According to al-Tūsī, he also studied the binomial expansion $(a + b)^n$ while extracting roots and derived a table for binomial coefficients for integral n from 1 to 12. Like al-Karajī, then, he appears to have known about the formation of binomial coefficients and to have had the essence of Pascal's triangle.

Al-Khayyām's *Risala,* which systematizes algebra and is the earliest known work satisfactorily to relate it to geometry rather than arithmetic, begins by tracing the terminology and techniques devised by ancient Greek geometers and prior Muslim algebraists for handling first- and second-degree equations. He observes that no work from the ancient Greeks on cubic equations, except for one problem from Archimedes, had come down to the Arabs. Arabic algebraists from Thābit to al-Māhānī, al-Bīrūnī, and al-Khāzin in the tenth century had made specific algebraic translations of third-degree problems, principally in their research in astronomy. Al-Khayyām enumerates all possible types of cubics in his organization of the field and solves them. Recognizing only positive coefficients, he classifies normal forms of twenty-five types of cubic equations. (Normal forms have positive coefficients.) Eleven of these he reduces to quadratics and solves by Euclidean methods. The fourteen types of irreducible cubics had to be solved by the intersection or contact points of conic sections used for constructing these equations. For cubics of the modern form $x^3 + b^2 x = c$ (or, as he writes, cube and sides equal a number), al-Khayyām uses a parabola and a semicircle, cubics of form $x^3 + ax^2 = c^2$ a hyperbola and a parabola, and cubics of form $x^3 + ax^2 + b^2 x = b^2 c$ an ellipse and a parabola. He solves one quartic equation, in modern symbols $(100 - x^2)(10 - x^2) = 8100$, by the intersection points of a circle and a hyperbola. Yet these geometrical figures have an auxiliary place in this theory. Demonstrations are algebraic.

In finding roots of equations, al-Khayyām does not go beyond sections of curves falling within our modern first quadrant. This limiting of roots to positive real numbers essentially emanates from al-Khwārizmī. Throughout the *Risala*, al-Khayyām cautions his reader that cubic equations of which the related cubic curves do not intersect have no root, while if the curves are tangential the equation has one root, and if the curves intersect in one or two points there are one or two roots. He did not recognize that under certain conditions cubics may have three positive roots. His classification of cubics and solutions with conics proved highly influential. Among others, Scipione del Ferro, René Descartes, and Isaac Newton drew on them.

The *Risala* thus skillfully follows in the tradition of inquiry of al-Khāzin and al-Haytham and represents the near culmination in medieval Islam of a form of geometric algebra that had detailed algebraic proofs. Al-Khayyām's grouping of cubic equations, together with prior Arabic studies of quartic equations and abstract algebraic computations, apparently prompted him to seek an arithmetical technique for obtaining roots of cubics similar to that for second-degree equations or, at least, a systematic array of solutions. Failing, he wrote, "It may be ... that men who come after us will succeed." Four centuries later the Italian Girolamo Cardano in his *Ars magna* fulfilled that hope.

Rashed Roshdi and Jan Hogendijk have shown that al-Khayyām's successor Sharaf al-Dīn al-Tūsī (fl. 1170), not to be confused with Nasīr al-Dīn, achieved the most advanced treatment of cubic equations in medieval Islam.[348] His 243-page *Treatise on Equations* covers linear, quadratic, and cubic equations. Recognizing only positive coefficients, Sharaf al-Dīn distinguishes eighteen types of cubic equations. He reduces to quadratics five types without a constant term and solves eight of the remaining thirteen types by al-Khayyām's geometrical constructions. Book II of Sharaf al-Dīn's *Treatise on Equations* presented for the first time the number of positive roots of cubic equations determined from the coefficients of the equation, and he recognizes that positive roots exist if and only if the discriminant is greater than zero. For the equation $x^3 + a = bx$, the discriminant is $b^3/27 - a^2/4$. The role of the discriminant is not yet generalized for reaching solutions by radicals. The methods of al-Khayyām and al-Tūsī were refined by al-Kāshī and discovered independently in Europe in the nineteenth century by Ruffini and Horner.

Al-Khayyām's other important mathematical work is his commentary on Euclid, the *Sharh ma ashkala min musadarat kitb Uqlīdīs* (*Explanation of the Difficulties in the Postulates of Euclid*). It strengthened the foundations of mathematics in two ways.

One way was to improve upon Euclid-Eudoxus' general theory of ratios and proportions in Book V, a topic long studied by Islamic mathematicians. Definition 5 of that book of the *Elements* defines ratios among commensurable quantities expressed now as a, b, c, and d, where b and $d \neq 0$, for $a/b > c/d$ and $a/b < c/d$, and the Euclidean algorithm in Book VII provides the greatest common divisor. The legitimacy that Book X of the *Elements* gives to incommensurable ratios is based on the theory of infinite, continued fractions. Al-Khayyām considered Abu Abdallah al-Māhānī's ninth century definition of proportion to be better and truer to an intuitive concept of ratio. By using infinite divisibility, al-Khayyām proves the two definitions to be equivalent. His proof is founded in a theorem relating the fourth proportional to the above three given magnitudes, a, b, c. He thus was the first to attempt a general demonstration of the theorem. Following al-Karajī, calculators who were his contemporaries, and land surveyors, he treats as numbers magnitudes in compound ratios. That includes treating as new kinds of numbers the ratio of the diagonal of a square to its side ($\sqrt{2}$) and the ratio of the circumference of a circle to its diameter (π). Thereby al-Khayyām extends arithmetical language to compound ratios and places irrationals and rationals on the same operational scale. His work completes the introduction of the positive real numbers in medieval Islam.

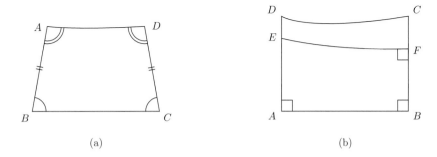

(a)   (b)

Versions of Thābit's quadrilateral.

The other major effort in al-Khayyām's commentary was to try to prove Euclid's fifth or parallel postulate, thereby denying it the status of postulate. Two of his Muslim predecessors, Thābit and al-Haytham, had attempted to prove the postulate by using an appeal to motion. Taking the dominant position of Muslim geometers, especially as found in al-Haytham's *Book on the Resolution of Doubt*, Al-Khayyām judges motion to be foreign to geometric proof. He begins with a principle equivalent to the fifth postulate, that "two converging straight lines intersect and it is impossible for two converging lines to diverge in the direction of convergence." Although he claims that this principle comes from Aristotle, it does not appear in Aristotle's known writings. Like Thābit and al-Haytham, al-Khayyām in constructing "Thābit's quadrilateral" drops two perpendiculars to one straight line. With them he perceives as the essential criterion of parallel lines not that they do not intersect but that, as *Elements* I.34 indicates, there is equidistance between them. He mentions Thābit only in passing. To prove Euclid's fifth postulate, al-Khayyām assumes that the two lower angles A and B in the quadrilateral are right angles and then explores three alternatives for the upper angles C and D: whether they are acute, falling short of two right angles, are obtuse, exceeding two right angles; or are equal to two right angles. If they are acute, extensions of the sides diverge; if obtuse, they converge. The upper angles must, therefore, be right angles and the two perpendiculars are everywhere equidistant. The conjectures on acute

and obtuse angles influenced Nasīr al-Dīn al-Tūsī in medieval Islam and Sac-
cheri in the eighteenth-century West. They are similar to the first theorems in
the non-Euclidean geometries of Lobachevsky, Bolyai, and Riemann.

In addition to astronomy, poetry, and mathematics, al-Khayyām contributed to philosophy, jurisprudence, history, and medicine. Disclaimer or no
in the *Rubaiyat*, he may be called a *Faylasuf*. He generally follows Avicenna's
brand of Arabic Aristotelianism modified by Neoplatonic elements. Yet he
departs from Platonic realism, developing what has later been called concep-
tualism. His contemporary Peter Abelard did the same in Latin Christendom.
For his theistic free thought, love of freedom and justice, philosophical poetry,
skepticism, and epicurean spirit, he has been anachronistically called the Persian Voltaire.

The Shi'ite *Faylasuf* Nasīr al-Dīn al-Tūsī (1201–1274), who revived major
accomplishments in mathematical science in eastern Islam during the thirteenth
century, had been educated at the Shi'ite school at Rus in Persia, studying
logic, metaphysics, natural philosophy, algebra, and elementary geometry. Afterward, he pursued advanced studies in Nishapur in philosophy, mathematics,
and medicine, especially Avicenna's *Canon*. While in Nishapur, he became
known as an excellent logician and astronomer, and his fame reached as far as
China. To continue his studies, al-Tūsī entered the service of the Ismaili rulers
of northern Persia. This isolated him from mathematical developments until
the Mongol monarch Hulegu, who esteemed astrology and astronomy, broke
Ismaili power in 1256 and brought al-Tūsī into his service as a scientific advisor
and director of endowments (*waqf*). Al-Tūsī became close to the monarch; he
even accompanied Hulegu on his sack of Baghdad in 1258.

After three years' labor, al-Tūsī gained the confidence of Hulegu and persuaded him to provide the finances for a major observatory with a large library
at Maragha in northwestern Persia. The library was to have books on all the
sciences and philosophy. Al-Tūsī became the director of the Maragha observatory and its team of astronomers. Under his leadership, the observatory became
a madrasa, a center for research and higher instruction in the sciences and philosophy. He guided the preparation of the *Ilkhani Zīj* (astronomical handbook)
of 1271, while he wrote a recension of Ptolemy's *Almagest* that exposes shortcomings of the Ptolemaic system and proposes revisions, such as eliminating
some planetary eccentricities. Al-Tūsī identified and encouraged apt students,
such as Qutb al-Din al-Shīrāzī (d. 1311), and had them participate in collaborative research.[349] His circle of researchers seem to have enjoyed themselves.
The translated title of al-Shirazi's study is *A Book I've Made, Don't Criticize
It, On Astronomy*.

The writings of the Maragha astronomers are only beginning to be translated, and we know little of their sources and lines of influence. Otto Neugebauer and Victor Roberts concur with Edward S. Kennedy, who discovered
their importance.[350] Kennedy proposes the term "Maragha school" for scholars
from al-Tūsī to ibn al-Shātir of Damascus (d. 1375). They produced substantial astronomical models that differ from Ptolemy's. Criticism of Ptolemaic

models from the *Almagest* and *Handy Tables* were not new in medieval Islam. The Maragha astronomers had probably closely studied al-Bitrūjī and ibn al-Haytham. Their criticism of Ptolemy, George Saliba argues, stemmed not from the generally good predictability that his planetary theory provides but from laws in the theory of planetary latitude along with observations that contradicted the Ptolemaic mathematical model for uniform circular motion in the heavens.[351] In Kennedy's estimation, ibn al-Shātir's *al-Zij al-Jadid* (*The New Astronomical Handbook*) offers a "pre-Copernican model," a precedent to Copernicus' use of a device now known as the "Tūsī Couple," which is depicted here.

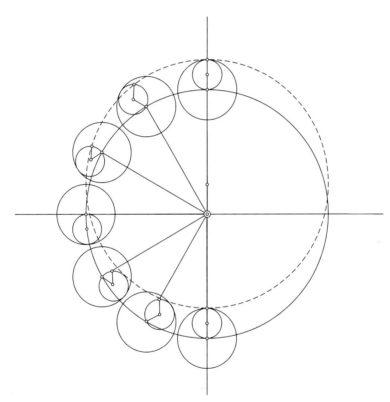

By having a smaller circle roll inside the circumference of a stationary circle, whose radius is twice that of the first, each point on the first lies on a straight line, the diameter of the second. The couple thus helps transform circular into linear motion and also the eccentric into the epicyclic model, thereby preserving the equant. Willy Hartner and Noel Swerdlow observe that the use of this "couple" could not be simply a coincidence.[352] Hartner's article on diagrams representing the "Couple" avers that for identical geometric points al-Tūsī and Copernicus have the same alphabetic references. The lines of transmission from ibn al-Shātir are not known, but possibly during his studies in Italian cities Copernicus had access to a Byzantine Greek manuscript on al-Shātir that is now housed in the Vatican Library.

As to al-Tūsī himself, mindful of the monarch's interests, he conducted experiments that included testing reactions of soldiers to sudden noises. Perhaps the early use of gunpowder prompted this experiment. To improve mathematical instruction, he wrote a series of recensions of texts by Euclid, Archimedes, Apollonius, and Menelaus besides that on the *Almagest*. These intermediate works became standard for teaching mathematics in eastern Islam. They are among the 150 treatises and letters that al-Tūsī wrote, 25 in Persian and the rest in Arabic. These show him to be the master of natural and occult science, philosophy, and theology and a composer of delicate poetry. For the quality and breadth of his writings and the scientific revival he generated, eastern Islam has judged al-Tūsī to be a leading *hakim* (wise man) comparable in influence to Avicenna. Avicenna was the greater of the two in medicine and narrowly in logic, al-Tūsī the leader in mathematical science.

Extending the work of al-Khayyām, al-Tūsī surpassed the Hellenistic Greeks in defining number. He defined number to include irrationals equally with rationals. His *Shakl al-qit* (*The Book of the Principle of Transversal*) states that every number is a ratio and in multiplying pairs of these ratios demonstrates the commutative property of numbers. His *Jawami al-Hisab* (*The Comprehensive Work on Computation with Board and Dust*) discusses numerical operations, extracting roots, al-Khayyām's work on binomial coefficients, and what is now known as Pascal's triangle. Through the *Jawami al-Hisab* and collaborative studies at Maragha, al-Tūsī advanced computational mathematics, especially with Indo-Arabic numerals, a trend that al-Kashī continued.

Al-Tūsī also went beyond the Greeks in treating parallels. He is among the mathematicians who attempted to find a proof of Euclid's fifth postulate. Possibly the foremost medieval Islamic student of this subject, he criticizes al-Khayyām's making equidistance the basic criterion of parallelism. Having probably only a partial knowledge of al-Khayyām's work, he finds it obscure. Al-Tūsī makes major use of the Thābit quadrilateral. In it, lines AB and CD are equal and perpendicular to BC and angles A and D are equal. Al-Tūsī drops ten successive perpendiculars and assumes the upper angle A to be obtuse. But this suggests that the successive perpendiculars are increasing, because in a triangle the greatest side is opposite the greatest angle. Similarly, al-Tūsī assumes the upper angle to be acute, which makes the successive perpendiculars shrink. Both assumptions thus contradict the original condition that AB and CD are equal. With an intuitive idea of straightness, al-Tūsī tacitly rejects the notion that two lines may converge for a time but inexorably diverge.

He and possibly his son Sadr al-Dīn, then, ingeniously sought to deduce the parallel postulate through a proof resting on the result that the angle sum of a triangle is 180° and its greatest side is opposite its greatest angle. This completes a circle of equivalences:

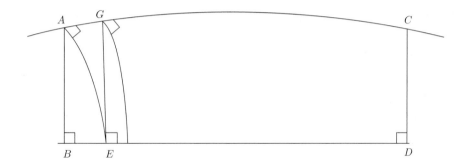

Through a point P not on line l
there is one and only one parallel line.

The curve equidistant from a
straight line is also straight.
$\Longleftarrow$
The sum of the angles
in a triangle is 180°.

It is in trigonometry that al-Tūsī extended the work of Abūal-Wafā and al-Bīrūnī and made his most original contribution to mathematics. His *Shakl al-Qita* separates trigonometry from astronomy and without recourse to Menelaus' theorem establishes it as an independent branch of mathematics. Completing earlier systems, he defines and uses all six trigonometric quantities, which simplifies trigonometry. The secant and cosecant had seldom been tabulated. Al-Tūsī defines the secant not by shadows but in this way: $\tan AB / \sec AB = \sin AB / R$, where R is the radius of the circle. His *Shakl al-Qita* also introduces the six cases for a right-angled spherical triangle. Given such a spherical right triangle ABC with legs a and b, hypotenuse c, and angle A at A, angle B at B, and angle C at C, then

$$\cos c = \cos a \cos b \qquad \cot A = \tan b \cot c$$
$$\cos c = \cot A \cot B \qquad \sin b = \sin \sin B$$
$$\cos A = \cos a \sin B \qquad \sin b = \tan \cot A$$

These six cases will yield the two chief formulas of spherical trigonometry:

$$\cos a = \cos b \cos c + \sin b \sin c \cos A, \qquad (1)$$

which expresses A by reference to the sides of the triangle, and

$$\cos A = \cos B \cos C + \sin B \sin C \cos a, \qquad (2)$$

which refers a to the angles. The *Shakl al-Qita* also gives the first clear description and proof of the relation now called the theorem of sines, which ibn Yūnus had known earlier.

Medieval Arabic mathematics appears to have declined after al-Tūsī. It had an occasional luminary, however, the most notable being the learned Persian

Jamshid al-Kāshī (fl. 1406–1437). Although the Arabs gave prominent calculators the nickname *al-Hasib* (the reckoner), al-Kāshī, who richly deserved the epithet, never received it.

Al-Kāshī first made his reputation in astronomy. After observing three lunar eclipses from his native town of Kāshān in 1406, he wrote on the dimensions of the cosmos in *Sullam al-Samā (The Stairway of Heaven)*. Eight years later he revised the astronomical tables of Nasir al-Dīn al-Tūsī, the *Khaqānī zīj*, dedicating this work to the khan Ulugh Bēg, a titular name meaning great prince. Ulugh Bēg ruled from his major city, Samarqand. The introduction of the *Zīj* spoke of al-Kāshī's years of poverty and the generosity of the khan that permitted him to complete this work. With his enlightened patronage of learning, Ulugh Bēg, who was a man of science, founded a *madrasa*, a school of higher learning, in the years from 1417 to 1420 in Samarqand. It differed from others of the time in stressing astronomy and the high level of understanding of all subjects taught there. After interviewing al-Kāshī to check his qualifications, Ulugh Bēg invited him to join the *madrasa*. Al-Kāshī was no longer an impoverished wandering scholar. He was the school's *maulana* (master), the most eminent among sixty men of science assembled there. Al-Kāshī, whom the khan considered "a second Ptolemy," helped plan, construct, and organize a three-story observatory, beginning in 1424, and aided in compiling for Ulugh Bēg trigonometric and planetary tables and a catalogue of coordinates of 1,018 fixed stars, the *Zīj-i Gurgāni*, which John Flamsteed was to use in the seventeenth century to compute his revised star tables. Archaeologists, beginning with V. L. Vyatkin, excavated the remains of the observatory from 1908 to 1948.[353] Vyatkin showed that the primary instrument at the observatory was not a quadrant but the Fakhrī sextant, a pre-telecopic sighting device having a radius of 40.04 meters. This made it the largest sextant of its time. A segment of its arc was located in a trench dug along the meridian line in a hill. Al-Kāshī's greatest mathematical achievements came in Samarqand, which remained for a quarter century until Ulugh Bēg's assassination in 1449 the chief center for the study of science in central Asia.

To make trigonometric tables for astronomical calculation more precise, al-Kāshī in two companion works made, for the first known time in the history of mathematics, a methodical presentation and use of finite decimal fractions. No record is known of the development of the theory of decimal fractions in the two and a half centuries that followed al-Samaw'al, but perhaps a successor will be found. The exposition of these fractions seems to have reached their apogee in medieval Islam in al-Kāshī's writings.

At the beginning of his computational *Risāla al-muhītiyya (The Treatise on the Circumference of a Circle)*, which appeared in 1423, al-Kāshī accurately approximates twice the ratio of the circumference of a circle to its radius, that is, 2π. His computations display a firm grasp of decimal fractions. Not satisfied with the Ptolemaic value of π as 3;8,30 (3.1466), which most Muslim astronomers accepted, he tackled the problem. He considers that number to be irrational. To obtain his better approximations, al-Kāshī employs a pinching method based on the traditional calculations of perimeters of inscribed and circumscribed regular polygons about the circle. He works with regular $3 \cdot 2^n$-sided polygons. The precision he seeks is to have an error in a circumference whose diameter equals 600,000 diameters of the Earth be less than the thickness of a horse's hair. He determines that this requires working with a regular polygon having having $3 \cdot 2^{28} = 805,306,368$ sides. The accuracy in each stage of his computation, including the refusal to let rounding errors accumulate, is impressive. He finds that the ratio 2π, initially in sexagesimal notation, is 6;16,59,28,1,34,51,46,14,50, which is correct to all of its places. This figure translates into the decimal fraction 6.2831853071795865, which is correct to sixteen decimal places. Al-Kāshī's purpose in seeking this close value of π was to be able to calculate precisely the circumference of the universe. No other mathematician approached his accuracy until Ludolf van Ceulen computed π to twenty and then to thirty-two places, publishing his results in 1615.

Miftāh al-Hisāb (*The Key to Arithmetic*), the companion of 1427 to the *Risāla,* methodically presents for the first time al-Kāshī's arithmetic of decimal fractions. A compendium in five books of algebra, arithmetic, and measurement, it describes the four arithmetical operations, a general method for extracting roots, and the calculation of areas of polygons and circle sections as well as volumes. Probably improving on al-Uqlīdisī's work on decimal fractions, al-Kāshī handles fractions to the ninth place and draws an analogy between operations with decimal and those with sexagesimal fractions, that is, he carries out operations on them the same as with integers. He was not the first to recognize that operations are equally valid for the two bases or any base. Al-Kāshī considered base ten simpler and more efficient than base 60.

The *Key to Arithmetic* is the summit achievement of medieval Islamic arithmetic. Persian *madrasas* were aware of it, perhaps transmitted through an intermediary, until the seventeenth century. But scientific calculators, particularly astronomers, persisted in using the sexagesimal system as well. After the fall of the Byzantine empire, al-Kāshī's younger colleague Alī Qūshjī moved to Constantinople. Al-Kāshī's thought on algebra and arithmetic appeared in Byzantine collections of problems. One Byzantine text containing such a collection was carried to Vienna in the West in 1562. Probably earlier transmissions to the West had occurred, albeit apparently sketchily with missing information. Western mathematicians did not achieve a comparable level of operations with decimal fractions until the work of Simon Stevin in the late sixteenth century.

Among al-Kāshī's contributions to algebra stated in a purely verbal form is his apparently independent derivation of an ingenious algorithm, which now bears the name Ruffini-Horner, that appears in Book IV of the *Key to Arithmetic.* To find the fifth root of 44,240,899,506,197, he divides it into cycles of five, because the perfect roots for powers of 10 are 10^0, 10^5, 10^{10}, ... The first integer in his solution has to be a number whose fifth power is not larger than 4,424. The number is 5 for the third cycle or hundred's place. Al-Kāshī estimates the next digit to be 4, or 540, but this is too big, and the answer is 3 or 530. He finds the next digit to be 6 and then adroitly finds the fractional part of his solution 21/414,237,740,281 using linear interpolation and the approximation $(n^k + r)^{1/k} \approx n + r/[(n+1)^k - n^k]$. In *Episodes in the Mathematics of Medieval Islam*, Berggren gives a step-by-step account of his computation.[354] Although al-Khayyām, who was probably the most imaginative in use of this algorithm before al-Kāshī, thought that he had originated it, the device can be traced at least to Abū al-Wafā's late tenth-century treatise *On Obtaining Cube and Fourth Roots*.

Al-Kāshī determined to $n = 9$ binomial coefficients for successive powers: in effect, Pascal's triangle to that point. This triangular arrangement of numbers had appeared earlier in the wiritings of al-Karajī, al-Khayyām, Nasir al-Dīn al-Tūsī, and Chinese algebraists. Al-Kāshī drew on prior Arabic writers, but apparently not the Chinese tradition. The modern symbolic representation of his binomial coefficients is $C(n, k)$, which means that from a set of n objects he counts ways of choosing k objects. From the binomial theorem for squares, he finds $(A+B)^2 - A^2 = \{C(2,2)B + C(2,1)A\}B$. He similarly finds identities for the cube of the binomial and higher powers. The triangle here, in which every number greater than 1 equals the sum of the two numbers to the left and right in the row above, results in our modern representation as

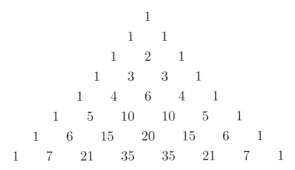

Written in response to increasing demands for greater precision in trigonometric tables, al-Kāshī's third major work, *Risāla al-watar w'al jaib* (*The Treatise on the Chord and Sine*), computes the sine of 1° correctly to ten sexagesimal places. The sine of 1° was to be the basis for his and Ulugh Bēg's tables. This work came long after Muslim astronomers had discovered Ptolemy's approximation of the chord of $l°$ in trigonometry, which rests on the inequality $\frac{2}{3}\,\mathrm{crd}(3°/2) < \mathrm{crd}(1°) < \frac{4}{3}\,\mathrm{crd}(3°/4)$. Ptolemy's method gives a value correct to two sexagesimal places. The best prior approximation, correct to four sexagesimal places, came from Abū al-Wafā and ibn Yūnus in the tenth century. From the eleventh century, Muslim mathematicians had known how to obtain a cubic equation of the type $px = q + x^3$ for the trisection of an angle. From this al-Kāshī proceeded to $x = (q + x^3)/p$ and devised his iterative method to compute ever closer roots. For x very small and positive, the root $x_1 = q/p$. This makes the next root $x_2 = (q + x_1^3)/p$ and so forth. Perhaps the interpolation procedures of ibn Yūnus inspired al-Kāshī's iterative method, which is based on this relationship: $\sin(3\theta) = 3\sin(\theta) - 0;0,4\sin(\theta)^3$. For $\theta = 1°$, this becomes $\sin(3°) = 3\sin(1°) - 0; 0, 4\sin(1°)^3$, and for a circle of radius 240 the sexagesimal approximation of $\sin(3°) \approx 3; 8, 24, 33, 59, 34, 28, 15$, or decimally about 3.140157374.

To compute $\sin(3°)$, al-Kāshī applies elementary Euclidean procedures to the chord of $72°$, which is the side of a regular pentagon inscribed in a circle, and the chord of $60°$, which is the side of an inscribed equilateral triangle. Next, the difference formula gives $\sin(12°) = \sin(72° - 60°)$, and then repeating the half-angle formulas gives in turn $\sin(6°)$ and $\sin(3°)$, all of which Ptolemy, Abū al-Wafā, and Ibn Yūnus knew. The sexagesimal value of $\sin 1°$ becomes the root of the equation $x = (900\sin 3° + x^3)/45 \cdot 60$. Substituting the value for $\sin(3°)$ in that relationship and letting $\sin(1°) = x$ in his iterative method, he reduces that to

$$x = \frac{x^3 + 47, 6; 8, 29, 53, 37, \ldots}{45, 0},$$

which is a cubic equation having one root equal to $\sin(1°)$.[355] Since the sexagesimal $\mathrm{crd}(1°)$ for a radius of 240 is near to 1, al-Kāshī writes $1; m, n, o, \ldots$, indicating by m, n, o succesive sexagesimal places.

$$1; m, n, o, \ldots = \frac{(1; m, n, o, \ldots)^3 + 47, 6; 8, 29, 53, \ldots}{45, 0}.$$

Subtracting 1 from both sides gives

$$; m, n, o, \ldots = \frac{(1; m, n, o, \ldots)^3 + 47, 6; 8, 29, 53, \ldots - 1}{45, 0},$$

which can be simplified to

$$; m, n, o, \ldots = \frac{(1; m, n, o, \ldots)^3 + 2, 6; 8, 29, \ldots}{45, 0}.$$

Since the two sides are equal place by place, al-Kāshī simply computes $[(1)^3 + 2, 6; 8, 29]/45 \approx; 2$, (49 or 50) to obtain m. The value of n similarly depends on m, so letting $1; m$ equal $1; 2$ and $1; 3$ gives $1; 2, 49, 39$ and $1; 2, 49, 43$, so $n = 49$. Successive approximations using the improved figure at each stage yield the sexagesimal value of $\sin(1)° \approx 1; 2, 49, 43, 11, 14, 44, 16, 26, 17, \ldots$. The decimal fraction for the modern sine $(1)°$ for radius 1 is approximately 0.017452406. All these figures are correct. In modern numerical analysis, Al-Kāshī's algorithm is known as fixed-point iteration. He made no attempt to prove its convergence in the neighborhood of 1, which was done much later with the mean-value theorem in differential calculus.

This account of medieval Islamic mathematics closes with the death of al-Kāshī in Samarqand. By that time the Muslim world had long suffered political disintegration and external assaults from the Seljuk Turks and European Crusaders in the West and from the Mongols in the East. As Arabic, Persian, and central Asian scholarship weakened, in Europe the Renaissance was beginning. Gaining knowledge of results of medieval Arabic and Persian learning amid a new round of discovery and partial assimilation of pagan learning from ancient Greece and Alexandria, European scholars working under new social, economic, and cultural conditions were to take the intellectual lead in mathematics.

12

Recovery and Expansion in Old Europe, 1000–1500

> [M]athematics ... reveals every genuine truth, for it
> knows every hidden secret, and bears the key to ev-
> ery subtlety of letters; whoever, then, has the effrontery
> to study physics while neglecting mathematics, should
> know from the start that he will never make his entry
> through the portals of wisdom.
> — Thomas Bradwardine,
> *Treatise on the Continuum*

In Latin Europe, three centuries of increasing stability beginning shortly before the year 1000 permitted rapid and sweeping advances in politics, economics, and culture.[356] Resurgent commerce spurred the adoption and development of more advanced computational techniques. A stream of translations from Arabic and Greek introduced algebra, brought back a complete Euclidean geometry, and, for the first time, put much of Archimedes' mathematics into Latin. By the early thirteenth century, two original mathematicians, Leonardo of Pisa and Jordanus de Nemore, were flourishing. During the two turbulent centuries following 1300, a separation grew between commercial and academic mathematics. Academic mathematics was increasingly associated with logical, philosophical, and theological investigations. It also profited from astronomers systematizing trigonometry.

12.1 Political Stabilization and Economic Development

After 950 religion and geography worked together to draw the northern European tribes into a unified Christendom. Tribal migrants, when protected by secure borders and united by common religion, became by degrees more settled and numerous. After the future German emperor Otto I in 955 defeated the Hungarian Magyars and Slavonic Wends, Scandinavia formed Europe's northern frontier. About the same time Poland, Bohemia, and Hungary protected the eastern flank. As pope, Gerbert oversaw the conversion of the Danes and Poles and presented the converted Hungarian king with the crown of St. Stephen. More important than mass conversions were social changes wrought by a greater interior assimilation of Christianity. Religion altered war and peace: there was a peace movement at home and "holy wars" abroad. Together these reorientations gave Europeans a common purpose and a respite from the worst destructiveness of feudal warfare. Relative peace and stability were preconditions for other internal developments.

Medieval Church leaders secured a series of protections for noncombatants and truces to restrict combat. Protecting their own interests, a council of bishops in Burgundy late in the tenth century decreed excommunication on knights who attacked monks, priests, church buildings, and peasants' fields. Cluny's abbot St. Odilo and Emperor Henry III, ruler from 1039 to 1056, supported this "peace of God" in French and German lands. Another council promulgated the "truce of God," which prohibited combat on Sundays. Later decrees added Thursdays, Fridays, Saturdays, and the sacred seasons of Lent and Advent. Partly through this truce, his ban against construction of private castles, and reintroduction of coinage to pay knights, William the Conqueror of England controlled his Norman-English barons.

The Crusades, a series of wars in the name of religion orchestrated by the papacy, drew opportunists, adventurers, and idealists to foreign lands.[357] Norman knights had earlier begun the military expansion by conquering Sicily and South Italy. Following the collapse of the caliphate of Cordoba in 1002, north Spanish Christians launched a *Reconquista*. In 1063, Pope Alexander II escalated the Iberian struggle to an international campaign to which French knights responded. The Crusader's Indulgence, a promise of full remission of sins to penitent soldiers at the moment of death, attracted lusty recruits, even as the Italian practice of carrying a cross into battle spread. This expansionist ethos along with frustration at the Seljuk Turkish control of the Holy Land and some concern for the security of Byzantium provoked military excursions into the Levant. Pilgrims had long cherished New Testament sites. Religious tourism, such as described in Chaucer's *Canterbury Tales*, contributed greatly to commerce and an exchange of knowledge. Weakened after the Turkish victory at Manzikert, Armenia, in 1071, the Byzantine emperor turned to the papacy for help. From Clermont in southern France, Urban II in 1095 proclaimed the First Crusade. Except for Germans, whose emperor was at war with the pope, people from all classes joined the effort. Jerusalem fell in 1099 and crusaders set up feudal kingdoms at Edessa and Antioch in Syria. Eight designated Crusades demarcated a nearly continuous flow of men and materiel until 1270. They provided knightly heroes and improvements in military technology, but were mostly military failures. Muslims regained Edessa in 1144 and Jerusalem in 1187.

The West, however, gained two enduring benefits from these actions: accelerated trade and sufficient contact with rich Byzantine and Islamic cultures to nurture an intellectual revival. From ports in the Levant, Venetians acquired trading concessions and the Genoese rivaled them there. Other Crusaders learned administration in the eastern kingdoms. Contact with Muslims at the frontiers helped diminish bigotries. Peter, the abbot of Cluny who visited the libraries of Spain, encouraged debates with educated Muslims and a translation of the *Qur'an* that was completed in 1143. Gerard of Wales (fl. 1180) refuted the ubiquitous slander that Muslims abstained from wine and pork because wild swine had eaten the drunken Muhammad. The knight-poet Wolfram von Eschenbach (fl. 1210) recognized "Muslims as God's handiwork." Encounters with Islamic and Byzantine learning opened a vast frontier to intellectually restless Latins.

From 1000 to 1300, the papacy reached its temporal zenith. Feudal conglomerates in France and England meanwhile moved toward centralized monarchies requiring a body of trained experts in legal, secretarial, and accounting arts.

Pontiffs made their offices more respected by pursuing ideals nurtured in the reformed monastery at Cluny, established in east-central France in 910, and Gorze in northwest Germany, begun about 933. Leo IX, who was pope between 1049 and 1054, recruited talented advisors, allied with Normans, and effectively used legates, councils, and correspondence. Papal control of the Latin Church, in which from 1059 only cardinals elected a pope, increased after 1130. By the mid-twelfth century the title Vicar of Christ had replaced the earlier, less imposing Vicar of Peter. Monastic and papal reformers attacked the abuses of Nicolaism or clerical concubinage and Simony, the sale of sacred things, each named after a disreputable Biblical figure. Reformers also challenged lay investiture, by which rulers controlled the selection of bishops and other high ecclesiastics. Emperor Henry IV's confrontation with Pope Gregory VII over investiture climaxed in a semblance of penitent submission as the emperor in 1077 knelt in the snow at Canossa in north Italy. Reforming popes and their allies made the papacy the center of spiritual authority in the West.

But Gregory VII, who also pressed papal claims for temporal sovereignty, died in exile. The power struggle continued between popes and German rulers. Emperors, popes, and lawyers disputed the ranking of spiritual and temporal leaders, while moderates held that each wielded a separate sword in defense of a complementary realm. The struggle exhausted even the talented emperor Frederick II, linguist, ornithologist, and patron of high culture, who ruled from 1215 to 1250. His Hohenstaufen dynasty was another casualty. A reconsolidation of feudal divisions undermined the Germanic empire, and Latin Christians split into imperial and papal factions.

In the contest between empire and papacy, French and English monarchs became stronger and emerged as new centers of cohesion. As Europe's strongest monarch, the Angevin-Norman Henry II, during his reign from 1154 to 1189, developed the Exchequer ("chessboard" for a means of reckoning of crown income similar to the abacus) and expanded the system of royal justice in England by deploying itinerant justices of the peace. In the rising kingdom, tensions grew between throne and altar. After Henry's complicity in the assassination of Thomas à Becket, archbishop of Canterbury, his episcopal successor Stephen

Langton inspired the Magna Carta. This charter of English constitutionalism stemmed the growing autocracy of the king, obliging him to respect the rights of the nobility and clergy but not the serf majority. In 1215 barons forced King John, notorious in tales of Robin Hood, to accept the charter. Subsequently, Edward I, whose rule stretched from 1272 to 1307, further stabilized the realm by holding regular parliaments (conversations) in assemblies with knights from shires and burghers from towns as representatives. Under the dynasty of Hugh Capet between 987 and 996, France was systematically consolidated and rose to cultural primacy in Europe. Capetians had the advantage of a two-century continuity of male heirs along with successful mergers through dynastic marriages, diplomatic skill, and military conquest. In the reign of St. Louis IX, extending from 1226 to 1270, the University of Paris, Gothic architecture, and chivalric style together won for France cultural hegemony in Europe. Louis's brother became Angevin king of Sicily and four popes came from French lands in the latter thirteenth century. Besides the English and Capetian monarchies, several lesser counties, duchies, and kingdoms such as Flanders, Savoy, and Bohemia organized around a central administration.

The incipient system of greater and lesser states indicates economic, demographic, and political advance. Beginning in the tenth century, Europe's economy quickened. Growth appeared first in Italian cities and spread to northern France and the Rhineland, the rest of German lands and, finally, to Scandinavia. The betterment of agricultural and manufacturing techniques encouraged economic development, as did an attendant increase in population that shifted part of the labor force from the fields to artisan and craft pursuits. Urban merchants expanded trade. Better nutrition and a drop in the average age for marrying, which lengthened the time span of fertility, supported population growth. Between 1086 and 1346, the estimated population of England tripled.

Improvement in agricultural production and therefore in nutrition resulted from bringing more land under more effective cultivation. The Cistercian order, founded in 1068, together with migrating peasants opened wilderness lands in central and western Europe. The three-field system—rotating fall and spring plantings with fallowings—known since Carolingian times spread erratically. So did the wheeled iron plow, heavy enough to aerate and drain heavy northern and steppe soils when pulled by shod horses in traction harness. Farmers bred better horses, mules, and sheep and introduced superior strains of cereals and beans to fields in some cases enriched by manuring. There is evidence of an advantageous climatic fluctuation. Crop yields—wheat, rye, barley, and oats were staples— varied greatly, but apparently increased fourfold in parts of northern France between Carolingian times and the eleventh century. Widespread use of water and windmills greatly reduced the need for servile labor and so freed laborers for work in urban crafts.

Agricultural surpluses and population growth made for a vigorous new urbanization. Town and city life centered on the local market called a *portus, vicus,* or *wic,* terms preserved in the endings of many town names. Some local markets evolved into long-distance outlets, where specialized products such

as Flemish textiles were traded. Old Roman cities, monastic or cathedral en-
closures, and forts became urban sites. A *ville* or *bourg* (fortification wall)
attracted merchants who constructed a *faubourg* (suburb) between the old wall
and the new. They took the name burghers. This bourgeois might clash in interest and outlook with the bishop residing in the old cathedral district (*cité*).

Between the eleventh century and 1348 by one estimate, western Europe's
town and city dwellers rose from less than two percent of the total population to about fifteen percent. Florence and Genoa each had more than 50,000
inhabitants. Paris, Bruges, and Ghent were comparable. Within these urban
centers merchants gained power. From the eleventh century Flemish and northern Italian merchants formed municipal corporations to guarantee their charter
rights and privileges. Increasingly, merchants joined themselves by oath and
organized cities into communes, which historian R. S. Lopez calls "governments
of, by, and for the merchants." Confident and energetic burghers achieved an
economic expansion marked by a construction boom and a reversal of the flow
of gold to the East.

As regional trade expanded along a corridor between north and south, the
Counts of Champagne in central France granted safe conduct to merchants en
route to transregional markets called Fairs. Such long-distance trading created
novel kinds of disputes, for whose resolution merchants developed laws and
Courts of the Fair. Sanctions imposed by courts for violations of merchant law
were soon recognized everywhere in a society that increasingly invited regula-
tion. In the economic surge, Italian city-states established merchant quarters in
commercial cities of the East. Later, Baltic merchants organized the Hanseatic
League.

Aggressive Italian merchants pioneered commercial techniques, such as
profit-sharing partnerships (*commenda*) between entrepreneurial investors and
traveling merchants. New instruments for credit, loans, and transfers of payment circumvented the Church's ban against usury. Improvements in transport
resulted in faster, safer, and more profitable commercial voyages. In the 1260s
the Venetians floated a round ship of about five hundred ton capacity, almost
three times that of the *Mayflower*. Genoese galleys, long ships with banks
of oars, made the first voyage to the north Atlantic, and before long overseas
(*oltremare*) would shift from the Levant to the New World.

Urban commerce fostered the growth of computational and practical mathematics. The monastic *computus*, like the prohibition of usury, became outmoded. From methods of commercial reckoning pioneered by Leonardo of Pisa,
Italian calculators developed an *arte della mercandatia*. Indo-Arabic numerals
appear in Venetian merchant journals of the late thirteenth century. In Genoa,
other Tuscan cities, and Venice, merchants reinvented a system of double-entry
bookkeeping. Chronological entries for debits and credits on facing pages or in
parallel columns date from the same time. Cross-referencing in journals separated ledgers that tracked individual clients or specific kinds of merchandise.
The system facilitated transfer of credits and debits and was a potential means
of rapid assessment of a business's current strength. Suspecting that Indo-

Arabic numerals would allow for fraudulent tampering, towns, among them Florence in 1299, enacted statues prohibiting their use. Yet they continued to win an acceptance that by the late sixteenth century had become general.

Shipping for increasing markets could send a craft for longer stretch- es without sight of land than previous navigational devices could handle, especially when clouds blotted out the stars. Relying on a "harbor finding art" from about 1250, pilots began dead-reckoning from one island, landmark, or designated point to another. They followed a compass heading while roughly logging the elapsed distance. A mariner's compass, adapted from an Oriental device, indicated the course to an accuracy of about five degrees. Its companion piece was the *portolano*, a new marine chart of landforms and distances drawn to scale. These port books originated as compilations of oceanographic data for particular ports and grew before 1270 to include tabulations of the entire Mediterranean. Crisscrossing radial spines from starlike compass rosettes displayed scaled distances and directional headings for sixteen or more compass bearings. These valuable charts included all major geographical features and their possessors kept them secret. To rectify zigzag courses, windblown navigators further referred to a set of traverse tables. Geographic knowledge and cartographic techniques advanced quickly: in 1320 a Genoese cartographer attempted to map the world, and the bulge of eastern Brazil showed on Andrea Bianco's map of 1448.

Builders of fortresses and cathedrals required a severely practical mathematics similar to that of merchants and navigators. Some scholars now proposed to link with theory the mathematical practice transmitted by oral tradition. The *Didacalicon* of Hugh of St. Victor (fl. 1125), canon at the Parisian cathedral school, paired the traditional liberal arts with seven mechanical ones, such as cloth making and commerce. This treatise divides geometry into a theoretical and a practical or instrument-using branch. Hugh's later *Practica Geometriae*, emphasizing simple agrimensorial instruments, adds the more complex astrolabe. Despite this work, Hugh lacked mastery of Euclidean geometry, a competence that remained absent from practical school manuals until the fourteenth century.

12.2   Role of Interactive Learning and Piety

The British medievalist R. W. Southern has argued that the primary catalyst for cultural and intellectual change came from a subtle shift of religious values, rather than economic improvement. He detects a shift toward a "new piety" and "new learning" visible shortly before the year 1000 as the fruit of a long germinating Christian educational program.[358] He traces to that learning and piety the peace movement and eleventh-century papal reforms. The new Franciscan order approved in 1208 and the Dominicans, founded in 1216, conspicuously manifested the compound. Friars appealed to an urban laity whose members worked at tasks requiring literacy and secretarial skills. Preaching friars supported by mendicant alms were competitors as well as co-workers with established secular clergy. The laity's sophistication and exposure to diverse, sometimes heterodox ideas increased the importance of preaching. Friars flocked to the new centers of higher learning to prepare for the challenge. By the mid-thirteenth century they were the dominant intellectual group in northern Europe.

New learning and piety encouraged curiosity about the physical world. This curiosity included a quest for precision in logic, law, and quantification. Recognition grew that quantification could impose order on commerce, trade, and some aspects of politics. The pursuit of better measurement in the accounting and navigational arts was akin to the regularization of trade and the plowman's labor, which came with the spread of standardized coins, weights, and measures. The English furlong of 220 yards set the unit for a plowed strip. In the *Doomsday Book* of 1085, William the Conqueror attempted to count all the landholdings and property of Norman England. Cardinal Henry of Susa (d. 1281) fixed the pope's dignity at $7,644\frac{1}{2}$ times that of the emperor. His multiple was Ptolemy's for the difference in brightness between sun and moon. Such efforts to quantify were common in the fourteenth century. Any significant advance in quantification, of course, required improved mathematical skills.

Recovery of mathematical learning, a component of the general expansion of scholarship, occurred in the twelfth and thirteenth centuries. Four phases demarcate the development of Latin academic mathematics. They began essentially with Latin translations from Greek and Arabic materials from about 1140 to the end of the century. By the first half of the thirteenth century the earliest original Latin mathematicians of the era appeared, Leonardo Fibonacci and Jordanus de Nemore. The systematic teaching of new mathematical techniques in the curriculum of new universities dates from about 1200 and continued for most of the century. From the mid-thirteenth century to about 1360, mathematics accommondated itself to the theological and philosophical interests of Latin Christian scholars. Practical mathematics continued for the most part within a vernacular tradition outside the universities. Scholarly discussions of relations between theory and practice were restrictively logical and bookish, providing only tenuous connections between them. Instrument making, however, offered a partial meeting ground between the two traditions. The academic Campanus of Novara (fl. 1296) invented a crude analogue computer for finding planetary longitudes. Other astronomers, such as Peter Nightingale (fl. 1290) in Paris and Denmark and the Norman Jean of Murs (fl. 1345), improved on the design. The invention of spectacles, reported in Venice in 1284, gives evidence of a revival and progress in optics, as the contemporaneous invention of the mechanical-astronomical clock bespeaks parallel progress in astronomy.

Translators dominated the formative period of the new learning. Limited in knowledge of their subjects, lacking dictionaries or grammars, and without adequate technical terms in medieval Latin, they attempted, with considerable success, to transpose the corpus of Greek and Arabic philosophy and sciences.

Reconquered Moorish Spain, center of a West Arabic culture, was a rich repository of manuscripts. Translation of Greek mathematics from Arabic to Latin was essential for transmitting ancient Greek learning. Translators from many lands migrated particularly to Toledo, recaptured in 1085. Translators sometimes worked in teams and followed an awkward two-step procedure. The Arabic text was read aloud and translated orally into Hebrew or Catalan and from that into Latin.[359] Passage through several languages meant additional errors and a reduction in resemblance to the Greek originals. Commitment by independent individuals, not central coordination, directed the translating enterprise, while ecclesiastical patronage, especially by two archbishops of Toledo, frequently supported translators. Sicily-Italy and the Crusader kingdoms were important secondary centers for translating Greek and Arabic works into Latin.

In the transfer of Arabic arithmetic and algebra, both in the first quadrivial subject, al-Khwārizmī exerted the greatest influence. Three renditions of his *On Calculation with Hindu Numerals*, the first before 1143, laid the foundations for algorithmic calculation by explaining the numerals, place-value notation, and arithmetical operations. His *Algebra* (see Chapter 11), Latinized by different hands, was fragmented. In 1145 Robert of Chester at Segovia translated its first section, on solutions for six types of linear and quadratic equations, binomial multiplications, and selected problems. Within the next quarter-century, Gerard of Cremona (1114–1187) translated the same section. The most prolific of the early translators from Arabic, Gerard made Latin versions of at least ninety Arabic texts. The second section of the *Algebra,* applying algebraic methods to mensurational problems, passed into Latin in two stages. Abraham bar Hayyam (fl. 1140) made an expanded version, which Plato of Tivoli translated into Latin as *Liber Embadorum (Book of Areas)* in 1145 at Barcelona. Gerard's translation of Abu Bakr's *Book of Measurements* duplicated much of Plato's translated materials.

Translators and editors gave more attention to Euclid than to any other mathematician. In the thirteenth and fourteenth centuries, over twenty Latin versions of the *Elements* circulated. The most popular edition, based on al-Hajjāj's Arabic rendering, departed considerably from the pristine Euclid. The widely traveled English monk Adelard of Bath (fl. 1126–1142), who prepared three different versions, was the first to put al-Hajjāj's text into Latin. Adelard probably obtained a copy of this Arabic text when, disguised as a Muslim student, he attended lectures in Cordoba. In 1259 the mathematical commentator Johannes Campanus of Novara expanded the shortest of his versions, now designated Adelard II, by furnishing proofs for propositions and adding supplemental material from Jordanus de Nemore. This became the standard edition. Meanwhile Gerard's translation introduced a second Arabic tradition drawn from redactions of the *Elements* by Ishaq ibn Hunayn and Thābit ibn Qurra. The Adelard-Campanus version of 1259, more popular than Gerard's but less faithful to the Euclidean text, spawned more than fifteen other versions. Even earlier, about 1160, an anonymous scribe had translated from the Greek the geometric, optical, and minor mathematical works of Euclid. His version went unnoticed.

Knowledge of more advanced Apollonian and especially Archimedean mathematics, assigned to geometry in the quadrivium, came piecemeal. Gerard translated Apollonius' *Conic Sections* and a version of Archimedes' *Dimension of the Circle*, which became relatively popular. His translation from Arabic to Latin of the *Discourses of the Sons of Moses* gives a related proof for squaring the circle. This short but important treatise contains a calculation of the value of π; Hero's formula for the area of a triangle with its first known Latin demonstration; ancient solutions to the problem of finding two mean proportionals between two given lines; the first known Latin trisection of an angle; and a method of approximating cubic roots to any degree of accuracy. Using a Greek source, the thirteenth-century English translator Johannes de Tinemue compiled a treatise *On Curved Surfaces* for determining curvilinear areas and volumes by the Archimedean method of exhaustion. In 1269, the Flemish Dominican William of Moerbeke translated from the Greek nine treatises by Archimedes. His highly literal renderings include the mechanical works that the Arabs had known only indirectly. After Moerbeke, Latin copies existed for all Archimedean works extant today except for the *Sand Reckoner, On Method,* the *Cattle Problem,* and *Stomachion.* Succeeding medieval mathematicians, however, used these higher materials sparingly and did not master them.

Scarcely less important than algebra and geometry for transforming European mathematical learning were translations in astronomy, the third branch of the quadrivium. Unlike algebra and geometry, which belonged to mathematics proper, the new astronomy was considered a mixed science, intermediate between physics and mathematics. Aristotle defines physics, mathematics, and metaphysics as the three major sciences. The main sciences progress from the most concrete to the most abstract levels, while subordinated sciences (astronomy, optics, and statics) are interposed between the lower levels. To understand the Ptolemaic and Arabic traditions in astronomy, students had to know about fractions, particularly sexagesimal ones. In studying spherics, advanced students became familiar with some trigonometry.

Adelard's translation of al-Khwārizmī's *Zīj* (*Astronomical Tables*) in 1126 introduced the West to sine tables and Hindu methods. The treatise, which fixes the calendar base on the Muslim *hegira* and the principal meridian at Arim (Indian Ujjain), was not at first usable in Europe. Eleventh- and twelfth-century astronomers adapted al-Zarquali's *Toledo Tables* to Marseilles, London, and Toulouse. Adaptations of these tables became the base of the *Alphonsine Tables*, a collaborative Muslim, Jewish, and Spanish project issued in Toledo in 1272. With sexagesimal refinements, these tables dominated European astronomy until Regiomontanus' more precise determinations.

A translation by the Moorish Christian John of Seville (fl. 1133–1142) of Masha'allah's *On the Composition and Use of the Astrolabe* extended the use of that instrument in calendrics and navigation. Herman of Carinthia's contemporaneous translation of Ptolemy's *Planisphere* introduced the Latin world to stereographic projection, which gave astrolabic reckoning a sound theoretical base. These new techniques, in which astrolabic observations yielded parameters for calculating conjunctions and eclipses, laid the foundations for a revival of scientific astronomy.

The key to constructing astronomical tables was the geometric model of planetary motions in Ptolemy's *Almagest*. While translation of that work from Greek to Latin at about 1160 by the anonymous translator of Euclid's *Elements* was mostly ignored, advanced students of astronomy and mathematics closely studied Gerard's version of about 1175, translated into Latin from Arabic. A report that Spanish Muslims possessed the *Almagest* may have sent Gerard to Toledo to begin his scholarly career. The small group of scholars who achieved technical mastery of the *Almagest* introduced mathematical astronomy in the Latin West.

Astronomical theory was indispensable for astrology and, again, translations from Arabic to Latin were the starting point. Abu Ma'shar's *Great Introduction* attracted wide attention. While astronomers investigated the causes and descriptions of celestial movements (*motus*), astrologers dealt with the companion effects of these movements. The science of astronomy, which included both pursuits, was employed in preparing the Christian calendar, but astrology was suspect since it seemed to encroach on divine omnipotence and human volition. Albertus Magnus composed a critical bibliography differentiating astronomical from dangerous astrological tracts. His *Speculum Astronomiae*, dated about 1260, displays an impressive array of Greek and Arabic technical works, as well as a growing theological antagonism to astrology. Aristotelian cosmologists intending to downplay the role of mathematics in celestial mechanics also resisted astrology. In the face of this formidable opposition astrologers continued to base their forecasts on mathematical models and their calculations were crucial in sustaining the revival of Ptolemaic mathematical astronomy.

Translators almost ignored music theory, the fourth quadrivial subject. But the omission was not a serious loss to the body of mathematical knowledge, since Boethian music centers on ratios and proportion already treated in his arithmetic. Inquiries on proportionality informed the revived study of Euclid and helped prepare for the interest in ratios in the fourteenth century. Monastic educators approached music more as a performing art than as a theory of acoustics and harmony. By the eleventh century monks had adapted Boethian monochords to liturgical plain chant: this permitted a sung accompaniment at intervals of a fourth or fifth below or above the melody. A century later the method of naming notes (after the first syllables of a hymn to St. John the Baptist) and of mensural notation was in use. The translations of new quadrivial materials set the stage for a rejuvenation of theoretical mathematics in the West.

12.3   Two Original Medieval Latin Mathematicians: Leonardo of Pisa and Jordanus de Nemore

From ancient Roman times to the twelfth century, Latin civilization had produced no mathematician of significant originality. Ending that barren situation, Leonardo of Pisa (ca. 1170 to after 1240) and Jordanus de Nemore (fl. 1220) inaugurated a second phase of Latin mathematics.

Leonardo, who belonged to a family named Bonacci, is today also known as Fibonacci, which he did not use. His father's name was not Bonacci and the name is not a shortened form of *filio Bonacci*. Leonardo's father was a career secretary who directed the Pisan customhouse in what is now Bougie, Algeria. He purposely brought his son there to be taught the art of calculation with new Indian numerals under a skilled teacher. This schooling, deepened through contacts with merchants and scholars and their books in the course of business travels to ports of Egypt, Byzantium, Syria, and Sicily and to Provence, gave young Leonardo a command of substantial portions of Arabic and ancient Greek mathematics.

At about 1200, Leonardo ended his early wanderings and returned to Pisa, where at intervals he published his mature mathematics. Italian teachers of the new commercial mathematics (*Maestri d'Abbaco*) rather quickly recognized its merit and utility. Leonardo's techniques, which they adopted and extended, brought their subject to a new level of sophistication. The northern Italian commercial setting and access to the learned court circle of the Hohenstaufen Frederick II in the Neapolitan and Sicilian south apparently provided intellectual stimulation to Leonardo, who had no known formal academic training. The Neapolitan court was a center for linguistic, philosophical, and mathematical inquiry. Michael Scot, Frederick's courtier, translator, and astrologer, was an intellectual associate of his. Leonardo probably resided for the rest of his life in Pisa, which in 1240 awarded him a pension for civil contributions.

Assessment of Leonardo's contributions to mathematics rest upon five extant writings: his *Liber Abbaci* (*Book of Counting*), which appeared in 1202 with an expanded version in 1228; a *Practica Geometriae* dated at 1220/1221; a *Flos* of 1225 responding to questions posed during an audience at the Norman court; an undated letter to the imperial philosopher Theodorus solving a traditional, indeterminate algebraic problem of a hundred birds; and the *Liber Quadratorum* (*Book of Squares*), a work of 1225. We lack his gloss on Book X of the *Elements*, which treats irrationals as numbers rather than geometric magnitudes. These writings show that Leonardo not only improved commercial arithmetics and theoretical geometry, but also made original contributions in number theory and indeterminate analysis in algebra.

The *Liber Abbaci*, a major starting point for the revival of theoretical mathematics in the Latin West, begins with Roman numerals and finger counting and proceeds to Indo-Arabic numerals. Its first section methodically covers computational operations with Arabic numerals and checks results by casting out nines. It also discusses proportions, approximations of square and cube roots, quadratic and indeterminate equations, and Archimedean mensurational problems. The second section treats calculation of wages, profits, interest, partnerships, and currency conversions. What Leonardo designates "abbacus" is not counting board reckoning but its replacement with Indo-Arabic numerals. Although al-Khwārizmī's arithmetic had brought notice of an early form of them, it was Leonardo's book that popularized them. As adopted, the new reckoning was called algorithmic (after al-Khwārizmī), rather than abbacusic as Leonardo had labeled it. Social inertia, however, encumbered algorithmic reckoning as did vestiges of the old sand box—cross outs, erasures, unintegrated subproducts, and subquotients. Traditionalists believed that it was easy to tamper with Indo-Arabic numerals, for example to change 6 to 0, and did not recognize the merit of decimal fractions within a place-value system. So while the new reckoning challenged, it did not displace the old Roman numeral system.

In addition to his computational instructions, Leonardo added a seemingly trivial but now famous rabbit problem, whose solution is the first recursive sequence known in mathematics in the West. The nineteenth-century number theorist Edouard Lucas was to give that solution the name of Fibonacci. The problem specifies that rabbit protoparents reach sexual maturity in one month. In successive months for a year, they breed a new male and female pair, as do all of their matured, paired descendants. The rabbits constitute a Fibonacci sequence $(1, 1, 2, 3, 5, 8, 13, 21, 34, 55, 89, 144, \ldots)$ in which each term is the sum of the preceding two. In modern form, any term $k_n = k_{n-1} + k_{n-2}$ for $n \geq 3$. Although in this sequence Leonardo probably had the gist of recursion, he lacked the means to supply a formula. Benefiting from advances in notation, Albert Girard would write the formula in 1634.

It took time for the remarkable nature of the Fibonacci sequence to become known. As you may have noted, every two successive terms are relatively prime. To prove this theorem, first assume the opposite. That is, there are some $n > 1$ that divide F_k and F_{k+1}. If so, n must also divide the difference $F_{k+1} - F_k = F_{k-1}$. This and the generating formula, $F_k - F_{k-1} = F_{k-2}$, invite the conclusion that n divides F_{k-2}. Working backward will show that $F_{k-3}, F_{k-4}, \ldots$, to F_1 are divisible by n. But F_1 is 1 and not divisible by $n > 1$. This contradiction invalidates the assumption and confirms the theorem. Another intriguing question about Fibonacci numbers is how many are prime. To the present we have no devices either for predicting exactly which Fibonacci numbers are primes or for establishing whether there are infinitely many.

Although Leonardo's major interest was algebra, his *Practica Geometriae* goes far beyond Hugh of St. Victor's. Drawing on the *Discourses of the Sons of Moses*, he gives classical solutions for duplicating the cube that also appear in Jordanus' treatise *On Triangles*. His mastery of Greek, Byzantine, and Arabic sources enabled him to provide original proofs with a rigor unknown in the West since the eclipse of Hellenistic science. The study is notable also in interweaving theoretical and commercial materials: with a standard chart for converting marketplace weights and measures, for example, juxtaposing an exposition of Euclidean irrationals from Book X of the *Elements*. Such linkages suggest that Leonardo respected the mathematical abilities of merchant readers.

Leonardo's audience at the Norman-Sicilian court in the presence of Emperor Frederick II prompted the *Flos*, three of whose fifteen challenge problems the courtier John of Palermo submitted. It contains both determinate and indeterminate problems. John was probably the foremost Western authority on conics. His translation of a short Arabic treatise demonstrating the asymptotic property of the hyperbola was the sole medieval Latin conical treatise not on optics. In modern terms, the first challenge problem was to solve these simultaneous equations: $x^2 + 5 = y^2$ and $x^2 - 5 = z^2$ or $y^2 - x^2 = x^2 - z^2 = 5$. Late in the *Flos* Leonardo offers the squares 961, 1,681, and 2,401, but does not reveal how he obtained these numbers. He appears to take advantage of a special class of numbers that he unveiled early in the *Flos*, for in the challenge problem he first replaces the number 5 with c for *congruum*. For even numbers $(a + b)$, he has $n = ab(a+b)(a-b)$ and for odd $(a+b)$, he gives $n = 4ab(a+b)(a-b)$. Each of these numbers is a *congruum* and Leonardo shows that each must be divisible by 24. He determines that for $x^2 + h$ and $x^2 - h$ to be simultaneously a square, h must be a *congruum*. Letting $a = 5$ and $b = 4$ yields $h = 720 = 5(12^2)$. Leonardo then has his original $y^2 - x^2 = x^2 + z^2 = 720$. Probably from a technique in Diophantus, Leonardo arrives at $2401 - 1681 = 1681 - 961$. These are $49^2 - 41^2 = 41^2 - 31^2$. With indeterminate equations, Leonardo generally seeks positive integral solutions but makes occasional use of zero and negative numbers.

Perhaps Leonardo's ability showed most in his obtaining a correct root without proof of a problem that can now be expressed as the cubic equation $x^3 + 2x^2 + 10x = 20$. He shows that its roots cannot be Euclidean, that is, an integer, a fraction of integers, or a number $a + \sqrt{b}$, for a and b rationals. The magnitude sought thus defies compass and straightedge construction and recalls Leonardo's proposal to treat irrationals as numbers rather than geometric magnitudes. He perhaps looked to the treatment of irrationals in *Elements* X. Although we do not know how he did it, Leonardo arrived at a six sexasgesimal place solution, 1;22,7,42,33,4,40, that is the most accurate approximation of an irrational to the sixteenth century. Perhaps he had learned what we call Horner's method from the medieval Arabs.

Leonardo's chief work, *Liber Quadratorum*, makes him the major number theorist in the West between Diophantus and Fermat. He determines in part Pythagorean triples of the form $[(a^2 - 1)/2]^2$ for odd a, which when added to a^2 equals $[(a^2 + 1)/2]^2$, with a corresponding $[(a/2)^2 + 1]^2$ for even a. He uses the Euclidean triples expressible as $2pq$, $p^2 - q^2$, and $p^2 + q^2$ and establishes still other triples using the Diophantine identity $(a^2 + b^2) \times (x^2 + y^2) = (ax + by)^2 + (bx - ay)^2 = (ay + bx)^2 + (by - ax)^2$. When translated into modern algebraic notations, Proposition 4 establishes these three identities now attributed to Lagrange:

1. If $A < B$ and $C < D$, then $(A^2 + B^2) \times (C^2 + D^2) = (BD \pm AC)^2 + (BC \mp AD)^2$,

2. If $A^2 + B^2 = M^2$, then $(A^2 + B^2) \times (C^2 + D^2) = (MC)^2 + (MD)^2$, and

3. If $C^2 + D^2 = N^2$, then $(A^2 + B^2) \times (C^2 + D^2) = (AN)^2 + (BN)^2$.

The second and third identities are simply statements of the distributive law. In the *Liber Quadratorum* Leonardo also advances a series of proofs of propositions in number theory that establish when it is not possible for a square to be a *congruum*, or for $x^2 + y^2$ and $x^2 - y^2$ to simultaneously be squares, or for $x^4 - y^4$ to be a square, and so forth.

Such work establishes Leonardo of Pisa as an original mathematician. Reintroducing Euclidean rigor to the Christian West after centuries of lax agrimensorial formulations, he also increased arithmetization in geometry. Readers of his works found complete and versatile supplemental proofs. Algebraic reasoning, especially in divisibility and factorization, was his forte. Although rhetorical formulation of problems limited Leonardo, he groped toward symbolism by sometimes letting a letter of the Latin alphabet stand for a general number. Academic mathematicians who succeeded Leonardo failed to pursue his advanced algebra and number theory, but Italian and German teachers of commercial arithmetics taught his computational algebra and practical mensuration. Eventually, theoretical algebra gained major status in the fifteenth century. It was called the *arte della cosa*, named from Fibonacci's *causa* (to signify an unknown). While Leonardo's algebra became an important source, his stature in mathematics was not fully recognized until the nineteenth century.

In the thirteenth century, Leonardo's only mathematical peer was Jordanus de Nemore. A library catalogue composed between 1246 and 1260 credits Jordanus with being the chief author of twelve treatises divided between mathematics and statics. The catalogue date fixes a rough termination for his activity and a single, internal reference in his *Elements [of the Science] of Weights* shows that it was written after *On Triangles*. His position resembles that of Leonardo in certain respects: the advanced work of both was insufficiently recognized by contemporaries; each was a "master," who cannot be traced to a university; neither seems to have been a cleric. Possibly Jordanus was Leonardo's junior, came from the north (perhaps Nemours, southwest of Paris) and lacked knowledge of the Pisan's writings but had absorbed common sources. There are no clues as to where and how Jordanus gained access to these. His major mathematical writings are *On Triangles* (or *Philotegni*), *On Given Numbers*, and an *Arithmetic* containing more than four hundred propositions. Those propositions are the standard source for late medieval theoretical arithmetic. Jordanus also composed more rudimentary works on algorisms and proportions in which Euclidean influence is pronounced.

Barnabus Hughes, who has edited *On Given Numbers*, credits Jordanus with reintroducing analysis in proving a proposition by reducing it to simpler, previously demonstrated propositions and basing the procedure on a concept of generalized number. (Credit for the beginning of modern algebraic analysis usually goes to François Viète in 1591.) This medieval treatise, Hughes observes, contains three steps essential for analytical algebra: construction of an equation that represents what is known and what is to be found, transformation of the equation into canonical form, and computation of numbers that satisfy the conditions specified. Hughes's English translation and algebraic notation support his claim more strongly than does Jordanus' Latin text.[360] Rhetorical expression clutters the work and its generalized number is ambiguous. Lacking operational symbols, Jordanus sometimes has to fit numbers to a particular case without reaching a general solution. This does not, however, detract from the merit of *De Numeris Datis,* which goes well beyond elementary algebra and points in the direction that the subject would follow.

Jordanus resembles and perhaps excels Leonardo in his mastery of geometry. Leonardo's *Practica* and Jordanus's *On Triangles,* which seem independent, share several advanced geometric demonstrations.

In Book IV, Jordanus addresses the three famous problems of antiquity. After criticizing two mechanical ways of trisecting an angle from the uncited *Sons of Moses*, he employs conic sections and a proposition from the *Optics* of Ibn al-Haytham. For squaring the circle he proposes not exhausting the area, but finding a third continuous proportional after the circumscribed square and its circle. His two solutions for finding mean proportionals between two given lines for duplicating the cube are from Archytas via the *Sons of Moses* and from Philo through Eutocius.

The short *Elementa super Ponderum* established Jordanus as the authority in the science of weights and theoretical statics. From its internal development and placement at the beginning of a Greek statical work, the treatise seems an attempt to reconstruct a series of lost Greek propositions. The reconstructed propositions give a theoretical foundation for weighing by the steelyard, a balance with sliding poise. Combining Aristotelian mechanics of motion with Archimedean statics, Jordanus produces his ingenious doctrine of positional gravity, allowing him to formulate a prototypical version of the principle of virtual work (input = output) and a geometric proof of the law of the lever. His proof is sound, but a double flaw in propositions leading to it muddles the treatise as a whole.

Subsequent treatises expand, modify, and correct the *Elements*. *On the Theory of Weights,* ascribed to Jordanus in some early versions, is the most sophisticated. A close reading of the text suggests that the author is an anonymous redactor, who is a close disciple of Jordanus. The author, whether Jordanus or a gifted follower, correctly uses for the first time in the development of mathematics the doctrine of positional gravity to analyze and state the law of the inclined plane. By projecting segments he shows, equivalently, that $F = W \sin \alpha$, where F is the force along the plane, W the free weight, and α the angle of inclination of the plane as illustrated here. If weight g : weight $h = XM : ZN$, then a force sufficient to lift h to M will also lift g to N.

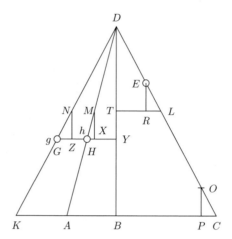

Positional gravity: Jordanus's inclined plane.

12.4   The Rise of Universities and Scholasticism

While Leonardo of Pisa pursued new mathematics in foreign places, others in
the Latin West were preparing universities that from about 1200 supervised
the burgeoning of European knowledge. Scholars in universities cast Latin
intellectual culture into the distinctive bookish form known as scholasticism
and launched an institutional phase of medieval mathematics. An outgrowth
of monastic and cathedral schools, the university was intimately connected to
the new learning, new piety, and new economic opportunities.

Commercial and cultural expansion required teachers. As a result, Italian
lay schools enlarged and multiplied while, in the north, urban schools at cathedrals or collegiate churches gradually replaced rural monastic schools. Teachers
who migrated, sometimes by invitation, to these schools were followed by pil-
grimlike students seeking ecclesiastical or governmental careers. Upon completing supervised study in the trivium and quadrivium, fledgling teachers received
certificates from a chief teacher (*scholasticus* or *ecolatre*). The most renowned
of the migrant teachers, such as the Breton Peter Abelard, who taught some
twenty future cardinals and fifty bishops, were known as expert dialectical logicians.

Better training in dialectics followed a pedagogical reorientation of the trivium away from its rhetorical core. Almost certainly Gerbert, who taught in
the monastic school at Rheims, was the first medieval Latin scholar to lecture
systematically on the full range of Boethian logic. The shift to Boethian logic
is apparent in the program of studies at cathedral schools of the late tenth
century. In the next century the humanist John of Salisbury (d. 1185), who
taught Latin literature at French cathedral schools, accurately observed that
Aristotle had awakened the Latin mind from sleep or death. As an indication of
its power, the encyclopedist Alexander Neckham (d. 1217) reported that when

pious persuasion failed a logical argument (from probabilities) had persuaded a dying student at Paris to accept the Christian doctrine of personal resurrection.

In the century after 1150, teaching masters and students developed their own organizational structure patterned on artisans' associations and the merchant guild, which completed the transformation of cathedral schools into universities. The new guild model broke with monastic precedent in presenting teaching masters as artisans who expected to be paid for their labor. Associations whose members swore to uphold a common interest were a pervasive medieval phenomenon going by various names, among them *societas* and *universitas*. These corporations acted as a single body. In recognition of this unity, jurists described the *universitas* as a "fictive person" in law, lack of a soul preventing it from claiming true personality. When teachers and students organized, a university resulted: the first reference to an educational *universitas* is the guild of masters and students at Paris in 1221. *Studium generale* is the medieval term that signified "university" in something like the modern sense.

Bologna and Paris epitomized the two types of medieval university. In 1158 Emperor Frederick Barbarossa granted law students at Bologna safe conduct and immunity from local laws. Subsequently, these students, most of them adult laymen, elected their rector and regulated salaries, schedules, and academic performances of the faculty. The faculty, however, retained control of examinations and admission into the guild of lawyers. Paris, which grew from an amalgamation of schools around Notre Dame cathedral, was chartered by King Philip Augustus in 1200. After a long struggle between the faculty and the Bishop's Chancellor over the right to grant degrees, the faculty triumphed. The *licentia docendi,* for which the candidate had to complete a prescribed curriculum in the remodeled trivium and quadrivium and pass a series of oral examinations, was a degree of admission to the guild of teachers throughout the Latin world and bestowed the title of master. The body of masters from arts, medicine, law, and theology elected a rector from the arts faculty as its chief officer. Masters regulated academic activities of the students, whose age for admission was about fifteen.

A bachelor of arts degree was more prestigious than a journeyman artisan's license. Granting of the bachelor's *bacca*, a symbolic teacher's rod whose judicious use Charlemagne had endorsed, meant that the recipient had undergone preparation to do supervised teaching as a graduate student—from *gradus*, a step leading to the teacher's pulpit. The baccalaureate was not a prerequisite for a master's program, which drew most entering students.

Instruction for the baccalaureate and the *licentia docendi* centered on the *lectio*, reading and commentary on a standard text, and the *disputatio*, formal oral argument. A Parisian master of arts degree required four to six years and for graduation a minimum age of twenty-one. Higher studies took longer, for example, theology, the "queen of the sciences," about twelve years beyond the master's degree and a minimum age of thirty-five.[361] Bologna granted a degree in Roman law, the "lucrative science," after eight years in higher studies, while Salerno's medical degree required six additional years.

Organized students, whose numbers in the thirteenth century reached about seven thousand at Paris, ten thousand at Bologna, and three thousand at Oxford, gained considerable power. Clashes between town and gown were common. When disputes went unresolved, one recourse for unsatisfied faculty and students was a migration. Migrants from Oxford, itself probably originating in an exodus of 1167 from Paris, founded Cambridge in 1209. Before 1212 the Parisian arts faculty had relocated from Notre Dame cathedral on an island in the Seine to its left bank. In 1222 students from Bologna settled at Padua. Crowded and poorly policed cities invited crime. At Paris students accused of lawbreaking had the privilege of being tried in ecclesiastical courts, where penalties were minimal. Of the shaved crown (*tonsura),* which was the mark of a student (*clericus*) and did not imply aspirations to the priesthood, King Philip Augustus observed that it was better protection than a knight's helmet. Adding to Philip's formal charter of 1200, Pope Gregory IX's decree *Parens Scientiarum* ("mother of the sciences") conferred special protection and privileges on Paris as well as giving that university an enduring sobriquet.

From writings of Church Fathers and Aristotle, *scholastici* or professors developed the Latin intellectual culture called scholasticism. They combined doctrinal content and pedagogical, dialectical method into a highly unified system. Their emphasis on scholarship as a cumulative tradition legitimated inquiry into all learned opinion. The twelfth-century master Bernard of Chartres compared his generation to dwarfs who see farther because they sit on the shoulder of giants. (Newton was later to use this metaphor, not original with Bernard, to describe himself.) Especially recognized were *auctores* (authorities) in the learned tradition. Schoolmen produced intricate textual commentaries on authoritative books and comprehensive *Summae* of universal knowledge. They embraced dialectics because it imposed order and organization on diverse and confusing materials.

Work about 1100 in Bologna that resurrected Justinian's *Corpus Iuris Civilis* was an early fruit of the new learning. Law and dialectics favored disputation, which quickly spread to other studies. Appearing at about 1122, Abelard's *Sic et non* (*Pro and Con*) introduced into theology the dialectical *questio,* which gave the subject a new orientation. Peter the Lombard's *Sententiae* (*Opinions,* Paris, ca. 1150) offered reasoned conclusions to Abelard's 168 unanswered questions and raised new ones. From it students learned scholastic reasoning. Two alternatives to the methodology of logic and law were also proposed. At Chartres, where Platonic thought was influential, Alan of Lisle (d. 1203) urged in *Theologicae Regulae* that theology be put into axioms on the model of geometry. As a second, but vanquished, alternative, John of Salisbury and others espoused a humanist program gracing with literary elegance the sensibilities and riches of the ancient world.

During the early thirteenth century, Robert Grosseteste (1168–1253) at Oxford carried beyond Alan's suggestion the notion of a mathematical approach to God and nature. He originated a "light metaphysics," which strongly influenced later Oxford scholars. As chancellor, bishop, and in completing from 1200 to 1209 the first Latin commentary on the *Posterior Analytics*, the center-piece of Aristotle's new logic, he had high standing. The *Posterior Analytics*, Book I, chapter 12, for example, provokes questions about the proper relations between mathematics and logic. Grosseteste doubts the primacy of logic. His light metaphysics in effect puts into geometric form an Augustinian rendering of Platonism that attributes human knowledge to an influx of divine illumi-nation, rather than to apprehension of ideas. Primordial light energy (*lux*) replicates its own image throughout space according to the laws of proportion and thereby gives the universe its extended "corporeity." The same principles that make intelligible the world, which on the first day of creation was filled with fundamental *lux*, also give intelligibility to Euclidean space. In examin-ing the geometrically patterned world, the philosopher attempts to go beyond mere recognition of facts of experience to an understanding of the reason for them. Since lines, angles, and figures express the causes of all natural events, mathematics is essential to that inquiry: notably *perspectiva*, geometrical op-tics applied to visual cognition. Grosseteste's Franciscan disciple, Roger Bacon (ca. 1219 to ca. 1292), who applied light metaphysics to the diffusion of divine grace, argued that mathematics underlies all disciplines, including logic, and that quantity is incomprehensible without it.

Neither light metaphysics nor Bacon's variant was generally accepted, but they influenced mathematically logical investigations in the fourteenth century at Oxford's Merton College, founded before 1264. The metaphysical orienta-tion of Bacon's program would probably have precluded substantive, original mathematics.[362]

At Paris, Aristotelianism of a legal and logical kind came to dominate study in the arts. The intellectual preeminence of Paris promised far-reaching effects of that victory. After initial resistance in 1210, other works of Aristotle joined his logic at the center of liberal arts education. Forty years later Paris would grant no master of arts degree unless the recipient had attended lectures on all available works of Aristotle, by then called simply "the Philosopher." Addition of these new branches of Aristotelianism refashioned the trivium and quadriv-ium. Under the new program natural philosophy included physics, theoretical astronomy, biology, psychology, and a prechemical science of matter. Two other philosophies, moral and metaphysical, completed the additions. The faculty of arts was redesignated as the faculty of arts and sciences.

The Italian Dominican and sometime Paris master Thomas Aquinas (1225–1274) brought to maturity the synthetic phase of Latin scholasticism. Within an era of faith, his *Summa Theologica*, premised on harmonious complementar-ity between reason and faith in revelation, was the most thorough and durable synthesis of Christianity and Aristotelianism ever achieved. It challenged the Augustinian theory of an identity of reason and faith that in practice subordi-nated reason. It was also contrary to an Aristotelian position, stemming from Arabic dialectical theologians, implying that faith and reason can coexist as contradictories. Thomas's principle of complementarity expanded the natural realm: studies of the entire natural domain are to proceed without recourse to supernatural explanations. Thomas identifies Aristotle's philosophical Prime Mover with the Trinitarian God of Christian revelation and reasons that the hierarchical order of the universe is intelligible and providential. Thomas could not, however, reconcile Aristotle's deterministic cosmology with divine omnipotence, a fundamental article of Christian faith.

In 1277 ecclesiastical authorities at Paris, apprehensive that a Christianized Aristotle was tantamount to a paganized Christianity, condemned 219 propositions of Aristotelian philosophical naturalism. The logician Pope John XXI acquiesced. While some condemnations were directed posthumously at Thomas, his reputation was later massively restored. The condemnations actually stimulated the development of natural philosophy by keeping open several questions about the physical universe, such as the motion of the Earth and the existence of a vacuum, otherwise foreclosed to orthodox Aristotelians. By Thomas's death Aristotelianism, now entering a critical as opposed to a synthetic phase, had been firmly implanted in the universities.

The third phase of medieval, academic mathematics, expansion of its study in the quadrivium, parallels the penetration and subsuming of the liberal arts curriculum by Aristotelian philosophy. During this institutional phase students recaptured lost computational techniques, theoretical geometry, and related mathematics needed for astronomy and perspective. The new reckoning had several Latin popularizations. The most notable were Alexander de Villa Dei's *Carmen de Algorismo* of about 1202, a versified short *computus* using Indo-Arabic numerals; a more ample *Algorismus vulgaris*, dated about 1240, by Oxford's Sacrobosco (John of Holywood); and an improved algorismus attributed to Jordanus. Leonardo's *Liber Abbaci* and Jordanus' *De Numeris Datis* were also available. For geometry, portions of the Latin *Elements* were assigned as a supplement to Hugh of St. Victor's practical manual. At thirteenth-century Oxford, where mathematics was more prominent than at Paris, a master had to affirm by oath that he had attended lectures on the first six books of the *Elements*. Campanus of Novara, editor of the definitive medieval Euclid, was possibly at Paris for a time.

For astronomy, students studied a relatively standardized compilation perhaps adapted from the Arabs. After 1250 the contents of the *corpus astronomicorum* included Sacrobosco's *Algorismus, Computus*, and *On the Sphere;* the calendar of Robert Grosseteste; an instrumental treatise on the old quadrant; the *Toledan Tables* for calculating celestial positions; the anonymous *Theoria Planetarum* on planetary kinematics of Hipparchus and Ptolemy; and four astronomical treatises by Thābit ibn Qurra. Paris and Bologna, joined by the Spanish university of Salamanca, founded in about 1227, were centers of astronomical studies. Amid old concerns about errors in the calendar and a new fascination with astrology, astronomical instruments multiplied along with predictions of celestial positions. Ptolemaic mathematical astronomy was indispensable for calendrics and astrology.

Scholastic compromise alleviated the worst disagreements between Ptolemaicists and Aristotelians. Explanations of the physical causes, nature, and purpose of celestial motions fell within the domain of natural philosophy, while the mathematical description of these motions belonged to astronomy. Aristotle's celestial mechanics attaches the planets to transparent, rotating spheres that transmit to the outermost sphere motion applied by the Prime Mover. By assumption, the system's center of movement coincides with Earth's center: all celestial velocities are uniform with respect to Earth. On the other hand, Ptolemy's kinematics of eccentric circles and epicycles violates the physical postulates of concentric spheres and uniform motions, but is superior to Aristotle's spheres for "saving the phenomena": for accurately describing and predicting planetary motions. Beginning with Roger Bacon, natural philosophers devised a compromise to retain Aristotle's spheres and generate Ptolemy's eccentric planetary orbits from their revolutions. In place of the strict concentricity of contiguous spheres, they proposed that planet-carrying spheres make contact at a single point with immediately enclosing spheres. This realignment, traceable to Alhazen, creates a subsystem of rolling, geocentric spheres replicating planetary motion eccentric relative to the Earth. Cosmologists insisted that the spheres were real. They also acknowledged that the compromise physico-mathematical description was simply hypothetical.

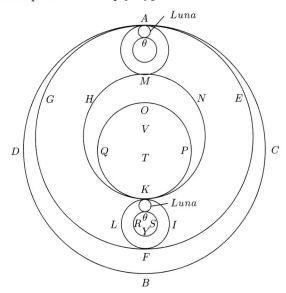

Bacon's planetary spheres.

Medieval optics and statics resemble astronomy in allowing distinct, sometimes divergent solutions in which mathematics figures prominently. To the science of perspective that Alhazen compounded of Euclid's and Ptolemy's geometric optics, the Greek atomists' physical theories of visual perception, and Galen's physiology of the eye, Bacon added St. Augustine and Grosseteste. The Silesian Witelo's *Perspectiva* of about 1270 adapted theorems from Alhazen, Theodosius, Eutocius, and Pappus to locate images on mirror surfaces.

Yet Witelo, unlike John of Palermo, did not grasp conics. He did not understand how parabolic construction depends upon the *latus rectum*, a straight
line drawn through the focus of a conic at right angles to the transverse diameter. In statics Jordanus reinvented theorems to establish equilibrium on
the steelyard. An anonymous fourteenth-century author at Paris discarded all
but two of these theorems in statics, for which he tailored new Archimedean
geometric proofs. The statical commentator's revision reflects a critical and analytical scholasticism that is less cosmological and succeeded the anthologies,
compoundings, and harmonization of the early period.

12.5   Style and Applications of Late Medieval Mathematical Thought

Strong scholasticizing tendencies that became more pronounced in critical Aristotelianism distinguish the fourth phase of medieval academic mathematics of
the late thirteenth century and the fourteenth. The logical and legal context
and the explication of authoritative texts continued. The major text was Euclid's *Elements*, but as an adjunct to the *Posterior Analytics*. This meant a
shift from mathematics in itself to its logical foundations. The *Elements* was
consulted less for its technical mathematics than as an exemplar of scientific
knowledge whose principles Aristotle had formulated.[363] Formal exposition became painstakingly didactic: the Latin Euclid was extensively reworked. This
scholastic exposition partly mended the broken Hellenistic tradition by reintroducing habits of critical thinking and proof. Commentators added lemmas
and alternative demonstrations; supplied missing steps in an argument, made
implicit assumptions or connections explicit, and gave directions for constructions. They examined the components of geometric demonstration, including
foundations, elements, processes, and relationships. In emphasizing the logical
structure of the *Elements*, teachers concentrated on processes of mathematical
thinking rather than results.

Among the fourteenth-century scholastic compilations that advanced mathematics was the Italian astrologer Dominicus de Clavasio's *Practica Geometriae* of 1346, which unified agrimensorial and theoretical geometries around
proportion. An anonymous monk and the Norman astronomer Jean of Murs
imaginatively devised what might be called *paralipomena* (Greek for "things
omitted"). After an unsuccessful search for proofs of two propositions of Apollonius, *Conics* I: 11 and I: 35, which Alhazen had reported and Witelo treated
unsatisfactorily, the unknown author reconstructs them. His proof restates a
rule that Archimedes had often used: that the *latus rectum* formed from a right-
angled cone equals a line double the axis of the parabola drawn as far as the
axis of the cone. The anonymous mathematician's *Speculi Almukefi Compositio*
(*Construction of the Parabolic Mirror*) is novel also in its use of trigonometric
lines for demonstrating the asymptotic property of a hyperbola and its firsthand

knowledge of iron metallurgical processes. Jean of Murs addressed Proposition I of Moerbeke-Archimedes' *Dimension of the Circle*, which proves the equality of areas for a circle and a right triangle whose including sides are equal to the circle's circumference and radius. By 1340 Jean had apparently realized that without further support this proposition begs the question, for it assumes the possibility of rectifying circumferences. Recognizing that Archimedes in Proposition 18 of *On Spirals* had demonstrated the construction, Jean culled that crucial proposition and others on which it rested and placed them before the unproved proposition.[364] His later *Art of Measurement* includes this *Circuli Quadratura*, the first known commentary to *On Spirals*. It shows a superior command of Archimedean geometry, Eutocius, and an algebraic mensurational treatise by Abu Bakr. Jean's better known *Quadripartitum Numerorum* contains arithmetic from al-Khwārizmī, algebra from Leonardo's *Flos*, and excerpts from medieval hydrostatics. He belongs in the front rank of fourteenth-century mathematicians.

The mathematics of that century was connected to movements, at once logical, philosophical, and theological, called nominalism at Oxford and terminism at Paris. Nominalists following William of Ockham (fl. 1335) turned from investigation of universal ideas to a close linguistic analysis of propositional statements. The larger meaning of this was a concentration on particulars as opposed to generalizations and the role of the individual and the human will as opposed to reflective, synthetic reasoning. The premise of God's unlimited freedom to work miracles encouraged nominalists to examine any imaginable hypothesis provided only that it contain no logical contradiction. Ockham's razor, a principle of parsimony, dictates the choice of the simplest among possibilities open to logic, language, and divine omnipotence.

By modifying Aristotle's doctrine of substantial forms, nominalists and terminists opened the way for a more quantitative analysis of change and motion than that previously pursued in the Latin West. A catalytic question in Peter Lombard's *Sentences* about increases and decreases in charity stimulated efforts at quantification. Mathematics of the infinite and the mathematical continuum became involved in clarifying such questions as how an infinite God relates to finite creatures or how members of different species, such as angels and flies, can be compared in any meaningful sense. In what John Murdoch has called their "frenzy to measure" everything possible,[365] logician-natural philosophers developed new quantitative languages with distinctive vocabularies and algorithms. The languages of *proportio* and *intensio et remisio* (increase and decrease) are particularly interesting.

Book V of Campanus's translation of the *Elements* shaped medieval treatments of ratio and proportion. Under Boethius' influence, however, the concept of proportion from Euclid's use of Eudoxus had been garbled, submerged, and recast, and Campanus's *Elements* is victim to the distortion. Ratio, defined as a relation between numbers, medieval mathematicians designated by specific and generic terms in a classificatory system and not, as now, by pairs of numbers, as 2 : 3 or 2/3. Late medieval mathematics defined, for instance, 5, 3 as a ratio of "greater inequality," specifically a "superbipartient third." Naming the precise relationship between the first and the second term gave the ratio its denomination. The lexicographic procedure of denominations isolated ratios from the arithmetic of fractions. Jordanus corrects some of that isolation by defining denomination as the number resulting from division. His definition permits him to treat some ratios as fractions. An example is *De Numeris Datis*, II: 2, 3. Still, into the late seventeenth century ratios continued to be referred to by their verbal denominations. This partial absorption of ratios into arithmetic after Jordanus was less important than a related expansion of medieval proportion theory relating ratios to one another. Book V of Campanus's Euclid had changed rather than transmitted Greek proportion theory. Definition 20 in Book VII of the original *Elements* contains a limited Pythagorean version of equality of proportions; the early Pythagorean form applies only to integers and their fractions, while Eudoxus' definition extends the theory of proportion beyond these rationals to include irrationals or incommensurable magnitudes. Campanus, who depended on distorted sources, adopts the limited Pythagorean version with a Boethian revision, using it to base equality of ratios on equality of their denominations.[366] Campanus was probably influenced by Jordanus, who had multiplied denominations by each other to compound a ratio of ratios. This opened the way for fourteenth-century scholastics to employ integers or fractions in lowest terms to denominate rational ratios that we treat as exponents.

Exploiting techniques from Euclid and Menelaus for compounding ratios, then, innovative Schoolmen made ratios subject to denomination by other ratios. The resulting "ratio of ratios" is the equivalent of exponents. Thomas Bradwardine at Merton College introduced this practice into natural philosophy and gave the first known exponential statement of a physical law (see below). Later in the century Nicole Oresme at Paris systematized the Mertonian materials into an *algorismus* in modern terms of rational and irrational exponents modeled on arithmetic operations with numbers.

Schoolmen likewise expanded proportionality theory to elucidate musicology. Medieval ratios governing musical intervals could be properly increased or decreased only when multiplied or divided by other ratios. The scholastic terms for such increase and decrease are composition (or addition) and decomposition (or subtraction). Mathematicians manipulated what are now exponents and roots by inserting geometric means between terms, which may be represented by the modern a_1 and a_n. Thus,

$$\text{if } a_1/a_2 = \ldots = a_{n-1}/a_n \quad \text{then}$$
$$a_1/a_n = a_1/a_2 \times a_2/a_3 \times \ldots \times a_{n-1}/a_n \quad \text{and}$$
$$a_1/a_n = (a_1/a_2)^{n-1}.$$

Similarly, a musical octave (*diapason*) was compounded out of intervals of the fifth (*diapente*) and the fourth (*diatesseron*): the compounded $(12/6) = (4/3)$ "added to" $(3/2)$. In the case of continuously proportional terms, this "addition" amounts to raising to powers, and "subtraction" is a matter of taking roots. Thus a "double ratio" is equivalent to squaring the ratio, while a subdupled ratio consists of extracting its square root.

12.6   The Merton School and the School of Paris

In *De proportionibus*, Thomas Bradwardine (ca. 1290–1349) at Oxford began in 1328 the representation of ratios, in his case for speeds, as what we identify as exponentials. In inventing a new "function" based on compounding ratios, he intended to remove an anomaly in scholastic interpretations of Aristotle's quantitative descriptions of motion. Aristotle postulated ratios relating forces (F_i), resistances (R_i), and velocities (V_i) of motion in *Physics* IV: V_i may be symbolized anachronistically as $V_i \propto F_i/R_i$, that is, $V_i = kF_i/R_i$. The formulation implies against all experience that when force and resistance are balanced ($F_i = R_i$) some value of movement ($V_i = k > 0$) occurs. In reformulating the Aristotelian generalization, Bradwardine asserted that speeds V_i change arithmetically, and the ratios of force to resistance (F_i/R_i), which determine the series of speeds, do so geometrically. In modern form, his dynamical law has $(F_1/R_1)^{V_2/V_1} = F_2/R_2$. In the case when $a = F_1/R_1$, $V = \log_a F/R$. Although physically erroneous, "Bradwardine's function" eliminated the anomaly (when $F/R = 0$, $\log a = 0$) and was the first exponential statement of a physical law. On strictly linguistic grounds this brilliant stroke seemed more faithful to the original text than the medieval alternatives. Beyond that it spurred further applications of mathematics to studies of motion, particularly instantaneous rather than actual or completed motion.

Bradwardine conducted his studies at Merton College, Oxford, beginning in the 1320s, and in the three following decades his followers, the Mertonians, by widely employing his language of proportions, became known as calculators. Applied to qualities and motions, the vocabulary of *intensio* and *remisio* measures variation across a given range (*latitudo*). It had originated in pharmacology; there numbered degrees (*gradi*) measured the body's health and the level of its humors.

Mertonians began by establishing more exact definitions for distributions of qualities or motions in a subject or across time. Their terminology classifies distributions as uniform or difform. The latter are subdivided into uniformly difform, varying at a constant rate, and difformly difform. Drawing again on the procedure of denomination, Mertonians devised rules so that a difform change when appropriately measured by the mean degree of its motion is equivalent to a uniform change. This technique had ramifications in physics. The Merton mean speed theorem in William Heytesbury's version gives a formula to measure the distance covered by a uniformly accelerating body. The distance is the same as would be traversed by another body moving at the constant velocity of the middle instant over the same period. Merton's mean speed theorem, which was developed in an arithmetical and logical fashion, likely influenced Galileo's formulation of the law of freely falling bodies, which he proved geometrically.

By treating degrees and latitudes as continua, the calculators reached a clearer understanding of infinitesimals and converging and diverging infinite series. This was rather a novelty, for the Schoolmen believed that the ancient Greeks had adopted a *horror infiniti*. By similarly applying the language of proportions to magnitudes broken into their proportional parts with corresponding changes in intensities, the Mertonians moved toward a concept of limit. Walter Burley's representative *On the First and Last Instant [of a Thing's Being or Non Being]* sharply distinguishes between extrinsic and intrinsic limits, insisting that a limit must be one or the other.

The most ingenious medieval Latin treatment of limits was in the *sophismata* (intricate logical puzzles) of Richard Kilvington and William Heytesbury. Besides considering complicated variants of Zeno's paradoxes involving footraces between Plato and Socrates and the instantaneous occupation of a given place by a moving body, they compared changes occurring at different rates with respect to limit points. A thought experiment by Richard Swineshead (fl. ca. 1350), called Calculator, measures the total heat of a body heated to degree 1 in the first half of an interval, then through degree 2 over the next quarter, degree 3 over the next eighth, and so on. This amounts to the infinite series $1/2 + 2/4 + 3/8 + \ldots + n/2^n$. Swineshead's verbal demonstration that the denomination of the heat of the whole is 2 degrees assumes that the sum of Zeno's geometric series $1/2 + 1/4 + 1/8 + \ldots + (1/2)^n$ equals 1. By regrouping his series, he obtains $1 +$ Zeno's series or 2. Swineshead then considers a body's heating and simultaneously expanding in every $2^{n\text{th}}$ proportional part, such that each proportional part expands twice as slowly as its predecessor. His denomination of the whole as infinity recognizes that the series $2^{2n-1} + 2^n$ diverges, because it cannot be summed. The complexity of his calculus of language made it inaccessible to Renaissance men of letters.

In fourteenth-century France the School of Paris succeeded that of Chartres. Two Parisian Schoolmen who theorized more about physical problems than the English and by 1350 had developed the Mertonian logicomathematical studies were Jean Buridan (ca. 1300 to ca. 1358) and Nicole Oresme (ca. 1320–1382).

Perhaps the examination of projectile motion in catapults and bows—cannon would come later—brought more attention to the subject of motive power or impetus. Buridan, like Bradwardine, initiated important lines of inquiry in physics, his novel, quantified doctrine of impetus resembling in certain aspects Newtonian inertial mechanics. His concept of impetus was quantity of matter times velocity, essentially, momentum. Buridan also clearly states a concept of limit: he divides the continuum of resistances that impede motions into movable and nonmovable classes at the point where the resisting force equals the moving forces.

Oresme was the most able fourteenth-century mathematician. A prominent churchman who became bishop of Lisieux, he was a royal advisor and pioneer monetary theorist. An authority on ancient Greek mathematics, he also translated into French works of Aristotle and perhaps Ptolemy. Today Oresme is known for his mathematical innovations.

Oresme's *Algorismus Proportionum* systematically introduces operational rules for multiplying and dividing ratios having integral and fractional exponents and provides steps toward more efficient notation. His *De Proportionibus proportionum* includes irrational exponents. In his system two ratios A and B can be related so that $A^n = B$. Exponent n is usually irrational. Oresme reformulates the terminology of Bradwardine, who had confused or left ambiguous arithmetic and exponential expressions. Oresme gives the terms "parts," "multiple," and "commensurable," all taken from the Pythagorean definition of proportion, a consistent exponential interpretation. Symbolically, A/B is part of C/D when $A/B = (C/D)^{1/n}$; A/B is commensurable with C/D when $A/B = (C/D)^{n/m}$ and n and m are integers. In Oresme's system $2/1$ is a *third part* of 8 and a *fifth part* of 32, that is, in our notation $(2/1)^3 = 8$ and $(2/1)^5 = 32$; $32/1$ is *five thirds part* of $8/1$: that is, $32/1 = (8/1)^{5/3}$.[367]

Oresme wrote *On the Ratio of Ratios* to refute the mathematical foundations of astrology. As teacher and later advisor to Charles V, he regarded astrologers at the French court as dangerous. This antiastrological treatise ends with an argument from probability that $A/B = (C/D)^{n/m}$, where n and m are integers. Consequently no two celestial bodies can ever precisely return to their original alignment in the sphere that marks a planetary conjunction: the exact alignments on which astrological predictions depend are mathematically impossible. The argument no more deterred astrologers than Zeno's did footracing.

On the Configuration of Qualities and Motions, appearing in the 1350s, is Oresme's second major contribution to medieval mathematics. To represent motions geometrically he plots by primitive graphing known as the latitude of forms, which are shown here, their intensities or velocities over extensions or time. In his graph the perpendicular is the intensity and the base line the time. Oresme relates areas under curves to distance traveled.

Enlarging on earlier usage but faithful to it, he develops a vocabulary. "Latitude" indicates intensity of a motion, "longitude" its extension. Scaled perpendiculars erected over a base line (= extension or time) show intensities point by point or speeds instant by instant. The entire figure gives the quantity of velocity while its *configuratio* indicates distribution of intensities. Rectangles represent uniform qualities or motions, right triangles show uniformly difform distributions from zero degree and trapezoids from some given degree. Stair steps and ellipses graph two varieties of difformly difform distributions. Oresme's geometric demonstration of the Merton mean speed theorem is simple. The right triangle BAC illustrating the distance covered in uniform acceleration from zero at B to a final velocity CFA is equal in area to the rectangle $BGFA$ representing uniform velocity at BG (part of the figure). Oresme's *configuratio* helped break down the bias against combining nonhomogeneous qualities in a multiplication: for example, velocity × time = length. It also foreshadowed analytical geometry in a limited way.

In *Questions on the Geometry of Euclid*, a mid-fourteenth-century work, Oresme surpassed Mertonian calculators in distinguishing convergent from

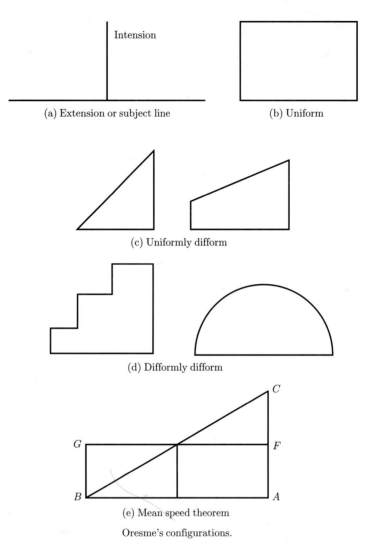

(a) Extension or subject line   (b) Uniform

(c) Uniformly difform

(d) Difformly difform

(e) Mean speed theorem

Oresme's configurations.

divergent series and in creating general summation procedures. He recognized that, if an infinite series is formed by adding to a given magnitude an infinite number of proportional parts in such a way that $a/b < 1$, then the series has a finite sum, while if $a/b > 1$, the series diverges. Oresme's summation of the geometric series $1 + 1/3 + 1/9 + 1/27 + \ldots + 1/3^{n-1} + \ldots = 3/2$ seems to imply a general procedure for summing all series of the form $1 + 1/m + 1/m^2 + 1/m^3 + \ldots + 1/m^n + \ldots$.

These late medieval treatments of infinite series were an important intellectual preparation for calculus. But not even the gifted Oresme could resolve the difficulties of the calculus of ratios that substituted for Eudoxean proportion. Unwieldy language unaided by adequate symbols made the medieval calculus esoteric and prey to copyist's error. Mediate and immediate denominations were ingenious circumlocutions but lacked rigor. Oresme himself shied away from applying ratios of ratios to cases of "minor inequality," that is, to a/b where $a < b$, lest he violate Euclid's axiom that a part is smaller than a whole. The mathematical materials were entwined in a web of scholastic lucubrations that were about to become obsolete.

12.7   The Fourteenth-Century Crisis and Recovery

The critical, skeptical tone of later scholasticism attended religious disarray in late thirteenth-century Europe. In 1277 Church authority condemned Aristotle, and for the moment western Europe lost confidence in the magisterium of that philosopher and the encyclopedic *Summa*. Beauvais's 160 foot Gothic cathedral, that summa of high medieval architecture, collapsed in 1284. In 1291 Muslims recaptured Acre north of Haifa, the last Crusader stronghold in the Middle East, and a new series of internal wars resumed.

Demographic disaster soon followed. At the century's turn Latin Europe experienced price inflation and severe fluctuations in weather: twenty bad harvests occurred between 1302 and 1348. Typhoid fever and animal diseases wasted the Low Countries. Chronic bad harvests forced people to eat seed grain and postpone marrying. That and perhaps amenorrhea caused by famine brought a sharp drop in the birth rate. Weakened by malnutrition, Europeans were susceptible to plague in bubonic and pneumonic forms. After the Genoese navy liberated the Straits of Gibraltar in 1291, Genoese and Venetian ships under improved rigging carried to Atlantic ports year-round cargoes—but soon bore black rats with infected lice as well. The Black Death, so named in the fifteenth century, reached Sicilian Messina from the Crimea on Genoese grain ships in October 1347 and thereafter struck recurrently until the seventeenth century. The estimated drop in European population between 1300 and 1450 ranges from one-half to two-thirds: plague killed perhaps a third. In 1349 Thomas Bradwardine, now Archbishop of Canterbury, was a victim.

In this crisis dynastic successions became uncertain and factional wars more likely. Failure of the Crusade increased at home an army of mercenaries. As expensive replacements for feudal vassals, they brought to warfare severe financial and disciplinary problems. Aided by Flemish, Gascon, and Burgundian factions, England fought France in the worst of these conflicts misnamed the Hundred Years' War: between about 1338 and 1453 occurred a sum of forty-four years of formal war. At issue were the Flemish textile industry, England's continental territories, and a disputed succession to the French throne. In the war's final phase, from 1415 to 1453, the French almost capitulated but rallied when Joan of Arc in 1429 broke the siege of Orleans.

Factionalism also divided the Latin Church. In response to unsettled political and economic conditions in the Italian Papal States, Clement V transferred the papacy to Avignon in southern France, where it remained from 1309 to 1377. Italian partisans, including the poets Dante and Petrarch, tax-weary subjects, and moralistic historians labeled it a "Babylonian Captivity" after the biblical exile of the tribes of Judah in 587 B.C. When the ailing Gregory XI restored the papacy to Rome, dissension worsened. At his death a Roman and an Avignonese claimant to the papacy emerged, and in 1378 the Great Schism, lasting until 1417, began. To resolve the dispute, reformists of the Conciliar Movement urged centering the Church government in a General Council of Bishops. The Council of Pisa in 1409 voted to depose both claimants and elected a third contender, which only compounded the problem of succession. The Council of Constance, beginning in 1414 and sitting for four years, deposed the Avignonese and Pisan claimants and elected a Roman as Pope Martin V. Until its fragmentation, the Conciliar Movement seemed to represent an emerging parliamentary counterforce to papal monarchy.

Theological controversy and reformist zeal brought the Church even more serious problems. At Oxford the Dominican John Wycliffe (d. 1384) questioned transubstantiation in the Eucharist along with clerical and papal authority. He sought to abolish pilgrimages, veneration of saints, and the church's ownership of property and to give responsibility to a laity provided with vernacular Bibles. English followers called Lollards (prayer mutterers) probably carried his doctrines to the Bohemian court, where they apparently influenced the theologian John Hus, rector of the University of Prague. Hus, summoned to the Council of Constance with an imperial guarantee of safe conduct, was nevertheless burned at the stake in 1415 after his condemnation. Outraged by this execution, Bohemian partisans seceded from Latin Christendom and from 1421 to 1436 defeated all foreign armies that intervened. A compact ending the war excluded Germans from Bohemian matters and recognized a Hussite Mass. In southeastern Europe, the fall of the Byzantine capital at Constantinople to the Ottoman Turks in 1453 increased the external peril from that quarter.

Yet even amid western Europe's troubles, commercial interdependence continued; and often one region's decline was offset by another's ascendancy. The commerce of southern Flanders, for example, passed to Brabant and Antwerp. Agricultural contraction forced landlords to be more productive in their use of land, labor, and implements. Shortages of skilled labor also prompted industrial employers to substitute capital in the form of better tools. Commercial instruments and navigational aids spread more widely and presaged another overseas expansion.

During the unsettled period despots, such as Gian Galeazzo Visconti (d. 1402) in Milan, took control of many Italian free cities. Milan, Venice, Florence, the Papal States, and Naples each became the center of a regional metropolitan state replacing the free communes. Genoa and Venice fought a series of costly trade wars. By 1443 Aragonese controlled Naples and Sicily. But this turmoil in the Italian peninsula also had a creative side. Italy became the cradle for humanism and the Renaissance (see Chapter 13).

An ex-Bolognese law student, Francesco Petrarca or Petrarch (1304–1374) castigated logical-legal scholasticism as cultural barbarism. He revived the Ciceronian rhetorical tradition of early imperial Rome. With his fellow Florentine Dante Alighieri (d. 1321) and Giovanni Boccaccio (d. 1375), he made Tuscan the literary dialect of the Italian peninsula. Petrarch, who wrote long letters to Cicero, portended the movement called humanism. Humanists oriented the curriculum on the trivium. They replaced logic with history and moral philosophy and exalted poetry and Latin literature. Italian humanists called Petrarch's new age the *Rinascita*, Renaissance or rebirth. Historians accept their term, which covers the period from the fourteenth into the sixteenth century, but question its aptness, its temporal and geographical boundaries, and especially its cultural break with a gothic "middle ages."

Florence was the home of humanism and center for its diffusion to other Italian courts and the rest of Europe. The threat that despotic Milan posed to Florentines, which became acute in the 1380s, probably helped to forge the distinctive characteristics of these republicans. The Florentine bourgeoisie exercised a degree of political power uncommon for the time. Its interests allied with high scholarship in defending the republic. Civic humanists encouraged study of the classics to build character: scholarly chancellors promoted civic responsibility to maintain the republic's strength. The Swiss historian Jakob Burkhardt was to argue that in place of medieval corporatism, these conditions nurtured self-confidence and independent personality. It was a concept of human dignity in keeping with that in the highborn Florentine Pico della Mirandola's oration of 1486. Leonardo Bruni, a chancellor (executive secretary of state) of the city, reintroduced the Ciceronian word *humanitas* to indicate the civilizing study of ancient learning. Light and landscape gave Florence a rare beauty enhanced, at mid century, by superlative artistic monuments, such as Filippo Brunelleschi's dome of the cathedral, Lorenzo Ghiberti's bronze doors for the baptistry, and Donatello's sculptures. Underwriting all this required wealth. Shrewd Florentines, who lacked access to major trade routes, took control of papal banking and tax collecting and by the end of the thirteenth century led Europe in international banking. Florence also manufactured fine wools.

The Medici banking family led the bourgeoisie in providing political stability and a remarkable patronage of art and scholarship. Cosimo de Medici, who came to power in 1434, and his grandson Lorenzo (d. 1492) were innovative diplomats in Italian balance of power politics. From Florence, humanism spread through Italy, favored by the density of the peninsula's urban population: more than one quarter in the fourteenth century.

Humanists with a taste for the classical Mediterranean past laid foundations for the modern disciplines of philology, archaeology, epigraphy, and numismatics. They encouraged translations from direct Greek sources. Familiarity with Latin classics was a prerequisite for civic virtue; study of Greek literature, especially the *Dialogues* of Plato and his Neoplatonist successors, deepened appreciation of beauty. Florentines flocked to learn from the Greek scholar Manuel Chrysoloras, who in 1397 occupied a new chair of Greek. A Platonic

Academy translating the Platonic corpus into Latin formed around Marsilio Ficino, engaged at Cosimo's request. Neoplatonic philosophy with its concept of resonance between the mind and the metaphysical realm suggested some transcendent spiritual force in humanity seeking perfect articulation in conduct and art. Since mathematics was more important to Plato than to Aristotle and since students now studied Greek mathematics directly and in context, mathematics benefited from the resumption of Platonic studies. Although Aristotle was no longer "the Philosopher," knowledge of him gained from examination of primary sources and new translations. At Padua, where Aristotelianism remained vigorous, peripatetic methodology subsequently contributed to the beginning of the scientific revolution in the late fifteenth century. Yet while humanist interest in Greek as well as Hebrew was hearty, it nowhere matched devotion to Latinity nor was it as strong as that for French imaginative literature.

Across the Alps scholarly humanists made common cause with Italian civic humanists. Both groups opposed scholasticism, and both collected manuscripts, often assisted by refugee Byzantine scholars. Acknowledged prince of the northern humanists was the itinerant scholar Desiderius Erasmus of Rotterdam (1466–1536). Erasmus, along with the German humanist Johann Reuchlin, had been educated in schools of the Brothers and Sisters of the Common Life, a new group dedicated to a communal devotional life and labor in the urban workplace. In *The Praise of Folly*, published in 1511, Erasmus calls for accuracy in the reading and interpretation of texts. Beyond that, the book is an appeal to tolerance, moderation, and reasoned doubt. The work of Erasmus and other northern humanists contributed to preparing the ground for the Protestant and Catholic Reformations, in which many of them took active roles. Printing was pivotal for the spread of both cultural and religious reforms.

Enthusiasm for their cause, however, flawed the humanist critique of scholasticism. Proclamations about the dignity of man glossed over the economic and social decline of women since the high Middle Ages. As a rule, scholastics had been more hospitable to natural philosophy than these activist litterateurs. Humanists, who complained about the baroque vocabulary and aridity of scholastic disputation, created their own formalisms and literary conceits, as in their florid dedications. Increased attention to historical context did not preclude historical slights and injustices. The preface to Erasmus' translation of the New Testament is condescending to women, Scots, and Irishmen. Nor did humanists strictly observe intellectual integrity. Church Fathers had found a biblical warrant to pirate pagan sources; humanists sometimes raided medieval sources while claiming their originality. Niccolo Tartaglia, who was outraged at the plagiarism of his algebra, did not hesitate to borrow extensively from Jordanus without credit.

Artists joined their scholarly humanist Italian compatriots in claiming the inauguration of a new age. Renaissance art, which developed focused linear perspective techniques, portraiture, naturalistic landscapes, and sculpture, makes up a major part of the West's artistic patrimony. Focused perspective painting holds a minor place in the history of mathematics. Dependent on angular measurements keyed to a vanishing point and a distant point, it achieves great geometrical clarity. But it is also static. Brunelleschi perhaps read Ptolemy's rules for spherical projection in the *Geography*, which Manuel Chrysoloras had brought to Florence. In any case, about 1425 he painted the first linear perspective scene since antiquity. Tomaso Masaccio in 1428 composed his *Trinity* as if the observer views the scene from a window sill at eye level of five feet ten inches. The treatise on painting by Leon Battista Alberti (ca. 1404–1472) gives precise, detailed instructions for drawing in linear perspective. Perspective painting is one of many instances during the Renaissance of a mating of the artistic and the scientific imagination.

During the second half of the fifteenth century while Italians produced a brilliant intellectual and artistic culture, voyages of discovery brought new prominence to the Atlantic states. The search for souls to be converted, the pursuit of new continents or legendary cities, the quest of trade routes, and the desire for governmental control of scarce resources prompted exploration that was both predatory and visionary.

Prince Henry the Navigator of Portugal, who organized astronomical and geographical expeditions, was the first to find profit in the Atlantic exploration. After capturing Ceuta in north Morocco in 1415, Prince Henry's ships edged down the west African coast. An interest in Indian spices, then a much broader category than now, grew out of success in trading African gold and, after the 1440s, slaves. In 1498 Vasco da Gama reached the Malabar coast of India, breaking the Muslim spice monopoly. His visit yielded a cargo of spices worth sixty times more than the cost of his expedition. In 1492 the Genoese Christopher Columbus, seeking in the service of Spain to reach the Orient by sailing west across the Atlantic, stumbled upon the Bahamas. The consequent revelation of a new continent sharpened the appetite for exploration. After the crew of the Portuguese Magellan (d. 1521) circumnavigated the globe, Spaniards realized both the vastness of the Pacific and the applicability of capitalism and technology to the precious metals of the Americas. Gold and silver flowing through Spain, Antwerp, and later Amsterdam contributed to scientific development, an armaments boom, and a steep inflation from 1475 to 1600. European diseases and enslaving of Indians in the mines had meanwhile devastated indigenous American populations within two generations.

Strong monarchs consolidated and expanded kingdoms along the Atlantic, while an empire emerged in eastern Europe. The union of Castile and Aragon in the marriage of Isabella to Ferdinand in 1469 and their policy of reducing the power of great nobles and ecclesiastics made for a strong and centralized Spanish kingdom. In France, where Louis XI during his reign from 1461 to 1483 used almost half the budget for a standing army, expansion came by invasion and absorption of Burgundian and Angevin territories. With his improved army, Louis XI, known as "the spider king" for his treachery, reduced aristocratic brigandage and probably laid the foundation for French absolutism. By balancing the budget without increasing taxes, the Tudor Henry VII restored the prestige of English royalty after the dynastic War of the Roses. He thus weakened the nobility and established order at the local level. All these monarchs cultivated political power through a kind of efficiency, opportunism, and secrecy that the exiled Florentine Niccolo Machiavelli in 1513 analyzed in *The Prince*. Ivan III , who was tsar from 1462 to 1505, made Muscovy independent of Mongols and Lithuanians and created a vast empire, using cannons and the river system to good advantage. Proclaiming Moscow the New Constantinople, Ivan attempted to reclaim Byzantine imperial and religious authority.

12.8   Late Fourteenth- and Fifteenth-Century Mathematics

Largely free of the animus against scholasticism that characterized much of humanist literature, mathematics in the late fourteenth century and the fifteenth advanced in a number of ways. Investigations in the logical form of mathematics continued, especially in some Italian universities, while curricula in the new central European universities emphasized mathematical preparation. Humanists uncovered lost Greek works and encouraged study of all ancient Greek mathematics. Expanding commerce stimulated wider reading and more efficient computation, particularly in Italy. Algebraic texts improved exposition and notation, and algebraic methods were used more often in solving geometric problems. And in attempts to align the calendar and refine astrological forecasting, astronomers freed trigonometry from spherical astronomy and launched its independent development in Europe. In the fifteenth century, mathematicians were increasingly sought in court circles, and instrument makers and cartographers put practical mathematics to proficient uses. Johann Müller (1436–1476) or Regiomontanus (the Latin rendering of Königsberg, his birthplace) was the leading mathematician-astronomer. He led in disseminating mathematics by the new print technology. Luca Pacioli's *Summa* of arithmetic and algebra gathered all these materials.

Study of Mertonian and Parisian scholastics proceeded at Padua, Bologna, Pavia, and Lisbon. The Italian connections had been formed early and continued into the scientific revolution: Giovanni di Casali at Bologna had anticipated Oresme in applying coordinates to kinematics. In the established universities, the quality of instruction in quadrivial mathematics began to decline. But new universities at Prague, founded in 1347, Vienna, refounded in 1383, and Krakow, established in 1397, addressed mathematical comprehension in their curricula and graduated able mathematicians. Some fresh mathematics surfaced, such as the first six books of the *Arithmetic* of Diophantus, which Regiomontanus discovered.

Manuscript libraries of ancient Greek authors and late sixteenth-century translations produced a nearly complete recovery of Greek higher mathematics. Pope Nicholas V, who planned the Vatican library, commissioned James of Cremona to retranslate the writings of Archimedes and three Eutocian commentaries. Cardinal Nicholas of Cusa (1401–1464), a major German philosopher active in reform of the calendar, tailored the Jacobus-Archimedean quadrature proofs, completed by 1453, to agree with his doctrine of "coincidence of opposites." He held a straight line and an infinite circle to be indistinguishable. Perhaps from Arabic sources, he computed an accurate value of π. Influenced by the scholastic calculus of ratios and Renaissance Neoplatonism, Cusa applied to mystical theology the paradoxes of infinity. His portrayal of the universe as having all points be the center suggests an infinite universe. The more able Regiomontanus not only followed the mathematics in Jacobus's manuscript but corrected its errors. His manuscript was the Latin source for the *editio princeps* of Archimedes, the definitive Latin translation completed in 1544.

At mid-*quattrrocento* humanists were avidly supporting manuscript copying. It is reported that Cosimo de' Medici employed forty-five scribes to produce two hundred volumes in twenty-two months. Although some modern historians question the accuracy of the data, such incentives encouraged engravers and other artisans to experiment. They investigated alloys for casting dies, additives to inks, adaptations of xylography (block printing), and rag paper manufacture, both of the latter originating in China. Copyists had replaced expensive vellum skins with paper made from linen rags. At Mainz about 1450 Johann Gutenberg (d. 1468), Johann Fust, and Peter Schofer organized an industry for reproducing books by printing with moveable type. Once the art of printing was thus invented, skilled German artisans quickly spread it. Although the first printed books were expensive and their quality generally poor, the printing industry grew phenomenally. By 1500 printers were working in more than a hundred locales and had produced at least six million copies of some 40,000 editions. Among the centers were Mainz, Strasbourg, Florence, Venice, Budapest, Antwerp, Krakow, and Paris. Religious books were most in demand, followed by ancient classics led by Cicero. The first printed book was Gutenberg's famous Bible on parchment. Luther later termed printing "God's highest act of grace." The coming of typography profoundly affected mathematics, as it did literature, religion and the other sciences.[368]

Although the first printed mathematical texts were not original, they contributed to learning by reducing copying errors. Not all scholars would have predicted that. Leone Battista Alberti warned that setting numerals in type would increase copying errors and urged authors to write out numbers in full, and Regiomontanus judged the proliferation of old errors through new print technology the single greatest threat to mathematical astronomy. Against convention, he set careful standards for proofreading before a final press run. Accurate, clearly drafted diagrams improved the learning of geometry.

The first printed arithmetic, known as the *Treviso Arithmetic* from its place of publication north of Venice in 1478, testified to the Italian lead in commercial arithmetic. Its anonymous author lists rules of computation for young people seeking to enter commerce. This tradition had long existed in Italian cities. A Florentine chronicle from the 1320s reports that between a thousand and twelve hundred children (from a population of about ninety thousand) were learning the *abacus* and *algorismus* in six schools. To Venice came Germans, especially Nurembergers, for instruction from the *maestri d'abaco*. Shortly after 1500, German *Rechenmeisters* and their commercial arithmetic texts became leaders. The *Treviso Arithmetic* also stimulated the publication of mathematical works in north Italian cities so that by 1500 two hundred printed mathematical texts appeared there, including Campanus of Novara's edition of Euclid's *Elements* with commentary, published in Venice in 1482.

While Italian *cossisti* circulated Leonardo of Pisa's algebraic methods, Nicolas Chuquet (fl. 1475), a Parisian physician in Lyons, wrote an original manuscript that improved numeration and notation. Except for a plagiarized section in 1520, it was not published until 1880 and then only the section Chuquet called "Triparty." Chuquet's mathematical style is at once simple and highly abstruse. Contrary to prevailing understanding, he takes "number" to include rational number, roots, and sums and differences of roots. He gives clear rules for decimal computation and divides large numbers into groupings of six figures. In British usage, to the number million he adds billion (10^{12}), trillion (10^{18}), and quadrillion (10^{24}). The terms for root or number and for second, third, and fourth roots and so forth replace geometric designations of square and cube roots. Like his predecessors, Chuquet considered negative numbers absurd. His apparently new notation facilitated performing operations with zero and negative numbers and addition and subtraction of exponents. Where we write 4, $5x$, $7x^3$, in "Triparty" he writes $.4.^0$, $.5'.$, $.7.^3$. Our $-12x$ and $-\sqrt[3]{12 \times x^3}$, he renders m $.12'.$ and m $R_x 3.12.^3$, where m stands for m[inus] or (-1) and R_x for r[adi]x or $\sqrt{}$.

Despite weakness in number theory in a premodern sense, a refusal to accept negative numbers, and a method inferior to Campanus's for division into mean and extreme ratios, the "Triparty" was insightful. It presaged and possibly influenced Rafael Bombelli's algebraic notation of 1572. Chuquet's book, like the work of Regiomontanus and Pacioli, confirms an increasing use of algebra for solving geometric problems. Historical misfortune distanced Chuquet from the mainstream of mathematical activity. Had the secondary author copied the entire text and not made misleading additions, the plagiarized version might have revealed his skill.

In 1460 a Greek emigré, Johannes Cardinal Bessarion, arrived in Vienna as papal legate. His coming shaped the careers of Georg Peuerbach (1423–1461) and his more talented student and Viennese collaborator Regiomontanus, whose early deaths—neither lived through a forty-first year—were a double misfortune for German mathematics. Bessarion's personal manuscript collection became the foundation of the Marciana library at Venice.

Bessarion pressed Peurbach for a more reliable translation of Ptolemy's *Almagest* than Gerard of Cremona's. He also urged a serviceable abridgement for students. Peurbach, thoroughly versed in the Gerard's version of the *Almagest* but not competent in Greek, began the abridgement. On his deathbed in 1461, he entrusted its completion to Regiomontanus, having finished only the first six books. Regiomontanus learned Greek thoroughly, in part from Bessarion, whom he accompanied to Italy. The *Epitome*, as it was called, of Ptolemy was completed before mid-1463 and printed in 1496. Regiomontanus was largely responsible for the book's accurate translation of Ptolemy and correction of his lunar theory. The *Epitome* would be the fundamental text for Ptolemaic astronomy until Kepler and Galileo. Their labors on it had convinced the two authors of the need for a comprehensive treatment of ratios between sides and angles in plane and spherical triangles. Reading it, a young student named Copernicus at Bologna was struck by errors in Ptolemy and worked on a new astronomy.

Association of the study of triangles with astronomy dates to Hipparchus, Menelaus, and Ptolemy, as well as to Indian and Arab sages. Indian scholars comparing half-chords to radii had been the first to recognize the sine relationship and subsequently obtained tangent lines. About 1126 Adelard of Bath, drawing on al-Khwārizmī, described tangents and drew up a table of sines. It was probably his work that first acquainted the West with sines and tangents. Following a misrendering of a phonetically transliterated Sanskrit term for chord in some Arabic sources, Plato of Tivoli apparently introduced *sinus* (fold, bosom) into Latin mathematical vocabulary. Adelard introduced the tangent called an erect or rotated shadow (*umbra rectus, umbra versus*) and Campanus prepared a table of tangents. Richard of Wallingford (fl. 1320), abbot of St. Albans, included them in his four part treatise on sines. In Provence Levi ben Gerson in 1321 was the first to state clearly the sine law in Latin. Put into symbols, it declares $a/\sin a = b/\sin b = c/\sin c$. Gerson proved this law, which was part of his procedures for solving right and general triangles. He utilized it, for example, when two sides and an opposite angle are known. Gerson also constructed a table of sines at intervals of $15'$. His sexagesimal expression of the sine of $15'$ is correct to seven decimal places. Thus a store of trigonometric materials existed, but scattered in astronomical manuscripts of limited circulation it did not form an intact corpus. The library sources of Peurbach and Regiomontanus had serious gaps, and no European yet knew the advanced East Arabic trigonometry by al-Tūsī and al-Kāshī.

Regiomontanus' greatest work, *De Triangulis Omnimodis (On Triangles of Every Kind)* in five books, completed in 1464 and printed in 1533, sums up the known work of his predecessors on chords and arcs and systematically studies problems related to plane and spherical triangles modeled on Euclid. Book II states the sine law in the language of ratios and gives the first algebraic solutions in Latin by quadratic equations for finding lengths of the sides of a triangle. After setting out the law of sines for spherical triangles in Book IV, Regiomontanus gives the law of cosines for spherical triangles its first practical formulation. Symbolically, $\cos A = (\cos a - \cos b \cos c)/\sin b \sin c$ or $\cos A = \cos b \cos c + \sin b \sin c \cos a$.

De Triangulis, which also treats the versed sine, does not use the tangent. In 1467, however, Regiomontanus in *Tabula Directionum*, first printed in 1490, computed a table of tangents. His earlier tables of sines had sine $90° = 6 \times 10^4$ or $6\ 10^6$ to agree with sexagesimal intervals of minutes and seconds. Setting sine $90° = 10^5$, he now recognizes the superiority of decimal notation. Calling tangents *numeri,* Regiomontanus with his decimal notation pioneered modern trigonometric tables.   But he did not admit negative values of cosines and tangents for obtuse angles and, like Chuquet, he rejected negative numbers. In practice *De Triangulis* separates trigonometry from astronomy. Another work by Regiomontanus, *Ephemerides Astronomicae,* which has no trigonometric tables, introduces a method of lunar distances for determining longitude at sea and was with Columbus on his fourth voyage to the new world.

Regiomontanus had settled in Nuremberg because of its instrument makers.   He made that city the site of the first European astronomical institute fitted with observatory and machine shop.   Since existing publishers avoided investing in scientific writings, particularly those requiring expensive diagrams, Regiomontanus also set up a printing press at his home in Nuremberg in 1471 and 1472 and projected publication of forty-seven works mainly in astronomy and mathematics. He sought to remove scribal and typographical errors from books and applied German letters that were simplified, rounded, and Latin-like for clear legibility. Regiomontanus's *Index* for his publishing project lists works by Ptolemy, Archimedes, Apollonius, Jordanus, Campanus-Euclid, Witelo, Peurbach, and himself. His death, likely from plague that spread after a Tiber flood, came while he was conferring with Vatican officials in Rome about reforming the ecclesiastical and Julian calendars. A sensational rumor asserted that the sons of George of Trebizond, whose commentary on Ptolemy's *Almagest* he planned to criticize extensively, had poisoned him.

Regiomontanus's *Index* is on a different level from the *Summa de Arithmetica, geometria, proportioni et proportionalita* of Luca Pacioli (1445–1517), completed in 1487 and printed in 1494. This six hundred page survey covers arithmetic, algebra in syncopated notation, elementary geometry, and Venetian double-entry bookkeeping, for which it became the international guide. Despite its vernacular Italian and its emphasis on problems, it belongs within the medieval encyclopedic tradition that prepared for the great expansion of mathematics in the next century. The *Summa*'s conclusion, that for cubic equations that today would be rendered $x^3 + mx^2 + nx = p$ a general solution is as impossible as squaring the circle, roughly ends medieval mathematics.

Between the pontificate of Gerbert and the encyclopedia of Pacioli, mathematics, like Europe, had changed greatly. A Franciscan friar, Pacioli was the major authority in capitalist bookkeeping methods and the first mathematician to have his portrait painted. Pacioli, who resided with Alberti and Leonardo da Vinci and taught mathematics at two Renaissance courts, lived at the edge of changes more sweeping than Scipione's algebra. In 1513 Machiavelli circulated the manuscript of his powerful argument that private and public moralities are incommensurable; in 1517, the year of Pacioli's death, Luther posted his reforming theses.

Part IV

The Transition to Modern Mathematics in Europe

13

The First Phase of the Scientific Revolution, ca. 1450–1600: Algebra and Geometry

> Tis all in pieces, all coherence gone
> All just supply, and all relation.
> . . .
> For of Meridians and Parallels,
> Man has weaved out a net, and this net thrown
> Upon the Heavens, and now they are his own.
> — John Donne, "Anatomy of the World:
> The First Anniversary"

Between roughly 1450 and 1600 Europeans moved unmistakably to the fore in both theoretical and practical mathematics. Algebra and to a lesser extent computational arithmetic, trigonometry, and geometry were now imaginatively developed.

Sixteenth-century algebra registered two seminal achievements. First came the discovery by Italian mathematicians of general arithmetical solutions for cubic and quartic equations by means of radicals. This broke the thousand-year dominance of geometric algebra, which solved these equations by using conic sections. In 1545 Girolamo Cardano published both arithmetical solutions in his *Ars magna*, which reoriented mathematics to the more powerful techniques of algebra. In the nascent theory of equations, syncopated algebra was inadequate to the formulation of problems and their solution. At this juncture algebraists, most of them Austrian, German, or Slavic, methodically improved algebraic notation. They were known as cossists, a name derived from the Italian *cosa* or "thing" that had come to stand for the unknown. Printing freed them from the relative isolation of earlier mathematicians, in which the development of notation had been fitful and extremely slow. Their transitional work culminated in the other major accomplishment, the pioneering symbolic algebra of François Viète. Finding roots of higher-order equations also broadened the realm of number by producing imaginary numbers.

Often in close connection with other fields, theoretical geometry evolved slowly during the sixteenth century. Studies of perspective by artists in addition to geographers and cartographers such as Gerardus Mercator produced the beginnings of projective geometry. A more important development is that zealous humanists rediscovered, reconstructed from fragments, and ably translated masterworks of Hellenistic higher geometry, especially by Apollonius and Archimedes. Among others, Francesco Maurolico and Frederigo Commandino carried out the recovery program. A Euclid revival also produced more exact translations of the *Elements* from Greek to Latin and vernacular languages. By 1600 print technology, which appears to have had a slower effect on mathematics than on natural philosophy, was making higher and Euclidean geometry more accessible. Eventually the introduction of proofreading corrected the bane of scribal drift.[369] This particular contribution of printing to mathematics, in which accurate transmission of the smallest detail is most important, went with the more general effects that print and paper had on the life of literature and scholarship: the standardization of vernacular languages, the creation of an intellectual community across Europe, and the movement away from the Scholastic premium on memorization and toward critical reading.

During the sixteenth century the nature of mathematics was changing and after 1550 its role in university education expanded. As in late ancient Alexandrian times, natural magic and Hermeticism were sources of inspiration, symbol, and methodology in mathematics and the natural sciences. Frances Yates, Allen Debus, and Robert Westman have demonstrated the significance of natural magic and mysticism in the sciences of the era. Yet as geometers considered what constitutes science and appealed more to experience and deductive reason, magic and astrology were losing intellectual respectability in mathematics. At the close of the sixteenth century, John Dee and Christoph Clavius argued for teaching mathematics beyond the first six books of Euclid as well as raising the status of the mathematics faculty at universities.

13.1   Practical Mathematics in High Renaissance and Reformation Europe

The acceleration in the evolution of mathematics in sixteenth-century Europe was part of larger historical developments: the High Renaissance,[370] the Reformation, and the beginnings of the scientific revolution. These shaped the style, direction, and application of mathematics.

Accompanying the growth in commercial capitalism and banking was a spate of mercantile arithmetics. Global exploration promoted and printing preserved more accurate star charts for navigation, and related trigonometric tables carried better values and were easier to use. In an age of warfare, mathematics was applied to gunnery and designing star-shaped forts for defense against cannon attacks. Ballistic experimentation on projectiles prompted a reexamination of free fall and the discovery of parabolic trajectories in place of Aristotelian abrupt angled falls. Equally important were applications of mathematics in shaping what in the eighteenth century would become the modern exact sciences. In mathematical astronomy, the new Copernican model appeared. In mechanics, Aristotelian thought was revised and attacked, and studies of impetus and centers of gravity employed the mathematically sophisticated Archimedean positions.

During the Italian and the later northern European High Renaissance and the contemporaneous Reformation, support for mathematics broadened. A growing audience of curiosi, some versed in Latin and others employing the vernacular, and embracing landed gentlemen, physicians, lawyers, *literati*, apothecaries, soldiers, and clerics, was attracted to mathematics for playful as well as practical purposes. They met in taverns or at noble courts to discuss natural philosophy. Some belonged to literary clubs that many European cities now supported, such as Padua's Academy of Science and Naples' *Oziosi*. The name *Oziosi*, or idlers, is humorous and mildly deprecatory, a custom followed among the new Italian societies. Before 1580 in Naples Giambattista della Porta (1535–1615) founded the Accademia dei Segreti (Academia Secretorum Naturae) that met at his house. His book *Magiae Naturalis* of 1580, which stresses experimentation and application in defining natural magic and conceives natural magic as likely the pinnacle of natural philosophy, synthesizes Neopolatonic and Hermetic ideas. After Porta was charged with practicing the black arts, the Inquisition closed his academy. Renaissance humanists showed renewed interest in the three classical problems of antiquity. To the practical and the theoretical objective of mathematics was joined a growing interest among Reformation religious reformers in algebra and arithmetic for cabalistic numerological clues to deeper Scriptural truths and prophecies, while natural philosophers appealed to numerology, magic squares, and harmonic ratios in their pursuit of fundamental relations in nature. Benefitting from spurts of prosperity, a few monarchs and nobles patronized mathematicians to add luster to their courts and brains to meet their ambitions. Cities employed a few for commercial architecture and for civil and military engineering projects. Patronage, such as that from the ninth Earl of Northumberland and in Prague the Holy Roman Emperor Rudolph II, remained the major support for original research in mathematics and natural philosophy.

An account of the sixteenth-century European context might begin with the crowning High Renaissance achievement in painting and sculpture, centered in the trade-rich Italian cities of Florence, Rome, and Venice.[371] The two-dimensional, formalized, and deeply religious art of the Middle Ages gave way to representations of classical beauty, the power of nature, and human dignity rendered in heroic human figures; and there was an increase in secular themes. Renaissance art reached its peak in the work of three masters: Leonardo da Vinci (1452–1519), Raphael of Urbino (1483–1520), and Michelangelo Buonarroti (1475–1564).

More than any of his contemporaries, Leonardo personified the Renaissance ideal of the universal man. Each of his nineteen surviving paintings innovatively conveys inner moods. The most celebrated are his *Last Supper, La Gioconda* or

Mona Lisa, and his self portrait. A self-taught genius, he contributed widely to science and technology as well as visual art. Believing it essential that artists who depict the human body know human anatomy directly and thoroughly, Leonardo obtained cadavers, dissected them, and in his notebooks made perhaps the first accurate drawings of bones and muscles. From his study of birds, this naturalist and vegetarian made sketches that place him in the tradition leading to the helicopter and the parachute. He made a model of an armored tank. A contribution to mechanics was his rejection of the physical possibility of perpetual motion. While Leonardo accepted theology as queen of the sciences, he came to believe that geometric rules underlie natural phenomena and applied this conviction to proportion and linear perspective, calling perspective "the rudder and guide rope of painting." His friend Luca Pacioli was probably his chief source of mathematical knowledge. Visual experience and accurate quantified data worked into geometric rules, he insists, provide a trustworthy means to distinguish true from false:

> The man who discredits the supreme certainty of mathematics is feeding on confusion and can never silence the contradictions of sophistical sciences, which lead to eternal quackery...for no human inquiry can be called science unless it pursues its path through mathematical exposition and demonstration.[372]

Leonardo ignored algebra and studied only geometry that is readily applicable to art. Attempting to square the circle, he inscribed regular polygons of three, six, eight, and twenty-four sides in it and he approximated the more difficult task of inscribing a heptagon in a circle. His notebooks confirm that the lunules of Hippocrates intrigued him. He rediscovered the simple proof that the area of the lunule constructed on the hypotenuse of an isosceles right triangle equals the triangle itself.

The irenic Raphael and melancholic Michelangelo possessed artistic genius comparable to Leonardo's. Raphael, who worked at the Sistine Chapel in Rome, painted tender madonnas and his greatest work, the fresco *School of Athens*, a powerful statement on the influence of Greek classicism. Centered on Plato and Aristotle in conversation and symbolizing the primacy of thought in human experience, *School of Athens* bespeaks the importance his age accorded to reading what the Renaissance called the Book of Divine Creation, or nature, which sixteenth-century scholars studied along with the other great book, the Scriptures.

Young Michelangelo sculpted several Pietas, reflecting the era's absorbing interest in death, and the eighteen-foot *David,* and he painted frescoes on the Sistine Chapel ceiling, including the *Creation of Adam*. His rendition of heroic anatomy on the Sistine Chapel ceiling, along with his working of perspective and of line and composition, was supreme. In his later years he altered the design of the great dome of St. Peter's Basilica.

To enhance their paintings, Renaissance artists perfected two techniques, chiaroscuro and linear perspective. Chiaroscuro blends light into shade to convey naturalness, and linear perspective adjusts the size of figures in a way to

School of Athens
(credit: Raphael. "The School of Athens." 1508. Stanze di Raffaello, Vatican Palace,
Vatican City)

give on a two-dimensional canvas the illusion of continuity and depth. The
word *perspective* comes from the Latin *perspectiva*, the medieval term for the
science of optics. In the Renaissance the science of optics became *perspectiva
communis* and the perspective construction of the artist *perspectiva artificialis*.
To construct mathematically correct perspective requires carrying out a conical
projection with the eye placed at the center of projection. Renaissance artists
took geometry to be the common ground between art and science and in developing linear perspective initially experimented with placing human figures in
circles, triangles, and ovals.

Three fifteenth-century scholar-artists, Filippo Brunelleschi (d. 1446), Leon
Battista Alberti (d. 1472), and Piero della Francesca (d. 1492), had led in
geometrizing artistic space. Searching for ideal proportion, each attempted to
discover the principle of optical recession to build an illusion of depth.

Perhaps his goldsmith's craft, which included some study of mathematics and astronomy, along with discussions among artists, astronomers, natural
philosophers, and geometers of perspective, prompted Brunelleschi to examine
the projection onto an astrolabe of a flat image of the heavens and to pursue
perspective construction. He probably invented the *construzione legittima* or

checkerboard grid for perspective. For the baptistry at Florence's cathedral and the Seignory palace, Brunelleschi constructed perspective pictures conveying an optical illusion of scenes in three dimensions in mirrors facing them when viewed through peep-holes or apertures at the back of the picture's panels. Brunelleschi's Old Sacristy of San Lorenzo and Pazzi Chapel in Florence probed secrets of optical recession.

Indebted to Vitruvius, Brunelleschi, and practical experience in Florentine studios, Alberti was the first to write directly on the cone of vision. His *Della pittura* on painting and patrons, completed in 1435 and printed initially in 1540, has a few pages on linear perspective. Attempting to establish perspective construction on the same academic level as quadrivium subjects, Alberti developed a pragmatic theory that all bodies emit rays in all directions that move rectilinearly and converge toward the eye, forming a visual pyramid or cone. In formulating this theory, he probably drew on antecedent intromission optics. Alberti described an image of a vertical picture plane and a horizontal ground plane with a single center-of-vision (V) vanishing point and equidistant parallel lines to the vanishing line. The horizontal plane may be likened to a square-titled pavement. Making the *construzione legittima* was tedious. Most had strings radiating from a centric point, commonly a nail in the wall.

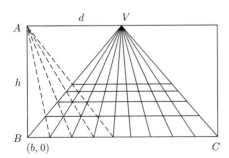 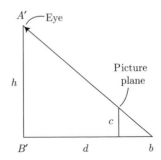

Of the three artists, Francesca was the most competent mathematician. Luca Pacioli was his student. Francesca's *De prospettiva pingendi* (*On Perspective in Painting*), which dates from about the 1460s, but was not printed until 1899, extends the checkerboard focused perspective from Alberti's plane figures to three dimensional planar objects, such as houses, wellheads, column capitals, and human heads. Using abacus and algebra to obtain precise perspective angulation with edges free of distortion, Francesca calculated distances between the parallel lines on the checkerboard. He achieved an almost cubist ordering of geometrical structures.

Near the end of *De perspettiva pingendi*, Francesca described a variant of a subject closely related to perspective, that is, perspectival anamorphoses. Anamorphoses refers to distorted images if pictures are viewed from a certain point or as a reflection in a curved mirror. Fostering the development were Renaissance wall and furniture decorations suggesting a hidden picture that reflects belief in a magical cosmos. Pictorial representations of elevated objects and columns and the determination of the visual angles to make these parallel also contributed. Leonardo da Vinci, Hans Holbein, in his representation of the skull in *The Ambassadors*, and Dürer's student Eric Schön's portraits employed amorphoses.

The talented Nuremberg artist and engraver Albrecht Dürer (1471–1528) was the principal High Renaissance contributor to perspective. After studying in Venice Euclid's *Elements*, Vitruvius' *De architectura*, Alberti's notion of plastic harmony, and Pacioli's *Divina Proportione*, Dürer made a series of woodcuts, entitled *The Life of the Virgin*, that display full mastery of techniques of linear perspective. Taking German artists up to the Italians in the abstract knowledge of perspective, he wrote *Vier Bücher von menschlicher Proportion (Four Books on Human Proportion)*, which was published posthumously in 1528, and the basic *Underweysung der Messung mit Zirckel und Richtscheyt in Linien, Ebnen und gantzen Corporen (Instruction on Measurement with the Compass and Ruler in Lines, Planes, and Whole Bodies)*, in four books published in 1525 teaching Italian methods of constructing plane curves and solids. Drawing largely upon terms and techniques employed by artisans,[373] Dürer also created a German vocabulary for perspective. His *Underweysung* draws such curves as the spiral of Archimedes, the logarithmic spiral, and conic sections. It constructs the five Platonic solids by a paper-folding technique still used today.

In the sixteenth century, perspective became increasingly important to the training of both artists and craftsmen, particularly architects and stage designers. Daniele Barbaro's *La practica della perspettiva*, which was printed in 1569, and Frederigo Commandino's comments on stereographic projection for astrolabes in his edition of Ptolemy's *Planispherium*, published in 1558, attest to this. But the creations of artists and craftsmen did not require a satisfactory deductive basis for perspective and none yet existed. The Italian mathematician Guidobaldo del Monte's late sixteenth century tract offered a comprehensive theory and rules of perspective based on a vanishing point. Fascinated with wonders and natural magic, the French Minim Jean François Niceron in his *La perspective curieuse*, published in 1638, was the first to define the close relation between perspective and anamorphoses, and French mathematician Gérard Desargues made the earliest recognition that properties of rays in the cone are invariant. A century later Brook Taylor and Johann Heinrich Lambert wrote the original definitive theoretical works on perspective.

The High Renaissance essentially ended in a political disaster for the autonomous Italian city-states. As Venice, Florence, the Papal States, Milan, and Naples fell to quarreling, new monarchies were emerging in transalpine Spain, France, and England. A series of French invasions between 1494 and 1516 left much of Italy in shambles. In addition, Turkish raiders plundered scores of Italian cities in the 1520s, and in 1527 Rome itself was sacked by the Spanish and Austrian army of Charles V.

During this period of foreign invasions, Baldassare Castiglione's *The Courtier*, composed between 1508 and 1528, outlined a secular education for an aspiring elite in state service. Transcending the medieval chivalric ideal, rooted in military virtue and loyalty, *The Courtier* prescribes a grounding in the liberal arts, including geometry and eloquence in speech. Of greater influence was the Florentine humanist Niccolo Machiavelli's *The Prince*, published in 1513, which founds the modern science of politics. In the interest of a unified Italy freed of foreign influence and awakened to the values of patriotism and war, Machiavelli calls for a ruler who will put power before more humane goals and, following "reasons of state," will not hesitate to deceive friends and destroy enemies in the national interest. Dictators of the next four centuries made *The Prince* what Bertrand Russell called "a handbook for gangsters," and the term *Machiavellian* has become a condemnatory epithet. But Machiavelli's work as a whole champions a republicanism and a civic virtue that are among the highest mundane ideals of the Renaissance and appropriate to the vigorous urban existence that nurtured much of the art and intellect of the age.

As the Renaissance faded in Italy, a new age of religious reformations spread across northern Europe. Protestantism from the time of Martin Luther (1483–1546) sundered western Christendom. A proliferation of churches—Lutheran, Zwinglian, Calvinist, Anglican, Anabaptist, and smaller sects—quickly fragmented Protestantism itself. Through pamphlets and books with powerful woodcut images, read privately and aloud to illiterate audiences, Protestantism made use of the newly invented printing press to propagate its teachings. The sales were massive. Luther himself had thirty works published by 1520, and his German translation of the Bible was crucial in the development of the German language. The initial reception of these publications was greatest among middle orders of society in northern cities, who were opposed to clerical exemptions from taxes and other civic obligations. Protestant reform thus made for the spread of printing and literacy and therefore the growth of regional communities of learning and, in time, the victory of the book in matters of authority.[374] The connections Catholic teaching and the Latin language had once made throughout the European West were to be replaced by networks of humane and scientific study. These had the indirect aid of a morality that some historians have ascribed to the Reformation. Max Weber's argument that the industrial capitalist order and the new sciences owe their beginnings chiefly to Protestantism is now largely discredited. But the Protestant ethic, demanding the discipline of work and thrift as a means to the mastery of the fallen and wayward self, has resonances with the modern project of mastery over the realm of abstract properties and the world of matter, with vast consequences for the evolution of mathematics.

Sixteenth-century Europe was also continuing to recover from two medieval crises: demographic and political. During the fourteenth century, between a third and a half of Europe's population had died from the combined bubonic, pneumonic, and septicemic plagues known as the Black Death. As it ebbed, an estimated sixty-nine million in 1500 grew by 1600 to a number calculated to be between ninety and ninety-five million. In Spain, France, and England new monarchs, using Machiavellian methods and control of the mining and metallurgy needed for cannon manufacture, struggled against powerful nobility, traditional political institutions, and local autonomies and formed dynastic states. These states developed a royal bureaucracy that labored to centralize administration. At court, vernacular languages were replacing Latin. Muscovy expanded and, after the disastrous loss of the Hungarians to the Ottoman Turks at the first Battle of Mohacs in 1526, the crowns of Hungary and Bohemia were incorporated into the Alpine Austrian monarchy. The new imperial Austria began to spread east and south along the Danube. For centuries kings and emperors would jockey for wealth, power, and glory in protracted wars.

For Europeans the sixteenth century was an age of unprecedented and predatory global expansion reaching to Brazil in the west and Japan in the east. After the Portuguese Prince Henry the Navigator commissioned the charting of the west coast of Africa, Vasco da Gama rounded Africa to reach India in 1498. This broke the Ottoman and Italian monopoly of the profitable spice trade with the Orient. In search of wealth, a route to India, and heathens to convert, Christopher Columbus journeyed across the unknown reaches of the Atlantic, but reached the Americas instead of India. From 1519 to 1522 Ferdinand Magellan's crew crossed the Atlantic and Pacific to circumnavigate the globe.

For a thousand years Ptolemy's world map had governed literate European thought about Earth's configuration. Its spherical compass enlarged the Mediterranean, diminished the southern hemisphere, and lacked the western hemisphere. Ptolemy's map was now obviously outdated. The work of the Flemish geographer Gerardus Mercator (d. 1594) superseded that ancient authority. After studying mathematics and astronomy with Gemma Frisius in Louvain and gaining a reputation for engraving, Mercator prepared the first modern maps of Europe and Britain. These he based partly on comparisons of information from travelers and his excellent editions of Ptolemy's *Geography*. In 1569 he designed a new world map for navigators with a projection now named for him. Mercator's stereographic projection thus addressed the practical problem of mapping the surface of the spherical Earth onto a plane. It is conformal, or angle preserving, so that it would be "similar in the small." Consequently, as a region grows smaller, it was meant to tend to an exact scale map. In the Mercator projection a scaled cylinder circumscribing the spherical Earth and tangent to it at the equator can be unrolled as a flat surface. Latitude and longitude lines appear as straight, mutually perpendicular lines with longitude lines equally spaced but converging on the poles. Mercator's map distortion increases toward the poles.

Population growth, new trade routes by sea, New World bullion, the political stability of monarchical states that were centralizing, and maturing capitalist institutions sustained a commercial revolution. Banking accompanied trade expansion, introducing checks and bills of exchange as substitutes for gold and silver. Circumventing the Church's condemnation of usury, banks in Florence and Venice exacted high-interest loans. Trade boomed in textiles, weapons, cutlery, glass, mining, printing, and shipping. A few people swiftly accumulated riches. The commercial activity encouraged numerous practical arithmetics in the skills for daily commercial activities and to serve as guides for merchant apprentices.

Since the thirteenth century, Italians had been writing practical arithmetics. Up to 1460, they numbered about three hundred.[375] Between 1460 and 1600 their production accelerated. Italian cities published some two-hundred practical arithmetics. Among the most popular were the Venetian Girolamo Taliente's *Opera che insegna* ..., published in 1515, and Francesco de Lagesio's *Libro de abaco*, which came out in 1517 or 1518. Only in the mid-twentieth century would historians rediscover these practical arithmetics and fill a notable gap in the historical record of Italian mathematics. Influenced by Leonardo of Pisa, some earlier authors of practical arithmetics included a chapter of rules of algebra and their application. These mundane manuscripts in the algebraic tradition before 1500 seeded the development of algebra from practical skill to abstract mathematics after 1500. They indicate a continuity of development in Italian algebra, rather than a sudden manifestation of Italian genius independent of past studies.

From the late fifteenth century, German and Slavic *Rechenmeisters*, teachers of arithmetic, had prepared extremely practical arithmetics, particularly in the Hansa trading towns fringing the Baltic. Their work, like that in Italian cities, was closely associated with algebra, the cossic art. After studying algebra and the Bamberg arithmetics in the 1480s, Johannes Widman (d. after 1498) of Bohemia and the University of Leipzig prepared in 1489 his *Behend und hüpsch Rechnung uff allen Kauffmanschaften*. The Trent *Algorismus* of 1475, the Bamberg arithmetics, and this work were the first printed arithmetics in German. The range and number of computations in Widman's *Rechnung* (*Reckoning*) put it beyond these predecessors. It also introduced in print the + and − signs to indicate surpluses and deficiencies in warehouse measures. These signs supplanted the Italian *p* for plus and *m* for minus. Work of the Saxon *Rechenmeister* Adam Ries (d. 1559) superseded that of Widman. Ries regarded arithmetic as the branch of mathematics of greatest use to art and trade. He was a mining official as well as a teacher in Annaberg, who constantly improved and revised his books. His *Rechenung nach der lenge, auff den Linihen und Feder* replaced older line counters, the abacus, and Roman numerals with algorithms using the feather pen and Indo-Arabic numerals. Although largely completed in 1525, it was not published until 1550, since the printing costs were great. Ries pioneered the use of Indo-Arabic numerals in German lands. His *Rechenung* was so clearly written and comprehensive that the name Ries became a synonym for arithmetic.

In Germany such work found encouragement. Unlike Medici bankers in Italy, German dukes and Fugger banking tycoons paid printing costs for commercial arithmetics. Luther's disciple Philipp Melanchthon advocated the improvement of mathematics education in German universities.

In commercial arithmetic and the algebra connected with it, western Europe lagged behind the central region until the mid-sixteenth century. While economic activity had expanded, early works there were generally poor.

In Spain a Dominican, Juan de Ortega, in 1512 and Juan Perez de Maya in 562 published commercial arithmetics. The *Livro de Algebra en Arithmetica y Geometria* of 1567 by the talented Portuguese cosmographer-royal Pedro Nuñez (d. 1577), who taught at the newly found University of Coimbra, was more theoretical.

Practical arithmetic followed a distinctive path in France. It had surfaced suddenly in the late fifteenth century. Its two likely sources were the Hebrew abacus tradition that flourished in southern France during the fifteenth century and Nicolas Chuquet's algebra. In calculating the Jewish religious calendar, Hebrew scholars now made sophisticated astronomical computations. In 1520 Étienne de la Roche, the only follower of Chuquet in the French kingdom, published his *L'arismetheque*. Based on both Chuquet's *Triparty* and Pacioli's *Summa*, it particularly popularized Chuquet's exponential notation. After de la Roche's *L'arismetheque* the pursuit of practical arithmetic in France temporarily ceased. The shift that Christian humanists brought to French intellectual life may help explain this. But while Christian humanists gave little attention to mathematics, the French humanists Jacques Lefevre d'Étaples (d. 1537) and Charles de Bouvelles (d. 1566) favored the abstract mathematics of Plato and Euclid. Oronce Finé (d. 1555), the first professor of mathematics at the College Royale, published in 1532 an *Arithmetica practica* with no practical problems. Following effective criticism that the logician and educational reformer Petrus Ramus (d. 1572) leveled at schools for neglecting mathematical studies altogether and pursuing abstractions unrelated to crafts or the marketplace, practical arithmetics reappeared and algebra blossomed in France after 1550.

Robert Recorde (d. 1558) composed in *The Ground of Artes*, published in 1543, England's most influential sixteenth-century practical arithmetic and in 1557 brought out a complementary book, entitled *The Whetstone of Witte*. Recorde, a physician, minor civil servant at the royal court, and briefly surveyor of the mines in Ireland, wrote clearly in the vernacular, which allowed craftsmen practitioners as well as scholars to study his texts. Like Pacioli, he broke with Latin authority. Following a dialogue form that leads the reader step by step to master techniques, the first edition of Recorde's *The Ground of Artes* treats whole numbers, basic operations with them, counter reckoning, and algorithms. The second enlarged edition of 1552 also includes fractions, progressions, alligations, and false position. *The Whetstone of Witte* is Recorde's only book to have but one edition, probably since London craftsmen had no immediate uses for its materials. It offers the second part of his arithmetic, which is taken from books on that subject in Euclid's *Elements* and elementary algebra through quadratic equations. Its algebraic section is based on the German cossic tradition, particularly the works of Johann Scheubel and Michael Stifel. *The Whetstone of Witte* uses their notation. *Whetstone* in the title refers directly to the cossic practice, since the Latin for whetstone is cos. Believing that nothing "can be more equal" than two parallel lines of equal length, Recorde introduces the equal sign, =, making it largely symbolic. He also employed negative coefficients, but rejected negative roots. In solving quadratic equations, in modern notation $x^2 = px - q$, having two positive rooots, he emphasized the relationships between coefficients and roots: $r_1 + r_2 = p$ and $r_1 r_2 = q$.

The Low Countries produced practical arithmetics in Latin, French, and Dutch. Most Latin and French arithmetics were printed in Antwerp. The Brussels physician Gemma Frisius wrote in 1540 *Arithmeticae practicae methodus facilis*, which went through fifty-nine editions. Until mid-century only a few practical arithmetics appeared in Dutch. The first printed in Dutch was Thomas Vander Nost's *The Way of Learning to Calculate According to the True Art of Algorisms in Whole and Broken Numbers*, published in Brussels in 1508 and in enlarged editions in 1510 and 1527. The second, printed in 1537 and again in 1545, was Gielis Vanden Hoecke's *An Extraordinary Book on the Noble Art of Arithmetic*. Many Dutch arithmetics appeared after 1550. They include the works of Creszfelt in 1557, Raets in 1566, Petri in 1567, Vander Gucht in 1569, Helmdyn in 1569, and Simon Stevin, who wrote the booklet *De Thiende* in 1585.

By 1600, then, many accurate commercial arithmetics were in print. Amid Europe's turbulent warfare, emphasis shifted partly to producing mathematical treatises related to military concerns such as firing tables for the artillery. Abstract algebras were also gaining in prominence.

13.2 A Northern Renaissance and the Early Scientific Revolution

A northern Renaissance, coinciding with the Reformation, contributed to the awakening of a new scientific mentality. Diffusion of Italian humanism stimulated northern thinkers, many of whom attended Italian universities, and after the French invasion of the Italian peninsula, Italian printing firms had a flourishing traffic in books to the north. In 1511 the philosophical and literary intellect Erasmus of Rotterdam completed his *Praise of Folly*. Though its chastising of superstition, sorcery, hate, violence, and social pretention differed in tone from the scientific mind in quest of cold precision, Erasmus's advocacy of tempered doubt concorded with the increasing refusal of scholars in the sciences and medicine to rely on texts by ancient authorities. The same is true of French scholars who gave Christian humanism a more secular emphasis than had Erasmus. In *Gargantua and Pantagruel*, serialized between 1532 and 1552, the Lyons physician François Rabelais ridicules scholasticism and superstition. His younger contemporary Michel de Montaigne (1533–1592) wrote elegant skeptical essays that challenge readers to confront the best of ancient thought. Such writings, along with the questioning temperament of such Renaissance humanists as Pico della Mirandola (d. 1494), Marsilio Ficino (d. 1499) of the Florentine Platonic Academy, and Petrus Ramus (d. 1572), embodied and promoted a major shift in European modes of thought. The greatest of the last plays of William Shakespeare (1564–1616), *The Tempest*, captures the imagination of the age, quickened by overseas discoveries and aware, as past eras had not been, that there might be such things as a brave new world. Symptomatic of that world was the scientific revolution that began in earnest in the mid-sixteenth century.

After two centuries of deepening assimilation of classical Greek, Hellenistic, and medieval Islamic texts, especially of Archimedes, and the recent explosion of print technology, European thinkers were poised to surpass the limits of knowledge in the sciences achieved mainly by ancient and medieval authorities (*auctores*), chief among them Ptolemy in astronomy, Aristotle in physics, and Galen and Avicenna in medicine. Although many scholars accepted ancient philosophy, more and more modern scholars with growing confidence began rejecting it, usually through revising and discarding scholastic translations and commentaries of works by time-honored ancient authorities that were viewed as adulterations and less often by directly assailing ancient thought.

Among the intellectual activities that, along with a new skeptical habit of mind, contributed to the determination to unlock the secrets (*arcanae*) was the Neoplatonic revival from Ficino through Francesco Patrizi's *A New Philosophy of the Universe*, published in 1591. The Dominican Tommaso Campanella's *Civitas Solis* of 1602 and, in 1620, his *De Sensu Rerum et Magia* (1620) also contributed. These texts worked to displace the Aristotelian logic of the scholastics, whose syllogisms confirmed existing knowledge but failed to press for the new knowledge that Renaissance humanists sought. They made use of deductive reason built on mathematical analogies and Euclidean axiomatics combined with increasing reliance on sensory observation and experience. Inspiring them was a panmathematical imagination drawn from Platonic, Neoplatonic, and Neopythagorean sources appropriated for quantified abstraction, complemented by a heightened sense of wonder and a pursuit of natural magic.[376] Together with the mechanist tradition from Archimedes and the atomists, who described nature by analogy to the machine, panmathematicism, through its early reduction of space and nature to geometry, challenged the supremacy of theology in the sciences. In this way and in seeking to expunge scholastic occult qualities from scientific discourse, panmathematics offered a new ontology. Still, not even all Neoplatonists agreed on the new relationship between mathematics and nature. Some, like the English magus John Dee, believed that mathematics is preparatory to the study of first principles in metaphysics. By the start of the seventeenth century, probably only Christoph Clavius and Johannes Kepler adhered to a Pythagorean and Platonic metaphysics that insisted on the centricity of mathematical reasoning and philosophy. With or without philosophical implications, the mathematical approach to nature, along with the study of the material base of the universe by atomists and alchemists, was essential to the new sciences.

A component of Renaissance and Reformation mathematics was occultist investigation related to number mysticism. Writings from the ancient Egyptian magus Hermes Trismegistus, particularly Ficino's translation of the *Hermetica*, as well as from Neoplatonism, especially speculation in Plato's *Timaeus*, and occult mathematics nourished the investigations of magic and mysticism. In general, the hermetic tradition refers to alchemy that embraces a belief in occult and manifest powers and in *prisca sapienta*. Hermetic thinkers perceived of man as a microcosm of the physical universe and to uncover deeper interrelationships in nature applied the analogy of microcosm to macrocosm. They also employed cabalistic numerology and magic squares to predict events and decipher greater truths from Scripture and other inspired sources. Although our knowledge of the extent of hermeticism remains unsure, it is clear that through the seventeenth century magic persisted as an inspirational source of mathematics. The notion of what constitutes scientific knowledge was evolving. Also important during the Reformation was an organic tradition, depicting all nature as animate.

While most scholars in the scientific revolution were trained at universities, the movement, especially at its inception, primarily occurred outside academe. It was communicated essentially through books, supplemented in time by scholarly articles. Hundreds of researchers in libraries, observatories, and their residences as well as in universities, aided by improvements in mensuration, proposed daring ideas and theories, made systematic observations and tests, discarded error, and elaborated theories more accurate and powerful than had their predecessors. They brought vast and what seem irreversible paradigm shifts in mathematical astronomy, mechanics, mathematics, anatomy, medicine, and chemical science. The conception of the universe that in 1500 dominated the European mind—geocentric, bounded, and hierarchically ordered toward teleological goals and vulnerable to demoniacal influences—had given way by 1700 to an infinite Copernican universe, free of hierarchy, Earth displaced from the center, and open to rational investigation. Herbert Butterfield has compared the scientific revolution to the rise of Christianity in importance.[377] While historians of early modern science agree with the significance of the achievement, Steven Shapin and many other scholars reject Butterfield's name and Thomas Kuhn's model for the age.[378] They seek a new master narrative that can join into one historiographic structure the diverse studies of early modern science. H. Floris Cohen, to the contrary, urges that, if refined, the concept is usable as a tool of historical interpretation.

During the sixteenth century, an age of striking developments in mathematical astronomy, the Polish astronomer and Catholic canon Nicolaus Copernicus (1473–1543) rejected the Ptolemaic-Apollonian geostatic model of the heavens. That model had planets move along epicycles situated on circles around the central Earth that was at rest. Astronomers computed stellar and planetary motions about it. After studies at Cracow and Bologna of canon law and astronomy and observations of the heavens made on the Baltic coast of Poland, Copernicus privately in his manuscript *Commentariolus*, completed by mid-1514, and publicly in *De Revolutionibus Orbium Coelestium* (*On Revolutions of the Heavenly Spheres*), published in 1543, proposed instead that the Earth rotates on its axis daily and revolves around the sun once every $365\frac{1}{4}$ days as it moves through the heavens, which he projected were far larger than the finite, closed Ptolemaic model. He thus eliminated the unique astronomical status of the Earth, making it one of several moving planets. At the center of the Copernican universe lies "the sun enthroned," the helostatic model.[379] Copernicus intended his reform of astronomy to simplify and improve the techniques for calculating positions of celestial bodies, but did not personally succeed in this. The size of his cosmos was "similar to the infinite," which allowed the concept of a center and explained his inability, having only naked eye observations to detect stellar parallax in an immense universe. Stellar parallax is the optical shift in the positions of stars observed from different points on a moving Earth's orbit. While opposed to the Aristotelian beliefs in a stationary Earth and immutable heavens, Copernican astronomy retained uniform celestial motions.

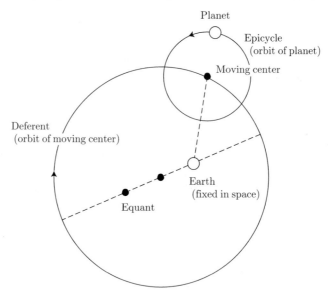

Ptolemaic system.

Copernicus's great work received a torrent of criticism. He called on critics to investigate his ideas "with both eyes open," knowing that a moving Earth and fixed sun ran counter to common sense and experience.[380] Aristotelians maintained that if Earth moved a strong wind would blow across its face. Scenting heresy, Protestant leaders, who accepted a literal interpretation of the Bible, railed against Copernicus. Citing Joshua 10:13, where Joshua commands the sun to stand still, Luther condemned Copernicus as a "fool" and "upstart astrologer," and Calvin called Copernicans "atheists."[381] Though Catholic opinion was generally hostile, Catholic authorities until the clash with Galileo tolerated Copernican astronomy as a speculation. The skeptic Montaigne denounced the vanity of the Copernican model, which endowed man "with godlike qualities ...."[382] Luther's major follower, Philipp Melanchthon (d. 1560), who wanted to "curb the imprudence of talents" of Copernicus, brought the astronomer Erasmus Reinhold to the University of Wittenberg.

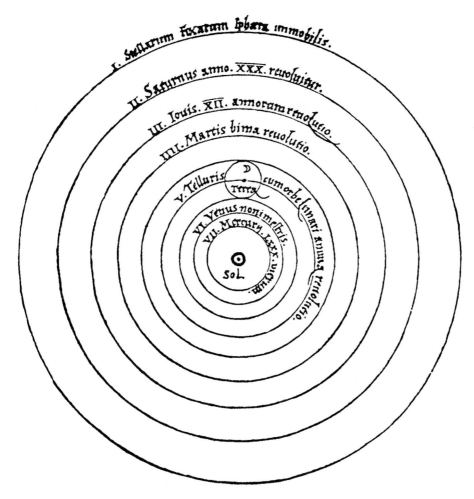

Copernican system

Yet Reinhold in his *Prutenic Tables* based planetary positions solely on the Copernican computational model.

The dispute did not simply pit the natural sciences against theology. The Apollonian-Ptolemaic model based on Aristotelian physics for a time made a respectable showing against Copernican astronomy, which lacked a supporting physics. The early Copernican model needed to be elaborated and refined. A supernova in Cassiopeia in 1572 and a large comet with a lengthy tail in 1577 convinced the Danish astronomer Tycho Brahe (1546–1601) of the inadequacy of the Aristotelian tenet that solid spheres comprise the unchanging heavens. Using his improved sighting devices, clocks, and celestial globes at Uraniborg castle and observatory as well as at the adjacent "Starburg" observatory, Tycho over decades made systematic and accurate observations of the planets and stars. He displayed computational prowess in recording them. Failing to detect stellar parallax with the naked eye and not accepting the great distances to the stars, Brahe attempted a compromise between Ptolemy and Copernicus. Tycho's treatise *On the Most Recent Phenomena of the Ethereal World*, published in 1588, places the fixed Earth at the center circled by the moon and sun. All other planets circle the sun. By contrast, the Dominican philosopher Giordano Bruno daringly extended Copernican astronomy. His universe is not heliostatic and closed but heliocentric and fully infinite. For heterodox speculations that human souls are material and Jesus not divine, in 1600 Bruno was burned at the stake in Rome to celebrate the centennial. The two natural philosophers crucial for gaining acceptance of Copernican astronomy and articulating it early in the seventeenth century were to be Johannes Kepler and Galileo Galilei.

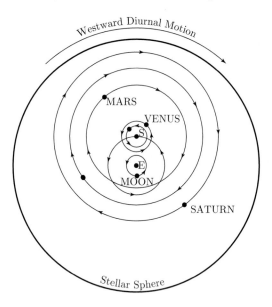

Tychonic system

   In anatomy and medicine the work of Andreas Vesalius (1514–1564), a Flemish physician, essentially challenged the thirteen-hundred-year dominance of Galen. In Paris, Vesalius studied Hippocrates, Galen, Dioscorides, and Avicenna, and he performed dexterous human dissections. In the medieval Latin West, religious leaders had severely limited dissections. At universities allowing them, teaching physicians who accepted Galenic anatomy consigned dissections to untutored barber-surgeons who tossed body parts to dogs around the dissecting table. Skillfully examining anatomic minutiae, Vesalius while at the University of Padua wrote *De humani corporis fabrica* (*On the Fabric of the Human Body*), published in 1543. Its remarkably accurate engravings, probably the work of John Stephen of Calcar, improved on Galen's descriptions of muscles, bones, blood vessels, and the brain. Its demonstration in book six that the mid-line septum of the heart is a solid partition contradicted the Galenic theory of the ebb and flow of two bloods (venous and arterial) passing through perforations in the septum. Followers of Galen heaped abuse on Vesalius as the "Luther of Anatomy."

Further research led to the overthrow of Galen's idea of ebb and flow of the blood. In the Spanish physician Michael Servetus's *Christianismi restitutio* (*On the Restoration of Christianity*), printed in 1553, for the first time a Western publication noted the pulmonary transit or lesser circulation, a loop from heart to lungs producing what we know as oxygenation of the blood. The University of Padua now produced the foremost research in cardiovascular physiology. In 1559, two centuries before the discovery of oxygen, Paduan Realdo Columbo in *De re anatomica* depicted the lungs as vitalizers of the blood, not Galenic furnaces consuming impurities. The starting point for Columbo's and subsequent investigations was the heartbeat subjected to mechanical analysis. After Andrea Cesalpino of Pisa demonstrated quantitatively in the 1570s that the capacity of the vascular system is inadequate to allow the constant generation of the blood in the liver and combustion in the lungs, as Galenists held, Girolamo Fabrici in Padua announced in 1603 his discovery of valves in the veins, which permit venous blood to travel only toward the heart. This research culminated in William Harvey's discovery of the general circulation of the blood. An English physician and student of Fabrici, Harvey (1578–1657) studied heartbeats or pulse in a series of vivisections from cold- to warm-blooded animals. In 1628 he announced his discovery in detail in *De motu cordis et sanguinis in animalibus* (*On the Motion of the Heart and Blood in Animals*).

Paracelsus (1493–1541) believed chemical science, not mathematics, to be the key to nature. Paracelsus was not the given name of this Swiss iconoclast. It likely refers to "surpassing Celsus," the great Roman physician. He was born Theophrastus Philippus Aureolus Bombastus von Hohenheim, and his tirades bequeathed our word "bombast." The most inflammatory of the leaders of the early scientific revolution, in Basel he publicly burned works of Aristotle, Galen, and Avicenna. He urged new observations in the medical sciences: direct investigations and observations, for example, in the tunnels of mines that would explain miners' diseases. Taking Galenic and Arabic herbal remedies not to be therapeutically exhaustive, he joined alchemy with medicine to create medical chemistry, distilling mercury, antimony, and arsenic medicines to harness their potencies and perhaps to attempt to detoxify them. The efficacy of these chemical medicines was synchronized with ingestions made during favorable configurations of cosmic sympathies from the stars reckoned by sidereal mathematics. Paracelsus rejected the Galenic theory that deficiencies or excesses in the balance of four humors cause illness, curable by purging or bloodletting. He pioneered instead a germ theory of disease. Paracelsus rejected the Aristotelian identification of four basic elements, instead proposing his three principles: idealized forms of salt, sulfur, and mercury. This proposal impelled a quest for a new definition of element and a new theory of matter as a basis for chemical science, to which Johann Joachim Becher's study of the composition of minerals and Johann Baptist van Helmont's of water as the chief element contributed.

13.3   Algebra: General Arithmetical Solutions for Third- and Fourth-Degree Equations

Algebra, the only branch of mathematics to register achievements comparable to accomplishments in mathematical astronomy, anatomy, and medical chemistry, by the end of the fifteenth century was under wide study by mathematicians in Italian commercial cities. They had come to believe it a better heuristic device than geometry. These scholars, unlike scholastics, consciously pursued discoveries. Many of the spate of commercial arithmetics appearing in Italian cities after 1460 included chapters on rules of algebra in the manner of Fibonacci, and together they nurtured advances in sixteenth-century algebraic theory. By the High Renaissance, practitioners had also learned from Indian, medieval Arab, and Persian sages, each of whom stressed arithmetical computations. Building on these traditions, Italian mathematicians placed algebra on an arithmetical rather than a geometrical basis, emphasizing equation solving.

The most important of Italian contributions was the discovery of the general solution of third- and fourth-degree equations by means of radicals.[383] In modern terms these equations can be written in this way:

$$x^3 + ax^2 + bx + c = 0 \qquad \text{(cubic)}$$
$$x^4 + ax^3 + bx^2 + cx + d = 0 \qquad \text{(quartic)}$$

Solving higher-order algebraic equations requires finding their roots or, put another way, all numbers x that satisfy them. Before 1500 in Europe, the coefficients and constants a, b, c, and d in such equations had to be positive and rational numbers. General solutions for first- and second-degree equations in positive rational numbers had been known in ancient Mesopotamia, but it was in Renaissance Italy that higher-degree equations yielded to general mathematical solutions.

The search had begun among palace and temple scribal mathematicians and school teachers in Old Babylon, and the ancient Greeks, particularly Diophantus, continued it. The Chinese pursued it separately. Invariably Hellenistic Greeks appealed to intersections of conic sections to solve particular cubics and to special cubics to obtain a general solution. Within medieval Islam, al-Khayyām classified normal forms of cubic equations with positive roots and offered geometric solutions. By the thirteenth century the Chinese, without the knowledge of the Latin West, had the equivalent of Pascal's triangle and approximation methods for solving cubics. At that time in the Latin West, John of Palermo challenged Leonardo of Pisa to solve the equation taken from Diophantus' *Arithmetica* VI.17 that in modern form is $x^3 + 2x^2 + 10x = 20$. Afterward Latin mathematicians sought to solve cubics using a cubic formula similar to the quadratic formula. In the West, Barnabus Hughes has shown, correct algebraic solutions of selected cubic equations date back as far as the work of Master Dardi of Pisa (fl. 1350). Fourteenth-century algebraists already had procedures for solving those cubics which are easily reduced to quadratics, but Master Dardi built on al-Khwārizmī's *Algebra* to create new procedures for other reducible cubics. He solved equations of the form $ax^3 + \sqrt{bx^3} = c$ by substituting $y^{2/3}$ for x and equations of type $ax^3 + bx^2 + cx = d$, where $a = 3b$, by substituting $y - b/3a$ for x. Only after 1500 were negative and complex numbers slowly admitted to the realm of number and new algebraic symbols added in the effort to render general solutions for third- and fourth-degree equations.

As interest mounted in algebra in late fifteenth-century commercial Italian cities, the search intensified. In 1463, the year that al-Khwārizmī's *Algebra* was translated into Italian, the mercantile arithmetic *Trattato di Praticha d'Arismetica* by Master Benedetto of Florence also appeared.[384] In 1494 the Franciscan friar Luca Pacioli, who at the beginning of the sixteenth century was to teach mathematics at the Universities of Pisa and Bologna and in 1509 would put into print his translation of the fifteen books of Euclid's *Elements*, published in Venice his *Summa de Arithmetica, Geometrica, Proportioni et Proportionalita*. The encyclopedic, six hundred-page *Summa* was more comprehensive than the practical arithmetics of the day. It is a two-part treatise: the first deals with practical and theoretical arithmetic, algebra, weights and measures, and the Venetian method of double entry bookkeeping; the second with Euclidean geometry. Borrowing freely from Euclid, Boethius, Leonardo of Pisa, Master Benedetto's *Trattato di Practicha d'Arismetrica*, and other mercantile arithmetics of the time, sometimes without acknowledgment, it was widely circulated and read. Written in Italian, it gave Pacioli's countrymen not schooled in Latin a source of mathematical studies.

More than any text since Leonardo's *Liber Abbaci* of 1202, the *Summa* promoted the growth of algebraic studies. Pacioli ingeniously solved linear and quadratic equations and those easily reducible to quadratics. Wide circulation of his treatise helped popularize its abbreviations and symbols, including the widely used *p* (*piu*) for addition, *m* (*meno*) for subtraction, *co, ce,* and *ae* for *cosa* (x), *censo* (x^2) *cece* (x^4), $--$ for =, and R_x (*radix*) for square root. He accepted only positive roots, holding negative quantities to be meaningless. In *Summa* I, dist. VIII, tractate 5, Pacioli, after unsuccessful attempts to solve third-degree equations by a single algebraic formula, judged the solution impossible.

Yet in the first or second decade of the sixteenth century Scipione dal Ferro (1465–1526), a businessman and, like Pacioli, a lecturer in arithmetic and geometry at Bologna, solved cubics of the simple type that have no quadratic term. In modern form, they are $x^3 + px = q$ and $x^3 + q = px$, where p and q are positive numbers. References in Tartaglia's *Quesiti*, Book IX, Question 25 and chapters I and XI of Cardano's *Ars magna* aid in placing the approximate time of discovery. There is no documented evidence that Pacioli encouraged dal Ferro's work or knew of it. How dal Ferro arrived at his solution is also unknown: no account of his discovery was printed or otherwise preserved. At Bologna, one of the oldest medieval universities, mathematics beyond the first books of Euclid had been made part of the curriculum by the late fourteenth century, and in 1450 Pope Nicholas V, in reorganizing the university, had added four positions in mathematics, which expanded to as many as eight by 1500. At the university, then, dal Ferro had access to a rich tradition that included the writings of Leonardo of Pisa, many classical and Arabic authors, the Tuscan *maestri d'abaco*, and Pacioli.

Retrospectively, we can see that the solution for the simplified type of cubics extends to third-degree equations the means for solving first- and second-degree equations. Comparison of equations of the first, second, and third degree:

$$x + a = 0, \tag{1}$$

$$x^2 + ax + b = 0, \tag{2}$$

$$x^3 + ax^2 + bx + c = 0, \tag{3}$$

indicates that equation (1) has a unique root $x = -a$ and equation (2) has two roots $x = -a/2 \pm \sqrt{a^2/4 - b}$, which differ unless the discriminant $a^2 - 4b = 0$. The latter is clear when equation (2) is recast as $(x + a/2)^2 = a^2/4 - b$. Extending this procedure to cubics reveals that $(x + a/3)^3 = x^3 + ax^2 + \ldots$, the dots signifying lower powers of x. Replacing x in equation (3) with $x - a/3$ gives the simpler but equivalent equation $x^3 + px + q = 0$, which dal Ferro solved.

What is transparent in our notation was opaque in dal Ferro's. More likely, his procedure derived from a careful study of the method for expressing sums of square and cube roots, a method that Italian algebraists extensively studied. He may have observed that the sum of square roots given as

$$x = \sqrt{a + \sqrt{b}} + \sqrt{a - \sqrt{b}} \tag{4}$$

can be squared and will satisfy this quadratic equation not having the first power of x:

$$x^2 = 2\sqrt{a^2 - b} + 2a \tag{5}$$

Analogously, the sum of cube roots given as

$$x = \sqrt[3]{a + \sqrt{b}} + \sqrt[3]{a - \sqrt{b}} \tag{6}$$

when cubed, will satisfy the reduced form of this cubic equation that is without the second power of x:

$$x^3 = 3\sqrt[3]{a^2 - b}x + 2a \tag{7}$$

Hence, equation (6) solves equation (7). Writing

$$a = q/2, b = q^2/4 - p^3/27 \quad \text{or} \quad p = 3\sqrt{a^2 - b}, q = 2a$$

gives cubic equation (7) and formula (6) the form $x^3 = px + q$, and the algebraist, if lucky, gets the formula

$$x = \sqrt[3]{p/2 + \sqrt{q^2/4 - p^3/27}} + \sqrt[3]{p/2 - \sqrt{q^2/4 - p^3/27}},$$

which is called Cardano's formula, because it was first made public in *Ars magna*. Cardano praised dal Ferro's discovery as a "beautiful and admirable accomplishment ... [which] surpasses all mortal ingenuity and human subtlety" while Bombelli called Scipione "a man uniquely gifted in his art."

Dal Ferro did not keep his solution strictly secret as was customary. Instead, he showed his manuscript to a disciple named Antonio Maria Fiore, to his son-in-law Annibale dalla Nave, and possibly to other close associates. The matter was dormant until 1535, when Fiore challenged Niccolo Tartaglia to solve various cubic equations. The outcome to the challenge showed that Tartaglia had discovered independently the algebraic rule for solving cubics, which he applied to cubics including the quadratic term as well.

As a youth, Niccolò Tartaglia (1499 or 1500 to 1557) had suffered much adversity. After his father Michela Fontana died about 1506, he lived in poverty. When the French army sacked his home town of Brescia in 1512, he received five serious head wounds from sabre cuts while he and others sought sanctuary in the cathedral, which French soldiers did not honor. Tartaglia was left for dead, but his mother nursed him back to health. A speech impediment resulting from a wound that cleft his jaw and palate gave him the nickname Tartaglia, the stammerer, which he later adopted as a surname in his publications. His formal education was limited to about fifteen days of elementary training in writing the alphabet to the letter k when he was fourteen. Unable to continue paying a tutor, he taught himself the rest of the alphabet, mathematics, and some Latin and Greek. As Tartaglia wrote in *Quaesiti* (Bk. VI, Question 8), he "had no other tutor, but only the constant company of a daughter of poverty called industry. I always worked from books of men who are dead."

Niccolò Tartaglia

After beginning his career by 1518 in Verona as an abacus instructor, Tartaglia headed a school in the Palazzo Massanti. In 1534 he moved to Venice to teach mathematics, remaining there for the rest of his life, except for an eighteen month return to Brescia in 1548–1549. In Venice he held the title of professor of mathematics and gave lessons at local churches. Most of his writings were published in that city. Since he earned only a minimal income, he and his family lived in reduced circumstances. He died impoverished and alone in Venice.

As early as 1530 a friend asked Tartaglia to solve two problems, which amount to cubics. He was to find a number whose cube added to three times its square makes 5. In modern notation, this can be expressed as $x^3 + 3x^2 = 5$. Tartaglia was also requested to find three numbers, the second exceeding the first by 2 and the third surpassing the second by 2 also, their product being 1000. The modern form is $x(x + 2)(x + 4) = 1,000$ or $x^3 + 6x^2 + 8x = 1,000$. Many contemporaries doubted that Tartaglia could solve third-degree equations and for a time he was unable to do so. In 1535, however, he announced that he could solve simple cubics arithmetically. Fiore probably believed that Tartaglia was bluffing—and with some reason. Following custom, Tartaglia had kept secret his method of solution. Not until the eighteenth century would mathematicians regularly publish their results. In the sixteenth century, oral *disputatio* of a scholastic kind rather than written publications decided the veracity of mathematical claims. The outcomes of these could influence academic appointments or carry monetary prizes. Thus Fiore challenged Tartaglia to a public contest. For this intellectual joust, Fiore and Tartaglia each posed thirty cubic equations that the other had to solve within a given time.

The contest proved no real match. Fiore, a mediocre mathematician, could solve not a single equation given him. He could apply dal Ferro's rule only to solve equations without the second-degree term. Consequently, he was undone when Tartaglia posed equations with a quadratic term, which Fiore could not reduce to the simplified form. The talented Tartaglia mastered all thirty equations that he had received. He could solve cubics of the form $x^3 + px = q$, where p and q are positive or negative numbers, as well as those of the type $x^3 + px^2 = q$. Among the equations Fiore proposed and Tartaglia's solutions were these in our notation:

$$x^3 + 9x^2 = 100 \qquad x = \sqrt[3]{24} - 2$$
$$x^3 + 3x^2 = 2 \qquad x = \sqrt[3]{3} - 1$$
$$x^3 + 6 = 7x^2 \qquad x = \sqrt[3]{15} + 3.$$

Contributing broadly after 1535 to pure and mixed mathematics while his reputation grew, Tartaglia pursued until the 1540s a vigorous study of cubics that allowed him to refine his still secret general solution. His *Nova Scientia* of 1537 began a sixteenth-century revival of the study of mechanics.[385] His appeal to suppositions in the form of axioms and postulates moved the science away from the verbal Aristotelian discursive examination of mechanics to more rigorous mathematical ground, which was later to bear fruit in the work of Galileo. Financial need, driving Tartaglia to military science, had motivated him to prepare *Nova Scientia* and the sections of *Quaesiti et Inventioni Diverse* of 1546 that developed the mixed mathematics associated with gunnery and warfare, that is, ballistics. *Nova Scientia* contains new firing tables for artillery and pioneers the study of projectile paths. Treating curved as well as straight motions and holding that projectile trajectory is everywhere a curved line broke into new territory. Both his *Nova Scientia* and *Quaesiti* reveal another side of Tartaglia: military projects seemed to raise ethical concerns that troubled him.

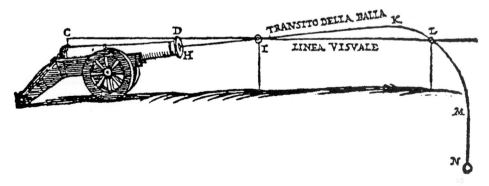

Quaesiti diagrams of trajectories

Tartaglia's last work, the two-volume *General trattato di numeri et misure*, published posthumously in Venice between 1556 and 1560, testifies to his mathematical inventiveness and knowledge. Modeled on Pacioli's *Summa* and intended to displace it, the study lucidly examines numerical operations, such as higher-root extractions. The *General trattato* introduces "Tartaglia's triangle," a configuration of coefficients of the binomial expansion similar to Pascal's triangle. While Tartaglia claimed that this array was his invention, it had appeared earlier. His *General trattato* reviews the geometry of the compass with fixed opening and introduces the theory of the surveyor's cross. This text completes a largely successful effort to outshine two mathematical rivals: Girolamo Cardano and Michael Stifel.

Tartaglia's career had its blemishes. In 1543, the year Copernicus and Vesalius published their seminal works, he completed the first Italian translation of Euclid's *Elements* and edited a thirteenth-century translation of works of Archimedes. He apparently claimed the rendering of Archimedes to be his own, when it was William of Moerbeke's. He took credit for solving the inclined plane problem in mechanics, an achievement which came from Jordanus de Nemore. Literary property and its trespass in the early days of printing were viewed more flexibly than after copyright laws. Still, there was a long tradition of giving credit to previous work.

Algebra, like the Italian cities, was beset with continuous strife. Into the early 1540s, Tartaglia sought to keep secret his solution of the cubic, wishing to publish it as the crowning achievement of a future book. Yet in March 1539 during a visit to Girolamo Cardano's house in Milan, after persistent inquiries, flattery, and a solemn oath not to divulge the coveted solution he outlined it for his host in obscure verse. Cardano was eager to learn the solution: in his *Practica arithmetica* of 1539, which was based on Pacioli's *Summa*, he had

agreed with Pacioli's statement that the cubic cannot be solved algebraically.
The twenty-five verses outlining the solution begin

> Quando che'l cubo con le case appresso
> Se occuaglia a qualche numero discreto
> Trovan dui altri, differenti in esso...
> [When the cube (x^3) and the things themselves
> (px) add up to some discrete number (q), then
> take two others, different from the first (u and v) ...]

No proof was provided, so Cardano set about the difficult task of finding one.

Checking rumors that Tartaglia was not the first to make the discovery,
Cardano and his disciple Ludovico Ferrari (1522–1565) went to Bologna to speak
with dalla Nave in 1543. They examined dal Ferro's papers and concluded that
his method and Tartaglia's were essentially the same. Afterward Cardano did
not feel bound to his pledge of secrecy to Tartaglia. He decided to publish in
Artis Magnae Sive de Regulis Algebraicis (*The Great Art of Solving Algebraic
Equations*), printed in Nuremberg in 1545, the general solution of the cubic
equation. This seminal work, which also contains the general solution of the
quartic, is known by the proud title *Ars magna*, setting algebra apart from the
lesser art of arithmetic.

Chapter XI of *Ars magna* recognizes sources. It begins with the background
to Cardano's solution:

> Scipio dal Ferro of Bologna well-nigh thirty years ago
> discovered this rule and handed it on to Antonio Maria
> Fior of Venice, whose contest with Niccolo Tartaglia
> of Brescia gave Niccolo occasion to discover it.
> [Tartaglia] gave it to me in response to my entreaties,
> though withholding the demonstration in [various]
> forms.[386]

This account and even more the ingenuity of the solution of cubic equations
by radicals, the generality of method, and the thorough demonstrations show
Cardano to be not a plagiarist but an impressive and original mathematician.
A crucial innovation, substituting, as more recent mathematicians would write
it, $x = y - a/3$ for x, allows Cardano to reduce cubic equations with a second-
degree term to the simplified form of the cubic, $y^3 + py = q$. In chapter XI
he solves by radicals a reduced form of the cubic, $x^3 + 6x = 20$. In modern
notation, Cardano's solution in several pages of prose essentially amounts to
this:

> Let $x = a - b$, where $a > b$, and then
> $$x^3 = (a - b)^3 = a^3 - 3a^2b + 3ab^2 - b^3$$
> $$= a^3 - b^3 - 3a^2b + 3ab^2$$
> $$= a^3 - b^3 - 3ab(a - b).$$

Adding px gives

$$x^3 + px = a^3 - b^3 - 3ab(a - b) + p(a - b) = q$$
$$= a^3 - b^3 - (3ab - p)(a - b) = q.$$

To determine a and b, eliminate the product of the two parentheses, which requires $3ab - p = 0$ or $3ab = p$. This leaves $a^3 - b^3 = q$. Substituting $b = p/3a$ yields

$$a^3 - (p/3a)^3 = q$$
$$a^6 - (p/3)^3 = a^3 q$$
$$a^6 - a^3 q = (p/3)^3.$$

Completing the square gives Cardano's rule:

$$a^6 - a^3 q + (q/2)^2 = (q/2)^2 + (p/3)^3,$$
$$(a^3 - q/2)^2 = (q/2)^2 + (p/3)^3,$$
$$a^3 - q/2 = \sqrt{(q/2)^2 + (p/3)^3},$$
$$a^3 = \sqrt{(q/2)^2 + (p/3)^3} + q/2,$$
$$a = \sqrt[3]{\sqrt{(q/2)^2 + (p/3)^3} + q/2},$$

which Cardano called the *binomium*. Similarly,

$$b = \sqrt[3]{\sqrt{(q/2)^2 + (p/3)^3} - q/2},$$

which he called the *apotome*. The solution for $x^3 + 6x = 20$, where $p = 6$ and $q = 20$ is then

$$x = a - b = \sqrt[3]{\sqrt{10^2 + (6/3)^3} + 20/2} - \sqrt[3]{\sqrt{10^2 + (6/3)^3} - 20/2}$$
$$= \sqrt[3]{\sqrt{108} + 10} - \sqrt[3]{\sqrt{108} - 10}.$$

While Cardano perceived that cubics might have three roots, he did not affirm that they must. Another limitation was his inability to resolve the *casus irreducibilis*, the irreducible case, in which all roots are real and distinct. To extract these three real roots by his rule, Cardano had to take a circuitous route, first approaching and then departing from the uncharted realm of complex numbers, which in modern notation have the form $a + \sqrt{-b}$, where a and b are integers and $b > 0$. When $a = 0$, these were considered imaginary numbers. This unusual process made extraction difficult without a well-formed notion of complex numbers. Cardano's solution for $x^3 = 15x + 4$, for example, is

$$x = \sqrt[3]{2 + \sqrt{-121}} + \sqrt[3]{2 - \sqrt{-121}},$$

but he believed that a negative number does not have a square root. He considered such solutions overly subtle and at first thought them useless. Work on solutions involving imaginary numbers meant, however, that the challenge to develop complex numbers could no longer be ignored. In 1572 Bombelli's *Algebra* presented the earliest demonstration of the reality and roots of the irreducible case, and Leonhard Euler would provide in 1732 the first complete exposition of how to employ Cardano's rule and obtain three real roots.

The general algebraic solution of the quartic equation followed quickly on that of the cubic and also appears in *Ars magna*. Zuanne da Coi had provided the impetus in 1540, when he challenged Cardano to solve the problem: "Divide 10 into three proportional parts, so that the product of the first and second is 6." If the three parts are w, x, and y, then $w + x + y = 10$, $wy = x^2$, and $wx = 6$, and when w and y are eliminated,

$$x^4 + 6x^2 + 36 = 60x \quad \text{or} \quad (x^2 + 6)^2 = 60x - 6x^2.$$

Cardano failed to solve this equation but assigned it to his student and secretary Ludovico Ferrari, a future professor of mathematics at Bologna. Ferrari, like his mentor, knew that substituting the quantity $(x + a/3)$ for x in cubics can reduce these equations to a simpler equivalent form without the quadratic term. He now sought to reduce quartic equations to equivalent cubics. In present-day notation, his method begins with adding a new unknown term y to $(x^2 + 6)^2 = 60x + 6x^2$ and completing the square on the left:

$$(x^2 + 6 + y)^2 = 6x^2 + 60x + y^2 + 12y + 2yx^2$$

or

$$(x^2 + 6 + y)^2 = (2y + 6)x^2 + 60x + (y^2 + 12y).$$

Next make the right side of the equation a perfect square. For this, eliminate the discriminant of the quadratic in x on the right side. This condition was well known. Meeting it requires this procedure:

$$60^2 = 4(2y + 6)(y^2 + 12y)$$

or

$$900 = (2y + 6)(y^2 + 12y), \text{ which yields the cubic}$$
$$450 = y^3 + 15y^2 + 36y.$$

Ferrari could then determine the value of y and upon placing it in, easily solve the equation that completes the square:

$$(x^2 + 6 + y)^2 = (2y + 6)x^2 + 60x + 12y + y^2.$$

Cardano credited Ferrari with solving this quartic and essentially followed his method to handle in chapter XXXIX of *Ars magna* several cases of quartics, such as $x^4 = cx^3 + n$ and $x^4 = bx^2 + cx^3 + nm$. Since he restricted quartics to those with positive coefficients, he had to consider at least twenty different cases. Cardano had devices other than Ferrari's method to handle special cases. Neither realized that Ferrari's method applies to all quartic equations, which in 1572 Rafael Bombelli first proved in his *Algebra*.

Its many examples lacking generalization and its view of number make *Ars magna* tedious for the modern reader. Equations such as $x^3 + px = q$, $x^3 + p = qx$, $x^3 = px^2 + qx + r$, and $x^3 + px^2 = qx + r$ in modern notation are solved in separate chapters with geometrical justification for Cardano's rule in each case. Cardano still regarded negative numbers as "fictitious" (*aestimationes vel fictae*). His initial encounter with imaginary numbers in solving problem 3 of chapter XXXVII, $5 + \sqrt{-15}$ and $5 - \sqrt{-15}$, led him to exclaim: "So progresses arithmetic subtly the end of which, as is said, is as refined as useless." Despite these misgivings, he continued to work with these and negative roots, which he eliminated from final results. Overcoming his own and his contemporaries' unease with imaginary numbers, in the section "De aleza regula" that he added to the 1570 edition of *Ars magna* Cardano extends his rule to them.

Cardano's syncopated notation, though it too looks awkward today, improved on the verbal stage of algebra. Here is how he wrote the equation $x^4 + 6x^2 + 36 = 60x$ and added $6x^2$ to each side:

1 *quad . quad .* p : 6 *quad .*p : 36 *aequalia* 60 *pos.*
+   6 *quad*   6 *quad*

―――――――――――――――――――――――――――――――――――――

1 *quad . quad .* p : 12 *quad . p* : 36 *aequalia* 6 *quad .* p. : 60 *pos.*

The solution to the reduced, equivalent cubic using his formulas is

R$_x$V *cubica* $287\frac{1}{2}$.p : R$_x$$80499\frac{1}{4}$.p : R$_x$V : *cubica* $287\frac{1}{2}$. m : R$_x$$80499\frac{1}{4}$. m.: 5,

which in modern symbols is

$$\sqrt[3]{287\frac{1}{2} + \sqrt{80,499\frac{1}{4}}} + \sqrt[3]{287\frac{1}{2} - \sqrt{80,499\frac{1}{4}}} - 5.$$

Note that Cardano lacked a single alphabetical symbol for the unknown, which he denotes by the abbreviation *pos (= positio)*. Similarly, he indicates powers by abbreviations for *quadratus, cubus,* and *quadratus quadratus*. R$_x$ represents the radical sign, V the root of the whole expression, p *(piu)* $+$ and m *(meno)* $-$.

Early typographic culture made for much confusion over proprietary and priority claims. In an age when mathematical discoveries were treated much as corporation research is today, Tartaglia was angered at a breach of an oath of secrecy and at being cheated of full recognition for his work. The publication of *Ars magna* prompted one of the most bitter and celebrated feuds in the history of mathematics.

Book IX of Tartaglia's *Quaesiti et Inventioni Diverse*, which appeared in 1546, reviews Cardano's attempts to obtain a general solution of cubics, giving a purported verbatim account of their meeting in Milan. Ferrari, who had been present at the meeting, protested this attack against Cardano, asserting instead that Cardano had never taken an oath of secrecy and had been magnanimous in mentioning Tartaglia in *Ars magna*. According to Ferrari, not only was Cardano innocent of plagiarism, but Tartaglia himself was a plagiarist. In a notice (*cartello*), Ferrari challenged him to a public disputation on arithmetic, algebra, geometry, and related areas, including optics. Tartaglia asked Cardano to fight his own battles, but did not draw him into the fray. Tartaglia termed Ferrari "Cardano's creature." After an insult-filled exchange of six *cartelli* by Ferrari and six *riposte* by Tartaglia, Tartaglia agreed to a public mathematical joust. Since both men taught mathematics publicly, the outcome could injure the reputation of either of them.

The first stage of the combat was by correspondence. In April and May 1547, each man proposed thirty-one algebra problems to be solved. Curiously, Tartaglia posed problems no more difficult than those in Pacioli's *Summa*. In July he solved twenty-six of Ferrari's problems, which led to cubics. In October, Ferrari solved Tartaglia's problems and criticized his opponent's solutions. Subsequently, both men agreed to a face-to-face debate on August 10, 1548, in Milan before a gathering of local dignitaries. The setting was not favorable to Tartaglia and he was later to argue that at the meeting he had not been allowed to state his case properly. Tartaglia had to leave on August 11 for Brescia before the debate ended. One report holds that he was lucky to escape with his life. It seems that Ferrari was perceived to be the victor, for Tartaglia lost his teaching position in Brescia and Ferrari was invited to lecture in Venice. The Milanese episode illustrates the breakdown of scholasticism in the degeneration of the *disputatio*, its most revered instrument of investigation.

Presenting Cardano's rule and noting relationships between roots and equation coefficients as well as between successive signs of terms and those of roots, *Ars magna* substantially created the theory of algebraic equations in Europe and strongly influenced its growth there. Its general solution of cubic and quartic equations led others to attempt over the next 280 years to solve higher-degree equations similarly. These attempts ended only with Paolo Ruffini's faulty proof in 1799 and Niels Abel's sound proof in 1824 of the impossibility of solving quintic or higher equations by root extraction except in a few special cases. Almost another half century elapsed after Abel before the group theory of Évariste Galois indicating which higher-degree equations admit to algebraic solution was well integrated into mathematics.

13.4   Cardano and Bombelli: Lives and Works

Tartaglia did not let die the feud of 1547. *General trattato . . .* accuses Cardano of perjury and questions his character. This book was long influential among geometers and helped put Cardano in a poor light. He probably deserves a better assessment.

Cardano, like his antagonist, lacked a stable personality. A man of passionate contrasts, part ancient, part modern, Cardano combined a powerful intellect and wide interests with childlike credulity, strong fears, and superstitions, all of which fed his fertile imagination. Magic and number mysticism, which he cherished, sparked his algebraic studies. Algebra, which he pursued more deeply than his contemporaries, was for him a magical talent as much as method. This interest in magic led him to become a formidable apologist for the occult arts and for the belief that various intelligences inhabit the universe. High temper, spirited quarrelsomeness, and perhaps a conceit that exceeds self-confidence characterized his personality, traits likely exacerbated by the difficulties of his life. In time he became vindictive. He imagined terrible punishments for his opponents, but did not carry them out.

Girolamo Cardano

The early life of Cardano had been harsh. The natural son of jurist Fazio Cardano, he was born and grew up in Pavia. Chronic illness scarred his youth, and his autobiography, *De vita propria liber*, claims scornful parents treated him miserably.[387] Yet his father, who belonged to Leonardo da Vinci's circle, urged him to study the classics, mathematics, and astrology. Young Cardano entered the University of Pavia in 1520 and graduated in medicine from Padua in 1526. He spent the next six years practicing medicine in nearby Saccolongo.

During his residence in Milan from 1534 to 1543, Cardano rose slowly from poverty to fame. The Milanese senator Francesco Sfondrato sponsored him to teach mathematics, astronomy, dialectics, Latin, and Greek at the "piattine" schools in the city, secular institutions named for their founder Tommaso Piatti (d. 1502). Cardano sought to practice medicine, but for many years the city's College of Physicians withheld admission. This denial was injurious, for mathematics was not yet a profession and did not provide a livelihood. In 1536 Cardano had the good fortune to gain Ferrari as a member of his household and as a student. Cardano's autobiography describes Ferrari as having "excelled as a youth all my pupils by the high degree of his learning." In 1539 Cardano displayed in his first book on mathematics, *Practica arithmetica*, a broad mastery of different methods of mnemonic calculation and confidence in handling linear and quadratic equations. The celebrated clash with Tartaglia over the solution of cubics began the same year. Only after persistent efforts did Cardano gain permission to practice medicine in Milan, and he soon became the city's most famous physician. By the early 1540s, only Vesalius surpassed him as a European authority on theoretical medicine, a situation that made many of Cardano's Milanese colleagues envious.

In 1543 Cardano returned to the University of Pavia, where he was technically a professor of medicine until 1560. He was an inspired teacher. Students flocked to his lectures, often trying to push in through the windows to a room in which every seat was filled. Publication in 1550 of his massive *De subtilitate* (*The Subtlety of Things*) and in 1557 its supplement, *De rerum varietate* (*The Various Realms*), enhanced his reputation beyond the bounds of medicine and mathematics. Written in elliptical Latin, they offer an animistic account of the natural sciences, both real and imaginary, of his time. These texts incorporate elements of cosmology, machine technology, astrology, alchemy, natural magic, superstition, and branches of the occult, even describing the influence of demons. Cardano reduced the four Aristotelian elements to three by holding that fire is a mode of matter. A fervent admirer of Archimedes, he studied the lever and inclined plane. His claim that mathematics had but limited application was sharply rebutted by the humanist Julius Caesar Scaliger. According to Pierre Duhem, similarities between Cardano's views and the thought of Leonardo da Vinci indicate that, in preparing these books, Cardano probably used Leonardo's manuscript notes. Whatever the source, Cardano introduced challenging ideas and spurred new investigations. The five editions of *De rerum varietate* and the eight editions of *De subtilitate* in the sixteenth century had widespread use.

From 1552 to 1559 Cardano traveled about western Europe. Noble and ecclesiastical leaders widely sought his medical services. He successfully treated the asthmatic Archbishop John Hamilton of St. Andrews, Scotland, in 1552 and then journeyed to the court of Edward VI in England. In debates with the medical faculty at the University of Paris, he defended Vesalian heresies. Cardano was at the height of his fame and prosperity in 1560, only to be banished from Pavia after several personal calamities. The worst was the trial and execution of his elder son, Giambattista, for poisoning his wife. Cardano was crushed; his medical practice declined; and his critics were more vocal. He now developed a passion for games of chance: cards, chess, dice, and knucklebones.

From 1562 to 1570, when the Inquisition imprisoned and tortured him, Cardano lived in moderate material comfort as a professor of medicine at the University of Bologna. Then, charged with heresy for casting a horoscope of Jesus and for his free thought, he was barred from teaching or publishing. After influential churchmen interceded to gain his release from prison, Cardano went to Rome, where Pope Pius V hired him as an astrologer. His final years were spent writing his autobiography, *De vita propria liber* (*The Book of My Life*) and various commentaries. These were the last of his more than two hundred treatises on mathematics, medicine, magic, mechanics, hydrodynamics, geology, philosophy, music, and religion. In 1576 he committed suicide. One suggestion is that this was to fulfill an astrological prediction of the date of his death.

For his contributions to algebra and to probability, Cardano has been adjudged the leading mathematician of his time. Consider his posthumously published *Liber de Ludo Aleae* (*Book on Games of Chance*). Probably written during his years in Bologna, it pioneered probability theory by basing games of chance not on the irreducible unpredictability known as pure luck but on rules and laws. *De ludo* was the first work to define probability as the ratio of favorable outcomes to all possible outcomes and gave an early form of Bernoulli's law of large numbers and the so-called power law. Cardano recognized that reaching the power law requires multiplying probabilities and not odds for independent events. After initial confusion over which to multiply, he generalized his findings for n trials to form the power law: if for n trials f is all possible outcomes and p the number of positive outcomes, the odds in favor of having positive outcomes are p^n to $f^n - p^n$. *De ludo* also solved problems relating to interrupted games of chance. But filled with classical allusions, pragmatic advice, strategies of competition, and his struggle to define terms better, it is difficult to follow and was not published until a century later in 1663, after Fermat and Pascal had conducted similar investigations and Huygens had published in 1656 his influential *Tractatus de ratiocinis in aleae ludo*, and so had little influence. Cardano's stature in mathematics seems greater to us than to his contemporaries and immediate successors.

Raphael Bombelli (1526–1572), from whom sixteenth-century algebra received its most comprehensive treatment and improved notation in Italy, was the son of a wool merchant. Becoming a hydraulic engineer and architect in the service of the bishop of Melfi, he helped reclaim land by desiccating marshes recognized for the wrong reason as threatening Rome with malaria. He also designed canals and repaired bridges over the Tiber. At mid-century he encountered the vigorous mathematical developments in northern and central Italian cities. Practical arithmetics were actively produced here. The dispute between Tartaglia and Ferrari over the solution of cubics was known through Tartaglia's *Quaesiti* and public circulation of the correspondence between the two antagonists. After carefully studying these materials, Bombelli concluded that none of his predecessors, except Cardano, had treated algebra in sufficient depth, while even Cardano lacked clarity in exposition. He therefore decided to write a treatise on algebra to succeed Cardano's *Ars magna* and to allow a student to master algebra without any other text.

As an engineer and architect, Bombelli had the leisure to prepare the manuscript, mostly between 1557 and 1560. While in Rome, he studied and translated part of a recently discovered codex of Diophantus' *Arithmetica* in the Vatican Library. The codex was marred by copying mistakes and numerical errors. Even so, Bombelli's translation of five of seven known books of the *Arithmetica* fired him with enthusiasm to complete his *Algebra*. The final version of his *Algebra*, which includes 143 problems from the *Arithmetica*, popularized that work in the Latin West. Bombelli neither thoroughly translated the

Arithmetica nor wrote a commentary. That achievement required the labors of Wilhelm Holzmann, who hellenized his name to be Xylander (d. 1576), and Claude Bachet (d. 1638).

Of the five books of Bombelli's *Algebra,* which present their subject systematically, only the first three were published in Venice in 1572. The last two remained in manuscript until rediscovered in 1923 and published six years later. Book I defines basic concepts, such as power and root, and illustrates fundamental algebraic operations. Book II introduces algebraic notation and solves equations of the first through the fourth degree. Only equations with positive coefficients are solved. After initially choosing for the manuscript version of Book III a series of practical problems, Bombelli replaced these in the printed version with abstract ones, including the 143 from Diophantus. His *Algebra* thereby runs counter to the tradition of applied problems that had informed practical arithmetics and Pacioli's *Summa.* Bombelli's *Algebra* is thus both a survey of earlier knowledge and an original contribution.

Because equations involving square and cube roots of negative (imaginary) numbers seem to lead the investigator back to the original equation, they had been termed irreducible, and Cardano believed that dal Ferro's rule could not solve those with three real roots. But Bombelli applies dal Ferro's rule to the equation $x^3 = 15x + 4$, and pursuing a circuitous path in and out of the domain of complex numbers, proves that the imaginary roots give a real value. Bombelli had "a wild thought in the judgment of many" that the two radicands of cube roots are related. We say that an imaginary root generates its conjugate: the radicals in the case Bombelli cites are $\sqrt[3]{2 + \sqrt{-121}}$ and $\sqrt[3]{2 - \sqrt{-121}}$. Near the end of Book I Bombelli introduces these two radicands as "another kind of cube root of an aggregate." Abbreviations of *piu di meno* and *meno di meno* denote square roots of a negative number so that *p. di m.* 5 stands for what in present-day notation is $+\sqrt{-5}$ or $+5i$ and *m. di m.* 5 the modern $-\sqrt{-5}$ or $-5i$. Beyond recognizing imaginary numbers, Bombelli sets forth correct rules for computing with them. His work eroded the notion of the mystical nature of imaginary numbers, but they were not to be fully accepted as complex numbers until the nineteenth century. Although bold with these numbers, Bombelli rejected negative roots.

Bombelli powerfully advanced algebraic notation. By drawing on past symbolism to indicate operations, he widened the acceptance of his *Algebra.* For operations, the traditional *p.* from *piu* denotes plus, and *m.* from *meno*, minus; and the conventional radical sign R_x followed by a *q (quadrata)* represents the square root ($R_x.q$); by *c (cubo)*, the cube root ($R_x.c$); by *qq (quadroquadrata)*, the fourth root ($R_x.qq$); and so forth. To collect two or more terms under the radical, Bombelli introduced a new symbol $\lfloor \quad \rfloor$. For example, $R_x \lfloor 4mR_x7 \rfloor$ represents $\sqrt{4 - \sqrt{7}}$. Here is the forerunner of the present bracket. Bombelli's chief improvements in notation, contained in Book II of the *Algebra,* relate to terms for the unknown x and its powers. Pacioli had referred to the unknown as *cosa* (thing, in Italian) or *res* (thing, in Latin); Bombelli called it *tanto* or *quantita*, which has a numerical meaning. Powers are called *dignita*, a term Tartaglia had introduced. Like Nicholas Chuquet, Bombelli placed exponents in a small semicircle, or later in time, in an arc above the numerical coefficient. The scheme for powers in Book II through the twelfth is:

Name of the *dignita*	Bombelli's notation	Modern notation
Tanto	1	x
Potenza	2	x^2
Cubo	3	x^3
...	...	...
Cubo di potenza di potenza	12	x^{12}

By this notation, $\overset{3}{2}$. p. $\overset{2}{5}$. m. $3 = 2x^3 + 5x^2 - 3$.

13.5   Algebra outside Italy and Cossic Arithmetic during the Sixteenth Century

Sixteenth-century mathematicians outside Italian cities also contributed to algebra and arithmetical computation. Algebraic results in midalpine central Europe before 1560 were less impressive than Italian achievements, but in computational arithmetic Austrian, German, and Slavic mathematicians began to emphasize decimal numeration and discovered a correlation between arithmetic and geometric series, which presaged logarithms. As part of a sudden appearance and blossoming of algebra in France, after 1550 dominance in the field passed there to François Viète, who generalized the theory of equations from Italian and midalpine masters.

The Silesian scholar Christoff Rudolff (d. ca. 1545), who ranks with Adam Ries and Michael Stifel as a leading mathematician in midalpine central Europe between 1500 and 1560, studied mathematics under Grammateus at the University of Vienna and then published in Strasbourg in 1525 the first comprehensive German text on algebra, or *Coss*, the German equivalent of the Italian *cosa* or unknown. This etymology suggests the diffusion of algebraic thought north out of Renaissance Italian cities.

Rudolff's *Coss*, which begins with a detailed explanation of place-value integers, occasionally employs decimal fractions and multiplies powers of numbers. Powers are represented by first letters of words, combinations of these, and special symbols, such as *zensus* or *z* for 2, *cubus* or *c* for 3, *zz* for 4, β for 5, *zc* for 6, and the like, which makes their multiplication like multiplying Roman numerals. This *Coss* substantially improves symbols. Following his Vienna teacher Heinrich Schrieber, Rudolff, in contrast to Johann Widman, uses the symbol + for addition rather than excesses and − for subtraction instead of deficiencies. Rudolff proposes a dot as a symbol for aggregation, for example his $\sqrt{.12} + \sqrt{105}$ in modern notation is $\sqrt{12 + \sqrt{105}}$. In place of the R_x symbol, he introduces the modern sign $\sqrt{}$ for square root, $\checkmark\checkmark$ for cube root, and $\checkmark\checkmark\checkmark$ for fourth root. These symbols add a diagonal stroke to forms used by earlier cossists. But Rudolff finds the last two symbols unwieldy. Since his work suggests the equivalent of setting $x^0 = 1$, an idea fundamental to logarithms, he came close to starting exponential arithmetic.

The second half of Rudolff's *Coss* gives basic rules (*cautelae*) that he utilizes for solving algebraic equations. His commercial, abstract, and recreational (*enigmata*) problems often lead to indeterminate first-degree equations and assume an eightfold classification of equations, rather than the twenty-four distinct types of earlier cossists or al-Khwārizmī's six basic types of quadratics. Rudolff gives the double root of the equivalent of $ax^2 + b = cx$. He recognizes rational, irrational and communicant roots, but like most authors of the time rejects negative numbers or zeros as roots. Roots having a common rational factor are communicant. The *Coss* ends with a discussion of three irreducible cubic equations and a drawing of a cube whose edge is $3 + \sqrt{2}$. Rudolff wanted thereby to stimulate more algebraic studies so that his successors would find a general solution for cubic equations.

Historians liken Rudolff's influence in Austrian and German lands to Leonardo of Pisa's in the Italian city states. Some contemporaries would have disagreed. They accused Rudolff of stealing from texts in the Vienna Library many of the 434 problems in his *Coss*, a charge Michael Stifel denied in his expanded 1553 edition of the *Coss*. In the eighteenth century, Stifel's edition was to be Leonhard Euler's first mathematics textbook.

An Augustinian monk, later an itinerant Lutheran pastor, and Reformation Germany's greatest cossist, Michael Stifel (ca. 1487–1567) immersed himself in the study of number mysticism to interpret prophecies in the books of Daniel and Revelation. For a cabalist interpretation of them, he developed a word calculus (*Wortrechnung*). He connected words and numbers, correlating twenty-three letters of the alphabet with the initial twenty-three triangular numbers. Applying his "holy arithmetic of numbers," Stifel arrived at the number 666 for the great malevolent beast of Revelation. By rearranging the Roman numerals in "Leo Decimus" to make DCLXVI, Stifel then equated that number with Pope Leo X, known for his avid hunting. The letter X seems to come from Leo X or the ten letters in Leo Decimus. Since the letter m stood for "mysterium," Stifel dropped it from his calculation.[388] Using his word calculus, Stifel predicted the ending of the world on October 3, 1533. When this prophesy failed to come true, Stifel was dismissed from his parish. After castigating him, Luther intervened to gain him another parish. Afterward, under the influence of the Wittenberg mathematics lecturer Jacob Milich, Stifel studied Euclid, Ptolemy, Boethius, Cusa, Jordanus, Ries, Dürer, and Rudolff and became an accomplished student of algebra at the University of Wittenberg.

Stifel wrote and had published four mathematical books. In 1544 appeared *Arithmetica Integra*, a survey of arithmetic and algebra that Milich had proposed. Its preface was by Philipp Melanchthon. The *Deutsche Arithmetice* of 1545 attempted to make the *Coss* more accessible to readers of the German vernacular by eliminating foreign words. In 1553 Stifel completed an expanded revision of Rudolff's *Coss* and a work on number mysticism, *Ein sehr wunder- barliche Wortrechnung.*

Arithmetica Integra, which epitomizes the effort among German cossists to replace verbal algebra with systematic notation and add original contributions, covers in its first book properties of number, proportions, the rule of three, and numerical relations in magic squares. In a table, this book impressively correlates a geometric and an arithmetic progression. The table was not entirely new, but Stifel surpassed Rudolff in correlating powers of 2 with their exponential numbers, including negatives:

-3	-2	-1	0	1	2	3	4
1/8	1/4	1/2	1	2	4	8	16

Stifel introduces the term exponents (*exponens*) for the upper series. The two series, he observes, can be expanded *ad infinitum,* and multiplication, division, and root extraction of the second row correspond to a simpler operation of addition or subtraction in the first. The table and accompanying text helped lay the foundations of logarithmic calculation.

The second book of *Arithmetica integra* deals with root extraction and irrational numbers. Apparently, Stifel independently discovered binomial coefficients and calculated their Pascal triangle to the seventeenth line. His book was the second in the Latin West to present the Pascal triangle. In 1527 Peter Apianus had given it to the ninth line. Having with great difficulty devised a general method to extract roots by using binomial coefficients, Stifel follows Euclid in rejecting irrational numbers as roots, but computes skillfully with them. In working with those of the form $\sqrt[n]{a} + \sqrt[n]{b}$, he draws on Book 10 of the *Elements.*

Solving equations is the subject of the third book. Combining into a single rule various cossic rules, Stifel seeks a standard form as essential to improving algebra. By letting a and b be either positive or negative he reduces to the form $x^2 = ax + b$ four separate cases of quadratic equations, and he arranges equations in descending powers of the unknown. In extracting roots of polynomials, Stifel surpassed his German predecessors, but he continued to ignore negative roots, dismissing negative numbers as absurd. Stifel sought a general arithmetic solution for cubic equations but had to await Cardano's *Ars magna.* Cardano's method appears in the 1553 edition of Rudolff's *Coss.*

Although his contributions to modern symbolic notation were minor, Stifel moved to its threshold. He represents the unknown with capital letters A, B, C, D, and F along with x in order to deal with more than one unknown. His *Arithmetica integra* indicates powers by a capital letter followed by the first letter of the word for a power, for example, Az for x^2, where $z = Zensus$ or *quadratum* and Ac, where $c = Cubus$. Stifel's later writings repeat the letter to get AAA for x^3, and $FFFF$ for x^4. His new method redesignated from the traditional R_x to $\sqrt{z}$ for square root the symbolic representation of roots. Later Stifel dropped the cossic sign z and used $\sqrt{.}$ and finally $\sqrt{}$ alone. Before the next step toward a general system occurred, an abstraction of algebraic operations had to separate them from their objects. Viète and Descartes were the first to achieve this.

Stifel influenced at least three generations of transalpine central and northwestern European mathematicians. His writings were a major source, for instance, of Robert Recorde's *Whetstone of Witte*. The last edition of *Arithmetica integra* was in 1586 and of the *Coss* in 1615. By 1620 Viète's more facile notation and methods and Napier's logarithms had opened new fields for research. Stifel's writings were thereby superseded and generally fell into disuse.

After 1550 algebraic studies suddenly appeared and quickly blossomed in France. The emphasis French humanists placed on abstract mathematics, along with Petrus Ramus's criticism of the absence in schools of mathematical studies connected to the marketplace, reinforced this movement. The first algebra published in France was by a German, Johann Scheubel, whose *Algebre compendiosa* came out in 1551. A series of algebras by French authors followed. Jacques Peletier's *L'algebre*, published in 1554, was based not on French forerunners but on Pacioli, Cardano, Stifel, Ries, Scheubel, and Pedro Nuñez. Peletier demonstrates how Stifel's symbols make algebra a more effective science. The *Logistica* of Jean Borel (Johannes Buteo), which was published in 1559, attempts to build algebra on the model of Greek geometry, particularly that in *Elements* II. Petrus Ramus's *Algebra* of 1560 urges greater inclusion of the "new" field of algebra in education. Inspired by Tartaglia and Nuñez, Guillaume Gosselin's *De arte magnae*, appearing in 1577, treats Diophantine equations. The scholar who in 1585 brought algebra to its highest level in France prior to François Viète (1540–1603) was the Dutch engineer Simon Stevin in *L'arithmetique*. This book cites Italian and German but no French authors and translates the first four books of Diophantus' *Arithmetica*. The greatest late sixteenth-century algebraist, however, was French. François Viète, generally known by the Latinized form of his name, Franciscus Vieta, generalized and extended contributions of his ancient, medieval Arabic, Italian, and midalpine predecessors.

The son of a lawyer and notary, Viète was trained for the law. Upon graduating from the University of Poitiers in 1560, he became a lawyer in his native city of Fontenay. Four years later he left the law and entered the service of the Huguenot Jean de Parthenay, tutoring Parthenay's daughter Catherine in cosmography and writing textbooks for her. In 1571 he was made an advocate in the Paris Parlement, the appellate court. While legal counselor to the Brittany Parlement at Rennes, a position he held from 1573 to 1579, Viète was asked to join the growing royal service. He advanced rapidly. In 1576, while he was still with the Brittany Parlement, King Henry III named him master of requests in Paris and three years later a member of the royal privy council.

François Viète

During the French wars of religion, Viète sympathized with the plight of the minority Protestant Huguenots. This placed him at odds with the Catholic League, which demanded religious unity in France. The League succeeded in having him banished from court in 1584. After opponents of the League attempted unsuccessfully to crush it, he fled Paris and moved to Tours. In 1589 Henry III recalled him to royal service as counselor at the Parlement there. In that year the monarch allied with the Protestant Henry Bourbon, who became king when the childless Henry III was assassinated. To strengthen his claim to the throne, Henry IV embraced the majority religion of Catholicism in 1593. Five years later this witty ruler issued the Edict of Nantes granting limited toleration to Huguenots. Viète remained in the king's service until poor health forced him to retire in 1602.

As his wealth increased, Viète had spent more of his leisure studying mathematical science and cryptography. When Spain in 1589 and 1590 militarily challenged Henry IV's accession to the French throne, Viète decoded intercepted letters intended for the Spanish emperor Philip II that contained a complex cipher of 413 characters. Philip believed that such decoding work could not be done by natural means. Reportedly, he complained to the Pope that the French were using black magic against him.

Viète had two periods of leisure that he devoted to highly fruitful investigations of trigonometry, geometry, algebra, and arithmetic. When absorbed in mathematics, he apparently sequestered himself in his study for days. During the first period, from 1564 to 1568, he closely linked mathematics to astronomy and cosmology. His manuscript "Harmonicum coeleste" offered geometric arguments supporting Ptolemaic over Copernican astronomy. This manuscript was intended as a preparatory work for his *Canon Mathematicus seu ad Triangula* (*Mathematical Laws Concerning Triangles*). It took eight years beginning in 1571 to print its two parts on trigonometric computations. Viète financed his own works, which escaped the oblivion that frequently follows such ventures. During the second and enforced period of leisure in the late 1580s, Viète virtually completed his chief book, *In Artem Analyticam Isagoge* (*Introduction to the Analytic Art*), which was printed in Tours in 1591. Analytic art (*ars analytica*) was the way he preferred to refer to algebra.

Prior to the mid-sixteenth century, algebra had been largely an appendage to geometry. Most of the surviving ancient Greek geometry was synthetic, and even Greek masters who used analysis, like Apollonius, followed it with a synthetic demonstration. Again, before 1500 mathematicians customarily used conic sections to solve cubic equations. But the intense study of quartic and quintic equations carried the disturbing suggestions that geometry, limited to three dimensions, is insufficient for convincing proofs. Viète pursued a new logic for algebra, intending to establish a general method and facile notation. In the preface to the *Isagoge*, Viète calls his discoveries "pure gold." They are "not alchemist's gold soon to go up in smoke, but the true metal, dug out from the mines where the dragons are standing watch."

A humanist skilled in gaining acceptance of new thought, Viète claimed to be restoring from antiquity a general method of analysis. He benefited from the skillful translation of ancient mathematical texts in his time. His *Isagoge* connects procedures in Diophantus' *Arithmetica* with theorems and problems in Book VII of Pappus' *Mathematical Collection*. More generally, it amalgamates with the numerical algebra of Diophantus, Cardano, Tartaglia, and perhaps Bombelli and Stifel, methods of analysis found in Euclid, Archimedes, Apollonius, and Pappus. Its new analysis for calculating magnitudes from canonically ordered equations or proportions exceeds these earlier methods.

Viète's *logistica speciosa* is analogous to geometric analysis and parallel to Diophantus' *logistica numerosa*, but more general. Diophantus had calculated only with numbers or equations with positive coefficients and thus his method is close to arithmetic. *Logistica speciosa* operates with letters representing species or classes of numbers and equations. What is done for a general case applies to an indefinite number of special cases. *Logistica speciosa* thereby generalizes arithmetic and numerical as opposed to geometrical algebra. It comes near to a modern definition of algebra.

The *Isagoge* divides the analytic art into three branches. Two are taken from Pappus: theorematic and problematic analysis. Viète calls theorematic analysis "zetetics" or truth seeking. Zetetics transforms problems into equations connecting the unknown with given knowns. Problematic analysis, or "poristics," investigates the truth of proposed theorems through symbolic manipulations in equations. To these branches, Viète adds rhetic analysis, or "exegetics." Exegetics solves equations derived by zetetics by determining the unknown through numerical calculations or geometric constructions.

Viète's notation is the clearest and most consistently applied of various attempts by sixteenth-century algebraists. To designate algebraic entities, he methodically adapts vowels and consonants of the alphabet. Capital consonants $(B, C, D, \dots)$ denote known quantities and capital vowels unknowns. This was the first unambiguous symbolic distinction between parameters and unknowns.

Its use was short-lived: in the 1630s Descartes modified Viète's notation by vowels and consonants, letting initial letters of the alphabet denote knowns and later letters unknowns. Descartes' notation continues in use. For the first power of the unknown, Viète later used the capital N from *numerus* rather than vowels. He also introduced letters both for unknowns and for general coefficients. With literal coefficients a, b, c in Cartesian notation, all quadratic equations can be expressed in the same general form as $ax^2 + bx + c = 0$. Earlier algebraists had separate classifications for equations, such as $2x^2 + 3x + 4 = 0$ and $x^2 - 5x + 7 = 0$, although the same method of solution applied to both. Viète adopts the German symbols for plus and minus and adds brackets for grouping terms.

Viète's notation keeps some of the awkwardness of earlier systems. Partly drawing on the past, he indicates powers by the word for each of its abbreviations: *A quadratum*, not x^2; *A cubus*, not x^3; .... The Latin *aequalis* would later give way to the double bar sign (=). Here is a typical equation from the *Zetetica*:

> *A cubus + A* in *quad.3 + A* in *B quad.3 + B cubo aequalia A + B cubo.*

In modern symbols this is $a^3 + 3a^2 + 3ab^2 + b^3 = (a + b)^3$.

The retention of some words and abbreviations hampered Viète's notation. His notation also fails to let the literal coefficients, a, b, c, ... , n, on which he performed operations, represent any real number beyond the rationals or any formative complex number. His *De Aequationum Recognitione et Emendatione* (*On the Recognition and Emendation of Equations*), completed in 1593 but not published until 1615, for example, leaves it "for the geometer rather than the arithmetician to set [irrationals] out accurately."[389] He rejects complex and negative numbers. It remained for Albert Girard (d. 1632) and Jan Hudde (d. 1704) to let literal coefficients represent positive and negative reals.

Franz van Schooten, who edited the collected works of Viète in 1646, improved on his notation. Van Schooten, like Stifel, denotes a power by the first letter of the word for the power: $Aq = a^2$, $Ac = a^3$, ... But he adopts the radical sign $\sqrt{}$, for square root, while Viète had employed the letter l.

One of the discoveries of pure gold is a general method in the *Isagoge* and *De emendatione* for solving both cubic and somewhat reducible quartic equations. This greatly improved the theory of equations, for Cardano and Bombelli had devised a separate method for each. It was probably by reducing equations that Viète reached his general method. To solve cubic equations, he removes the next highest term, x^2, as had Cardano. In today's notation, let $x = y - b/3$ to transform $x^3 + bx^2 + cx + d = 0$ to $y^3 + py + q = 0$. Substituting $y = z - p/3z$ and multiplyiing each term by z^3 gives $z^6 + qz^3 - (p/3)^3 = 0$, a quadratic in z^3 consequently, $z^3 = -q/2 + \sqrt{R}$, where $R = (p/3)^3 + (q/2)^2$. Among the equations Viète solves is $x^3 + 3B^2x = 2z^3$. To obtain $y^6 + 2z^3y^3 = B^6$, a quadratic in y^3, he has $y^2 + yx = B^2$ and solves for x and y. Viète arrives at this value for B^2, since it is assumed to be a rectangle whose smaller side is y and larger side is x. This means that $(B^2 - y^2)/y = x$. To solve irreducible cubics, he employs trigonometric substitutions. To solve quartic equations, Viète also begins by reducing them, removing their x^3 term. Having $x = y - a/4$ yields depressed quartics of the form $y^4 + a^2y^2 = c^4 - b^3y$. Viète then moves the y^2 term to the right-hand side of the equation, and adds to each side $x^2y^2 + y^4/4$, which gives the result

$$(x^2 + y^2/2)^2 = x^2(y^2 - a^2) - b^3x + y^4/4 + c^4.$$

Next the value of y is chosen to make the right-hand side a perfect square. The required condition is $y^6 - a^2y^4 + 4c^4y^2 = 4a^2c^4 + b^6$, a cubic in y^2. After substituting this value of y, Viète takes the square root of both sides of the equation. This yields two quadratic equations for y, which can be solved.

   De emendatione contains another algebraic novelty. Chapter Fourteen gives formulas relating coefficients of an equation to its roots. Viète's formulas show that if the coefficient of the second term of a quadratic equation is minus the sum of two numbers whose product is the third term, then those numbers are the roots of the equation. His rejection of negative roots hindered his fully grasping this relationship. The *Invention Nouvelle en l'Algebre (New Invention in Algebra)* of 1629 by Albert Girard, who accepted negative and complex roots and coefficients, contains the first statement of the relations between roots and coefficients.

   In solving equations, Viète retains the limiting ancient Greek principle of homogeneity, which holds that the first power represents line segments, the second power areas, and the third power solids. He thus uses algebra chiefly to solve geometric problems. But in drawing on the higher geometry of Apollonius and Pappus rather than elementary sections of Euclid, Viète surpassed his predecessors. Relating the new algebra to the higher geometry of antiquity was a crucial development. Descartes and Fermat would soon interpret the powers of unknowns without geometric constraints.

   In 1593 Viète applied trigonometry to solve a forty-fifth degree equation in x. That year a Belgian mathematician, Adriaan van Roomen (d. 1615), in *Ideae Mathematicae pars Prima*, a book on the calculation of trigonometric chords and the quadrature of the circle, had challenged all geometers to solve this problem:

$$x^{45} - 45x^{43} + 95x^{41} - 12,300x^{39} + \ldots - 3,795x^3 + 45 = K.$$

When the Spanish Netherlands ambassador boasted to Henry IV that no Frenchman could solve this problem, the king summoned Viète. His faith was not misplaced.

   Van Roomen's problem, as Viète quickly recognized, reduces to this: when "given the chord of an arc of a circle, find the chord of the 45th part of that arc." Having recently found multiple-angle formulas for sine and cosine by cleverly manipulating right triangles and related mathematical identities, he recognized that van Roomen's problem can be solved by assuming that $K = 2\sin(45\theta)$.

Viète could express this in powers of $\sin \theta$. From

$$\sin n\theta = n \sin \theta - n(n^2 - 1^2) \sin^3 \theta / 3! + n(n^2 - 1^2)(n^2 - 3^2) \sin^3 \theta / 5! + \ldots,$$

he found that

$$2 \sin(45\theta) = 2(45 \sin \theta) - 45(45^2 - 1) \sin^3 \theta / 3! + 45(45^2 - 3^2) \sin^5 \theta / 5! + \ldots.$$

If $x = 2 \sin \theta$, the right-hand side becomes $45x - 3,795x^3 + 95,634x^5 - \ldots - 45x^{43} + x^{45}$. Viète saw at once that $45 = 5 \times 3 \times 3$. He could thus solve the resulting equation by breaking it into a fifth-degree or quintic equation $(5z - 5z^3 + z^5 = y)$ and two cubics $(3x - x^3 = a$ and $3y - y^3 = x)$. He rapidly found solutions for these. Legend holds that within minutes he gave the king two answers written in pencil. The next day he had a complete set of twenty-three positive roots. These appeared in 1595 in his *Responsum*. Viète did not possible solutions, for the rest involved negative sines, which were unintelligible to him.

Although it has often been said that mathematics was chiefly a hobby for Viète and in the *Responsum* he states that he does "not profess to be a mathematician," his reference to the field as "ars nostra" in the preface to *Isagoge*, his substantial mathematical publications, and his thorough efforts to solve problems such as Roomen's argue that mathematics was his intellectual passion and primary interest. He sets standards so high that only a dedicated practitioner could embrace them. It was not sufficient to solve van Roomen's problem; Viète attempted to prove his answer using conic sections from Apollonius and others. In 1600 he had a rigorous geometric proof that so impressed Roomen that he journeyed to Fontenay to meet Viète. The two men became good friends.

De Numerosa Potestatum adds a method for calculating roots of higher-order equations in a manner not unlike the method of ordinary root extraction later devised by Ruffini and Horner. Viète approximates the first digit of a root, or r, and then determines the next lower place-value digit by following a repetition of the process that begins with substituting the first approximation in the equation. He computes its difference from the value given and divides that difference by an amount equal in modern notation to $|(f(r+s_i) - f(r)| - s_i^n$ where, for an equation of standard form $f(x) = 0$, n is the degree and s_i is the unit for the denomination of the root's next digit: for example, a first guess of 10 for the ten's position means that s_i is 1, since that is the next lowest place value. To find the root of the equation $x^5 - 5x^3 + 500x = 7,905,504$, Viète, upon assuming that $r = 20$ and $s_i = 1$, computes $7,905,504 - (r+1)^5 + 5(r+1)^3 - 500(r+1)$, which is 3,857,208, and calculates for the divisor 878,295. The quotient gives 4 as the next digit, which is in the unit's place. The desired root is $20 + 4 = 24$. This method, in use in the 1680s, was so laborious that one geometer described its computations as "work unfit for a Christian."

The *logistica speciosa* and systematic notation in Viète's writings of the 1590s began symbolic algebra in western Europe. After the *Isagoge*, his *Zeteticorum Libri Quinque (Five Books of Zetetics)*, which was published in 1593, demonstrated the power and range of the new algebra. Viète next wrote *De*

Numerosa Potestatam ... (*The Numerical Resolution of Powers* ...), which Marin Getaldič edited and published in Paris in 1600, and *De Emendatione*. This series of writings established Viète's mathematical reputation.

Viète's influence on algebra and mathematics in general is often underestimated, perhaps because he had few disciples. Wide circulation of his writings did not compensate for their lack of clarity. Perhaps in part, but only so, because Viéte's methods and notation were new, many readers found them unintelligible. One attempted translator complained that "it would take a second Viète in order to translate the first." A better gauge of Viète's importance than the number of disciples is a comparison of the achievement of mathematicians who started from his work with the accomplishment of others who labored within a different tradition. The first successor generation included a Croatian, Marin Getaldič and the French exile Albert Girard, and a second generation gathered Pierre Fermat and Gilles Roberval, all outstanding mathematicians. Contributions of Fermat and Roberval to algebra and the origins of calculus overshadowed the accomplishments of Galileo's disciples Bonaventura Cavalieri and Evangelista Torricelli. With Viète and his successors, France displaced Italian cities at the forefront of theoretical mathematics.

13.6   Viète on the Three Classical Problems and Gregorian Calendar Reform

In 1592 the Italo-French classical scholar Joseph Justus Scaliger (d. 1609), addressing the three famous ancient problems, claimed to have trisected an angle, duplicated a cube, and squared the circle by means of ruler and compass only, a claim Oronce Finé had made a half-century earlier.[390] During public lectures at Tours, Viète proved erroneous Scaliger's assertions without mentioning his name. He trisected an angle by the verging process and doubled the cube by constructing two mean proportionals between two line segments. Squaring the circle or, more precisely, determining the value of π gained the attention of European scholars through the seventeenth century. Viète asserted correctly that the lunules of Hippocrates can be squared but not the circle. By inscribing regular polygons of 4, 8, 16, ... sides in a circle having a unit radius, he found π correct to ten decimal places, which suggests that he did not know al-Kāshī's superior approximation. Viète's calculations apparently extended to a regular polygon of $6 \times (2^{16}) = 393,216$ sides.

Book VIII of *Variorum de rebus mathematicis responsorum* ..., published in 1593, which contains these results, also presents in its eighteenth chapter one of the first precise theoretical expressions for π as an infinite product. In modern notation, Viète gives twice the inverse or

$$2/\pi = \cos(90°/2) \cdot \cos(90°/4) \cdot \cos(90°/8) \cdot \ldots$$

$$= \sqrt{1/2} \times \sqrt{1/2 + 1/2\sqrt{1/2}} \times \sqrt{1/2 + 1/2\sqrt{1/2 + 1/2\sqrt{1/2}}} \ldots$$

The infinite product can be derived either by inscribing in a circle regular polygons with indefinitely increasing numbers of sides or by using recursive trigonometric series. To obtain his infinite product, Viète simply inscribes a square in a unit circle and then applies the recursive trigonometric formula $a_{2n} = a_n \sec \alpha/n$, where n is the number of sides and a_n the inscribed regular polygon's area. Here n is allowed to grow ever larger. In this way, Viète computes the area of a unit circle to be

$$2 \times 1/(\sqrt{\frac{1}{2}} \times \sqrt{\frac{1}{2} + \frac{1}{2}\sqrt{\frac{1}{2}}} \times \sqrt{\frac{1}{2} + \frac{1}{2}\sqrt{\frac{1}{2} + \frac{1}{2}\sqrt{\frac{1}{2}}} \cdots}).$$

From this result, he derives his value for π. This work takes into arithmetic, algebra, and trigonometry the mathematics of the infinitely large and infinitely small, fields previously dominated by geometry. While not entirely new, the quest for infinite products began the end for the Archimedean exhaustion method of squaring the circle.

Lacking sufficient theoretical training, Scaliger was not swayed by Viète's reprimand. In *Cyclometrica Elementa Duo* and in *Mesolabium* in 1594, he repeated his erroneous claim for squaring the circle by drawing an inscribed quadrilateral having its diameter and two diagonals stand in arithmetical proportion. Scaliger continued that Archimedes' approximation of π in the *Dimension of the Circle* was incorrect. Among the several mathematicians who rose to refute Scaliger, Viète used the Archimedean spiral to offer two approximations of a segment of a circle but again denied that the circle can be squared in the manner that Scaliger proposed. Two others arrived at better approximations of π: Adriaan van Roomen in *Archimedis Circuli Dimensionem* of 1597 found a value correct to sixteen decimal places, and Ludolph van Ceulen (d. 1610) pushed it to thirty-five. His value, first published in Willebrord Snell's *Cyclometricus* of 1621, so impressed contemporaries that π became known as the "Ludolphine constant."

Beginning in 1594, Viète had a quarrel with the German Jesuit Christoph Clavius (1537–1612) over calendar reform.[391] It was not to his credit. Proposed in 46 B.C., the early version of the Julian calendar of $365\frac{1}{4}$ days had the solar year eleven minutes too long. The precise hours in a day, moreover, were not firmly established until the early sixteenth century. In the precession of the new moon, the early Julian calendar produced every millennium a growing error at the least of almost three and a quarter days that affected the dating of Easter. For centuries, the exact date of Easter was a source of discord among Christians. Exodus 12:2 and 6 set the Jewish Passover on the fourteenth day of the lunar month Nisan; Scriptures put the Resurrection three days later. But early Christians also placed the Resurrection on the first day of the week or a Sunday in the solar calendar. Each dating had support among early Christians. From Bede's *De temporum ratione* of 725, Dionysian reckoning gained wide acceptance for dating Easter. It equates nineteen solar years with 235 lunations or lunar months. Each lunation was twenty-nine or thirty days. When the lunar and the solar year differed by more than thirty days, calendar alignment required adding an extra lunation. In the eleventh and twelfth centuries, *compotus*, the study and collation of lunar and solar movements, flourished in England.[392] Gerland, Roger of Hereford, and later Robert Grosseteste applied this work to increasing the precision of the existing Julian calendar. At the time, astronomers attempted to determine the length of the tropical year, which they initially believed to be variable. Obtaining it requires determining when the ecliptic or sun's path intersects the celestial equatorial coordinates in two equinoctial points or equinoxes and accounting in computations for the precession of the equinoxes. By the late middle ages the tropical year was set at 365.243, which is longer than the actual value of 365.242.

Efforts to relate the Julian, tropical, and solar years compounded the drift in the calendar. By 1500 the total error amounted to one day for every 128.6 years, and not the 134 years assumed in the *Alphonsine Tables*.[393] Today, a working rule for computing the number of years in this increasing drift is $130.1 - Y/1000$, where Y equals the year A.D. Not such long-term astronomical motivations but improving the calendar as a determinant of uniform liturgical worship whose cycles are set out in breviaries and missals prompted Pope Gregory XIII to establish a congregation in Rome in 1580 to reform it.

Clavius, who was shifting from Ptolemaic geocentric astronomy to the Copernican heliostatic model, along with other astronomers completed the proposed reform in 1582. The Julian calendar after Scaliger was in error by eleven days and Pope Gregory XIII in the apostolic letter, or papal bull, *Inter Gravissimas* agreed to drop the days between October 4 and October 15, 1582, and in the future to make into ordinary rather than leap years three of four centennial years, those congruent with 100, 200, and 300 but not 400. Thus began the Gregorian calendar that soon appeared in astronomical literature in the writings of Kepler.[394]

The Gregorian calendar initially roused vehement opposition among natural philosophers. Among them were Viète, who had wanted merely to rectify errors in the Julian calendar, and Tobias Müller. Two years after Clavius turned down proposed corrections submitted by him and Pierre Mettayer in 1600, Viète and Pierre's son Jean went so far as to publish a libelous attack against him. Acceptance of the new calendar in Protestant regions was clouded by its association with Catholicism.

13.7   Theoretical Geometry from 1500 to 1600

During the sixteenth century, European mathematicians constructed a foundation for renewing theoretical geometry superior to any other since antiquity. Humanist translations and restorations of antique texts following on the scholastic recovery of mathematical literacy were crucial to this renewal. Most translations of higher geometry from ancient Greek sources and Arabic and medieval translations were made into Latin, the language of learning. When different versions of a text existed, painstaking comparisons and paleographic investigations permitted more exact translations. The Italian scholars Francesco Maurolico (1494–1575) and Frederigo Commandino (1509–1575) set high standards for these and added elucidating annotations and commentaries. Among the authors retranslated or restored were Apollonius, Archimedes, Euclid, Pappus, and al-Haytham. Through medieval and prior Renaissance times, little attention in Europe had gone to the higher geometry of Apollonius and Archimedes, and the *Mathematical Collection* of Pappus was virtually unknown. A parallel revival of Euclid produced a more accurate Latin *Elements*, along with vernacular translations and critiques of Euclid's proof theory. Collectively, these works provided a solid basis for better mathematical textbooks, chiefly by Christoph Clavius, and for progress in geometry.

The Benedictine mathematician, cosmographer, engineer, historian, and optician Francesco Maurolico, whose parents were Greek refugees from Constantinople, published in 1558 a compendium, entitled *Gnomonica*, that translated the *Sphaerics* of Theodosius of Bythnia and another of Menelaus of Alexandria, the *De sphaerica mobile* of Autolycus of Pitane, the *De Habilitationibus* of Theodosius, the *Phaenomena* of Euclid, and trigonometric tables. In his critical, six-volume *Histoire de l'astronomie*, published between 1817 and 1827, the French mathematician Jean-Baptiste Joseph Delambre credits Maurolico with introducing the secant function to the Latin West, but Copernicus had possessed it earlier. The *Gnomonica* proved that the tip of a gnomon's shadow traces a conic section.

Conic sections drew Maurolico, who first translated from Greek into Latin Books I through IV of Apollonius's *Conics*. Reconstruction of the lost last four books of *Conics* became a vogue among mathematicians and stimulated the development of higher geometry through the seventeenth century. From brief references in chapter prefaces, Maurolico attempted to reconstruct Book V on normals to conic sections, or maxima and minima, and Book VI. His restoration was completed in 1547. Unlike Apollonius, Maurolico presents conics as sections of a cone cut by a plane rather than as plane curves, the standard approach prior to Descartes. Maurolico, in his edition of all known mathematical writings of Archimedes, skillfully determines centers of gravity of a pyramid and a paraboloid of revolution in mechanics. His collected works suggest a competent revival of the study of ancient Greek curves, such as Archimedes' spiral and Diocles' cissoid. But Maurolico's influence on higher geometry in the late sixteenth century was limited, because his translation and reconstruction of *Conics* was not printed until 1659 and his Archimedean compilation came out in 1685, a century after his death. The 1685 compilation was based on the *Archimedes opera*, brought out in 1679 by Giovanni Borelli of the Accademia del Cimento in Tuscany.

Maurolico's chief work, his two-book *Arithmetic*, completed in 1557 and published posthumously as part of his *Opuscula Mathematica* in 1575, advanced proof theory and the theory of numbers. It appeared as the work of the ancient Diophantus was first being translated into Latin, preparing the way for the research of Bachet de Méziriac and Fermat. The first book comprehensively covers figurate numbers from the Pythagoreans, Nicomachus, and Boethius. Maurolico's treatment of polygonal numbers is more complete than that of Diophantus. The American mathematician Leonard Eugene Dickson has noted the ingenuity in Maurolico's proof that every polygonal number is hexagonal and thus triangular. The second book of the *Arithmetic* deals with irrationals and is original in treating Euclid's *Elements* X as arithmetic. Maurolico achieved greater generality than his Greek predecessors. His search for generality led to a formula relating the faces, vertices, and edges of regular polyhedra, a formula later found by Descartes and Euler. By frequently appealing to reasoning by recursion in proving theorems, Book I shows that Maurolico followed an early form of the mathematical principle of complete induction, for example in his proof of the binomial theorem for $n = 3, 4$. Apparently, the fourteenth-century studies of number by Levi Ben Gerson employing it were unknown to him. In his *Traité du triangle arithmétique*, completed in 1654, Blaise Pascal would first clearly state and systematically apply the principle of complete induction.

A lesser mathematician than Maurolico, Frederigo Commandino was the foremost mid-sixteenth century translator of ancient Greek geometric classics. While physician to the Duke of Urbino and his cardinal brother-in-law in Rome, Commandino worked assiduously at this task for twenty years. His translations are based on a Greek manuscript in Venice, a printed Greek text, and a Latin translation that had appeared earlier in 1544 in Basel. His first two volumes were published in 1558. It is the earliest collection of five treatises by Archimedes: *Dimension of a Circle* that included Eutocius' commentary, *Quadrature of the Parabola, Spirals, Conoids and Spheroids*, and *Sand Reckoner*. Commandino's edition of Ptolemy's *Planisphere* together with a commentary was printed in 1558. Onto the plane of the equator of the celestial sphere, Ptolemy had stereographically projected circles on that sphere. Commandino's edition of Ptolemy's *Analemma* appeared in 1562, and afterward he corrected a Latin translation of portions of Archimedes' *Floating Bodies,* which quickened interest in related conics. In 1566 he provided the leading translation of the first four books of the *Conics* of Apollonius, to which he added Eutocius' commentary. After the English physician, mathematician, and occultist John Dee (1527–1608) gave him a Latin translation manuscript of an Arabic rendering of Euclid's *On Divisions* [of Figures], he published in 1570 a new translation and a treatise generalizing its arguments. In 1572 appeared Commandino's translation of Euclid's *Elements* from Greek into Latin, adding extensive notes. His translation of Books III through VII of the *Mathematical Collection* of Pappus was published posthumously in 1588. Primarily through Commandino's efforts, Western scholars had available by 1600 most of the principal surviving ancient Greek texts on higher geometry.

In addition to translating, annotating, and restoring antique higher-geometry texts, sixteenth-century scholars made improved translations of Euclid's *Elements.* They also wrote commentaries on its books.

Bartolomeo Zamberti in 1505 had made a Latin translation from a Theonine

Greek text without medieval Latin and Arabic sources. He thereby arranged the *Elements* more accurately than the medieval Campanus version had done and filled in some omitted parts. Campanus' partisans, notably Luca Pacioli, responded by correcting erroneous text and drawings that had appeared in their version, while Jacques Lefevre d'Étaples in preparing the first Latin edition of the *Elements* in France used those of Campanus and Zamberti. Proofs devised by Campanus followed those of Theon that now appeared in Zamberti. The German theologian Simon Grynaeus took the next step in restoring the original *Elements*. In 1553 Grynaeus published in Basel the first complete Greek text (*editio princeps*) of the *Elements*, based on Theon. All other Greek editions before 1700 were partial.

At mid-century a revival of Euclid was underway. More than the *editio princeps*, the Campanus and Zamberti versions influenced editors of Latin Euclids. In France, Oronce Finé based on Zamberti his six-book Latin version of the *Elements*, which was published in 1536. Finé inserted into Zamberti's translation some missing propositions from Book VI. After Pierre Mondore edited Book X (the later books were still not well understood), the Horace scholar Jacques Peletier issued in 1557 another six-book edition, *In Euclidis Elementa Demonstrationum*, based on Campanus' version from Arabic to Latin. Peletier criticized Euclid's method of superposition for proving congruence theorems: superposition, Peletier observed, implies motion. In 1566 François de Candale prepared a Latin *Elements* of the traditional fifteen books based on those of Campanus and Zamberti. The Campanus and Zamberti versions strongly influenced vernacular renderings, which brought the *Elements* to a larger audience. In 1543 Tartaglia had first translated it into Italian. His edition draws heavily on Campanus but adds missing propositions and premises from Zamberti. Johann Scheubel's translation of Books VII through IX, which appeared in 1558, and the rendering by Wilhelm Holzmann or Xylander of Books I through VI, published in 1562, were the first German editions of the *Elements*. From 1564 to 1566, Pierre Forcadel made the first French translation of Books I through IX. In 1570 Henry Billingsley completed the first translation into English of the entire *Elements*, using Campanus and Zamberti as sources. John Dee, who had studied Euclid at Cambridge and taught the *Elements* in Paris in 1550, wrote its influential preface,[395] extensively edited Billingsley's *Elements* in the 1560s, and added many annotations, alternate proofs, and corollaries.[396]

Closely reflecting the scope and discussion of mathematics given in Proclus' *Commentary*, Dee's *Preface* first surveys arithmetic and geometry, the two principal fields of mathematics, and then classifies and describes the many mathematical arts and their practical derivatives whose application extends from the supernatural realm to the natural sensible world. In establishing through mathematics the structure of astronomy and astrology and the causal relations within each, Dee mainly follows Neoplatonic philosophy, but also conforms to the nature of Aristotelian science, which he rejects in mechanics. Among the mathematical arts he includes archemastrie, astrology, numerology in theology, thaumaturgics or the study of marvels, and Archimedean wonderworking. This is in keeping with a time in which varieties of such magic persisted; in Renaissance Neoplatonism, natural magic, Hermetic analogies between microcosm and macrocosm, the numerology of the cabala, optical divination, and the detection of mystical relations and weights in Reformation efforts to unlock the secrets of nature. But the chief purpose of the *Preface,* Dee explains, is to promote a mathematics that will better meet the "sundry purposes in the Common Wealth."[397] Sections on architecture, astronomy, gunnery, horometry, judicial proportions, music, navigation, and zography (painting, drawing, and engraving) attest to this goal. Appealing to such sources as Archytas, Archimedes, Hero, Vitruvius, and Roger Bacon, Dee endeavors to make comprehensible to his vernacular audience the mathematical base of various technologies. An apparent object here is to demystify technology in its power, for example, to produce special effects in plays.[398] The *Elements* of Billingsley and Dee along with the other vernacular translations, including the first into Spanish by Rodrigo Zamorano completed in 1576, were intended primarily for people unschooled in Latin: curiosi, builders, goldsmiths, merchants, metalworkers, mining and military engineers, painters, and other common folk.

After a spate of mid-century critics pointed out errors in Latin Euclids, Commandino's translation of the *Elements* from Greek into Latin in 1572 superseded the renderings by Campanus and Zamberti. Commandino, the principal editor of Euclid during the Renaissance and Reformation, criticized his Latin predecessors for straying from the Greek source. The 1533 *editio princeps* of Grynaeus and a second Greek manuscript had allowed a more critical assessment of the principal Greek source at hand. Commandino discovered more scholia.

By 1572 these critical translations and a more accurate Latin *Elements* provided a sound base for better mathematical textbooks. At the Collegio Romano, Clavius gained a renown that began with his magisterial Latin fifteen-book *Elements*, published in 1574. Clavius's edition of the *Elements* was not so much a new translation as a collection of materials culled from previous editions with his annotations. It illuminates the continuing presence of Campanus, the new influence of Commandino, and the widening transmission of the *Elements.* Clavius gives directions for constructions and lengthy *excursi* on debatable issues. His Euclid includes 486 original propositions plus an additional 671 to elucidate them. This masterful schooltext rapidly became the standard and made Clavius known as "the Euclid of the sixteenth century." His influence extended beyond Europe to China. As noted in Chapter Ten, his student Matteo Ricci and Chinese literati translated his first six books of the *Elements* into Chinese.

On the following page is a chart of the filiations of the chief editions of Euclid's *Elements* during the sixteenth century.

Until Peyrard at the start of the nineteenth century discovered what is generally considered the pristine Euclid, the improvement in the Campanus version and new editions by Zamberti, Grynaeus, and especially Commandino were

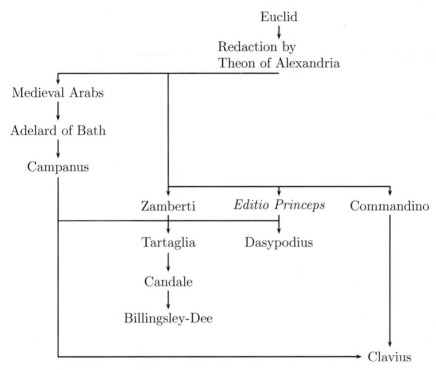

largely taken to be the canonical *Elements*.

   A few late sixteenth-century editors of vernacular Euclids called for the expansion of mathematics in college curricula and the attainment of a higher cognitive status for it. In advocating educational reform, John Dee, for example, recognized that Latin and Scholastic faculty might view their translations as a threat. Praising universities as "storehouse treasures of all sciences and arts," he maintained that increased attention to the usefulness of mathematics could make universities more highly "regarded and esteemed."[399] At the Jesuit Collegio Romano, Clavius labored to elevate from a preparatory art to an advanced level in philosophy the place of mathematics in the curriculum, comparable in status to metaphysics and physics. In the 1580s he was complaining of disrespect for mathematics teachers and their subject at Jesuit colleges, which suggests strong opposition to his effort. The Coimbra commentaries on Aristotle depict mathematics as mainly calculation, and Jesuit philosophers, such as Benito Pereira, argued that the pure mathematics of geometry and arithmetic lack the causal connections necessary for a true science. Appealing mainly to the authority of Plato and Aristotle, Clavius contended that mathematics has the logical methodology and experiential basis essential for the study of astronomy, cosmology, natural philosophy, and optics and that these are universal: mathematics thus meets Aristotle's criteria for *scientia*.[400] The mathematical disciplines gained greater attention at the Collegio Romano, when

CHRISTOPHORVS CLAVIVS BAMBERGENSIS E
SOCIETATE IESV ÆTATIS SVAE ANNO L XIX·

Christopher Clavius

in 1581 Pope Gregory XIII raised it to a university, now known as the Gregorian University, and provided new buildings, books, and astronomical instruments. In 1599 Clavius in the revision of the general curriculum, *ratio studiorum*, at Jesuit colleges won a greater role for mathematics. Citing hundreds of passages from Plato, Aristotle, Euclid, and Proclus, Joseph Blancanus in *Aristotelis loca mathematica* ..., published in 1615, supported his mentor Clavius by defending the perfection of geometric demonstrations and the excellence of knowledge that mathematics supplies.[401]

In addition to the *Elements*, Clavius prepared a series of large textbooks covering other fields of mathematics: *Epitome Practicae*, which was published in 1583, the *Astrolabium* of 1593, the *Geometria Practica* of 1604, *Triangula Sphaerica* in 1611, and *Algebra*. These were mostly practical. Clavius's *Algebra* follows Stifel in style of exposition and notation. It introduces the German + and − into Italian cities and parentheses for collecting terms. *Astrolabium*, *Geometria Practica*, and *Triangula Sphaerica* contain almost all the trigonometry of the time. The *Astrolabium* pioneers the use of the decimal point and gives the pre-Napierian *prosthaphaeresis* method of computation and rules for solving spherical triangles. Wide use of these mathematical manuals of Clavius at Catholic universities, particularly those of the Jesuits, as the standard introduction to the subject made him the mathematics preceptor of Catholic Europe.

14

Transformation of Mathematics, ca. 1600–1660: I

Keplerius, an alumnus of earth,
makes assault on the heavens,
Seek not for ladders: the earth itself takes flight.
— J. Seussis

[T]his grand book [of nature], the universe, ... stands
continually open to our gaze. But the book cannot be
understood unless one first learns to comprehend the
language and read the letters in which it is composed. It
is written in the language of mathematics, and its characters are triangles, circles, and other geometric figures
without which it is humanly impossible to understand
a single word of it; without these, one wanders about in
a dark labyrinth.
— Galileo Galilei, *The Assayer* (1623)

Roughly from 1600 to 1660, the application and influence of mathematics
expanded greatly in Europe, and its serious pursuit quickened. A surer mastery
and reconstruction of the higher geometry of Apollonius' *Conics*, Archimedes'
writings, and Pappus' *Mathematical Collection* contributed, along with continuing study of medieval Arabic thinkers. So did two potent traditions. Renaissance Platonic and Pythagorean panmathematicism conceived of the cosmos
as constructed and ordered on geometric principles and numerical harmonies.
The mechanical tradition depicted nature as a huge machine, the model for
which was the mechanical clock, and searched for hidden mechanisms underlying phenomena.[402] Building on these traditions and an Aristotelian epistemology, European scholars revitalized quantitative physico-mathematics and
applied it to the study of nature. A catalyst for this was the belief in natural, as opposed to diabolical, magic that numbers were a lever that activates
hidden powers in the universe and makes possible reaching a higher sphere of
understanding. The new physico-mathematics, however, was strictly quantitative, divested alike of weakening numerology and older verbalism. Able to

447

achieve greater precision, it contributed strongly to the overthrow of the Aristotelian natural philosophy of essences that was qualitative and had yielded only probable results.[403] European savants who introduced general symbols that facilitated the development of physico-mathematics had the printing press to promote its diffusion. By coupling mathematical demonstration to physical phenomena, it claimed a certitude comparable to that of logico-mathematics. Its model profoundly affected the form and content of the high sciences of mathematical astronomy, mechanics, and optics as well as the mixed mathematics of pneumatics, ballistics, and hydraulics. Kepler's three planetary laws, Galileo's kinematics, and Descartes' geometric physics are among its achievements. Young Kepler followed a Neoplatonic route to compute more accurate ratios for the distances separating the known planets from the sun. In *Mysterium Cosmographicum*, published in 1597, he circumscribed each of the five nested regular polyhedra with shells traversed by orbiting planets.

At the beginning of a rapid succession of mathematical achievements unparalleled since the Hellenistic third century B.C., the natural sciences and commerce adopted decimal fractions. Independently of each other, meanwhile, John Napier and Joost Bürgi invented logarithms. These accomplishments moved Europeans ahead of computing mathematicians elsewhere. After Thomas Harriot and Albert Girard codified algebra, René Descartes and Pierre Fermat imaginatively coupled geometry with trigonometry and the theory of equations to create analytic geometry, thereby subjecting geometry to algebra. At mid-century, Gérard Desargues and Blaise Pascal invented projective geometry, while Marin Mersenne, Pascal, and above all Fermat revived number theory. Pascal and Christian Huygens also laid the foundations of classical mathematical probability. French savants now dominated mathematical research.

Prominent by mid-century was the pursuit of methods for solving a surge of problems mainly in six classes, then largely separate, that set the groundwork for calculus. European geometers and natural philosophers sought to find the instantaneous rate of change in dynamics. Study of constant acceleration added attention to finding the slopes of curves and the areas under them. European geometers set out to determine tangents and normals to a curve at any point, which relate to its slope, as well as extreme values of curves. They sought to resolve quadrature and cubature problems, originating in antiquity, that require computing precisely curvilinear areas and volumes. Here they turned from Archimedean exhaustion techniques to the infinitary methods associated with Kepler's infinitesimals and Cavalieri's indivisibles. They located centers of gravity and initiated successful rectification—calculating ratios between straight lines and segments of curves. Gilles Roberval, for example, probably rectified the cycloid, as did Christopher Wren, who demonstrated that an arc length of a general cycloid is eight times its generating circle's radius.[404] All these problems involve curves, which were the principal analytical subject of the time. The acceptance in geometry and arithmetic of infinitary techniques, not simply as heuristic expedients, as in ancient Greece,[405] but for developing formal proofs reflects the ubiquitous role of approximation in mathematics.

Studies of Archimedes and dissemination of his exhaustion methods are the apparent reason seventeenth-century geometers advanced apogogic—fastidious and rigorous—exposition.

In content, style, and notation, by 1660 Europeans had moved mathematics to the edge of modernity. A transmutation, Martha Ornstein writes, was occurring in seventeenth-century science, and the same can be said for mathematics.

14.1   Crises and State Making: A Synopsis

From 1600 to 1660, armies increasing in size, with more potent weapons and better training, threatened older empires and kingdoms. As these crumbled, the royal state grew in power, especially in the west. Since their introduction in the fifteenth century, cannon and gunpowder had become widely available in Europe. This weaponry challenged regimes already weakened and helped a patchwork of rival royal states surpass them. Controlling the production of cannon and often of mortars, kings gained great power over their states. These new monarchies were not national states as understood today. They were dynastic conglomerations gained through war, inheritance, and marriage alliances. Within them centralizing administrations were slowly developing at the expense of traditional rights and privileges held by corporate bodies, among them the Church and towns, and by individuals, such as princes, nobles, and provincial officials. The term "absolutism" often applied to these monarchies is too sweeping. Though kings were believed to rule by divine right, so ordained by God and proclaimed in Romans xiii.1, they were circumscribed by Christian morality. On one side Machiavelli's *The Prince* of 1513, the French jurist Jean Bodin's *Colloquium of the Seven*, completed in 1588 or 1593, and the English political theorist Thomas Hobbes' *Leviathan*, published in 1651, advocated an expansion of sovereign power. An opposing body of literature existed on the right of resistance to tyrants. At this time empires were waning. Only Muscovy in the East and the Ottomans in the Balkans escaped, while imperial Austria in central Europe barely survived.

Within and among empires and dynastic states, political uncertainty came of unstable boundaries, inheritance counterclaims, and the clustering of geographical units disparate in culture, customs, religion, and language. Generally working in concert with the great nobility, rulers pursued wealth, power, and glory through brutal, expensive wars that became a permanent feature of early modern Europe. Efforts of monarchs to compel religious uniformity, such as the Catholic French crown's actions against Huguenots, further strained tensions. Religious antagonisms provoked such conflicts that these defined the seventeenth century to 1660 as the era of wars of religion.

Inclement weather worsened the misery. The most terrible winters would come after mid-century, but even before frequent cold periods ruined harvests and produced famine. Europe was now also experiencing the worst bubonic plague in three hundred years. Primarily in clogged towns, the plague joined with smallpox, influenza, and typhus to take huge tolls. In these circumstances, Europe's population fell from an estimated ninety-five or a hundred million in 1600 to eighty or ninety million in 1660.

The Thirty Years' War conducted in central Europe intermittently from 1618 to 1648 joined political passion and religious fervor in one of the cruelest episodes in European history to 1914.[406] Austria's Catholic Habsburg rulers struggled to subject separatist Protestant Bohemian nobles to the Holy Roman Emperor and to restore traditional Catholicism in regions where Lutheranism and Calvinism were spreading. Historians define four phases of this war, the Bohemian, Danish, Swedish, and General, that together put a million men under arms. For the time, armies were enormous. Many were raised by entrepreneurial businessmen seeking the profits of war. Unprofessional and undisciplined soldiers far from home, along with mercenaries, many of them outlaws, devastated many towns, villages, and industries in the main war zones of south Germany, Bohemia, and Pomerania. An estimated half of the population there died directly in battle or by infection and social dislocation.[407] The hero of Hans Grimmelshausen's near autobiography, *The Adventures of Simplicissimus the German*, composed in the 1660s, bears classic—if slightly exaggerated—witness to the woes of the war. War badly damaged the social and economic structure that had supported Tycho Brahe briefly and Johannes Kepler at Rudolph II's imperial court in Prague. In that city a modicum of scientific research continued, often in isolation, in the work of Joost Bürgi, Marcus Marci, and Gregory of St. Vincent, who wrote the huge pedestrian tome *Opus Geometricum*, published in 1647, under the misleading title *Problema Austriacum*. By the Austrian problem Gregory meant squaring the circle. Its solution he thought to be less remarkable than achieving social justice in the empire.[408] Through a geometric summation procedure applied to Cavalieri's method of indivisibles, Gregory mistakenly believed that he had squared the circle. But geometers doubted this and he lost stature among them in 1651 when Huygens found an error in the calculations.

In 1648, a series of agreements known as the Peace of Westphalia ended hostilities in the Holy Roman Empire. This settlement reasserted religious stability on the principle established at the Peace of Augsburg of 1550 authorizing each ruler to determine the religion of his subjects. In imperial courts, Calvinist princes now gained equal status with their Lutheran and Catholic counterparts. This agreement officially acknowledged the split in Latin Christianity, but the Catholic Church retained a position of primacy. The intense religious divisions were to prolong for two centuries a fragmented Germany consisting of more than three hundred political subdivisions. The treaty diminished the influence of Austrian Habsburg rulers in north German states and placed dynastic interests above religious allegiance. It is therefore often considered a secular constitution.

Consisting of Aragon and Castile, Spain had acquired economic and military hegemony in the sixteenth century, but a reversal of geopolitical roles transpired in the seventeenth. Spanish expansion in Europe and across the

Atlantic overextended its once formidable army. Castileans opposed new taxes at the same time that Dutch and English pirates raided the Spanish silver fleet from America. Rebellions in Catalonia and Portugal helped to drain the treasury, and so did the protracted war in the Low Countries, where the Protestant Dutch republic's fight for independence against Catholic Spain again flared up in 1621 and ended in Dutch victory in 1648. By the 1630s Spain's main enemy was not Holland but Catholic France. The French army defeated the Spanish at the Battle of Rocroi in 1635 during the Thirty Years War and in 1659 forced the Spanish to sign the humiliating Treaty of the Pyrenees. By then, economic and military dominance had shifted decisively from Mediterranean lands to states on the north Atlantic and Baltic trade routes. Spain's sixteenth-century enemies, England, France, and Holland, were now in the forefront.

The seafaring and trading states of England and the Dutch oligarchic republic differed in developing away from absolute royal states. A vital middle order of society invigorated by growing commerce fought to remove the many difficulties blocking constitutionalism and representative government. England's Stuart monarchs struggled with Parliament. Charles I, whose reign began in 1625, imposed forced loans, collected illegal taxes, such as ship money, and billeted troops in private English homes. In response the House of Commons issued the Petition of Right of 1628, forbidding these practices without its consent. Religious divisions heightened the problem facing the monarchy. Charles's support of the *Book of Common Prayer* and the episcopal system antagonized powerful English Puritans. The elaborate rituals of that system reminded many Englishmen of Roman Catholicism. Constitutional crisis turned to civil war that began in 1642. Oliver Cromwell and the Parliament's New Model Army defeated the king, who was beheaded in 1649. A few years of a republican Commonwealth gave way to a protectorate under Cromwell. Shortly after his death, parliament, tired of political uncertainty, called to the throne Charles II, son of the executed monarch. From 1600 to 1660, the economy of the Dutch republic, which was composed of seven provinces in a federal structure, surpassed all others in Europe. Amsterdam's bourse or stock exchange was the foremost in Europe and Dutch bankers for the first time gained social respect. Holland reclaimed fertile land from the sea. Yet even its prosperous economy and defeat of the Spanish did not free Holland from anxiety, for just to the south were the armies of the ambitious French king.

As the sixteenth century had been Spain's, most of the seventeenth belonged to France. It was to become the model absolutist state. Its ruthless king Louis XIII, reigning from 1610 to 1643, and his first minister Armand Jean du Plessis, Cardinal de Richelieu (1585–1642) crushed the possibility of representative government and Protestant resistance. Richelieu, best known in learning for founding the Sorbonne, was building the modern centralized state. The principle of *raison d'état* placed the interest of the country above moral constraint and picking intendants assured a body of provincial noble officials loyal to the king. Crop failures, increased taxes, and the behavior of the brutal police of Louis XIII brought peasant uprisings. From 1649 to 1652 office-holding nobles of the sword, who demanded that taxes for the war against Spain await approval of the Paris parlement, a judicial not a legislative body, essentially an appellate court, rose during the Fronde against the regency established for the youthful Louis XIV. The result was near anarchy. The French crown soon defeated the divided *frondeurs*.

The political crises in France and England in the middle years of the seventeenth century aggravated the distress that the Thirty Years War brought to Europe in general. Historians refer to the mid-seventeenth century as a particular time of troubles. An English preacher speaking the perennial language of his profession accurately declared to parliament in 1643, "These days are days of shaking, and this shaking is universal." And so it was on battlefields and in starving villages.

14.2   Early Baroque Art and Thought

Despite the troubles, the era from about 1590 to 1750 produced a distinct culture, which historians have named the age of the baroque. The term "baroque" may derive from the Portuguese *barocco,* meaning *rough* or *irregular* in the sense of flamboyant rather than uncouth. It applies to European painting, sculpture, architecture, literature, and thought. For most historians it exemplifies the tenacity of Europe's republic of letters underwritten by an expansion of secular and religious patronage. From courts and noble households, a wealthier and growing bourgeoisie, and thriving towns came an increase in patronage, partly in the form of new academies and enlarged artists' studios.

The visual arts continued the Renaissance efflorescence, now with a refined geometric cast. Well-trained painters and sculptors faced no major technical problem in representation. In painting, the baroque style, characterized by vivid colors and climactic scenes, such as moments of martyrdom or religious ecstasy, replaced the classic austere balance. A forehint of a culture devoted to scientific projects and a requisite ethos of order and self-disciplined work was a contrasting baroque depiction of solid bourgeois domesticity.

Leadership in painting, etching, and drawing shifted from Italian cities to Catholic Flanders under Austrian rule and to Calvinist Holland to its north. One of the two principal early baroque masters was the learned Peter Paul Rubens (d. 1640). Born near Cologne, he primarily resided in Antwerp and went on diplomatic missions to Spain and Italian cities, where he made hundreds of drawings of high Renaissance art and sculptures. In Flanders, his stylistic authority equaled that of Michelangelo a century before. At his home and factory in Antwerp, he and his students made altarpieces and decorative panels and painted canvases. Rubens is known for his supreme brushwork, his fleshy female nudes, and his poignant crucifixions, such as in *Raising of the Cross* of 1609–1610. His style is to spiral into the pictures, moving down into the shadows. The foremost Dutch master was Rembrandt van Rijn (d. 1669) in the burgeoning port of Amsterdam. He signed his artwork with his first name only and that is how he is known to the present. Rembrandt's conception of art as a form of reasoned organization, which accords with Descartes' rational cosmology and Spinoza's geometrical ethics, appealed to prosperous Dutch merchants, bankers, and guilds. One of his first commissioned pictures was *The Anatomy Lesson of Dr. Nicolaes Tulp* in 1632, showing a public dissection of the corpse of an executed prisoner conducted before leaders of the surgeons' guild. The picture breaks with the prior military presentation of a dissection and gives minute details of the facial expressions of the audience and the features of the cadaver. As a group portrait, it belongs to a genre then gaining in popularity. Such work captures a spirit of civic pride, showing guilds, not princes and aristocrats, as leaders of Dutch society.

The Anatomy Lesson of Dr. Tulp
The Hague, Mauritshuis

Art critics remember Rembrandt for his naturalism, effective lighting, and honest detail, such as in *Three Trees*. The detailed scientific interest in anatomy, the artistic depiction of its study, and the representation of the world in solid dimensions, all this goes more with the observational, empirical side of the scientific revolution than with its abstract and theoretical component. Among major artists elsewhere were Spain's court painter Diego Velázquez (d. 1660), who created sweeping dramatic scenes such as the surrender of Breda, and Prague's Karl Skreta (d. 1674), whose warm chiaroscuros synthesized Roman and Bolognese naturalism.

In architecture and sculpture, Italian cities generally retained leadership. At the papal court in Rome, Giovanni Lorenzo Bernini (d. 1680) directed the sculptural program for the nave at St. Peter's and constructed its famous spiral-columned baldachinno or canopy as well as the spectacular *Fountain of the Four Rivers*.[409] In Piedmont the Theatine priest Guarino Guarini (d. 1683) demonstrated a geometric sensibility in positioning square designs beneath convex surfaces. His domes consist of a fabric of receding tunnels formed by superimposed segmented arches that filter light through intervening spaces. He took this idea from Islamic flying arches, applying it in the dome of the Cathedral of Turin.

Many men of letters continued to appeal to biblical authority. But a critical spirit, both rationalistic and empiricist, was also evident. Critical reasoning about the physical world conflicted with traditional, faith-based knowledge. Besides the new critical studies was the radically skeptical legacy of ancient pyrrhonism and a recoil from the excesses of witch trials. Literature in England and France expressed the competing themes.

In England, the brooding clergyman John Donne (d. 1631) voiced the prevailing position among men of letters before mid-century when he worried that "the new philosophy calls all in doubt" and in *Ignatius his Conclave* condemned Copernicus and his views. In France, the austere Christian moralist and mathematician Blaise Pascal (1623–1662), whose satirical *Provincial Letters*, published in 1656 and 1657, helped create modern French prose, was distressed at the separation in intellectual circles between established religion and Cartesian rationalism. In his *Pensées*, published posthumously in 1670, Pascal agrees with Galileo that the underlying structure of the physical world is given in handwriting of the three branches of mathematics: mechanics (motion), arithmetic (number), and geometry (space). Fathoming the inexhaustible diversity of the firmament beyond the intelligible underlying structure, however, depends on faith. It is an error to make the study of science attempt to explicate the most profound truth and to treat God as merely a filip. "The God of the Christians is not simply the author of geometrical truth," Pascal writes. "I cannot forgive Descartes for eliminating the analogy between God and the world—that is, the '*imagines*' and '*vestigia Deo in mundo*'—and making God simply a watchmaker."

Taken aback at the destruction of the Thirty Years War, northern European scholars systematically studied questions of international relations; while the political crises of the 1640s brought an absolutist response. The Dutch diplomat and jurist Hugo Grotius (d. 1645), who was a disciple of Cicero and Seneca, in 1625 denounced aggressive wars in *De Iure Belli ac Pacis* ("Of the Law of War and Peace"). The first thorough treatise founding international law on man's nature and his place in the universe, it argues that while people may choose their form of government, when the sovereign is established they must give unconditional obedience. The German jurist Samuel von Pufendorf's *Of the Law of Nature and Nations* of 1672 further argues legal principles founded in the concept of natural law and claims that only defensive wars are justified.

After experiencing the English Civil War, the English scholar Thomas Hobbes, who was the most original political theorist after Jean Bodin, thundered for absolutism. In exchange for security the people would have to obey the sovereign. Hobbes' general ideas of order sprang from his participation in the research in Mersenne's circle on mathematics and natural philosophy, especially in optics. Hobbes defines geometry, "the only science ... God [hath] bestow[ed] on mankind," as the model for civil order. From the laws of motion and the necessity of restraining powerful forces in nature he concludes that social forces need a system of social constraints. Absolute rule, he insisted, alone can prevent an anarchic state of nature having constant war that made "the life of man solitary, poor, nasty, brutish, and short."

14.3   A Second Phase of the Scientific Revolution

Nurturing a geometric spirit of order, clarity, and precision, scholars and artisans produced during the first half of the seventeenth century what historians have termed a second phase of the scientific revolution or, more recently, a period of transformation in early modern science. Natural philosophers scrutinized the roles of religion and natural magic; from scholastic universals and the notion of unusual marvels and wonders moved to the study of regular particulars; and accepted as fact information from new instruments, rejecting the contention that the telescope conveyed merely optical illusion. Bacon's induction and Descartes' scientific rationalism provided new methodologies, and Cartesian displaced essentialist Aristotelian science. In reducing results to numbers and representing these in charts, diagrams, and graphs rather than contending with the ambiguity of words, scholars brought to science greater clearness of expression and exactitude. Quantifiers in the new sciences were urged to cultivate higher standards of impartiality, in part from practices in the law. Since the acquisition and publication of knowledge remained precarious, geometers and natural philosophers developed networks of correspondence extending across Europe to convey new findings accurately and to pose playful and occasionally arrogant riddles and challenge problems. Refining sixteenth-century models of academies in such fields as art, poetry, and natural knowledge in environments of sociability and civility, they promoted collaborative research. Initially private, such practices led to academies that were nodes of transmission of reliable information and sources of funding. In turn, the academies improved research techniques. This interworking of subjects, motivations, and transformed methods and practices, Lorraine Daston states, comprises the moral economy of the sciences during the early seventeenth century.[410]

Physico-mathematics and New Methods: Highlights

The second phase of the scientific revolution began with an increased attention to mathematics together with a telling attack on Ptolemaic-Apollonian astronomy and its supporting Aristotelian celestial physics. For at least a century, astronomy had fostered novelty. The leaders of the attack, Johannes Kepler and Galileo Galilei , were to become militantly Copernican. In creating a new mathematical structure for astronomy, Kepler discovered his first two planetary laws: the planets revolve round the sun in ellipses with the sun at one focus, and each planet moves with varying speeds during the course of its orbit, such that for equal times a line drawn from sun to planet or radius vector sweeps over equal areas.

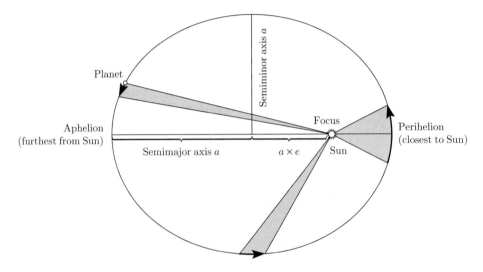

A planet's elliptical orbit

In *Astronomia nova ...* (*New Astronomy Studied Through Causes, or Celestial Physics Derived from the Investigations of the Motion of the Planet Mars Founded on Observations of the Noble Tycho Brahe*), published in 1609, Kepler (1571–1630) announced and demonstrated these laws. Since the sun comprises 99.85 percent of the mass of our solar system, it effectively controls planetary motion. Although the other planets cause small perturbations that over time adjust the ellipses' size, shape, and inclination, Kepler's laws accurately describe planetary paths. Within a half century, Aristotelian celestial physics, which projected circular and uniform motion in the heavens, was to be obsolescent. In 1610 Galileo's *Sidereus Nuncius* (Starry Messenger) appeared. Among its telescopic discoveries was the mountains on the moon, which contradicted the Aristotelian notion of perfectly regular crystalline spheres in the heavens.

At the University of Tübingen, which he attended from 1587 to 1594 on a scholarship funded by the Duke of Württemberg, the frail and myopic Kepler had begun to show his promise and the direction of his research. While preparing to be a Lutheran theologian, he was influenced by Michael Maestlin, who cautiously taught Copernican astronomy. In 1594, before concluding his theological education, Kepler accepted the position of provincial mathematician and teacher in the Lutheran school, the *Landschaftsschule*, in Graz. This capital of the Austrian province of Styria also had a Catholic university. Kepler was to spend his entire career in imperial Austrian cities buffeted by forces of the Counter-Reformation and the Thirty Years War. In Graz up to 1600, Kepler started defining the boundaries of astronomy.

Johannes Kepler

Late sixteenth-century wars and plagues had stimulated scholarly as well as popular interest in astrology. Common astrology as then practiced Kepler dismissed as the "foolish little daughter of respectable astronomy." He insisted that the beliefs of common astrology arose from superstition, and he attributed any of its right predictions to luck. He turned to respectable astronomy to complement his small school salary, involving preparation of an annual almanac containing astrological forecasts. His first almanac, for 1595, correctly foretold a bitterly harsh winter and an invasion by the Turks. Later he labored to reform astrology by basing the zodiac on archetypal geometric forms, geometric ratios, and musical harmonies of the spheres that he believed underlay the firmament. He saw the zodiac as an extension of the soul and thus knowable to it.

Almost alone in his day, Kepler held to a Neoplatonic and Pythagorean metaphysics that recognized the centrality of mathematics in philosophy. His first book, *Mysterium Cosmographicum* (*The Secret of the Universe*) of 1597, which supports Copernican over Ptolemaic astronomy, imaginatively fashions mathematical astronomy and seeks to raise it from the status of an Aristotelian mixed science or mere hypothesis to a certain science fundamental to cosmology. Displaying, according to Kepler's biographer Max Caspar, the skill of a profound geometer and artist, the study appeals to an aesthetic sense of the exquisite regularity and order in the cosmos, a perception drawn mainly from Plato's *Timaeus* and the Tübingen theologian Jakob Heerbrand's discovery in the beauty of the universe of an argument for the reality of God.[411] After making extensive computations, he assumed that geometric relations rather than numbers underlie the construction of the physical universe. Properties of numbers are only accidental. Finding the straight line and circle alone insufficient to describe the universe, Kepler's fertile imagination in seeking to determine the ratios of planetary spacings from the sun fixed upon the Platonic polyhedral hypothesis. His model encases the orbit of each planet in a thick sphere circumscribing each nested polyhedra. The thickness accommodates planetary eccentricities. The five regular polyhedra schema allows for the six known planets.

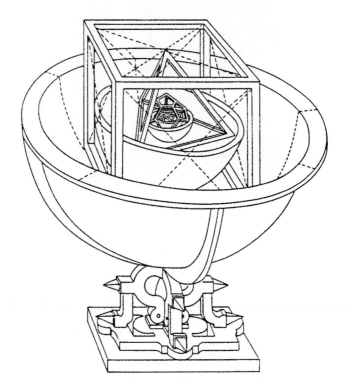

The nested polyhedra and planetary spheres of Kepler

Kepler recognized that his new hypothesis would face critics: there had been debate since antiquity about this epistemological status in astronomy. In *Scholae mathematicae*, published in 1569, the French anti-Aristotelian Peter Ramus had argued that Aristotelian astronomers had corrupted a perfect *prisca astronomia* by importing fallacious hypotheses, such as epicycles and uniform motion. In the same vein, the Imperial Austrian mathematician Nicholas Ursus had expressed a widespread view that any hypothesis in astronomy was *aitemata* or false and spurious. To confirm his hypothesis, Kepler employed strict logic, good observations, and massive calculations. Erasmus Reinhold's *Prutenic Tables* that had displaced the earlier Alphonsine tables and that in part revised and expanded those in Copernicus's *De Revolutionibus* were available to him. At the time, the *Prutenic Tables* were the standard tool for astronomical computations. Kepler had access to writings of Copernicus and Rheticus and knew the work of Tycho Brahe. He also depended on Maestlin's computations of positions of Mercury. Planetary eccentricities obtained by their predecessors were doubtful, but Kepler reduced the error to five percent. For a first text on cosmography, he accepted this approximation.

The *Mysterium Cosmographicum,* which suggests Kepler's views on the appropriate domain and application of mathematics, was well received. He sent a copy to Tycho Brahe, who was impressed with the author's ability but wanted the inclusion of planetary eccentricities mapped by his own better observations. Tycho deepened Kepler's respect for accumulating ever more precise data. Two copies of the *Mysterium* went to Venice. Galileo, who was probably still unknown to Kepler, received a copy. Apparently he sent a perfunctory note of appreciation.

The insistence on reliable computations and a geometric base that informed Kepler's cosmology from the beginning animates his commentary on other writings of his era.[412] He censures Jean Bodin for a misapplication of mathematics in connecting political state systems to proportions—equating democracy with arithmetical proportion, aristocracy with geometric proportion, and monarchy with the harmonic. Kepler's *Harmonies of the World* of 1619 and an *Apologia* attached to a reissue of the *Mysterium* challenge the scheme of divine harmonies in the influential Mosaic philosophy of the Oxford physician Robert Fludd that attempted to refute Copernican astronomy with an alchemical cosmology. Among the English Fludd's writings amalgamate all strains of the occult—alchemical, cabala, Paracelsian, and Neoplatonic. Hating the Aristotelian corpus and asserting the close connection between religion and science, he looked for wisdom in the Scriptures and the Book of Nature. Fludd took true alchemy, not mathematics, to be the key to understanding the universe. But his description of the universe is a mix of divine musical harmonies, microcosm and macrocosm correspondences, number mysticism, and occult, as opposed to manifest, powers in nature.[413] In *Metaphysical, Physical and Technical History of the Macro and Microcosm,* published from 1617 to 1621, Fludd responded scathingly to Kepler, who replied in his *Apologia.* Kepler agreed with Fludd's application of music to astronomy but found his calculations wrong, since they were not based on sufficient observations. Kepler also rejected the Oxonian's view that the properties and relationships among numbers are the highest field of mathematics and that occult sources are superior to modern experiments and observations. Kepler's mathematical description of the universe clashed with Fludd's hermetic—qualitative, alchemical, and symbolic—representation of concentric circles with the sun the center of the heavens, based on the *Vulgate's* Psalms 18:5, and Earth the center of the universe.

Rumors of torture of Protestants circulated in Graz, and Kepler, unable to obtain a faculty post at Tübingen, moved in 1600 to the court of Emperor Rudolph II in Prague, where he remained until 1612. At the Rudolphine court, he found an exhilarating freedom to pursue his research. Tycho Brahe, who had just moved to Prague as an imperial mathematician, had Kepler concentrate on the study of Mars. Apparently he wanted Kepler to defend his compromise in astronomy. But his younger colleague remained a Copernican. When Tycho died in 1601, Kepler became an imperial mathematician and, following a lengthy legal fight with the eminent Dane's kin, gained control of Tycho's treasury of superior astronomical observations.

Proceeding methodically, Kepler made seminal contributions to optics as well. Like earlier astronomers, he knew that optical refraction in the atmosphere of light from celestial bodies reduces the accuracy of astronomical observations. His book *Ad Vitellionem Paralipomena* (*Appendix to Vitello*) or *Astronomiae pars Optica*, published in 1604, refers to the Pole Vitello's thirteenth-century treatise on perspective. Kepler's appendix criticizes refractive theories of Alhazen, Giambattista della Porta, and Vitello, as well as della Porta's mixture in *Magia Naturalis* of 1558 of the natural and the incredible. This clash of ideas Kepler likened to a war, and he pleaded with Rudolph II to dispatch sufficient funds to allow him "to withstand the siege." His revision of Tycho Brahe's refractive tables clarified the consequences of the law of refraction, but it remained for Willebrord Snel about 1621 and René Descartes in his *Dioptrics* of 1637 to formulate the law of refraction: for all angles of deviation α, the ratio of the sines $\sin \beta / \sin \alpha$ is constant.

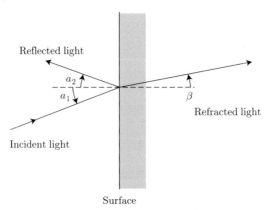

Reflection and refraction at the surface between two media. (In this example, air is assumed to be on the left of the surface, glass on the right.)

In optics, Descartes was to consider Kepler his teacher. Roughly six months after receiving a copy of Galileo's *Sidereus Nuncius* that reported its author's rich telescopic discoveries, Kepler completed his *Dioptrics*. It examines the optics of lenses and presents the theory of new telescopes with a concave eyepiece, like Galileo's, and a more effective convex eyepiece. In 1615 the Jesuit

Christoph Scheiner first constructed the convex model, which is today called the Kepler telescope. It took time for natural philosophers to master the optics in Kepler's obscure *Dioptrics*. By applying his theory to lenses, they were to go beyond craftsmen in constructing better telescopes and microscopes. Kepler's *Ad Vitellionem Paralipomena* and *Dioptrics* founded modern geometrical optics and removed a major obstacle in observational astronomy. He was a pioneer in wearing eyeglasses. His personality mixed characteristics of his times with more modern traits. Observing a supernova in 1604, Kepler believed that it portended a mass migration to the New World.

After 1604 Kepler was approaching his first two planetary laws of elliptical orbits and varying speeds, though not without setbacks. To his conviction regarding underlying geometrical relations and his polyhedral hypothesis was added the idea that the sun emits a magnetic force, or soul in his terminology, that attracts the planets, thereby influencing their paths. The magnetic force diminishes linearly according to the distance from the sun $(1/d)$. Kepler was replacing with a dynamical astronomy the ancient geometrical descriptions of the universe. His painstaking analysis of Brahe's systematic observations of the path of Mars, extensive triangulations of its positions vis-à-vis Earth, and a confusing labyrinth of computations brought him to his first two planetary laws.

As early as 1603 Kepler had conjectured an oval orbit. The ovoid approximates an ellipse. But defective figures for the longitude of Mars, computational mistakes, in some cases in simple divisions, and an inability to connect the oval orbit with the hypothesis of the sun's magnetism forced Kepler briefly to set this idea aside. Then in early 1605, recognizing that observations on the departure of Mars's path from a circle were inadequate, he began a renewed assault on the problem that included revised triangulations of the Martian position and at least seventy repetitions of computations that finally reduced to an acceptable $2'$ the error in observing the path of Mars. Even so, a mutual canceling out of computational errors was to lead him finally "as if by miracle" to the second law. Arthur Koestler portrays Kepler as a sleepwalker stumbling on the correct answer, but the surprises are only a portion of his process of discovery.[414]

Kepler's magnificent folio *Astronomia Nova* offers an animistic and mechanical account of the universe. The Earth and sun, for example, are huge organisms, and Earth's breathing partly affects the tides. Although *Astronomia Nova* was written in the prescribed Latin academic fashion, its mathematical structures were breaking with the past. Filled with studies of curves and intricate computations, it was too mathematical for most scholars of his day. Its preface confesses, "It is a hard task to write mathematical and above all astronomical books for few can understand them. I myself who am considered a mathematician become tired when reading my own work." The emphasis on newness, to which the title attests, was to spread in the sciences, as Bacon's *New Atlantis*, Galileo's *Two New Sciences*, and Robert Boyle's *New Experiments Physico-mechanical Touching the Spring of the Air* of 1660 demonstrate.

When Rudolph II died in 1612, Kepler was permitted to move to Linz, the capital of Upper Austria, and to retain his title of imperial mathematician. He had wanted to return to Tübingen but, citing his Calvinist leanings, the Lutheran clergy blocked him. In Linz he taught at the *Bürgerschule*.

Tired of "the treadmill of mathematical calculation," Kepler put aside astronomy until 1616. His *Stereometria*, whose infinitesimal method is a geometrical integration that presages calculus, was published in 1615. Three years later Kepler completed his *Harmonice Mundi* (*Harmony of the World*), which contains with little comment his third planetary law. It holds, in modern form $p = a^{\frac{3}{2}}$, that the squares of the periods of revolution (p^2) of any two planets are as the cubes of their mean distance from the sun (a^3).[415] Kepler's seven-book compendium *Epitome Astronomiae Copernicanae*, published from 1618 to 1621, explains its derivation. Book IV bases it on the relation Period = Path Length $L \times$ Matter M of the Planet)/(Driving Force $S \times$ Volume V of the Planet) and the proportionality of L/S to distance a and that of M/V to $1/\sqrt{a}$, together with archetypal principles. Its title notwithstanding, the *Epitome*, which contains Kepler's three planetary laws, is an introduction to his astronomy. In 1619, soon after its first trial of Galileo, the Catholic Church placed the *Epitome* on the *Index of Prohibited Books*. Even so, to 1650 it was apparently the most widely read astronomical text in Europe. Until the 1650s, when better lenses for telescopes improved observational astronomy, Aristotelians and the Tychonic camp, including the Dane Christian Severin or Longomontanus, opposed Kepler's planetary laws, and even some Copernicans, like Philip van Lansberge, whose 1632 astronomical tables are based on epicyclic theory, did not accept them completely.

As imperial mathematician Kepler had to complete Tycho's planetary tables. His Rudolphine tables were to contain 120 folio pages of precepts and 119 pages of planetary, lunar, and solar tables. Kepler employed Napier's logarithms for computing planetary positions. While prior tables had erred as much as $5°$, Kepler's fell within $\pm 10'$, an improvement by a factor of 30. Printing the Rudolphine tables was difficult. Emperor Ferdinand II owed Kepler 6,300 gulden in back pay and was not collecting taxes imposed on Nuremberg and other cities for these funds. As the Counter-Reformation entered Linz during the Thirty Years War, most Lutheran printers had to leave the city and Bavarian soldiers billeted in remaining printing shops. In the summer of 1626 Lutheran peasants set fire to parts of Linz, including buildings with presses. The Rudolphine tables were finally printed in Ulm in 1627. The city of Linz has recently commemorated its illustrious resident Kepler by naming its university after him.

The career of Galileo Galilei (1564–1642), whom most historians of science define as the foremost scientist of his time, began at the University of Pisa in 1589. Three years later he moved to the University of Padua in the Venetian republic. The University of Padua was so distinguished in medicine and natural philosophy that it is often perceived to be at the center of the scientific revolution.

Galileo Galilei

Galileo's main research was not in mathematical astronomy. But in attempting to unify astronomy, mechanics, and mathematics, he became a powerful partisan of Copernican astronomy. This and his novel effort to draw a line between the authority of religion and that in the realm of what was called the *Book of Nature* led to clashes with leaders of the Catholic Church. Writing his books in Italian rather than Latin, he broke with the language of the universities and the Church, reaching a larger audience in Italian cities. Deeply influenced by the "divine Archimedes," Galileo radically transformed mechanics. He extended geometrization from statics to kinematics. He improved on the experimental methods of Renaissance technicians and analyzed facts of experience by panmathematical reason. Platonic mathematization alone, he believed, could transcend phenomena, allowing him to apprehend their essence as the empiricism of the time could not.[416] This combination of experiment and better quantitative measurement, joined with mathematical laws and numerical relationships for portraying the regularity in the rational universe, brought fruitful results and displaced the traditional Scholastic logical and verbal approach to mechanics that rested more on texts than on facts. For these achievements, many historians and physicists consider Galileo the founder of modern mechanics and experimental physics.

Historians of science have long disputed the relative influences of various sources on Galileo's mechanics. Pierre Duhem emphasizes Nicole Oresme and William Wallace, the fifteenth-century Paduan Aristotelians. For both Galileo's kinematics is not a rupture from medieval Latin science.[417] Alexandre Koyré denies in his *Études galiléennes* of 1940 that Galileo reaches his conclusions on motion from pioneering empirical testing. Such tests would require machined inclined planes, spherical hard bodies, and a timing device beyond any then available. Rather he deduced it in the fashion of the scholastics from thought experiments, but his mathematical approach did break with the past. Still-man Drake shows to the contrary that existing laboratory equipment permitted Galileo to measure time and motion more precisely than his predecessors. Despite deviation between his measurements and the law, Drake proposes, experiments and not abstract mathematics substantially forced him to rethink his position on accelerated motion. Beyond this, Galileo invented new concepts for measuring velocity. Drake finds him not an heir to the medieval Christian tradition.[418]

In 1604 Galileo stated the law of free fall: ratios of spaces traveled from rest are as the square of elapsed time, in modern symbols $s = \frac{1}{2}at^2$. Apparently he had been studying not mean values but instantaneously varying velocities, comparing infinite sets of them. For equal times, comparisons of these infinite sets may have suggested roughly a 1, 3, 5, ... progression in spaces traversed along an inclined plane. If so, he likely assumed these odd numbers to be exact. At first he misidentified the significant variable as distance rather than time. Only later did he confirm the relation between time, velocity, and distance following Oresme's fourteenth-century way of deriving the mean-degree theorem, called Merton's rule. Oresme's velocity-time graph had points on a horizontal indicate instants of time and lengths of perpendiculars the velocity of a uniformly accelerated body. The end points of the perpendicular segments lie on a straight line. Oresme was able thereby to verify geometrically Merton's rule that at midpoint the velocity is half that at the terminal point. The geometrical diagram here shows that the ratio of area covered in the first half to that in the second half is 1 : 3. Dividing the time into three equal parts, the distance traversed as shown by areas has a ratio of 1 : 3 : 5. For four equal subdivisions, the ratio is 1 : 3 : 5 : 7. Thus Galileo saw that the distances covered are indeed as odd numbers. Since the sum of n odd numbers is n^2, this gave the basis for Galileo's law of free fall.

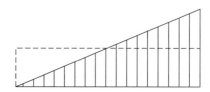

A note stating the correct version given in his *Discorsi* of 1638 apparently dates from 1609. In testing this law, Galileo conducted experiments that involved rolling a metal ball down an inclined plane. He did not, as is sometimes said, drop weights from the leaning tower of Pisa as another test.

After viewing the supernova of 1604, Galileo criticized the Aristotelian concept of the unchanging heavens. He was not diverted primarily from mechanics to astronomy until 1609, when news reached Venice of the Dutch lens maker Hans Lippershey's invention of the telescope. Spurred by that information and probably also an Italian spyglass from the 1590s, Galileo assisted by skilled artisans through trial and error quickly made a nine-power telescope. By the end of 1609 they had fashioned a thirty-two power instrument having a plano-convex objective and a plano-concave eyepiece. Transforming into a powerful research instrument what had been considered a curiosity of natural magic, he turned the telescope to the heavens with startling results.

In his *Sidereus Nuncius* of 1610, his first published work, Galileo at forty-six hastily recited his discoveries. *Sidereus Nuncius,* a twenty-four page booklet written in Italian prose, made Galileo a celebrity. His discoveries begin with the mountains on the moon. Comparing the radii of Earth and moon led him to believe that the lunar mountains are four times the height of any on Earth. Overturning the Aristotelian concept of smooth crystalline spheres in the heavens, the investigation revealed resemblances between moon and Earth, challenging the Aristotelian separation of celestial from terrestrial physics that Newton would later eliminate. Galileo also discovered four new satellite bodies near Jupiter. This contradicted the anti-Copernican view that only Earth's moon is preempted from orbiting the sun. Even Kepler, who wondered whether with so many moons Jupiter might be nobler than the Earth, had doubts about these until Galileo sent him a telescope. Galileo's observation that many stars appear the same in size and brightness through the telescope as with the naked eye suggested to him that the universe is much larger than the small closed Ptolemaic cosmos.

In 1610 Galileo moved to Florence as first philosopher and mathematician to the grand duke Cosimo II de' Medici, a position that left more time for his research. Continuing his astronomical observations, he discovered that Venus has phases like the moon, which suggested that it revolves about the sun. Lacking journals, he followed the practice of his day of keeping this discovery secret but communicating the gist of it by an anagram to establish his priority. The letters of the anagram can be recast as *Cynthiae figuras aemulatur mater amorum,* that is, the mother of love imitates Cynthia's shapes. Galileo's first decisively Copernican tract, the *Letter on Sunspots* of 1613, showed that even the divine sun is changing. It also demolished Scheiner's belief that sunspots are tiny planets, gaining the enmity of that influential Jesuit astronomer. This tract was daring. Official astronomy held to the Tychonic compromise. Kepler had just proclaimed elliptical planetary orbits, but Galileo to the end of his career defended circularity in the heavens. The advance of science here is clearly not linear. At universities, Aristotelians rejected Copernican astronomy, since it could not be reconciled with their physics. Citing deviations from the literal truth of Scripture, Protestant leaders had rejected Copernican astronomy, but Catholic churchmen long tolerated it as an adventurous mathematical hypothesis and not physical reality. But that was changing. The support of Tommaso Campanella, imprisoned for heresy in Naples, was likely a further liability to Galileo, whose *Letter to the Grand Duchess* of 1615 pleads for freedom in scientific research. The Dominicans and Robert Cardinal Bellarmine, the Church's leading theologian, now viewed Copernican theory as "very dangerous" to faith. They too held that it suggested the falsity of Scripture. In 1616 Bellarmine summoned Galileo to appear before the Inquisition and cautioned him not to hold or teach Copernican astronomy, except as a mathematical supposition. In this year that Cervantes and Shakespeare died, the Catholic Church first placed Copernicus's *De revolutionibus* on the *Index*, where it stayed until 1822.

Galileo turned to the study of comets, the foundation of mechanics, and correspondence with Francesco Cavalieri on indivisibles. Responding to an attack by the Jesuit Orazio Grassi, his *Il Saggiatore* (*The Assayer*), published by the Lincean Academy in 1623, dismisses Grassi's view of the circular orbits of comets and proclaims mathematics the language of the sciences. Galileo's predecessors had not employed the two-book metaphor, the Book of Nature and the Scriptures, with a line of competency drawn between them. This breviary for scientific inquiry with its confession of faith in mathematics presents mathematics, not theology, as Galileo's final authority in the sciences. Also troubling to the ecclesiastical officials in *Il Saggiatore* was an atomistic theory of matter, which seemed to subvert the doctrine of transubstantiation. But in 1623 Galileo's friend Maffeo Cardinal Barberini, who had defended him in correspondence of 1616, was elected Pope Urban VIII. The next year he granted Galileo permission to publish on Copernican astronomy, but only if treated as hypothetical; otherwise it would supposedly limit divine omnipotence. While working on this treatise for the next six years, Galileo briefly questioned the accuracy of Tycho's observations, which brought Kepler to the defense of Tycho. In 1632 Galileo published his *Dialogo* (*Dialogue on the Two Chief Systems of the World*). The dialogue, which follows the Socratic format, is animated and plausible. It has three participants: Salviati, a supporter of Copernicus; Simplicio, an obstinate, dull supporter of Aristotle and Ptolemy; and Sagredo, an educated layman. Although the preface and closing argument are anti-Copernican, the papal position is put in the mouth of the discredited Simplicio and thus carries no weight.

Angered, Urban VIII had the Inquisition begin proceedings on charges of heresy and the breaking of a pledge made in 1616 to Bellarmine, now deceased. The ill, sixty-eight year old Galileo was ordered to Rome. If he did not come peacefully, he was to be brought in chains. The grand duke of Tuscany did not require Galileo to go, but he went to the trial. After a threat of torture and observing the means of torture, on his knees Galileo recanted "his errors and [the] heresy" of Copernican theory that are "contrary to the Holy Church." An apocryphal story has it that upon rising he whispered of the Earth *Eppur si muove*. Not all historians view this condemnation as simply an obscurantist church's attack on freedom of thought. Pietro Redondi suggests that, besides being hostile to Galileo's atomic doctrine, the Pope was acting on his alliance with Spain by criticizing a defender of France.[419] Mario Biagioli depicts the trial as a case of the downfall of a court favorite. The trial, lasting twenty days, and the recantation did briefly retard the study of heliostatic astronomy among Catholic scholars, making even Mersenne in France and Descartes in the sanctuary of the Dutch republic guarded in expressing support for it.

Galileo's trial seems a distraction from the major challenges facing the Vatican. Threats internally from Protestant dismemberment in northern Europe and externally from Turkish incursions against Austria preoccupied Roman authorities. These situations dictated a strategy of containment. The power politics of the Italian peninsula, in which the Papal States were a leading participant, reinforced this agenda. In less threatening times Galileo's challenge might have been an opportunity for expansive rethinking: in stress-filled, contractive times it seemed to undermine stability and engender confusion. That potent misrendering helped relegate it to a matter to be settled by inquisitorial functionaries following routine procedures.

Galileo spent the final eight years of his life under house arrest at his farm in Arcetri outside Florence. Not allowed to attend Mass or consult physicians, he was blind for his last four years. Within a year he completed his major work, *Discorsi* (*Discourses and Mathematical Demonstrations Concerning the Two New Sciences*). The manuscript had to be smuggled out of Florence and was published in Leiden in 1638. The new sciences are the engineering science of the strength of materials and elements of the mathematical analysis of kinematics. Galileo's continuity physics treats the infinite and the infinitely small. He considers paradoxes of the infinite not as puzzles to be solved, as Aristotle had attempted for Zeno's, but as realities, and the mathematical instant as a physical entity. This distinguishes his work from finite medieval physics. Demonstrating great insight, he sets the squares, 1, 4, 16, 25, 36, 49, ... , into one-to-one correspondence with the integers. His last discourse treats the parabolic trajectory of projectiles.

By the 1620s ferment in the high sciences, mathematics, and medicine was intensifying a search for a methodology that might legitimate natural philosophy and elevate it above the level of hypothesis. William Harvey's *De Motu Cordis* ... , published in 1628, for example, was transforming medicine. His discovery of the general circulation of the blood, building partly on a graduated series of studies of the heartbeat from single-chamber to four-chambered hearts, was displacing the Galenic theory of the ebb and flow of the blood even before the new microscope made possible the discovery of the capillaries.[420] European thinkers contended with the obscurity of the occult and the skepticism of a pyrrhonism that held knowledge to be relative to properties of the perceiver and encompassing circumstances. Two new methods that now appeared were the induction propounded by the English statesman and jurist Francis Bacon (1561–1626), Lord Chancellor to James I, and the scientific rationalism of the French philosopher René Descartes (1596–1650).

Bacon was averse to reverence for old authority in the sciences as well as to natural magic. His *Novum organum* (*New Logic*), which appeared as parts of the omnibus volume *Great Instauration* in 1620, rejects Aristotle's method. Scholastic logic and its syllogisms, he argues, are acceptable for civil business and opinion but "not nearly subtle enough to deal with nature." A practical man, Bacon had profited from reading Vanoccio Biringuccio's *Pirotechnia* of 1540 on metallurgy and Georgius Agricola's *De Re Metallica* of 1556 concerning metallurgy and mining. These books describe tools, natural processes, and applications of knowledge. Conceiving of the new sciences as grounded in sensory experience, Bacon searched for efficient natural rather than final causes. His universe is mechanistic and filled with the actions of corpuscles, essentially impenetrable inert and active atoms. An active power in atoms suggests a continuing influence of Renaissance animism.

For investigating the structure of the physical world, Bacon proposes a form of induction leading to the "invention not of arguments but of arts." Induction, he writes, is not simply collecting raw physical facts, resembling the work of the ants, nor is it speculating in abstraction, suggesting the work of the spider that spins its web from itself; it is instead like the activity of the bee that "gathers its materials from the flowers of the garden and the field but transforms and digests it by a power of its own." The work of the active interpretive mind complements observation. Histories or extensive collections of data are not to be simple gatherings of unrelated materials. Bacon intends his methodological rules to improve research, filling a role not unlike that of the ruler and compass for constructions in geometry. Sound investigators must first purge the major types of intellectual prejudice that aphorisms 36 through 38 of *Novum organum* call the four "idols." Catalogued in reference to their societal and linguistic settings, these are the anthropomorphism of the tribe; the personal and environmental prejudice of the cave, an allusion from Plato's *Republic;* deceptions from the emptiness or ambiguity of words of the marketplace; and the dogmatic philosophical systems of the theater. Bacon wants knowledge of the natural sciences to be cooperative, cumulative, and public. Free and open communication is fundamental to their advancement.

Like Pliny, Bacon believed that the greatest virtue of the sciences lay in their usefulness. He saw his role as a propagandist for the sciences and a canvasser of their achievements, in his words "a bell-ringer ...[calling] others to church," but he knew little about their actual substance. Bacon treated mathematics simply as a tool, rejected the "absurd opinions" of both Ptolemaic and Copernican theory, misunderstood Galileo, and underestimated the work of Harvey.

On the continent Descartes, who epitomized the *ésprit géométrique* of the age, inaugurated modern rationalism. Since the nineteenth-century neo-Kantian Kuno Fischer, philosophers have mainly interpreted Descartes as beginning an era in which the argument between Platonism and Aristotelianism gives way to a competition between the epistemologies driven by skepticism: the rationalist and the empiricist.[421] Descartes' ontology divides the world into two fundamental entities, thought or mind, which is unextended thinking substance, and matter or extension together with its properties. This opposition constitutes the celebrated Cartesian dualism, developed and sharpened beyond that of the Stoics in antiquity who had a separation of the two by degrees. Descartes concurs with panmathematicism in that only mathematical laws satisfactorily describe the material world, being accessible through deduction coupled with instantaneous intuitions and producing clear and distinct ideas. He begins his method by what is today called Cartesian doubt: doubting everything, including his own existence. In this search he determines that he cannot doubt the existence of doubt itself and reaches the incontrovertible truth *cogito ergo sum*

("I think, therefore I exist"). Written in elegant French and published in 1637, Descartes' great work, *Discours de la méthode*, offers his program for "conducting the reason and seeking truth in the sciences." Its ahistorical treatment offers not a propectus for a scientific revolution, but a program for how results are processed and presented.[422] In his unalloyed rationalism, mathematics is the purest cognitive exercise; among the natural sciences only mathematics achieves certainty in proofs.

Descartes' criterion of clear and distinct ideas allows him to export his generalized method from mathematics to all the sciences. *Discours* 2 presents its four fundamental rules:

> [N]ever to accept anything as true that I did not know to be evidently so: that is to say, carefully to avoid precipitancy and prejudice, and to include in my judgments nothing more than what presented itself so clearly and so distinctly to my mind that I might have no occasion to place it in doubt.
> [T]o divide each of the difficulties that I was examining into as many parts as might be possible and necessary in order best to solve it.
> [T]o conduct my thoughts in an orderly way, beginning with the simplest objects and the easiest to know, in order to climb gradually, as by degrees, as far as the knowledge of the most complex, and even supposing some order among those objects which do not precede each other naturally.
> [E]verywhere to make such complete enumerations and such general reviews that I would be sure to have omitted nothing.[423]

In starting with the simple and general and developing a long chain of mathematical reasoning that proceeds from causation to specific particulars, Cartesian rationalism reverses the Baconian order of inquiry. While allowing conjectures or hypotheses as a basic research tool, Descartes recognizes that introductory principles can be probable or false. A thorough and accurate enumeration of results through *expériences*, that is, observation and experiment, is required to establish and refine principles. This induction is a lower but second root of his method. He based his own research on optics, sounds, the rainbow, and the circulation of blood in fish and rabbits on experiential procedures.

Although Cartesian rationalism soon triumphed on the continent, it had powerful opposing forces in an age when belief in witchcraft was pervasive. But witchcraft trials may also have generated a backlash of skepticism and support for rationalism. Although Descartes adamantly denied that his method and natural philosophy encouraged theological unorthodoxy, application of his mind and matter dualism led others to materialism and some to question the very reality of God. Descartes argued that an infinitely perfect God must exist on grounds that we conceive the idea of such a being and that the content of our ideas requires an adequate cause. The adequate source of this idea has to be the infinite mind.

The Inquisition's trial and condemnation of Galileo for supporting Copernican astronomy in the *Dialogue,* making him into "a criminal," and burning copies of this book in Rome particularly shocked Descartes.[424]   At the end of 1633, he had just completed the manuscript of *Le monde (The World),* which also espouses Copernican astronomy, when he heard the news. He wrote to Mersenne three times asking who in the Church supported the condemnation and its standing outside Italy. Although he resided in Protestant Holland and publishers there and in Paris faced no Inquisition, he delayed publication of *Le monde,* as he reported in *Discours vi,* rather than offer it in a mutilated form. Descartes continued to develop his ideas in physics and astronomy, seeking to avoid censure from misunderstanding. His *Le monde* appeared in 1641.

Following upon it, Descartes' magisterial *Principia Philosophia* (hereafter *Principles of Philosophy),* published in 1644, systematically presents his distinctive, comprehensive science that was to supplant Aristotle's.   Written not in French but Latin, the language more likely to gain acceptance from universities and the Catholic Church, and composed while he resided near Leiden and had access to its libraries, the *Principles* explains the material world in terms of an all-embracing corpuscular mechanics. It asserts that "the rules of mechanics ... are those of nature."

Descartes attempts to banish from that world the antipathies, appetites, spirits, sympathies, and occult qualities of the scholastics. These either have no existence or are the measurable effect of some purely corporeal cause. Corporeal substance has to be extended in space, and extension can be understood only through shape and motion. Matter in motion is thus the fundamental palpable reality. Descartes recognizes three types of material corpuscles. The finest comprise the ether; the second grade, the sun, other stars, and the medium for transmitting light; and the coarsest, the planets and comets. Descartes' brittle corpuscles resemble classical atoms in being impenetrable but differ in being continuously divisible, exceeding the fineness of matter for the medium of light. Reduced to its simplest form, Cartesian mechanics resembles impacting billiard balls as bodies act on one another.   Action by contact, or impulsion, is the only admissible force.   Cartesian mechanics preserves an invariant "quantity of motion," which approaches the modern concept of momentum (mv, where m = quantity of matter and v = velocity). Under the Cartesian law of inertia, motion acquires a status of reality equal to that of rest, and the distinction in status between rectilinear and circular motion vanishes. Both concepts are major departures from Aristotelian physics. Descartes criticizes Galileo for concentrating on particular as opposed to general cases and for granting existence to the state of nothingness, the vacuum, which Descartes does not admit. He has space filled with a primal matter or ether that comprises a plenum.

Descartes continues the Aristotelian separation between terrestrial and celestial mechanics. Essential to his mechanistic cosmology is the theory of *tourbillons* or vortices, which are huge whirlpools of the ether. The heavens consist of an indefinite number of contiguous vortices, each centering on a star. Celestial motions such as planetary orbits, along with the tides, terrestrial gravity, and magnetism, depend on it.

Terrestrial gravity, for example, is a resolution between roughly centrifugal force propelling bodies away from the center and the downward-spiraling ether revolving around Earth. Bodies with a lesser centrifugal force are pushed downward, just as fluids force bodies that have lesser specific gravity to rise. In prescribing uniform and circular motion in the heavens, the initial Cartesian vortex

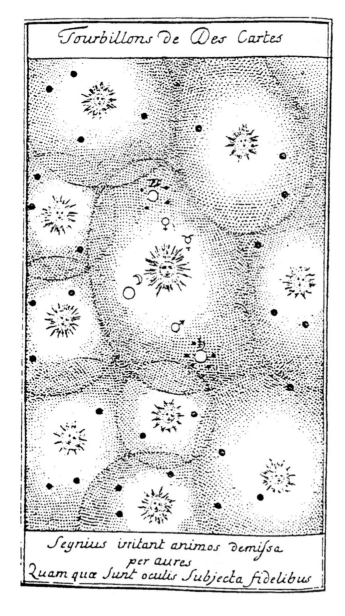

Tourbillons in the Cartesian cosmology

theory clearly contradicted many known laws, including Kepler's second for elliptical planetary orbits and varying speeds of planets. Having available the accurate laws of physics from Huygens, Mariotte, Wallis, and Wren, Nicolas de Malebranche (1638–1715) enjoyed a clear advantage over Descartes, who had depended upon *a priori* principles in founding his mechanics. He devoted himself to synthesizing and reforming Cartesian science, even as he grafted on new thought. Malebranche defended the Cartesian conservation of momentum principle against criticisms by Huygens and Leibniz. His *Research on Truth*, published in 1674, also attempted to make the vortex theory more plausible. Its illustration 16 adds to the large tourbillons an infinite number of little, perfectly elastic tourbillons whose size can be infinitely small. All large tourbillons are comprised of these. Planets and satellites travel along these complex vortices, which by impinging on each other can produce ellipses, a position Pierre Mazière defended in winning the prize of the Paris Academy in 1726. Malebranche also has all matter carried along in small vortices and explains as consequences chemical reactions, fire, and thunder.

Descartes, who wished to geometrize terrestrial and celestial physics, did so by an almost entirely speculative system requiring continuous mechanical contacts. His analytic geometry, moreover, is limited to finite series and lacks a more powerful mathematical technique, such as calculus, to exploit infinite series and make increasingly precise approximations. Physics had not yet expunged the vagueness of more or less.

In pursuit of a universalized natural philosophy, Descartes' *Principles of Philosophy* extends his mechanical program to the nature and movement of animals. He considered them unthinking machines and extended the comparison in some degree to human beings, whom he describes as "earthen machines." As background study, Descartes had made extensive anatomical dissections on animals at butcher shops. Of course the soul, which connects humans with God, and the mind remain outside corpuscular mechanics.

The new Cartesian science had its critics. Catholic theologians claimed that its reduction of matter to extension undercut the doctrinal transubstantiation. It thereby supported Protestant challenges as to how Christ became present in the eucharist. There was also popular ridicule for the desensitized Cartesian view of animals. Yet Cartesian science quickly gained dominance in western Europe. Superceding Aristotelian essentialist physics, in modified form it was to enjoy primacy there until the end of the 1730s. Jacques Rouhault's masterful *Traité de physique* (*Treatise on Physics*) of 1671, a synthesis of Cartesian science, became the leading natural philosophy textbook of its era. Malebranche's circle in Paris accepted Rouhault's book as faithful to Cartesian science. Against Leibniz, who proposed the alternative conservation of *vis viva* (roughly twice kinetic energy), Malebranche insisted that momentum, the Cartesian measure of force, is conserved. His redesigned small and large vortices better explained celestial motions, and he defined light as a vibration in the ether. The peak of Cartesian influence in the sciences came with the publication of Bernard Fontenelle's *Entrétiens sur la pluralité des mondes*

(*Conversations on the Plurality of Worlds*) in 1686.[425] Its five evenings' presentations address Copernican astronomy, the work of Galileo and Kepler, and the Cartesian vortex cosmology. The *Entrétiens* is the most famous and most widely translated work of Fontenelle, who steadily revised it and later became secretary of the Paris Academy of Sciences. But a potent rival appeared in 1687 when Newton published his *Principia mathematica*.

Patronage, Academies, and Universities

Knowledge of the organization and financing of the sciences and their fit within a seventeenth-century cultural and intellectual setting are crucial to understanding their development. Studies by Mario Biagioli, Mordechai Feingold, Paula Findlen, Margaret Jacob, Alexander Keller, Pamela Long, David Lux, Robert Merton, Bruce Moran, and Charles Webster, insisting on the socially derivative aspect of scientific activity at that time, have been instrumental in moving historiography to this position, which is still being refined.[426]

During the sixteenth century, the main centers of scientific research shifted away from universities, which were losing governmental support and faced the Counter-Reformation, to the courts of princes, great landed nobles, and cardinals. Urban merchants, trading and banking oligarchs, military captains, and publishing houses also supported this realignment of resources, and new academies and projects were created. To the 1660s, personal patronage funded most of these activities.[427] Among at least seven hundred learned societies founded in the sixteenth century, only a small minority concentrated on the sciences.[428] The individuals and small groups conducting research there probably numbered no more than a few thousand people.

Patronage enabled a new style of research and dictated its terms and topics. At late medieval universities, natural philosophers had concentrated rigorously on the logic and language of known knowledge. Their work appears as more of a textual and library study than field work. John Murdoch describes it as a discourse in natural philosophy that leaves out nature.[429] In contrast, princely courts and academies adopted for the sciences the sixteenth-century motif of *venatio*, a hunt to discover by design or accident secrets and uses of nature.[430] *Venatio* with its glamorous trappings fitted the self-image of Renaissance princes. Tasks at courts were to glorify and legitimize rulers, great nobles, and urban patricians. *Praxis*, military and political action, and *techne*, or constructive activities, came to be closely affiliated. The built environment of new palaces, churches, painting, and sculpture expressed the authority of the state. At courts, natural philosophers collaborated closely with personnel previously considered ancillary—craftsmen, engineers, geometers, and technicians. A major task was developing armaments and military techniques. In this, a rationalization of craft practices contributed to the incorporation of aspects of the mechanical arts into mechanics, and attention to precision measurement refined mechanical engineering.[431] Courts also included magicians and alchemists informed by hermetic philosophy. Central European princes avidly patronized alchemy, which as a primary objective sought to transmute base metals into gold and silver. Alchemical research helped give legitimacy to the empiricism of the laboratory.[432]

The Holy Roman Emperor Rudolph II, who reigned from 1576 to 1612, was the foremost patron and protector of the sciences. Amid the religious stress of the Counter-Reformation, he inclined to humanism. After moving his court from Vienna to Prague in 1582, he set about making it an artistic and cultural capital comparable in splendor to major Italian cities.[433] Rudolph especially had his ambassadors to Madrid and Rome identify art works on the market and applied to Fugger bankers for credit to purchase them. He collected paintings by Breughel, Corregio, and Titian, as well as by da Vinci and Raphael, which at that time every royal collection prized. Rudolph displayed these at Hradĉany palace, whose great rooms included an ethnographic museum, a library of rare books, and a *Kunstkammer*. *Kunstkammern*, housing collections of instruments, technical contrivances, and curiosities, were popular among Austrian, German, and Italian princes and great nobles. Rudolph's included a table inlaid with precious gems representing animals, flowers, and trees that conveyed such an illusion of reality that it was considered a wonder of the world. He also had a zoo on the palace grounds. Rudolph had an observatory built, and over time his court invited Nicholas R. Ursus, Tycho Brahe, and Johannes Kepler there. Their assignments included astrology, which the monarch considered a means of attuning the mind to that of God. Goldsmiths were attracted to Rudolph's court, which housed alchemical laboratories. Apparently, he dabbled in searches for the philosopher's stone that could produce the basic substance of the universe. For hosting many astrologers, alchemists, cabalists, magicians, prophetic mystics, and Paracelsian physicians, Rudolph is sometimes depicted as a Dr. Faustus and his court a breeding ground for the occult arts.[434] This seems somewhat anachronistic, since the lines between astrology and astronomy or natural magic and magnetism, for example, were not yet clearly drawn. Two other principal science patrons were Prince Maurice (d. 1625) of Holland, whose court included Simon Stevin and young René Descartes, and Henry Percy, the ninth, "wizard earl" of Northumberland, who supported such mathematicians as Thomas Harriot.

Across central and western Europe, the sixteenth-century Platonic revival had given rise to academies, set up by private initiative, that disputed the Aristotelian thought of the universities and the Church. Up to 1800, more than 2,500 academies were established.[435] The first of these cultivated societies had a broad intellectual base, but later academies were likely to specialize in art, poetry, or natural science. Members were educated people having some leisure—landed gentlemen, lawyers, physicians, merchants, apothecaries, teachers, clerics, and soldiers. The patron host might encourage science meetings to include a dissection, experiment, or astronomical observation. These structures stimulated and helped sustain the scientific enterprise. Since scientific publication was neither easy or extensive, a few intelligencers in science academies sought to build a correspondence network and transmission nodes.

From about 1623 to his death, the Minim monk Marin Mersenne (1588–1648) had the most famous meetings in France concentrating on the mathematical sciences. Among the regular company carefully selected to convene at the Minim convent near the Place des Vosges were to be Claude Henry, Claude Mydorge, Étienne and Blaise Pascal, Gilles Personne de Roberval, René Descartes, Pierre Gassendi, and Thomas Hobbes. In 1635 Mersenne organized these meetings to form the *Academia Parisiensis*. Its character was not that of a restrictive academy but more of a salon, having an informal atmosphere and wine served during discussions. Developing a network of at least seventy-eight correspondents across Europe and reaching to Turkey and Tunisia, he was essentially the secretary of the early republic of science. Letters received, some nearly illegible, were carefully and clearly copied. They and criticisms were sent to others. This was crucial when few had opportunities for encounters with distant colleagues and no scientific journals existed. Mersenne's circle helped to introduce and spread in France the mechanics and mathematics of Galileo and his disciples. Mersenne himself provided Descartes, who resided at various locations in Holland, the latest information on meteorological phenomena such as parhelia, on musical consonance, and on Galileo's *Dialogue* and trial. Accurately reproducing the letters of Descartes was not simple. His handwriting is unclear, while his spelling and punctuation occasionally are careless.[436] The circle supported Descartes' rational method and new system of the world and promoted Pascal's study of the cycloid. In his book *La verité des sciences* of 1625, Mersenne defended against pyrrhonian sceptics' criticisms and paradoxes the certainty and utility of mathematical truths. His argument followed the stance of Clavius. Numbers, he asserted, arise in the intellect and ultimately the intellect conferring existence on them is God. Numbers are then connected to the natural world by way of logic in a necessary union.

In 1603 the eighteen-year old Marquis Frederico Cesi and three friends founded in Rome the *Accademia dei Lincei*, whose members engaged in a lynx-eyed hunt for knowledge. The emblem of the lynx was derived from Giambattista della Porta's book *Phytognomonica* of 1588. Cesi advocated freedom of scientific inquiry and, in place of the customary hoarding, a collegial publication of findings. When named a member in 1610, the aged Porta was the star of the *Lincei*. But the next year the *Lincei* selected Galileo a member. In part through his studies with the telescope, a name given at a *Lincei* meeting, Galileo imparted to this *Accademia* a new direction and overshadowed Porta. Upon Cesi's death in 1630, the institution collapsed.

In 1657 the junior Prince Leopold de' Medici of Tuscany began in Florence the *Accademia del Cimento*, or Academy of Experiments. It is often considered largely the archteype for project research and the modern science academy.[437] The court academy and printing house were merged, and laboratories were built in the academy rooms at the Pitti Palace. The cream of Galileo's students, namely Vincenzo Viviani and Giovanni Alfonso Borelli, extended his program in mechanics and Evangelista Torricelli's experiments on the vacuum. Not they or the other eight members but Prince Leopold, the Cimento's supervisor, decided when it would meet and which of the recommended problems to study. Acting as princely servants, the members made many contributions to instrumentation, including refining Torricelli's barometer. The new instruments improved their research on atmospheric pressue and thermometry, showing, for example, that the temperature at which the freezing of water occurs is constant. The Cimento's motto was "provando e riprovando," test and retest. Selected experiments of members, also including Francesco Redi and Nicholas Steno, were published in 1667 in the illustrated volume *Saggi di Naturali Esperienze*, which retrospectively gave the name to the Cimento. An encomium to the richness of Medicean science, the *Saggi* gives no names of authors and holds experiments to be credible by the presence of the prince. Collective standards for certifying knowledge were missing. Court etiquette required there be no disputes before the prince, so the sanitized *Saggi* makes no mention of the disagreements and colorful insults among members that appear in their correspondence.[438] In 1662 the Cimento's activities slowed and, after Prince Leopold became a cardinal, his and the other Medici's interest in it waned as clerical animosity to the sciences grew worse. Consequently, in 1667 after a brilliant decade of work, it closed.

In England John Wilkins, the Puritan brother-in-law of Oliver Cromwell, was from 1648 patron to an experimental science club at Oxford. Strongly influenced by Baconian programs, it investigated natural philosophy, chemical science, and even the most abstruse mathematics. Before 1645, a voluntary group had met to promote experimental science at Gresham College in London. In 1648 the London group migrated to Oxford. Three years later the experimental science club set out rules for admission, contributions for instruments, and meeting times. Its members combatted Aristotelian science and, among other things, set about cataloguing the books in the Bodleian Library. John Wallis headed the experimental science club. Among its other members would be Robert Boyle, Robert Hooke, Henry Oldenburg, and William Petty. Wilkins and Oldenburg knew Mersenne's circle and Habert de Montmor's Academy in Paris, which succeeded it under Pierre Gassendi. They also had Italian links to the *Accademia del Cimento.* A separate club may have continued to meet in London through the 1650s. Historians have depicted the Oxford and London clubs as invisible colleges, but the Baconian emphasis on useful, collaborative application over academic study seems more germane in mid-seventeenth century England. In 1658–1659, the Oxford group returned to London. In November 1660 a dozen members met at Gresham College and resolved to seek more regular ways of debating and meetings. The Royal Society of London was aborning. The club's Puritan members deftly secured as patron the restored monarch Charles II, who chartered the Royal Society of London in 1663.[439]

The presence of early seventeenth-century science academies did not mean that a scientific community or communicating network was being formed. Expressions of mutual admiration notwithstanding, jealousy, discord, and clever combat divided academies. But work on shared problems and techniques did produce some congealing. The financing of academies came from personal patronage that entailed paternalism and dependence until after 1660. By then royal state funding permitted some independence.

Few early seventeenth-century universities, whose curricula centered on philosophy, religion, law, and medicine, made innovations in the sciences. Still natural philosophers and geometers acquired higher formal education there and belonged to a community of scholars whose pedagogy, investigations of method, and research interests occasionally fostered advances in science. Padua was the setting for Galileo's mechanics, fundamental physiological research from Vesalius to Fabricius culminating in Harvey's discovery of the circulation of the blood, and important botanical studies. Paduan Aristotelianism was a heterodox derivation from that endorsed elsewhere by Catholic authorities. At the small and confessional German universities, patronage devolved from the prince to the bureaucracy associated with centralized state building. The Melanchthon circle at Wittenberg developed and elaborated the Prutenic Tables in astronomy. Among the few German curricular reforms in the sciences, in 1609 Marburg introduced Paracelsian chemical medicine as a discipline.[440] Alchemical and Paracelsian thought also found a sympathetic audience at the court of Landgraf Moritz of Hessen-Kassel. Moritz devised recipes to be tested in the chemical laboratory. Giessen, Heidelberg, and Jena accepted chemical medicine and granted academic status to chemical pharmacy. Dutch Leiden and the French College de France occasionally contributed to mathematics. Oxford added the Savilian Professor of Geometry in 1619 and strengthened its Bodleian Library, Oxford's and Cambridge's move into the forefront of the mathematical sciences did not come until after John Wallis was appointed Savilian Professor in 1649. Meanwhile Gresham College, which the London merchant Sir Thomas Gresham founded in 1597, was compiling a solid record of accomplishment. Three of its seven professorships were in the sciences. These faculty resided in Gresham's mansion house and were obliged to give public lectures. The college mainly pursued practical mathematics. Part of the impetus for it came from the need among London sea captains for better astronomical and navigational charts along with improved cartography and geography. Early sixteenth-century Gresham faculty, who included Henry Briggs and William Oughtred, were to study William Gilbert's *De Magnete,* fashion and popularize logarithms, and invent the slide rule. John Wallis was a member of its faculty in the 1640s and Christopher Wren was from 1657 to 1661.

14.4   Computational Arithmetic in Western Europe from 1570 to 1630

Motivations

The "miraculous power of modern calculation," according to Florian Cajori writing before the general use of computers, rests on three computational inventions: Indo-Arabic decimal numerals, decimal fractions, and logarithms.[441] The general adoption of decimal fractions and the invention of logarithms took place in western Europe near the start of the seventeenth century. Both came partly of the effort to simplify laborious trigonometric computations involving the large numbers demanded in astronomy and thereby to make tables more accurate. The need to improve navigational charts for growing transoceanic trade also promoted these computations. Astronomers, weighing the level of precision of the dominant Ptolemaic and Apollonian geostatic model against that of the Copernican heliocentric system, had reason to welcome a means of calculation that was simpler, less arduous, and less susceptible to error.

Viète, whose computations pave the way for a general adoption of decimal fractions and the invention of logarithms, became convinced on the basis of geometric arguments that Ptolemy's geocentric model was more reliable than Copernican theory. It was probably astronomical debate occasioned by the new Copernican model that prompted him to make his earliest extensive trigonometric calculations. The first part of his *Canon Mathematicus, Seu ad Triangula cum Appendicibus*, which appeared in 1571, and the second, published in 1579, generalized the more analytic approach to trigonometry known as goniometry. Its planned last two parts on astronomy went unpublished. Viète employed and mostly superseded methods from Ptolemy's *Almagest* as refined in al-Zarqālī's *Canones sive Regulae super Tabulas Astronomiae*, which Gerard of Cremona translated, and in al-Tūsī's *Shakl al-Gita*. He also drew on John of Gmunden's *Tractatus de Sinibus, Chordis, et Arcubus*, which had influenced Georg Joachim Rheticus, a disciple of Copernicus, along with the writings of Regiomontanus.

From the first two parts of the *Canon Mathematicus*, which gave the trigonometry preparatory to "Harmonicum Coeleste," through *Apollonius Gallus*, published in 1600, Viète labored to compute the six trigonometric lines more finely than at the fifteen-degree intervals that his predecessors had managed. For right triangles and spherical triangles, which he like the ancients divides into two right angles, he succeeds in constructing these lines degree by degree. Viète then divides the radius of his unit circle into sixty parts to achieve minute by minute tables for them.

Reinforcing a practice developing among his contemporaries, Viète's *Canon* advocates decimal rather than sexagesimal fractions. Following Regiomontanus, he writes:

> Sexagesimal and sixties are to be used sparingly or never in mathematics, and thousandths and thousands, hundreths and hundreds, tenths and tens, and similar progressions, ascending and descending,... frequently or exclusively.[442]

The second part of the *Canon* puts the fractional part of decimal fractions in smaller print than the integral portion and separates the two parts by a vertical line. Seeking to avoid fractions insofar as possible, Viète chooses a sufficiently large hypotenuse for sine tables. Regiomontanus had a hypotenuse of 10^7 parts; Viète's has 10^5 parts, a fundamental figure in the number system of Apollonius.

Recognizing that the total sine and two other elements determine a spherical right triangle and aided by tables of the six trigonometric lines for angles to the nearest minute, Viète systematically computes right, oblique plane, and spherical triangles. His *Canon* appears to be the first published book in western Europe to do so. Its third of six tables gives multiplications of sides of right triangles, and the fourth provides quotients from dividing these by the Egyptian year, day, and hour along with major subdivisions. Viète excels in multiplication and division, as do his rivals of the time Joost Bürgi and Adriaan van Roomen. By using an arc of a great circle perpendicular to one side, Viète subdivides spherical triangles into two right triangles to compute them, as had the ancients and Regiomontanus.

Searching for new analytic relationships and simpler computations, Viète in the *Canon* derives several trigonometric identities. These include $\sin\theta = \sin(60° + \theta) - \sin(60° - \theta)$. In modern notation, Viète's derivation is

$$\sin(60° + \theta) = \sin 60° \cos\theta + \cos 60° \sin\theta$$
$$\sin(60° - \theta) = \sin 60° \cos\theta - \cos 60° \sin\theta$$
$$\sin(60° + \theta) - \sin(60° - \theta) = \sin 60° \cos\theta + \cos 60° \sin\theta$$
$$- \sin 60° \cos\theta + \cos 60° \sin\theta$$
$$= 2\cos 60° \sin\theta$$
$$= 2(\tfrac{1}{2})\sin\theta = \sin\theta.$$

Viète also finds that $\sin\alpha + \sin\beta = 2\sin\frac{(\alpha+\beta)}{2}\cos\frac{(\alpha-\beta)}{2}$. Letting $\frac{(\alpha+\beta)}{2} = A$ and $\frac{(\alpha-\beta)}{2} = B$ gives the important rules $\sin(A+B)+\sin(A-B) = 2\sin A \cos B$ and similarly $\sin(A + B) - \sin(A - B) = 2\cos A \sin B$. Putting angles α and β above the radius OD in the accompanying drawing and proceeding in an analogous manner will yield $\cos(A + B) + \cos(A - B) = 2\cos A \cos B$ and $\cos(A - B) - \cos(A + B) = 2\sin A \sin B$. Viète also expresses $\cos n\theta$ as a function of $\cos\theta$ for $n = 1, 2, \ldots, 9$.

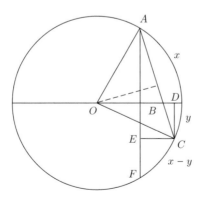

Viète's derivation of trigonometric identities for simplifying computations was part of the development of *prosthaphaeresis*. At leading, late sixteenth-century astronomical observatories, most notably at Tycho Brahe's Uraniborg castle and the adjacent Sytjaerneborg observatory on the island of Hven in the

Danish Sound and later in Prague, *prosthaphaeresis* was the principal method for shortening lengthy multiplications. Trigonometric identities reduce multiplications to addition and division to subtraction.[443] Since it is easier to add and subtract than to multiply and divide, this method made the computation of trigonometric tables quicker and more accurate.

Stevin's Work and Notation for Decimal Fractions

The northward shift of ascendancy in commerce, art, and letters from Italian cities is background for Simon Stevin (1548–1620), the next major mathematician to advance computational arithmetic. From 1582, he lived at Leiden, where he was a leading partisan of decimal fractions.[444] Stevin also contributed to trigonometry, statics, and hydrostatics. In the Dutch republic under its governor Prince Maurice of Nassau, learning flowered.

Stevin was born in Bruges in the Spanish Netherlands, which is now Belgium. Beginning as a cashier in the merchant houses of Bruges and Antwerp and then traveling through Prussia, Poland, Denmark, and Norway between 1571 and 1577, he arrived in Leiden in 1582. Its prosperity and intellectual vitality were attractive and even more so the escape it offered from the persecution of Protestants in the Catholic Spanish Netherlands. In moving to Leiden, Stevin had joined the flow of migrants during the Dutch war against Spain that lasted from 1568 to 1648.

A career closely tied to commercial, administrative, and military affairs in the Dutch republic led Stevin to write primarily in Flemish. In 1583 he had enrolled at the University of Leiden and studied the native language. He admired the power of its simple and modest words and its ability to shape new terms in the natural sciences. Stevin came to consider Flemish, rather than Latin, the language from which others had arisen. His writing in Flemish allowed him to reach a wide audience in the Spanish Netherlands as well as Holland. Already in 1582 he had completed *Tafeln van Interest*, published in Antwerp, which made public rules for computing single and compound interest. Its tables, the invention of Jean Trenchant, allow for rapid computation of annuities and discounts. Banking houses had kept such tables secret. They soon came into common use in the Dutch republic. Stevin's *Bookkeeping . . . in the Italian Manner,* appearing in 1608, introduced into Holland the superior Venetian double-entry system.

After 1600 Stevin's career advanced rapidly. That year he began teaching mathematics at the Leiden School of Engineering and working as an engineer. About this time he built a sail-rigged land carriage. With a proper wind, his carriage carried twenty-six passengers along the shore faster than any pulled by galloping horses. In 1603 Prince Maurice appointed him army intendant and inspector of Holland's dikes.

Stevin advised Prince Maurice on military and navigational matters and oversaw education for his army. After extensive examinations of camps, military equipment, and sieges, he recommended that fortifications be defended with modern artillery rather than conventional small firearms. Ancient Roman fortifications described by Polybius were his starting point for these studies. Stevin designed sluice gates to flood areas and repulse invaders, a technique crucial in defending Holland against the Spanish and later the French.

For the seafaring Dutch with their thriving commerce and their overseas colonies, improved navigation was vital. Stevin developed a theory of tides founded primarily on available data and proposed a scheme for charting longitude based on a global survey of the magnetic needle's deviation from the astronomical meridian. The proposal proved inadequate when it was shown that secular variations also affect magnetism. In the eighteenth century, John Harrison's development of the maritime spring chronometer was to solve more simply the problem of determining longitude.

Stevin was among the first unconditionally to support the Copernican system. His *De Hemelloop* of 1608 calls it "the true theory." In mathematics, he drew upon the discovery, reconstruction, and assimilation of major writings of antique science, the work especially of Archimedes, Apollonius, Diophantus, and Euclid. The Dutch jurist Hugo Grotius said of Stevin that he viewed ancient sources as part of a first age of wisdom and sought to effect a second such age.

After completing his mercantile tables, Stevin in 1583 published *Problemata Geometria*. Strongly influenced by Euclid and Archimedes, it includes a discussion of how to inscribe regular and semiregular solids in a sphere. Like Dürer, Stevin flattens the solid surfaces by projection of sections onto a plane. His chief book on mathematics, *The Arithmetic of Simon Stevin of Bruges*, appeared in French in 1585. Appended to it was Stevin's translation of Books V and VI of Diophantus' *Arithmetica*, building on the translations of the first four books by Wilhelm Xylander and Bachet de Méziriac. Stevin also appended his now-famous *La Disme*, which with its Flemish original, the twenty-nine page pamphlet *De Thiende* (*The Art of Tenths*), introduced decimal fractions into western Europe for comprehensive use.

The timing was most opportune. Earlier in the century Adam Ries, Christoph Rudolff, and Michael Stifel had advocated an increasing use of decimal numeration. Published in 1541, Regiomontanus's decimal trigonometric tables with $\sin 90° = 6 \times 10^6$ further promoted its employment. Perhaps the Persian al-Kāshī's *Key to Arithmetic* of 1427 most influenced Stevin. In Delft, which was near to where Stevin wrote his pamphlet, accounts of the Turkish decimal method derived from al-Kāshī were being read. Also close to Delft after 1579 worked the Leiden botanist Justus Lipsius, who while in Vienna in 1564 had looked through more than 240 manuscripts in Greek from Constantinople, perhaps al-Kāshī's among them.[445] Just before Stevin, authors had computed with both sexagesimal and decimal fractions, but generally limited the latter to the sciences. Stevin explicitly sought also to employ them broadly in the "affairs of men": in business, commerce, surveying, and tapestry measuring. The preface of *De Thiende* declares his intention to "teach the [comparatively] easy performance of all reckonings ... addition, subtraction, multiplication, and division" with pen and with decimal fractions in the same way as integers.

It was probably a reading of Diophantus that persuaded Stevin to treat decimal numbers as unbroken: in computations, to treat the integer and the fractional part the same. But after reading Stifel and Bombelli, he arrived at an unwieldy circled numerical notation. The symbols 0, 1, 2, 3, ... represent a geometric progression by tenths: $(1/10)^0$, $(1/10)^1$, $(1/10)^2$, $(1/10)^3$, ... Stevin calls 1 "prime," 2 "second," and 3 "third," instead of tenths, hundreths, and thousandths. His symbols and terms seem an analogy to sexagesimal notation, where $1/60 = 1'$ and $1'/60 = 1''$. In Stevin's notation

$$32.57 \quad \text{is} \quad 32 \quad 0 \quad 5 \quad 1 \quad 7 \quad 2.$$

Stevin's arithmetic operations with these numbers are in principle the same as today's. Here, for example, is how Stevin multiplies 32.57 by 89.46.

$$
\begin{array}{r}
0\,1\,2 \\
3\,2\,5\,7 \\
8\,9\,4\,6 \\
\hline
1\,9\,5\,4\,2 \\
1\,3\,0\,2\,8 \\
2\,9\,3\,1\,3 \\
2\,6\,0\,5\,6 \\
\hline
2\,9\,1\,3\,7\,1\,2\,2 \\
\underline{0\,1\,2\,3\,4} \quad \text{or } 2913.7122.
\end{array}
$$

In division, subtracting the last underlined figures in his divisor from the last in his dividend locates the separatrix between the integer and the fraction portion of the answer. When these figures are 2 and 5, the answer has three places after the separatrix.

By consistently using decimal fractions and his system of operations, Stevin believed, computers, whether clerks or astronomers, could become proficient in operating with them and thereby save time and labor. He asked surveyors to test his decimal fractions. After mastering his system, they expressed satisfaction.

Support for decimal fractions spread quickly, especially after the invention of logarithms. Within fifty years they were widely adopted in Europe. Stevin's clumsy notation briefly retarded that process. He had overlooked the redundancy in his system, which fails to separate fractions from integers by a single symbol. In 1592 the Bolognese mapmaker Giovanni Antonio Magini (1555–1617) corrected that fault by introducing the decimal point in his *Tabula Tetragonica*, a table of squares of integers. This table allowed for computing as differences between two squares the products of two factors. Magini was a Ptolemaic astronomer who was to endorse libels against the young Copernican Galileo. In his *Astrolabium*, published in 1592, Christoph Clavius used the decimal point in sine tables. In *The Art of Tenths*, Robert Norman's English translation in 1608, Stevin's circles are replaced with parentheses around numbers. The effort of Viète's *Canon Mathematicus* of 1579 and *Algebra* of 1608 and more so of Kepler and John Napier were crucial in gaining an acceptance of decimal fractions. Napier also effectively addressed their notation. The English translation in 1616 of his *Canon Mirificus* by Edward Wright with the author's approval has a point as a general separator in numbers but not between units and tenths. In *Rabdologia*, Napier in 1617 wavered between a decimal comma and a point to separate the unit from tenths. His *Constructio* of 1619 asserts that "whatever occurs after the point is a fraction." Bonaventura Cavalieri, Magini's successor in Bologna in 1629, also adopted that symbolism.

Other notation still competed with it. In *Arithmetica Logarithmica*, Gresham professor Henry Briggs in 1624 underlined the fractional part, that is, 57<u>462</u>, as an example, for 57.462. Albert Girard's *Invention Nouvelle* of 1629 employs the decimal comma. In *Clavis Mathematicae*, William Oughtred, an admirer of Stevin, in 1631 separated the integer from the fractional part by a slash, as had Rudolff, and underlined the fractional part, so that 83/<u>461</u> is 83.461. Not until the eighteenth century was the decimal point the standard notation.

In the final two sections of *De Thiende*, Stevin urges that trades and states adopt a decimal system of currency, weights and measures, and degrees of arc. But the implementation of such a scheme began only two centuries later with the introduction of the metric system during the French Revolution.

Stevin's *Arithmetic* also recasts the origins of numbers and enlarges their domain. The ancient Greeks had believed that unity generates numbers but is not itself a number. Proceeding from this assumption, Euclid defined numbers as collections of units. Through the ages geometers debated whether unity is a number. Even in his polemic with Ferrari during the 1540s, Tartaglia had hedged by calling unity a potential number. Stevin speaks with certitude. Unity is a number, because all four arithmetical operations can be performed on it and it can be continuously subdivided into as many parts as a computer wishes. Stevin defines number as "the quantity of each thing" and argues that zero, viewed as a geometrical point, generates numbers. Thus, numbers comprise a geometrical continuum.

Stevin asserts that the many distinctions of classes of incommensurable lines in Euclid's Book X are unnecessary. While he distinguishes between commensurable and incommensurable number pairs, the incommensurable lacking a common divisor (he denies that incommensurable numbers) or in more recent terminology irrationals, are "absurd ... or inexplicable," and, like Ptolemy and medieval Islamic algebraists, he works extensively with them. He finds no difference in nature between the integer 2 and the square root of 2, because decimal fractions can represent the square root as accurately as desired. So all numbers, whether discrete or continuous and geometric, have the same nature, a view rejected by Stevin's contemporaries but vindicated by nineteenth-century number theorists who completed the task of embedding discrete numbers into a linear continuum. Stevin does not accept all numbers. He rejects the new imaginary numbers, which weakens his efforts to solve cubic and quartic equations. It took another century before Leonhard Euler made imaginary or complex numbers more precise.

Stevin's principal contribution to algebra in the *Arithmetic* is simplifying rules for solving quadratic equations. Cardano, Stifel, Viète, and others believed that quadratic equations consist of three irreducible types. Stevin reduces them to one type: $ax^2 + bx + c = 0$ in the symbolism of Viète and Descartes. He does so by accepting negative numbers and equating two processes: subtracting a positive number and adding a negative number.

In the treatise *De Deursichtige*, Stevin treats perspective mathematically. Given an object and a perspective sketch of it, he determines the location of the observer's eye. This solves the inverse problem of perspective. Rather than adhering strictly to ancient Greek *reductio ad absurdum* proofs in quadrature problems, he proposes a clumsy forerunner of integration.

After the *Arithmetic*, Stevin published the theoretical *Algebraic Appendix* and *Practice of Geometry* in 1594 but was turning from mathematics to physics and state projects. In 1586 in the *Art of Weighing*, the first systematic work on hydrostatics since Archimedes, and in *Scientific Thought* nineteen years later Stevin's theoretical bent remained evident. The *Art of Weighing*, which appeared a decade before Galileo's Pisan research, reports dropping lead spheres of different weights from the same height to measure their times of fall, and he describes his solution of the problem of the equilibrium of heavy bodies on an inclined plane. His famous diagram "wreath of spheres," or *clootcrans*, accompanies his law of the inclined plane. The *clootcrans* has two inclined planes, *AB* and *BC*. The length of *AB* is twice that of *BC*. Disregarding friction, the wreath of spheres is draped around the triangle *ABC*. On its own, the wreath will not start to rotate. The pull of the four spheres on the left equals that of the two on the right. Stevin has the force of gravity vary inversely as the length of the inclined plane. The *clootcrans* diagram bears the confident inscription *Wonder en es, gheen wonder* (What appears a wonder is not a wonder.) It was chosen recently as the colophon for Charles Gillispie's *Dictionary of Scientific Biography*.

Stevin's wreath of spheres

Stevin's writings, all dedicated to Prince Maurice, were collected and published in five volumes in Dutch in 1608. Albert Girard edited Stevin's *Arithmetic* in 1625 and his edition of *Les oeuvres de Simon Stevin* came out in 1634.

The Inventions and Transmission of Logarithms

The Scot John Napier or Neper (1550–1617), first inventor of logarithms for positive numbers, was the eighth laird of Merchiston, an estate near Edinburgh. From 1563 to sometime before 1566, he studied at St. Andrews, giving special attention to religion and the Apocalypse. Afterward Napier apparently traveled to the continent to expand his learning. He returned to Scotland in 1571 and married Elizabeth Stirling the next year. After she died in 1579, he married Agnes Chisolm. Napier fathered a dozen children. In 1608 his family moved to Merchiston Castle, where the penurious Napier lived for the rest of his life. His logarithms were the cardinal computational accomplishment of the Reformation.

A fervid Scottish Presbyterian, Napier championed the cause of John Knox (d. 1572). The St. Bartholomew's Day massacre of Calvinists in Paris in 1572 must have intensified his opposition to Catholicism. The threatened invasion of Scotland by the Armada of Phillip II of Catholic Spain in 1588 prompted Napier to enter public service in the Edinburgh Presbytery. In 1593 he completed a tract entitled *A Plaine Discovery of the Whole Revelation of Saint John: Set Downe in Two Treatises* intended for the Scottish king, James VI, the future James I of England. Identifying the Pope as the Antichrist, it enjoined the king to "purge his house, family, and court of all Papists, Atheists, and Newtrals." God, he announced, had ordained 6,000 years to elapse between the creation of Earth and its destruction. Uncertain about the precise date of creation, Napier maintained that the world would end sometime between 1688 and 1700. *A Plaine Discovery*, which ran through twenty-one English editions and had many translations, also investigated symbolism presented in the Apocalypse and annexed oracles of Sybilla that concurred with that biblical book.[446]

John Napier

As a landowner, Napier concentrated on improving crops and cattle. In their tillage, he and his sons experimented with different manures and discovered how to apply common salt, or perhaps saltpeter, to the enrichment of soil consistency. His eldest son, Archibald, was granted a monopoly for this method. Napier shared the zest of his contemporaries for technical invention. In 1597 he invented a hydraulic screw and revolving axis that improved on Archimedes' for controlling water levels in coal pits. His designs for the defense of Protestant Scotland included diagrams for burning mirrors, reminiscent of a contrivance that legend attributed to Archimedes, that could set afire enemy ships at a distance and an artillery piece capable of destroying everything in an arc. He even wrote about a device "for sayling under water" that had antecedents in Leonardo. Philip II never attacked and these military engines were probably not constructed. Napier's contemporaries considered the application of mathematical processes for building mechanical devices to be mathematical magic.

Napier addressed two mathematical problems important to his age. In a search for hidden scriptural prophecies, he employed mystical numerology. To improve the accuracy of mathematical astronomy charts being used in global exploration, he endeavored to simplify computations in spherical trigonometry. Only the leisure available to a landed aristocrat and a personal devotion to study allowed him to spend twenty years from about 1594 pondering geometric and arithmetic sequences of numbers.

To that end, Napier read and quoted from the underlying theory in Euclid and studied sequences in Archimedes' work. He examined the research of sixteenth-century mathematicians on the correspondence between terms in an arithmetic and a geometric progression and Stifel's statement in *Arithmetica Integra* of the basic laws applying to them. His recognition of a correspondence between the two sets of terms in these progressions, as in Stifel's table for powers of 2 and exponents, was crucial to inventing logarithms. While Stifel and many others had perceived the correspondence, gaps between successive terms with a base such as 2 were too large to make accurate interpolations between the two sequences. Stifel's mathematics also lacked index notation.

In the mid-1590s John Craig, physician to James VI, informed Napier that in their laborious efforts to improve the accuracy of trigonometric computations with large numbers for astronomical tables and to reduce the time spent on them, Tycho Brahe and his island of Hven assistants were freely using the conversion process of *prosthaphaeresis*. Thereupon his study of the correspondence between arithmetic and geometric sequences deepened. Although *prosthaphaeresis* did not contribute directly to his construction of logarithms, its aim was the same: to employ trigonometric rules for reducing to addition the multiplications of sines or cosines and to subtraction their divisions, as well as for carrying out the extraction of roots. Tycho's group, for example, used Ibn Yunus's identity

$$2\cos A \cos B = \cos(A + B) + \cos(A - B)$$

to reduce to the addition of two trigonometric half chords the multiplication of two others. Here is how an astronomer applying that identity would multiply 80,765 times 97,803:

He will first rewrite this problem in notation equivalent to $(0.80765 \times 0.97803) \times 10^{10}$ and consult a table of cosines. He then computes $\cos A = 0.80765/2 = 0.40383$ and $\cos B = 0.97803$. Using radian measures, where $360° = 2\pi$, he will find that $A = 1.155$ and $B = 0.210$. It is simple to compute $A + B = 1.365$ and $A - B = 0.945$. From tables, next read $\cos(A + B) = 0.208$ and $\cos(A - B) = 0.588$. Adding gives $\cos(A - B) + \cos(A + B) = 0.796$. This means that $80,765 \times 97,803 = 7.96 \times 10^9$. Five-place trigonometric tables give the more accurate 7.90×10^9. A still more accurate value is 7.899059×10^9.

Napier sent his preliminary logarithmic results to Tycho for approval, possibly as early as 1594. Tycho, in turn, mentioned them to Kepler, who wrote in a letter that "some Scotsman" was attempting to simplify computations. Napier worked extensively with a trigonometric identity well known in his time:

$$2 \sin A \sin B = \cos(A - B) - \cos(A + B).$$

Since he addressed spherical trigonometry, his tables were restricted to logarithms of natural sines and tangents. He followed the sine tables of Erasmus Reinhold in the *Prutenic Tables*.

Whatever the origins of Napier's idea of logarithms, in 1614 he announced his invention in *Mirifici Logarithmorum Canonis Descriptio* (*A Description of the Marvelous Laws of Logarithms*), which contains a sine table of angles for each minute of arc. Keep in mind that sines were not yet numerical ratios but semichord lengths of arcs on a circle, and a number large enough, such as 10^7, to avoid fractions was customarily selected for the radius. The method and theory behind constructing logarithmic tables followed in *Mirifici logarithmorum canonis constructio* ("A Construction of the Marvelous Laws of Logarithms"), published in 1619. Napier had initially called this invention "artificial numbers," a phrase still used in the *Constructio*, but from joining the Greek words *logos* (in this usage, ratio) and *arithmos* (number), he coined the compound term "logarithm." It refers to the common ratio of the two sequences of numbers. The concept of indices that are not integers was still obscure, and not until 1728 would Euler define the logarithm of a number as the exponent to which a certain number, called the base, must be raised to give the original number.

Although the essential concept for logarithms—knowledge of a correspondence in terms between a geometric and an arithmetical sequence—clearly preceded Napier, he shows a mastery of mathematics in crafting three other basic elements. To keep close together the geometric progression of powers of a given number, he chooses that number to be very near to 1, namely $1 - 10^{-7} = 0.9999999$. But this bunches the geometric progression too closely, and so to balance that and not have to introduce fractions before reaching seven significant figures, he multiplies each of the integral powers of that number by 10^7. Here he follows Regiomontanus by making the whole sine or radius of the circle 10^7 units. Napier also brilliantly introduces into logarithms the concept of continuity by founding the correspondence between a geometric and an arithmetical progression on two particles moving continuously along two parallel lines over time.

In Napier's kinematic model, two particles P and Q begin with the same velocity on parallel lines. As the drawing here illustrates, particle Q uniformly maintains that velocity along a line of indefinite length, $A'Z'$, giving increases in distance over successive very short time intervals of 10^{-7}. At the same time, the velocity of particle P diminishes at each point, such that it is proportional to the distance remaining to the line's terminus point of Z.

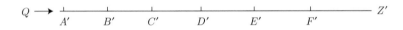

In compiling his clumsy logarithmic tables, Napier starts with successive multiplications by 10^7 of integral powers of his selected number. That number arises from AB nearly equaling 1 and the distance BZ or velocity at B being $10^{-7} - 1$. In the next interval of 10^{-7}, Q reaches C' and P arrives at C, making $BC = 10^{-7}(10^7 - 1) = 1 - 10^{-7}$. The distance CZ is thus $10^7 - 1 - (1 - 10^{-7})$ or $10^7(1 - 10^{-7})^2$, or P's velocity at C. Distances DZ, EZ, etc. are thus $10^7(1 - 10^{-7})^3$, $10^7(1 - 10^{-7})^4$, ... Particle P's motion along AZ reflects the decreasing geometric progression for sines, while the distances on $A'Z'$ comprise an arithmetic series. Napier's logarithms thus increase arithmetically, while modern logarithms increase geometrically.

Without exponents and having virtually no modern notation, Napier lacked important tools of modern analysis. His propositions on continuous functions are thus intuitive, but he clearly grasped functional relations between continuous variables. Basing inequalities on geometric means, he computes bounds within which logarithms must lie. Arithmetic means of these bounds give him values accurate to seven significant figures. Employing ingenious interpolations as well, Napier tabulates sines and tangents at one-minute intervals from zero to 90 degrees.

What exactly are Napier's logarithms? Missing the concept of a base, his tables depend on integral powers of the number $(1-10^{-7})$. Because $(1-10^{-7})^{10^7}$ closely approximates $\lim_{n\to\infty}(1 - 1/n)^n$ or $1/e$ Napier's logarithms divided by 10^7 virtually have a base of $1/e$. The transcendental number e was yet to be discovered. Initially, his hyperbolic form of logarithms amounts to Nap.log $y = 10^7 \log_{1/e}(y/10^7)$. His logarithm of the whole sine of 10^7 was 0, of $10^7(1-10^{-7})$ was 1, and so forth. Later he had the logarithm of 1 equal to 0. Lacking a base, Napier called his logarithms "equidistant comparisons of proportional numbers." Unlike modern logarithms, his logarithm of a product does not equal the sum of its logarithms, nor does that of a quotient equal the difference between them. Dividing the sum or difference by 10^7, however, gives the desired result. Completion of the form of natural logarithms awaited Euler, who made e their base. Still, natural logarithms are sometimes called Napierian.

It is less known that in his pursuit of simplified computations Napier invented crude calculating devices, including movable, engraved rods of wood or ivory, from which evidently, with some confusion outside mathematical circles, they are known as Napier's bones. In *The Fortunes of Nigel*, Sir Walter Scott either jokingly or ignorantly has his clockmaker David Ramsay swear "by the [skeletal] bones of the immortal Napier!"[447] Inscribed on Napier's bones are multiplications of digits by 1 through 9. Essentially they are a mechanical multiplication table and ancestor of the slide rule. To multiply 238×9, move the rods 2, 3, and 8 to the ninth position. This gives $\boxed{\frac{1}{8}\,\frac{2}{7}\,\frac{7}{2}}$. Adding diagonally, which is the lattice method of multiplication of the medieval Arabs and Indians, yields 2,142. For the calculation of square and cube roots, there were separate bones. Napier's own description of his bones appears in *Rabdologia*, a 1617 publication notable also for its last section's expression of numbers in the binary scale. Napier's bones, in vogue for many years, presaged interest in building mechanical devices that could free astronomers, mathematicians, and tax officials from the drudgery of extensive computations.

Astronomers and mathematicians, engaged in tedious and time-consuming trigonometric computations involving large numbers that were subject to "slippery errors," welcomed Napier's logarithms quickly and enthusiastically. Soon after the *Descriptio* appeared in 1614, computers in trigonometry widely embraced this work. Logarithms facilitated accuracy in their astronomical calculations and permitted enormous savings of time and energy. As Pierre-Simon de Laplace was to remark two centuries later, "[Logarithms] by shortening labors [on computations] doubled the life of astronomers."

In the summer of 1616, Henry Briggs (1561–1631), the first professor of geometry at the new Gresham College in London, traveled to Edinburgh to meet Napier. Logarithms had fully absorbed his teaching and research for at least a year, and he wanted to learn "by what engine of wit or ingenuity ... [Napier had come] first to think of this most excellent help in astronomy." During the month-long visit, Napier informed him of his discarding the hyperbolic form of logarithm and proposed that the logarithm of 1, that is 10^0, equals 0 and that $\log R$ or the whole radius equals 10^{10}. Even though he had earlier declared in correspondence to Napier and stated at Gresham that $\log 1 = 0$ "would be much more convenient," Briggs' *Arithmetica logarithmica,* published in 1624, relates his alteration of the basic form to $\log 10 = 1$. Briggs, then, worked with a fixed base of 10. So were born Briggsian or common logarithms.

Briggs's construction of fourteen place tables in base 10 begins by taking successive square roots of 10. Finding $10^{\frac{1}{2}}$ or $\sqrt{10} = 3.162\ldots$, his $3.162 = 0.5000$ and $10^{\frac{3}{4}} = 5.623\ldots$ giving $\log 5.623 = 0.7500$. By 1617 he had compiled tables of common logarithms from 1 to 1,000. They appeared in London as *Logarithmorum Chilias Prima*. Briggs's *Arithmetica Logarithmica*, which appeared in 1624, gives 30,000 decimal logarithms from 1 to 20,000 and from 90,000 to 100,000 as well as ways to find the intermediate values. For the second edition of the *Arithmetica* in 1628, the Dutch mathematician Adriaan Vlacq (d. 1666 or early 1667) provided the intermediate 70,000 chiliads to ten decimal places. The Briggs-Vlacq tables were not superseded until twenty-place tables were computed between 1924 and 1949 in Britain.

Briggs greatly influenced the development of logarithms. His definition incorporating a fixed base prevailed throughout the seventeenth century. His *Arithmetica* also introduced the names "characteristic" for the portion of the logarithm before the decimal point and "mantissa," a Latin term for appendix, for the part after the decimal point. The nomenclature is still in use.

Gresham College faculty spread knowledge of logarithms and applied them to critical problems. In lectures there and after 1620 at Oxford, Briggs promoted Napier's invention. The Gresham group around Briggs—Henry Gellibrand, William Gilbert, Edmund Gunter, William Oughtred, and Edward Wright—pioneered applying logarithms to trigonometric problems related to navigation and mechanics. Wright's first English translation of Napier's *Descriptio*, published in 1616, added a few natural logarithms. Edmund Gunter (d. 1626) invented a sector with sine, tangent, logarithmic, and meridional scales. For two centuries British navigators were to use improved versions of it to solve plane, spherical, and nautical triangles. Gunter's sector, invented before 1623, apparently preceded the rectilinear slide rule, which William Oughtred had invented by 1621. Oughtred and his student Richard Delamain bitterly quarreled over priority for the circular slide rule. Both seem to deserve credit for it, but Delamain published first in 1630. Neither had a runner for the slide rule. Isaac Newton suggested that in 1675, but it would not be added until almost a century later. In 1657, Oughtred in his thirty-six page *Trigonometria* gave sine, tangent, and secant tables to seven decimal places and logarithmic tables to seven places.

On the continent logarithms were accepted just as rapidly and tables were promptly developed. In 1618 Benjamin Ursinus published the first continental work on them. It excerpts Briggs's *Logarithmorum* shortened from seven to five significant figures. Initially, through correspondence Johannes Kepler made logarithms known throughout imperial Austrian and north German lands. Dissatisfied with Napier's tables and impatient with waiting for his colleague Bürgi's, he calculated his own, the *Centrobarya*, that had base e. Kepler dedicated to Napier his *Ephemeris* of 1620. On the Italian peninsula Bonaventura Cavalieri extolled logarithms, while Edward Wingate brought them into vogue in France. Scientific circles in Prague did not have to learn of them through Napier's work. In an instance of the kind of coincidental discovery that often happens in the sciences, they had been invented independently by the Liechtenstein instrument maker Joost Bürgi (1552–1632).

Bürgi, an important figure in his own right, became about 1603 court watchmaker in the imperial cultural center assembled by Rudolph II in Prague. He also assisted Kepler and did for him computations related to astronomy. In order to compute sine tables more easily and accurately, from the mid-1580s Bürgi had worked at the Kassel observatory to improve *prosthaphaeresis*. At the same time, he learned in outline from other writings of the ratios of geometric and arithmetic progressions given in Stifel's *Arithmetica Integra*. Bürgi's system, like Napier's, is not based on exponents, and his logarithm of a quotient is not the difference, nor is that of a product the sum of the logarithms of the constituent numbers. But while Napier's approach is kinematic and in a geometric progression employs a number slightly less than 1, Bürgi is algebraic and has a number slightly greater than 1: $1 + 10^{-4}$ or 1.0001. His geometric progression goes to 1,000,000,000 and his corresponding arithmetical progression to 230,270,022. His system tabulates a column of "black" numbers multiplied by 10^8 and "red" numbers or logarithms by 10^5. When divided by 10^5, his logarithms agree to four significant figures with natural logarithms having base e. His tables are arranged antilogarithmically, however, which would have made difficult their spread and use had they been known.

When Kepler departed in 1612, Bürgi was isolated in Prague. After the death of Rudolph II the city disintegrated as a flourishing cultural center. The start of the Thirty Years War in Prague in 1618 and the defeat of the Protestant Bohemian nobles at the Battle of White Mountain outside the city in 1620 worsened the situation there. Bürgi did not publish his *Arithmetische und geometrische Progress-Tabulen, sambt grundlichen Unterricht* ... until that year, and few copies of the promised accompanying handwritten instructions were saved. His discovery apparently went almost entirely unnoticed outside Prague.

How mathematicians treat their creations and how other mathematicians and society respond to them vary over time. The case of Napier is unusual in the rapid acceptance of his invention among peers. Since Cardano, secrecy was ending in mathematics and the practice increasing of making public new creations. Napier's *Descriptio* and *Constructio* had made logarithms accessible to mathematicians. Napier's work would have long-range importance in astronomy, navigation, banking, and even number theory. In reaching his neat three-page proof that π is transcendental, soon after the laborious proof in 1882 by Carl Louis Ferdinand Lindemann, David Hilbert was to connect the logarithm and trigonometric functions. Within the field of mathematics, these proofs of π's transcendental nature resolved definitively the ancient problem of the squaring of the circle, since all constructible numbers must be algebraic, and it invigorated research on other fundamental problems. But Napier's aristocratic society, which rewarded landowning and commerce, had given little financial recognition to intellectual creations. For his invention Napier received no monetary recompense.

The response of Napier's contemporaries in an age of witchcraft was also hardly reassuring. His imagination and ingenuity in mathematics combined with his retiring nature and study were capable of rousing suspicions among witch hunters. The prestige he enjoyed as a Protestant divine and author of

A Plaine Discovery offered some protection. Even so, hints of unholy doings surfaced. This mentality did not indicate simply folk or regional backwardness. Cardano and Viète had been suspected of practicing the dark arts. For softness in persecuting witches, in Scotland James VI's demonology treatise admonished Reginald Scot, who had detected in witch hunts ignorance, roguery, and false accusations.

14.5   Early Seventeenth-Century Algebra: Harriot, Girard, and Getaldić

In early seventeenth-century algebra, western European thinkers codified and extended major achievements of the preceding century. Remember that Italian masters of that earlier time, most notably Cardano, had discovered general solutions by radicals of cubic and quartic equations and fashioned a theory of polynomial equations. Before Viète, the absence of an effective notation had hindered algebraists. In the 1590s, his introduction of adequate symbols, including capital vowels for unknowns, completed the freeing of algebra from the Euclidean geometric style of proof and made it easier to manipulate equations. Only for all positive roots had Viète stated the relationship between coefficients and roots of equations up to quintics. A general statement was still missing. As the chief locus of mathematical learning was shifting from Italian cities to France, two prominent algebraists emerged further north, one in England and the other in Holland. Following on Viète's program, a third also labored in the Dubrovnik republic on the Adriatic.

The demanding life of Thomas Harriot (ca. 1560–1621), who began a continuous line of English algebraists, included adventure, broad interest in the sciences, and support from noble patrons. In the service of Walter Raleigh, he participated in a cartographic and ethnographic expedition in 1585 to the part of Virginia now known as North Carolina. This seems to make him the first European mathematician to visit the Americas. After returning to England, Harriot published *A Briefe and True Report of the New-Found Land of Virginia* and joined a circle that Shakespeare called the "School of Night." A central figure was the older mathematical magus John Dee, whom he surpassed. After 1600 Harriot corresponded with Kepler on optics and astronomy. This was his only known contact with a natural philosopher of comparable rank. After leaving Raleigh's service about 1598, Harriot received an annual pension of £300 from Henry, Earl of Northumberland. When the earl was confined in the Tower of London after the Gunpowder Plot of 1605, Harriot was imprisoned briefly but was judged not to have been implicated.

Harriot examined many major scientific problems of his day. He improved the theory of the rainbow by measuring the refraction of light rays in liquids and glass. He studied patterns of rainfall and wind. He read Kepler's *Astronomia Nova* of 1609 and corrected minor errors in its computations. Kepler's book forced him reluctantly to accept Copernican astronomy. When news reached him of Galileo's telescopic discoveries, Harriot and his assistant built an improved 30**x** telescope. From their diligent telescopic observations from 1610 to 1613, they made detailed maps of the moon and drew sundisks with sunspots. In *De Reflexione Corporum Rotundorum*, an unpublished tract of 1619, he detailed his geometrical theory of elastic bodies, which has formulas for the rebounding of two bodies, a subject neglected through the Renaissance, but given some prominence by the game of billiards among the late sixteenth century well-to-do.[448] Likely his patron Northumberland's bowling greens also made this an important field of inquiry for Harriot. His results were akin to those found later for the conservation of the quantity of motion, or roughly momentum, by Descartes and the conservation of *vis viva*, or roughly twice kinetic energy, by Leibniz. Continuing the work of Tartaglia, Harriot strove to make ballistics a mathematical science. In a thorough study, he proved that ballistic trajectories must be tilted parabolas. For theory and gunnery, his tilted parabolic shapes and ranges exceeded results from the Italian upright parabolas.[449] Harriot's mathematical deduction is more elegant than that of James Gregory later. For algebra, his reputation rests principally on his *Artis Analyticae Praxis* (*The Practice of the Art of Analysis*), published posthumously in 1631. That it was not published in his lifetime may largely be the result of his poor health from cancer.

That book's attempt to treat linear motion through examples from linear through quintic equations that were carefully worked out extends Viète's new algebra. Harriot decomposes equations into simple factors, such as $x^2 - 5x + 6 = (x-3)(x-2)$, but depicts negative roots as useless and imaginary roots as impossible. Apparently his reason for rejecting those roots here is that the *Praxis* was intended for mathematical amateurs, for elsewhere he studies imaginary roots, which he calls "noetical." This ambiguity suggests that Harriot only partly grasped the relationship between roots and coefficients. But he does discuss this relationship more clearly than Viète had.

The notation in the *Praxis* simplifies Viète's and makes additions. Simplification comes of uniformly treating all algebraic equations and the use of small letters instead of capitals. Vowels represent unknowns and consonants knowns. Harriot's representation of powers is superior. He repeats the lower case of the unknowns in this way: $a^2 = aa$, $a^3 = aaa$, ... His employment of Robert Recorde's equality sign helped make it standard. His equation *aaa – 3.bba = 2.ccc* in modern notation is $x^3 - 3b^2x = 2c^3$. Other new signs are $<$ for "less than" and $>$ for "greater than." Although the amount of new symbolism in *Praxis* was moderate, at least compared with that in William Oughtred's *Clavis Mathematicus*, which also appeared in 1631, the number of new notations and symbolic shorthand in the first edition of the *Praxis* unfamiliar to printers was enough to make for many errors. It took decades for the new notation to be accepted

A generation younger than Harriot, the French-born mathematician Albert Girard (1595–1632) was a Huguenot and therefore liable to come to the attention of French authorities as a heretic. For Protestants life in Lorraine was precarious, and from St. Mihiel in that province young Girard moved to the Dutch republic, probably before 1616. There he pursued as his major work the elaboration of Viète's algebra.

In Holland, Girard attained a scholarly reputation but not financial success. Before becoming an engineer in the army of Prince Frederick Henry of Nassau, he probably studied at the University of Leiden. As his ability in theoretical mathematics gained wide respect Constantine Huygens praised him in 1629 as a "*vir stupendus*" and Pierre Gassendi traveled from France to meet him. But when Girard's patron died, he was left isolated and could find no regular financial support. In the preface to his edition of *Les Oeuvres de Simon Stevin*, which came out posthumously in 1634, Girard complains that he was helpless without a patron and ruined. In its dedication, his widow states that she was left poor with eleven orphans to raise.

Girard's main contributions to algebra appeared in 1629 in the theoretical *Invention nouvelle en l'algebre* (*A New Invention in Algebra*). Girard, who believed that Viète "surpasses all his predecessors in algebra," in his thorough study of cubic and quartic equations ably develops Viète's *logistica speciosa*, which he calls "literal algebra." His central concept is the relation between roots and coefficients. Building on it, Girard concludes but does not prove that "every algebraic equation ... admits of as many solutions as the denomination [power] of the highest quantity indicates." This is the fundamental theorem of classical algebra and Girard's principal result in the field. He also preferred arranging equations in alternating order of decreasing degree on each side of an equation. This means that he rewrites $x^4 = 3x^3 + 6x^2 - 3x - 12$ as $x^4 - 6x^2 + 12 = 3x^3 - 3x$. This arrangement helps him state the relation between coefficients and roots, and Girard asserts that the sum of the roots in an equation must equal the coefficient of the second highest power, or 3 from x^3 in this case. Aided by trigonometric tables and iterations, Girard presents a geometric method more elegant than Cardano's rule to solve cubic equations that have three real roots. Like Pappus and Viète, he reduces these cubic equations to trisecting the arc of a circle with a hyperbola. The figure indicates the three roots.

Exceeding Viète in number theory, Girard accepts negative solutions to equations and recognizes imaginary roots. He explains negative solutions geometrically in this way:

> The minus solution is explained in geometry by moving backward;
> the minus sign moves back when the plus advances.

Girard shows that a fourth-degree problem arising from Pappus has two negative and two positive solutions. For representing roots he, like Stevin, prefers fractional exponents. Girard correctly asserts that the "numerator is the power and the denominator the root." As an alternative to fractional exponents for designating higher roots, he proposes notation that improves on Stevin's and is still in use. He represents the cube root not as $\sqrt{3}$ but as $\sqrt[3]{}$.

Separately continuing Viète's program of restoring treatises of Apollonius and elaborating algebraic analysis was Marin Getaldić (or Ghetaldi, 1568–1626). Born into a patrician family in Dubrovnik, as a peripatetic scholar he traveled near the turn of the century to Rome, where he spoke with Clavius, and to England, Flanders, and France. In Paris he met and worked with Viète. Getaldić had previously been studying Archimedes and parabolic surfaces for the construction of burning mirrors. Taken with his ability, Viète entrusted Getaldić with completing his reconstruction of Apollonius' *On Tangencies* and *On Inclinations.*[450] While Euclid emphasizes the method of synthesis, Apollonius gives greater attention to the method of analysis, starting with relations. The only extant information on these treatises came from complicated statements relating to Apollonius' problems in the preface to Book VII of Pappus' *Mathematical Collection.* These statements in outline form were open to different interpretations. Viète's *Apollonius Gallus* of 1600 reconstructs *On Tangencies.* It proposes ten problems, all involving circles, and solves them. But after reading Commandino's translation of Pappus, Getaldić believed that the treatise contained six more problems, which in the *Supplementum Apollonii Galli* of 1607 he restores and proves. The five problems of Apollonius' *On Inclinations* appeared in Getaldić's two-volume *Apollonius Redivivus* of 1607 and 1613. Paul Guldin and Galileo received copies, and in 1612 Alexander Anderson wrote *Supplemetum Apollonius Redivivus.* Samuel Horsley's edition of Apollonius for Oxford in 1770 observes that, while in presenting solutions and proofs Getaldić employed the geometric method of synthesis, he must have reached them by the algebraic method, which his posthumous *De Resolutione et Compositione Mathematica* of 1630 articulates.

The early influence of Getaldić is not clear. His residence after 1603 in Dubrovnik placed him outside the mainstream of western and central European mathematics, with which he had few contacts after 1609. In Dubrovnik Getaldić held legal positions and served as a diplomat on missions to Constantinople. His astronomical observations and experiments with burning mirrors raised charges locally that he was a sorcerer and magician. The cave where he conducted experiments was called "the magician's den."

The elaboration of Viète's algebra by Harriot and Girard improved the manipulation of algebraic equations, and Getaldić helped restore Apollonius' method of analysis. This work added essential parts of the foundation on which analytic or coordinate geometry was to be built.

15

Transformation of Mathematics, ca. 1600–1660: II: To the Edge of Modernity

> For the human mind has within it a sort of spark of
> the divine, in which the first seeds of useful ways of
> thinking are sown, seeds which, however neglected and
> stifled by studies which impede them, often bear fruit of
> their own accord. This is our experience in the simplest
> of sciences, arithmetic and geometry ....
> — René Descartes, *Rules for Direction of the Mind*
>   (1628)

> So let arithmetic reclaim the doctrine of whole numbers
> as a patrimony all its own.
> — Pierre de Fermat

15.1   A New Pentecost

The creation of the rudiments of analytic geometry, a term not invented until
the nineteenth century, is the chief mathematical achievement of the seventeenth century up to 1660. Infinitesimal calculus, the other principal accomplishment of the full century, came later. Analytic geometry offered a novel and
powerful method that for the first time applied algebraic techniques to solving
problems of Euclidean geometry. Before Viète, properties of curves had been
discovered by geometrical construction. In modern notation, the fundamental
principle of analytic geometry is that every algebraic equation in two unknowns,
$F(x, y) = 0$, defines a curve in space, a principle necessary though not sufficient
in itself for the invention of the new field. The marriage of algebra and geometry, two disciplines previously largely separate, has been indispensable to later
mathematical research.

In the West, the groundings of analytic geometry had developed over a long
time. Ancient geometers, notably Apollonius in *Conics*, described curves by

equations but in verbal form. In the *Elements*, Euclid and in medieval Islam al-Khwārizmī and al-Khayyām developed geometric algebra. By increasing interest in the study of curves, Renaissance and early Reformation painting and the high sciences strengthened the influence on mathematicians of Apollonius's *Conics*, the more so after Commandino's translation of 1566. The more exact translations and reconstructions that sixteenth-century humanists achieved of Hellenistic mathematical classics, especially the higher geometry of Archimedes, Apollonius, and Pappus, and a growing mastery of these writings also revived interest in the exemplary rigor of Archimedes, particularly as exhibited in his methods of exhaustion. In their proofs, Renaissance and Reformation geometers endeavored to meet that standard and sought more; they wanted a heuristic tool for invention, for which Aristotle and ancient geometry were insufficient help. They searched Archimedes' *Quadrature of the Parabola*, Apollonius' *Conics*, *On Contacts*, and *On Tangencies*, Diophantus' *Arithmetica*, the writings of Hero and Proclus, and above all Pappus's *Mathematical Collection* for analytic clues that might serve as a method of discovery.

As the sixteenth century ended, two other elements required for analytic geometry were emerging: a new symbolism and greater skill in manipulating algebraic equations. Algebraic transformations allow the discovery of unsuspected properties of curves. The language of algebra, moreover, is more conducive to discoveries, and algebra can simplify many proofs that are difficult by geometric methods. Viète's symbolic algebra of the 1590s and likely Getaldič's determination of equations that correspond to various geometric constructions began in Europe the intimate connection between algebra and geometry. By the 1620s, algebraists were systematically solving third- and fourth-degree equations. Viète had started with determinate equations in one unknown, but grew increasingly interested in indeterminate equations and systems in two or more unknowns: equations having infinite solutions.

During the 1630s René Descartes and Pierre de Fermat pioneered, as Michael Mahoney argues, in exploiting the heuristic potential of a combination of an expanding knowledge of curves, the application of algebra to solving determinate geometric problems, and the treatment of indeterminate equations with continuous variables in two unknowns.[451] They invented this complex novelty almost simultaneously and apparently took complementary paths to its founding.

The Gentleman from Poitou and His Géométrie

Descartes came from the middle order of French society. The families of both of his parents, Joachim and Jeanne, née Brochard, belonged to the *noblesse de robe*. His father was a lawyer and judge. In social status, this order was below the landed nobility and above the bourgeoisie. Young René, who exhibited an extraordinary curiosity, the elder Descartes called his "little philosopher." Probably in 1606, at the age of ten, he was sent to the Jesuit Royal College at La Flèche. Its rector, his cousin Étienne Charlet, whom Descartes considered a second father, permitted the frail boy to spend mornings in bed meditating and writing. The curriculum began with three years of instruction in the codified humanistic trivium of grammar, rhetoric, and dialectic together with the quadrivium. The second three years added metaphysics, natural philosophy, and ethics. The dialectic centered on Aristotle's *Organon* and its recent Paduan and Coimbran commentators as well as Aquinas and his critics. But the Aristotelian dialectic built on the syllogism and scholastic logic often led to quibbling, and this did not satisfy Descartes. Oratory and the "ravishing delicacy and sweetness" of poetry with its "enthusiasm and power of imagination" drew him more.[452] Apparently he had not yet grasped mathematics as primary to his search for fundamental truths.

Currently, scholars are debating the place of mathematics at early Jesuit colleges and their influence on the development of mathematics.[453] At La Flèche mathematics included the medieval quadrivium and the subordinate sciences of astrology, geodesy, perspective, practical arithmetic, and mechanics. Students read commentaries of Clavius, who stressed utility and novelties, such as Galileo's discovery of the Medician stars—the moons of Jupiter. Mathematical instruction at La Flèche also addressed civil and military architecture, since its noble pupils would become military officers. In its curriculum mathematics was marginal. Descartes' profound interest in the subject was to emerge later in Holland.

Leaving La Flèche in 1615, Descartes went to Paris and then to the University of Poitiers, where in 1616 he received a law degree. After serving in 1618 as an unpaid officer in the Dutch army of Prince Maurice of Nassau, he was briefly an officer in the Bavarian Catholic League military and the Imperial Austrian army. Little of his movements and research from 1621 to 1625 can be established with confidence. No documentary record survives. We know that in 1623 Descartes was in Perron to sell land comprising a third of his mother's estate. Proceeds from this sale provided the income that permitted him to live the life of a gentleman with simple tastes. In 1624 and early 1625 he traveled to Basel, Innsbruck, Venice, and Tuscany, but did not meet Galileo in Florence. From the middle of 1625 to mid-1628, Descartes resided in Paris, where he interacted extensively with Father Mersenne and Claude Mydorge, translator of Euclid into Latin and student of conic sections. Showing increasing interest in mathematics, he compiled in 1626 a three-volume *Synopsis mathematica* on pure and applied mathematics. Descartes now studied the telescope and had the assistance of the glass cutter Jean Ferrier and Mydorge in examining the causes of refraction in different media.[454] By 1628 he had discovered the sine condition of refraction. In today's symbols, if ABH is the angle of incidence i and GBI the angle of refraction r, then in the same bodies $\sin i : \sin r$ is constant in all refractions. Mydorge, who independently discovered this condition mathematically, along with Mersenne, circulated Descartes' ideas widely within the Parisian intelligentsia, so that he was swarmed with visitors.

In late 1628 or early 1629 Descartes' formative intellectual years were complete. Taking only a Bible and Aquinas's *Summa*, he left France for the Netherlands. Its commercial prosperity, degree of religious toleration, and liberal

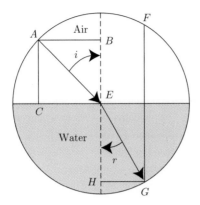

Sine condition of refraction.

support of the new sciences, especially Copernican astronomy, may be among his reasons. Most important was his desire to retreat and pursue his research without interruption. During his twenty-year stay in Holland, a combination of that desire and an occasional effort to be near a friend such as Henri Reneri, who was introducing Cartesian science at Utrecht, brought him to move eighteen times. Personal grieving interrupted Descartes' studies in 1640, when his five-year-old daughter Francine died of a fever. Francine's mother was Descartes' servant Hélène. He never married.

René Descartes

Descartes visited France in 1644 and again in 1647, complaining of insults from Calvinist divines at Utrecht, adherents of a conservative brand of scholastic philosophy who accused him of atheism. During the 1647 visit, Descartes saddened by the death of Mydorge and the illness of Mersenne, met with Gassendi, Hobbes, and the young prodigy Blaise Pascal. He also spoke with his staunch mathematical adversary Gilles Roberval—"that monster Rob," Descartes called him—a leading theorist of infinitesimals. Mersenne, having discussed Torricelli's recent theory of a vacuum, had met with the Pascal family in Rouen to review his failure to repeat accurately Torricelli's experiments. Mersenne lacked adequate glass tubes, so the Pascals had the excellent glassmakers of Rouen prepare instruments for Blaise. Descartes, convinced that a vacuum could not exist, perhaps suggested that Pascal conduct a barometric experiment on atmospheric pressure at different altitudes. In 1648 since Pascal was in frail health, having had purges and bleedingss, his brother-in-law, Florin Périer, who was a judge, carried crude barometers to the 4,000 foot summit of Puy de Dôme near Clermont. Six tests during which the barometer reading fell from 26 inches at the base to 23 inches at the summit confirmed that atmospheric air had weight and suggested the existence of a vacuum, but the Cartesians attacked the study. When Louis XV granted Descartes a pension of 3,000 livres in 1648, he returned to France, possibly intending to stay. But a fiscal crisis leading to the civil war known as the Fronde had Paris in turmoil, and he retreated to Holland. The next year he accepted Swedish Queen Christina's provisional patronage in contrast to what he considered the persecution to which he was subject in Holland. In January 1650 Descartes was required to give her lessons on passions, ethics, and perhaps theology. The lessons were to be at 5 A.M. three days a week. During Sweden's severe winter, he contracted pneumonia and died in Stockholm. Seventeen years later his body, most likely with a substitute skull, was returned to Paris, where it now lies in the church of St. Germain-des-Pres.

The reputation of Descartes in mathematics rests primarily on his invention of analytic geometry, which he published in 1637 in the three-book *Géométrie*. It appeared as an appendix to his *Discours*. He was not chiefly a mathematician and, except for a few fragments and letters, this is his only original work in the field. But its impact was profound. John Stuart Mill was to call analytic geometry "the greatest single step ever made in the progress of the exact sciences."

How long had the idea of it germinated in Descartes' mind before the publication of *Géométrie*? A story has it that observing a fly crawling on a ceiling near a corner turned him to imagining how to express the path of the fly according to distances from adjacent walls. This tale is as fictional, as pleasant, and as useful pedagogically as the tale of Newton and the apple. What years of pains and reflections does it encapsulate?

The first inkling seemingly appeared with Descartes' efforts to solve a challenge problem in mathematics set out on a placard in Breda, Holland, in November 1618. By chance he met Isaac Beeckman, who was reading the placard. Descartes, who was only beginning to learn Flemish and eager for gamesmanship, asked Beeckman to translate into Latin the posted problem. The two men quickly became good friends. Beeckman agreed to instruct the young man from Poitou in physical science and mathematics. Beeckman kept a journal that supplies information on their interactions. Up to early 1619 he set out exercises in mathematics, mechanics, hydrostatics, and acoustics. In mathematics

Descartes considered, for example, whether chains hanging between two nails describe a conic section. In mechanics he applied to such problems as free fall Beeckman's corpuscular natural philosophy. Eschewing occult and immaterial causes, Beeckman reduced phenomena to the speed and direction of corpuscles on an atomic level. This tutoring and collaboration, it appears, developed the mathematical skills of Descartes and his intense interest in the subject. His study of the architecture of the proportional compass recognized that representing the roots of irreducible cubic equations is equivalent to trisecting an angle and the roots of an irreducible quartic equation to duplicating a cube. Afterward in March 1619, Descartes wrote to his mentor that he had caught a glimpse of the "foundations of a marvellous [new mathematical] science."[455]

Possibly influenced by the Belgian mathematician Adriaan Roomen's *Apologia pro Archimede* of 1597, Descartes pursued through the summer and fall a *mathesis universalis* whose method, he believed, lay hidden in the writings of such ancients as Diophantus and Pappus. "A kind of pernicious cunning," he wrote, had hidden it from moderns.[456] Hermetic tracts by Raymond Lull and Cornelius Agrippa on the grandeur of the art of finding the truth, or *ars inveniendi veritatem*, probably influenced Descartes briefly, along with the secret key to knowledge propounded by the mysterious Rosicrucians, whose religious movement had swept across the German states. Four years later in Paris he would deny being a Rosicrucian.

Following travels to Denmark and German states in 1619, Descartes had joined the Bavarian Catholic League army soon after the opening of the Thirty Years War. His first comprehensive biographer Adrien Baillet (d. 1706) relates that during the search for his new mathematics Descartes scorned even his usual solitary walks. Exhausting himself, he found its beginnings, and in winter quarters in a village near Ulm experienced on November 10 a feeling of intense joy followed by delirium. That night he had three successive dreams in what scholars call the episode of the stove-heated room. These dreams suggest that he experienced an intellectual Pentecost during which the spirit of truth descended on him and shifted his principal attention from *mathesis universalis* to method, a critical step leading to analytic geometry.[457]

By 1619, perhaps from a reading of Pappus, Descartes was refining or perhaps misconstruing the two ancient classes of curves: geometrical and mechanical. Descartes distinguished between them by the traditional instrumental and kinematic criteria. Geometrical curves were constructed by the simple, continuous motion of the hand or a machine with a straightedge and a mesolabe compass. They were precise and exact. From antiquity, the simpler of these curves were called "plane" and those involving conic sections, "solid." Descartes now also studied the class of curves more complex than the straight line, circle, and conics: what the ancients grouped together as mechanical curves. He was probably the first to distinguish two types of mechanical curves, in Leibniz's later terminology the algebraic and the transcendental. Algebraic curves, such as the cissoid and the conchoid, he found can be exactly drawn, while others, such as the quadratrix and the spiral, are generated kinematically by two simultaneous motions whose relation is not precisely determined. Since algebraic curves had an exactness of construction, Descartes accepted them as geometrical. The rejection of all mechanical curves from proper geometry in the past, Descartes believed, created too great an exclusion. By removing the stigma from some mechanical curves, he was expanding the domain of elementary geometry.

Another nine years elapsed before Descartes embarked on his continuous work culminating in analytic geometry, so his studies of 1619 alone do not give him priority over Fermat. From the quasi-pictorial model for proofs of mathematics, his methodology was shifting by 1628 to the criterion of clarity and distinctness of ideas that metaphysics claims to guarantee. In mechanically generating from conics a hierarchy of increasingly complex curves, he was to argue that these could be "conceived as clearly and distinctly as ... the circle, or at least ... the conic sections."[458] These curves were needed for solving geometrically equations higher than the fourth degree. One curve that was to appear often in his *Géométrie* is the trident $x^3 - 2ax^2 - a^2x + 2a^3 = axy$, which gives a Pappus locus and a means to solve higher-degree equations. By 1630 Descartes was judging equations to be more important than simplicity in the methods of construction of curves. Still, the instrumental along with the algebraic criteria were to persist in his work. Though it was to cause him great difficulty, he was reluctant to accept the occasional incompatibility of these criteria for classifying curves.

Descartes holds that a curve is acceptable geometrically only if each of its points is the intersection of two straight lines, each proceeding parallel to a fixed axis with commensurable velocities. These lines relate to what is today called an axis of coordinates and an origin. Descartes defines the distances as x and y but does not provide the second or y-axis. His axes are positioned at a given angle that allows the lines to meet. They comprise an oblique coordinate system. The thirty-two figures in Descartes' *Géométrie* do not explicitly use rectangular coordinates, and he restricts curves to the first quadrant. So he has only the embryo of our coordinate system. Not he but Leibniz introduced the word "coordinate."

Book II of *Géométrie* states that a single algebraic equation in two variables must define all points of geometrical curves. From the modern standpoint, Descartes mistakenly stresses determinate rather than indeterminate equations. Finding the curves of these equations is called a locus problem. Solving them requires obtaining a locus of points that meets certain conditions. A simple case is the locus of points at a fixed distance r from a given point in a plane. The answer is a circle, which today is represented by the equation $x^2 + y^2 = r^2$.

In 1630 Descartes enrolled as a mathematics student at the University of Leiden. Its professor of mathematics, Jacobus Golius, was studying the *Conics* of Apollonius. Addressing the three-line problem, which seeks the locus of a point in a plane moving such that the product of the distances from two set lines in a given direction is proportional to the distance squared from the third, Book III of *Conics* proves that the locus is a conic section. Apollonius stops short of four lines but assumes that there too the solution is a conic. In the

part of Book VII of his *Mathematical Collection* that is a commentary on the
Conics, Pappus states the four-line locus problem and generalizes it to six lines
and more. Given four lines in a plane, it requires finding the locus of a point
moving such that the product of the distances from two set lines is constantly
proportional to the product of the distances from the remaining two. Despite
the persistent efforts of ancient geometers, the four-line problem had remained
unsolved until late in 1631, when Golius sent it as a challenge to Descartes and
Mydorge among others. That winter Descartes gave the accompanying pictorial
representation and solved the problem algebraically as a quadratrix. His pointwise construction of the quadratrix involves repeated bisections of curves and
not a relation that exactly locates each point. By Descartes' definition, it was
thus not a legitimate geometric curve. Instead, the quadratrix is a mechanical
curve. A purely geometric solution of the four-line problem was first achieved
by Blaise Pascal and first published by Newton in his *Principia mathematica*.

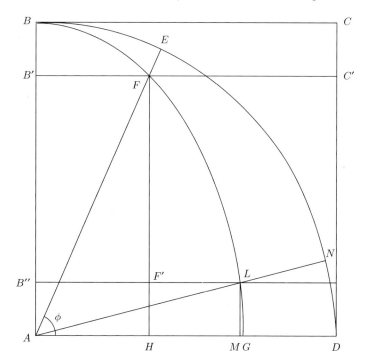

Descartes' resolution of Pappus' four-line locus problem and his generalizations of it and the three-line locus problem are central to his creation of analytic
geometry. The details of his solution appear at the end of Book I and in Book II
of the *Géométrie*. In treating the problem algebraically, Descartes expresses relations among the lines as two variables only, x and y. This radically broke
with the past. From Greek antiquity through Viète, in what is known as the
principle of homogeneity, geometers had depicted the multiplication of two lines
as an area (initially aa in Cartesian notation) and three lines as a volume (aaa).

The product of four or more lines had no dimensional interpretation. Pappus' attempt to overcome by compound ratios that limitation within the four-line locus problem had failed. Descartes' introduction of line segments that are to function arbitrarily as unit lengths or 1, to which he refers all other lengths in proportion theory, surmounts the dimension obstacle by divorcing numbers from representation by line segments. Thus he makes the line segment a/b satisfy the proportion $a/b : 1 = a : b$; $a \times b$, the proportion $1 : a = b : ab$; and $\sqrt{a}$, that of $1 : \sqrt{a} = \sqrt{a} : a$. For Descartes, a^2 is simply the fourth term in the proportion $1 : a = a : a^2$. He constructs $1 : a = a : a^2$ by the similar triangles, beginning with a triangle having sides 1 and a. The side corresponding to 1 is a, and the other side is a^2. From Euclid V on ratios depending on homogeneous quantities and from the three relations $1 : a = a : a^2 = a^2 : a^3 = \ldots$, he linked a^3 to the unit magnitude. Apparently, Descartes did little if any original research in mathematics after this four-line locus proof in 1632.

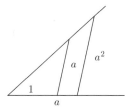

Consolidating work from various earlier projects, Descartes wrote in 1636 his *Géométrie*. Its three books refine his mathematical thought and pieces it together in a general though not systematic order. The first book examines problems from determinate and indeterminate equations that can be constructed "by means of circles and straight lines only," and the second, "the nature of curves." The third on "solid and supersolid" constructions of roots of determinate equations principally offers an elementary theory of equations. The two other appendices, *Météors* on atmospheric phenomena, such as snow crystals, the cause of lightning, and the rainbow, and *Dioptriquse* on properties of light, Descartes had already completed. Probably the knowledge that Fermat had by the mid-1630s worked out the fundamentals of analytic geometry and might appear first in print prompted him to hurry to complete the *Géométrie*. He had failed to reach a satisfactory agreement with the publisher Elzeviers, but their neighbor Jan Maire was to print 3,000 copies. In April or May of 1637, as Fermat's manuscript *Ad locos planos et solidos isagoge (Introduction to Plane and Solid Loci)* was circulating in Paris, Mersenne by coincidence was applying on behalf of Descartes for a privilege from the king to publish and distribute the *Discours*. The chancellor's secretary Jean Beaugrand, the censor asked to review the privilege request in Paris, had been Fermat's mentor. Descartes, who was antagonistic to Fermat—in a letter to Mersenne, he refers to Fermat's mathematics as dung[459]—received a copy of the *Introduction* but claimed to have returned it unread. The *Géométrie* was to be a *tour de force* eclipsing Fermat's unpublished *Introduction*.

The first book of the *Géométrie* opens with the assertion that in ordinary geometry ruler and compass suffice for all constructions. These constructions are equivalent to obtaining the roots of a second-degree algebraic equation. To this point the reader might expect Descartes to review rather exhaustive traditional materials from Euclid. But instead he explains how new algebraic procedures from "the arithmetic calculus" relate "to the operations of geometry" and can solve geometric problems. Beyond the simple adding and subtracting of line segments, his solution of the four-line locus problem applies nondimensional methods to the more difficult task of multiplying or dividing of them and extracting a root. The first book also gives Descartes' new symbolism, which greatly improves on the old cossist notation and refines that of Viète partly by skillfully employing the letters of the alphabet. Where Viète had the capital consonants, B, C, ... , Z, for knowns and capital vowels for unknowns, Descartes had the lowercase typography of the first letters of the alphabet a, b, c, and so on, stand for known constants and the last, x, y and z, for unknowns, a convention that remains in use to the present. Descartes pioneered the use of the same letter for a negative and a positive quantity, and he gave negative roots of equations a status equal to that of positive roots. He also excised from algebraic equations verbal remnants, such as the words "square" and "cube," and replaced Viète's words for powers, such as *quadratus* and *cubus*, by introducing our symbol system of numerical superscripts for powers, such as a^2, a^3, a^4, ... , but did not extend this notational advance to fractional powers.

A major innovation in Book II of *Géométrie* is its purely algebraic method for constructing the normal to any point of a curve having a known equation. In modern notation, given the equation of a curve, $f(x,y) = 0$, and a point P on it with coordinates (x_0, y_0), Descartes draws the normal from P to a point $C(x_1, 0)$ on the x-axis using a clumsy circle technique. The equation of a circle with center C and passing through a point P on the circumference is $(x - x_0)^2 + y^2 = (x_0 - x_1)^2 + y_0^2$. Generally the circle having center C and radius PC intersects the curve at two points.Discovering that when these two points coincide the line PC is the normal, Descartes reduces the normal construction problem to finding an equal double root of a quadratic equation.

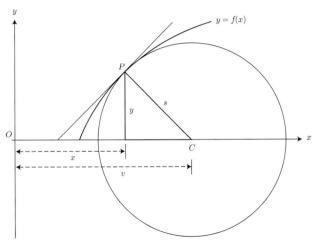

Since the tangent is the perpendicular through the point on the curve to the normal line, Descartes' method indirectly provides for many curves a means of drawing the tangent to points on them. Among these curves are his favorite, the folium, which he proposed to Mersenne in 1638. Its equation is $x^3 + y^3 = 3axy$.

The single folium drawn in the first quadrant is a leaf, but constructing it in each quadrant gives a four-leaf clover. Except for Apollonius' studies and Archimedes' efforts to draw the tangent to his spiral, geometry had not pursued before 1600 the tangent construction problem. But this problem was to be critical for determining slopes of curves and rates of change in differential calculus. Descartes maintained that optical studies had drawn his attention to the subject.

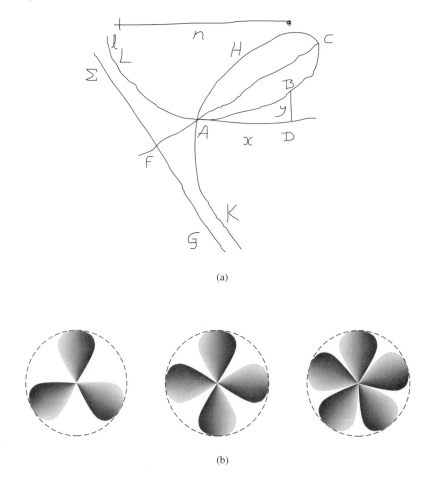

(a)

(b)

For an example of Descartes' method of drawing tangents to curves, consider the parabola $y^2 = 4x$ at point $(1, 2)$. He considers the circle equation $(x - x_2)^2 + y^2 = (1 - x_2)^2 + 4$. Eliminating y gives $(x - x_2)^2 + 4x = (1 - x_2)^2 + 4$ or $x^2 + 2x(2 - x_2) + (2x_2 - 5) = 0$. Obtaining two equal roots requires that the discriminant vanish: $(2 - x_2)^2 - (2x_2 - 5) = 0$ or $x_2 = 3$. The circle's center is $(3, 0)$ and the normal and the tangent may be drawn to the point $(1, 2)$. As the algebra becomes more complex, the problem grows quite daunting. There are better methods for constructing the tangent, and two others appeared almost simultaneously. From his method of maxima and minima, Fermat had a streamlined procedure that he announced in the unpublished memoir *Methodus*, dating from 1629, and sent to Roberval and Mersenne in late 1637. Although Descartes was to criticize Fermat's findings as paralogistic and the result of trial and error, he followed Fermat's lead in improving his own tangent construction method by revising the circle procedure so that the tangent became the secant's limiting position. Descartes and his partisans argued that their procedure was superior.

Among other endeavors in Book III, which addresses the elementary theory of equations, are a method for depressing the degree of a third degree or higher equation having a known root, a procedure for removing the second term, and Descartes' remarkable rule of signs. This rule asserts that the maximum number of both positive and negative roots of an equation may equal either the variation in the signs of the successive coefficients or that number with a positive integer subtracted. Thus, $x^3 + 2x^2 - 4x + 2 = 0$ may have at most two real, positive roots, which Descartes calls "true" roots, or none. His rule allows this equation at most one false, or negative, root. For any degree of an equation that exceeds the number of positive and negative roots, Descartes accepts imaginary roots, but realizes that his rule may not indicate how many imaginary roots there are. John Wallis's *Algebra* of 1685 was the first work explicitly to point out that the rule is a mistake for imaginary roots and that Descartes through inadvertence or error had not made this clear. In his *Arithmetica universalis* of 1707, Isaac Newton formulated the rule more precisely.

The dominant interpretation given by Henk Bos, Carl Boyer, Emily Grosholz, and Timothy Lenoir is that in the Cartesian system algebra is but a tool in the service of geometric construction.[460] Even when Descartes writes the equation, Bos argues, roots have to be constructed geometrically. Descartes' belief that algebraic equations lack the clear and distinct ideas of geometry supports this position. So does the indication that Ramist methods of invention, a Renaissance ordering system of mnemonics, and hermetic constructions of knowledge from a set of mental loci influenced him. E. Giusti maintains to the contrary that algebraic equations form the core of Descartes' program, in which the construction of roots geometrically or by machine has a role more rhetorical than scientific. Stephen Gaukroger finds the *Géométrie* to intermingle with new algebraic methods the old geometrical or instrumental criterion without giving priority to either. He presents Descartes as wrestling mightily with efforts to reconcile the two for classifying curves. In *Dioptrique* and Book II of *Géométrie*, Gaukroger observes, to achieve pictorial clarity Descartes adds a third representation in string constructions. In designing flower beds in the shape of an ellipse or hyperbola, gardeners had long employed string constructions. To draw an ellipse, begin with two stakes H and I. Connect together in a third point B the end points of lines tied to them. Around stakes H and I move the taut string end from B. Descartes' generalization of the gardener's method produced what is now known as the "ovals of Descartes" that he applied in optics. These instrumental constructions have unresolved ratios between straight lines and curves, so they are not proper geometry. Descartes' concluding statement in Book II, "I think that I have omitted nothing essential to an understanding of curved lines," exceeds rhetorical hyperbole.

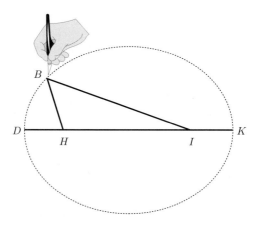

Its lack of clarity and general omission of detailed proofs make the *Géométrie* difficult to read and grasp. Most likely Descartes, who had expended great effort in attempting to master anew the materials for the *Géométrie*, did not again deeply immerse himself in original mathematical research in writing it. This more likely accounts for the obscurity of the text than Descartes' explanation that he intentionally omitted much "to give others the pleasure of discovering [it] for themselves."

Descartes' *Géométrie* quickly provoked controversy. In 1631 Beaugrand had published a new edition with commentary of Viète's *In artem analyticem isagoge*, which had been out of print, and issued the first publication of Viète's *Ad logisticem speciosam notae priores*. Mersenne assisted with the annotations. While Descartes claimed to have depended solely on his own resources, Beaugrand wrote in letters to Mersenne and two anonymous pamphlets that he detected borrowings from Viète and Harriot. Apparently, he was the first to make this charge of plagiarism, which temporarily damaged Descartes' reputation. A more substantive criticism was directed at his circle method of finding a tangent to a curve. Drawing a tangent was now a major problem. Among others, Roberval and Fermat and later Mercator, Sluse, Wallis, and Barrow developed methods for this. Roberval's and Fermat's methods came from their study of extrema and infinitesimals. Since infinitesimals lacked intuitive clarity, Descartes rejected them and the possible superiority of the associated tangent methods of Roberval and Fermat. He treated critics of his *Géométrie* with disdain, calling Roberval "less than a rational animal," Fermat "a braggart," and

Thomas Hobbes "contemptible."[461] Descartes' mathematical accomplishment continued to be questioned into the late seventeenth century. Isaac Barrow judged Cavalieri's methods of indivisibles to be superior, and Gottfried Leibniz asserted that "those well versed in analysis and geometry know that Descartes has found nothing of consequence in algebra."

Perhaps influenced by his friend Johannes Faulhaber, the author of *Numerus figuratis* of 1614, and Kepler's *Harmonice mundi*, Descartes sometime between 1620 and 1623 wrote *De solidorum elementis*. *De solidorum*, which is not connected with his other mathematical studies, attempts to extend his results from plane figures to polyhedra, concentrating on Archimedes' semiregular cases, and it examines figurate numbers. Selected polygonal numbers Descartes reduces to triangular numbers. The number 20, for example, is a triangular number composed of lines of 1, 3, 6, and 10 dots. Descartes also offers a sketchy, intuitive algebraic proof that there can be no more than the five Platonic or regular solids and calculates polygonal numbers for them and eleven semiregular polyhedra. Most scholars studying *De solidorum* investigate whether it contains the germ of Euler's formula for convex polyhedra, $v - e + f = 2$, where v is the number of vertices, e that of edges, and f the number of faces. Descartes finds that solids have twice as many plane angles as sides and that the total of plane angles is $2\phi + 2\alpha - 4$, where ϕ is the number of faces and α the number of solid angles. But he does not put together these two observations. Attributing Euler's formula to him appears mistaken.

"The Learned Councillor from Toulouse"

Strongly influenced by Viète's program and working within a circle of correspondents, Pierre de Fermat (1601–1665) independently invented analytic geometry. Through a method of maxima and minima, along with summation techniques largely from Archimedes for obtaining the quadrature of curves, including upper and lower bounds, he also developed the beginnings of differential and integral calculus. Fermat also made major discoveries—he called them his reveries or musings—in algebra, in part in elimination theory and the handling of Diophantine equations; in number theory; and in the founding of probability, which appears in his correspondence with Blaise Pascal and Christian Huygens. Throughout his work, Fermat sought general methods and posed novel questions. Pascal called him "the greatest mathematician in all of Europe" during the mid-seventeenth century.[462]

Fermat was born into a bourgeois family steadily rising in wealth. His father, Dominique, was a successful leather merchant; his mother, Claire de Long, was from a high *famille de robe* of jurists. In school Pierre became well versed in Greek and Latin, as well as Spanish and Italian. He was praised for the poetry he wrote in different languages. He was later to collect ancient manuscripts and was sought after to emend Greek texts. The primary path to upward social mobility for those of Fermat's social background was the law. By 1631 he had received his baccalaureate in civil law from the University of

Orleans. In June of that year he married his fourth cousin Louise de Long. The marriage produced two sons and three daughters.

Fermat's professional career was divided among three cities. After 1631 he mainly resided in Toulouse, where he was councillor to its parlement.[463] As a member of the *noblesse de robe* he was now able to add the "de" to his name. He frequently presided over the *conseil generale* of nearby Beaumont, his birthplace, and from 1632 belonged to the commission of the *Chambre de l'Édit* of Nantes, which had jurisdiction over the tense relations between Catholics and the Protestant Huguenots. The high rate of mortality among magistrates may largely account for his rapid rise to *conseiller aux enquêtes* for the Toulouse parlement in 1638, to its highest or criminal court in 1642, and to spokesman for its Grand Chamber in 1648. For his preoccupation with mathematical research and his insufficient and confused records in legal cases, political critics considered him naive. Fermat apparently enjoyed good health until a plague struck Toulouse in 1653. After recovering he fell ill again in 1660 and was unable to travel as he had planned to meet Pascal, also fragile, near Clermont and Christiaan Huygens in Paris. Though only thirty-seven, Pascal was aged, being able to walk only with a cane and unable to ride a horse or to travel some distance in a carriage. Fermat's farthest journey from Toulouse was to Bordeaux. Probably because of his seniority and his evenhandedness, the staunch Catholic Fermat occasionally presided over the *Chambre de l'Édit* of the Huguenot stronghold of Castres. While acting in that capacity, he died there in 1665.

Not yet a profession mathematics lacked opportunities for employment or advance as well as specialized journals following a widely accepted set of standards. Its practitioners pursued a hodgepodge of activities and diverse styles. Still lacking even a common name in France, theoretical mathematicians preferred to be called geometers. Classical geometers looked to models and methods from Greek antiquity and saw themselves as mainly recovering and refining past thought. The term *mathematicus* retained its ancient meaning of astrologer or astronomer. Mystics and artists also drew on mathematics. In working on fortifications, commerce, harbor design, mapmaking, and navigational charts, the small group of applied mathematicians relied on problem-solving skills and models mostly from ancient Greece and Alexandria. Fermat was to pursue mathematics as a sideline financially. But he developed a passion for it and took pride in a growing reputation arising from his innovations and his contributions to style and method.

Terse and modest, Fermat seems to have sought generally to avoid the scientific disputes that were endemic in the seventeenth-century. Yet he enjoyed the intellectual exchanges that pitted him against Descartes and Wallis. Except for planning in the mid-1650s to write a book on number theory and a more ambitious project to collaborate with Blaise Pascal and Pierre Carcavi in preparing the bulk of his mathematics for publication, neither of which came to fruition, Fermat to the consternation of his friends refused to publish.[464] This has been attributed to a wish for anonymity, but he shared results by circulating manuscripts that drew admiration. More likely he was unwilling or lacked the time to perfect materials for publication. During an age unclear about models and standards, Fermat found writing final proofs most difficult. In number theory, it bordered on paralysis. Mathematical illiteracy of typesetters was another obstacle to publication. This meant that an author or a colleague familiar with his style and notation had to supervise directly the preparation of a text by the printer. As always books could be sharply criticized. Such obstacles, to be sure, did not keep mathematicians from publishing, but compounded the difficulties for anyone with temperamental reticence or competing obligations.

Aside from a few reminiscences in letters to Mersenne and Roberval in 1636 and 1637, almost no record survives on what drew Fermat's profound interest to mathematics. It probably arose during a visit to Bordeaux in the late 1620s before he took up legal studies at Orleans. In Bordeaux apparently he first encountered the writings of Viète. In France Ramus and Viète had encouraged the restoration of ancient mathematical treatises, particularly work on geometrical analysis by Archimedes, Apollonius, and Pappus, and they had presented algebra as analytic in nature. In the late 1620s and the next decade, Viète's Bordeaux followers were collecting his writings and diffusing his methods. Fermat now became friends with Étienne d'Espagnet, a jurist with a taste for mathematics who possessed some of Viète's writings. He also became acquainted with the mathematician Pierre Prades and the philologist François Philon. He may have met Jean Beaugrand by 1630, a year before Beaugrand's edition with commentary on Viète's *Introduction to the Analytic Art* and his previously unpublished *Notae Priores*. Mersenne added annotations. At any rate, Fermat corresponded with Beaugrand, who probably introduced him to Viète's work in depth. Fermat adopted Viète's symbolic algebra as a research tool and treated as equal in importance to traditional geometry his conception of algebra as the "analytic art." He saw himself as continuing Viète's tradition. Beaugrand, who praised the merits of Fermat's thought, also occasionally claimed as his own some of Fermat's work. By the time Beaugrand died in 1640, relations between the two men had cooled.

The conceptual development of Fermat's research leading to the analytic geometry in his watershed *Introduction* is rather complex but started about 1628.[465] Commencing with his study of Viète's symbolic algebra uniting geometry and arithmetic, he proceeded to attempt to recover lost ancient analytical works cited in Pappus' Book VII and accepted Pappus' goal of seeking generality in mathematics. Generalization he found in Viète's reduction analysis. By 1630 Fermat had restored 16 of 147 theorems in Apollonius' two-book *Plane Loci* on the construction of locus problems. He was not an antiquarian who simply wanted to reproduce faithfully a past work. Fermat retraced Apollonius' rigorous synthesis constructions, but was beginning verbally to shift his attention to the correspondence between geometric loci and the solutions to indeterminate algebraic equations in two unknowns. This differs from Descartes' concentration on geometrically constructing roots of determinate equations having one unknown. After completing work on the *Plane Loci*, Fermat investigated the *symptomata* or defining properties of conic sections. Final proofs for *Plane*

Loci, particularly for II.5, confirm an *ad hoc* shift to the algebraization of locus problems about 1635. Theorem II.5 states:

> if, from any number of given points, straight lines are drawn to a point, and if the sum of the squares of the lines is equal to a given area, the point lies on a circumference given in position.

A systematic change in direction was crucial to composing Fermat's *Introduction*. To reach his analytic geometry he was still in need of his uniaxial system of graphing for sketching solutions.

Among sources other than Viète and Apollonius for Fermat's work were Euclid's *Porisms*, which he attempted to restore, Diophantus' *Arithmetica*, of which he was beginning a long study, and Pappus' *Collection*. A comment in Pappus on the singularity of extrema and his Bordeaux colleagues' investigations had prompted him to write around 1629 a *Method for Determining Maxima and Minima and Tangents to Curved Lines*. This treatise, prepared before he expressed curves as equations, lacked the clarity of his *Introduction*. Editor Cornelis de Waard conjectures that by 1637 Fermat had completed a larger amended *Method*, but his correspondence does not mention it and Michael Mahoney dates that manuscript to the spring of 1638.[466]

In contrast to Descartes, who had proposed only a few new mechanical curves, Fermat in the *Method* defines many analytically. In modern notation, he considers equations of the form $y = x^n$, for $n > 0$, which are now known as the "parabolas of Fermat"; for $n < 0$, the "hyperbolas of Fermat"; and $r^n = a\theta$, now known as the "spirals of Fermat." He also proposes the cubic curve known today as the witch of Agnesi. In 1636 Carcavi probably took the original version of the *Method* to Paris for Roberval to examine.

Fermat's amended *Method* gives a procedure for finding an extreme-value solution of any polynomial equation, values at the top or bottom of any smooth curve. Its procedure depends on adequality or counterfactual equality. Fermat changes a variable A by a very slight, that is, a geometric infinitesimal, amount E, obtaining $A + E$, and then examines the neighborhood, including when $A = A + E$ or $E = 0$. This procedure was to become fundamental in infinitary calculus. In modern notation $A = x$ and $E = \Delta x$. To find the maximum value of $bA^2 - A^3$, begin with $bA^2 - A^3 = M^3$. Substituting $(A + E)$ for A gives $b(A + E)^2 - (A + E)^3 = M^3$ or $2bAE + bE^2 - 3A^2E - 3AE^2 - E^3 = 0$. M is a maximum when there is a repeated root, $A = A + E$. Dividing by E and then setting $E = 0$ gives $2bA - 3A^2 = 0$ or $A = 2b/3$ and $M^3 = 4b^2/27$. This provides the abscissas of a polynomial at maximum and minimum points.

When the slope at a point is zero, the point is either a maximum, a minimum, or a point of inflection. Near a maximum point of a curve, the slope goes progressively from positive to zero to negative, and near a minimum from positive to negative values. The anomalous zero, when the slope on either side has the same sign, gives an inflection point. Fermat establishes the nature of what we call the second derivative criterion. In modern notation, for a maximum

$f''(x) < 0$ and for a minimum $f''(x) > 0$. But lacking a general overview here, he fails to investigate the condition $f''(x) = 0$ that gives points of inflection.

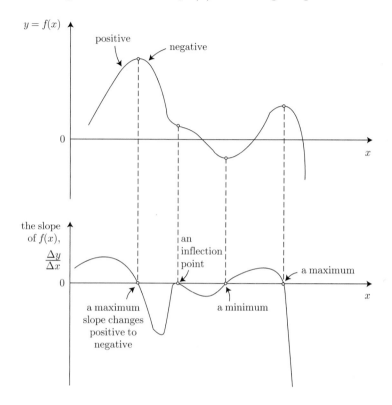

Most mathematicians were impressed with the power of Fermat's method, but a few reflective and philosophical mathematicians, such as Thomas Hobbes, questioned how E could not be zero and at the same time equal zero. The limit concept was needed. In modern notation, Fermat's method closely approximates $\lim_{E \to 0}\{f(A + E) - f(A)\}/E$, or the first derivative of an algebraic equation. The first derivative gives the slope of the curve at any point $x = A$. Fermat was nearing but does not yet have the process of differentiation, for he lacks a reason for neglecting E at the end of his computation and he pursues geometric problems using infinitesimals, a concept Descartes criticizes, rather than variables and functions.

In 1636 Fermat's first creative period in mathematics produced the algebraic geometry of his *Introduction*. Since he refers in letters to completing his *Introduction* after his *Plane Loci*, that is after 1635, and since he sent it to Paris in late 1637, he probably composed it early in 1636, perhaps shortly before he started to correspond with Mersenne and Roberval that spring. Although the analytic geometries of Descartes in print and of Fermat in circulated manuscript appeared almost simultaneously, again in creating them they had independently pursued inverses of the fundamental principle of the field.

After restoring Apollonius' *Plane Loci* and examining Pappus' *Collection* and Viète's reduction analysis, Fermat reformulated problems as generally as possible, derived an indeterminate equation, and then obtained the locus of the problem, while Descartes largely began geometrically with the locus. In 1636 Fermat added the device that located position in his analytic geometry. This was his uniaxial system: a single axis and moving ordinate. Like Descartes, he did not have a coordinate system. His system divides second degree indeterminate equations into seven canonical forms. Working only in the first quadrant, he has $x^2 = y^2$ and every linear equation, for example, $Dx - Ey = 0$, represent a straight line. By translating axes, he reduces to the form $xy = k^2$, an axial hyperbola, equations of form $xy + a^2 = bx + cy$. He demonstrates $a^2 \pm x^2 = by$ to be a parabola; $a^2 + x^2 = ky^2$, an equilateral hyperbola, giving both branches, $a^2 - x^2 = ky^2$, an ellipse, and $x^2 + y^2 + 2ax + 2by = c^2$, a circle. Fermat shows that each curve is uniquely determined by the coefficients. His presentation differs from Descartes in employing Viète's notation and algebra, while Descartes' gives his superior new notation and a new theory of equations.

The *Introduction*'s tangent algorithm is a major corollary to Fermat's method of maxima and minima. Starting with a curve, in his case the generalized parabola $y = x^n$, OP' in the drawing, and a point $P = (a, b)$ on it, he constructs a tangent to the curve at P.

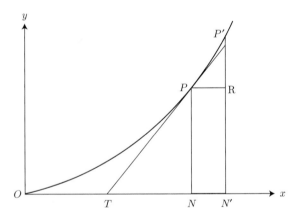

The tangent line segment must cut the curve at only one point. Seventeenth-century geometers knew from Apollonius how to find the length now termed the subtangent, or TN here. This length on the x-axis from the ordinate line gives the point, T, from which the tangent line segment can be drawn to P. Fermat thus already knew the correct answer. His task was to devise a better method to obtain it. Clearly, as the curve proceeds from point to point, its slope varies. Drawing a tangent requires computing the slope of the curve. In modern notation, the slope of a curve $y = f(x)$ approximately equals $\Delta y/\Delta x$. It is critical to make two points, P and P', very close. He shrinks the distance between these points, his E, to an infinitesimal, or essentially Δx. Here E equals PT. In this fashion Fermat obtains a precise answer. Counterfactually, he assumes that the ordinate of the curve $f(a-s)$ in this drawing adequals, that is, nearly equals, the ordinate to the tangent point. This gives the similar triangles $b/t \sim f(a-s)(t-s)$. Allowing s to go to 0 as in the method of maxima and minima, he correctly finds $t = f(a)/f'(a)$, which he applied to many curves.

Fermat's quadratures, together with his methods of maxima and minima and of tangents that obtain the first derivative using infinitesimals, led Lagrange, Laplace, and Fourier to regard him as "the true inventor of the differential calculus.[467] But he did not develop and rigorously prove the inverse relationship and lacked the limit concept.

Controversy and the Early Transmission of Analytic Geometry

In the spring of 1638 a bitter quarrel erupted between Descartes and Fermat, centering on dioptrics along with Fermat's method of maxima and minima and his tangent rule. It was to involve most mathematicians in Paris.

Apparently, in April or May of 1637 Beaugrand obtained, perhaps by stealing, a copy of Descartes' *Dioptrique* and *Géométrie*, yet to be released. The laxness of Mersenne, who had been entrusted with copies, may partly explain the acquisition. Beaugrand, whose *Geostatics* Descartes' had harshly judged, sought to reduce the effect of Descartes' publication by circulating it for criticism before it appeared on the market. Fermat, unfamiliar with the temperament of Descartes, did not respond until Mersenne asked for comments. Fermat began by questioning the derivation of the Cartesian laws of reflection and refraction as well as the belief in the instantaneous transmission of light.[468] Innocently, he became part of what Descartes believed was a conspiracy within the Paris establishment, largely the Roberval faction within Mersenne's circle, to destroy the *Discours* and its essays. In December of 1637 Fermat's brief and vague *Method* and the *Introduction*, possibly Roberval's copies, were delivered to Mersenne to convey to Descartes, who still knew little of Fermat. Perhaps this was done to show the independence of Fermat's research. In early 1638 Descartes set out to demolish the Toulouse upstart. The dispute over Fermat's method of maxima and minima and his tangent construction taught an erring Descartes little, but pushed Fermat to a deeper level of understanding.

Skirting the issue of plagiarism, Descartes asserted that Fermat's method of maxima and minima was implicit in the *Géométrie* but was less effective, because it did not have Descartes' more general method, applicable to both plane and solid loci. He found that substituting the name of each conic section for parabola in Fermat's tangent algorithm yields the same result, an indefensible consequence. He also showed that Fermat's is not a purely mechanical procedure. The hypercritical Descartes opined that intuition was the appropriate avenue from Fermat's method to the tangent rule, but that Fermat had fallen on the rule "by accident," a lucky result of trial and error (*à tâtons*), a criticism that particularly stung Fermat. Descartes challenged Fermat to find the tangent to the curve $x^3 + y^3 = pxy$ and dismissively predicted that he would fail to do so. But he extended his rule to apply to this curve and all others generated by polynomials $p(x, y) = 0$. Fermat's tangent algorithm lacked a rigororous foundation, but possessed correctness and greater mathematical generality, virtues that Descartes chose to ignore.

Descartes was initially upset that Fermat refused to let his letters on the controversy be published and open to public rebuttal. The intervention of Roberval and Étienne Pascal on the side of Fermat hardened Descartes' position. Roberval resolved vectorially into a vertical and a horizontal component the velocity of a point moving along a curve. The tangent had to lie on the diagonal of the rectangle that the two velocity vectors form. The components' procedure was not new with Roberval. Archimedes, Galileo, and Descartes had employed it, and Newton would later apply it for differentiation. In March Descartes wrote seeking the opinions of Hardy, Mydorge, and Desargues on what he perceived as failures in Fermat's methods, Desargues suggesting in response that the dispute rested on a simple misunderstanding. Revising his tangent rule, Fermat noted that Descartes was incorrect in seeking for the increment $A + E$ an operation different from that for the decrement $A - E$: both, Fermat observed, yield the same result.

By the summer of 1638, Descartes saw that Fermat was not part of the adversarial Roberval and Pascal faction, recognized their agreement that a maximum or minimum appears at repeated roots, and admitted that Fermat's methods were essentially right. But the intense polemic clouded Fermat's victory and mathematical reputation and increased the hatred of Roberval for Descartes.

Probably in 1639 or 1640, after reading the *Géométrie* with its advanced theory of equations, Fermat provided in *Analytic Investigation of the Method of Maxima and Minima* a better justification for his original method of maxima and minima. Central to it is the role of coefficients in manipulating an equation in search of a repeated root. Its claim that extreme values are unique is now known to be untrue. Since Fermat believed that he had already justified the tangent algorithm and its relation to his method, the *Analytic Investigation* does not address these topics.

Analytic geometry gained acceptance rather rapidly and was extensively developed. Descartes' decision to publish in French rather than Latin, the language of scholars, and to leave gaps in his highly abstruse *Géométrie,* along with Fermat's aversion to publishing, briefly slowed the process. Crucial to the transmission of analytic geometry and its establishment within the university mathematical curriculum was that Descartes' *Géométrie* quickly found a capable commentator, Franz van Schooten the Younger (ca. 1615–1660), a professor of mathematics at Leiden. In 1637, when Descartes was in Leiden to oversee the printing of his *Discours*, van Schooten had met him and quickly recognized the merit of the *Géométrie*. He began an intensive study of it, going to Paris, where Mersenne's circle also made available manuscripts of Viète and Fermat.

The Elsevier publishing house commissioned van Schooten to gather the writings of Viète for publication, but in light of Descartes' unfavorable comments declined to follow his recommendation to publish Fermat's collected works as well.

Van Schooten's first Latin edition of the *Géométrie* of 1649 contains a lengthy introduction and commentary by the French mathematical jurist Florimond Debeaune that includes Fermat's approach to loci, along with better figures, systematic order, and further explanatory commentary that made it didactically successful. Printed from 1659 to 1661, an enlarged two-volume second edition, fully twice the size of the original, expands van Schooten's explanatory commentary and adds Fermat's tangent method, Jan de Witt's study of conic sections, Christiaan Huygens's study of the parabola, and Hendrik van Heuraet's curve rectification. Volume I contains the rule that Jan Hudde, van Schooten's student, who became a burgomaster of Amsterdam, devised for determining double roots: the solution that Descartes' tangent method had sought. But, following Fermat, Hudde holds that near an extreme value, that is, a maximum or minimum point, a function has a double zero. His rule states that, if a polynomial equation has two equal roots and the equation is multiplied term by term by an arithmetic progression whose largest term is the degree of the equation, the resulting equation will have one of the roots. Consider the equation $x^3 - 4x^2 + 4x = 0$ and multiply its coefficients by the arithmetical progression 3, 2, 1, 0 to obtain $3x^3 - 8x^2 + 4x = 0$. Since the two equations have a common root of 2, it is the double root of the original. When Hudde claimed priority for extreme-value methods, Huygens had to defend Fermat's title to the honor. A third Latin edition of the *Géométrie* appeared in 1683; a fourth, in 1695, contains remarks by Jakob Bernoulli.

Van Schooten's Latin editions of the *Géométrie* were to be the major source of analytic geometry for young leading mathematicians. Wallis learned the subject from the 1649 edition, Newton from that and the two-volume second issue, and Euler from the 1695 edition. Apparently, van Schooten's editions were the chief foundation for the form and advance of analytic geometry until Leonhard Euler's *Analysin Infinitorum* of 1748 put the subject into its modern form.

Principally through the work of van Schooten as well as texts by Jan de Witt and John Wallis, described as the first two textbooks on analytic geometry, the subject became popular among mathematicians at mid-century. Their texts completed the arithmetizing of conic sections initiated by Descartes. Van Schooten had taught de Witt the mathematics of Descartes and Fermat. At least three years before assuming the post of grand pensionary or prime minister of Holland in 1653, de Witt composed his two-book *Elementa curvarum linearum* (*Elements of Curves*). Book I of his *Elementa*, which variously defines the parabola, hyperbola, and ellipse kinematically as plane loci and planimetrically, is mostly synthetic. Introducing the word "directrix," it includes definitions of the ratio of focus to directrix. Its geometric theory, drawing on the initial books of Apollonius' *Conics*, meets Cartesian criteria for construction. Like Descartes, de Witt ignores negative coordinates, displaying drawings only in the first quadrant. Rather than synthesize as in Book I of his *Elementa*, Book II systematically employs coordinates and the new method of analytic geometry to treat conic sections. It extends Fermat's algebraic methodology and joins it to modern Cartesian notation. De Witt represents a straight line as a first degree equation. For each conic he reduces more complicated equations to a normal form. He considers, for example, the equation $y^2 + 2bxy/a + 2cy = bx - b^2x^2/a^2 - c^2$. Substituting $z = y + bx/a + c$, he arrives at $z^2 = 2bcx/a + bx$. For $d = 2bc/a + b$ this reduces further to $z^2 = dx$. De Witt knew that if the second-degree terms are a perfect square, the equation represents a parabola. Since his *Elementa* was not published until 1659 in Amsterdam, John Wallis, whose *Tractatus de sectionibus conicis* appeared in print in 1655, raised the question of priority for having the initial analytic geometry text, claiming that de Witt's work palely simulated his. Although neglecting logical rigor, Wallis profitably replaced geometric with numerical concepts wherever possible, even for proportions. He assumes that second-degree equations define conics as loci of points in a coordinate system without any reference to the cone and from these deduces the common properties of the three curves. This work placed Wallis closer to Fermat than to Descartes.

Questions remain to the present about the influence of Fermat's ideas in the early dissemination of analytic geometry. Scholars long assumed that it was minor, since he did not publish. This practice may have deprived his name from the actual influence, but van Schooten's addition of Fermat's tangent method to his Latin editions of the *Géométrie* and the examinations by de Witt and Wallis of Fermat's mathematical ideas argue to the contrary. The February 1665 issue of the *Journal des sçavans* announcing Fermat's death catalogues his mathematical works and gives him priority over Descartes for analytical geometry. The eulogist, probably Carcavi, pleaded that his works be published, but to no avail.

Although at his death the bulk of Fermat's mathematics, mainly in correspondence with Carcavi, Frénicle, Huygens, Pascal, Wallis, and others, remained in manuscript, Wallis published a few of Fermat's letters on number theory in his *Commercium epistolicum* of 1658 and the Jesuit Jacques de Billy's thirty-six page essay *Doctrinae analyticae inventum novum* (*New Invention of the Analytic Doctrine*) was condensed from a series of Fermat's letters. Billy quotes from and explains Fermat's *Appendix to ... Bachet's Dissertation on Double Equations in the Style of Diophantus* that greatly extends and makes more effective the traditional, ancient method of double equations. In response to a problem from Frenicle, Fermat in 1644 had devised an iterated analysis by degrees that finds a root for a first equation, let us say x, and makes it part of the solution, say $x + m$, that allows the transformation of the original equations to higher double equations. Fermat was the first to recognize that the difference of the original and higher equations are only numbers and squares, that is, neighboring species, and when they can be neglected. He was justifiably proud of his new method. Not only was it important to algebra but also to giving results and making breakthroughs in number theory that are akin to those moderns obtain by systematically employing group theoretic properties.[469]   All but one of his letters to Billy and the *Appendix* are now lost. Billy's *Doctrinae* is the introduction to the 1670 edition of Diophantus's *Arithmetica* by Fermat's son (Clément)-Samuel. Samuel's edition includes the full notes that his father wrote in the margins of his copy of Bachet's *Diophantus*. Billy, who took the role of popularizer of Fermat's ideas, along with others introduced his work into secondary and university mathematical curricula. He did this at the College de Clermont that became the Lycée Louis le Grand. In addition, Samuel published in *Varia opera mathematica* in 1679 a small sample of his father's correspondence and scattered memoirs, and that year Fermat's *Introduction* was published posthumously. All of this suggests that Fermat's ideas had a demonstrable early impact. After Huygens, Newton, and Leibniz quickly surpassed his work in analysis though, his writings in this area had little to offer leading researchers and his influence waned.

15.2   The Invention of Projective Geometry

In creating projective geometry, the French architect and military engineer Gérard Desargues (1591–1661), a member of the Mersenne circle, sought to rationalize and unify under new synthetic rather than analytic procedures of perspective and anamorphosis a range of ad hoc graphical techniques that Renaissance artists and architects had proposed. Anamorphosis is a pictorial rendition by which the picture looks undistorted when viewed from an extremity, such as from its right edge or bottom. By introducing a pencil of rays and sheaf of planes and by discarding the rigid figures of Euclid for mobile ones, Desargues improved on these techniques. Incorporating projective methods that run counter to Euclid, Desargues hypothesizes that parallel lines if "produced to infinity in both directions" meet. This is like the concept of the vanishing point in perspective drawings. Desargues also considers varying transformations of lengths and angles.

Having mastered the geometry of Apollonius and sharing the knowledge that an oblique view of a circle gives an ellipse, Desargues in his pamphlet *Brouillon projet ...* (*Rough Sketch of What Happens When a Cone Meets a Plane*), published in 1639, explores properties of conics that under projection remain invariant.[470] By central projection from a vertex at infinity, he simultaneously relates these figures to the circle, which is a plane section of the three-dimensional cone and cylinder. He concludes that projectively all conics are equivalent to a circle. Postulating, like Kepler, that lines have a point at infinity, Desargues differentiates among each conic section as to how many of those points each has. In their plane, projections of circles that are ellipses do not meet, that is, have no points at infinity, while that of the parabola touch in one and that of the hyperbola cuts the infinite line at two points. Desargues found these results by tilting each conic section to arrive at his perspectives.

His work unifies the theory of conics. Since at infinity the straight line does not differ from any other line, his projective geometry eliminates its privileged status. Although his projective views lay outside the realm of ancient Greek geometry, Desargues utilizes classical geometry to express them and establish their advantages. At the time when he was formulating his ideas, the method in Descartes' analytical geometry remained obscure.[471]

Two principal theorems for founding projective geometry appeared and received elaboration not in Desargues' *Brouillon projet* but in the treatise *Manière universelle de Sieur Desargues, pour pratiquer la perspective . . .* (*Universal Method of Mr. Desargues for Applying Perspective . . .*), published in 1648 by his chief disciple, the engraver Abraham Bosse (1602–1676). Rendered into modern terms, its first theorem, considering two triangles related by projection from a common point, holds that the intersection points from pairs of corresponding sides of these two triangles must be collinear. To prove this, the *Manière universelle* applies Menelaus' theorem. This theorem now bears Desargues' name. Here is an illustration.

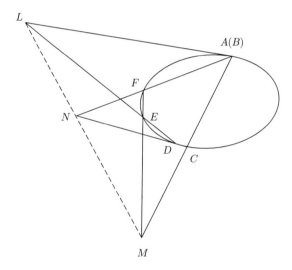

The other theorem holds the invariance in projection not of length and angle but of cross ratios. In this figure, on the line AD the cross ratio for points A, B, C, and D is $CA/CB = DA/DB$.

The *Brouillon projet* and its projective geometry otherwise were largely ignored by the scientific and the artisanal community. Desargues printed only fifty copies, sending most of them to scientific colleagues. His pamphlet's highly original vocabulary, including terms borrowed from botany and biology, its unfamiliar notation, and the author's refusal to use any Cartesian symbols made it difficult for draftsmen, geometers, mechanics, and mathematical practitioners to appreciate its contribution. In 1640 Desargues published an essay under the same title offering techniques for stonecutting, sundial construction, and gnomonics. Since he attacked procedures governed by powerful trade guilds,

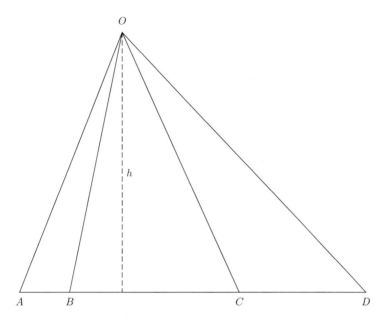

they opposed him. In the resulting polemic, the stonecutter J. Curabelle issued an attack that Bosse judged mediocre, erroneous, and plagiaristic. After about 1644 Desargues turned from polemics mainly to work as an architect and engineer in Lyons and Paris. The complex staircases that he designed in a few houses required exact stonecutting graphical procedures. Desargues also installed a water system using new epicycloidal wheels at a chateau that by 1671 was acquired by Charles Perrault, who wrote "Cinderella."

In a scientific community just accepting analytic geometry and approaching infinitary analysis, this new synthetic version of geometry found little enthusiasm. After Desargues criticized Beaugrand's geostatics, Beaugrand claimed that demonstrations from Apollonius were superior to those of the *Brouillon projet*. Descartes, a correspondent of Desargues since 1637, described his text as "a beautiful invention," but maintained that conics could be expressed more easily with algebra. The *Brouillon projet* was soon lost until a handwritten copy by Philippe de la Hire was discovered in 1864 in a Paris library.

Only the young Blaise Pascal (1623–1662) fully recognized the significance of Desargues' work. Early in 1639 his father Étienne had brought his brilliant fifteen-year-old son Blaise to participate in the weekly meetings of Mersenne's Parisian academy. There Blaise met Desargues and in June discovered the important theorem that he called his mystic hexagon (*mysterium hexagrammicum*): if a hexagon is inscribed in a conic, then the three points of intersection of opposite sides (*P*, *Q*, and *R* in the accompanying drawing) are collinear.

This property geometrically formulates the condition when six points in one plane belong to a conic. In 1640 Pascal printed it in a one-page handbill titled *Essai pour les coniques* that was circulated to very few for discussion at Mersenne's academy. There Pascal stated "that I owe the little that I have

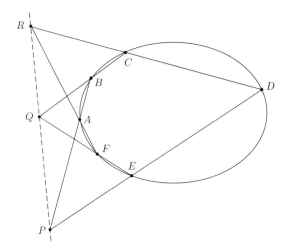

found on this subject [projective geometry] to [Desargues'] writings. The *Essai* outlined a major treatise projected to deduce from a single proposition, Pascal's mystic hexagon theorem, all propositions in the *Conics* of Apollonius. Except for Descartes, who lacked interest "in the work of [this] boy," his handbill was warmly received. By December Pascal had deduced most of Apollonius' theorems. But at that time he had to join his father in Rouen, discontinuing his contacts with Mersenne's academy that had prompted and nurtured the flowering of his mathematical ideas. In 1641 his health began to worsen. Afterward he only worked occasionally on his treatise. In 1648 he reported having geometrically solved Pappus' four-line problem, on which he and Descartes had independently been working.

By 1654 Pascal had put in order and completed his large manuscript on projective geometry. Mersenne had assumed that it would have more than four hundred propositions on conics and apply projective methods to solve geometric problems. Only Leibniz, who in Paris in 1676 was shown a copy by Pascal's Perier heirs, has left much information about it.[472] In a summary, he divided it into six sections, beginning with the projective generation of conics, tangents, and secants and including a section on the three- and four-line locus and another on the *magna problema*, placing a set conic on a set cone of revolution. Apparently, Pascal lacked good notation and substantial experience in symbolic algebra. Leibniz preserved a large excerpt of this elegant work, which sets forth more explicitly than had Desargues the ideas of projective geometry. Leibniz urged that this treatise be published within a volume of Pascal's writings on conics and tangents. It never was.

Apparently, Leibniz wanted it published before two anticipated expositions of La Hire. Although a follower of Desargues who carried some of his language, such as "knots" for points on the axis, into analytic geometry, La Hire in the *Nouveaux élémens des sections coniques* of 1679 preferred Cartesian methods. But his principal work, *Sectionaes Conicae* of 1685, returned to treating quadrilaterals, conjugate diameters, normals, poles, tangents and properties of conics by synthetic projective methods. In general, his results are no better than those of Desargues and Pascal. La Hire attempted to demonstrate that projective methods were superior to the those of Apollonius and Descartes. Pascal had thought that projective geometry might rank at least as the equal of Cartesian analytic geometry. The eclipse of his *Essai* together with the obscurity and two centuries' loss of Desargues' *Brouillon projet*, however, deprived it of opportunity for development. One immediate impact of the projective geometry of Desargues, Pascal, and La Hire was to foster elaboration of the concepts of continuous change in geometric figures, transformation, and invariance.

15.3   Quadratures in Retrospect: Multiple Roots of Integral Calculus

Infinitesimals and Indivisibles

By the late sixteenth century, European geometers were gaining through the translations and research of Regiomontanus, Commandino, Maurolico, and others a mastery of exhaustion methods and proofs taken mainly from Eudoxus and Archimedes. Not based on a simple theory, exhaustion proofs were highly complex, sometimes involving convexity lemmas. Some investigators sought to refine and extend the Archimedean model, making it more powerful in comparing convex curves. But others found exhaustion proofs prolix and looked for simpler, more direct methods. This largely illusory seventeenth-century perception of prolixity relates to the logical roughness of proofs. A logical symbolism revealing the general power of the exhaustion method and enshrining its value as a proof form came later. Among most geometers an unwillingness to short-circuit the cumbrous double *reductio ad absurdum* from each particular proof contributed to the misconception. Although praising exhaustion proofs for their logical rigor, geometers generally believed that the ancients had hidden in them their methods of discovery. This fashionable belief encouraged a few to recover this subterranean process of discovery in its elegant proofs. They hoped formal justification carried telltale signs of a link between heuristics and formalism. Commercial pressure for more accurate and less complex computations of volumes of curvilinear solids, particularly of barrels for liquid goods, moved in the same direction.

Improvement of methods required a deeper understanding of the arithmetic of the infinite. In determining the pressure on a vertical square wall of a vessel of water, Simon Stevin mentally subdivided the wall into thin horizontal strips, increasing these strips from four to over a thousand. The result given in his supplemental demonstration by numbers does not depend on the reversal of inequalities in exhaustion and is closer to the fundamental concept of the limit of the magnitude vanishing. But he did not recognize that his sequences formed infinite series.[473] In *De centro gravitatis solidorum*, published in 1604, Luca Valerio, a mathematics professor at the Sapienza in Rome and a correspondent of

Galileo, in finding the center of gravity of a segment of a parabola added lemmas that eliminated the need for repetitions of the *reductio ad absurdum*. Struggling with the limitations of the methods of exhaustion, Kepler, Bonaventura Cavalieri (ca. 1598–1647), and Gregory of St. Vincent (1584–1667) developed infinitary techniques of geometric integration that presaged calculus. Their methods employed infinitely small elements or infinitesimals and indivisibles.

Kepler's *Doliometria*, or *Stereometria doliorum vinariorum* of 1615, is a practical treatise on curvilinear measurement. He had installed new wine casks in his house and found that wine merchants crudely measured volumes of barrels with a gauging rod extended along the diagonal length without adequately considering their shape. On review of Archimedean stereometry, the mensurational tabulations for wine barrels in Linz, and general applications, Kepler discards Archimedean procedures and their modifications by Stevin and Valerio.[474] Inspired by the thought of Nicholas of Cusa and Giordano Bruno on the infinite, he assumes instead that curvilinear solids are aggregates of an infinite number of extremely thin circular laminae or similar cross sectional forms. The *Doliometria's* first problem determines the area of a circle by regarding it as a regular polygon having an infinite number of sides. The base of each infinitesimal triangle is located on the circumference and its vertex is the center of the circle. The area of a circle obtained by summing the areas of the inscribed triangles is thus one-half the circumference times the radius. Kepler similarly divides a sphere into an infinite number of narrow cones whose common vertex is the center of the sphere and computes its volume to be one-third the surface area times the radius. He demonstrates that the cube is the largest right parallelepiped having square bases that can be inscribed in a sphere. From wine barrel tables for selecting optimal proportions, Kepler shows that when nearing maximum volume the amount of volume change stemming from dimensional change grows smaller. His infinitary technique is intuitive. For justification, he often appeals to the law of continuity.

Kepler's atomistic technique, which Archimedes would have considered merely heuristic, is less rigorous than the exhaustion methods, but more efficacious in setting up problems. Although it put him outside the bounds of ancient Greek geometry, this work pleased Linz's wine merchants by showing that the simple gauging rod imposed as a standard closely approximates the volume of Austrian wine barrels. Kepler hoped to profit from sales of the *Doliometria*, the first publication in Linz, but few copies were bought. This led him to edit hastily a reordered popular German version of it entitled *Messekunst Archimedis*.

Kepler's infinitesimal method quickly came under assault. In *Vindicae Archimedis* of 1616, the Scottish mathematician Alexander Anderson, a student of Viète, criticized it for lacking in rigor. By the late 1620s, two methods of indivisibles, the collective and the distributive, for comparing two geometric figures that Cavalieri, under the influence of Kepler and Galileo, had developed drew most of the attention of geometers.

A member of the small Jesuat order, Cavalieri had been a pupil in Pisa of the Benedictine Benedetto Castelli, who put him in touch with Castelli's own teacher, Galileo. Between 1619 and 1641 the two exchanged a hundred letters. When Cavalieri failed in his first attempt to obtain the chair of mathematics in Bologna in 1619, he claimed that the reason was the unpopularity of the Jesuats in Rome. (This order should not be confused with the Jesuits.) Probably on Galileo's recommendation, he succeeded to that post in 1629. His contract was renewed every three years until he died in 1647. Cavalieri wrote eight books on mathematics, optics, and astronomy, which included astrology. These books include a logarithmic table.

In the seven-hundred page *Geometria indivisibilis continuorum nova quadam ratione promota* (*Geometry Advanced in a New Way by the Indivisibles of the Continua*) of 1635 and the *Exercitationes geometricae sex* of 1647, Cavalieri presents in detail his two methods of indivisibles. These verbose and difficult books made him a leading mathematician. The *Geometria's* first six books elaborate the collective method, and the seventh, the distributive. Cavalieri saw merits to both. The collective method was most important for geometrical investigations and finding centers of gravity, since it was more general, applying to figures of equal or unequal height, while the distributive compares the finite features of aggregates, thereby supposedly providing a foundation "free from the concept of infinity" that applies to aggregates of indivisibles. Cavalieri's distributive method does, however, accept nonfinitistic techniques. Primarily a defense of the indivisibilist procedure, the *Exercitationes* expand upon both methods. Theorem 1 of book VII bases these methods on a new principle now named for Cavalieri:

> If between the same parallels any two plane figures are constructed, and if in them, any straight line being drawn equidistant from the parallels, the included portions of any of these lines are equal, the plane figures are also equal to one another; and if between the same parallel planes any solid figures are constructed, and if in them, any planes being drawn equidistant from the parallel planes, the included plane figures out of any one of the planes so drawn are equal, the solid figures are likewise equal to one another.[475]

Theorem 2 extends this result to plane figures in constant ratio, and theorem 3 gives the analogue result for solids.

Unlike Kepler, who except for a few lapses divided geometrical figures into infinitesimals of the same dimension, Cavalieri has indivisibles of a lower dimension make up a figure. He assumes that a line is composed of "all the points," a plane of "all the lines," and a solid of "all the planes." "All" he defines as an infinite number. Thus, equidistant and parallel infinitely thin line segments compose an area; equidistant and parallel infinitely thin plane sections, a volume. Critics, such as the Jesuits Paul Guldin and André Tacquet, raised technical objections, noting that Cavalieri failed to provide postulates justifying the rules of operation between these collections, whether they can be added or compared, thus possibly giving rise to paradoxes, and objected to his use of superposition and his notion of infinity and the continuum.[476] Non-Averroistic Aristotelians, the group opposing the Paduan Aristotelians, and many anti-Aristotelians rejected the decomposition of the continuum into indivisibles, for this implied atomism. Paolo Mancosu argues tellingly that Cavalieri and his critics desired to attain certain and evident mathematical proofs that eliminated contradiction and met the criteria of Aristotle's *Posterior Analytics*. Generally Cavalieri's critics stood closer to the ancients than he did.[477]

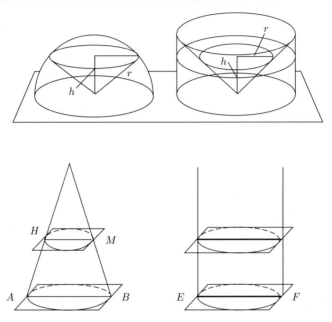

Although Cavalieri's principle has deficiencies in rigor and obscurity about the limit process, his methods of indivisibles were important methods of discovery. Their generality made computations of cubatures or solids for pyramids, cones, and cylinders simpler and quicker than ancient techniques that applied only to specific cases. The characteristic trend in seventeenth-century geometry was away from special constructions and toward general theorems. Cavalieri's methods also were to supplant the Archimedean methods of exhaustion, which obtained upper and lower bounds by inscribing and circumscribing curvilinear figures with regular rectilinear figures whose sides progressively increase. Cavalieri's determination of the area of an ellipse and the volume of a sphere demonstrates this novelty. Discarding the ancients' series of rectilinear approximating figures, Cavalieri systematically employs thin rectangles to approximate the area under curves. He and other seventeenth-century geometers follow Stevin in this practice. The heights of his rectangles increase slightly at each step to approximate the height of the curve. This requires normalizing the associated series. As the width of rectangles grows ever smaller and their number greater, pinching in more closely to the curve, the sum of their areas comes increasingly close to filling the space under the curve. Here Cavalieri replaces the indirect proof of the ancients with his method depending on infinitely many, infinitely thin rectangles. Essentially applying this summation technique, he computes the area under a parabola to be what in modern notation is equivalent to $\int_0^a x^n dx = a^{n+1}/(n+1)$ for $n = 1$ through 9. It remained for Roberval and Fermat to generalize this for all positive and fractional n. Summing squares of the integers as had the ancient Babylonians, Cavalieri determined that the volume of a cone compared with that of a cylinder having the same base and height is $1/3$ for large n.

Although his *Geometria* was not published until 1635, Cavalieri was composing it from 1621 to 1627, when he completed a near final draft. Teaching responsibilities and occasional ill health are the reasons he gave for the delay in publishing. Kirsti Andersen argues that a more pressing reason was his attempt to have Galileo endorse his methods before publication. The endorsement never came. As Paolo Mancosu observes, Galileo rejected Cavalieri's attempts in the method of indivisibles to measure infinite collections. Galileo's spokesman Salviati stated that for infinities "it cannot be said that one is greater, or less than, or equal to, another."[478]

Without naming Cavalieri, Galileo's *Dialogo* and correspondence of October 1634 propose two paradoxes that arise from what he saw as the perils of Cavalieri's attempts to measure infinite collections and the dangers of analogy in transferring equality for n-dimensional entities to $(n-1)$-dimensional counterparts. These are the paradox of the bowl, or soup dish, and the paradox of concentric circles. For the bowl, Galileo begins with two equal surfaces and two equal bodies, having the surfaces as bases. He was to show that one surface and body, continually lessened, so that the remaining parts are equal, ends in a long line, containing infinitely many points, while the other two become a single point. Galileo first draws a semicircle AFB, having center C, and has rectangle $ADEB$ circumscribing it.[479] He next draws line segments CD and CE, along with fixed radius CF perpendicular to AD and DE. When the figure rotates around the axis CF, semicircle AFB generates a hemisphere, rectangle $ADEB$ a cylinder, and triangle CDE a cone. Removing the cylinder and hemisphere leaves a bowl. Raising the cutting plane always produces cuts in the cone equal to cuts in the bowl. But the bowl ends in a circumference of a circle and the

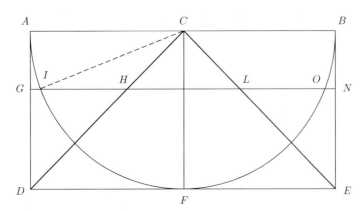

cone in a point, which gives the paradox. Since rectangle $ADEB$ can be made as large as desired, the circumference of all circles may equal a single point.

Neither Galileo nor his disciple Torricelli could resolve this situation or the similar concentric circles' paradox arising from compounding the continuum from indivisibles. Still, here and in his pairing of natural numbers with their subset of squares, he brought attention to the one-to-one concept. Although he rejects one-to-one correspondence for measuring the infinite realm, it was to be crucial in Cantor's defining of infinite sets and in distinguishing their sizes.

The Jesuit Gregory of St. Vincent's *Opus geometricum quadraturae circuli et sectionum coni* (*Geometrical Work on the Squaring of the Circle and of Conic Sections*), consisting of more than 1,250 folio pages, makes him an independent pioneer of infinitesimal analysis. A student of Clavius in Rome, he resided in Louvain, Rome, and Prague, where he located in 1628 as part of the Counter-Reformation in the Austrian monarchy. Although not published until 1647, his *Opus geometricum* was probably mostly written between 1621 and 1628, when he extracted the theory of conics from Commandino's editions of Archimedes, Apollonius, and Pappus and devised his method of infinitesimals. After Gregory suffered a stroke in 1628, his writing slowed and in 1631 during the Thirty Years War he had to flee Prague ahead of the Swedes without his papers. His colleague Rodrigo de Arriaga saved them, but the papers did not catch up with Gregory until 1641.

Gregory's poorly organized *Opus geometricum* covers theorems on the circle and triangles, Zeno's paradox of Achilles and the tortoise, sums of geometric series,[480] the conic sections, and the relation between the parabola and the Archimedean spiral. In an attempt to determine when Achilles will overtake the tortoise, Gregory employs proportion and geometric series. His appeal to sensory experience more than purely abstract mathematical reasoning deterred him from the use of limits of infinite series, so important in the formulation of infinitesimal calculus. He was perhaps the first though to propose explicitly a curve as a terminus, not an approximation, of an inscribed or circumscribed figure. This shows an appreciation of a geometrical limit concept. For calculating areas intercepted under a rectangular hyperbola between successive ordinates, Gregory inscribes and circumscribes rectangles but does so in a geometric progression. Expressed in modern form, $\int_a^b dx/x = \log b - \log a$.

Not Gregory but the Belgian Jesuit Alfonso Antonio de Sarasa (d. 1667) recognized the relationship between the logarithmic function and this summing process for the hyperbola. Put another way, his conclusion upon reading Gregory's *Opus* in 1649 was that the area under the portion of a hyperbola between two asymptotes has the logarithmic property $L(\alpha\beta) = L(\alpha) + L(\beta)$. In grasping the potential role of the logarithmic function, integral calculus precedes differential calculus. Efforts to calculate these areas also prompted by the 1660s the emergence of the power series methods so essential to calculus.

Applying in book X of his *Opus* his summation process from book VII for cubature of volumes, Gregory contends that he has squared the circle. He applies infinitely many, infinitely thin triangles to fill an n-sided polygon. Following

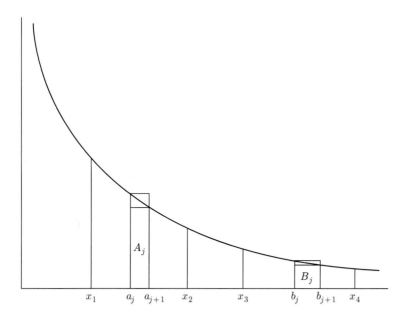

Gregory of St. Vincent and the area under the hyperbola $xy = 1$

Nicholas of Cusa, he substitutes for n an infinite number of sides. In 1651 Huygens uncovered the error in Gregory's mistaken assumption that in constructing geometrical solids from plane figures $X_2/Y_2 = (X_1/Y_1)^n$. Gregory's claim to be a circle-squarer reportedly caused many mathematicians to treat him with disdain until Huygens and Leibniz rehabilitated his reputation.

The French Triumvirate

Combining extensions of Archimedean geometry with number theory, the French triumvirate of Roberval, Fermat, and Pascal at mid-century prepared the ground for the introduction of infinitesimal analysis. The near simultaneous appearance of infinitesimal procedures suggests widespread interest.

In addition to Archimedes, whom he acknowledged, Roberval seems committed to improving on Kepler, Valerio, and Cavalieri. For Archimedes' methods of exhaustion, he substitutes infinitary methods. Although he credits Cavalieri with considering surfaces as composed not of lines but of small bits of surface, and solids of small bits of solids, that advance seems to be original. Arithmetically dividing figures into sections continually decreasing in magnitude was a more exact procedure than deploying Cavalieri's geometric fixed indivisibles of a lower degree. The decrease proceeds, not unlike a procedure by Stevin, to produce an infinite series, but there is no limit. Roberval's division process is critical for formulating the definite integral, which represents the area under a curve occurring between two fixed points on the x-axis. Since the presence of higher-order differences for squares, cubes, and so forth of the infinitely small does not substantially change the result, he neglects them. His finding, the equivalent of $\int_0^a x^n dx = a^{n+1}/(n+1)$ for positive integers n, came perhaps at Fermat's suggestion. This finding partly reflects the ancient association of numbers with geometric magnitudes, as espoused by the Pythagoreans and Nicomachus.

Roberval connects with integers the small lines composing a line segment. In the case of measuring the area of a square, he employs successive right isosceles triangles with sides consisting of 4, 5, 6, ... indivisibles or points. The number of points for the triangle of 4 is 10 or $1/2(4)^2 + 1/2(4)$. Since these triangles constitute half the square, Roberval finds what in our notation is $\int_0^a x\, dx = a^2/2$. In using such infinitesimal components as parallelograms, pyramids, and cylinders, he is imaginative. Since his *Traité des indivisibles* remained unpublished until 1693, Roberval's influence was chiefly through correspondence with the Pascals and Fermat.

From 1637 to 1657 fellow mathematicians heard almost nothing from Fermat. Even so, his responses in 1643 to questions raised by Cavalieri and letters to Torricelli in 1646 show that he remained active in studying quadratures. When John Wallis's *Arithmetica Infinitorum* appeared in 1657, Fermat wrote that he had obtained the same results earlier. His reply to Wallis in what is now known as the *Treatise on Quadratures* probably took a year to complete. The first part contains a direct procedure for computing the quadrature of all regular and higher parabolas, $x^p y^q = k$ (q and p integers and $q \neq p$); the second, a reduction analysis giving the quadrature of any curve. Fermat applies to sums of powers of integers Roberval's generalization of Archimedes' inequalities to discover that $1^m + 2^m + \ldots + n^m > (n^{m+1})/(m+1) > 1^m + 2^m + \ldots + (n-1)^m$. This formula gave him sums of encasing rectangles constructed on subintervals. These were the basis for computing the areas bounded by parabolas and other curves $y = px^m$ for positive and negative m, except for -1. Most likely Fermat had devised this method by 1640.

His initial procedure dissatisfied Fermat, since it did not apply to fractional hyperbolas and parabolas. He recognized that it is sufficient to divide the area under a finite curve into an infinite number of equal rectilinear figures, but saw that dealing with infinite lengths, especially for higher hyperbolas, posed a dilemma. He proposed dividing the x-axis into a progression of adequal, or nearly equal, segments, which he was to call his logarithmic method. Breaking decisively with the method of exhaustion, adequality transforms the method of quadratures. Its infinite converging series of adequal rectilinear figures "go to nothing." After circumscribing adequal rectangles, Fermat found no need to calculate inscribing others as well. Each progression of adequal rectangles offers a limit. The sum is the curvilinear area. With adequality, Fermat calculated as $(m/[n-m])x_0 y_0$ the area under hyperbolas to the right of x_0 and quickly saw that his new procedure also applied to parabolas, giving the area from 0 to x_0 as $(m/[n+m])x_0 y_0$. Since he did not publish his *Treatise on Quadratures* and his son Samuel was sloppy in editing it, his influence in altering the method of quadratures is unclear.

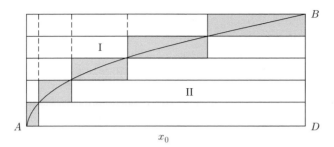

It is perhaps surprising that Fermat did not recognize the inverse relationship between tangent construction and his method of quadratures for parabolas and hyperbolas. In addressing various problems with each, he relied on clever geometric transformations. His determination of centers of gravity, which depended on his method of maxima and minima, likewise did not bring out this connection, nor did the style of his quadratures make clear the path to integration. The record again suggests that calling him the inventor of calculus is unwarranted.

Blaise Pascal, the third major French contributor to the method of indivisibles, was a Christian apologist and a master of French prose, as well as a prominent geometer and natural philosopher. Pascal epitomized the tension felt between religion and Cartesian rationalism, whose exemplar was Euclidean geometry. Leading the last part of his life in a state of Augustinian piety together with Jansenist self-denial, including renunciation of his property, Blaise rejected Hobbesian materialism and Spinoza's pantheism. Nor did he accept the effort of some of his contemporaries, including Cambridge Platonists such as Henry More, to harmonize religion and scientific rationalism by proving the existence of God from the works of nature while creating a scientific theism. Instead Pascal, who portrays man as a feeble but thinking reed, dichotomizes Cartesian reason, which engenders the sciences and whose province is the brain, from the portions of religion and philosophy springing from faith and the heart. Even in the physical sciences he recognizes limits to mathematics and its applications. Preoccupied by interest in grace late in life, he proposed a deeper-rooted base than Cartesian certitude. "The God of the Christians," Pascal insisted, "is not simply the author of geometrical truth." His primary achievements in mathematics were in geometry and number theory. Besides helping to create projective geometry, he refined the geometry of indivisibles, discovered properties of new curves, and computed the binomial coefficients for positive integers in his arithmetical triangle. He also contributed to the origins of classical probability. More than innovations, a systematic ordering along with a clarification of ideas offered vaguely by predecessors principally characterize his mathematical work.

Born in Clermont Ferrand, Pascal was the only son of Étienne Pascal, a judge, and Antoinette Begon. After Blaise's mother died when he was three, Étienne raised the children alone, giving them home schooling even after the family moved to Paris in 1632. At home Blaise mastered Greek and Latin and studied history, philosophy, and theology. His father required him to think for himself. A year after writing a treatise on sound at the age of eleven, Blaise began to study geometry on his own. To that point his father had excluded mathematics, possibly afraid that an intense pursuit of it would strain his son's fragile health. Reportedly through a game whose pieces were "coins," "three-cornered hats," "tables," and "sticks," representing circles, triangles, rectangles, and lines, Blaise reinvented the equivalent of Euclid's thirty-second theorem in Book I of the *Elements*: the angles of the three-cornered hat are equal to one-half the angles of the table. It is often assumed that he discovered the preceding thirty-one theorems as well, but this is doubtful.

Blaise Pascal

In 1635 Étienne, recognizing his son's precocity, introduced him to Mersenne's Parisian Academy. Blaise took an active part in the scientific discussions each Saturday. There he met Carcavi and Mydorge and sometimes Desargues, Gassendi, and Roberval. Under their tutelage and scholarly exchanges, his mathematical research began to blossom. Late in 1639 his father was appointed *intendant* in Rouen, however, and his frail teen-age son had to leave Paris.

Blaise recognized the hard work, drudgery, and need for precision in his father's calculations for accounting and for assessing and collecting taxes. To free the mind from computations by pen or counter and to assure accuracy, he designed and produced from 1642 to 1643 a clockwork device that was an arithmetic calculator, the Pascaline, which his contemporaries called Pascal's wheel. It successfully reduced addition and subtraction to a simple movement of wheels and shafts. Since the Pascaline had eight wheels with ten cogs each, representing the numbers from zero to nine, it could handle columns of eight figures. Possible slippage in the wheels made it reliable only through five figures. For addition Pascal slowly moved the wheels forward and for subtraction, backward. A complete revolution of a wheel or pivot through ten numbers produced a movement of only one unit in the next higher- or lower-order wheel.

This was the basic idea in his planning. The instrument was fourteen by five by three and one-half inches, roughly the size of a shoe box.

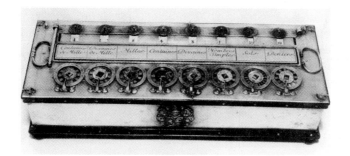

The Pascaline

By the time he recruited the most skillful artisans and finished building this first essentially digital calculator, Pascal had already earned a mathematical reputation. The Pascaline suffered most from the friction of its gears. In manufacturing fifty copies, Pascal experimented with different designs and materials, including wood, brass, copper, ebony, and ivory. He gained a royal privilege or monopoly for their sale, but the price was set so high, a hundred livres, that most could not afford it. Only a few were sold, so production was ended in 1652. Although its concept was to influence Leibniz, it had no immediate successors. Still, many of Pascal's contemporaries considered this calculator his foremost mathematical achievement. Eight copies have survived to the present.

In *Augustinus*, the Catholic theologian Cornelius Jansen (d. 1638) supported the Augustinian view that man is totally sinful and that God alone confers grace and salvation by eternal predestination. Jansen rejected any role for free will in salvation as incompatible with the utter and absolute gratuity of divine grace. His views were close to Calvinism. The Jesuits championed the theology that good works are part of the process of salvation. The introduction of the Pascal family to Jansenism by their uncles in 1646 drew them more deeply into religion. Blaise's first conversion, as he referred to it, was mainly doctrinal. It soon brought him into this quarrel within the Catholic Church.

Although religion was now foremost to Pascal, he continued to study natural philosophy extensively. Repeating through 1647 Torricelli's experiments with quicksilver and with water and wine in glass tubing, he authoritatively confirmed against Cartesian objections the existence of the vacuum and the weight and pressure of air. In performing earlier tests, Mersenne and Gassendi had displayed experimental versatility. But lacking sufficient glass tubes, their tests were inconclusive. Pascal had to defend himself against Jesuit accusations that he was taking credit for Torricelli's research. He published a twenty-page account of his findings, entitled *Great Experiment Concerning the Equilibrium of Fluids*. Although his barometric observations continued to 1651, Pascal by late 1647 turned more to investigating hydrostatics. He synthesized and extended the work of Stevin, Galileo and Torricelli by following a rigorous experimental method. By early 1654, it seems, his study of physics ended. His larger two works, *Treatises on the Equilibrium of Liquids and the Heaviness of the Mass of Air*, only appeared posthumously in 1663. This delay reduced Pascal's influence, since he had competition from Otto von Guericke and Robert Boyle on the vacuum and the compressibility of air. As Leon Brunschvig and Pierre Boutroux assert, Pascal considered the results of these experiments to follow from general principles and not to be sources of new hypotheses.[481] The 1663 treatises give Pascal's law that a fluid in a vessel exerts pressure equally in all directions and explain equilibrium of fluids in vessels through a multiplying of forces.

As his fragile health worsened in 1647, Pascal returned to Paris. Bleedings and purgings contributed to, rather than alleviated, the internal ills which exhausted him. Four years later his physicians advised him to set aside his research and as a diversion to frequent the drawing rooms of polite society. His friendship with Artus Gouffier, duc de Roannez, whom he had known as a child, was renewed. The duke, who was almost the same age, was taken with their discussions of God, life, and science. He presented Pascal to his circle and introduced him to the courtier and soldier Antoine Gombaud, Chevalier de Méré, who was a linguist and classical scholar dedicated to Plato. Despite his Jansenistic leanings, Pascal enjoyed dancing and playing backgammon, piquet, and various dice games at gambling places. After the death of his father Étienne in 1651, Pascal was alone in Paris and his profane activities apparently increased.

Developing "an extreme aversion to the [follies and] beguilements of the world," Pascal experienced on the night of November 23, 1654, a second conversion, mystical in nature. In what has been called either a religious illumination or a hallucination, he saw flames, suggestive perhaps of Moses' burning bush, and believed that he was conversing with God for two hours. Pascal resolved to retire from the political world and devote himself entirely to religion. But his writings disprove the interpretation that he totally set aside mathematics from 1654 to 1658. As he was to write to Fermat, geometry remained the "highest exercise of the mind," even as he concentrated on matters of the heart.[482] In 1655 Pascal moved to the Jansenist stronghold of Port Royal, near Paris. His *Lettres provinciales* (*Letters Written to a Provincial by One of His Friends*) of 1656 and 1657, composed under the pseudonym Louis de Montalte, eloquently champion Jansenism, defending it from censure directed against its partisan Antoine Arnauld at the Sorbonne and assailing its enemies the Jesuits, particularly for their casuistry, which as he portrays it comes out to be situational ethics to eliminate or reduce the gravity of sinful acts. In each *provinciale*, Pascal strove for the highest degree of clarity, rewriting the eighteenth, for example, thirteen times. The elegance, lightness, and wit of the eighteen satirical pamphlets comprising the *Lettres* captivated Paris and made them popular in England. He found some support in mainstream Catholicism.

The influential French orator bishop Jacques Bossuet spoke strongly against clerical casuistical excess and in 1679 Pope Innocent XI condemned it. The *provinciales's* analysis of ancient Pyrhonnism and contemporary rationalistic deism helped to persuade Pierre Bayle of the dangers to religion from methodical skepticism. The method of the *Lettres* together with their commentary on justice would contribute to Voltaire's denunciation of religious superstition.

In 1660 life became more difficult for the Jansenists in France. The French State Council burned the *Lettres* of the unknown Montalte, and Louis XIV closed Port Royal and allowed the enforcement of the papal bull *Ad sacram sedem* that banned Jansenist theology. Pascal spent his final years working quietly on a great book of Christian apologetics on grace, the truth of Christian religion, and the nature and fate of man. Fragments discovered upon his death were published under different titles, most often the inappropriate *Pensées* in 1670.

In mathematics, the year 1654 was most productive for Pascal. His first work on indivisibles, *Potestatum numericarum summa*, was completed. It provided a new approach that makes possible the summation of numerical powers, or essentially the integral of x^n, a result that Roberval knew. Rather than proceeding by way of geometric propositions, Pascal follows a hint from Roberval about figurate numbers that he discerned in his *Traité du triangle arithmétique*, also of 1654.

1	1	1	1	1	1	1	1	...
1	2	3	4	5	6	7	8	...
1	3	6	10	15	21	28	36	...
1	4	10	20	35	56	84	...	
1	5	15	35	70	126	...		
1	6	21	56	126	...			
1	7	28	84	...				
1	8	36	...					
1	9	...						
1	...							

Our modern form of this triangle rotates his version by 45°. Moritz Cantor notes the importance of figurate numbers to Pascal and his contemporaries, who took the integers in the first column of the arithmetic triangle to be the number of points making up a line; those in second column, the sums of the integers in the previous column to the position of the integer, give lines; and the integers in the third column after 1 (3, 6, 10, ...), the binomial coefficients for $m = 2$, are triangular. [483] The number m of the first column is zero. The fourth column integers after 1 (4, 10, 20, ...), the coefficients for $m = 3$, are pyramidal, those for $m = 4$ are triangulotriangular, and so forth.

But for $m > 3$, seventeenth-century geometric techniques failed, so Pascal had to proceed by analogy. One point, he contends, adds nothing to a line, nor does a single line increase an area. Thus, in considering ratios of squares or cubes, Pascal treats as zero all the terms of lower dimension. Sometimes

Cavalieri is credited with making this refinement, but his indivisibles were fixed. Pascal explicitly compares indivisibles to zero. These two concepts became central to differential calculus. Against the logical objection to omitting terms of lower dimension, Pascal appeals to a higher cognition surpassing reason. In computing the total lines of a triangle to be half the area of the longest line's square, the total of squares of lines to be one-third the cube, and so on, he has in modern symbols $\int_0^a x^n dx = a^{n+1}/(n+1)$ for all integral n except for -1.

Pascal's study of indivisibles continues in the fragment "De l'esprit géométrique," a segment of an introduction to geometry for Jansenist schools written in 1657. Breaking with the Aristotelians, who treat the infinitely large only as potentiality, it considers the infinitely large and the infinitely small as complementary actualities. Generalizing Roberval's work, Pascal's *Traité des sinus du quart de cercle* (*Treatise on the Sines of a Quadrant of a Circle*) of 1659 derives formulas for what amount to definite integrals of powers of the sine by employing infinitesimal rectangles. In addition to balancing infinitesimals and Archimedes' mechanical method in summing ordinates, Pascal's *Traité* is perhaps most notable for introducing the differential or characteristic triangle to construct tangents. Forerunners of these triangles whose sides can be made infinitely small had appeared in diagrams from Snell in 1624 and from Roberval's *Traité des indivisibles* and Torricelli's *De infinitis hyperbolis*. Fermat, Huygens, Hudde, and Heurat were also close to this insight. In 1672 young Leibniz coined the expression that English renders as "characteristic triangle" and saw it as the breakthrough to show that the drawing of tangents is the inverse of finding areas under curves, which led to his invention of calculus.

Leibniz was amazed that Pascal had not taken the step. Pascal, he wrote to Jakob Bernoulli in April of 1703, had acted as though he were blindfolded. Carl Boyer traces Pascal's failure to progress into calculus to his preference for classical geometry,[484] while René Taton blames Pascal's refusal to accept Cartesian symbolism, along with the formalization it allowed, and his preoccupation with religious issues, which interrupted his mathematical research.[485] Although Pascal lacks notation, Nicholas Bourbaki praises the power of his clear and precise language. As soon as the basic techniques of calculus are known, his statements are quickly transcribable into its formulas.[486]

15.4   A Cornucopia of Curves, the Helen of Geometers, and Rectification

The ancient Greeks had devised few curves. Their reconstructions and a firmer grasp of the higher geometry of Apollonius and Archimedes, and especially the new procedures of analytic geometry, allowed European mathematicians to generate many. Communication of problems and results, chiefly from Mersenne's circle, made them widely available, and the geometry of indivisibles opened fresh paths and made for a convergence of mathematical interests in the mid-seventeenth century. Conic sections and Archimedes' spiral were starting points. Eager to generalize, Fermat had introduced the higher parabolas $x^m = 2ay^n$, comparing their lengths with those of arcs of the higher spirals $r = a\theta$ that he and Roberval had originated. Next to conic sections, the cycloid was the most widely studied curve. The cycloid is the curve traced out by a fixed point, say a nail, on a circle rolling along a straight line. Because it provoked frequent bitter quarrels over the priority for discovering its properties or because it was the solution to challenge problems, Jean Montucla called it "the apple of discord" and, elaborating the same Greek myth, other mathematicians have dubbed it "the Helen of geometers." Late in the century, for example, it was the solution to the Bernoullis' brachistochrone problem, finding the curve of quickest descent.

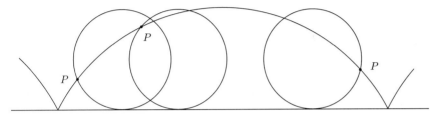

Perhaps from contacts with Galileo, Mersenne had in 1615 brought this curve to the attention of his circle and correspondents. In 1628, when Roberval visited him in Paris, Mersenne recommended examining it. Roberval initially named the cycloid after the Greek word for wheel, *trochoid*. Working in the geometry of infinitesimals, he succeeded by late 1636 in determining the center of gravity of both the cycloid, $x = r\theta$, and its partner curve: $y = r(1 - \cos\frac{x}{r})$, if r is the circle's radius. He does not recognize the partner curve as basically a cosine curve. Partitioning the arch of a cycloid into infinitesimally narrow vertical rectangles each having width ϵ, according to Cavalieri's principle, and a height equal to the mean value of the function in each interval, he first finds that the area between the cycloid and its companion curve equals half the generating circle. Using the same principle, he finds that the companion curve bisects a rectangle, whose area is half the circumference of the generating circle times the diameter or $2\pi r^2$. This makes the area under the companion curve equal to πr^2. The area under a half-arch of the cycloid is, therefore, three halves that of the generating circle, or three times that of its generating circle under a complete arch. By reducing the problem to that of integrating the sine, Roberval apparently also rectified the cycloid. Until 1635 a dictum of Aristotle, repeated in Book II of Descartes' *Géométrie*, had led mathematicians to believe it impossible to obtain a rectilinear length equal to a portion of any given algebraic curve. For a time Roberval kept his discoveries secret, perhaps in order to use the knowledge in his successful triennial competitions for the Ramus chair at the College Royal in Paris, which he held from 1634.

A correspondent of Carcavi and Mersenne and a pupil of Cavalieri, whom he surpassed in the method of indivisibles, Evangelista Torricelli (1608–1647)

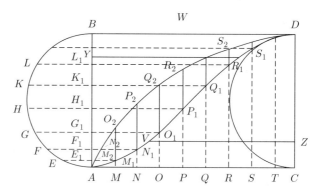

Roberval's computation of the area that a cycloid bounds.

independently made these cycloidal and other conic findings probably in Rome. An appointee of the Medicis to the Tuscan post of mathematician and philosopher that Galileo had held, Torricelli was prominent in Florentine culture: a witty conversationalist, an author of comedies that have not survived, and a participant in the founding of an academy. In the third section of his *De sphaera et solidos sphaeralibus*, compiled about 1641 which became part of his *Opera geometrica* three years later, he published his findings on the cycloid, including his computation that the quadrature under an arc of this curve is triple the area of its generating circle. This result he attributes to Galileo, whom he had visited in Arcetri. To an infinitely long solid generated by rotating a hyperbola $xy = k^2$ about the y-axis, Torricelli applied Cavalieri's principle extended to curved indivisibles as a heuristic, along with exhaustion methods taken from Archimedes and Galileo for demonstration, to make and confirm the striking discovery that the volume of the long solid equals that of the cylinder with altitude k/a and radius equal to the hyperbola's semidiameter, $AS = \sqrt{2k}$. In 1645 he also demonstrated that the length of a logarithmic or equiangular spiral $r = ae^{b\theta}$ winding backward around the pole O from $\theta = 0$ equals the polar tangent PT's length, which again proved wrong the Cartesian rejection of the rectification of algebraic curves.

The next year Roberval accused Torricelli of plagiarizing him on the cycloid and Fermat on maxima and minima. Roberval, who deserves priority for his work on the properties of the cycloid, had not published. While preparing all of his letters with French mathematicians for publication to establish his independence and possible priority, Torricelli died suddenly, probably of typhoid fever, at the age of thirty-nine.

Pascal was convinced that he had perfected the method of indivisibles and had broadened its application to solve problems concerning the cycloid or *roulette;* to compute generally quadratures or areas under curves, such as parabolas and spirals, and cubatures of volumes; and to determine rectifications and centers of gravity. In June 1658 he proposed as a challenge to mathematicians six problems on the cycloid that he had just solved. In issuing the challenge he used the pseudonym Amos Dettonville, an anagram for Louis de

Montalte. Dettonville offered a prize of 600 francs for the correct answers and superior demonstrations for these problems and set a deadline of October 1. His questions appeared in a series of three circulars or pamphlets.

Historians and mathematicians have long been intrigued by the psychological factors that inspired Pascal to embark on this affair of the cycloid. Having completed his *provinciales*, he was in 1658 working on what became his *Pensées*. He was absorbed in the Bible and the *Imitation of Christ*, translating parts of the *Vulgate* into French, and reading diverse authors, including Saints Theresa and Augustine, Arnauld, Mersenne, Epictetus, Seneca, Montaigne, and possibly Giordano Bruno and Girolamo Cardano. The house near Porte-Saint-Michel containing his study was located between two tennis courts. The cries of players must have occasionally diverted his attention outside to the trajectories of their tennis balls. Perhaps such volleys helped sustain or revive his early interest in geometrical curves.

A letter of Pascal's sister Gilberte discounts such a background by reporting a sudden immersion into the cycloid. As his health deteriorated in the spring of 1658, she says, he was beset one night by a toothache that kept him awake. To take his mind off the pain, he thought about an unsolved problem on the cycloid that Mersenne had posed. Now he was able to obtain a first result for this curve, to be followed by a second, and the second by a third. Pascal, who was then suffering from insomnia, does not refer to the toothache, writing merely that an unexpected occasion gave him the opportunity to think again on geometry, which he had set aside for a time.[487] Geometers had heatedly debated problems involving the cycloid for two decades. As Henri Poincaré and Jacques Hadamard construe the situation, Pascal's mind had worked on related problems subconsciously until they rose to a conscious level in a seemingly sudden burst. Soon Pascal was able to tell Roannez that he had solved six cycloid problems. Dazzled at his friend's genius, Roannez insisted that he publish the results. The challenge pamphlets became the vehicle. Roannez believed that the publication of this geometric treatise would increase the prestige of Jansenism and provide favorable publicity for Pascal's planned book against atheism.

After beginning the contest, knowing only the germane work of Stevin, Cavalieri, Gregory of St. Vincent, and Tacquet, Pascal gained knowledge of most of the unpublished research of Roberval and Fermat on the cycloid. This caused him to modify the challenge questions in later circulars and to revise his solutions and his assessment of the priority controversy of the 1640s between Roberval and Torricelli over cycloid rectification. Previously, Pascal had regarded Torricelli with great respect. But his brief *History of the Cycloid* of October 1658 harshly accuses the deceased Torricelli of stealing Roberval's ideas from papers sent to Galileo. Pascal's giving to Roberval sole credit for cycloid rectification angered Italian mathematicians.

Many leading mathematicians participated in Pascal's cycloid competition. By October, Christiaan Huygens had solved four of the problems, while John Wallis and Antoine de Lalouvère, who taught at the Jesuit college in Toulouse, claimed to have solutions for all six. The pseudonymous Dettonville, who dominated the awards' committee, found Wallis's work incomplete and refused to grant an extension of the time limit. He accused Lalouvère of stealing Roberval's ideas. Lalouvère did not realize that Dettonville, the chief judge in the competition, was Pascal. The prize went to the unknown Dettonville for having achieved the best solutions. These were published in a fourth circular in January 1659.

That year Pascal quickly published his four circulars as *Lettres de Amos Dettonville.* The problems he posed did not submit to elementary solutions. Using Archimedes' ideal balance and static moments, he finds centers of gravity of arbitrary segments of cycloids. Pascal proves that the arch of a generalized cycloid equals the semicircumference of an ellipse. The *Lettres de Amos Dettonville* convey his deepening understanding of infinitesimals in computing the area of any segment of a cycloid up to a line parallel to the base and the volumes of solids of revolution generated by rotating these segments. Pascal implicitly employs the characteristic triangle. The *Lettres de Amos Dettonville,* along with his *Treatise on the Sines of a Quadrant of a Circle,* moved Pascal remarkably close to constructing infinitesimal calculus.

In 1660 the competition finalist Lalouvère published his own book on the cycloid, *Veterum geometria promota in septem de cycloide libris.* More than for his findings, it is known today for being the first book dedicated to his fellow Toulousan Fermat and for containing as an appendix the only major work of Fermat published in his lifetime, *Geometrical Dissertation on the Comparison of Curved Lines with Straight Lines.* The *Geometrical Dissertation* shows that Fermat's quadrature methods apply directly to the problem of curve rectification. The author did not give his permission for its printing. This treatise appeared under the pseudonymous acronym M. P. E. A. S.

15.5   Number: Love and Theory

Mid-seventeenth century mathematicians also pioneered number theory. A persisting interest in number-related recreations and, more so, successful searches for new types of numbers in algebra from Cardano through Viète were stimuli. Advances were more directly tied to the new algebra, prompting the rediscovery and mastery of six books of Diophantus' *Arithmetica,* which provided a foundation for number theoretical studies. A new type of mathematician for the time was attracted as much to arithmetic as to geometry. André Weil divides them into number lovers, such as Mersenne, Frenicle de Bessy, and Viscount William Brouncker, who provided closure or polish to proofs and extended theories, and original thinkers, among them Carcavi, Pascal, and above all Fermat. The ferment was not universal: John Wallis criticized number theory as wearying computation. A lively dialogue consisting of challenge problems and exchanges of work in progress ensued through correspondence.

Again, the recovery of six books of Diophantus' *Arithmetica* began with Bombelli, who found a codex of the *Arithmetica* in the Vatican Library and incorporated 143 problems from it in his *Algebra* of 1572. Bombelli was a popularizer of Diophantus, not a translator. In 1575 Wilhelm Holzmann, hellenized as Xylander, published his Latin translation and the Greek original. It is not surprising, given the number of scribal omissions and copying and computational errors, that he did not reach comprehension of many passages. Regiomontanus considered printing the *Arithmetica* but did not. Building on Bombelli and Xylander, the French poet, philologist, and mathematician Claude-Gaspar Bachet de Méziriac (1581–1638) in 1621 made a more graceful Latin translation, *Diophanti Alexandrini arithmeticorum libri sex*, accompanied by the Greek original and a lengthy commentary. Bachet, known mainly for his humanistic studies, was upon its founding in 1634 made a member of the Académie Française. His only other writing in mathematics, the recreational *Problemes plaisans et delectables qui se font par les nombres* of 1612, includes card tricks, puzzles, and a procedure for constructing magic squares of odd order. Bachet was reluctant to admit his debt to Xylander, whom the classicist T. L. Heath gives most of the credit for the effective communication of Diophantus. This interpretation goes contrary to observations by Fermat and Weil that Bachet corrects many defects in Xylander's translation, fills in missing proofs, and makes sense of many corrupted passages.[488] Deliberate emphasis on decomposing integers into sums of squares displays mathematical acumen. Bachet's *Arithmetica* was both a reliable text and a mathematically sound translation.

From the late sixteenth century, mathematicians pondered the ancient Greek division of numbers into three classes: abundant, perfect, and submultiple, along with amicable numbers and magic squares. Such interest accorded with the revival of classical themes generally and drew on curiosity about the relations of numbers to music and to mystical meanings in magic and astrology. Some mathematical effort went into the making of horoscopes and talismans. Abundant numbers have a value greater than the sum of what were called their aliquot parts. The origin of the name "aliquot part" is uncertain, but it had come to mean any proper divisor, including 1, but not the number n itself. The plural "aliquot parts" referred mainly to their sum. Perfect numbers, which were also then called aliquot parts, have a value equal to the sum (s) of these proper divisors: $s(n_i) = n$; for example, $1 + 2 + 3 = 6$. Mathematicians also began to look at submultiple numbers that are solutions of an integral multiple, $s(n_i) = a \times n$, where a is 2 or 3 or another small positive integer. For example, the sum of 672 and its divisors is 1,344 or 2×672. Amicable numbers are pairs n, m satisfying $s(n_i) = m$ and $s(m_i) = n$.

Perfect and amicable numbers had been studied occasionally since the ancient Pythagoreans discovered them. Euclid IX.36 explains that $2^{n-1}(2^n - 1)$ is perfect if the second factor is prime and $n > 1$. All of the thirty-six perfect numbers known to the present are even.[489] We do not know whether any odd ones exist. Nicomachus' *Introductio Arithmetica* lists four perfect numbers: 6, 28, 496, and 8,128. These meet the Euclidean form for $n = 2, 3, 5,$ and 7, that is, $2(2^2 - 1) = 2 \times 3 = 6$, $2^2(2^3 - 1) = 4 \times 7 = 28$, $2^4(2^5 - 1) = 16 \times 31 = 496$, and $2^6(2^7 - 1) = 64 \times 127 = 8{,}128$. In *The City of God*, St. Augustine finds felicitous accord between six as a perfect number and its choice by God for the number of days to create the world. Whether a transmission of knowledge from medieval Islam relating to number theory occurred is not yet known. A western manuscript of 1456 correctly records the fifth perfect number, 33,550,336. Until Hudalrichus Regius factored $2^{11} - 1 = 2{,}047 = 23 \times 89$, sixteenth-century mathematicians in the Latin West wrongly assumed that $M_n = 2^n - 1$ is prime for every prime n. The symbol M stands for what is now called a Mersenne number and its subscript n gives the power to which 2 is raised. Regius also gave the fifth perfect number. The *Treatise on Perfect Numbers* of 1603 by Pietro Antonio Cataldi (d. 1626) of Bologna, who is best known for refining Hero's method for extracting roots by continued fractions, lists factors of numbers up to 800 and primes to 750. Cataldi observes that $2^n - 1$ is composite if n is composite and confirms that $2^{13} - 1$, $2^{17} - 1$, and $2^{19} - 1$ are prime. He thus states that $2^n - 1$ is prime for $n = 2, 5, 7, 13, 17$, and 19 and conjectures that it is also prime for $n = 23, 29, 31$, and 37. In correspondence Mersenne, Fermat, Brûlart de Saint Martin, Frenicle de Bessy, Descartes, Jumeau de Sainte-Croix, Roberval, and other Parisian mathematicians discussed perfect and amicable numbers, notably on the problem of perfect numbers, which was crucial by this time, determining which of the numbers $2^n - 1$ are prime.

The preface to Mersenne's *Universal Harmony* of 1636 testifies to Fermat's acknowledged prowess in computing aliquot parts, for he discovered the amicable numbers 17,296 and 18,416 and that the submultiples 672 and 120 are half the sum of the aliquot parts of those corresponding numbers. Fermat's amicable pair were the first friendly numbers discovered in the West since the Pythagorean 220 and 284. Possibly, Fermat computed these from a rule that he devised that Thābit ibn Qurra in medieval Islam had known. In 1638 Descartes sent Mersenne the third pair of amicable numbers, 9,363,548 and 9,457,506, and a rule for obtaining such numbers: $2^{n+1}(18 \times 2^n - 1)$, if the second factor is prime, and $2^{n+1}(3 \times 2^n - 1)(6 \times 2^n - 1)$, if the second and third factors are prime. A year later Descartes claimed to have mastered these number problems.

In March 1640 Frenicle had Mersenne forward a letter challenging Fermat to find a twenty-digit perfect number and its successor.[490] Fermat had kept secret his methods for rapidly computing large numbers, and Frenicle was apparently trying to draw them out. On the mistaken assumption that each dizaine, the decimal interval from 10^n to 10^{n+1}, contains at least one perfect number, a letter by Fermat probably written in June of 1640 to Mersenne observes that the challenge requires determining whether any number between 10^{20} and 10^{22} is prime. Putting this into the Euclidean perfect number form gives $10^{20} < 2^{n-1}(2^n - 1) < 10^{22}$, which means that $34 \leq n \leq 37$. As Fermat explains, this cannot be prime from $n = 34$ to 36, since $2^n - 1$ cannot be prime unless n is prime. The question, then, will be whether $2^{37} - 1$ is a Euclidean prime or a submultiple. This number lies in territory where Eratosthenes' sieve cannot practically be applied. Apparently Frenicle had decomposed $2^{37} - 1$ and set the problem as a trap. Initially Fermat thought the number was prime, but a discovery that he made in 1640 permitted him to factor it.

Fermat's June letter to Mersenne states what he calls "the fundamental proposition of aliquot parts," which we know as his little theorem to separate it from his last or great theorem.[491] In modern notation, it asserts that for a and p prime, $a^p \equiv a \pmod{p}$. Fermat's little theorem has two classical formulations. Not having Gauss's symbolism "congruent modulo p," he might say the equivalent: $a^{p-1} - 1$ is divisible by prime p for any integer a or that for any number a and a prime p, p divides $a^p - a$. As a case of the first classical form, consider that since 11 is prime $4^{10} - 1$ must be divisible by 11. Simple calculation shows that $4^{10} - 1 = 1{,}048{,}575$ and that $1{,}048{,}575 = 11 \times 95{,}325$. Fermat's letter states the decisive case of the theorem for Euclidean primes: when $a = 2$, if p is prime, then p divides $2^{p-1} - 1$. Probably he found this result experimentally by mathematical induction. One type of mathematical induction is sometimes known as Fermatian or incomplete induction in contrast to the more expansive Cantorian or Baconian complete induction. Neither the letter to Mersenne nor that of October 18, 1640, to Frenicle containing the general form of the little theorem supplies a proof. If he "did not fear being too long," states his letter to Frenicle, he would have sent the proof that he has.[492] Characteristically, he left proofs for others.

By 1636 Fermat had developed "a very beautiful proposition" allowing him to compute the binomial coefficients for $(x + y)^p$, designated here by the modern symbol $\binom{p}{k}$, denoting the number in the pth column and kth row of the arithmetical triangle, in which the initial column and row are numbered zero. Among others, Weil believes that he noticed that p divides all the coefficients. If so, he very likely proceeded to prove his little theorem additively by replacing $a = 2$ by $(1 + 1)$. Thus, $2^p = (1 + 1)^p = 1^p + \binom{p}{1} + \binom{p}{2} + \ldots + \binom{p}{p-1} + 1$. Transposing this gives $(1 + 1)^p - 1^p - 1 = \binom{p}{1} + \binom{p}{2} + \ldots + \binom{p}{p-1}$. Since the factor prime p divides all numerators on the right side of the equation, where the formula $\binom{p}{k} = \frac{p!}{(p-k)!k!}$ was perhaps derived partly from Viète, it must divide the amount on the left side. Applying the multinomial formula, Leibniz in an unpublished treatise by 1680 proved Fermat's little theorem for $a > 2$, and Euler, by extending induction in a 1742 St. Petersburg *Commentarii* article, first published a proof. Around 1750 Euler demonstrated it more generally multiplicatively by using the second formulation of the theorem.

From the consequences of his little theorem, Fermat was able to identify all possible prime factors for a given number. The number posed was $2^{37} - 1 = 137{,}438{,}953{,}471$. Although Fermat lacks a technique for deciding primality directly, he could check all possible factors one by one. From these checks, he was to find that $2^{37} - 1 = 223 \times 616{,}318{,}177$.[493] He does not say how, but he apparently arrives at ways for limiting the number of possible divisors. Perhaps Fermat assumed that q is a prime divisor of $2^p - 1$. In modern symbols this means that $2^{q-1} \equiv 1 \pmod{p}$, and by assumption he has $2^p \equiv 1 \pmod{q}$. It is then possible to show that $p \mid (q - 1)$. In this case, p is prime and thus must be a smallest power. Since q is even here, it follows that $2p \mid (q-1)$. This gives $(q-1) = 2pk$, for some k, and so the possible prime divisor q for $2^{37} - 1$ must be odd and sought among the primes of the form $q = 2pk + 1$. For $p = 37$, the first candidate is $2(2 \times 37) + 1 = 149$, which does not work. The next is $3(2 \times 37) + 1 = 223$, which is a divisor. The prime number factorings of $616{,}318{,}177 + 1 = 2 \times 7^3 \times 898{,}423$; $898{,}423 + 1 = 2^2 \times 112{,}303$; $112{,}303 + 1 = 2^4 \times 7019$; $7^8 - 1 = 2^6 \times 3 \times 5^2 \times 1{,}201$; and $3^9 - 1 = 2 \times 13 \times 757$ confirm Fermat's decomposition, which disproves part of Pietro Cataldi's conjecture.[494]

In a letter to Frenicle, Fermat raises another important question for aliquot parts: when is $2^m + 1$ prime? It cannot be, he says, when m has an odd divisor greater than 1. He conjectures that the number is prime for $m = 2^r$ up to $r = 6$. For r no matter how large, integers of this form are today called Fermat numbers. These numbers quickly grow very large. Those up to $r = 6$ are 3, 5, 257, and 65,537 together with 4,294,967,297 and 18,446,744,073,709,551,617. Fermat may not have attempted to factor $2^{32} + 1$ or possibly made an error in his computations that he did not check. His method of decomposing $2^{37} - 1$ suggests that this number must have a prime divisor of form $2pk + 1$ or $64k + 1$, which would make 193, 257, 479, 577, 641, and so on possible divisors. Euler's proof in 1736, almost a century later, that 641 is a divisor shows Fermat's supposition to be false. So far, the only primes of the form $2^m + 1$, where $m = 2^\gamma$, is known are 3, 5, 17, 257, and 65,537.

The preface to Mersenne's *Cogitata Physico-Mathematica* of 1644, when Fermat's and Roberval's interest had turned from perfect and amicable numbers to diophantine analysis and sums of squares, asserts that when $n = 2, 3, 5, 7, 13,$ 17, and 19, M_n is known to be prime. He proposes the same is true for $n = 31,$ 67, 127, and 257. For all other primes less than 257, the preface suggests, M_n is composite. Although part of the conjecture is incorrect, numbers of the form $2^n - 1$ are today called Mersenne numbers and designated by M_n. The primes among these are called Mersenne primes, M_p. In his studies Mersenne did not go beyond M_{19}. Extensive tables of primes and composites, of course, did not yet exist, and techniques for decomposing M_{31} and larger M_p were not yet available. It was to be another century before Euler proved that M_{31} is prime. Two more centuries were required to determine five mistakes in Mersenne's listing. M_{61}, M_{89}, and M_{107} are prime, while M_{67} and M_{257} are composite.

Drawn to its beauty and subtlety, Fermat sought to restore number theory to the status that Plato had accorded it. It was probably his favorite field of mathematics. When Fermat began his studies by 1636, he had as guides little more than Euclid's *Elements* VI–IX, the algebraic methods and problems of Viète's *Zetetica* that ably analyzed much of Diophantus's *Arithmetica*,[495] and Bachet's *Diophantus*. Bachet's *Diophantus* probably sparked his initial interest in number theory. Although it long fascinated him, he differed with Diophantus by searching for integral rather than the more general rational solutions to equations. Analogies with other fields had not yet been developed. Fermat investigated not only perfect and amicable numbers but also figurate numbers, Pythagorean triads, and diophantine equations, as well as sums of squares, cubes, and figurate numbers. The twin classical concepts that integers are either divisible or prime guided his research. At the time number theory lacked an active research tradition, and so few mathematicians outside of Toulouse, Paris, Leiden, London, and Oxford responded to Fermat. Even so, working within a circle of correspondents, he was to give new direction to the venture.

Soon after beginning his number theoretical studies, Fermat found that "the ordinary methods found in the books ... [are] inadequate to proving ... difficult theorems."[496] His brief "Relation des nouvelles découvertes en la science des nombres," sent through his friend Carcavi to Christiaan Huygens in 1659 (a copy was rediscovered in 1879 among Huygens' manuscripts) says that he had in short order "discovered [but he did not say how] a singular method ... which I called the *infinite descent*."[497] This method became his principal "route for arriving at theorems."[498] It is a form of *reductio ad absurdum*.

The "Relation" notes Fermat's efforts to prove the theorem that primes of form $4n + 1$, in modern notation $p \equiv 1 \pmod 4$, have the property of splitting uniquely into two integral squares. In 1640 Frenicle had observed that Bachet, in attempting to restore defective propositions in Diophantus' *Arithmetica* and go beyond them, calculated particular figures satisfying Diophantus III.19 on numbers that are sums of squares, but was unable to attain a general rule. Frenicle challenged Fermat to provide the missing general procedure. In a hurriedly written Christmas letter to Mersenne, Fermat announced most of it.[499] Cases, for example, are for $n = 4$, $17 = 16+1$, and for $n = 7$, $29 = 25+4$. By infinite descent Fermat shows that if primes $q = 4n + 1$ have the desired property of being the sum of two squares, then some prime $p < q$, p of the form $4m + 1$, must have that same property. Ultimately, the smallest prime comes from the form $4m + 1$ given in the theorem and will lack the property of being the sum of two squares. In this case the process, if continued long enough, leads to the smallest prime of 5, which equals $2^2 + 1^2$. By contradicting the original assumption that primes $4m + 1$ are not all the sums of two squares, Fermat has theoretically proved that they are. Weil observes that Fermat recognized the need for a mathematical fact, such as an identity on divisibility, to reach conclusions about reducing factors. Though he lacked that fact and thus a full proof, his work "except for minor details" has the gist of Euler's first proof of this theorem in 1754, which contains a decomposition lemma.[500] Fermat recognized that his method of infinite descent was more useful for negative propositions or disproofs. In his words, "the trick and slant for arriving at" affirmative propositions or formal proofs made it very difficult to apply for that purpose.[501]

In a letter of 1640 to Mersenne, Fermat had without proof generalized the right triangle relationship 3, 4, 5 by conjecturing that $4n + 1$ primes are the hypotenuse of only one right triangle with integral arms: the square is the hypotenuse of only two, for $5^2\{15, 20, 25\}$ and $\{7, 24, 25\}$; the cube of only three, for $5^3\{35, 120, 125\}$, $\{44, 117, 125\}$, and $\{75, 100, 125\}$. In these cases $25^2 = 15^2 + 20^2 = 7^2 + 24^2$ and $35^2 + 120^2 = 44^2 + 117^2 = 75^2 + 100^2$.

From 1643 to 1654 Fermat did not correspond with the Parisian mathematicians. They construed him as a righteous Archimedes posing for latter day Conons problems that had no solution. In 1654, he responded to questions about probability that Pascal had raised in a letter to him. Since the questions about interrupted games of chance that are discussed below partly involve number theory, it is not surprising that Fermat attempted to revive his interest in the subject. But unable to summon any interest in number theory, Pascal curtly rebuffed Fermat writing, "Look elsewhere."[502]

Two years later the English political operative Kenelm Digby visited Fermat in Toulouse and brought a copy of John Wallis's *Arithmetic of Infinities.* Joseph Hoffmann surmises that Fermat, on a review of its theory of quadratures and integral power series, decided to contest its completeness and assert the superiority of his mathematics of the late 1630s.[503] Wallis's results, such as those on binomial coefficients that Newton was to generalize, employed incomplete induction, or heuristic arguments. In a letter, Fermat criticized Wallis's method, which can confirm special cases but fails to achieve universally true results. Perhaps angered at the harsh tone of this criticism, Wallis wrote to Digby that he would not respond to it.[504]

In 1657 Fermat sent challenge problems to Wallis by name, intending them also for Brouncker, Frenicle, van Schooten, Huygens, and "all other [mathematicians] in Europe."[505] The problems asked the recipient to prove that no cube is the sum of two cubes and that $x^2 + 2 = y^3$ has but one positive integral solution and $x^2 + 4 = y^3$ has two. The computed solution to the first equation is $x = 5$, $y = 3$; the answers to the second are $x = 2$, $y = 2$ and $x = 11$, $y = 5$. Fermat also wanted a proof that adding 1 to square powers of 2 always gives a prime and urged arithmeticians to find a cube that added to its aliquot parts gives a square. The number 343 equals 7^3. Adding 343 to its aliquot parts of 1, 7, and 49 gives 400 or 20^2. An ingenious problem solver, who had devised the new techniques of infinite descent and a general solution of double equations, apparently Fermat wanted not so much to satisfy his vanity as to spur the development of a foundation or general canon of techniques for a new arithmetic. He was to be disappointed. Tempers flared among the contestants. Troubled by Fermat's negative propositions in general, Wallis called for stronger positive solutions. Taking a position reminiscent of Descartes, he opined that Fermat's findings might partly be the result of accident and luck.[506] Perhaps it was number theory in general that offended him. Wallis's *Commercium Epistolicum,* his correspondence with Fermat published in 1658, disparages the subject, depicting its negative propositions as common and familiar.[507] E. T. Bell contends that Wallis's studies of number propositions, often resorting to tables of test values, required less theoretical understanding than those of Fermat and led him to a lesser opinion of number theory.[508] When Frenicle, who pressed the English for integral solutions, solved the challenge problems, a cultural nationalism entered the dispute.

In proceeding from specific solutions to an appreciation of the structure of problems, Fermat's correspondence with Wallis and a letter of February 1657 to Frenicle largely singles out the diophantine equation, $x^2 - Ny^2 = 1$ that has unlimited solutions for N positive and not a perfect square. He likely did not know Archimedes' bovine problem or the method of the Indian sages, who loved large numbers, for obtaining roots of the equivalent of such equations. He had read Theon of Smyrna on side and diagonal numbers that involve the case $x^2 - 2y^2 = \pm 1$. The first solution from Brouncker and Wallis, essentially following the pattern of the Euclidean algorithm for finding a greatest common divisor, was fractional. Believing that even a novice could provide a fractional solution for this equation, Fermat requested a solution giving integers, which today are referred to as units of real quadratic fields. This insistence on integers rather than rational solutions suggests that a more correct term for diophantine analysis is Fermatian analysis. Brouncker's response to Fermat assumes that $x^2 - Ny^2 = \pm 1$ has a solution (x, y) and that possible smaller pairs of solutions exist. Brouncker and Wallis confirm this for the numerical cases that they study, including $N = 13$. Perhaps mischievously Fermat inquired about the cases for the small numbers $N = 61$ and $N = 109$. He must have known that the solutions for $N = 61$ involve ten digits and that for $N = 109$, fifteen.[509] Fermat understood that Brouncker and Wallis had not gone far enough to test when their process terminates or repeats itself. Brouncker also gave answers to such equations using continued fractions and what equates to $x = 2r/r^2 - a$, $y = r^2 + a/r^2 - a$ for r any integer. Fermat's attempt in this correspondence to make number theory a prominent branch of mathematics did not meet with success.

A letter of 1654 from Fermat to Pascal and a later letter to Huygens claim that by using his method of infinite descent he invented a general rule that gives all solutions to these diophantine equations. As usual, he does not reveal this rule. The English mathematician John Pell (d. 1685) probably copied the equation $x^2 - Ny^2 = 1$ from Wallis's correspondence. It is not covered in the notes that he added to Brouncker's translation of J. H. Rahn's *Introduction to Algebra* of 1659. In an August 1730 letter to Christian Goldbach and again in Chapter 7 of his *Algebra* of 1770, Euler mistakenly attributes a peculiar method for this clever equation to "a learned Englishman named Pell."[510] Although this method actually came from Brouncker, Euler's designation of "Pell's equation" is convenient and not ambiguous, since it is the only concept associated with Pell. Consequently, the name remains in use to the present.

Building upon Diophantus's *Arithmetica* Book II, problem eight, to divide a given square number into two squares, Fermat had posited in the margin of his copy of Bachet in 1637:

> No cube can be split into two cubes, nor any biquadrate [fourth power] into two biquadrates, nor generally any power beyond the second into two of the same kind.[511]

On page 61 of the 1670 *Diophantus*, Fermat's son Samuel put this observation in Latin between problems eight and nine of Book II. Translated into modern symbols the comment means that except for trivial solutions where $z = x$ or y

and the other variable is zero, for $n > 2$ the diophantine equation $x^n + y^n = z^n$ has no solution among the integers. Since it does not reduce to determining points having positive rational coordinates on curves of genus 0 or 1, which he primarily considered, Weil believes that Fermat must have approached this problem with trepidation. The concept of genus for classifying curves given here stems from Alfred Clebsch in the nineteenth century. Algebraic curves of genus 0 have at least one obvious rational point, and those of genus 1 are elliptic and have a visible rational point and rational pair.[512] Young Fermat probably discovered this using his own method of infinite descent and perhaps thought that he had a proof. The older Fermat knew better, but confident in his conjecture he added to the remark, "For this, I have discovered a truly remarkable proof which the margin is too narrow to hold." The long search for a proof of the impossibility of integer solutions for the $x^n + y^n = z^n$ for $n > 2$ has in popular mathematics garnered for this equation the title Fermat's last or great theorem, even though it is not his last or his most important discovery.

The conditions and techniques required for general solvability of this elementary diophantine equation, seemingly intuitively true, were long missing. The path to its resolution attracted the attention of leading mathematicians. For $n = 3$, Euler provided in 1753 a flawed proof, and Gauss followed with a complete one, while Dirichlet partially proved it for the more difficult case of $n = 7$. Kummer added the condition that primes must be regular and essentially solved the equation for regular primes less than 61.[513] Before World War I the Wolfskehl prize in Göttingen offered 100,000 marks for a general proof of FLT. Only after the development of a complete theory of complex numbers and the evolution of modern algebraic geometry was a general proof achieved, that of Andrew Wiles of Princeton in 1994.

15.6   The Formation of Classical Probability

Mid-seventeenth century philosophical, religious, literary, legal, scientific, and medical texts abound in references to different forms of the notions of uncertainty and probability. Belonging to mixed mathematics, along with such fields as pneumatics and hydrodynamics, the investigation of classical probability, known in French as the *art de conjecturer* or *analyse des hasards* and lasting to about the 1830s, was initially the sum of its applications. Theory was not yet separate from practice. At first alchemists, astrologers, physicians, and practitioners of the low sciences that were undemonstrative also appealed to a crude probability, for example, in correlating fevers with disease or sightings of comets with the deaths of monarchs.

Scholarly debate continues over the sources of classical probability and the conceptual change that ushered it in. Ian Hacking traces this development to a mid-seventeenth century merging of two constituents.[514] A traditional epistemological component deriving from ancient texts that provided a measure for determining degrees of reasonableness of opinion without statistical background became linked to a novel concept that natural evidence could be quantitatively evaluated based on the frequency of occurrence. Philosophers sought a calculus of reason based on a set of computational rules against which doubts and ambiguities could be tested and inferences converted into judgments. Uneasy with the paths either to total doubt via Pyrrhonian skepticism or to absolute certitude via Cartesian methodical doubt, they sought a middle way. Their mission was to achieve a level of moral certitude appealing to the intelligent, sensible man. The novel computational component paralleled, so Hacking argues, the new aleatory law for fair contracts, a legal field developed among others by Grotius and Pufendorf. Aleatory law, the name taken from the Latin term for dice, includes any contract in which chance plays a part, such as the delivery of goods after shipment. But Hacking's argument deriving aleatory law from a theory of statistical regularity or the computation of odds from observed frequencies is questionable. The initial intent of aleatory legislation was not to lay down norms for setting odds. Rather it was to devise procedures to ensure maximal reciprocity when compensations were awarded in equity settlements. Also dubious is Hacking's historical claim tracing the rise of mathematical probability to a sudden shift in semiotics, the study of the use of signs in human communication. The signatory status of testimony, he argues, changed quite abruptly about 1650. In a novel turn, testimony, legal or otherwise, was considered connecting a word sign to the situation or event, which it depicted on a hitherto unrecognized basis, that of evidence. Unlike the case of natural and conventional signs, which linked a magical formula to an event, its effectiveness thus depended on the frequency of occurrence. Daniel Garber and Sandy Zabell reject the abrupt shift.[515] While inherently efficacious signs frequently appear, for example, in ancient divinations and portents, another semiotic stream connecting probability to relative frequency appears as early as Cicero and Quintillian and resurfaces in late medieval canon law and the philosophy of Nicole Oresme.[516]

Gottfried Leibniz, who in his letters cites the "intelligent but ... half comprehending" work of Chevalier de Méré on the subject,[517] suggests that early quantitative probability stemmed mainly from the illicit pastime of gambling, an institutionalized form of risk taking, which the Catholic Church condemned but clearly could not prevent.[518] Gambling was popular among the nobility. Siméon-Denis Poisson at the beginning of his *Recherches sur la probabilité ...* of 1837 posits a similar interpretation. While all pioneer probabilists from Cardano to Huygens solved gambling problems, Hacking and Lorraine Daston consider this argument incomplete and misleading. Most countries, except for England, considered life insurance and annuities a form of gambling until the eighteenth century and declared them illegal.[519] Prior to the work of Jan Hudde and John de Witt on pensions in the 1670s, neither was on sound actuarial footing, and life insurance lacked mathematical certainty: John Graunt provided the first extensive mortality tables.

In addition to philosophy, political economy, and psychology, Lorraine Daston finds jurisprudence crucial in influencing the creation of mathematical probability.[520] She cites the work of jurists on aleatory contracts, along with their response to the Church's proscription against usury. Lawyers attempted, for example, to equate to services rendered the interest on investments on a merchant-shipping expedition. Legal examinations into the reliability of causes and testimony, the credibility of witnesses, and the nature of evidence, as well as the preparation of sentencing guidelines, also figure in Daston's interpretation.

Commencing with an exchange of letters from spring to October 1654 between Pascal and Fermat and culminating in Jakob Bernoulli's *Ars Conjectandi* (*Art of Conjecturing*) of 1713, often considered its founding work, mathematical probability theory rapidly advanced. Early in this period, Pascal, Fermat, and Huygens, primarily in his *Tractatus de ratiociniis in ludo aleae* (*Treatise on Calculations in Games of Chance*) of 1657, proposed new mathematical lemmas, propositions, and proofs. These applied to a range of practical probabilistic problems not originally considered. Huygens followed by one year the posthumous publication of Cardano's treatise of the same name.

When his associate de Méré proclaimed a scandal in the arithmetic of probabilities because some gambling problems produced contradictory results, Pascal in early 1654 solicited advice from his father's old friend Fermat. In their brief correspondence, Pascal was to compose three letters and Fermat four.[521]

While Pascal's first letter is lost, Fermat's reply shows that both define the chance of success as the ratio of favorable results to all possible results. Although neither writer cites sources, by 1637 Mersenne's circle knew practically all of their problems, except for its "problem of points," and were seeking solutions. These problems seek to determine the number of decisions needed to give two or more players an equal chance to win and how to divide stakes accordingly in prematurely ended games of chance. Pascal's July 29 letter raises a seeming contradiction in computing the probability of obtaining a double six upon throwing two dice.

Fermat's response to Pascal's lost letter and the start of Pascal's second indicate that de Méré had studied the problem of obtaining a six on the toss of one dice. Apparently, de Méré reasoned that the probability is $1/6$, and so that of not getting a six is $5/6$. For n throws the second value becomes $(5/6)^n$. Thus, the probability that six will appear at least once is $1 - (5/6)^n$. To find the smallest number of trials that allow the player more than a 50-50 chance of success here requires computing $1 - (5/6)^n > 1/2$. For four throws, $1 - (5/6)^4 = 1 - 625/1,296 = 671/1,296$, which correctly gives the odds which de Méré has computed as 671 to 625. Cardano had found the same odds.

Looking next at the problem of double sixes on a toss of two dice, De Méré proposes that n/N is constant, where n is the number of trials needed to guarantee a 50-50 draw in obtaining all sixes and N is the total number of outcomes from a single toss of the dice. Since $n = 4$, or $4/6$, for a better than even chance of tossing a single die for six, he assumed the same odds for the fewest throws of two dice needed for obtaining two sixes. In this case there is a total $N = 36$, and the n should thus equal 24, or $24/36$. But this is not correct. Pascal, whose letters describe de Méré as a savant combining heart and imagination, but judges him not to have advanced beyond calculation to higher theory in mathematics, quickly finds the error and dispatches de Méré's so-called scandal.[522] The next game must be considered as well. Reasoning recursively, Pascal lists all possibilities for the two dice in a six-by-six matrix, beginning the first line with $(1,1)(1,2)$ and ending the sixth line with $(6,5)$ and $(6,6)$. This matrix shows that there are thirty-five throws that do not have a double six. The crucial figure is thus $(35/36)^n$. For twenty-four tosses, $1 - (35/36)^{24} = 0.04914$, which is less than one-half. Patient recursive observation and correct computation indicate that, since $1-(35/36)^{25} = 0.5055$, twenty-five throws are required for a better than even chance of a double six.

Fermat and Pascal independently discovered the formula for binomial coefficients, so important in number theory and algebra, and were to apply it to resolve interrupted games of chance. Fermat had informed Roberval in 1636 and Mersenne in 1638 of finding this "very beautiful proposition."[523] Its basis seems to be Fermat's recursion formula relating triangular $(m = 2)$, pyramidal $(m = 3)$, and triangulotriangular $(m = 4)$ numbers, and so forth, in modern notation becoming $\binom{n+m}{m+1} = \binom{n+m-1}{m} + \binom{n+m-1}{m+1}$.[524] [The modern notation of $\binom{n}{r}$ modifies Euler's 1778 notation of $\binom{n}{r}$]. Apparently Fermat did not know that Peter Apian in his *Rechnung* of 1527 had introduced to the Latin West the triangular arrangement of these coefficients and that Tartaglia and Cardano had generated small tables of them. Possibly Fermat knew of the computation of the relationship to the seventeenth line that Stifel had provided in his *Arithmetica Integra* of 1544; more likely, he was familiar with Viète's similar calculation in *Angular Section* of 1615. Fermat and Pascal undoubtedly were aware of Mersenne's having in 1636 or 1637 extended Cardano's table to $n = 36$. [525] Not having learned of Fermat's discovery, Pascal made it in 1654 and revealed it in his famous *Traité du triangle arithmétique*, which proves relations between binomial, combinatorial, and figurate numbers.[526] This treatise was not released until 1665, but Fermat received a copy, so it is considered an integral part of their correspondence.

Pascal's form of the arithmetical triangle neatly arranges and computes the numbers. The first row are all ones; the second row, sums of the numbers up to it in the first row; the third row, the sums of the second; and so on. Each upward sloping diagonal of the triangle gives the coefficients for expansions of the binomial $(x + y)^n$ for $n = 1, 2, \ldots$. Further binomial coefficients are easy to compute without appeal to the binomial theorem. In each cell of right triangular sets, the sum of the two figures on the hypotenuse is the third.[527] Put another way, each number is the sum of the number immediately to its left and the number directly above it. In row 7, for example, $28 = 7 + 21$. The modern form of the arithmetical triangle shifts Pascal's representation $45°$. Employing the binomial coefficients, Pascal solves the "problem of points." Beginning with the number of points that each player lacks to win, he sums them and selects the diagonal whose total elements is the sum. The case in which player A lacks three points to win and B four gives $n = 3 + 4 = 7$. The diagonal with seven entries is 1, 6, 15, 20, 15, 6, 1. The first player's stake is the sum of all terms

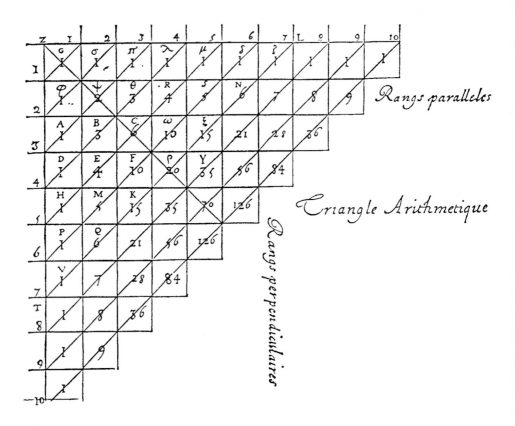

on the diagonal or 42, while that of B is the sum of the terms whose number is the first person's lack, or $1 + 6 + 15 = 22$. The ratio for dividing the stakes is then $42/22 = 21/11$.

In addition to the recursion solution for particular cases, Pascal—to be followed in August by Fermat—gave a general solution by applying combinatorial methods, a particular strength of Fermat. Pascal circulated Fermat's solution in Paris, where Roberval initially doubted Fermat's methods as counterfactual. Fermat required players to terminate their game not at the break even point, or $n - 1$ plays, but at its full outcome, or $2n - 2$. At first Pascal shared Roberval's reservations, but Fermat's August 29 letter convinced him that computation based on play until full outcome is necessary for order. The total outcomes in the preceding problem have $n - 1 = 6$ and $2^6 = 64$. The induction method shows that the proper division of stakes in the next unplayed game is the proportion of $\sum_{r=0}^{n-2} \binom{n}{r}$ to 2^6. A's stake equals the numerator of the binomial sum to the total, or $42/64$, and B's likewise is $22/64$. The stake division is $42/64 :$ $22/64$, which again equals $42/22 = 21/11$.

The clarity of Pascal's presentation of the arithmetical triangle and the

nineteen consequences he derived from it in his *Traité* have assured that it
be known as Pascal's triangle. The twelfth consequence is the most significant for its proof about ratios that formulates the demonstrative procedure of
mathematical induction. Incomplete induction had been known since Euclid,
and Francesco Maurolico's *Arithmeticorum Libri Duo* of 1575 had reasoning
by recursion. Pascal offered the first explicit and complete statement of mathematical induction. He did not name his procedure mathematical induction.
That name was coined by Augustus de Morgan in his article "Induction" that
appeared in the *Penny Cyclopedia* in 1838. This procedure is an invaluable tool
to mathematicians. Pascal bases mathematical induction on two lemmas that
apply to propositions covering an infinity of cases. The first lemma establishes
some proposition $P(n)$ as true (meaningful) for a smallest $n = n_1$. The second
shows that if the assertion $P(k)$ holds true then the next assertion $P(k + 1)$
must also be valid. Applying logic and mathematical facts, Pascal establishes
that the two lemmas hold.

The Port Royal *Logic* of 1662 by Antoine Arnauld and Pierre Nicole, whose
fourth book sought to quantify the degress of frequency of contingent events
and the occurrence of good or evil based on testimony rather than miracles,
and Pascal's *Pensées* report another prominent element in classical probability,
Pascal's wager.[528] This matter of Christian apologetics, whether God is or is
not, was originally written on two pages filled with erasures and corrections.
Reason, Pascal holds, cannot provide a valid proof of God's reality. He proposes
a wager to settle the issue in a prudential way. Hacking neatly casts his wager as
an exercise in decision theory. When an outcome is uncertain, the theory seeks
to discover the best decision. Decision problems require an exhaustive account
of all possible states: in this case, God is or is not. These problems also must
list all possible actions: here the two choices are to lead a pious life or a worldly
and libertine life. In deciding the best course of action, Pascal examines four
combinations. Assuming the simplest combination, that God exists and the
pious life is to be pursued can lead to the infinite blessings of salvation. Even in
the two instances conjecturing that there is no God, a finite chance, no matter
how small, exists that the opposite is true. Pascal's probability distribution
decisively argues for the pursuit of the pious life. Following the short pleasures
of the world he considers a wrong judgment. Among others, John Locke, who
translated the Port Royal *Logic* into English, gave a version of Pascal's wager.
His appears in his *Essay Concerning Human Understanding* of 1704.[529]

After meeting Roberval and Gassendi in Paris in 1655 and Fermat the next
year and solving probability problems that Fermat had sent him, the Dutch
mathematical physicist Christiaan Huygens (1629–1695) wrote his brief *Tractatus de ratiociniis in ludo aleae*. It was published in 1657 at the end of the
Exercitationum Mathematicarum of Franz van Schooten, who had taught him
mathematics at Leiden, immersing him in the methods of Viète, Descartes, and
Fermat.

Huygens's *De ratiociniis* consists of three propositions and eleven problems
on games of chance.[530] Although in the Dutch original he did not use the

equivalent of the word *expectation*, either he or van Schooten introduced it in the Latin translation. Apparently influenced by Dutch commerce and accounting methods, Huygens wrote initially about the "value [or price] of a chance." The value or expectation generalized the arithmetical mean. Proposition one, for example, introduces a lottery in which two players have an equal chance to win and the winner will pay the loser amount a. If each player has put in the same amount x, this will leave the winner $2x - a$. If $2x - a = b$, the lottery will be $x = (a + b)/2$. This last amount is Huygens's value. Proposition three generalizes this result. If two players each invest a stake of $(a + b)/2$ and the first has p chances to win a and the second q chances to win b, reaching odds of $1/2$ or better requires computing $(pa + qb)/(p + q)$. Problem eleven analyzes de Méré's problem of obtaining a double six in throwing two dice. Huygens asserts that a player has one chance of winning a and thirty-five chances of losing, that is, winning 0: so the value for winning is $(1/36)a$. Moving through thirty-six throws, he finds the value of his chance of winning is $(1a + \{35/36\}a)/36$ or $71a/1{,}296$. The first pair of tosses gives 71 out of 1,296 chances of winning and 1,225 of not. Continuing this argument, the value for attaining a double six in four series of throws is $(71a + 1{,}225\{71/1{,}296\})a/1{,}296$, but this is far less than the $(1/2)a$ sought. Computations show a small disadvantage at twenty-four throws and the reverse at twenty-five.

Although Huygens adopts the numerically cumbersome recursion method of Pascal and eschews combinatorics, the *De ratiociniis* was well received. Its first problem, for example, has two players A and B using dice and A winning upon throwing six points and B for seven points. A has the first throw, B the next two, and so on, until one or the other wins. It seeks the ratio of A's chances to B's. Huygens gives the answer of 10,355 to 12,276. The only published tract on probability to 1713, the *De ratiociniis* underwent several editions. Jakob Bernoulli's *Ars Conjectandi* superceded it but did not end its influence, since Bernoulli begins his text with a complete, annotated edition of Huygens's treatise.

16

The Apex of the Scientific Revolution I: Setting and Laureates

> Freely I stepped into the void, the first.
> — Huygens, quoting from Horace, in asserting
> his innovation in the study of centrifugal
> force, *De vi centrifuga*, in 1659

> There could be only one Newton, there was only one
> world to discover.
> — Joseph-Louis Lagrange

In 1660 European mathematics consisted of a swiftly expanding mass of concepts, various methods and techniques of solution with increasing attention to the infinite, unsolved problems, and tantalizing paradoxes, often intertwined in a disordered way. In addressing these, a dynamic nexus formed made up of genius and a selective and occasionally transposing synthesis of Cartesian geometry, the arithmetic of infinities, power series expansions, including for trigonometric relations and their inverses, and computations with infinitesimals of areas under curves and volumes of solids, joined with tensions from critical issues in mechanics, astronomy, navigation, artillery, commmercial contracts, games, and demographics that often shaped major problems and areas for research. From all this, leading European mathematicians produced by 1730 rich and diverse technical achievements. Foremost among them were Isaac Newton, who invented fluxional calculus, and Gottfried Leibniz, apparently along with Jakob and Johann Bernoulli, who created a similar differential version. This powerful new instrument allowed natural philosophers to uncover properties of force and to begin to describe more exactly the motions of the moon, planets, comets, and tides.[531] Centering on infinite series, differential and integral calculus also permitted mathematicians to solve problems of instantaneous acceleration and motion in physics that finite ancient and Cartesian geometries could not. Other branches of mathematics now had piecemeal developments. To algebra, Leibniz added determinants, while in number theory he first summed $\pi/4$ as an infinite series from his study of the quadrature of the circle, and the Bernoullis derived their numbers. Jakob Bernoulli's law of large numbers significantly forwarded probability theory.

Contributing to the quality and extent of this accomplishment was a change in the vital repository of primary writings. No longer were Hellenistic Greek sources the main starting point for mathematical research. Aside from providing guiding methods, these ancient texts were increasingly consigned to antiquarians and translators. Mathematical investigators now looked principally to the works of modern thinkers, such as Viète, Kepler, Descartes, Fermat, Pascal, and Wallis. Mathematics profited also from a broadened institutional setting. Most major mathematicians now spent much of their career in universities. But new scientific bodies chartered by monarchs, above all the Royal Society of London and the Paris Academy of Sciences, gave additional status and resources to the sciences. These state institutions became significant loci of mathematical studies, and their journals together with the new *Acta eruditorum* were essential in the diffusion of scientific knowledge.

In most of western Europe, Cartesian thought dominated the republic of letters. Together with the accompanying spread of a geometric spirit and the growing success of mathematics aided by better instruments in more precisely measuring and describing physical phenomena, it encouraged a few savants to believe that mathematics could grapple with everything. Baruch Spinoza's application of the axiomatic method in his *Ethics* of 1677 was a foretaste. So was the claim in 1699 of Bernard Fontenelle, secretary of the Paris Academy of Sciences, that "a work on ethics, politics, criticism, and, perhaps, even rhetoric will be better, other things being equal, if done by a geometer."[532]

16.1   An Age of Absolutism

This groundwork of modern mathematics was laid during an age of political absolutism from 1660 to 1730. Along with high intellectual creativity went crises and suffering, as continental European kings extended their power by centralizing royal states. This coexistence of progress with wretchedness testifies to the vitality of Europe's republic of letters when supported by increasing patronage and new state science institutions.

The scourge of epidemics reappeared: London's 1665 bubonic plague killed an estimated 100,000 people. In 1720 plague claimed another 90,000 in Provence, including half the population of Marseilles. Bubonic plague was blamed on corrupt air. It was not yet known that fleas borne by black rats, gerbils, and squirrels were the cause. Rat baffles on ship lines had a commercial—not hygienic—function. They were intended simply to safeguard goods from damage. Quarantine or isolation, under development in the west, was still haphazard in the east. Among endemic diseases were smallpox, dysentery, and typhus. A climate that had turned very cool and damp, remembered as a little ice age, was catastrophic for the near-subsistence farming that preceded crop rotation and the adoption of the high-yield potato. Entire village populations became wandering beggars foraging for food. In the 1690s death was widespread, and in the winter of 1709–1710, during the last great famine of the period, Parisians died in the streets from cold and hunger.

Frequent wars increased the misery. At the close of this era alone, three major conflicts raged. In the west the War of the Spanish Succession from 1702 to 1714 pitted France against an alliance led by Austria and England. In the northeast in the Great Northern War from 1700 to 1720, Petrine Russia fought Sweden. And in the southeast during the Austro-Turkish War from 1716 to 1718, Prince Eugen and the Imperial Austrian army drove the Ottoman Turkish forces beyond Belgrade in the Balkans. No longer religious antagonisms but a quest for prestige and territory among absolute rulers, coupled with a growing global commercial rivalry, shaped warfare in Europe. Since the existing diplomacy could not limit war, European leaders adopted the Renaissance concept of equilibrium, rendered as balance of power, to offset a superior power by an alliance of lesser ones.

In response to the chaos of the 1640s and the costs of widespread warfare, European sovereigns and their ruling elites seeking stability fashioned two types of state building—absolutist and constitutional. Increasingly centralized, most states had a growing and efficient bureaucracy that could extract more taxes. Nearly all on the continent were absolute. In these agrarian states ambitious rulers were unchecked by a vigorous and widespread urban populace accompanied by strong representative institutions. Generally they won the support of the nobility for a pyramidal structure in society, the king granting privileges at the expense of a subservient peasantry at the bottom. The nobility still competed with the monarch for power, and lesser nobles quarreled with greater magnates. An alternative constitutional model developed in England, the Dutch republic, and Sweden. In these countries, strong towns and representative bodies, together with merchant burghers enjoying the independence that comes with the exercise of commerce and industry, obstructed absolutism. These states also depended on relatively small armies.

For three decades after the Parliament restored the Stuart monarchy in 1660, trade and commerce in the maritime state and constitutional exemplar Britain continued to expand. The British merchant fleet doubled the tonnage it hauled; New Amsterdam in North America became New York; and Lloyd's of London began insuring ships sailing to the Americas. Whether Whig or Tory controlled Parliament, the chief goal of British policy was to gain commercial advantage. France probed areas south of Quebec and in 1699 set up a colony in Louisiana, but the Peace of Utrecht of 1713 countered French expansion, giving Britain some of Canada. Parliament meanwhile struggled with Stuart kings over royal prerogatives, opposing new taxes and fearing an effort to restore Catholicism as the state religion. Protestant fears increased with the failure of James II to censure Louis XIV for revoking in 1685 the Edict of Nantes, which had guaranteed toleration to French Calvinist Huguenots. The birth in 1687 of James II's son, who was to be raised a Catholic, provoked a constitutional crisis. During the next year's Glorious Revolution, Parliament again triumphed, inviting William and Mary to be the new monarchs and passing a Bill of Rights reasserting parliamentary financial authority, recognizing tolerance for Protestant dissenters, and guaranteeing the rights of property owners. John Locke's *Two Treatises of Government* of 1690 holds that the rule of kings rests on contracts between the monarch and the people. While the people surrender some liberty to gain security, the rights of life, liberty, and property remain innate, founded in nature. Rejecting patriarchal models of authority of his time, Locke argues that the king must respect these laws of nature. A ruler who breaks the trust the people have given him to maintain natural rights can be removed. The argument was later to appear in the American Declaration of Independence.

The federated Dutch republic was the most densely populated state in Europe, except for a few on the Italian peninsula, with half of its population of almost two million residing in cities. The country's richest land, recently reclaimed from the sea, had produced an agricultural surplus, while the banks and warehouses of Amsterdam surpassed all others in Europe, and the city rivaled London as an international port. Dutch merchants traded with Poland for grain, Sweden for iron ore, Africa for slaves, and the west and east Indies for spices. Yet the republic, in 1660 the leading commercial and naval power in Europe, was practically a British satellite seven decades later. As powerful merchant families gained tighter control over profitable governmental posts, social mobility lessened. Under the dominance of the stern Calvinist Dutch Reformed Church, a country once relatively tolerant in religion turned to persecuting dissenters and homosexuals. French invasions, Spanish embargoes, and British commercial rivalry put the Netherlands at peril. The high taxation required for resistance to the French military incursions combined with high wages to undercut the sale of Dutch goods. The Dutch republic's vitality weakened and its economic primacy ended.

In France the handsome and energetic Louis XIV established the model for absolute rule. Declaring himself without equal, *Nec pluribus impar*, and selecting for his emblem the sun, representing the virtues of firmness, equity, and benevolence, he is remembered as the Sun King. The early phase of absolutism was subsequently named Ludovican for him. Coming to the throne as a youth in 1643, Louis in 1661 assumed personal responsibility for government and reigned until 1715. He justified a shrewd pursuit of state centralization by Bishop Jacques Bossuet's revived theory of the divine right of kings, which held that royal authority comes from God alone (but not, as is sometimes thought, that kings rule without constraints.) Louis took the suggestion of the controller general Jean-Baptiste Colbert that monumental buildings show the greatness of the prince. Twelve miles west of Paris, Louis in disregard of costs converted the hunting lodge of Versailles into a magnificent palace complex. He moved the royal court there and rarely visited Paris. Amid the architectural grandeur of Versailles, wigged courtiers followed catechized etiquette. The degree of a bow or the size of a wig precisely signaled a person's hierarchical rank. Notwithstanding the pomp, the theater, the operas, and the hunting, the courtiers suffered physical torments in damp, freezing apartments and perhaps psychological ones as well struggling to offer constant flattery to the king and his retainers. Drawing about 10,000 great and lesser nobles, officials, and servants to lavish Versailles or nearby, Louis XIV cut adrift from its lands the old nobility of the sword and domesticated it. He further weakened the second estate by selling noble offices and installing members of the middle order of society as ministers of his Council and *intendants* (administrators), who resided in the towns of the royal provinces.

Absolutist France relied financially on mercantilism. Created by Colbert (1619–1683), who managed public administration, public works, and taxation, it mistakenly perceived and sought economic self-sufficiency through establishing royal monopolies, assuring that exports exceed imports, and accumulating gold reserves. More effectively collecting from tax farmers, Colbert doubled crown revenues. Louis XIV created the war ministry, and Camille le Tellier Louvois, known as the king's "evil genius," transformed a poorly organized army into the most powerful in Europe, the best armed, trained, and administered, growing rapidly from 30,000 in 1659 to 97,000 in 1666, to 360,000 in 1710.[533] Louvois subjected it to civilian control. As suppliers of munitions, clothing, and equipment, manufacturers had an interest in supporting the established order.

As the motto "one king, one law, and one faith" suggests, Louis came to desire religious conformity. Probably under the influence of Jesuit confessors and advisers, he suppressed the Jansenists, the movement to which Pascal had belonged: these Catholic puritans with Augustinian views of grace that God freely imparts to humans, were depicted by their enemies as "Calvinists who go to mass." In 1685 Louis revoked the Edict of Nantes, Henry IV's grant of toleration to French Huguenots. Numbering about a million and three fourths, they comprised almost ten percent of the population. Repressive laws closing churches accompanied the revocation and dragoons, mounted infantry, participated in efforts at forced conversion. Though the king forbade the Huguenots to leave France, about 200,000 emigrated to England, Holland, Prussia, and South Africa, many of them merchants or skilled craftsmen.

Economic surpluses accumulated in the 1670s, largely from exports of linen, cotton, woolen cloth, and luxury goods, especially silk and tapestries, as well as funds from improved tax collection were spent mostly on war. Intent on gaining territory as a sign of his greatness, the Sun King embarked on a series of large-scale aggressions to achieve what geography and his policies defined as natural frontiers. By this calculation, France's southern boundary was the Pyrenees mountains. Reaching the Rhine boundary to the north and northeast involved France in the Dutch War from 1672 to 1678, when the Netherlands opened dikes and by flooding prevented French conquest, and from 1688 to 1697 the War of the League of Augsburg, also known as the War of the Palatine Succession. Undisputed hegemony, which encouraged expansionist thrusts did not guarantee success. While no single power could withstand France, other states formed wary defensive alliances. Although by the mid-1690s the tide of battle had begun to shift against the French, Louis could not resist the temptation of Spain, when its king died without heir. During the War of the Spanish Succession at the end of Louis's reign, a Grand Alliance of armies commanded by Austria's Prince Eugen and England's Duke of Marlborough defeated the French outside France. The war was fought not only in Europe but in America, Asia, and wherever the powers had colonies and trading stations. During the harsh winter of 1709 and 1710, military losses and famine brought France to the verge of collapse. The costs of the war almost bankrupted the French crown. While the treaties of Utrecht in 1713 and Rastatt in 1714 recognized Philip V, Louis' grandson, as king of Spain, Philip had to reject any claim to the French throne. These treaties gave the Spanish Netherlands and northern Italy to Austria. Britain gained recognition for the Protestant succession in the monarchy, expansion of its colonial domain in North America, and the *asiento*, control over the import of black slaves into the Spanish empire.

Quilted with large estates populated by legally bound serfs and having no overseas empires, the states of central Europe were also changing in makeup. The Peace of Westphalia had deprived the Austrian Habsburgs as Holy Roman Emperors of some of their influence in north German affairs. Afterward, closely supported by roughly a hundred wealthy noble families in Vienna, they consolidated their rule in Austria and Bohemia, reconquered Hungary from the Ottomans, and extended their control in the Balkans beyond Croatia. In imperial Austria the Counter-Reformation peaked in the lavish building of monumental baroque palaces, monasteries, and churches, such as Fischer von Erlach's Belvedere in Vienna. In addition to their being an ideological statement, these structures probably served as a means of tax evasion. In the region, meanwhile, two new powers, Brandenburg-Prussia and Russia, were emerging.

After Westphalia the Hohenzollern rulers merged the three noncontiguous areas of Brandenburg, which were poor in natural resources and soil. Acceptance by most Prussian Junkers, or nobles, of royal authority in exchange for privileges cleared the way for Frederick William, who reigned from 1640 to 1688. Known as the Great Elector, he created a permanent standing army to crush opposition from rebellious nobles, to gain concessions from his small towns, and to acquire territory. Afterward he recruited the sons of the nobility to lead his army and bureaucracy. Within the Great Elector's centralized administrative bureaucracy, perhaps the first civil service in early modern Europe, the General Directory over Finance, War, and Royal Domains was especially effective, amassing a full treasury by following the cameralist doctrines. To improve the economy, the cameralists retained the protections and manufactures of mercantilism, but were to base the economic order on the land and the peasant, setting forth the idea of agriculture as an elastic source of taxable wealth. The directory collected twice as much revenue as the French crown. Frederick William also attracted skilled immigrants, including 20,000 French Huguenots, who went to Berlin after 1685. Below the Junkers, the burghers were powerless and the peasantry had lost its freedom.

After a troubled early seventeenth century, the new Romanov tsars in Russia gradually subdued the boyars, or great magnate nobles, and crushed the near endemic peasant uprisings. Generally, they had the support of the Orthodox Church. Before the reign of Peter the Great (1672–1725), Russia had been remote and isolated on the eastern periphery of Europe, and western Europe knew little of it. Impressed with western technology, military strength, and efficient administration, this tsar, nearly seven feet tall and dressed as a workman, took a supposedly incognito trip west, during which he may have met Newton in London. Afterward Peter introduced a program of forced modernization. He centralized his government by reorganizing civil administration into a series of colleges or bureaus headed by ministers, named from a cadre of nobles, and by establishing in 1711 a nine-man Senate following a Swedish model to guide the bureaucracy. The main task of the bureaucracy was to collect taxes for the army. More than two-thirds of state revenues were spent on the first standing Russian army, which Peter began. He built an elegant new capital, St. Petersburg, as a "window on the Baltic Sea." His government acquired craftsmen from the west, opened schools, and founded the St. Petersburg Academy of Sciences. His turn to the west angered old noble families, and his efforts to subordinate the Orthodox church to the state ired its traditional leaders. Together these affronts to tradition produced four uprisings that he subdued. Russia defeated two powers that had hemmed it in, the Ottoman empire, now badly decayed, and Sweden. During the Great Northern War, it supplanted Sweden as the dominant power on the Baltic and moved closer to western and central European affairs.

16.2   The Visual Arts, Literature, and Method

After the mid-century magnificence of a Rembrandt, Europe's late baroque visual arts achieved an afterglow expressed through the open market sale of pictures in the Netherlands of a host of "little masters." Among them was the prolific Jacob van Ruisdael (d. 1682), whose subjects were biblical scenes, daily and family life, landscapes, shipping, parlor games, skating, and churches. These pictures that adorned the walls of palaces, chateaux, and small houses reflected the prosperity of the Dutch state and the importance of the household. As noted above, the French military invasions of Holland failed, but a French cultural invasion triumphed, turning Dutch artists and leaders to French classicism. After Jan Vermeer (d. 1675), Dutch painting began to languish. Departing from historic themes, the French painter Jean-Antoine Watteau (d. 1721) idealized courtship in his graceful portrayal of *fêtes gallantes,* festivals held in parks that were attended by beautifully dressed young men and women.

As a display of authority, absolute monarchs along with popes supported monumental architecture. In this regard, baroque Rome, a city of fountains, was foremost. Much of its recent design was due to Bernini, who had drafted the original plans for the nave of St. Peter's, although Bernini himself went to Louis XIV's Versailles in 1665.[534] A single disaster and one genius prompted the flowering of English architecture. The disaster was the great fire of 1666 that destroyed most of central London. The genius was the young professor of astronomy at Oxford, Christopher Wren (1632–1723). From 1672 to 1675 Wren redesigned Saint Paul's Cathedral. Its complete rebuilding was finished in 1710. The interior of St. Paul's, which has long Gothic spaces as the clergy desired, stresses linear arches, circles, and frames.

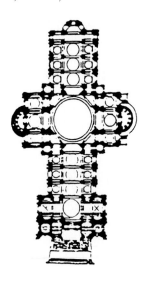

Interior of St. Paul's cathedral

The new sciences affected English literature, as did absolutism and aristocratic culture its French counterpart. John Milton, whose *Areopagitica* of 1664 is an eloquent defense of freedom of the press, considered scientific studies a cognate to the Puritan world view. From the Ptolemaic model of the universe given in *Paradise Lost*, Milton shifted in his later *Paradise Regained* to a Copernican cosmos. But many English authors still opposed the new sciences. Walter Charleton decried "the dark lanthorne of reason." John Bunyan in 1666 in *Grace Abounding* placed divine designs and interventions beyond the limits of human reason. In France, drama dominated. Expensive theatrical productions in Paris and at Versailles enjoyed royal funding. The two leading playwrights were Jean-Baptiste Poquelin (d. 1673), whose adopted name was Molière, and the Jansenist Jean Racine (d. 1699), who had ancient Greece as his muse. Though he fulsomely praised Louis XIV, the lonely Molière ridiculed aristocratic and ecclesiastical pretensions, criticized the bourgeois for pedestrian behavior, and in *Tartuffe* lambasted hypocrisy. Tragedian Racine in *Iphigénie* and *Phédre* explores the human character and hidden passions.

Although historians disagree on the interactions between espoused methodologies and actual natural knowledge practices, questions of method were taken seriously during the late seventeenth century. Correct method, it was believed, is the guide to knowledge of the natural world, to justification of this knowledge, and to making it powerful. Questions of proper method increased tensions between religious faith and scientific reason. While most late baroque men of letters reserved their principal loyalty for religion, skepticism and evolving critical rational methods influenced biblical criticism and more so the conduct of investigations of the physical world. The primary new rationalism was that of Leibniz. Scientific practitioners and philosophical partisans debated whether the most secure knowledge comes from deductive reason, critical empiricism, to which Locke was a notable contributor, or Newton's coupling of critical empiricism with infinitary analysis in his new mathematical science of rational mechanics.

Prominent in the debate between faith and reason was the French Huguenot refugee Pierre Bayle (1647–1706). While residing in Louis XIV's France, he briefly converted to Catholicism and then returned to Calvinism. His position thereupon became untenable in France on the verge of the revocation of the Edict of Nantes. In Holland in 1682, Bayle issued his *Diverse Thoughts on the Comet of 1680*. Condemning superstition and holding that the comet was a natural event and not a portent of divine wrath, the tract nonetheless questions the adequacy of all natural philosophy, ancient and modern. Bayle's *Historical and Critical Dictionary*, published in 1697 and revised and expanded in 1702, extends the range of his work beyond that of Baruch Spinoza (d. 1677) and Richard Simon (d. 1712), who developed principles of biblical criticism to reveal interpolated tales of miracles.[535] Building on a critical method, a genial skepticism from the ancients, Montaigne, and Christian thinkers, and more accurate factual accounts, his *Dictionary* demonstrates crimes of major figures in the bible and profane history, criticizes errors and inconsistencies in past accounts of them, advocates toleration as a public necessity, holds that an atheist can be a virtuous citizen, and takes on Descartes, Spinoza, Leibniz, and Locke. The *Dictionary* objects, for example, to Leibniz's concept that a preestablished harmony between mind and matter enables learning. Yet were it not mechanical and poorly founded, Bayle decided, it would be preferable to Cartesian occasionalism. Descartes had placed in the pineal gland the physical communication nexus between the two human components. Occasionalism contended that God alone, by employing the soul or body as merely an occasion for cause or effect, becomes responsible for all fundamental causal activity. Judging Cartesian certainty as humanly unattainable, Bayle observes that reason will lead the inquirer astray and ultimately produce only perplexities. The secure guide to truth is religious faith and an acceptance of revelation. Bayle's articles "Manicheans" and "Paulicians" especially assert the primacy of faith. Yet both Catholic and Protestant dogmatists were infuriated at his religious views, and the *Dictionary* was banned in France.

Bayle's *Dictionary* became a summa of skepticism. Undermining the intellectual foundations of late seventeenth-century western Europe, his influential critical method and his stress on factual accuracy made Bayle a progenitor of Europe's eighteenth-century Enlightenment, putting him in the company of Leibniz, Locke, and Newton. Incomplete, too digressive, and obscene, his *Dictionary* was outmoded by mid-century in the face of the thoroughness and scientific scholarship of Denis Diderot's famous *Encyclopédie*, but Enlightenment philosophes considered it an arsenal for their movement.

The German savant Gottfried Wilhelm Leibniz (1646–1716) was the leading contributor to continental rationalism. His *Ultimate Origination of the Universe* of 1697 conceives of a Platonic universe informed by reason, as exemplified by Euclidean geometry. This universe, no matter how arbitrary its operations may appear, is emphatically intelligible. Every truth is analytic. Only God can know with certainty, since that requires infinite analysis by a single mind. Along with the dominant Cartesian scientific rationalism, a complex variety of Aristotelian and other ancient thought persisted in Leibniz's time. While holding Descartes together with Galileo in the highest esteem in natural philosophy, Leibniz rejects as an atheistic neglect of divine providence Descartes' argument in *Principles of Philosophy* (III, 47) that corpuscular matter takes "on all the forms of which it is capable." Likewise wanting, according to Leibniz, is Descartes' rationalism articulated in the four rules of the *Discourse on Method*.

At the core of Leibniz's *philosophia perrenis* is a logic based on three axiomatic principles. The principle of the predicate in notion has a germinal form in the unpolished *Dissertatio de arte combinatoria (Dissertation on the Combinatorial Art)* of his youth, which was published in 1666, and its earliest precise formulation in section eight of his *Discourse on Metaphysics*, written in 1685. It maintains that for every true proposition the subject contains or can be reduced to the concept of the predicate to the minutest detail. Adopted from the thirteenth-century Catalan encyclopedist Ramon Lull, this principle, posited for a logic of invention, reverses traditional Aristotelian syllogistic reasoning that confirms. From the classical expression *nihil est sine ratione*, Leibniz derives what we know as the principle of sufficient reason. In the *Discourse* and his correspondence with Arnauld of 1686, he begins to infer from the omnipredicate his new principle as a cosmological generalization. Leibniz also called the root of his principle that nothing happens in the universe without a reason the common axiom. Holding to the concept of the freedom of the will, he explains that reason might incline to an act without necessitating it. Free will, states Leibniz, is one of two labyrinths that may lead reason astray. The other is the continuum. Leibniz treats the principle of sufficient reason in different fashions, sometimes suggesting that it entails teleology but mostly presenting it as a criterion for a logical grounding. Newton and the Bernoullis subsequently elaborated this principle. The other logical premise in Leibniz's theory of judgment equates identity with noncontradiction, which underlies consistency.

Leibniz's epistemology and natural philosophy are grounded in his theory of primal substance or monads, which were initially metaphysical points of energy. At its most fundamental level, his plenum universe consists only of animate monads.[536] Infinite aggregates and the results of them make up extended spatiotemporal matter. The monads exist in a strictly hierarchical great chain of being, differentiated by a psychological gradation of the relative clarity or confusion of their perceptions. There is no reciprocal influence among the windowless, that is, completely opaque, monads or between them and the composite physical world. Interaction between them is apparent rather than real. Yet, while Leibniz recognized that at the level of a human organism sensory experience is fraught with a host of possible deceptions, he, no less than Locke and the British empiricists, accepted the results of experience and experiment as sound. Phenomena from the material universe are communicated to the mind mediated, as it were, in current terminology by a master computer program in which the mind and matter act in a harmony preestablished by God. Descartes' and Leibniz's analogy for this harmony was the workings of two perfectly synchronized clocks. Subjecting experiential and experimental data to an organizing discipline of reason prompts an *ars inveniendi* to improve their precision. The process attains contingent truths of fact of which one may be certain.

Although Leibniz wanted to publish the *Discourse on Metaphysics* and the *Correspondence with Arnauld* that the two conducted from 1685 to 1690, neither was to appear until the mid-nineteenth century. Leibniz's article "Système nouveau de la nature et de la communication des substances" ("New System of the Nature and Communication of Substances") in the *Journal des sçavans* for 1695 introduced in print his logic and metaphysics. This article and its clarifications largely made his reputation in philosophy.

Initially through correspondence, Leibniz disputed the methodology and pessimism of Bayle's revised *Historical Dictionary*. For his friend Queen Sophie Charlotte in Berlin, he had begun from 1698 to 1705 a spirited critique of Bayle in what became the tract *Theodicy* on divine providence, human freedom, and the origin of evil. At Berlin's royal court Bayle's thought was popular. Leibniz completed his manuscript in 1707 there, but news of the death of his opponent in 1706 caused him to hold back this tract from the press. At the urging of friends, however, the first edition appeared in 1710 in French to reach a large audience.

Although the *Theodicy* was published anonymously in Amsterdam, the author was obvious. The second edition in 1712 carried Leibniz's name. The cardinal problem of the *Theodicy* is reconciling the freedom of the will with the preestablished harmony. It asserts that the preestablished harmony is natural, flowing from the principle that nature is governed by attaining what is best. Thus, it is not arbitrary and mechanical. Bayle had erroneously compared the soul to a passive atom, rather than conceiving it as a animate monad possessing composite tendencies. The *Theodicy* also objects to Bayle's skepticism about reason and his subordinating it to religious faith as a guide to truth. Leibniz argues that revelation and truths gained through the light of reason cannot be contradictory. When the eternal and necessary truths of reason pose an insurmountable obstacle to presumed religious revelation, reason is primary and the article of supposed revelation is false. Rebuking Bayle's pessimism, Leibniz proposes the law of metaphysical optimism: from among the infinite number of possible worlds God has selected for existence the best of all possible worlds for attaining the greatest perfection and the greatest harmony. Greatest perfection, a concept which he borrows from Aristotelian metaphysics, is his *summum bonum*. It requires a maximum of variety within a superior order that makes possible the variety. The *Theodicy* remained popular during the Enlightenment until 1759, when Voltaire in the jeremiad *Candide* ridiculed a naive version of the law of optimism that accepted whatever is as being right.

In *An Essay Concerning Human Understanding* of 1690, John Locke (1632–1704), weary of the medieval Aristotelian philosophy taught at Oxford, set out the roots of modern critical empiricism.[537] In its introductory epistle, he states that he looked for guidance in method to the moderns Boyle, Sydenham, "the great Huygens," and "the incomparable Mr. Newton."[538] They, like Galileo and Bacon, recommended skepticism, rejecting the authority of ancient and medieval texts, and directly inspecting the Book of Nature. Locke's epistemology rejects innate ideas and holds that we learn only through the senses and reflection. Human beings are born in ignorance and their passive mind at birth is a blank tablet—Aristotle's *tabula rasa*—upon which the senses write. The *Essay* also proposes another metaphor less familiar today: that the mind is an empty cabinet that the senses furnish. With this in mind, natural philosophers have to study primary qualities of solidity, extension, figure, motion, and number that are essential elements of things and secondary qualities of color, sound, taste, and feel that are in our mind. Primary qualities are best investigated, the argument goes, by use of the senses through disciplined observation and experiment. Lacking a means of testing the authenticity of signals reaching the mind through the senses, which might be a screen as easily as a conduit, Locke's theory of knowledge does not guarantee the attainment of the clear and certain truths that he claims for his method.

Locke's epistemology and critical empiricism put him at odds with both Bayle and Leibniz. In the dialogue *Nouveaux essais sur l'entendement* (*New Essays On Human Understanding*), Leibniz responded.[539] These essays, in which Philalèthe speaks for Locke and Théophile for Leibniz, describe Locke's sensationalist methodology as narrow, applying only to contingent truths and not to the necessary truths of mathematics, metaphysics, and ethics. Leibniz calls a fiction the notion of the *tabula rasa* and insists on the validity of innate ideas. Following Plato, Leibniz sees the mind as active and endowed with a store of innate ideas. He distinguishes between conscious perception and apperception. Our mind is not aware of the numerous and minute insensible apperceptions whose gradual coherence by degree results in conscious perceptions. Without such a process we would have to think constantly about an infinity of things. To explain this chain of cohesion of thought, Leibniz appeals to the law of continuity, holding to the adage *natura non facit saltum*, nature does not leap. Ideas originate not in sensation but reflection. And Leibniz proposes in a conciliatory way that Locke's different notion of reflection may uncover innate ideas and truths. Leibniz praises Locke's rejection of the occult qualities set forth by the scholastics, but is troubled by his acceptance of Newton's centripetal force since, in the fashion of an occult quality, it acts in a place where it is not present. Acting at a distance serves to erode the distinction between the natural and the miraculous or suffers from a deficient causality.

Apparently, Locke saw extracts from the *New Essays* manuscript but disdained them, possibly finding their ideas reactionary, problematic, or a paradox. When Leibniz finished the manuscript in 1708, he learned of Locke's death. Averse to refuting recently deceased authors and believing the limitations of Locke's epistemology were clear, he did not send the manuscript to press. The *New Essays* were published posthumously in 1768 in Leipzig and Amsterdam as part of a German cultural renascence. They appeared as part of a six-volume edition of Leibniz's collected works edited by Louis Dutens.

At the Royal Society of London in the 1660s and 1670s, the experimental program of Robert Boyle dominated. It expressed great confidence in systematic practice-based programs without aspirations to inquiries into causes and effects that could yield the certainty of geometrical demonstrations. Their results were probable and provisional. The early Royal Society considered claims to apodictic certainty in natural philosophy a return to the older dogmatism that it sought to expunge. But in the *Principia mathematica* Newton built rational mechanics on a powerful, new mathematical method. The ancients had two mechanics: the rational based on demonstrations and the practical, that is, the manual arts. Adopting a widely held late seventeenth-century view, probably taken mainly from Isaac Barrow, Newton conceived of mathematics as "founded in [deriving from] mechanics."[540] Human artificers actively construct mathematical objects, such as the straight line, circle, and other curves, reaching differing degrees of accuracy in their work. Geometry, which is considered exact, is simply the perfectly mechanical and part of a higher universal mechanics. In the preface to the *Principia*, Newton rejects occult qualities and Aristotelian substantial forms and insists on subjecting natural phenomena to infinitary mathematical laws.

Newton's causal, infinitary mathematical method, which was central to the eighteenth-century evolution of the modern exact sciences, has both theoretical and experiential components. Distinguishing descriptive mathematical laws from behavior resulting from physical properties of matter, he analyzes forces and motion as derivative effects and from them reaches causes, thereby reversing the starting point of Kepler. Besides satisfying idealized conditions of theoretical cogency, Newton's mathematicized physical constructs must fit increasingly exacting observational and experimental conditions and results in measuring phenomena of the physical world.[541] He thereby sought to obtain a deeper level of mathematical principles for natural philosophy, rather than those of arbitrary systems. Newton's rational mechanics thus approaches the certainty of mathematics. It combines a hierarchy of mathematical and physical causes investigated partially and described by means of Euclidean and Cartesian geometries and fluxional calculus with rigorous experimental verification that is currently within his reach and refinements required by forthcoming greater accuracy. Although he does not draw attention to fluxional calculus in the *Principia,* some of the results were derived by applying it.

In the 1670s, Newton had asserted that the science of light and colors as well as optics depends on physical principles and mathematical demonstration.[542] Extending the boundaries of mathematics slightly beyond astronomy, geography, mechanics, navigation, and optics, he was arguing that sufficient evidence exists in the science of colors to make it a mathematical science and thus subject to mathematical reasoning. To prove its propositions, he employed mixed mathematics and the geometry of light rays.

Discussion in the *Principia* attempts to firm up the distinction between hypotheses and theory. Hypotheses are physical conjectures whose lack of general mathematical descriptions makes them less well founded than theories. The *Principia* aims to present theories, and especially in Book II, although it depends partly on unstated assumptions, urges the elimination of hypotheses. Query 28 of Newton's *Opticks* states that "the main business of natural philosophy is to argue from phenomena without feigning hypotheses, and to deduce causes from effects ...."[543] By the word feigning, he probably did not have today's meaning, faking, but framing in mind.

Newton's mathematical method largely brings to fruition and allows him to exploit suggestive proposals from Kepler's faulty celestial dynamics in *Astronomia nova* and Galileo's kinematics in the *Dialogo* and the *Discorsi*. With them, mechanics shed the perfect accuracy of theoretical geometry and got beyond an arithmetic that merely quantifies. Newton's method of analysis and synthesis leans toward analysis, but without a background in advanced mathematics it was difficult to follow. Locke, who apparently wrote the summarizing review of the *Principia* for the Dutch journal *Bibliothèque universelle*, was unable to understand its mathematical propositions. To grasp its physics without the mathematics, he worked with Huygens and sought to meet Newton.[544] They did so in 1689. Each considered the other his intellectual peer. Newton, who had shunned correspondence in the past, embraced an exchange of letters with Locke on alchemy, critical empiricism, and what he viewed as trinitarian corruptions in the scriptures.

A second obstacle to the acceptance of Newton's mechanics, in addition to its abstruse mathematical formulation, was its concept of the force of attraction acting at a distance that was central to the *Principia*. Seeking to avoid misunderstanding, including the stigma of occultism, Newton represents this centripetal force as a manifest quality on grounds that any opaque cause whose effects can be measured can be so categorized. This denies Leibniz's charge that gravitational attraction requires a miraculous or nonrational cause or is, as Leibniz and the Cartesians claimed, an occult quality, a category that Descartes had expunged from natural philosophy. Centering at the Paris Academy of Sciences, the quarrel between Cartesians and Newtonians over the nature of attraction was to continue for more than a half-century, a debate that Bernard Cohen has found to demonstrate the revolutionary nature of Newton's method.[545] Appealing to the writings of the French astronomer Alexis Clairaut, Cohen defines revolutionary as epoch-making, which signifies an event or circumstance inaugurating a new age or the first major step in a revolution.

16.3   Enhanced Organizational Base and New Journals

As components of an expanding institutional base after 1660, universities offered improved education in methods and significant problems, and a few provided a reflective setting with a well-stocked library, including major recent works. While philosophy, the humanities, and religion dominated the curriculum and mathematics was a lesser subject, a handful of universities in Catholic Paris and Protestant England, Scotland, Holland, and Switzerland added positions in the field. At these institutions mathematics flourished. At Paris's Collège Royal (now the Collège de France) Gilles Personne de Roberval was the Gassendi professor of mathematics from 1655 to 1675, and Jean Gallois, an opponent of the new calculus, taught geometry. Appointed to the new professorship of mathematics at the Collège Mazarin in 1688, Pierre Varignon moved to the Collège Royal in 1704. Up to 1660 the standard mathematical course in English universities had been the elementary portions of Euclid's *Elements,* taught as a structure for rational thought. From mid-century a few English mathematicians studied on the continent. Others examined continental literature and distinguished themselves in resolving the challenge problems that they encountered. At Oxford the Savilian professors John Wallis and Christopher Wren and at Cambridge the new Lucasian professors Isaac Barrow and Isaac Newton did so. In Scotland, James Gregory, before moving to Edinburgh, occupied the new chair of mathematics at St. Andrews from 1668 to 1674. At Switzerland's small University of Basel, Jakob Bernoulli was professor of mathematics from 1687 to 1705, and his brother Johannes succeeded him.

In most of Catholic Europe, caught up in the Counter-Reformation, mathematics did not fare well. Neither Italian or Austrian universities stood out in the field. The University of Vienna was building an important collection of mathematics books and astronomical instruments that Johannes von Gmunden had begun in the mid-fifteenth century.[546] But in imperial Austria the strengths developed in the teaching of mathematics before the Thirty Years War slackened as the Jesuits, who controlled education, gave a lesser place to mathematics in the curriculum, and the monarchy reimposed older restrictive practices on mathematics teachers. University mathematical faculty were not allowed to initiate courses, could only read verbatim from approved texts, and were assigned courses by the drawing of lots, even when the assignments were not in the professor's field.[547]

London was, with Paris, a leader in founding a state science body that fostered the growth of mathematics. The Royal Society of London was chartered in 1662. Preceding it was the Florentine Accademia del Cimento, which by 1660 had been providing a distinguished record of scientific achievement and an organizational model. The Royal Society evolved from smaller groups pursuing the new natural philosophy that challenged older Aristotelian learning. The Merton thesis offers a sociological explanation of the new situation by closely correlating seventeenth-century English scientific growth with Puritanism. In 1649 John Wilkins's club that included Wallis had divided, meeting at Wadham College in Oxford and at Gresham College in London. An Anglican who was loyal to the Commonwealth, Wilkins was trusted and supported by Cromwell and the Puritans, even though he rejected scriptural authority over the natural sciences. In 1660 these clubs founded the Royal Society.[548] Antonia Fraser finds the Merton thesis incomplete for the chartering of the Royal Society. Not Puritan leaders but the Royalist Robert Moray brought the proposal to Charles II, who subsequently sent questions on inventions and laboratory experiments, some of which Charles himself performed.[549] Although English universities blocked it from granting degrees, the Royal Society soon surpassed them in research extending the frontiers of the mathematical sciences.

In Paris the king's chief minister and a perceptive scientific amateur Colbert proposed in 1666 the founding of the Paris Academy of Sciences as a general resource for the glory and utility of the state.[550] For guidance Colbert looked to the Tuscan and London models and to the fluid, private scientific groups in Paris. These included Rohault's Cartesian Wednesdays, Henri-Louis de Montmor's Academy where visual demonstration replaced verbal argument, Charles Perrault's proposed General Academy, and the Compagnie des Sciences et Artes, which supported utilitarian projects and worked with reports of travelers and linguists.[551] The Paris Academy consisted of two sections: the mathematical or exact sciences and the physical and experimental, including anatomy, botany, chemical science, physics, and mechanical inventions. Wednesdays were for discussions of the mathematical sciences and Saturdays for physical topics. Like the Royal Society, the Paris Academy was quickly ahead of the universities in mathematical research.

In the mid-eighteenth century, state science academies in Berlin and St. Petersburg became influential. In 1701, largely through the efforts of Gottfried Leibniz, the Brandenburg Society of the Sciences was founded in Berlin, then a small military town. With no university and little royal or noble appreciation, it remained moribund until renovated in the 1740s. In 1725 the St. Petersburg Academy of Sciences started as a European rather than a Russian institution. Its members came from the Germanies, Switzerland, and France. Chief among them during its first five years were the senior members Jakob Hermann and Joseph Delisle and the younger Daniel Bernoulli and Leonhard Euler.

In the pursuit and acquisition of royal financial support for mathematics and natural philosophy, the four academies were not unprecedented. What made this situation distinctive were the level of state funding and the growing recognition of these academies as instruments of state power. One indicator of the change was that each academy also had a new observatory.

Not yet fully collaborative, both the Royal Society and the Paris Academy aimed nevertheless to compile comprehensive collections, observations, and experiments and to consolidate these into great storehouses of knowledge. These extensive bodies of information were to lay the evidentiary groundwork for building theories. Neither set a limit on the range of its physical research, though the Paris Academy had more teamwork.

The early Royal Society principally followed the inductive procedures of its leading member, Robert Boyle, and the Paris Academy, adopted essentially the method of Francis Bacon. The early Royal Society was larger, having one hundred fellows from the nobility, gentry, professions, and universities, many with only a passing interest in science. Their weekly meetings lasted from three to five hours. The early *Philosophical Transactions* give a major place to mathematical subjects, such as the nature and observed positions of comets, the satellites of Jupiter, sunspots, lunar motion, and optical lenses, along with rules of motion and acceleration. They also consider such diverse materials as the vacuum, plants sensitive to touch, ant eggs, animals, trades, coffee, and beer. Fellows described particular, observed experiences rather than resorting to Aristotelian universal statements. They challenged theories and devised new apparatus for experiments and observations. A small group, including Christopher Wren, John Flamsteed, Edmond Halley, and Robert Hooke, concentrated on mechanics and astronomy. In their methodology, Boyle's suspicion of mathematics did not go unchallenged. At this time Isaac Barrow accorded mathematics a central place in the sciences. Barrow's view and Boylean induction were later to converge. A prominent critic of Boylean procedures at the early Royal Society was Thomas Hobbes, who was not named a fellow. Although Hobbes gave systematic collecting a role in natural history, in accordance with a criterion from Aristotle he said that collecting fell short of natural philosophy because it yields no secure knowledge of causes.[552]

The mathematical sciences favored by Colbert dominated the early Paris Academy. Accordingly, astronomers, geometers, and mechanists were twenty-two of its first thirty-six members. They included Christiaan Huygens, Pierre Carcavi, Edme Marriotte, and the Italian astronomer Gian Domenico Cassini (1625–1712). Huygens and Cassini, who were effectively directors, received handsome stipends of 6,000 livres. The imposing Paris Observatory, designed by Perrault in 1667 and completed in 1672, confirmed the Crown's commitment to the sciences. It was to become Europe's center of excellence in astronomy. There astronomers studied eclipses, sunspots, the parallax of Mars, and the satellites of Jupiter and Saturn, and measured arcs of meridian, labored to determine longitude more accurately, and assisted with cartography. Like the Royal Society, the Paris Academy opposed the ill temper, pointless debates, and authoritarianism in universities that hindered progress in the sciences, insisting instead during meetings on civility, good order, and conduct becoming to a civic gentleman of intellectual curiosity. Clearly questions of the merits of new ideas, claims to priority, egos, and dyspeptic tempers produced continuing combative exchanges among scholars in the sciences.

In funding, the Royal Society and the Paris Academy starkly contrasted. Chronically short of income, Charles II gave the Royal Society privileges but no funds. The Society's privileges included the right to appoint a printer, to correspond abroad, to build a college, and to issue warrants for bodies of executed criminals for anatomical studies. Fellows who lacked personal fortunes were obliged to hold other positions. Until the great plague they continued to meet at Gresham College. In 1664 Charles II founded the Royal Observatory at Greenwich, designed by Wren, but the astronomer royal John Flamsteed had to outfit it with his own equipment and gifts from friends.

In Paris, the Crown funded the regular members, who were called pensioners, provided housing, and equipped laboratories, but did not support student assistants. Pensioners called themselves "the company." Most came from Paris or other cities in the north of France, where literacy was high. Scientific as well as professional talent flowed to the capital from the provinces. Colbert employed large pensions to draw from abroad a few eminent scholars, such as Huygens and Cassini. The state equipped the Royal Observatory and funded expeditions to the Canary Islands, Cayenne, and China. Under Colbert, who to 1683 spent an average of 87,700 livres a year on the Academy, it thrived, but under Louvois, who to 1691 annually spent on average only 26,484 livres, it waned.[553]

Improved instruments were another crucial component to advancing research in the mathematical sciences. Better barometers, clocks, compound microscopes, air pumps, telescopes, glass tubes, and surveying equipment were now produced. In Paris, Berlin, and St. Petersburg the royal state purchased these. In England, the Royal Society's first curator of experiments, Robert Hooke, added the spring-driven watch, the wheel barometer, and the cross-hair sight and screw adjustment for reading settings directly from the telescope. He typically demonstrated these to edify and entertain the fellows at meetings of the Royal Society. For so skillfully enhancing practically all scientific instrumentation and effectively bringing it to bear on scientific research, the French astronomer Joseph-Jérôme Lalande in *Traité d'astronomie* of 1764 called Hooke "the Newton of mechanics."

More open publishing practices now developed that lifted much of the secrecy that had inhibited the communication of the latest scientific findings and underlay the move from private to public science. Up to 1660 no journals in mathematics, the sciences, or substantial related materials existed. Books and correspondence transmitted scientific results. But the mails were unreliable. Without specialized journals, European geometers and natural philosophers had no reliable means for rapidly communicating scientific results, setting standards, fashioning public science, and building a scientific community. Three important scientific journals appeared: the *Philosophical Transactions* of the Royal Society, the *Journal des sçavans*, and *Acta eruditorum*. They were a slightly delayed response, since neither the Royal Society nor the Paris Academy had possessed a journal at its inception. The Royal Society was more open about communicating results. The voluminous correspondence seeking new information by its intelligencer Henry Oldenburg, together with the reports of fellows and a desire for notices of new books, was responsible for beginning the *Philosophical Transactions* in 1665. The Paris Academy had initially been closed to outsiders, conducting its research in secret. But in 1667 it started to share announcements about its work in the *Journal des sçavans*. This weekly, founded in 1665 in time to include an obituary of *"ce grand homme"* Fermat, became a model for the sciences by presenting for specialists and the lay reading public detailed results of experiments and matters of general scientific interest. Even after the Paris Academy adopted it, the *Journal* had to suspend operations occasionally for lack of funds. Having enough separately printed reports, in 1676 the Academy began to issue collections of them in a *Recueil de plusiers traitez de mathématique de l'académie*. From 1701 its annual *Histoire* of articles appeared. In 1681 Otto Mencke of Leipzig discussed with Leibniz founding the journal *Acta eruditorum Lipsienium* (*Proceedings of the Scholars of Leipzig*) to keep German scholars abreast of new books and to publish learned articles in the sciences. Begun the next year, it regularly ran Leibniz's contributions under his initials G. G. L. Geometers and natural philosophers could also submit papers to a few new philosophical journals, such as the *Bibliothèque universelle*, the *Mémoires des Trevoux*, Pierre Bayle's *Nouvelles de la république des lettres* in the Netherlands, later renamed the *Histoire des ouvrages des savans*, Wilhelm Tentzel of Hamburg's *Monatliche Unterredungen*, and Modena's *Giornale de' Letterati*.

16.4   Selective, Transmuting Synthesis and Genius: Highlights

During the century after 1660, mechanics, celestial dynamics, and mathematics were to be the sciences *par excellence*, followed by optics, matter theory, and nascent energy theory. Cartesian science, as refined by Malebranche and Rohault and professed in Holland, France, and England, preempted varieties of academic Aristotelian and scholastic thought. It encompassed a corpuscular theory of matter, laws of impact of hard bodies, and a vortex cosmology in a plenum universe. Despite this initial success, Huygens and Leibniz quickly rejected the laws of Cartesian mechanics, in part questioning the Cartesian principle of the conservation of momentum. Cartesian science was further undermined, but not yet toppled, at the Paris Observatory, where the Danish astronomer Ole Rømer (1644–1710) calculated the finite speed of light. Even with his rather accurate approximation, the traditional assumption from Aristotle to Descartes of the instantaneous transmission of light persisted. Galileo's *Discorsi* had raised the question of whether the speed of light is finite. Besides Cartesian thought, major scientific efforts were directed toward refining demonstrations of Kepler's three laws or a close approximation, building on Galileo's program in mechanics, and probing in the theory of matter the implications of Pierre Gassendi's revived Democritan atomism. The effort produced Newton's *Principia Mathematica*, which rejected the Cartesian vortex cosmology, extended the work of Kepler and Galileo, and gave new direction to mechanics, celestial dynamics, and mathematics. On another front, the Bernoullis investigated elastic collisions and the foundations of physics. Late in the era, Newton's emissary Samuel Clarke debated Leibniz over whether space, time, and motion are absolute or relative, and a polemic between Cartesians and Newtonians, centering at the Paris Academy, was growing.

Pendulums, Monads, and the Speed of Light

Christiaan Huygens, who until Newton's *Principia mathematica* would be considered the foremost natural philosopher and mathematician in Europe, was a member of a prominent, erudite Dutch family. He was introduced as a youth to Cartesian geometry and physics. His father Constantijn corresponded with Descartes, who was often a guest at his house. From 1645 to 1647 Christiaan studied at Leiden under Franz van Schooten, who taught ancient mathematics, the modern methods of Viète, Descartes, and Fermat, and Cartesian physics. Van Schooten recommended that he study as well Girard's edition of Simon Stevin's mathematics. Against Girard and Juan Caramuel Lobkowitz, who held distances of fall to be proportional to times elapsed, young Huygens confirmed in 1646 Galileo's times-squared law of acceleration. He showed that distance fallen forms "the arithmetical sequence 1, 3, 5, 7, 9 ... and thus the total distance fallen is as 1, 4, 9, 16 ...."[554]. Soon after he acquired a copy of Galileo's *Discorsi*, containing that law and material on the parabolic paths of projectiles. Young Huygens had planned to write a treatise on his discoveries on motion, but upon reading the *Discorsi* declined, saying "I did not wish to write the *Iliad* after Homer had."[555]

The father wrote to Mersenne of his son's confirmation of Galileo's acceleration law. Mersenne began to suggest to Christiaan topics for investigation, such as Torricelli's work on the cycloid, a vibrating body's center of oscillation, the motion of a nonideal pendulum, and an exact measurement of the rate of free fall using a pendulum, a topic he did not turn to for a dozen years. Christiaan asserted that following the precepts of Galileo and questions of Mersenne offered "a better way to the study of motion."

Christiaan Huygens

From 1650 to 1666, Huygens worked at his family home in The Hague. While displaying universal interest in the sciences, he contributed especially to mathematics, where he rigorously improved a geometric form of infinitesimal mathematics and applied it more widely to the study of nature. His adept geometric insights yielded fruitful techniques and results, such as reducing the ratio of two auxiliary parabolas to the same ratio with a circle and finding that the cycloid is the isochronous curve for the pendulum clock. Huygens' discovery of complex and fruitful geometric relationships, many of which algebraic representations would have masked, was invaluable for the creation of calculus. The claim that his strictly geometric focus probably blocked him from making innovations that Leibniz and Newton were soon to formulate seems inadequate. Huygens also fundamentally advanced mechanics, chronometry, astronomy, hydrostatics, and geometrical optics. In Paris he met Gassendi and Roberval in 1655 and Desargues and Pascal in 1660–1661. During a visit in 1661 to London's Gresham College, he met Brouncker, Oldenburg, Wallis, and Wren, giving an account of his book *Systema Saturnium* and his pendulum clock, and he followed Boyle's air pump experiments. In both Paris and London he encouraged founding permanent state science institutions. Not surprisingly, in 1663 the Royal Society elected him a member and Colbert granted him 1,200 livres for his mathematical achievements and invention of the pendulum clock. In an audience with the minister and king, Huygens pledged to dedicate to Louis his forthcoming *Horologium oscillatorium*, which he planned to release after sea trials of his clocks. Three years later the Paris Academy of Sciences offered him membership, an ample pension of 6,000 livres, and living quarters in the Royal Library.

Protected by Colbert, Huygens was until 1681 essentially the director and premier member of the Paris Academy. He organized its extensive Baconian program of research in physics and collaborated closely with research in astronomy. In 1681 ill health forced his return to the Netherlands. The death of Colbert and the worsening position of Protestants in France caused him to remain to his death in Holland, mostly in The Hague. At a meeting in 1689 of the Royal Society, which Newton attended, Huygens delivered a paper on his pulse theory of light and another on the causes of gravity, planning to publish both soon. He and Newton met twice more to discuss optics, colors, experiments made with thin films, and probably the *Principia*. Huygens promised to send a copy of his forthcoming *Treatise on Light*.

Meanwhile, in 1652 Huygens had begun investigating in mechanics the laws of collisions of elastic bodies. In his characteristic search to quantify precisely a problem and to structure results, especially from a study of Archimedes, Galileo, and Mersenne's *Harmonie universelle,* along with his own experiments, he perhaps grew dissatisfied with the verbal and qualitative Cartesian physics, if he had ever been that accepting of it. From more accurate experiments, Huygens concluded that Descartes' *Principles of Philosophy* is filled with "fictions as so many truths," substituting a "pleasant romance for a true history." Huygens's *De motu corporum ex percussione* of 1656 refutes the seven Cartesian *a priori* rules on the collision of bodies. Lagrange's *Mécanique analytique* of 1788 agrees that Huygens's main goal was to build on Galileo.[556]

Huygens, who derived the laws of impact from Galileo's law of inertia, agreed with the Cartesians that motion can be defined only relatively to given points, a view that Newton was later to reject. Huygens discovered the centrifugal tendency (or disjoining *conatus)* and its laws. Centrifugal force is the outward force exhibited in the pull on a rapidly rotated body, as when a bucket of water is swung around on a rope in circles. It is the force that flings the bucket in a straight path when the rope is released. Huygens considered this physical swirl to be opposite to the centripetal tendency for heavy bodies, and he fitted it into a Cartesian vortex explanation of gravity. His assumption that a mechanical system's center of gravity cannot rise until forced suggests recognition of the conservation of energy. The Royal Society asked Huygens together with Wallis and Wren to clarify the laws of impact. In 1668 he presented without proof the laws for elastic collisions. Searching for an entity for a conservation law, he rejected Cartesian *quantitas motus*, or roughly momentum, that is, quantity of matter times the absolute value of the velocity, $m|v|$. Probably he considered it far removed from an intuitive conception of motion and rejected its quantification. From his study of impacts in elastic collisions, Huygens settled instead on the conservation of what Leibniz was to name *vis viva*, mv^2. But Huygens disagreed with Leibniz's claim that *vis viva* is, like attraction, a fundamental force in nature. His discoveries in mechanics appeared in his major writing, *Horologium oscillatorium (Pendulum Clocks, or Geometric Proofs Relating to the Motion of Pendula Adapted to Clocks)* of 1673. Its dedication to Louis XIV improved his support in Paris, but drew ire from the Dutch republic, which was at war with France.

Skilled in grinding lenses and reducing optical aberrations, Huygens along with his brother had constructed from 1655 the best telescopes for prescribed lengths. Using a 92**X** telescope, the most powerful of its day, he studied Saturn, which through weaker telescopes appeared to be three joined stars. Huygens discovered a satellite or moon of Saturn, later named Titan, and explained that the "arms" of Saturn are not adjacent stars but a ring. In 1659 Huygens dedicated his *Systema Saturnium* to the founder of the Accademia del Cimento, Leopold d'Medici. When Honoré Fabri, a corresponding member, disputed its findings,[557] Leopold asked which position was more tenable. Cimento members built mechanical models but published no conclusion. After five years of debate, Fabri accepted Huygens's explanation.

Among the scholars with whom Huygens now increasingly corresponded on curves and the pendulum were Gregory of St. Vincent, Sluse, van Schooten, and Wallis. Employing the swings of a pendulum as regulator, Huygens in the winter of 1656–1657 made a more accurate mechanical clock. This differed from the existing cogwheel clocks in which balances controlled movement. The sources of motive power of cogwheel clocks were often unreliable. Huygens's clockwork applied Galileo's discovery of isochronicity of oscillations of pendulums of fixed length; that is, irrespective of the their arc length pendulums have a near uniform rate of swing even as they are dampened. Within a short period of time this mechanical phenomenon thus repeats itself. Through an armature that Huygens cleverly constructed with slanted teeth for nudging the pendulum rhythmically, feedback from the oscillation determined time intervals. The curve traced by the pendulum bob is generally called the isochrone, but sometimes the tautochrone after Leibniz defined the isochrone as the semicubical parabola. Huygens discovered an error in Galileo: oscillations in circular pendulums are only isochronous for small angles of deviation from the vertical. To correct this, he introduced curved plates that effectively shortened the pendulum's length as it swung widely. His pendulum model became popular: it was employed, for example, in the tower clock at Utrecht.

Seeking to make it a precision instrument, Huygens steadily perfected the clock. He initially kept stable the amplitude of swing by making it small. But even a minor disturbance could stop such a clock. To have a tautochronous clock keeping accurate time, Huygens had to eliminate the period depending upon amplitude. This led him to investigate by infinitesimal geometry paths of the pendulum bob. Treating them either as a semicircle or as a parabola at its lowest point was unsatisfactory. From his study of normals and tangents to curves and participation in Pascal's cycloid competition in 1658, Huygens determined that the path of the bob that describes an exact condition rather than a rough approximation is the cycloid. He rigorously proved this his "most fortunate finding" in an Archimedean fashion. Thus, to construct an isochronous pendulum, he had to find what cheeks or side supports of the pendulum string will cause it to move the bob along in a cycloidal path. Cheeks consisting of the intersection of arcs from two cycloids, he conjectured, will do so, keeping it an isochrone, a theorem proved by Christopher Wren.

Huygens set out to prove that his isochronous pendulum clocks would make it possible "to measure longitude even if this is to be done at sea," a consideration of paramount importance in seafaring Holland and carrying possible economic benefits.[558] To this point, the longitude problem had seemed intractable. A first sea trial using the pendulum clock by his brother Lodewijk on a trip to Spain in 1660 failed during a severe storm that damaged five Dutch merchant marine ships. Huygens realized that not the pendulum clock but its cousin, the cycloid clock, had both the accuracy and allowed for large amplitude at sea. Earlier timepieces, such as Brahe's astronomical clock that had to be adjusted daily by the blow of a hammer, were not sufficiently accurate. For two years, Huygens extensively refined the cycloid clock to make it seaworthy. To keep it from being affected by rough seas, he built an apparatus to support the undisturbed vertical suspension of clocks on ships, and he wrote an instructional pamphlet for pilots on determining longitude by using clocks. Forced into a partnership with the British Alexander Bruce and supported by the Royal Society, Huygens obtained a patent for the new marine clocks. A sea trial in 1662 again failed. One clock crashed to the deck of a tossing ship and another stopped completely. Huygens was becoming depressed. But sea trials from London to Lisbon in 1663, and another by the English fleet to the south

Atlantic, including the Cape of Good Hope, ending two years later showed that cycloidal clocks worked well. The British captain Richard Holmes made exaggerated claims for discovering an island using clocks.[559] Under Moray, the Royal Society investigated whether the length of the cycloidal pendulum clock should be a universal measure. But Hooke criticized Huygens' measure of the constant of free fall from gravitational attraction by employing the swing of the pendulum, improving upon Riccioli's 15 feet/second, and the belief in the easy synchronicity of clocks, while British scholars and reports frrom French vessels supported by the Paris Academy showed the pendulum clock to be too slow for measuring longitude.[560]

Propositions 7 through 9 of Part III of the *Horologium* treat the rectification of various types of curves, a topic of all major mathematicians of the period. Before the mid-seventeenth century, rectification had been thought impossible. Aristotle and Averroes had denied that a rational proportion could exist between a curved and a straight line. But Archimedes' axiom for convex plane figures in *On the Sphere and Cylinder* provided a base for later rectifications: "that which is included is the lesser."[561] Huygens's work on rectifications largely exceeded the boundaries of Cartesian geometry and opened up a thorough study of transcendental problems. Shortly after 1657, William Neil, Hendrik van Heuraet, and Fermat rectified the semicubical parabola, $y^3 = ax^2$, and Wren the cycloid in the Pascal competition. In his search for evolutes, the envelope of normals to a curve, Huygens had found the evolute of a parabola to be the semicubical parabola. Fellows of the Royal Society challenged Huygens's claim in the *Horologium* to priority for rectification, defending the claim of Neil of Gresham College, by then deceased, and observing that Wren's results were published in 1660. Wallis now broke off correspondence with Huygens.

Chonometry also produced heated arguments. Although the last part of *Horologium oscillatorium* gives without proofs thirteen propositions on centrifugal force intended to justify theoretically a second timepiece, the cycloid clock, the isochronous path of whose bob follows from his theory of evolutes, the book principally examines and popularizes single and compound pendulum clocks, along with balance springs, rather than pendulums as regulators. It introduces the concept of the center of oscillation. Huygens makes two assumptions in computing the center of oscillation. He assumes that when it is solely influenced by gravity a system's center of gravity cannot rise and also that without friction a system's center of gravity that descends from a given height will, when parts are directed upward, return to the initial height. Galileo's students claimed priority for the pendulum clock for their teacher and his son Vincenzio Galileo, for himself. Galileo had suggested using pendulums *per se* to keep time. Among others, Mersenne and Riccioli and before them Leonardo da Vinci had made a similar proposal. Historians Leopold Defossez and Bert Hall defend Huygens' priority, observing that he had to overcome significant conceptual and technical hurdles in transposing the theoretical pendulum into a reliable physical mechanism.[562]

In 1673 and 1674 Huygens designed his first balance-spring clock, which brought another priority dispute. Colbert agreed to have the clock made secretly and granted a patent. But the clockmaker, Isaac Thuret, claimed partial credit for the invention. Huygens had to threaten legal action. In January of 1674/75 he sent an anagram to Oldenburg of the Royal Society to establish his priority. After Thuret revealed the secret inner workings of the balance spring clock, Huygens also sent a letter and brief description scheduled for publication in the February 1675 edition of the *Journal des sçavans*.[563] Although he would not show its inner workings, Hooke had a clock constructed and alleged that Oldenburg and the Royal Society had circulated to Huygens his earlier ideas on clocks. It was true that Oldenburg had transmitted a few allusions to them. The Royal Society's President Brouncker asked Huygens to submit a balance-spring clock. Huygens had three made. One for Louis XIV and another for the Dutch stadholder William III worked well. But the third, sent to England, kept stopping. Hooke increased his attacks on Huygens and likened Oldenburg to a French spy. Huygens and his diplomat father vigorously protested these allegations. The Royal Society reprimanded Hooke and in the *Philosophical Transactions* for November 1676 supported its secretary Oldenburg.[564]

In the autumn of 1672 Huygens, probably at his Royal Library apartments, met the young Saxon scholar and diplomat Leibniz, for whom Oldenburg had sent a letter of commendation.[565] Leibniz wanted to discuss the principles of motion and his work on summing infinite series. Huygens asked him to read Wallis's *Arithmetica infinitorum* and Gregory of St. Vincent's *Opus Geometricum* and posed a challenge problem of summing the reciprocals of triangular numbers: $S_n = 1 + 1/3 + 1/6 + 1/10 + 1/15 + \ldots$. Huygens had obtained the answer by systematically summing the reciprocals of other polygonal numbers. But Leibniz cleverly divided S_n by 2 and regrouped his terms, so that $S_n/2 = (1 - 1/2) + (1/2 - 1/3) + (1/3 - 1/4) + \ldots = 1 - 1/2 + 1/2 - 1/3 + 1/3 + \ldots = 1$, and so; S_n for n going to infinity is 2. He wrote up for the *Journal des sçavans* his results with lemmas, but from late 1672 to 1674 it suspended operations. If Huygens had any questions about Leibniz's ability, this experience convinced him otherwise.

Born in the Protestant city of Leipzig in 1646, two years before the end of the Thirty Years War, the universal genius Leibniz was to be a progenitor of modern German idealism and a principal architect of the modern sciences. He made original contributions to fields ranging from mathematics, metaphysics, history, jurisprudence, and linguistics to physics, psychology, and theology. As systematized and extended by his disciple Christian Wolff, his rational philosophical and scientific ideas dominated German intellectual life until displaced by Immanuel Kant's critical philosophy in the 1780s.

The son of Friedrich Leibniz, a professor of moral philosophy, and Catharina, née Schmuck, Gottfried was a precocious youth, attending Nicolai school from 1653 to 1661, but was largely self-taught. Studying in the library of his father, who died in 1652, by 1661 he was fluent in Latin and able to stammer in Greek. This linguistic preparation enabled him to read classical and Hellenistic Greek and Roman literature and the early Church fathers. Aristotelian logic and Lutheran theology now fired his imagination.

In 1661 Leibniz entered the University of Leipzig, where Jacob Thomasius was his advisor and teacher. Thomasius disdained both academic scholasticism and the modern thinking of Bacon, Hobbes, Galileo, Gassendi, Grotius, and Descartes. The original Aristotle he held to be superior. In the summer of 1663 at Jena, Erhard Weigel persuaded Leibniz to seek to reconcile Aristotelian and modern or anti-Aristotelian thought on the basis of a Euclidean axiomatic model. Subsequently Leibniz closely studied Plato and perhaps even more Plotinus, working to accommodate them with modern thinkers. After receiving the bachelor's degree at Leipzig in 1664, he worked on a master's degree. Two years later he applied to enter studies for the degree of doctor of law. Probably since he had not completed a five year law internship, he was refused admission. In October 1666 Leibniz left Leipzig and enrolled at the University of Altdorf, near Nuremberg. Altdorf granted him the doctorate the following February for his dissertation *On Difficult Cases in Law*. Desiring to devise major reforms in the law, politics, religion, and the sciences that could not be accomplished within the confines of a university, he declined a post at Altdorf.

Like Cicero and Francis Bacon, Leibniz became a courtier. Shy but friendly, a voracious reader, didactic, ever curious, and filled with personal memories and political plans, he embarked on a busy adult life. After serving as secretary to a Nuremberg alchemical society, Leibniz in 1667 was appointed judge in the high court of appeal of the elector-archbishop Johann Philip von Schonborn of Mainz. He labored to codify Roman civil law and apply the result to German states. When Louis XIV planned to attack the Germans, they proposed by custom a diversionary tactic. Leibniz hoped to entice Louis to capture trade with the Orient and build a Suez canal. Sent to Paris, the young diplomat never gained an audience with the king. But from 1672 to 1676 he resided in Paris, mastering the French language, the second among the aristocracy and the republic of letters across Europe, studying with Huygens, learning from Nicolas Malebranche's circle the latest developments in Cartesian thought, and corresponding about logic and monads with the difficult Antoine Arnauld. After building a calculating machine that improved upon Pascal's, he was unanimously elected in 1673 a fellow of the Royal Society. His calculator, which multiplied by repeated addition and divided by repeated subtraction, was intended to free men from losing "hours like slaves in the labor of calculation." Leibniz was also now inventing differential calculus. Even with this achievement, no academic or governmental position existed for Leibniz in Catholic Paris in 1675. This was probably in part because he was a Lutheran. In addition, the Paris Academy had two prominent foreign members, Huygens and Cassini, and apparently there was some opposition to adding a third to a leadership position.

In the summer of 1676 Leibniz's years of academic preparation were completed and he embarked on a new phase: the development of his mature ideas and a defense of them in spirited polemics. That October he reluctantly accepted from the Duke of Hanover the position of privy counselor of justice, librarian, and minister without portfolio. On his way to Brunswick, he stopped in Holland. In Amsterdam he met the microscopist Jan Swammerdam and the burgomaster and mathematician Jan Hudde (1629–1704). A disciple of Schooten, Hudde had devised two rules to find double roots of equations, one of which slightly alters Fermat's theorem on maxima and minima. Along with Mengoli, he also used infinite series to obtain the quadrature of the hyperbola. Leibniz was then improving upon Mercator's work on that quadrature. Proceeding by boat to Delft, he met Antoni van Leeuwenhoek, who had discovered a microscopic world filled with minute organisms or "living atoms." This work influenced Leibniz's evolving concept of animate monads. In The Hague, he met Spinoza, who was completing the *Ethics*. Leibniz found statements in Spinoza's pantheism that he judged false or contradictory. He argued that Spinoza supported immortality and providence only in words, not in conviction.

Leibniz was to develop his metaphysics and cosmology, which rest upon his monadic doctrine, a theory of primal substance. In the late seventeenth century, two major theories of elemental matter existed: Descartes' concept of corpuscular extension and Gassendi's of passive, impenetrable, and indivisible atoms. From a study of cohesion in the 1670s, Leibniz concluded that primal substance could not be Gassendi's atoms with their hooks and eyelets. By the time of his *Discourse on Metaphysics*, he was rejecting as materialistic the concept of mere extension as the basic constituent of matter. Radically revising the view of what ultimately constitutes matter, Leibniz praised Plato for correctly maintaining that primal substance must be active. The *Sophist*, *Phaedo*, particularly Cebes' description of the enduring force in the soul, and perhaps Plotinus' panpsychistic *Enneads* were instructive. Monads possessing principles of action, that is, directive entelechies, were Leibniz's primal substance.[566] The term "monad" he likely borrowed from the Cambridge Platonists Henry More and Ralph Cudworth.

Leibniz's doctrine of monads appears in his "New System of the Nature and Communication of Substances" and is revised in the *Monadology*, written in Vienna in 1714.[567] In the earlier work, the monads are metaphysical points of energy; in the latter, geometric points of energy without parts. In either case they comprise the continuum, which Leibniz views as a labyrinth. At its deepest level, the physical world consists only of animate monads. Material bodies have no place in Leibniz's fundamental ontology. Aggregates and results of monads make up extended bodies that move according to the laws of mechanics. That spatial extension results from geometric point monads poses a dilemma for Leibniz. He observes that the least physical minima still consist of an infinity of monads. Georg Cantor's later concept of a nondenumerable infinity might have given greater assurance. Allowing animate monads a formative role in extension produces the system of panorganicism, whereby the universe is a plenum and nature is filled everywhere with organic creatures.

Building on the doctrine of monads, Leibniz attempted to found a new science of dynamics. It explicitly opposed Cartesian physics. After exploring Descartes' laws of reflection and refraction in early pieces in *Acta eruditorum*, in the article "Brevis Demonstratis Erroris ..." ("Short Demonstration of a

Remarkable Error in Descartes"), Leibniz in 1686 struck at the principle of the conservation of momentum that lay at the heart of Cartesian physics. This essay asserts that God conserves another force in the universe. Descartes had estimated the force of bodies in motion to be the quantity of matter, or roughly mass, times the absolute value of the velocity. From Leibniz's study of pendulums and the force required to raise a body to a certain height comes the conclusion, in accord with his mentor Huygens, that the actual measure of force conserved is mv^2, which Leibniz names living force. He goes beyond Huygens in asserting that mv^2 is not simply a mensurational convenience but a fundamental force in the universe. Its conservation law unifies his formative analytical mechanics. Motion, Leibniz contends, arises from the appetites or strivings of monads. A momentary state, it arises from derivative forces, the two chief ones being dead force and living force. So began a sixty year dispute, before Jean d'Alembert showed that both sides are right. Newton's force (F) acting over time gives $mv = Ft$, and acting over space it gives $mv^2 = 2Fs$. After carefully reading and annotating Newton's *Principia* and criticizing its concept of attraction acting at a distance,[568] Leibniz wrote in 1695 the fullest account of his new dynamics in *Specimen dynamicum* and in 1714 gave a short version in *Principles of Nature and Grace*. Leibniz's physics based on energy and the mathematical continuum was to contrast sharply with Newton's atomistic model.[569]

Spending the remainder of his career from 1676 in Hanoverian service, Leibniz had begun in part by corresponding with Bishop Bossuet and Leopold I in an effort known as irenicism to reunify the Christian churches. In this many Lutheran and Calvinist clergy opposed him. Leibniz built up the natural sciences in Hanover, purchasing for the ducal collection such private libraries as Martin Fogel's 3,600 specialized volumes, and discussing with Herman Brand the discovery of phosphorous from distilled urine. When he exposed a deception that alleged to turn sand into gold, he gained the enmity of the alchemist Johann Joachim Becher. For a prediction that a carriage could some day travel from Hanover to Amsterdam in six hours, Becher ridiculed Leibniz. In politics, Leibniz confirmed primogeniture as the rule for succession and gained in 1692 the elevation of Hanover from a duchy to an electorate of the Holy Roman Emperor. The Hanovers were the ninth elector.

His growing reputation notwithstanding, Leibniz's position in Hanover eroded under each new ruler. Duke Ernst August succeeded Johann Friedrich in 1679, and the illiterate Georg Ludwig became elector in 1698. They considered Leibniz's work in mathematics and natural philosophy a distraction from his official duties. In 1679 Leibniz unsuccessfully sought positions as imperial librarian in Vienna or a pensioner of the Paris Academy of Sciences. Leibniz believed that French and Dutch historians, particularly Jean Mabillon, a critical student of primary sources, had made history and genealogy a science. He proposed composing an accurate family history of the Hanovers, tracing them back possibly to the ancient Romans. A thorough investigation of legal precedents would improve public law, and documented knowledge of the Hanover genealogy, unlike self-serving histories of past courtiers, could be crucial in adjudicating the family's claims to lands. In 1685 Ernst August commissioned Leibniz to be historian of the Hanoverian House of Brunswick-Luneburg. Preparing the *Brunswick scriptores* was his main court assignment thereafter. Armed with letters of introduction, he undertook in 1687 a grand tour of southern Germany, Austria, and northern Italy seeking primary sources and artifacts, including inscriptions in cemeteries. In 1691 Leibniz also became director of the Wolfenbüttel Library near Hanover. He required the compiling of an alphabetical listing of all its authors, which was completed in 1699.

Gottfried Wilhelm Leibniz

The energetic polymath Leibniz also endeavored to construct a strong state institutional framework for the sciences in central Europe and Russia. To Leopold I in Vienna and to Peter the Great of Russia, whom he met in Brandenburg, Leibniz proposed a royal science society. But the most promising locale was Berlin, whose learned Queen Sophie Charlotte had been his student. In 1697 she authorized construction of an observatory, partly to address calendar reform. When Georg Ludwig became elector the next year, he refused his mother's request to make Leibniz vice chancellor and attempted to curb his travels. Only after several requests from his sister did Georg Ludwig permit Leibniz to go to Berlin. The court chaplain Daniel Jablonski had drawn up a plan for the electoral science society. Based on the Florentine, London, and Paris models, it was to pursue abstract learning or *curiosa* but give more attention than they to *utilia* improving agriculture, manufacturing, industry, and commerce. Leibniz, who was elected a foreign member of the Paris Academy in February 1700, revised the plans. He objected to naming the new society an academy, since that smacked of scholasticism. The Royal Brandenburg Society of Sciences, later renamed the Berlin Academy, was founded on July 11, the Hohenzollern elector's birthday. A grand masquerade was held. Leibniz was invited to attend as an astrologer wearing a tall peaked hat, but declined the honor. He was named president for life.

To keep in the forefront of research in law, history, mathematics, natural philosophy, philosophy, and politics, Leibniz engaged in a massive correspondence and traveled widely. After 1700 he exchanged letters with Bayle and Locke on method and Joachim Bouvet on the Jesuit mission to China. Among his scientific correspondents in Groningen was Johann Bernoulli, in the German states Christian Wolff, and in Paris Pierre Varignon, who discussed planetary motions and the calculus dispute between Joseph Saurin and Michel Rolle. From Basel Jakob Bernoulli corresponded with him on calculus and Jakob Hermann on algebra. Leibniz also asked Fontenelle, the secretary of the Paris Academy, about corrections in Kepler's *Rudolphine Tables* and Cassini's and de la Hire's modifications of Kepler's ellipses for planetary orbits, and he described the superiority of his mathematics over Cartesian geometry and the work of his own collaborator on calculus, Walther von Tschirnhaus. In 1701 Fontenelle invited a paper for the new annual *Histoire* of the Paris Academy. Leibniz had worked with the Court mathematician in Berlin, Philippe Naudé, who had taken great pains to construct a table in binary notation of the natural numbers up to 1023. His "Essay on a New Science of Numbers," which proposed that binary numbers are better for representing transcendental numbers than are terms in infinite series of rational fractions, was primarily a work in number theory, examining periods in its table as subjects of algebraic operations.

As relations between Hanover and Brandenburg-Prussia worsened, the members of the Brandenburg Society of Sciences elected a director in 1710, effectively removing Leibniz as president. In the simmering, decade-long quarrel over who invented calculus, the Royal Society recognized in 1713 the priority of its president Newton. More troubling that year was Roger Cotes's preface to the second edition of Newton's *Principia*, associating the Cartesian and Leibnizian plenum with fatalism and atheism, a charge that Leibniz set out to refute. In 1713 Leibniz was in Vienna discussing the close of the War of the Spanish Succession. The young Emperor Charles VI agreed to found an imperial science society and nominated him for president. To Prince Eugen, Leibniz sent a report on the importance of the telescope, microscope, and magnetic compass to making progress in knowledge and of the printing press to its transmission. The prince endorsed the idea of the new institution. Based on the Brandenburg Society, the imperial society was to have an observatory, laboratories, botanical gardens, and a zoo. A calendar monopoly and a tax on stamped paper on all except the poor were to finance it. Leibniz urged the establishment of a commission to found it. But the War of the Spanish Succession had depleted the Habsburg treasury of Charles VI and the Jesuits opposed the idea.

When letters arrived for Leibniz to return to Hanover, he wrote that Vienna was experiencing a plague and he did not want to carry that home with him. Suddenly in August 1714 Leibniz received news that Queen Anne of England had died and that Georg Ludwig had become George I. Hoping to accompany George to England, Leibniz set out for Hanover, stopping only briefly in Dresden and Leipzig. He arrived on 14 September, but George, the new Prince of Wales, and most of the court had departed three days earlier. George and Newton opposed his petition to come to London as the historiographer of England, giving him a status equal to the master of the mint. The first volume of the *Brunswick scriptores*, covering the period from 768 to 1024, was not yet completed. George ordered Leibniz to stay in Brunswick to make up for time spent in Vienna and to work on the *Scriptores*. Within a month of his return, he finished volume one and had it printed mainly at his own expense. He wanted to embark on the planned second volume, but since he could not receive funding for his research assistants and was personally bearing the costs of the project, he delayed proceeding, writing, "I will leave the Bavarian Guelfs resting in their old documents, until I am richer."[570] The first volume of the *Brunswick scriptores* was to win Edward Gibbon's praise of him as a master of the history of the middle ages and, in 1959, Friedrich Meinecke's citation in *The Origins of Historicism* of the *Scriptores* as being among the first histories to grasp the full richness and diversity of the human experience.

During his last two years, Leibniz was in near isolation in Hanover. Still, his correspondence continued unabated. From 1711 a leading theme in his correspondence with mathematicians and natural philosophers had been the calculus priority quarrel with Newton and his partisans. Leibniz had wanted to widen the debate into philosophy. Cotes' preface to the second edition of Newton's *Principia* provided the occasion. For Leibniz the universe was a plenum filled with animate monads. For Cotes plenum explanations suggested some necessity in nature and its adherents faced the danger of sinking "into the mire of the infamous herd who dream that all things are governed by fate and not providence."[571] Leibniz and others took this as an assault on Hanoverian Lutherans by British Protestants. In May 1715, soon after receipt of Cotes's preface, Leibniz wrote to Caroline, Princess of Wales, who was from Hanover. She asked Samuel Clarke, the king's chaplain and a friend of Newton, to clarify this matter. In their correspondence, Leibniz claimed that the principle of sufficient reason properly applied led to freedom, not necessity, and returned to the familiar charge that attraction is an occult quality or perpetual miracle.[572] He also criticized Newton's thought in Queries 28 and 31 of the *Opticks* that space is the sensorium of God, or a sensory organ to perceive the world. Leibniz considered this a suggestion that objects exist independent of God. This notion and John Locke's materialistic philosophy he held to be dangers contributing to the decline of natural religion in England, a charge that incensed Newton. Clarke replied that Newton meant the sensorium figuratively, not literally. The Leibniz-Clarke correspondence also addressed the nature of space, time, and motion. It set the framework for the debate over them that lasted throughout the eighteenth century. In the scholium to Definition 8 and Book III of the *Principia,* Newton had distinguished between space and time being absolute or relative. He accepted their being absolute and real, while Leibniz argued, partly on epistemological grounds, that they are relative. For Leibniz, space was an order of coexisting phenomena and time an order of events in succession. For

Clarke, experiments demonstrated the necessity of distinguishing absolute from relative motion and that space and time are quantities, not simply relations. Until the invention of non-Euclidean geometries more than a century later, Leibniz's relative space and time were difficult to defend in physics.

Leibniz's house in Hanover, as shown in a drawing made in 1828

Although his arthritis and gout were tolerable, Leibniz's health was weakening in 1716. After a week-long illness, he died on the night of November 14. His last conversation concerned a report that an iron nail had been turned into gold. A school choir, a few close friends, his nephew, and his secretary, Johann Georg Eckhart, attended the funeral. Since he had declined to request a Lutheran pastor in his final hours and had not often attended church or taken communion, the Lutheran clergy boycotted it. So did the nobility, probably since the new king had expressed displeasure with him. The logarithmic spiral, Johann Bernoulli's favorite curve, was inscribed on the lid of his casket, along with the phrase *Inclinata resurget* ("Though bent, he will rise."). Only the Paris Academy in the *Éloge* by Fontenelle in 1717 lauded his work.

As young Leibniz was preparing to depart from Paris in 1676, Rømer made a crucial advance in the study of light by boldly claiming that its speed is finite.

This contradicted the ancient and prevailing Cartesian theories of its instantaneous transmission. Rømer based his assertion on systematic observations made at the Royal Observatory in Paris of occultations and eclipses of Jupiter's moon Io, deriving a first approximation of 221,200 kilometers (140,000 miles) per second, which is on the correct order of magnitude. The modern value is 279,792.5 kilometers (186,282 miles) per second.

Among the circumstances that made possible Rømer's discovery was the construction of superior telescopes by Parisian lensmakers, by 1670, of 150**X** magnification. These powerful but delicate instruments had been installed in the Royal Observatory associated with the Paris Academy of Sciences. Working skillfully with these, Cassini discovered the two rings of Saturn and accurately measured the distance from the sun to Earth. Previously thought to be 8 million kilometers, he computed it to be 146 million kilometers, which is almost the modern value of 149.6 million. Since Galileo, the Medician stars had been widely studied. Eclipse tables for them helped in determining longitude. The reappearance of Io from around Jupiter to our sight puzzled Cassini, for it was sometimes a little early or a little late. He could not explain this discrepancy by the degree of accuracy of observations or otherwise. Cassini had Huygens's latest pendulum clock to confirm the time. After eight months of observations of eclipses of Jupiter's satellites from Tycho Brahe's observatory of Uraniborg in 1671, Jean Picard had invited Rømer to the Paris Observatory.

Rømer noticed that the intervals between Jovian satellite eclipses increase as the distance between Earth and Jupiter lengthen. The greatest difference in the interval for the reappearance of the Jovian moons is twenty-two minutes when the two planets are farthest apart. In September 1676 Rømer, assuming that light travels at a finite speed and employing an estimate based on its traversing Cassini's average diameter of Earth's orbit, predicted that the reemergence of Io in November would be ten minutes late. Observations at the Paris Observatory, in which Cassini took part, found him correct to within nearly a second. Rømer had calculated that light took twenty-two minutes to cross the diameter of Earth's orbit. Today it is known to take sixteen minutes and forty seconds. His assumption and the doubling of Cassini's astronomical unit of 146 million kilometers for the average diameter explain the degree of error in reaching the speed of 221,200 kilometers per second.

Even after the November observation, Rømer's conjecture of the finite speed of light was considered fantastic and Cartesian theory prevailed. Cassini remained doubtful of Rømer's explanation. Only Halley accepted it. Not until James Bradley discovered stellar aberration in 1729 and deduced that it takes light from the sun eight minutes and twelve seconds to reach Earth, offering a better approximation of the speed of light, was Rømer's hypothesis to be widely accepted.

Newton's Prisms and Silent Face

Early on Christmas day 1642 (O.S.) Isaac Newton was born in the manor house of the hamlet of Woolsthorpe in Lincolnshire. By the old calendar, it was still the year of Galileo's death. In the Netherlands, Descartes was active in his research. The combined work of these two men together with that of Newton was to comprise much of the core of the scientific revolution. Newton's mother, Hanna Ayscough, whose family brought formal learning into the lineage, was already a widow. The father, also named Isaac, who came from a successful yeoman family, had died six months after the marriage. Newton was to take the date of his birth and his being a posthumous child as signs of his having a prophetic power to restore higher knowledge that humanity had possessed before the time of Moses but had lost afterward. Family legend had it that he was a premature, weakly baby who could have fitted into a quart pot at birth.

When Isaac was three, his mother married the prosperous sixty-three year old clergyman Barnabas Smith, who rebuilt the Woolsthorpe house and left Isaac there to be raised by his Ayscough grandparents for eight years. The departure of his mother must have been disturbing. Growing up in relative isolation without playmates, Newton was a lonely child given to meditation. This loneliness may partly account for his tortured, neurotic adult personality. The English Civil War had begun and people feared raiding parties from the forces of the contenders, Cromwell's parliament and Charles I. When Smith died in 1653, Newton's mother returned to Woolsthorpe. Two years later she sent Isaac to Grantham to attend grammar school, which emphasized Latin and the Bible. His precocity was emerging. Miss Storer, a younger friend of Newton, was to describe the adolescent Isaac as "a sober, silent, thinking lad . . . known scarce to play with the boys abroad, at their silly amusements."[573] He proficiently constructed mechanical models, including an operating windmill, set aloft fiery kites that frightened neighbors, built a sundial, and on the walls of his room made accurate drawings of birds, ships, plants, and famous men. In 1659 his mother recalled him to Woolsthorpe, intending to make him a farmer. His reaction was angry and peevish. The next year his Uncle William and the Grantham schoolmaster persuaded his mother to let him return to grammar school to prepare for possibly entering the nearby university, Cambridge.

In 1661 Newton entered Trinity College, Cambridge. His mother did not give full financial support, so he matriculated as a sizar who did chores for wealthier students. Apparently, his menial status intensified his alienation. In Cambridge's conventional Aristotelian curriculum of the first two years, Newton did not distinguish himself. A notebook he kept indicates that, after examining a book on judicial astrology in 1663, he sought to know more about the sciences and turned to reading reformed or anti-Aristotelian philosophy. His readings of Descartes' *Dioptriques, Géométrie,* and *Principia philosophia,* Galileo's *Dialogo,* though not his *Discorsi,* and Walter Charleton's translation of the atomism of Pierre Gassendi, to which he inclined, along with Thomas Hobbes, Robert Boyle, Kenelm Digby, Joseph Glanville, Henry More, and John Wallis, stirred his imagination. So did his study of the astronomical observations of Jeremy Horrox, possibly from Johannes Hevelius's *Mercurius in Sole Visus* of 1662, and reading of Thomas Streete's *Astronomica Carolina* of 1661, which was one of his introductions to Kepler. He learned the second or area law elsewhere, possibly from Giambattista Riccioli's *Almagestum novum astronomiam* of 1651.

Newton's notebook questions suggest that Cambridge institutionally contributed little in moving him to the new sciences and gaining mastery over them. In this, he was principally self-taught. Newton received a scholarship in 1664 and was awarded the baccalaureate degree the next year when the bubonic plague was descending on England.

As the pestilence spread, universities were closed. Some students went with tutors to neighboring villages, but from June 1665 to March 1666 and again from June 1666 to late April 1667 Newton was in Woolsthorpe.

Almost fifty years later, Newton was to remember discovering during those two plague years the general binomial theorem. And in a crucial experiment with a prism, he decomposed white light into a spectrum of colors and attempted to refract one of the colored rays a second time with another prism.

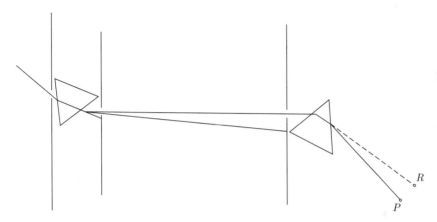

Newton's *experimentum crucis.*

If the traditional theory that white light is the primitive state of light and that it is homogeneous were true, the second refraction should prompt another color change. But Newton believed that white light is instead made up of a mixture of various colored rays. If that is correct, the second refraction should produce no color change. It produced none. His first prism dispersion showed further that in the visible range the refractive index for red light is the smallest and that for violet light the largest.

Newton also then created the embryo of his fluxional calculus, developing the method of Gregory and Slusius for determining the slope of tangents and discovering that computing the areas under curves is its inverse method. That connection between the two is the central relationship of calculus. As Newton would recall, he also conceived of attraction as extending to the moon and calculated that the inverse-square law of attraction, not the prevailing Cartesian physics, allows "pretty nearly" for planetary and lunar orbits.[574]

Such recollections from about 1714, when Newton was older, his thinking unclear, and his remembrances colored by emotions and priority disputes are partial, sometimes inconsistent, and inadequate on their own. They must be confirmed by the record of his mathematical writings, edited by Derek White-side. Apparently, Newton employed a value equivalent to 60 miles for an arc of meridian from Snell, not having the far more accurate 69.1 miles published by Jean Picard in *Mesure de la terre* in 1671. Assuming that the moon is $60r$ from Earth, where $r =$ Earth's radius, and the mean period of the moon is 27 days, 7 hours, and 43 minutes, the circumference of the orbit of the moon using Picard's figure is 69.1 miles/degree $\times 60 \times 360° \times 5{,}280$ ft/mile. Working with right triangles inscribed in a circle gives the fall of the moon per minute as $(69.1 \times 60 \times 360 \times 5280)^2 \pi / (69.1 \times 60 \times 360 \times 5{,}280)(39{,}343\text{min})^2 = 16$ ft. This computation supports the inverse-square law, which indicates that the moon's fall per minute is $(1/2)\{(32.2 \text{ ft/sec}^2 \times (60 \text{ sec/min})^2\}/(60)^2 = 16.1$ ft. But the figure of 60 miles for an arc of meridian gives an incorrect value for the radius of Earth and distorts a proof that homogeneous spheres attract not from the surface but as if all mass is concentrated at the center. This included Newton's famous Moon-apple thought experiment of 1666. He was not sure that the idea of having all force concentrated at the center was credible until he proved this in 1685 and listed that proposition as No. 71 of Book I of the *Principia*. Newton also lacked a precise measure of free fall per second. This awaited Huygens's pendulum studies. At best his 1666 calculation for centripetal force gave a rough approximation. The decisive calculation came later in 1684.

In 1665 and 1666 Newton thus laid the groundwork for his revolutionary scientific achievement. This was remarkable, but not a sudden break with the past. By 1664 Newton had rejected academic Aristotelianism and was moving beyond his mentors. In Woolsthorpe, the freedom from academic assignments and the psychological lift at being home, some historians have suggested, facilitated intense reflection. Yet during a visit to Cambridge in the late spring of 1666 Newton wrote two tracts on physics, one of which outlines his method of fluxions. His extraordinary scientific program had moved beyond the inchoate state of his notebooks from before 1665, but his physics took another twenty years of patient, intermittent revision and insight and at least as late as 1696 he was still developing his fluxional calculus. Richard Westfall refers to this two-year period of extremely fruitful scientific creativity as *anni mirabiles*.

Returning to Trinity in April 1667, Newton desired a permanent position of senior fellow. The chances appeared slim. Royal influence in these selections was growing. The alternative for Newton was to become a village vicar. In 1668 he completed his master of arts degree, which assured his advance to major fellow. Except for the brief time in that position, when he socialized and occasionally went to taverns in the evenings with friends, such as his roommate John Wickins, at Cambridge he was a quirky scholar, unkempt, known to eat sparingly and alone, skip meals, study to two or three each morning, and occasionally forget to sleep. To guard his research time and avoid campus distractions, Newton subsequently chose to be isolated, seldom leaving his chamber. On occasion transported in thought when he was figuring out a mistake or had a new idea, he was known to return hastily from a walk to jot down his new results and to write standing at this desk, not taking the time to pull up a chair.[575] Stories of his absent-mindedness were rife. Even so, his colleagues held him in awe. He was a compulsively precise scholar who had a gift for thinking long on a problem to the exclusion of all else.

Isaac Newton at the age of forty-six, as drawn by Sir Godfrey Kneller

Apparently seeking a higher position at court and considering himself primarily a divine, Isaac Barrow resigned his position of Lucasian professor of mathematics and natural philosophy in 1669, which barred him from politics and administrative posts, and became royal chaplain. Impressed especially with the *De analysi* manuscript, Barrow, the first to hold the chair endowed by Henry Lucas, a member of Parliament, had the young and obscure Newton appointed to succeed him. Newton was to occupy the post to 1701. The method and content of Barrow's optical lectures, along with a reading of Descartes' *Dioptrique*, Robert Boyle's *Experiments and Considerations Touching Colors* of 1664, and especially Robert Hooke's *Micrographia*, published in 1665, inspired Newton to study optics, the theory of colors, and dioptrics, as well as the telescope and microscope, and to grind lenses. In January 1670 he gave his initial Lucasian lecture. His topic was the properties of light. Whiteside believes that he was nervous in speaking to a large audience.[576]

Dissatisfied with Hooke's theory of light as pulses and colors as confused impressions, Newton proposed instead that light is corpuscular, that is, a stream of moving particles, and set out to express colors through quantitative laws, particularly by the degrees of refrangibility, as he rotated the prisms in his crucial experiment.[577] His reports to the Royal Society in the early 1670s on the manipulations in his optical experiments were sparse and sketchy, since he considered a detailed historical narration too tedious. From his prismatic studies, Newton developed a small reflecting telescope that magnified forty times and eliminated colors from the image.

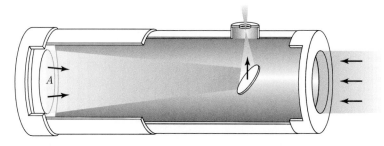

Newton's telescope

Although he had completed his optical investigations by 1670, Newton was unable to publish his new theory of light and colors until 1672 in the *Philosophical Transactions*. The Great Fire of London of 1666 had depressed the book trade, making prohibitive the issuing of scientific and technical works promising few sales. Heavy subsidies from the Royal Society were essential to scientific publication. The 1672 paper, Newton's first, founded the science of spectroscopy. Shortly before it was published, he was elected a Fellow of the Royal Society, the highest scientific honor in Britain, for the portable reflecting telescope that he had designed and constructed. London artisans did not have the skill to copy it. Newton had wrestled with the nature of light in private for eight years and apparently expected general approbation for the new theory in his paper. He was mistaken. Considering himself the authority on optics, Hooke wrote a lengthy critique implying that he, not Newton, had conducted the prismatic experiments and that Newton misunderstood them.

Newton's paper required clarification. The Royal Society received a dozen letters filled with criticisms and questions. Among these, the French Jesuit Ignace Pardies asked whether the differing angles of incidence in the sun rays invalidated the paper's crucial experiment. Huygens saw Newton's theory as ingenious and a "highly probable hypothesis" but inquired whether establishing it requires going beyond two colors, yellow and blue. Newton bridled at the word "hypothesis" to describe his theory: he believed that all questions about it could be answered by solid experiments. The anonymity that he had enjoyed to 1672 was lost. He received as gifts books by Boyle and Huygens, and in Paris Leibniz read the article. In London the Royal Society's secretary Henry Oldenburg asked Newton to write an extensive treatise on light and colors.

Newton responded angrily to Hooke's patronizing comments. The pattern for their future relationship was set. He refused to publish his tract on light before Hooke's death. Apparently, Collins also failed to obtain a subsidy for its publication. Although mostly completed by 1675, Newton's *Opticks* was to appear only after that event almost thirty years later. The controversy had an acid effect on him. Writing to Leibniz, Newton reproached himself for his "imprudence for parting with so substantial a blessing as my quiet to run after a shadow."

In August of 1684 came the now famous visit of Edmond Halley. At a meeting of the Royal Society the previous January, Halley, Wren, and Hooke had considered the major problem before natural philosophy, that of deriving from the principles of dynamics Kepler's three laws of planetary motion or variants of these. All had known since the previous decade the inverse-square law of attraction, which holds that the force of attraction between two bodies is the reciprocal of the square of their distance from each other, but had failed to demonstrate from it the laws of celestial motion. In January Halley and Wren admitted having failed in their attempt, while Hooke claimed he was keeping his solution secret. Recognizing Newton as an expert mathematician who might have addressed this problem, Halley communicated to him a challenge problem posed by Wren involving theoretically demonstrating the orbits of the solar planets. The prize for its solution was to be a book costing up to forty shillings. Halley inquired what, under the inverse-square law, would be the exact shapes of the planetary ovals. Ellipses, Newton immediately responded. Halley asked how he knew and Newton replied, "I have calculated it."[578] When Halley asked for the calculation, Newton said that he could not find it among his papers. Westfall believes that this was a charade. Newton had the paper, which still exists, but in a hasty drawing had confused an ellipse's axes with conjugate diameters. Perhaps knowing that he occasionally made errors, Newton adhered to results required by Kepler's laws. He promised to reconstruct his lost paper and send it to Halley.

So began what Whiteside describes as Newton's second *annus mirabilis*.[579] Halley's challenge seized his imagination and from August 1684 to the spring of 1686 he was almost totally absorbed in preparing his *Principia mathematica*, probably the greatest achievement in the history of science. During the decade preceding 1684, theological studies and alchemical experiments had dominated Newton's studies. These did not end, but his intellectual life was now redirected to the grandeur of a new dynamics. The timing was right. Newton had developed the beginnings of fluxional calculus and after almost twenty years of research had definitely rejected the Cartesian vortex cosmology.

With Edward Paget, a Trinity fellow, Newton in November sent Halley a nine-page treatise entitled *De motu corporum in gyrum* ("On the Motion of Bodies in Orbit"). Consisting of four theorems and seven problems starting with two definitions and two hypotheses, it was far more than Halley had expected. Its first definition introduces into mechanics a new term, centripetal force, by which "a body is impelled or attracted" towards a point considered a center. Newton consciously coined this term in parallel to Huygens's centrifugal force, fleeing the center. His second definition asserts that motion in a straight line, in addition to Aristotelian rest, is the natural state, and the second hypothesis has rectilinear motion proceeding uniformly to infinity by a body's inherent force unless hindered by an external obstruction. From his definitions and hypotheses, the concept of centrifugal force generated by planetary rotation, and a version of the mathematics of infinitesimals that assumes that results for finite triangles hold for infinitesimal ones, *De motu* demonstrates Kepler's three laws in limited circumstances. The pull of the inverse-square centripetal force against the rectilinear directions of the planets produces a conic orbit, which for bodies having less than a certain velocity is elliptical.

In *De motu* and its revisions Newton clearly set for himself the daunting task of developing a new quantitative science of dynamics to cap Galileo's kinematics. Revisions redefine inherent force as the force of inertia and put it in matter, not a body. They consider space and time to be absolute and reject Cartesian relativism as atheistic. *De motu*'s study of a projectile traversing a resisting medium also presages the wider horizon of a general dynamics. Halley immediately saw the treatise to be revolutionary and returned quickly to Cambridge to confer with Newton and have him agree to publish his new dynamics. Receiving a request from Newton for data to make his demonstrations more precise, the astronomer royal John Flamsteed likewise found *De Motu* extraordinary. Newton's queries about the comet of 1680–1681, the velocities of Jupiter and Saturn and the periods of their satellites, and observations of the tides broadened his investigations, suggesting that all conform to the same mechanical principles as do planetary orbits.

The *Principia mathematica* was to consist of three books. By April 1686 Newton had completed an expansion of the manuscript of Book I and was composing Books II and III. As promised, Halley gained quick approval from the Royal Society council for its publication. At this point, Hooke wrote that Newton had taken from their correspondence of 1679 the concept of the inverse-square law of attraction. In his anger, Newton threatened to withhold Book III from publication. But calmed by Halley, he submitted Book II and the revision of Book III in November. Halley edited the manuscript, polishing its mathematics, and it was published in the summer of 1687. British natural philosophers and geometers greeted it with rapture.

An older practical mechanics and a *scientia media* mixing mathematics and natural philosophy gave way in the *Principia mathematica* to a program for mathematicizing natural philosophy. From it, modern mathematical physics was substantially to develop. Its full title, *Philosophiae naturalis principia mathematica,* read in the reverse order in English, *The Mathematical Principles of Natural Philosophy*, thus indicates what is the great achievement of this landmark in the history of science. Written in a classical Euclidean format, it consists of almost two hundred propositions that are theorems or problems, many corollaries and lemmas for establishing results, and scholia for comments and interpolations. Newton begins with eight definitions of such basic concepts as momentum, mass that begins distinguishing it from weight, and innate, impressed, and centripetal forces. These improve on their predecessors. Building on these definitions, his three axioms or laws of motion, starting with inertia, and the inverse-square law of attraction, Newton proves all the main properties of moving bodies. Occasionally he employs bodies as mass points. Motion joins

Aristotelian rest as a natural state. Newton's *Principia* unifies terrestrial and celestial mechanics, which from Aristotle to Descartes and Leibniz had been separate. Much of the material in prior treatises on mechanics had been vague and filled with special cases. Newton's writing is clear and concise.

In creating a new science of dynamics from first principles, the *Principia* combines critical synthesis and products of Newton's own genius. Newton himself rejected the term "dynamics," which Leibniz had borrowed from the Greek. The difficulty of the task that Newton faced is suggested in his discarding of several drafts and the *Principia*'s several editions. The major source of his reputation is the first half of Book I, treating the motion of one or two bodies in a vacuum. In systematizing the science of dynamics, Newton selects, marshals, perfects, and formalizes contributions to mechanics from the previous century, especially Kepler's three laws of planetary motion, Galileo's kinematics, and Huygens's laws of elastic collisions, along with Gassendan atomism. He refutes the Cartesian vortex cosmology and plenum. His dynamics provides the first satisfactory theoretical justification of Copernican and Keplerian astronomy. The second half of Book I of the *Principia* on the three-body problem for the moon and comets shows his unmatched command of the core of mechanics. Although lacking series of differential equations for systems of more than two free bodies, his method obtains approximate results not surpassed until Clairaut, d'Alembert, and Euler obtained the first of these equations sixty years later. In Book II on the motion of bodies in resisting media, for example air or water, his method fails and much of it is incorrect. This book especially deals with pendulums. Book III on astronomy formulates new concepts and seeks numerical predictions to compare with measured results. While Newton worked in local time and space, the *Principia* presents from the opening scholium a framework of absolute time, space, and motion. Space was the sensorium of God as guarantor of its unchallengeable, absolute status.

As Lucasian professor, Newton had only to deliver up to twenty lectures annually in addition to his research. He taught but one semester each year, thus submitting ten lectures for the archives. Apparently, Newton had little success in his teaching. So few students attended his classes and fewer still understood him that he later reported that often "for want of hearers [he] read to the walls."[580]   Late in his academic career Newton believed that he had made the mistake of attempting to teach advanced mathematics to students who lacked proper preparation.

His confidence buoyed by the response to the *Principia* that placed him at the forefront of natural philosophy, Newton began to look beyond the cloistered academic world of Cambridge. His Arian views in religion clashed with Cambridge's Trinitarian orthodoxy. Having supported the Glorious Revolution, Newton was narrowly elected a Whig member of Parliament in 1689. In 1691 he planned a major treatise on fluxional mathematics comparable to the *Principia*, but stopped with his tract *De quadratura curvarum* ("On the Quadrature of Curves"). But in the autumn of 1693 Newton suffered from a breakdown bringing sleeplessness, memory loss, and paranoia. Excessive study and a fire that destroyed some of his papers have been suggested as causes. But Newton was deeply engaged in 1692 in correspondence with Boyle and Locke on alchemy and was conducting alchemical experiments with red earth and mercury. Newton's conditions in 1693 are similar to the effects of chronic mercury poisoning.[581] His paranoia turned him briefly against his friends Samuel Pepys and John Locke: he accused Locke of endeavoring "to embroil me with women."[582] With the breakdown, Newton's creative research mostly ended. The acuity of his genius remained throughout the decade though, enabling him, for example, to solve quickly the Bernoullis' brachistochrone problem in 1697.

In 1695 the British government, lacking experts in economics, appointed Newton master of the mint. Rising costs for building the navy and army in opposition to Louis XIV's expansions had brought the country to the verge of bankruptcy. Newton carried out a minting of new coins, introducing milled-edged coins to keep them from being clipped undetected around their rims, and established precise standards of purity for alloys to avoid adulteration. He was reelected to Parliament in 1701, but defeated in 1706. Because his nemesis Hooke attended the weekly meetings of the Royal Society, Newton usually did not. The death of Hooke in March 1703 opened the way for his election to its presidency, which occurred at its annual meeting that November. Two years later Queen Anne knighted him.

In his final years, Newton was occupied with a number of scientific and humanistic studies and more deeply with theology. In 1704, his second major book, *Opticks,* was published. Despite hectoring by Wallis for its earlier release, he had waited until a year after Hooke's death. The *Opticks* contains his corpuscular theory of light, his nutshell theory of matter projecting a nearly vacuous universe, his impenetrable, passive atoms, and a critical empirical methodology. Angered at Flamsteed, who he thought had not supplied enough lunar and cometary observations, Newton engaged in an ugly effort to wrest from him the preparation of a star catalogue.[583] From 1711 to 1712 he guided the deliberations of the Royal Society committee over whether he or Leibniz deserved priority for the creation of calculus. He continued the investigations of alchemy, hermetic philosophy, and theology that he had begun in the 1670s. Working as an historian and Arian heretic, Newton prepared *The Chronology of Ancient Kingdoms,* which confirmed dating by astronomy, and *Observations upon the Prophecies, Daniel and the Apocalypse of St. John the Divine,* both of which appeared posthumously in 1728. After five years of declining health ending with the pain of a gall stone, Newton died on March 20, 1727, and was buried in Westminster Abbey, an honor reserved mostly for royalty.

By the 1750s, Newtonian theory had been transmitted across Europe, and the general explanation of its dynamics involving gravitational attraction of the motion of planets, the tides, our moon, comets, and falling stones was steadily being verified. Telescopic observations and the formulation of differential equations precisely describing these phenomena, especially in lunar theory, were crucial. Also confirmed through measurements of arcs of meridian in Lapland and Peru was Newton's mathematical demonstration in the *Principia* that Earth, being subject to centripetal and centrifugal forces, must have an equatorial bulge and polar flattening. By 1750, the variety, exactness, and predictive power of his science had brought Newton nearly apotheosis among natural philosophers and the larger literate public in Europe. Alexander Pope's famous couplet, submitted for inscription in Westminster Abbey, but rejected, catches the sentiment:

> Nature and Nature's Laws lay hid in night
> God said, Let Newton be, and all was light.[584]

And for understanding the development of Newton's thought and its place in the sciences, there is Newton's own reflection on his life's work written shortly before his death:

> I don't know what I may seem to the world, but, as to myself, I seem to have been only like a boy playing on the sea shore, and diverting myself in now and then finding a smoother pebble or a prettier shell than ordinary, whilst the great ocean of truth lay all undiscovered before me.[585]

17

The Apex of the Scientific Revolution II: Calculus to Probability

> Other very learned men have sought in many devious ways what someone versed in this calculus can accomplish in these lines as by magic.
> — Gottfried Wilhelm Leibniz, 1684

> This [brachistochrone] problem seems to be one of the most curious and beautiful that has ever been proposed, and I would very much like to apply my efforts to it, but for this it would be necessary that you reduce it to pure mathematics, since physics bothers me....
> — L'Hôpital in a letter of June 15, 1696, to Johann Bernoulli

When mathematicians, exploiting the processes of Cavalieri's geometry of indivisibles and several infinitesimal methods, joined the twin procedures of Descartes' construction of tangents and Wallis's infinitary tabulation of quadratures, calculus emerged from its intuitive cocoon. The appearance of calculus, together with analytic geometry, marks the beginning of modern mathematics. In addition to Fermat, Pascal, and Roberval in France, Torricelli in Italian cities was pursuing its groundwork paths. In making Cavalieri's two indivisible techniques more flexible, Torricelli's *Opera geometrica* of 1644 brought greater clarity and rigor to them. He sought to uncover a superior heuristic for discovering new propositions known to the ancients, but kept secret. After van Schooten died in 1660 and Huygens departed for Paris six years later, mathematical contributions in the Low Countries slackened, but René-François de Sluse (d. 1685) at Liège generalized the tangent drawing methods of Fermat and Descartes. The *Philosophical Transactions* for 1673 published his results. Meanwhile in Britain, John Wallis, Nicolaus Mercator, James Gregory, and Isaac Barrow, turned to the study of quadratures, tangents, and infinite series, subjects then gaining increasing attention. The differential versions of calculus of Newton, Leibniz, and the Bernoullis, founded in good measure on these investigations, were not fully formed: fundamental concepts remained to be clarified and no systematic structure of basic rules and procedures existed. Criticism and controversy were to culminate in the bitter quarrel between the Newtonian camp in Britain and the adherents of Leibniz on the continent over priority for the invention.

17.1   Major Sources in Britain after 1650

After studying Torricelli's *Opera geometrica*, as well as works by Descartes, Harriot, Oughtred, Stevin, and Roberval, John Wallis labored to replace the indivisibles in infinite geometric progression with an infinite arithmetical sequence. The arithmetical technique permitted computation of areas under curves, including conic sections, and cubatures of cones and cylinders, and embodied a rough notion of limit.

Wallis's transition from geometric procedures and lines to algebraic methods and numerical expressions appears in the *De sectionibus conicis tractatus* of 1655. This book, coupled with de Witt's independent *Elementa curvarum*, essentially completed the arithmetization of conic sections that Descartes had started. Their advance made it possible to prove the properties of conic sections by methods of plane coordinate geometry alone. Wallis's *De sectionibus* also introduced for infinity the "love knot" symbol ∞, now standard, and in place of infinitely small rectangles conceptualized lines or *non quanta*. For these infinitely small quantities, he wrote $1/\infty = 0$. In this he was bolder than Fermat, who had not referred to what is deemed E as an infinitesimal. But Wallis dangerously assumed that rules that hold for finite procedures do so also for infinite ones.

John Wallis

Further reading of Stevin and Roberval persuaded Wallis in the *Arithmetic of Infinities* to adjust his method so as to approximate arithmetical limits more nearly. In formalizing his tabulations, he concluded what in modern symbols is $\int_0^a x^m \, dx = a^{m+1}/(m+1)$, for all m except the transition value of -1. This is the definite integral computed between the limits 0 and a. Apparently unaware that this proposition had existed during the previous two decades in various forms, Wallis arrived at it from his calculations paralleling Cavalieri's indivisibles.

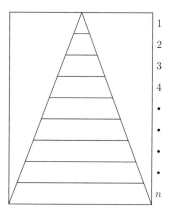

To compare with squares of indivisibles inscribed in a triangle the squares inscribed in a rectangle, Cavalieri had assumed that the triangle has an infinity of indivisibles or abscissas whose lengths are 0, 1, 2, 3, and so forth. Plainly (the area of the triangle)/(the area of the rectangle) $= (0+1+2+3+\ldots+n)/(n+n+n+\ldots+n) = 1/2$. Wallis examines the ratio of the squares for these two figures: $(0^2+1^2)/(1^2+1^2) = 1/2 = 1/3+1/6$. For three figures the ratio of the squares is $(0^2+1^2+2^2)/(2^2+2^2+2^2) = 5/12 = 1/3+1/12$. For four figures it is $(0^2+1^2+2^2+3^2)/(3^2+3^2+3^2+3^2) = 7/18 = 1/3+1/18$. In modern symbols, Wallis proceeds to what is $(0^2+1^2+2^2+\ldots+n^2)/(n^2+n^2+n^2+\ldots+n^2) = 1/3+1/6n$. If the sequence continues to infinity $1/6n$ becomes zero, and the ratio of squares, which in modern symbols is $\int_0^1 x^2 \, dx$, is exactly the limiting value of $1/3$. The analogous ratio for the third power, Wallis concludes, is $1/4$; for the fourth, $1/5$; for the fifth, $1/6$; and so on. He formalizes his results so that for m a positive integer the quadrature under the curves $y = x^m$ is $x^{m+1}/(m+1)$. For $\int y \, dx$, having no specified limits, Wallis had the rule for the indefinite integral. But this integral remains undetermined, since integration simply adds a constant. For example, $\int x^2 \, dx = x^3/3 + C$. By substituting y^k for x in $(1-x^{1/k})$, he transforms $\int_0^1 (1-x^{1/k}) \, dx$ into the normal beta integral. Following suggestions in Oresme and Stevin, Wallis further extends the use of exponential notation by apparently assuming that what is true for fractional values of exponent m in x^m holds as well for negative and irrational values, but gives no proof. Interpolations, iterations, and incomplete induction based on what was known to be valid for a few instances are essential to his computations. His justifications of interpolation were questionable. One of them depends on the principle of continuity: the assumption that what is known true for end values is also true for the intermediate values. Wallis's inclusion of helped counter the persistent view from antiquity that they are not strictly legitimate numbers.

Wallis quickly extended his results to compute the quadrature under curves of the form $y = x^p(a^n - x^n)^m$, where m, n, and p are positive integers. The curves $y = (1 - x^2)^n$, where $n = 1, 2, 3, \ldots$, particularly interested him. After determining their quadrature, he sought to "square the circle," which means to compute the area under the curve $y = (1 - x^2)^{\frac{1}{2}}$. In modern notation, the area of the quadrant of a circle is $\int_0^1 (1 - x^2)^{\frac{1}{2}} \, dx$, which equals $\pi/4$. Reaching this result for a curve having a fractional exponent was more difficult. It required expanding $(1 - x^2)^{\frac{1}{2}}$ into a power series, a series of powers of a variable. If c_0, c_1, c_2, $\ldots$ are a sequence of real coefficients and x is the independent variable, then the power series is $\sum_{n=0}^{\infty} c_n x^n = c_0 + c_1 x + c_2 x^2 + \ldots$. Lacking the deeper insight and sounder intuition of Newton, Wallis failed to obtain the power series for exponent $n = 1/2$. To no avail, he sought help from Gregory of Saint Vincent's *Opus geometricum*. Knowing that $\pi/4$ is the area under $(1 - x^2)^{\frac{1}{2}}$ from 0 to 1 and employing a bold, ingenious procedure of interpolations from the computed areas of $(1 - x)^n$ for $n = 1, 2, 3, \ldots$, he discovered instead of an infinite series his famous infinite product for its reciprocal, $4/\pi = (3/2)(3/4)(5/4)(5/6)(7/6)(7/8)(9/8) \ldots$, now known as Wallis's formula.[586]   Wallis asked Brouncker to derive an alternative expression. Brouncker was able to transform Wallis's product into a continued fraction, determining that $4/\pi = 1 + 1^2/(2 + \{3^2\}/(2 + [5^2\}/(2 + \{7^2\})]/2 + \ldots) \ldots$. From this sequence, he was able to compute π correctly to ten decimal places. Impressed by this result, Wallis devoted the final two pages of his *Arithmetic of Infinities* to a theory of continued fractions.

Finding his rigor wanting, Huygens and Fermat questioned Wallis's computational procedures. To test them, Fermat issued challenge problems to all English mathematicians, such as obtaining a cube that gives a square when added to all of its aliquot parts. An example is $7^3 + 7^2 + 7^1 + 7^0 = (20)^2$. Wallis's responses, combining some trial and error with perceptive guessing, appeared in the *Commercium epistolicum* of 1658. Wallis never achieved the deeper number theoretical insights of Fermat. Hobbes, who claimed to have obtained a quadrature of the circle, also criticized the *Arithmetic of Infinities*, calling it a "scurvy book." Its arithmetization was, he believed, absurd; it was merely a "scab of symbols".[587]   J. F. Scott believes that Wallis's opposition to Hobbes's *Leviathan* helps account for the vitriolic exchanges between the two men.

Early in his career, Nicolaus Mercator (1620–1687) calculates logarithms and espoused Keplerian astronomy. His *Trigonometria sphaericum logarithmica* of 1651 computed logarithms of the sine, cosine, tangent, and cotangent lengths. Two years later Mercator moved from Denmark to London, where he worked as a mathematical tutor, becoming friends with John Collins and John Pell. His *Hypothesis astronomica nova*, published in 1664, explains Kepler's planetary laws and added a another link to its mystical components: the ratio of the apsides, the maximum and minimum distances of the planet from the sun, must be the golden section of $(\sqrt{5}-1)/2$. Possibly Newton first learned Kepler's laws from this source. Principally for his invention in 1666 of a marine chronometer on a model developed by Huygens, Mercator was elected a fellow of the Royal Society.

In 1668 a supplement to Mercator's *Logarithmotechnia* (*Logarithmic Procedures*) presented power series expansions of logarithms, its author's greatest achievement in mathematics. The modern natural logarithm of x is the integral of x^{-1} for $x \neq 0$. Oresme and the Merton college group had worked with $1/(1-x) \approx 1+x+x^2+x^3+\ldots+x^{n-1}+x^n$. For $\mid x \mid < l$ and arbitrarily large n, the last terms approach zero. Wallis had provided a rule for the indefinite integral of x^m together with long division to give the power series of the integral of $1/(1-x)$, in later notation $\int dx/(1-x) = \int[1 + x + x^2 + x^3 \ldots]\,dx = x + x^2/2 + x^3/3 + x^4/4 + \ldots$. But the model of the quadrature of the hyperbola by Gregory of St. Vincent and de Sarasa prompted Mercator's work. They had computed the area under a rectangular hyperbola $y = 1/(1 + x)$ over the interval $[0, 1]$ and knew that this equals $\ln(1 + x)$, where ln stands for natural logarithm. To compute the area, Mercator divided that interval into n subintervals, or length x/n. In modern symbols, he established that $\ln(1+x) = \int_0^x dx/(1+x) = x - x^2/2 + x^3/3 - x^4/4 + \ldots$. From Pietro Mengoli he adopted the name "natural logarithms" for values from this series.

Among others working on the quadrature of the hyperbola were Hudde and Newton. Both, having independently arrived at the same sum earlier, had not published. Newton's results were given later in *De analysi*. Gregory's *Vera quadratura* had just made it a major problem for discussion. Apparently, ten years earlier Brouncker, seeking to find common logarithms to base 10, had achieved a quadrature of the rectangular hyperbola. In order not to lose priority, he published his results in the *Philosophical Transactions* for April 1668. In a square of $0 \leq x \leq 1$ and $0 \leq y \leq 1$, Brouncker constructed hyperbola $y = 1/(1 + x)$ and repeatedly bisected the area under it.

He found the hyperbola area or $\log 2$ to be greater than $1/1\cdot2+(1/3\cdot4)+(1/5\cdot6+1/7\cdot8)+(1/9\cdot10+1/11\cdot12+1/13\cdot14+1/15\cdot16)+\ldots$ and the remaining area of $(1-\log 2)$ to exceed $1/2\cdot3+(1/4\cdot5+1/6\cdot7)+(1/8\cdot9+1/10\cdot11+1/12\cdot13+1/14\cdot15)$. Finding series solving the problem of $\log 2$ are convergent, he computed it to be 0.69314709. This result allowed Brouncker to show 2.30 proportional to $\log 10$. But his generalized procedures led to poorly convergent, unfruitful expansions.

James Gregory (1638–1675) of Scotland was the grandson of Alexander Anderson, who had edited Viète's writings. In 1662, Gregory traveled to London, becoming friends with Robert Moray, the interim president of the Royal Society, and John Collins, its librarian. From 1664 to 1667, he studied geometry, astronomy, and mechanics at Padua under Stefano degli Angeli, a student of Cavalieri and partisan of Torricelli, and Gabriele Manfredi. Gregory devoted a lifetime of research to the binomial theorem, algebraic equations, number theory, and infinitesimal methods for the quadrature of conic sections and general spirals.

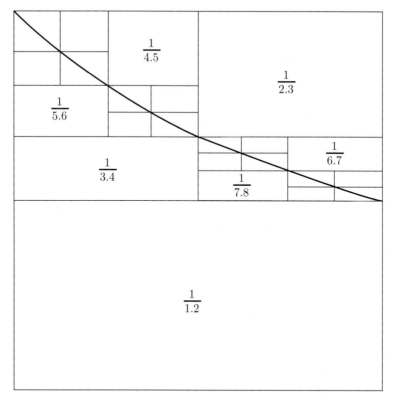

Gregory's *Vera circuli et hyperbolae quadratura* was published in Padua in 1667. It generalizes to conics Archimedes' algorithm of an unbounded double sequence for measuring the area of a circle. Gregory's method begins by inscribing the circle within a triangle and circumscribing it around a rectangle. He then increases indefinitely the number of sides of both until the difference between them is arbitrarily small. Similar bisections of bounded arcs produce a sequence of *mixtilinea* (i_n, I_n) that he proves is the geometric mean of i_n, I_n and the harmonic mean of i_{n+1}, I_{n+1}. Adding a generalization of the inequalities of Snell and Huygens for central conics, $(i_0 = 2I_0) > I > (4i_1 - i_0)$, the *Vera quadratura* calculates sectors of circles and hyperbolas to fifteen places and concludes that π cannot be algebraic. Gregory's method is clearly ingenious, but Huygens attacked it as unoriginal and ineffective, and he asserted that π is algebraic. Proof of the transcendence of π took two more centuries.

In 1668 Gregory also published in Padua his eclectic compendium *Geometriae pars universalis* (*The Universal Part of Geometry*). It provides a geometrical analysis of the work of his time on centers of gravity, rectification, tangent problems, quadrature, and cubatures of solids and surfaces of revolution. The *Geometriae pars universalis* borrows a geometric tangent method from Fermat, Hendrik van Heuraet, and Wren in his treatise on the cycloid. The work presents a differential triangle and simplifies related demonstrations. Knowing the area under a curve from a fixed point to x, Gregory was able to draw a new curve to ordinate x. He was a step away from constructing its tangent and on the verge of the fundamental theorem breakthrough. But neither here nor in generalizing the rectification of the semicubical parabola (problem 6) did he prove the fundamental theorem of calculus. If he grasped the significance of the inverse relationship between quadrature and the method of tangents, he did not mention it. Gregory discovered Taylor's series and geometrically deduced Mercator's series, $\pm \ln(1 \pm x)$. In Padua he probably learned that the arctan x gives the area under curve $y = 1/(1+x^2)$. By long division, $1/(1+x^2) = 1 - x^2 + x^4 - x^6 + \dots$. From rules of Cavalieri and Wallis, Gregory obtained what in modern notation is $\int_0^1 dx/(1 + x^2) = \arctan x = x - x^3/3 + x^5/5 - x^7/7 + x^9/9 - \dots$, now known as Gregory's series. About two hundred years earlier, Keralan scholars in southern India had discovered it. Late in 1668 *Exercitationes geometricae*, Gregory's riposte to Huygens, appeared in London. It is mainly an appendix to *Vera quadratura*.

Impressed with Gregory's work, the Royal Society elected him a fellow in 1668. On Robert Moray's recommendation, he became professor of mathematics at St. Andrews the next year. Collins, who transmitted copies of letters sent to the Royal Society from Barrow, Huygens, Sluse, and Newton, was to be his chief contact with mathematical and scientific advances. At St. Andrews, Gregory established the first public observatory in Britain, traveling to consult with Flamsteed on equipment. Essentially forced out for urging a curriculum reform including adding to the offerings in mathematics and science, in 1674 Gregory accepted the newly endowed chair of mathematics at Edinburgh. While observing Jupiter's satellites the next year, he suffered a blinding stroke, dying a few days later.

Isaac Barrow (1630–1677), who probably came as close to calculus as any other mathematician, including Fermat and Wallis, was a royalist. Consequently, Cromwell removed his name from the list for Regius professor of Greek at Cambridge in 1655. After the restoration in 1660, Barrow took holy orders and was named Regius professor. Two years later he became concurrently Gresham professor of geometry in London and was elected a fellow of the Royal Society. In 1663 Barrow was appointed to the Lucasian chair of mathematics at Trinity College, Cambridge, a position from which he resigned in 1669. Only two chairs of divinity paid larger stipends than its annual about £100. From the manuscript *De Analysi* and a discussion with him of Mercator's *Logarithmotechnia*, Barrow had learned of Newton's brilliance, and he effectively had the obscure young man appointed his successor. Within a year the ambitious Barrow was named royal chaplain in London. The king admired his scholarship and wit. Barrow was named in 1673 to the post that he had wanted, master of Trinity College, and two years later became vice chancellor.

In mathematics as generally in politics, Barrow was conservative, preferring ancient Greek geometry to algebra and its formalisms, which he subordinated to logic with the status of mere tools. In general, his position was at opposites to Wallis's. Barrow's commitment to geometry and his undervaluation of algebra are cited as reasons for his not making the analytical breakthrough to calculus.

Isaac Barrow

Barrow's first publication, *Euclidis elementorum libri XV*, was printed probably in 1654. As a book tailored to the historic quadrivium, it emphasizes the deductive, axiomatic structure of the *Elements*, rather than its geometric content. In the years to 1675, Barrow, who was a master of Greek and Arabic, meticulously edited an epitome of Euclid's *Data*, Archimedes' *On the Sphere and Cylinder*, the first four books of Apollonius' *Conics*, and three books of Theodosius' *Sphaerics*. The most widely read of his translations was the pocket size *Elements*, known as Barrow's *Euclid*.

Barrow's two chief publications in mathematics are the *Lectiones mathematicae* and the *Lectiones geometricae*. The *Lectiones mathematicae*, the first Lucasian series, were given from 1664 to 1666 and printed posthumously in 1683. Among their topics are mathematical deduction, proportions and incommensurables, and infinity and the infinitesimal. While contending like Aristotle in *Posterior Analytics* that the logical structure of the physical sciences is the same as that of mathematics, Barrow departs from scholasticism by insisting that the emerging procedures of physico-mathematics are superior to Aristotelian essentialist natural philosophy.[588] Hypotheses in physics must be verified not by submission to an Aristotelian test of ongoing observation, but by a few decisive experiments, if possible only one. Robert Kargon interprets Barrow in this regard as rejecting hypothetical physics, rather than endorsing Archimedean methods.[589] Edited by Newton and Collins, the *Lectiones geometricae*, which appeared in print in 1670, is a study of higher geometry. In an unusal twist, the several printers of its editions to 1674 went bankrupt.[590] Having read Mersenne and Torricelli and accepting a kinematic view of geometry, Barrow speaks of the uniform fluent of time and aggregations of the steady motion of points and lines as generating curves. His investigation of Cavalieri, Descartes, Gregory of St. Vincent, Pascal, Fermat, Torricelli, Roberval, Schooten, Hudde, Huygens, Wallis, Wren, and most of all James Gregory brought him systematically to seek to generalize quadrature, rectification, and tangent procedures.

Descartes' and perhaps Fermat's tangent algorithms that underpin Barrow's compendium on tangents, together with his revision of propositions six and seven from Gregory's *Geometriae pars universalis* and Wallis's account in *De cycloide* of William Neil's rectification, are crucial to bringing him close to the fundamental theorem of calculus.

Proposition 11 near the end of lecture X and proposition 19 of lecture XI together comprise the fundamental theorem of calculus. Barrow in proposition 11 offers a general "method for finding tangents [to a point P on a curve] by calculation." It is not just any curve but the graph of an area accumulation "function." In modern notation, Barrow is showing that $d/dx \int_a^x f(t)\, dt = f(x)$. Let the modern $f(x, y) = 0$ give his graph. Between P and a neighboring point Q the method draws the hypotenuse of right triangle PQR. The tangent crosses the x-axis at point J and PR extended crosses the x-axis at M, and the resulting triangles PTM and PQR are nearly similar. Barrow lets $PR = a$ and $QR = e$. If the coordinates of P are (x, y), those of Q are $(x - e, y - a)$. In place of Fermat's single E, Barrow makes his right triangle indefinitely small, that is, a differential triangle. In this case a may be roughly likened to $\triangle y$ and e to $\triangle x$. Barrow's task is to find the ratio a/e, which is the slope of the tangent. Although apparently not knowing Fermat's work, since his name does not appear, Barrow proceeds in the same fashion. He "rejected all terms" in which there is no a or e ... or in which a and e are above the first power, or are multiplied together. Proposition 19 proves the second inverse part of the fundamental theorem. From infinitesimal rectangles under specific curves formed from what is today called the derivative function, Barrow computes the area under them. His finding in modern notation is $\int_a^b f'(x)\, dx = f(b) - f(a)$.

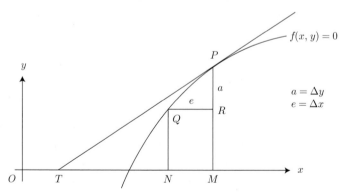

On the advice of Newton, Barrow elsewhere discusses his technique of algebraically determining the tangent for a special Lamé curve, $x^3 + y^3 = r^3$. The substitution is $(x - e)^3 + (y - a)^3 = r^3$, which gives $x^3 - 3x^3e + 3xe^3 - e^3 + y^3 - 3y^3a + 3ay^2 - a^3 = r^3$. Subtracting $x^3 + y^3 = r^3$ gives $3x^2e + 3xe^2 - e^3 - 3y^2a + 3ay^2 - a^3 = 0$. Removing terms having powers of a and e above the first leaves $3x^2e + 3y^2a = 0$ and $3x^2e = -3y^2a$. Dividing by 3 yields $x^2e = -y^2a$ or $x^2/y^2 = a/e$. This procedure, which Barrow does not prove, ignores Fermat's idea of adequality or pseudo-equality. To be made rigorous, it requires only a theory of limits. Among the curves to which Barrow applies this procedure are the folium of Descartes ($x^3 + y^3 = rxy$), the kappa curve ($x^2(x^2 + y^2) = r^2y^2$), and the quadratrix ($y = (r - x)\tan x/2r$).

Although his tangent method improves on Fermat and closely resembles modern differentiation, Barrow does not give special recognition to propositions 11 and 19 as being important. They are just propositions in a list. Nor does Barrow identify as fundamental the inverse relationship between tangent problems and quadrature. His work is with infinitesimals and geometrical problems, not symbols and functions. His a and e do not have the implications of our $\triangle x$ and $\triangle y$, and he cannot justify neglecting higher powers of them, which requires a grasp of limits. His method can reduce tangent problems to the calculation of quadratures but not the reverse.

By the mid-1660s, European geometers using techniques of indivisibles and infinitesimals had succeeded in computing quadratures of many curves, cubatures of curvilinear figures, and rectifications of curves. They had developed adept means of constructing tangents that are basic to differentiation, were starting to grapple with the theory of limits, and were close to recognizing the fundamental theorem of calculus. To invent calculus, Newton and Leibniz still had to provide a clear formulation of its fundamental concepts and rules, the beginning of a systematic organization of these rules, and a rich symbolism.

17.2   Newton's Forging of Fluxional Calculus

Newton made deep and complex contributions across mathematics, advancing algebra, calculus, infinite series, methods of computation and approximation, finite differences, classical and analytic geometry, and probability. He published no definitive tome in mathematics comparable to his *Principia mathematica* in mechanics or his *Opticks*. Instead he transmitted his mathematical ideas through correspondence, the circulation of manuscripts, and a few published pieces. Among these publications are the *De analysi per aequationes numero terminorum infinitas* (*On Analysis by Equations Unlimited in Their Number of Terms*), composed in 1669 and privately circulated but not printed until 1711; the "De quadratura curvarum" of 1691, which appeared in truncated forms in Wallis's Latin *Algebra* of 1693 and as an appendix to *Opticks*; *Universal Arithmetick* (the term means algebra), published in 1707; and the *Methodus differentialis* of 1711. Newton's *Method of Fluxions and Differentials*, a translation by John Colson, was published posthumously in 1736. From 1959 to 1977 *The Correspondence of Isaac Newton* came out in sections. H. W. Turnbull edited the first three volumes, J. F. Scott the fourth, and A. Rupert Hall and Laura Tilling the last three volumes.[591] A more critical resource for preparation of this study than his letters is Derek T. Whiteside's annotated *The Mathematical Papers of Isaac Newton,* published in eight sizable volumes from 1967 to 1981.

Newton's foremost mathematical achievement was the creation and application of fluxional calculus. The calculus, which treats the mathematics of continuity, is today part of mathematical analysis. Newton's invention depended on perceiving the interlocking structures of Cartesian analytical geometry, infinitesimal analysis, and the expansion of functions as infinite series, together with developing the techniques connecting them. It was to take more than two hundred years for the fundamentals of calculus to evolve satisfactorily, arriving at a rigorous understanding of such properties of infinite series as convergence, continuity beyond the nonmathematical concept of smoothness, and primitives. In modern symbols, if the derivative of $F(x)$ is the function $f(x)$, then $F(x)$ is its primitive. Thus, primitives are essentially antiderivatives.

In the process of inventing infinitesimal calculus and recasting its foundations to the fluxional version, Newton began in 1664 as a mathematical apprentice, reading extensively, taking notes, and gaining a firm grasp of the existing core of mathematics.[592] His two principal formative influences were Descartes and Wallis. Shortly before Christmas of 1664 he purchased van Schooten's 1659 Latin edition of Descartes' *Geometry* with annotations and Wallis's *Arithmetic of Infinities*. The *Arithmetic* was part of Wallis's first *Opera omnia*, prepared in collaboration with Brouncker in 1656 and 1657.[593] Newton also read Wallis's *Commercium epistolicum*, including letters of Fermat on number theory, and his *Tractatus duo* of 1659, examining the cycloid and cissoid. He mastered Descartes and Wallis without the assistance of any formal instruction. Reading these texts in two- and three-page segments, Newton did not proceed until he understood each portion in depth. Apparently, Descartes, van Schooten's supplemental authors, and Wallis revealed the most promising lines of development in mathematics. Newton examined as well Viète's *Opera mathematica*, edited by van Schooten in 1646. For an adequate arithmetical symbolism, Newton depended on the third edition of Oughtred's small *Clavis mathematicae*, and for the algebraic he turned to Descartes. His studies of elementary scholastic logic and primarily of Barrow's *Euclid*, which he read at least twice, guided him in deductive procedures and the nature of axiomatic proof. Newton annotated Books V, VII, and X on proportions and number theory. He read no other ancient Greek geometer than Euclid. John Conduitt has him at first consider the *Elements* "a trifling book," finding its first theorems obvious, but changing his assessment upon proceeding to its more difficult theorems.[594] By late 1665, within a scant year of beginning these intense studies, Newton had become a master of mathematics, taking details from texts of others but inspired chiefly by his fertile mind. By the next year he was the peer, if not the superior, of Gregory and Huygens.

In the winter of 1664–1665, Newton at twenty two was investigating Wallis's *Arithmetic of Infinities*, particularly Propositions 118 through 121 on the quadrature of curves. In 1654 Pascal had computed for positive integral $n = 1$, 2, 3, ... the coefficients for the expansion of $(a + b)^n$. His rule is $(a + b)^n = a^n + (n/1)a^{n-1}b + (n(n-1)/1 \cdot 2)(a^{n-2}b^2) + \ldots + b^n$. Newton probably did not yet know that Pascal had shown in his triangle that the coefficient $_nC_r$ in each row is the sum of its two superiors, the term to the left and the other to the right in the row above, that can be stated as $(n-1)C_{r-1} + (n-1)C_r$. He noticed instead that Wallis had obtained as upper limits of a sequence of sums the sums of areas under a sequence of curves or integrals.[595] Working with the curves $y = (1 - t^2)^n$, Wallis found in present-day notation:

$$\int_0^x (1 - t^2)\, dt = x - \frac{1}{3}x^3,$$

$$\int_0^x (1 - t^2)^2\, dt = x - \frac{2}{3}x^3 + \frac{1}{5}x^5,$$

$$\int_0^x (1 - t^2)^3\, dt = x - \frac{3}{3}x^3 + \frac{3}{5}x^5 - \frac{1}{7}x^7,$$

$$\int_0^x (1 - t^2)^4\, dt = x - \frac{4}{3}x^3 + \frac{6}{5}x^5 - \frac{4}{7}x^7 + \frac{1}{9}x^9, \text{ and so forth.}$$

But Wallis had been unable to compute the areas of circles, that is, when the exponent n equals $1/2$.

In a breakthrough, Newton replaced by a free variable x Wallis's fixed upper bound in integrals for areas. He observed that each sequence has x as the initial term, subsequent x's increase in odd powers, 3, 5, 7, 9, ..., with algebraic signs alternating. He further observed that the second term is $\frac{0}{3}x^3$ or $\frac{1}{3}x^3$ or $\frac{2}{3}x^3$ or $\frac{3}{3}x^3$, depending on whether n is 0, 1, 2, 3, and so forth. Reading down the x^3 terms thus gives an arithmetical progression. By analogy using Wallis's intercalation principle, Newton concluded that the first two terms of $\int_0^x (1 - t^2)^{\frac{1}{2}}\, dt$ are $x - \frac{1/2}{3}x^3$. An algorithmic cascade, $(P + PQ)^{m/n} = P^{m/n} + (m/n)AQ + [(m-n)/2n]BQ + [(m-2n)/3n]CQ + \ldots$, where $P + PQ$ is the quantity whose root, power, or root of a power is to be ascertained, and $A = P^{m/n}$, $B = mAQ/n$, $C = (m-n)BQ/2n$, similarly reveals the next values to be $-\frac{1/8}{5}x^5 - \frac{1/16}{7}x^7 - \frac{5/128}{9} - \ldots$. In modern notation Newton showed that $\int_0^x (1 - t^2)^n\, dt = x - \binom{n}{1}1/3x^3 + \binom{n}{2}1/5x^5 - \ldots$, where

$$\binom{n}{p} = \frac{n(n-1)\ldots(n-p+1)}{1 \cdot 2 \cdot 3 \cdot 4 \cdot 5 \cdots p}.$$

The denominator is thus $p!$. Apparently, Newton did not initially recognize that this integral form had the coefficients in the binomial expansion. Upon closer examination he observed that

$$\frac{1}{8} = \frac{1}{2} \cdot \frac{1}{4} = \frac{1}{2}\left(\frac{1}{2} - 1\right)/2;$$

$$\frac{1}{16} = \frac{1}{2} \cdot \frac{1}{4} \cdot \frac{3}{6} = \frac{1}{2} \cdot \left(\frac{1}{2} - 1\right)\left(\frac{1}{2} - 2\right)/1 \cdot 2 \cdot 3;$$

and

$$\frac{5}{128} = \frac{1}{2} \cdot \frac{1}{4} \cdot \frac{3}{6} \cdot \frac{5}{8} = \frac{1}{2}\left(\frac{1}{2} - 1\right)\left(\frac{1}{2} - 2\right)\left(\frac{1}{2} - 3\right)/1 \cdot 2 \cdot 3 \cdot 4.$$

These are simply the sequence of binomial coefficients: $\binom{n}{1}$, $\binom{n}{2}$, $\binom{n}{3}$, $\binom{n}{4}$. Newton now either read Pascal, whose treatise on the arithmetical triangle had just been published, or rediscovered that triangle on his own. Proceeding from a quadrature problem, Newton discovered through these computations that Pascal's rule holds for $n = 1/2$. He soon determined that the fractions need not have a denominator of 2. He also understood that by interpolation he could obtain the result from the binomial cascade by deriving $(1 - x^2)^{\frac{1}{2}} = 1 - \frac{1}{2}x^2 - \frac{1}{8}x^4 - \frac{1}{16}x^6 - \frac{5}{128}x^8 - \ldots$ and integrating the series term by term to compute the area.

By Wallisian intercalation and extrapolation coupled with perceptive guesses by analogy, Newton had succeeded in computing the binomial coefficients for fractional and negative exponents n, thereby generalizing the binomial theorem. Previously, the coefficients had been known only for positive integral exponents n. Newton knew that his interpolation method was dangerous. His guesses were occasionally wrong, but he quickly saw and corrected errors. Newton did not publish his findings on the binomial theorem. They became known through the private circulation of his *De analysi* after 1669 and were published in Wallis's *Algebra* of 1685. Nor did Newton provide a rigorous proof of the binomial theorem for real n. Niels Abel would provide that a century and a half later.

Newton saw that his generalized binomial theorem "much shortened" the extraction of roots.[596] His verification that $(1 - x)^{\frac{1}{2}} = 1 - (1/2)x - (1/8)x - (1/16)x - (5/12)x - \ldots$ allowed him to find a decimal approximation of $\sqrt{7}$ in this fashion. $7 = 9(7/9) = 9(1 - 2/9)$, and hence $\sqrt{7} = \sqrt{9(1 - 2/9)} = 3\sqrt{1 - 2/9}$. Substituting $2/9$ for x in the first six terms of the preceding expansion gives $\sqrt{7} \approx 3(1 - 1/9 - 1/162 - 1/1458 - 5/52488 - 7/472392) \approx 2.64576\ldots$. His result from only six numerical terms differs from the true value of $\sqrt{7}$ by merely 0.00001. Greater accuracy could be obtained by extending the binomial to a greater number of terms. Newton could apply the same technique for approximating cube, fourth, and higher roots. The most remarkable features of Newton's binomial procedure for computing roots, William Dunham asserts, are that it exactly indicates which fractions to employ, and that it generates results mechanically without requiring any ingenuity on the part of the calculator.[597]

Among the first to emphasize the importance of infinite power series expansions, Newton was to make them central to his calculus. His generalized binomial expansion was only a beginning. Reaching it showed analysis with infinite series to have the same inner consistency as finite algebraic operations and to share their general laws. Converging infinite series were no longer to be merely alternative approximating devices. Momentum had been gathering for determining power series for areas under curves and zones of circles, lengths of arcs of conic sections, and trigonometric functions and their inverses. An instance is Kenelm Digby's asking John Wallis to compute the area under the hyperbola $y^2 = 1 + x^2$. By the autumn of 1665 while at Boothby to avoid the plague, Newton had found that it equals $x - x^2/2 + x^3/3 - x^4/4 + x^5/5 - x^6/6 + \ldots$,

Mercator's series for $\ln(1 + x)$. Letting $x = 1$, for example, gives $\ln 2 = 1 - 1/2 + 1/3 - 1/4 + 1/5 - + \ldots$. Newton enthusiastically computed the area under the hyperbola accurately to fifty-two decimal places. Employing a version of binomial theorem, he and Gregory independently interpolated logarithms. Subsequently, in *De analysi* Newton circulated his finding to assure that he received credit for this work. Descartes' algebra to describe curves had been limited to finite expressions. By expanding functions into infinite series and operating with them, Newton increased the power of mathematics. One advantage of infinite power series expansions is that they can simply represent transcendental curves.

Essential to a mature calculus is a satisfactory foundation. Newton was exploring elements relating to this matter. From the beginning, the concept of infinitesimals as a basis for his method of tangents and the discontinuity of indivisibles had troubled him. During 1666 Newton was working intermittently on mathematics. In a forty-seven page fluxional tract of October 1666, he offered an alternative to infinitesimals and Wallis's static summations by resolving problems by motion.[598] His new kinematic approach considered variables to be continuously flowing quantities. Continuously moving lines sweep out areas and, following defined conditions, continuously moving points trace curves. The status of infinitesimals that are greater than zero but smaller than any quantity was unclear. Newton summarizes tables of integrals for simple functions and gives examples of differentiation. He proves his fluxional algorithm and applies it to resolve tangent problems. Probably from Sluse's rule for constructing tangents that in modern notation equals $dy/dx = (\partial f/\partial x)/(\partial f/\partial y)$, Newton defined what he later termed a fluxion as a beginning of flowing quantities or velocity of rate of change. In his later dot notation, the fluxion for $x^2 - y^2 + 1 = 0$ is $2xx - 2yy$. Newton's fluxion is a differential that in modern notation is $(\partial f/\partial x)(dx/dt) + (\partial f/\partial y)(dy/dt) = 0$. It is not the derivative, but the ratio of y to x is. Recognizing that the operation of quadrature or integration is the inverse of differentiation gives the fundamental theorem of calculus, Newton reduces all geometrical problems to the classes of these two operations. His central inversion theorem of 1666 essentially uses the new functions of the modern derivative and the indefinite integral: in now current symbols, $(d/dx)\int_a^x f(x)\, dx = f(x)$ and $\int_a^x F'(t)\, dt = F(x) - F(a)$.

This formative 1666 tract on fluxional calculus ends Newton's first highly creative mathematical period. Afterward he periodically revised the foundations of fluxional calculus, extended it to equations and problems that for a time were recalcitrant, and sought to assemble his ideas on it into a collective unity. Newton had known that he was talented, but the October 1666 tract so far surpassed work by any other mathematician of the time that he recognized the extraordinary nature of his genius and his responsibility assiduously to pursue his scientific work. About fifty years later, he was to write of the plague years 1665 and 1666, "in those days I was in the prime of my age for invention and minded mathematics and [natural] philosophy more than at any time since."[599] Other topics, especially optics, mechanics, and alchemy, now

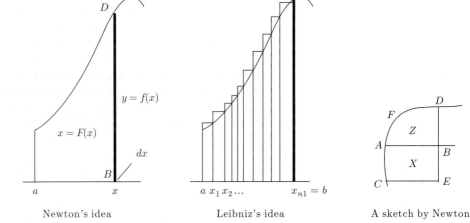

Newton's idea   Leibniz's idea   A sketch by Newton

Comparison of Newton's and Leibniz's ideas of primitives

competed for Newton's time. For thirty months he put mathematics aside.

Most likely a desire to establish his priority for logarithmic series after reading Mercator's *Logarithmotechnia* and entreaties from Barrow, to whom it is dedicated, prompted Newton to write in 1669 what has become known as *De analysi*. His first methodic treatise on fluxional calculus, it seems a collection of ideas that he had been pursuing while chiefly occupied with optics. Its first of three rules gives the processes of differentiation and its inverse of integration in Newton's terms of abscissas and ordinates to a curve. By use of the infinitesimally small time interval o, abscissas and ordinates vanish. But this procedure is not justified, for Newton simply neglects terms containing o and its powers as a final step in computing a series expansions with them. Rule 1 gives Mercator's series $1/(1 + x^2)$, which Hudde had come upon in 1656 and Newton in 1665. This ratio equals $1 - x^2 + x^4 - x^6 + \dots$. Integrating these term by term gives $x - x^3/3 + x^5/5 - x^7/7 + \dots$, the arctan x series. It was known that the closer that radian angle x approaches to 0, the closer x approximates $\sin x$. Expanding $x = (1 - t^2)^{\frac{1}{2}}$, in modern notation Newton's result is $\sin x = x - x^3/3! + x^5/5! - x^7/7! + x^9/9! - \dots$. He also discovered that the $\arcsin x = +\frac{1}{2}x^3/3 + \frac{1}{2}(\frac{3}{4})x^5/5 + \frac{1}{2}(\frac{3}{4})(\frac{5}{6})x^7 + \dots$. This and a reversal of Mercator's logarithmic series and successive approximations yielded the exponential series $x = 1 + y + y/2 + y/6 + \dots$. The derivative of the exponential e^x is the same as the original function. Newton applied his principles to the rectification and quadrature of algebraic curves, including the circle and hyperbola $y = a/(b + x)$, and many mechanical curves, such as the cycloid and quadratrix. Combining his central inverse processes of integration and differentiation opened, he recognized, a nearly limitless expansion of mathematical territory.

Although meant to assure that Newton received credit for his work independently of Mercator, *De Analysi* was only circulated privately, mainly through Barrow and Collins, so until its publication in 1711 it remained semisecret. It

Definition. Consider a right-angled triangle with hypotenuse 1. If x denotes the length of the leg opposite the angle, arcsin x is the length of the arc. The values arccos x and arctan x are defined analogously.

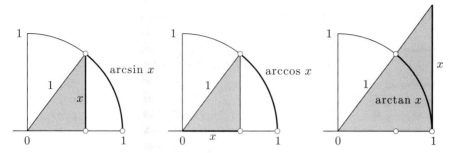

Definition of arcsin x, arccos x, and arctan x.

is questionable whether Barrow, Collins, and the few others who saw the manuscript recognized the potential of its new methods. Even Leibniz, to whom Collins showed the manuscript during a visit to London in 1676, seemingly passed over its infinitesimal procedures, believing that they repeated earlier work.

In the winter of 1670–71, Newton began work on a monograph *De methodus serierum et fluxionum* (*A Treatise on the Method of [Infinite] Series and Fluxions*), revising earlier studies, that was intended to support his claim to priority for calculus. It consists of about 140 manuscript pages. After editing Barrow's *Lectiones geometricae*, Newton introduces the terminology *fluent* for the sum (an integral) of a quantity, say x or y, that varies or flows, and *fluxion* for the differential or velocity as rate of change. His concept of a *moment*, the amount the fluent quantity increases over an infinitely small time, sustains an element of infinitesimals. The status of moments seems to have troubled Newton, for they are evanescent, vanishing without explanation. He also introduces his notation of pricked or dot letters, such as $\dot{x}$, for the fluxion and small o's for moments. It is helpful to consider the moments o as varying, but they could be fixed. Among examples that Newton gives to illustrate his new method, which is similar to modern differentiation for finding the slope of the tangent, is the equation $x^3 - ax^2 + axy - y^3 = 0$. Substituting $x + \dot{x}o$ for x, and $y + \dot{y}o$ for y, gives

$$x^3 + 3x^2\dot{x}o + 3x\dot{x}^2o^2 + \dot{x}^3o^3 - (ax^2 + 2ax\dot{x}o + a\dot{x}^2o^2)$$
$$+ (axy + ax\dot{y}o + a\dot{x}yo + a\dot{x}\dot{y}o^2) - (y^3 + 3y^2\dot{y}o + 3y\dot{y}^2o^2 + \dot{y}^3o^3) = 0.$$

Expunging $x^3 - ax^2 + axy - y^3 = 0$ and dividing the result by an infinitely small o, Newton obtains $3x^2\dot{x} + 3x\dot{x}^2o + \dot{x}^3o^2 - 2ax\dot{x} - a\dot{x}^2o + ax\dot{y} + a\dot{x}y + a\dot{x}\dot{y}o - 3y^2\dot{y} - 3y^2\dot{y} - 3\dot{y}^2yo - 3\dot{y}^3o^2 = 0$. Since o is infinitely small, terms "multiplied by it will be nothing in respect to the rest: I therefore reject them." Blotting out terms that are multiples of o and o^2 and subtracting equals yields $3x^2\dot{x} - 2ax\dot{x} + a\dot{x}y + ax\dot{y} - 3y^2\dot{y} = 0$. This gives the slope of the tangent

$\dot{y}/\dot{x} = (3x^2 - 2ax + ay)/(3y^2 - ax)$.

While *De methodus serierum* demonstrates a breadth of vision and a mastery of technical detail and insists that the method of fluxions is well founded on the geometry of the ancients, Newton knew that many fundamental concepts had to be clarified, methods made more comprehensive, and concise symbolism devised. A systematic structure for calculus including these was far distant. Newton was particularly displeased with the vanishing infinitesimal moments, writing that "in mathematics the minutest errors are not to be neglected." In the winter of 1671 he completed more than half of the manuscript of *De methodus serierum*, but made an author's typical lament that it was growing longer than expected. He wrote to Collins at the Royal Society that he would possibly finish it in the summer of 1672, but he never did. Collins believed that Newton's interest had now shifted away from mathematics, that he had become more "intent upon Chimicall Studies and practices, and both he and Dr. Barrow &c [are] beginning to think math[emati]call Speculations to grow at least nice and dry, if not somewhat barren."[600]

Seeking to eliminate from calculus the concept of infinitesimals, Newton was to achieve a more satisfactory foundation in limiting processes. His new method of what he calls "prime and ultimate ratios" or "first and last ratios" appears in Section One, Book One of the *Principia*. Book One essentially gives the derivative as we do today, $\lim_{h \to 0}\{[f(x+h) - f(x)]/h\}$. Depending on geometric and kinematic intuition, it seems the closest of the new calculus methods to Archimedean exhaustion.

In the summer of 1691 Newton was drawn again to pure and applied mathematics. Leibniz had begun publishing on the differential calculus in 1684 and made no mention of the work of Newton. In light of their 1676 correspondence, Newton was apparently nursing a sense of grievance at this neglect. In London he now met David Gregory. Recognizing the power of his favor, Gregory shamelessly flattered Newton, who recommended him for the Savilian chair of astronomy at Oxford. Newton made this choice over Halley, whom he was later to support for the Savilian chair of geometry. In the fall Gregory asked Newton for a letter on his binomial expansion. Newton, seeing that Gregory lacked a good account of his method of quadrature, told him that his manuscript of 1671 was no longer acceptable. In defense of his priority over Leibniz and probably in reaction to Wallis's report that continental mathematicians were employing Newton's "notions (of fluxions) ... (under) the name of Leibniz's calculus differentialis," Newton set about composing a larger treatise known as *De quadratura curvarum*.[601] In working with infinite converging series, it expands on his method of "prime and ultimate ratios" and applies his fluxional method to solve a set of problems of mechanics that Leibniz was considering. Its dot notation and the letter Q as an alternative to Leibniz's $\int$ for summa seem a direct competition with Leibniz's notation. Newton claimed that from 1676 he had been able to solve the problems in *De quadratura*, but the solutions did not appear until the 1690s. Newton's circle of friends in London, including Nicholas Fatio de Duillier, expected a priority dispute, but Newton's interest in the subject rapidly declined. To avoid the debate and embarrassment that publications would cause, he withheld *De quadratura* from the press.

17.3   The Creation of Leibniz's Differential Calculus

On the continent during this same period, Leibniz made original and extensive contributions to mathematics, which stood as the touchstone of his natural philosophy. Chief among them were symbolic logic and calculus.

Influenced by Ramon Lull, Leibniz pursued symbolic logic as part of a search for a universal characteristic or language of thought. Believing that all complex ideas arise from a small number of simple ideas, Leibniz sought an alphabet of human thought in a code like hieroglyphics or Chinese script. He associated each simple idea with a prime number and complex ideas with products of primes. The grammar and syntax of his alphabet of thought were to be analogous to methods of arithmetical and algebraic computation. Reasoning by means of his universal characteristic was supposed to be free of Aristotelian syllogistic errors and to foster discovery of new truths. Even before going to Paris, Leibniz realized that this project required the preparation of an encyclopedia classifying and analyzing complex thoughts. Possibly the magnitude of that undertaking, given the number of his other involvements, kept him from publishing on symbolic logic. In the nineteenth century George Boole, Gottlob Frege, and Giuseppe Peano would independently take up the project. Otherwise, Leibniz treated algebra as part of combinatorics and was one of the founders of determinant theory for helping to solve equations.[602] He pioneered topology, which he called *analysis situs*, and developed complex and binary numbers.

A letter of April 1703 from Leibniz to Jakob Bernoulli declares that his research leading to calculus did not begin until his arrival in Paris in 1672.[603] His only prior mathematical training had been in lessons from Johann Kühn on Euclid's *Elements* in the summer of 1661, and Kühn probably dealt only with commentaries on its first books. As a college student, Leibniz also read on his own the beginning algebra in Johann Lantz's *Institutiones arithmeticae* of 1616 and apparently portions of Clavius' *Geometria practica* and Cavalieri's *Geometria indivisibilibus*. He came by chance across Cavalieri's book and Vincent Léotaud's *Elements of Curves* of 1654 in Nuremberg, but did not yet understand the profundity of Cavalieri's ideas. As he remarks in the letter to Bernoulli, he was then "about to swim without corks," an alternative to the more common metaphor "I tried to run before I could even walk."[604] In Paris Leibniz was to exchange ideas with the city's leading mathematicians. Encouraged by Pierre de Carcavi, the royal librarian, he designed his new calculating machine and had a working model by 1673. Leibniz had met Huygens in the fall of 1672 and apparently passed on information that in combinatorics he had examined successions of differences, such as $b_1 = a_1 - a_2$ and $b_2 = a_2 - a_3$. Huygens recommended that he read Wallis's *Arithmetica* and Gregory of St. Vincent's *Opus geometricum*. Leibniz quickly generalized Gregory's method of summing and applied it to summing differences between successive forms in a geometric series. Challenged by Huygens, Leibniz summed the nettlesome series in modern symbols $S_n = \sum_{n=1}^{\infty} 2/n(n+1)$. Its terms, recall, are the reciprocals of triangular numbers, $1, 3, 6, 10, \ldots$. Leibniz observed that since $2/n(n+1) = 2\{1/n - 1/(n+1)\}$, each two successive terms of the series can be expressed as a difference. For the first n terms, the sum is simply $2\{1/1 - 1/(n+1)\}$. For n going to infinity, the sum is $2(1 - 1/\infty) = 2$. Leibniz's clever regrouping of terms and rapid computation impressed Huygens, who had obtained the same answer.

Making an analogy to Pascal's arithmetical triangle and employing his concept of differences among triangular reciprocals, Leibniz devised a method for generating a triangle of fractions that he named the "harmonic triangle." Its first row contains the inverses of the integers, the harmonic series. Each of its terms can be obtained by subtracting the element immediately above it from the term to the left in the prior row. For example $1/2 - 1/6 = 1/3$ and $1/12 - 1/30 = 1/20$. In Pascal's arithmetical triangle, each element not in the first column is the sum of elements in the row above and to the left.

$$
\begin{array}{ccccccccc}
1/1 & 1/2 & 1/3 & 1/4 & 1/5 & 1/6 & 1/7 & \cdots \\
 & 1/2 & 1/6 & 1/12 & 1/20 & 1/30 & 1/42 & \cdots \\
 & & 1/3 & 1/12 & 1/30 & 1/60 & 1/105 & \cdots \\
 & & & 1/4 & 1/20 & 1/60 & 1/140 & \cdots \\
 & & & & 1/5 & 1/30 & 1/105 & \cdots \\
 & & & & & 1/6 & 1/42 & \cdots \\
 & & & & & & 1/7 & \cdots
\end{array}
$$

In the harmonic triangle each element not in the first row is the sum of all elements in the next row and to the right. The number of terms in each row is infinite. The second row, for example, equals 1, that is, the sum of the reciprocals of triangular numbers. The sum of the third row, $1/3 + 1/12 + 1/30 + 1/\{n(n+1)(n+2)/2\}$ is $1/2$, which is one-third the sum of the reciprocals of pyramidal numbers, $\{n(n+1)((n+2)\}/(1\cdot2\cdot3)$. The fourth row is one-fourth of the reciprocals of four-dimensional figurate numbers. Except for the terms comprising the first row—the harmonic series—Leibniz thus found that all series in the harmonic triangle converge. His results were not novel, but he perceived that "the operations of summing sequences and of taking their difference sequences are in a sense inverse to each other."[605] From this arithmetical study, he soon extrapolated a geometrical interpretation. Infinitely many equidistant ordinates or infinitely many infinitesimal rectangles summed give the quadrature under curves, while the subtraction of successive ordinates approximates the local slope of the tangent.

In January and February of 1673, Leibniz visited London on a peace mission. Previous exchanges of information had given him little knowledge of the work of British mathematicians, which he perhaps undervalued. For two years he had corresponded with Oldenburg on philosophy and mechanics, in part seeking comments from Wallis and Mercator on his tract *Hypothesis physica nova* of 1671. Leibniz brought to London his wooden calculating machine that performed the four basic arithmetical operations. In a letter Huygens described it as promising. Leibniz demonstrated his machine at the Royal Society, where Hooke inspected it. Leibniz's model was superior to a British calculator by Samuel Morland that could multiply and divide only with the help of Napier's bones. At a dinner at his home, Robert Boyle introduced Leibniz to the rather morose John Pell, reputedly one of Britain's two leading mathematicians, the other being Wallis. On mathematics Collins was to correspond mostly with Gregory, Newton, and Pell. Pell's major research had been completed in the Cromwellian period.[606] Leibniz described to Pell his method of differences and interpolation for generating the harmonic triangle and his summations. Pell was not receptive. Without speaking to the method, the acerbic Pell defined the results as a mere offshoot of the latest literature, which amounted to a veiled charge of plagiarism. He and Oldenburg noted that the summations of figurate numbers could be found in Gabriel Mouton's report on the work of the Catholic canon François Regnaud in Lyons and in Mengoli. Pell wrote to Collins that Mengoli had also shown that the harmonic series diverges and, referring to Newton's logarithmic approximations, that the English had obtained partial sums for the series.[607]

Leibniz claimed originality for his method, but his rough handling by Pell and other fellows of the Royal Society made him painfully aware that his mathematical preparation was inadequate. He had not yet read Mouton or Mengoli and had only skimmed through Pascal's work on the arithmetical triangle. Occasionally and fatefully, throughout his career Leibniz did not cite fully his sources. In England the issue of plagiarism was to continue to haunt him. Adding to the attack by Pell, Hooke belittled Leibniz's calculator and called for improvements, which Leibniz had already been planning. Primarily for his calculator, Leibniz was in April unanimously elected a fellow of the Royal Society. Oldenburg led the support for him.

Embarrassed by the brush with Pell, Leibniz on his return to Paris decided to study higher mathematics intensely. His period of what he called "proud ignorance" was about to end. He met Jacques Ozanam and worked with him on problems of Diophantine analysis and number theory. Delivering a letter from Oldenburg to Huygens gave Leibniz the opportunity to reestablish that crucial contact. The *Horologium oscillatorium* had just been published. Huygens gave Leibniz a copy and explained that his pendulum studies derived from Archimedes' methods for centers of gravity.[608]. As Leibniz related in a letter to Tschirnhaus, Huygens laughed off his comment "that a straight line drawn through the center of gravity must always divide an area into two equal parts." Huygens now agreed to mentor Leibniz in higher mathematics and physics. For the next three years, he did so partly through correspondence. Thinking Leibniz to be more than a beginner in mathematics, Huygens started him reading Pascal's *Letters to Dettonville*, which was in the Royal Library.

According to Joseph Hoffmann, during 1673 and early 1674 Leibniz progressed from an amateur to a nearly mature mathematician.[609] Huygens recommended works by Descartes, Fabri, Gregory of St. Vincent, James Gregory, Paul Guldin, Mercator, Sluse, and perhaps la Hire.[610] Leibniz borrowed these books from the Royal Library and searched for others by French and Jesuit authors but found few. The only major author missing was Fermat, but Huygens was familiar with his work and must have filled that gap. Studying with Roberval, Leibniz now stressing van Schooten's second edition of Descartes' *Geometria* was able to identify oversights in it. He worked through many propositions of Gregory's difficult *Geometriae pars universalis*, which yielded him in his search for a simplified tangent method during the summer of 1673 one example. Leibniz also began to explore the quadrature of curves. In the second, expanded edition of Sluse's *Mesolabum*, whose index gave the latest developments with infinitesimals, he found a tangent and a quadrature method. Leibniz, who corresponded with Sluse, adopted his tangent method and learned from him how to construct the normal to a conic. Sluse's tangent method was the most general, applying to all geometric curves. In 1673 Leibniz investigated other trigonometric series, obtaining the expansions of $\sin x$, $\cos x$, and $\arctan x$. To this point, English authors seem to have had little influence on Leibniz. While in London he had purchased a copy of Barrow's *Lectiones geometricae*. Hofmann argues that tackling this book requires a firm grounding in Euclid's *Elements*, which Leibniz had not yet seriously studied.[611] Possession of the book has created confusion over Barrow's influence. Ehrenfried Walter von Tschirnhaus (1651–1708) asserted that Leibniz had taken from Barrow his basic ideas for the calculus followed by Jakob Bernoulli in the *Acta eruditorum* for January 1691.[612] Leibniz denied this. Hofmann traces to a later period the marginal notes in the *Lectiones geometricae* that Leibniz made. As will be seen shortly, Hofmann and Whiteside argue that Barrow had little or no impact.

As he read Pascal's *Letters to Dettonville* and particularly the treatise on the quadrature of sine curves, Leibniz would remember, a light suddenly burst upon him. By 1674 he was perceiving that the operation of quadrature, that is, the summing of equidistant ordinates or infinitely thin rectangles, is the inverse of finding the ratio of differences of successive ordinates and abscissas that approximate the slope of tangents. Making these approximations exact requires that the differences be infinitely small. Leibniz wrestled with the notion of the infinitely small, seeking, but never achieving, a logically rigorous explanation of it. Pascal's *Letters* suggested to him a major idea that connected the summing with the differencing processes. It was the differential triangle, which he named the characteristic triangle. In his later notation, this infinitesimal right triangle has hypotenuse ds connecting a tangent and nearby point and sides dx and dy. It is similar to the triangle whose sides are ordinate y, tangent τ, and subtangent t. Among others, Fermat, Torricelli, Roberval, and Barrow had earlier known the characteristic triangle.

By the summer of 1674 Leibniz had found an expert watchmaker named Olivier, who made an improved version of his calculating machine. This project,

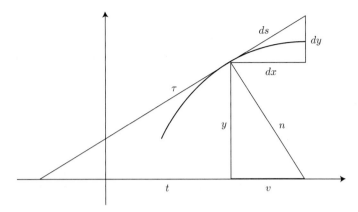

by his accounting, cost him a small fortune. In his apartments Leibniz showed friends the new calculator and word of it spread among scientific circles in Paris. Étienne Périer, Pascal's eldest nephew and his executor, who was visiting the city, asked to see the machine that bettered his uncle's in multiplication and division. Leibniz, who deeply admired Pascal's writings, agreed.[613] He praised Périer for bringing out a second edition of the *Pensées*. Périer was taken by Leibniz's graciousness, charm, and mathematical ability. Regretting that much of Pascal's work remained unprinted, Périer agreed to share with Leibniz the unpublished mathematical papers, including the writings on projective geometry, if he would accept responsibility for their safety. Leibniz was elated soon to have access to all of Pascal's mathematical work.

In July of 1674 Leibniz wrote to Oldenburg of the stir that his calculating machine was causing in Paris and of his discovery of a most important theorem in infinitesimal geometry, which is known as Leibniz's transmutation rule or theorem. His first application of this theorem was to calculate the rational quadrature of a segment of a cycloid. To Huygens, who had earlier computed a quadrature of a cycloidal zone and now recognized the originality of Leibniz's method, he submitted a report of his findings, *Accessio ad arithmeticam infinitorum*. The theorem is so general and fertile that it enabled Leibniz to confirm all known quadratures and to propose a foundation for Wallis's arithmetic of infinities. Having been occupied chiefly with political and literary tasks set by his patrons, Leibniz informed Oldenburg that he had obtained his rule more from intuition than from laborious study.

The basic idea of the transmutation theorem is simple. To determine the areas of curvilinear figures, Cavalieri had divided them into infinitesimal rectangles of the same width. Leibniz now evaluated the areas under curves in two ways, utilizing both infinitesimal rectangles and, from Desargues and Pascal, infinitesimal triangles.

Consider the curve OPA, having two neighboring points P and P'. For a succession of the point pairs, P and P', the combined area of all the small triangles added to that of $\triangle OAB$ equals the area of $OPAB$. Leibniz follows the procedure of drawing a tangent PT, according to Sluse's rule, and the lines OS, OF, and SQQ', as indicated in the illustration. Drawing a characteristic

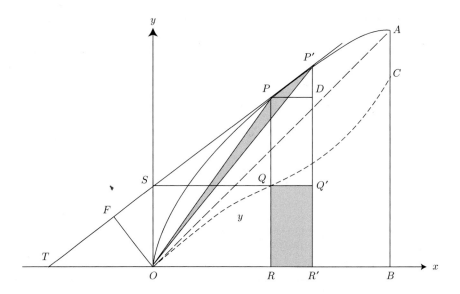

triangle, whose hypotenuse goes from tangent point P to P', gives $PP'/PD = OS/OF$. The result is that $\triangle OPP' = \frac{1}{2}$ rectangle $RQQ'R'$. Thus, the area of $OPAB$ = rectangle $RQQ'R' + \triangle OAB$ = area $OQCB + \triangle OAB$. As point P proceeds from O to A, a corresponding point Q, as shown, traces a trans-muted curve. Leibniz thus effectively reduces to the quadrature of a simpler curve that of another given curve. In modern notation, for $z = y - x\,dy/dx$, his transmutation rule gives $\int_0^{x_0} y\,dx = \frac{1}{2}\int_0^{x_0} z\,dx + x_0 y_0$, which reduces to $\int_0^{x_0} y\,dx = x_0 y_0 - \int_0^{x_0} x(dy/dx)\,dx$, integration by parts. Besides applying his rule to the rational cycloid segment, Leibniz employed it to obtain the quadra-tures of higher parabolas and a circle.

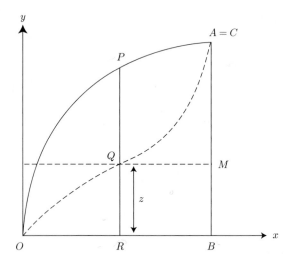

Huygens apparently saw nothing new in principle in Leibniz's theorem and considered the cycloid quadrature evident, though surprising. He added improvements in pencil to the *Accessio* and wrote a complimentary November letter to Leibniz, noting that he and Gregory had squared the cissoid.[614] Baffled by Leibniz's quadrature of the circle, Huygens requested further information. By a clever change of axes for the transmuted curve of the circle of radius a, the new curve represented by $y^2 = 2ax - x^2$, Leibniz had found that area $OQMB = \int_0^z (a - x)\, dx = az - \frac{2}{3}z^3/a + \frac{2}{5}z^5/a^3 - \ldots$. For $z = a$, this becomes area $OQCB = a^2(1 - 2/3 + 2/5 - 2/7 + \ldots)$. From his transmutation theorem, Leibniz knew that area $OPAB = \frac{1}{4}\pi a^2 = \frac{1}{2}$area $OCQB + \frac{1}{2}a^2$. Inserting into this last equation his result for the area $OQCB$ yields the formula for the arithmetical quadrature of the circle, the series $\pi/4 = 1 - 1/3 + 1/5 - 1/7 + 1/9 - \ldots$, which is the series for arctangent(1). Leibniz's method avoids the difficult extrapolations of Newton, and his series converge more rapidly. Among them is the alternative $\pi/8 = 1/(1 \cdot 3) + 1/(5 \cdot 7) + 1(9 \cdot 11) + \ldots$. Leibniz's investigation of the quadrature of the hyperbola produced $\frac{1}{4}\log 2 = 1/(2 \cdot 4) + 1/(6 \cdot 8) + 1/(10 \cdot 12) + \ldots$. His results for two basic transcendental problems, the quadratures of the circle and the hyperbola, probably convinced Leibniz that deeper connections existed between them.

Lacking crucial parts of Leibniz's correspondence, James M. Child has equated with Barrow's quadratures in polar coordinates Leibniz's transmutation given in Cartesian coordinates. He concludes that Leibniz's *Accessio* was indebted to Barrow's *Lectiones geometricae*, arguing that it introduced him to tangent constructions and influenced Sluse.[615] Holding Leibniz's method to be too close to that of Barrow to be accidental, J. F. Scott concurs.[616] Hofmann disagrees with both scholars.[617] The rectangular coordinates in which Leibniz initially presented his transformations Hofmann finds to be unrelated to Barrow's method. Hofmann surmises that Sluse was more influential. Leibniz was probably present at the Royal Society meeting of January 1673 when Sluse's letter on his tangent method was read. Leibniz, moreover, was the intermediary for the correspondence of Oldenburg and Collins with Sluse. Hofmann believes that Leibniz did not seriously study Barrow's book until Tschirnhaus mentioned it in 1675. Again, the *Lectiones geometricae* largely generalizes quadrature, tangent, and rectification procedures known on the continent and in England, above all there from the work of Gregory. Whiteside concludes that those procedures were the more likely source for Leibniz.[618]

In 1675 Leibniz was deeply engaged in studying geometrical loci, graphical techniques for solving equations, and trigonometric functions and their inverses. In Paris he worked with Roberval on solving algebraic equations and drew tangents to Archimedes' spiral. In March he wrote to Oldenburg and Collins inquiring whether British geometers could rectify the hyperbola and ellipse. Collins provided results from Newton's binomial series and Gregory's $\sin \alpha$, $\tan \alpha$, and $\arctan \alpha$ series, but without describing their methods. At the end of September Tschirnhaus, who had attended Leiden University, joined Leibniz's circle of friends in Paris. During his visit to England, Tschirnhaus had met

Wallis and had spoken with Oldenburg and Collins. Through him the three sent a report containing recently obtained mathematical results by the English using infinitesimal methods.[619] The report was intended for Huygens and Leibniz. In return Collins sought information on unpublished related work by French mathematicians.

In a series of notes dating from October and November of 1675, Leibniz invented the basic elements of his version of differential calculus. This he accomplished as he moved toward the algebraization of infinitesimal methods to free them of their opacity and wordiness. Hastily prepared, the notes contain errors, which is likely Leibniz's reason for not publishing them. They include his developing an excellent notation and rules for operations with them. Quite likely Huygens, who had a notation for first-order infinitesimals that he did not publish, provided guidance. For infinitely small intervals, Leibniz gives the sum of the x's as ult.x, and the sum of the very thin rectangles xw as omn.xw.

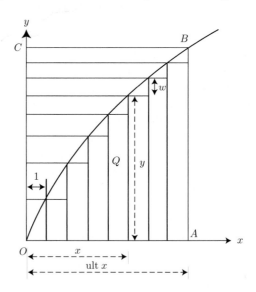

Leibniz discovered the relation omn.xw =$x$ult.x,omn.w $-$ omn.omn.w. Since $w = a/x$ is a rectangular hyperbola, omn.a/x gives its quadrature, which is a logarithm. Clarifying his ideas, Leibniz had l equal the xth successive difference in a sequence of infinitely small differences, such that omn.xl = x,omn.$l$$-$omn.omn.$l$. Letting $l = x$ yields omn.x^2 = x omn.x $-$ omn.omn.x. Leibniz showed that omn.x = $x^2/2$. Thus, omn.$x^2/2$ = $x \times x^2/2$ $-$ omn.$x^2/2$ and omn.$x^2/2$ = $x^3/3$. Pursuing rules for operations with the notation omn., Leibniz on October 29 introduced an important symbol, the elongated script s, $\int$, to stand for *summa*, and had $\int l$ =omn.l, the sum of all l's. But for several years he was inconsistent in its use. The November 11 manuscript, titled "Examples of the Inverse Method of Tangents," considers differentiation, recognizing its differences as the reciprocal of the sums of quadrature. The work employs $\int$ for summing rectangles or areas and $x \mid d$ and then dx for differences between x's. Having recognized that neither $\int$ nor dx involves changing dimensions, for example, if x is a line, then dx is a line segment infinitesimal in length, Leibniz continued to seek rules for operating with arbitrarily small dx. Part of his search involved extrapolating rules of geometry to infinitesimals, such as in the sides of infinitesimal triangles. The meaning of the ratio dx/dy is unclear. It is not yet the modern derivative.

Leibniz, a masterful notation builder, worked steadily to perfect his symbols. He soon introduced $\log x$, d^{-1} for $\int$, d^{-n} for the nth integration of summation, and d^n for the nth derivative of a product. In modern notation, the nth derivative is $(xy)^n = x^n y^0 + n x^{n-1} y^1 + \ldots + n x^1 y^{n-1}$. Leibniz's apt notation calls to mind a line from Goethe's *Faust*: "War es ein Gott, der diese Zeichen schreib?" (Was it a god, who wrote these symbols?)

In his new analytical method, Leibniz recognized that Sluse's tangent rule had to be superceded. Applying his transmutation rule, he almost effortlessly determined quadratures of parabolas and hyperbolas that Wallis had found with difficulty by induction. Claude Dechales, author of the *Cursus mathematicus* of 1674, challenged Leibniz to compute the portion of a circular cone lying between its base and a plane parallel to its axis.[620] Using his quadratures of the circle and the hyperbola, Leibniz completed the computation in one night. These quantities he defined by converging infinite series.

Among the unsatisfactory results in Leibniz's early labors to formulate the basic rules for his calculus was his first assumption, put in his later notation, that $d(xy) = dx\,dy$. But he quickly determined what are $(dx/dt)(dy/dt) \neq d(xy)/dt$ and $d(y/x)/dt \neq (dy/dt)/(dx/dt)$. At the end of 1675, Leibniz wrote to Gallois that while his arithmetical quadratures were not geometrical, they were more than mechanical and approximations, for they were exact. He cited the greater generality and unifying power of his transmutation rule.

Leibniz's *Calculus tangentium differentialis*, an unpublished manuscript dated November 1676, fixes on dx/dy as the best way to determine a tangent and correctly gives the chain rule for differentiating a function. Here, as throughout his work, he emphasizes a crude concept of function and its variables.

Work in England was now coming more to Leibniz's attention. At the end of 1675 Collins had sent him some of Newton's results on infinite series without the author's permission and without proofs. At that time, the Danish geometer Georg Mohr was visiting Paris. Leibniz did not meet him until the spring of the next year. Mohr showed materials that he had obtained directly from Collins on Newton's sine and arcsin series. Their elegance struck Leibniz. In a letter dated May 12 he asked Oldenburg to have Collins send more details and proofs of these and Newton's tangent method. Engaged in completing his proof for the arithmetical quadrature of the circle, Leibniz saw Newton's demonstrations as critical. He offered to send his complete quadrature demonstration to England in return as a present. Since Oldenburg had just sent him information on Newton's work on April 12, some historians have wondered why Leibniz made the request.[621] But Oldenburg did not understand the mathematics, and likely Leibniz wanted a response from a mathematician. Since Collins was ill in the autumn of 1676, Oldenburg asked Newton to reply. It was a tense time at the Royal Society, for the debate between Huygens, Oldenburg, and Hooke over spring watches was in progress, and Hooke among others was subjecting Newton to sharp criticisms of his theory of colors.

Meanwhile, in 1676 a rift grew between Tschirnhaus and Leibniz. Tschirnhaus believed that Descartes' analytic geometry was most important and that Leibniz's differential calculus with its infinitesimals was not part of pure mathematics. Leibniz knew calculus was something new and significant, not a minor extension of analytic geometry. Both scholars sought to correspond with Newton, who was dismissive of Tschirnhaus. Urged by Oldenburg, Newton wrote two letters to Leibniz, sending on October 26 the first, which he labeled the *Epistola prior*, and on November 14 the second, the *Epistola posterior*. Their tone suggests a poor impression of Leibniz.

Drawing on materials from *De analysi* and *De methodis,* the *Epistola prior* presents Newton's binomial theorem and applications. Leibniz quickly recognized Newton as the foremost mathematical genius in Britain. The *Epistola prior*, he wrote, was "filled with more remarkable ideas about analysis than many thick volumes."[622] Leibniz sent a response describing his transmutation method and posing specific questions about infinite series. In October he visited London for ten days and spoke to Collins, who showed him Newton's *De analysi* and *Historiola* on tangents, as well as Gregory's method of maxima and minima. Leibniz took notes on infinite series, but not on the embryonic fluxional calculus. Neither Leibniz nor Collins, who quickly realized the extent of his indiscretion, informed Newton, who would not learn of it until over forty years later.

The *Epistola posterior* emphasizes infinite series, but conceals in anagrams consisting of a mixture of letters and numbers two critical passages about fluxional calculus. Newton believed that these could not be decoded but if necessary could confirm his discovery. One anagram is

$$6accdæ13eff7i3/9n4o4qrr4s8t12vx.$$

It counts the number of each letter in the Latin sentence "*Data aequatione quotcunque fluentes quantitates involvente, fluxiones invenire et vice versa.*" The English translation is "Given an equation involving any number of fluent quantities, to find the fluxions and vice versa."[623]

Upon receipt of the *Epistola posterior* on July 11, 1677, Leibniz penned a reply praising Newton and giving correct rules for differentiating products of two functions, $d(xy) = x\,dy + y\,dx$; quotients, $d(x/y) = (y\,dx - x\,dy)/y^2$; and powers or roots, $dx = nx^{n-1}\,dx$. Since dx and dy are infinitely small, Leibniz in reaching his results disregarded multiples or powers of them. For example, the least difference from xy becomes $(x+dx)(y+dy) - xy$. Dropping $dx\,dy$ as incomparably smaller than the other terms gives $d(xy) = x\,dy + y\,dx$. Leibniz's reply also poses acute questions. He had made good progress since

November 1675. But the correspondence between Leibniz and Newton was not to continue. In September Oldenburg died, removing the crucial connection. Newton's paranoia was growing in his debate over colors with Hooke. In 1677 neither he nor Leibniz published on calculus, which might have established priority.

Although primarily occupied with judicial, diplomatic, and library tasks, Leibniz continued to pursue basic procedures in calculus and to clarify its fundamental concepts. By 1680 he had completed his unpublished *Elementa calculi novi pro differentiis et summis, tangentibus, et quadratrus*, which reviews the steps he had made toward calculus. It considers dx to be the differences among abscissas, and dy, the differences among ordinates. Leibniz treats differences and sums as inverses, such that "$\int dx = x$ and ... $d \int x = dx$." For Leibniz, curves are infinitangular polygons comprised of infinitely many rectilinear segments. Tangents extend these segments. The *Elementa* also offers, in an imagery similar to that of Newton later, Leibniz's concept of momentary increments, and it relates in detail Leibniz's tangent method.

In the October 1684 issue of the monthly *Acta eruditorum*, Leibniz published the first article on calculus. It commenced his series in the *Acta* on the subject that was to go beyond 1695. The shortened title of this six-page paper is *Nova methodus*. Its full title in English translation is "A New Method for Maxima and Minima as well as Tangents, which is Impeded by neither Fractional nor Irrational Quantities, and a Remarkable Type of Calculus for Them." Apparently, Leibniz's recognition that infinitesimals are quite susceptible to logical inconsistencies delayed his publication on calculus. To counter such critics, he defines a first-order differential as a finite assignable line segment, rather than an infinitely small quantity. For dx a given quantity, Leibniz represents dy by $dy : dx = y :$ subtangent. It is his most satisfactory definition of the term, but failing to give an expression for the subtangent, it is incomplete.

The *Nova methodus* synthesizes Leibniz's newer findings. It presents without proofs his rules from 1677 for the differentials of differences, sums, products, quotients, and powers or roots. Leibniz also offers conditions, $dx = 0$, for maxima and minima, as well as conditions for concavity, convexity, and inflection points. Points of inflection require second-order differentials $d\,dx = 0$, where neither x nor $dx = 0$. In emphasizing the inverse relationship between differentiation and integration, Leibniz like Newton includes transcendental curves. This is largely a departure from Descartes, who had excluded some of these mechanical curves from pure geometry. This is perhaps the first time that the word "transcendental" appears in its modern meaning of non-algebraic. Leibniz's definition of the tangent to any point of a curve, including the cycloid, as "a line that connects two points of the curve at an infinitely small distance" is not satisfactory. The *Nova methodus* applies its new methods to solve Fermat's optical refraction problem of determining the path a light ray follows between two different media, a topic the *Acta* had covered two years before. It obtains Snell's law of sines from this and reduces a curve proposed by Florimond De

Beaune, which passes through the origin of Cartesian coordinates and satisfies $dy/dx = (x-y)/a$, to a logarithmic curve. Descartes had been unable to obtain De Beaune's curve as a whole.

Boldly, Leibniz claims that "our methods are of astonishing and unequaled facility" and are "only the beginning of a much more sublime geometry, pertaining to even the most difficult and most beautiful problems of applied mathematics, which without our differential calculus or something similar no one could attack with any such ease."[624] Coining the name "differential calculus," the 1684 paper lacks adequate definitions and sufficient explanation to convey it skillfully. In leaving gaps, it is reminiscent of Descartes' *Géométrie*. Even Leibniz's most gifted disciples, the Swiss brothers Jakob and Johann Bernoulli described this article as "an enigma rather than an explication."[625]

In a second memoir in *Acta eruditorum* in 1686, Leibniz tackled the inverse of the tangent problem, integral calculus, setting out its fundamental rules. The first published work on that subject, it does not use the name. Instead, it retains for the inverse of differential calculus, *calculus summatorius*. In his *Autobiography* Johann Bernoulli claims having first coined the name "integration" for the inverse.[626]   In an article of 1690 in the *Acta* identifying the isochrone curve, Jakob Bernoulli first employed the name "integral" in its modern sense. A year later Leibniz adopted the name "integral calculus" to replace *calculus summatorius*.

Titled in English translation "On Recondite Geometry and the Analysis of Indivisibles and Infinities," the 1686 paper stresses the reciprocal relationship between differentiation and integration and the importance of transcendental, along with algebraic, curves. Leibniz gives as the equation for the cycloid $y = \sqrt{2x - x^2} + \int dx/(\sqrt{2x - x^2})$, His method and notation allow him to handle curves that Newton's method of infinite series could not. Perhaps caught up in his algorithms, Leibniz was to accept in a letter to *Acta Eruditorum* in 1713 Guido Grandi of Pisa's assertion that the infinite series $1 - 1 + 1 - 1 + \ldots$ equal $1/2$. Grandi related it to the mystery of creation, for it can be rewritten as $0 + 0 + 0 + = 1/2$, giving *creation ex nihilo*. Leibniz resolves this paradox by holding that the sum of the series is 0 for an even number of terms and 1 for an odd number. But the legitimacy of his position was disputed, since it was not thought possible to describe infinite extension as composed of an even or odd number of members. He is less circumspect than Newton in working with divergent series.

Leibniz's computation of the simple integral for π employing the derivatives of trigonometric functions suggests the superiority of his method and notation to Newton's. While Newton had approximated $\pi/4$ by extrapolations between series, Leibniz, in finding that $\pi/4 = \int_0^1 dx/(1 + x^2)$, worked with the simple integral for $\tan \alpha = \sin \alpha / \cos \alpha$. Applying the chain rule to the derivative of the tangent gives $d \tan \alpha/d\alpha = \frac{d}{d\alpha}(\sin \alpha / \cos \alpha) = (\sin \alpha / \cos^2 \alpha)d \cos \alpha/d\alpha + (1/\cos \alpha)d \sin \alpha/d\alpha = (\sin^2 \alpha + \cos^2 \alpha)/\cos^2 \alpha = 1/\cos^2 \alpha = \sec^2 \alpha$. The identity $1 + \tan^2 \alpha = 1 + \sin^2 \alpha/\cos^2 \alpha = (\cos^2 \alpha + \sin^2 \alpha)/\cos^2 \alpha = \sec^2 \alpha$. Thus the integral of $1/(1 + x^2)$, where $x = \tan \alpha$, is

$$\int_0^1 dx/(1 + x^2) = \int_0^{\pi/4} \frac{\sec^2 \alpha \, d\alpha}{\sec^2 \alpha} = \int_0^{\pi/4} d\alpha = \pi/4.$$

Among other contributions to calculus and algebra, Leibniz generalized the binomial theorem to the multinomial theorem. This gives series expansions of such expressions as $(x + y + z)^n$.

From the beginnings of calculus, the concept of the infinitely small had troubled mathematicians. Remember that Newton quickly discarded infinitesimals for prime and ultimate ratios. In developing higher-order differentials and the rules for canceling them, which are fundamental to his new calculus, Leibniz went in the opposite direction. While mathematicians accepted the results of Newton and Leibniz, they found them obscure in methods and lacking in good definitions. In 1694 the Dutch mathematician Bernard Nieuwentijdt attacked the new calculus for these reasons. He held Leibniz's higher-order differentials to be no clearer than Newton's evanescent increments or Barrow's letting quantities a and e both equal zero. Nieuwentijdt called for a justification of each of these without the use of infinitesimals, but failed in his attempt. Leibniz responded in the 1695 *Acta eruditorum*, calling his critics "overprecise" and likening them to ancient skeptics. The differential he defined as "less than any given quantity."[627] It was the counterpart to his monads and a Janus-faced tool in computation. Neglect of the differential he compared to Archimedes' assumption that two quantities differing by less than any given amount are equal. The new calculus was, he thought, better adapted to discovery than Archimedes' methods of exhaustion. For justifying differentials, Leibniz refused to appeal to the metaphysics of the continuum. He relied on geometric intuition, on the law of continuity as a postulate, possibly from Kepler, and even more on analogies, such as Hobbes's proposal that the *conatus* is to motion as a point to space or Leibniz's own declaration that a point is to Earth as Earth to the heavens. Two decades later in his *Historia et origo calculi differentialis* Leibniz was to depict differentials as amphibians swimming between existence and nonexistence. Critics continued to ask when they are assignable and unassignable. In calculus, rigorous definitions and foundations did not emerge until the nineteenth century.

17.4   The Bernoullis of Basel

Although groundbreaking, Leibniz's *Acta eruditorum* papers of 1684 and 1686 on calculus were so sketchy and difficult as to be for nearly all readers barely intelligible. On the continent, differential calculus spread mainly not from these memoirs, but from the articles and letters of Jakob Bernoulli (1654–1705), who had reportedly mastered the Leibnizian calculus in a few days of study in 1687 and started publishing on it three years later, and his brother Johann (1667–1748).[628] These two are also known by the familiar French versions of their names, Jacques and Jean. The Bernoullis are one of the most remarkable families in the history of science. From 1650 to 1900 they would produce eleven prominent mathematicians and physicists, the most distinguished being Johann's son Daniel. Like Leibniz, the two brothers who founded this astonishing dynasty confined most of their publications to articles in scholarly journals. They worked to clarify fundamental concepts, to discover deeper connections between problems, to shift emphasis from geometric curves to algebraic functions, and to set out branches. From 1693 they did so partly through Johann's extensive correspondence with Leibniz.

Jakob (I) Bernoulli, the first mathematician in the family, was the fifth child and eldest son of Nikolaus Bernoulli, a prosperous druggist and town magistrate, and Margaretha Schönauer. Their Calvinist Bernoulli forebears had fled before the fury of Catholic persecution in the Spanish Netherlands in 1583, moving in 1622 from Amsterdam to Basel. Five years after receiving a master of arts in philosophy from the small University of Basel in 1671, Jakob in accordance with his father's wishes obtained a licentiate in theology. But as his diary *Meditations* shows, during a stay from 1677 to 1679 in France he studied more the rational method and science of Descartes and his disciples, especially Nicolas Malebranche. Probably the appearance of the great comet in 1680 clinched his shift from theology to mathematics. Jakob now adopted as his motto *Invito patre, sidera verso* ("Against the will of my father, I study the stars"). In 1681 he published a pamphlet on the great comet. For rejecting the notion that comets were signs of divine displeasure and proposing laws that govern their paths, it drew criticism from theologians. Appearing six years before Newton's *Principia*, his laws based on collisions between particles in the ether were faulty. After a second educational trip in 1681 and 1682, when he met Jan Hudde in Holland and Robert Boyle and Robert Hooke in England, Jakob conducted experiments in mechanics in Basel and began lecturing and sending reports to the *Journal des sçavans* and *Acta eruditorum*. In 1687 the University of Basel named him professor of mathematics, a position that he held to his death.

Jakob's contributions to calculus were not chiefly through formulating theories but by adroitly analyzing significant problems. Obstinate, self-willed, insecure, irritable, and vindictive, he was quick to clash with other scholars. Perhaps from as early as the mid-1680s, the rivalry between the slower but more penetrating Jakob and his argumentative and arrogant younger brother Johann, who gained more attention, particularly angered him.

Jakob Bernoulli's *Meditations* and his *Theory of Series*, completed from 1682 to 1704, indicate that between 1682 and 1689 he was familiarizing himself with the basic concepts and questions of what became the new calculus. The *Theory of Series* consists of five dissertations, which are made up of sixty consecutive propositions. In 1682 Leibniz's conclusions for the series for $\pi/4$ and $\log 2$ drew Bernoulli's attention. The next year he set forth the problem of continuous compounding interest, which requires finding $\lim_{n \to \infty}(1 + 1/n)^n$, the exponential series. Expanding by use of the binomial theorem the expression $(1 + 1/n^n)$ gives $(1 + 1/n)^n < 1 + 1/1 + 1/1 \cdot 2 + \ldots + 1/1 \cdot 2 \ldots n < 1 + 1 + 1/2 + 1/2^2 + 1/2^3 + 1/2^{n-1} < 3$, so its limit must lie between 2 and 3. Part Two of Jakob's *Ars Conjectandi* derives the exponential series using Bernoullian numbers. In Euler's notation, the exponential function is e^x. Next to π, e is the most famous number.[629] It was soon computed as nearly $2.7182818\ldots$. From 1684 Jakob examined the writings of John Wallis and Isaac Barrow on mathematics, optics, and mechanics, which brought him to problems in infinitesimal geometry. By 1687 he was working closely with Johannes. Along with Paul Euler, the father of Leonhard, Johann resided at his house while a student at the University of Basel. That year the Bernoulli brothers started delving into articles by Leibniz and Tschirnhaus in *Acta eruditorum* suggesting the differential calculus and its application to mechanics. Mastering Leibniz's method in 1687, Jakob believed that it simply formalized Barrow's account of infinitesimals.

From the late 1680s, as other mathematicians were sharply increasing the number of curves, Huygens and Leibniz were seeking the shapes of the transcendental curves called the catenary, the isochrone, and the tractrix. The isochrone is the plane curve that traces the fall of an object near the Earth's surface having a uniform velocity of descent from one point to another not directly under it. In 1690 Jakob showed that the required curve is a semicubical parabola and posed as a counterproblem the catenary, the shape of a chain loaded by elastic weights uniformly distributed between two points: for example, the cable of a suspension bridge. To the naked eye, it seems that this curve could be a parabola, which is algebraic, as Galileo had thought. But this is not the case. Only the new differential calculus could show the catenary to be transcendental. Jakob could not resolve this problem, but correct solutions by Huygens, Leibniz, and Johann Bernoulli appeared in *Acta eruditorum* in 1691. In a letter of October 9, 1690, Huygens stated that this problem offered a test of the quality of Leibniz's new calculus algorithm, while Johann Bernoulli claimed solving the problem during an entire night of work on it.[630] By different paths, the three found that the curve must satisfy the differential equation $dy/dx = c/a$, where $c = $ the constant and $a = $ the arc length. These problems in mechanics made the Bernoullis more aware of the full power of the new calculus.

In Proposition 43 of his *Theory of Series*, Jakob turned to logarithmic series, restoring to the center of research a subject that Leibniz had not pursued since 1682. Jakob probably discovered that logarithms are the inverse of the exponential function, which is $d(\ln x)/dx = 1/x$.[631] Since Euler was later the first to make e the base of natural logarithms, it is likely that study of the area under the hyperbola $y = 1/x$ led to the number e. The Bernoulli brothers recreated expressions of $\sin n\theta$ and $\cos n\theta$ built on $\sin \theta$ and $\cos \theta$, and Johann knew the inverse relationship between trigonometric functions and imaginary logarithms. Although their concept of functions was vague, they uncovered connections among three of the most important functions in calculus: trigonometric, logarithmic, and exponential. In the 1694 *Acta eruditorum*, Jakob introduced polar coordinates, which locate a point P from its distance r from a set point O, called the *pole*, usually the *origin* in a coordinate system, and an angle θ between line OP and a reference line, usually the x-axis. This gives polar coordinates (r, θ) for P.

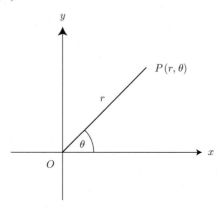

Polar coordinates.

Finding that by transforming curves into these polar coordinates he could investigate many new curves, Jakob proposed in 1694 the curve known as the "lemniscate of Bernoulli," whose polar equation is $r^2 = a^2 \cos 2\theta$. It resembles a figure eight on its side or a knotted ribbon, that is, a *lemniscus*. Jakob was more drawn to a curve identified by Harriot and rediscovered and rectified by Torricelli, the logarithmic or equiangular spiral. Its modern polar equation is $r = e^{a\theta}$. For the constant a positive, the spiral turns counterclockwise. In connection with his brother's private instruction in 1692 of Guillaume François Antoine, Marquis de L'Hôpital (1661–1704), the grand seigneur of French mathematics, Jakob examined the logarithmic spiral, discovering several of its striking characteristics: that its evolute, polar pedal curve, and its caustic, or the envelope of rays refracted along it, trace an equal logarithmic spiral. An evolute is the locus of the points of the changing centers of curvature of the original curve as we proceed along it. The logarithmic spiral thus remains invariant under most geometric transformations. Jakob called it the *spira mirabilis* and ordered that it be engraved on his tombstone.

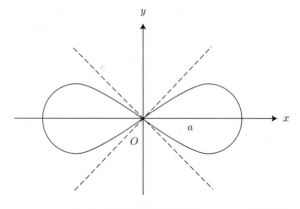

Lemniscate of Bernoulli.

In his work on calculus Jakob Bernoulli gave more attention than his contemporaries to the divergence and convergence of infinite series. Unaware of earlier work on the harmonic series, in proposition 16 of his *Theory of Series* he credited Johannes with discovering its divergence. In 1689 Jakob began with Leibniz's demonstration that the series of the reciprocals of triangular numbers, $S = 1 + 1/3 + 1/6 + 1/10 + \ldots$, converges and proceeded to show that the infinite series of reciprocals of square numbers, in modern notation $\zeta(2) = \sum_{n-1}^{\infty} n^{-2} = 1 + 1/2^2 + 1/3^2 + \ldots$, also does. Through a comparison term to term, Jakob saw that $1/4 < 1/3$, $1/9 < 1/6$, $1/16 < 1/10$, $\ldots$, or that $1/n < 1/n(n + 1)/2$. This meant that the sum of the series of reciprocals of square numbers must be less than 2, and so the series converges. An appendix to Jakob's *Ars conjectandi* similarly proves that the series $1/\sqrt{1} + 1/\sqrt{2} + 1/\sqrt{3} + 1/\sqrt{4} \ldots$ diverges by comparing it term by term with the harmonic series. Seeking better criteria for convergence, Jakob in proposition 59 of his *Theory of Series* improved on a procedure by the Swiss engineer John Christopher Fatio de Duiller that Euler was to develop into his series transformation. Euler wrote the series $\sum_{n=0}^{\infty} b_n$ as $\sum_{n=0}^{\infty} (-1)^n a_n$. By several algebraic steps, he demonstrated that $\sum_{n=0}^{\infty} (-1)^n a_n = \sum_{n=0}^{\infty} \triangle^n a_0 / 2^{n+1}$, where $\triangle^n$ is the nth finite difference. This offers the benefit of transforming a converging series into another that more rapidly converges, as well as some divergent into convergent series.

Assuming continuity in the operations of nature, Jakob Bernoulli contributed to continuum mechanics with fruitful consequences for calculus. He discovered the *elastica*, the curve of a thin elastic rod being compressed at each end, showing that it satisfies the differential equation $ds = dx/\sqrt{1 - x^2}$. In geometrically interpreting its integral, he introduced his lemniscate curve. Jakob's lemniscate integral is an early elliptic integral. It was later discovered that these integrals are closely connected with Diophantine algebra, which is important to the whole of Leibnizian analysis. Jakob's studies of elasticity, which are prominent among the last five propositions of the *Theory of Series,* along with investigations of arcs and amplitude vibrations of simple pendulums, helped lead Euler to formulate the theory of elliptic functions, which Niels Abel discovered have the significant characteristic of double periodicity.

Johann (I), the tenth child in the Bernoulli family, was seemingly destined for a career in the family business. After proving a poor salesman, he was allowed to enroll at the University of Basel in 1683. He received the master of arts degree two years later and embarked on the study of medicine. In 1694 he completed his doctoral dissertation on the movement of muscles. The mechanics and mathematics of the Italian iatromechanist Giovanni Borelli, a member of the Accademia del Cimento, influenced him. Johann also attended his brother's lectures and privately studied mathematics with him. Johann's solution in 1691 of the catenary, a feat he shared with Huygens and Leibniz, quickly placed him among the first rank of European mathematicians.

After teaching Leibnizian calculus at Geneva in 1691, Johann moved to Paris, where he won acceptance in Malebranche's mathematical circle for his method of determining the radius of curvature of a curve using the equation $p = dx/ds : d^2y/ds^2$. He met L'Hôpital, whom he agreed to teach for an ample remuneration the secrets of infinitesimal calculus directly in Paris and through correspondence after his return to Basel. Chiefly for its intellectual results, the contract has been called "the most extraordinary ... in the history of mathematics."[632] The lessons and correspondence are the basis for L'Hôpital's *Analyse des infiniment petits*, the first textbook on differential calculus. In 1692 Johann was introduced to Pierre Varignon, who became his close friend and disciple. The two subsequently embarked on a voluminous correspondence.

The competition between the Bernoulli brothers was intensifying. From 1692 to 1695 they independently solved the *velaria* problem, the shape of a sail filled by the wind, an issue important for positioning a sail boat on the Rhine. The differential equation $(dx/ds)^3 = a\,d^2y/dx^2$ describes the shape. When Jakob called his brother his pupil on the *velaria* problem, who could only report what his teacher taught him, Johann was incensed.

In the 1694 *Acta eruditorum* article "Effectiones omnium quadraturam ...," Johann employed a method that is nearly our integration by parts repetitively for computing integrals and from that derived, without recognizing its full importance, a most general series that is a slightly modified Taylor series.[633] His error term is an integral. Newton had earlier found a similar series and Taylor later did so. In modern notation where x is a constant and z is a variable between 0 and 1, what Johann found is $f(x) = f(x - zx) + zxf'(x - zx)/1! + z^2x^2f''(x - zx)/2! + \ldots$. With Leibniz he devised partial differentiation, which they kept secret for more than a decade. Partial differentiation extended differential calculus from one to two and three independent variables. This requires an altered methodology. In obtaining the partial derivative, now written $\partial f(x, y)/\partial x$ of $f(x, y)$ with respect to x, they held y constant so that $\partial f(x, y)/\partial x = \lim_{\triangle x \to 0}\{f(x + \triangle x, y) - f(x, y)\}/\triangle x$. Keeping x constant gives the partial derivative $\partial f(x, y)/\partial y$. This operation also holds for higher derivatives.

Since his brother held the only mathematics position at the University of Basel, Johann knew that he must look elsewhere for employment. In 1695 he rejected a position at Halle, but accepted a professorship at Groningen in Holland, for which Huygens had recommended him. In the June 1696 issue of *Acta eruditorum*, Johann challenged the mathematical community to solve the brachistochrone problem, finding the curve of quickest descent by a particle from one point to another not directly beneath it, and set a time limit of six months. News that Jakob agreed with the mistaken notion from Proposition 36 of Galileo's *Discorsi* that the circle is the quickest path probably prompted the public contest. Since six months seemed too short a time, Leibniz urged Johann to reissue the brachistochrone and another problem in a flyer addressed to "the shrewdest mathematicians of all the world" with an extended deadline of Easter. In September Leibniz wrote that he had solved the brachistochrone problem the day it arrived, but he never submitted his solution. Leibniz also sent a letter to Jakob Bernoulli asking him to try to solve that problem. He correctly predicted that there would be five solutions: besides himself, Newton, the Bernoulli brothers, and L'Hôpital would provide these. Johann agreed to the extension of the competition. The flyer was sent across Europe. Among others the *Philosophical Transactions*, the *Journal des sçavans*, Wallis, and Newton were to receive copies. By December no satisfactory answer had been returned.

Since Newton had remained silent on the brachistochrone problem through 1696, Bernoulli assumed that it had stumped him. Bernoulli, moreover, had just accused Newton of taking from Leibniz's *Acta* papers the calculus methods in the segment of "De quadratura" published in Wallis's *Opera*. For whatever reason, when he received the brachistochrone problem from France on January 29, 1696/1697, Newton took it as a personal challenge. He had just come home at 4:00 P.M., tired from working on recoinage at the Tower. He wrote the date of receipt on the problem sheet and decided not to go to bed until he had solved both problems.[634] He had the solutions by 4:00 A.M. Newton anonymously forwarded them to Charles Montague, the president of the Royal Society, and they were sent to Bernoulli. From the authority and skill in the paper, Bernoulli immediately knew its author to be Newton *ex ungue leonem* ("as the lion is recognized from his claw"). Newton and in a more rambling manner Leibniz had found the solution to be the differential equation for the common cycloid. Johann was astonished that the curve of solution was the same as that of the isochrone of Huygens. He called the cycloid "the fateful curve of the seventeenth century." It was also referred to as the "Helen of Geometers," since it was the solution to several challenge problems.

From 1693, Johann had been studying Huygens's wave theory of light and was intrigued with Fermat's principle that light always follows the path of least time, rather than of least distance, which is a straight line. Fermat's principle supported his belief that "nature tends always to proceed in the simplest way." Using his acute intuition, Johann began by reducing the brachistochrone problem from a mechanical to an optical problem, in effect a matter of layers of wave fronts. From his analogy to Fermat's principle, the law of refraction, $v/\sin \alpha = K$, he arrived at the solution of the differential equation of a one-parameter family of cycloids.[635] In this case, v is the velocity or speed of light and equals $\sqrt{2gy}$, where g is the acceleration from gravity and y the distance of fall along the y-axis. Since from Fermat's principle $\sin \alpha = 1/\sqrt{1+y'^2}$, $K = \sqrt{1+y'^2} \cdot \sqrt{2gy}$ or $dx = \sqrt{y\,dy/(c-y)}$. Substituting $y = c \cdot \sin^2 u = c/2 - (c/2)\cos 2u$ leads to the solution $x - x_0 = cu - (c/2)\sin 2u$, which gives the cycloid.

In his article "Solutio Problematum Fraternorum ..." in *Acta eruditorum* for May 1697, Jakob extended his early, more mathematical solution with differential equations to attack a new class of curves and challenged his brother to do likewise. The new curves had to satisfy isoperimetric conditions. The ancient isoperimetric problem had required finding among closed curves the perimeter that encloses the greatest area. The answer is the circle. In ordinary calculus, the researcher may seek a value x that maximizes or minimizes a function $y = f(x)$. Jakob here pursued curves that give extremal values, maxima and minima, of definite integrals. This could be applied, for example, to determining the lowest center of gravity. Johannes failed to grasp the variational complexity of the problem. His solution given in the *Histoire* of the Paris Academy for 1697 was incomplete and its differential equation too low by one order. Jakob subjected it to scathing criticism. This completed breaking the relationship between the two men. In his March 1701 *Acta eruditorum* paper "Analysis magni problematis isoperimetrici" ("Analysis of the Large Isoperimetrical Problem"), Jakob asserted that the study of his new class of curves requires a second degree of freedom among variables and called for a general method. Not until 1718 did Johann admit that his brother, in allowing two ordinates to vary, had the better method.[636] The work of the Bernoulli brothers on the brachistochrone and isoperimetric curves laid the ground for the semigeometric stage of the calculus of variations developed by Euler.

In the 1697 *Acta* article " Principia calculi exponentialum seu percurrentium," Johann began founding exponential calculus, the calculus of the exponential function, e^x. He examined simple exponential curves, $y = a^x$, and general exponentials, such as $y = x^x$. Johann computed the quadrature under the curve $y = x^x$ from $x = 0$ to $x = 1$ to equal $1/1^1 - 1/2^2 + 1/3^2 - 1/4^2 + \ldots$. To reach this result, he expanded the equivalent of $x = e^{x \ln x}$ into the exponential series and integrated term by term.

Despite attractive positions offered by the Universities of Leiden and Utrecht, Johann Bernoulli upon Jakob's death in 1705 succeeded to the chair of mathematics in Basel, and he remained there. Family concerns drew him to his home city. Leibniz attempted to find him a more prestigious post elsewhere. In recommending him in 1714 to succeed Jakob Hermann as professor of mathematics at Padua, Leibniz called him "a luminary of our century."[637]

Johann Bernoulli, who investigated central problems and issues in theoretical and applied physics, had discovered before October 1712 an error in Proposition 10 of Book 2 of Newton's *Principia*. The proposition addressed the movement of a body in a resisting medium influenced by gravitational attraction. The numerical value of $1/2$ that Newton had computed in it was too small. Johann mistakenly attributed the error to confusion over higher differentials. His nephew Nicholas went to London to inform Newton of the mistake. It took Newton four days to make the correction. The first edition's numerical value of $1/2$ should be $3/4$. To arrive at $3/4$, the correct result, Newton increased resistance by $3/2$. The young Trinity fellow Roger Cotes, who was named in 1709 to oversee the second edition of the *Principia*, put these revised figures in to the new edition. Newton thought that Bernoulli's finding of this error might weaken his claims against Leibniz for priority in calculus. It probably did with the pugnacious Bernoulli, a partisan of Leibniz. His book on fluid dynamics, published in 1714, held that the reason the Cartesians rejected Leibniz's *vis viva* was that they were confused in their treatment of momentum. In his *Hydraulica* of 1732, which clarifies the internal pressure of fluids on tubes,

Johann unsuccessfully attempts to make the conservation of *vis viva* a basis for fluid dynamics.

In competition for the prestigious biennial prizes at the Paris Academy in the early 1730s, Johann employed the Cartesian vortex cosmology, which challenged Newton's law of gravitational attraction and its action at a distance, to explain the inclination of planetary orbits in relation to the solar equator. In a losing effort to that of Colin Maclaurin on the time required for colliding elastic spheres to become rigid, he introduced the modern delta function. Maclaurin argued that these spheres become inelastic at the limit of absolute hardness. But Bernoulli contended that hardness does not affect the scattering of the force on the spheres. As that force tends to infinity, the time of contact and spherical deformation approaches zero, so the integral for that arbitrarily small interval remains finite. From their applications of differential calculus to problems in mechanics, their study of probability, and Leibniz's principle of sufficient reason, Jakob and Johann had proposed the law of determinism in classical physics: "If all events from now to eternity were constantly observed, it would be found that everything occurs for a definite reason."

By the 1720s Johann Bernoulli was the mathematical preceptor of Europe. Leibniz had died and Newton was essentially inactive in mathematics. The pinnacle of Johann's teaching career was tutoring Euler in 1725 and working with him for the next two years. Among his contemporaries, Leibniz and Euler were his only heroes. As a member of the Basel school board, he attempted with limited success to reform the teaching of mathematics at its Latin school.

17.5   More Quarrels and Early Articulation

Resistance at the Paris Academy

In the 1690s the new calculus was being transmitted to France. In the winter of 1691 and 1692, Johann Bernoulli joined the mathematical circle in Paris centered on the charismatic Father Malebranche, including L. R. Byzance, L'Hôpital, Pierre de Montmort, Charles Reyneau, and Pierre Varignon. They were making an intense effort to understand calculus. Bernoulli agreed to tutor L'Hôpital on the secrets of differential and integral calculus. In 1692 L'Hôpital wrote to Leibniz that he had acquired his *Acta* articles, apparently in 1688, and had understood the method of tangents, but could not grasp the inverse procedure.[638] Again, the result of Bernoulli's teaching was L'Hôpital's *Analyse des infiniment petits pour l'intelligence des lignes courbes* of 1696. It argues that the nearly infinitely small and the nearly infinitely large exist and presents Leibnizian calculus as reliable. L'Hôpital's book begins with definitions of a variable as a continuously increasing or decreasing quantity and a differential as the infinitely small difference by which a variable changes. Differentials are thus variable. The *Analyse* treats the construction of tangents, maxima and minima, and applications of calculus to mechanics, along with higher-order differentials, points of inflection, and evolutes. The ninth chapter gives what is now known as L'Hôpital's rule: for differentiable functions $f(x)$ and $g(x)$, if $f(a) = 0$ and $g(a) = 0$ at $x = a$ and it is known that $\lim_{x \to a} f'(x)/g'(x)$ exists, then $\lim_{x \to a} f(x)/g(x) = \lim_{x \to a} f'(x)g'(x)$. The preface by Fontenelle, who upon the renewal of the Paris Academy in 1699 was to be appointed its secretary, examines the concept of infinity and reviews the study of geometric curves from Archimedes to Newton and Leibniz.

L'Hôpital's *Analyse* provided the elementary introduction to Leibnizian calculus that mathematicians had awaited for a decade. It dominated the field for a half-century. After L'Hôpital's death in 1704, a controversy erupted over whose achievement it was. Although admitting a debt to Leibniz and "the young professor at Groningen," L'Hôpital had not stated that the book was mainly the work of Johann Bernoulli. As praise for the book grew, Bernoulli made that claim. Not until his manuscripts on calculus from the early 1690s were discovered in Basel in 1921 would the issue be resolved in his favor.

Pierre Varignon

Appealing to the work of Newton, especially Book I, Lemma XI of the *Principia*, and Leibniz, who had asserted that the proportional size of a differential might be compared to that of a grain of sand to Earth,[639] Varignon now became a major exponent of the continuous variability of differentials. L'Hôpital worried that Leibniz's support for the concept of the infinitely small was weakening. In a letter of February 1702 to Varignon largely published in the *Journal des sçavans*, Leibniz depicted infinitesimals "if not as real, as well-founded fictitious entities" or as "ideal notions," in modern terms as operators, vital to reaching exact results by its algorithms.[640] The final results thus justified calculus pragmatically. Leibniz urged Varignon to look more to its applications than to philosophical foundations.[641]

Even after L'Hôpital's *Analyse* was greeted with strong approval, Leibnizian calculus was attacked at the Paris Academy. From 1700 to 1707 the Paris academicians were deeply divided over the foundations and admissibility of the methods of the new Leibnizian calculus.[642] Varignon, one of three pensionary geometers, led what became a group of adherents to its infinitesimalist form codified by L'Hôpital. The other two were Abbé Jean Gallois, an admirer of ancient Greek geometry, and his disciple the Cartesian algebraist Michel Rolle, both of whom criticized the existing foundations and infinitesimal methods of differential calculus. Rolle found L'Hôpital's calculus lacking in rigorous logical and metaphysical foundations. He was dubious about both the existence of higher-order differentials, the idea that the sum of a quantity and its differential is the original quantity, and the treatment of differentials as both nonzeroes and absolute zeroes. These contradicted the Archimedean postulate and Euclid's fifth axiom. Rolle argued that there were only finite quantities and zeros with no in-between zone inhabited by amphibians. The differential algorithm led to error and failed to provide all the maxima and minima of the selected curve for $a^{1/3}(y-b) = (x^2 - 2ax + a^2 - b^2)^{2/3}$, while an older rule from Hudde did. Adhering to the Leibnizian formalism, Varignon depicted the differential dx as a process and showed that Rolle had made mistakes with the differential algorithm and the sketching of the curve. Rolle did not recognize, moreover, that Hudde's rule gives intersection points as well as maxima and minima.[643]

In late 1701 the first phase of the dispute was ending. Although Varignon wanted to let the learned public judge, the academy forbade any public mention of the quarrel. Following academy procedures, a commission consisting of Cassini, la Hire, and Gouye was appointed to resolve the argument. It sided with Rolle but did not issue a judgment.

In 1702 the dispute went public. That year Gallois' group suffered a partial setback when the French government transferred editorship of the *Journal des sçavans* from him, hired specialist editors by fields, and named Fontenelle as book review editor. It did retain Gallois' associate Abbé Jean-Paul Bignon as a director. Joseph Saurin (1659–1737), the new mathematical editor, who had moved to Paris from Geneva and was a friend of the Bernoullis, began elaborating calculus. Rolle challenged whether multiple singularities of curves, having dy/dx assume the form $0/0$, could be obtained from the differential method of tangents. In 1702 and 1703, Saurin showed that L'Hôpital's *Analyse* and rule handle such indeterminate expressions, and he accused Rolle of plagiarizing L'Hôpital's method by simply using variants of his notation. The debate was becoming more personal and political.

A telling moment in the quarrel came with Fontenelle's eulogy for L'Hôpital in the *Histoire* of the academy for 1704. Up to this point, Fontenelle had not publicly taken sides. Using Leibnizian terminology, his eulogy now described differential calculus as "sublime geometry." Passing quickly over foundational problems, Fontenelle contended that its opponents did not understand calculus. He was engaged in working on a "true metaphysics" for it. Fontenelle dismissed as invective Rolle's attacks on Saurin.

A commission appointed in 1705 did not decide on the merits of calculus but praised the "good heart" of Saurin for proper exchanges. Leibniz and

Johann Bernoulli wrote asking for a definitive statement. In 1707 the Paris Academy elected Saurin a senior pensionary geometer. The Academy's reproaches and the efforts of Varignon to reconcile infinitesimals with Euclidean geometry prompted Rolle to withdraw from the fray. Rolle's departure coupled with the death of Gallois in 1707 assured that Leibnizian calculus was generally accepted at the Paris Academy.

Parisian academicians continued into the 1720s to argue over the applications of calculus and the lack of clarity and rigor in its foundations. In the 1709 *Mémoires*, Saurin wrote on the cycloid as the curve of quickest descent, supporting Johann Bernoulli's differential method. That year, in response to Huygens's criticism of the Cartesian vortex explanation of gravity, he defended Cartesian vortices, arguing that even though the vortex rotates with great speed, a very attenuated ether is all but nonresistant to objects on Earth. Newton's gravitational attraction, especially its idea of action at a distance, he rejected as threatening to return mechanics to "the ancient shadows of Peripateticism." In 1723 and 1725 Saurin corrected misunderstandings about extreme values in multivalued curves and L'Hôpital's theorem. Shortly before his death in 1722, Varignon was planning to explore the foundations of calculus in his courses at the Collège Royal. In 1725 his commentary on L'Hôpital's *Analyse* was published posthumously. Fontenelle's *Élemens de la géométrie de l'infini* of 1727 states that the resistance to calculus has ended. The work distinguishes between mathematical infinity and metaphysical infinity and treats infinitesimals as reciprocals of orders of infinity. Like Wallis and Johann Bernoulli, it holds that arithmetical studies freed of geometric ratios are fundamental to calculus, which reflects the shift of emphasis on the continent from geometric curves in calculus to algebraic equations.

Priority and the Royal Society

At the turn of the eighteenth century, critics observed that although the new calculus could lead to accurate results, it represented approximation more than precision, since the foundations of calculus, particularly the transition from infinitesimal to zero, were vague, changing, and not well organized. Leibniz's notation masked the logical basis of his calculus. An adequate distinction between fluxional and Leibnizian calculus was missing. In these circumstances, an aggressive cultural nationalism that divided German and British scholars combined with Newton's paranoia and the partisanship of his friends in London to produce a pathological moment in the history of mathematics, a dispute over priority for the invention of calculus. The nearly simultaneous appearance of two forms of differential calculus, coupled with the general assumption within Europe's republic of letters that a scientific theory can have only one originator, made the quarrel nearly unresolvable.

In 1690, Leibniz and Newton were each eager to establish his originality in inventing calculus. That July Leibniz wrote to Huygens:

I do not know Sir if you have seen in the Leipzig *Acta* a method of calculation that I propose in order to subject to analysis that which M. Descartes himself had excepted from it .... By means of this calculus I presume to draw tangents and to solve problems of maxima and minima, even when the equations are much complicated with roots and fractions ... and by the same method I make the curves that M. Descartes called mechanical submit to analysis.[644]

Although he had seen the articles, Huygens found them obscure. He asked Leibniz to explain clearly his methods and sent some challenge problems. Leibniz easily solved all problems. Huygens judged his methods "good and useful" but not as important as Leibniz claimed. Up to 1690, Newton had not published his fluxional calculus. He now set out to write a complete exposition of it, entitled "De quadratura," but ceased work on it in 1691, reportedly not wishing to stir resentments by its publication.

Newton's London circle of friends and Wallis now took up the priority issue. Newton permitted the Swiss engineer Nicholas Fatio de Duiller, who became known as "Newton's ape," to see his unpublished papers. After examining them, Fatio was convinced that Leibniz was the second inventor. In Holland, he professed to Huygens to understand all of Leibnizian calculus. In December 1691, he wrote from London to Huygens:

Mr. Newton is beyond question the first Author of the differential calculus and ... he knew it as well or more perfectly than Mr. Leibniz yet knows it, or before the latter had even the least idea of it, which idea itself did not come to him, it seems, before Mr. Newton wrote to him [in 1676] on the subject .... I cannot cease to wonder that Leibniz takes no notice of this in the Leipzig *Acta*.[645]

Fatio now claimed to have independently developed the equivalent of Leibnizian calculus, and thus to be the equal of Leibniz and the Bernoullis as a mathematician. The next February, he repeated that Newton's letters of 1676 explained differential calculus. Impressed with "De quadratura," Fatio asserted that Newton had a "perfect original" of the subject and Leibniz only "a botched and very imperfect copy."[646] The combative Wallis, known for his defense of English priorities, beset Newton with inquiries about Leibniz and sent Fatio's assessments abroad. He asked Newton for materials to insert in the preface to the first volume of his *Opera omnia*, especially concerning the two letters to Leibniz.

Soon after returning from his historical research trip to Italian cities and Vienna in 1690, Leibniz published three articles in *Acta eruditorum:* on optics, motion in a resisting medium, and planetary motion. The last provided a foundation for Kepler's laws that expanded Leibniz's arguments with Newton. This case reversed the publication chronology that existed in the quarrel over calculus, for Newton's work in the *Philosophical Transactions* on optics and his

Principia mathematica had preceded Leibniz's relevant three memoirs. Leibniz admitted Newton's influence on him but denied having seen the *Principia*. Actually he had. These *Acta* articles seem intended for claims not of priority but of independence and possibly equivalence in obtaining Kepler's ellipses, which Leibniz accomplished by assuming a harmonic circulation of ether, following a technique closer to Kepler's.[647] Leibniz claimed to have largely finished these articles fifteen years earlier in Paris, refraining from publishing them until he obtained a closer agreement between his theory and new observations. The scientific community was developing criteria for judging priority claims or the timing of unpublished research. A second, independent scholar had to corroborate these assertions. Huygens asked Leibniz who in Paris could testify to the time he had composed the papers in question, but there was no one. Among the British, this seemed another instance of his plagiarism. Later Newton would speak of having his ideas stolen and "annoyingly" repeated.[648]

In March 1693 Leibniz attempted to open a philosophical correspondence with Newton. It was to be part of a triangular exchange that included Wallis. All seemed amicable. Leibniz praised Newton for his work on infinite series, the *Principia*, and optics, which Huygens had related to him.[649] He exhorted Newton to provide the finishing touches on calculus and to continue his mathematical treatment of nature. He tried to excuse the long gap in their correspondence by saying that he had not wished to burden Newton. While Newton undoubtedly did not think that Leibniz had gained ideas from the unintelligible anagrams of his second 1676 letter, he was probably angry at Leibniz's silence about that correspondence in his articles in *Acta eruditorum* for 1684 and 1686. Although this failure to give credit to Newton is understandable, since Newton had not yet published on the method of fluxions, it was neither generous nor judicious of Leibniz. This failure to cite Newton would lead him to nurse a grievance. Probably his illness caused him not to respond to Leibniz until October and then briefly. After emphasizing that he cherished friendship with "one of the chief geometers of the century," Newton wrote what could be taken as a veiled threat about revealing in Wallis's forthcoming *Algebra* his own methods concealed in the 1676 anagrams. Leibniz dropped this effort to resume correspondence.

Wallis's preface to volume I of his *Opera* in 1694 asserts that Newton alone deserved priority for differential calculus. Receiving a copy from Huygens in September, Leibniz was relieved that Wallis had said little about Newton's inverse method of tangents and had not attacked his technique. After reading volume 2 of Wallis's *Opera,* containing portions of "De quadatura," Johann Bernoulli wrote to Leibniz that Newton might have derived its methods from Leibniz's articles in the *Acta*. This probably reflected continental thought of Leibniz's priority. In 1695 Huygens died and Leibniz succeeded him as the doyen of continental mathematicians. That year Newton agreed to let Wallis publish the two *Epistolae*. They appeared in volume 3 of Wallis's *Opera*, published in 1699. Newton explained both anagrams in the *Epistola posterior* and claimed that he had created the methods of "De Quadratura" in the 1670s, but these seem to date from closer to the 1690s. Though volume 3 of Wallis's *Opera* did not circulate widely on the continent and was not treated as inflammatory, the basis of the dispute between Newton and Leibniz that had been known to few was now made public. The reviewer of this *Opera* in the *Acta*, probably Leibniz, announced that it testified to Leibniz's early genius. In a letter to Thomas Burnet, he expressed satisfaction "with Mr. Newton, but not with Mr. Wallis who treats me a little coldly in his [preface] ... through an amusing affectation of attributing everything to his own nation."[650]

In 1699 the tone of the dispute was becoming shriller. In a paper in the *Acta*, Leibniz declared that Fatio and Newton possessed his calculus. This angered Fatio. In *Lineae brevissimi descensus investigatio geometrica duplex* (*A Twofold Geometrical Investigation of the Line of Briefest Descent*), Fatio insisted that he had independently invented calculus and in no way had a debt to Leibniz. He judged Newton to be the "first and by many years the most senior inventor of the calculus" and, as a pointed insult, added "whether Leibniz, the second inventor borrowed anything from him, I prefer that the judgment be not mine, but theirs who have seen Newton's letters and other manuscripts."[651] Leibniz probably thought that Newton had a hand in this attack, not realizing that Newton and Fatio had fallen out six years before. In another article in *Acta eruditorum* in 1699, Leibniz incorrectly asserted that only in 1684 had he known that Newton had a method of tangents and criticized as showing the weakness of fluxional calculus David Gregory's efforts to demonstrate the catenary.

In 1702 George Cheyne (1671–1743) came from Scotland to London, seeking to publish his *Fluxionum methodus inversa* (*The Inverse Method of Fluxions*), which is based on the work of Newton and David Gregory. He met Newton, who offered to pay for the printing, but Cheyne declined that arrangement. Newton, who refused to see Cheyne again, was upset with the errors in the book and its assertion that Newton's fluxional calculus coincided with Leibniz's. Though calling the book remarkable, Johann Bernoulli observed that Cheyne's xenophobia left no credit for continental mathematicians.

After thirty years of delay and revisions, Newton decided he must publish in detail on his fluxional calculus. The complete "De quadratura" was appended to his *Opticks* in 1704. The anonymous review in the *Acta* in January 1705, written by Leibniz, transposed Newton's fluxions into Leibnizian differences. Since Newton's publications were later, a reader of the review might conclude that he was the plagiarist. In an article honoring Jakob Bernoulli in the *Acta* a year later, Jakob Hermann cited "the great discovery of the century, the Leibnizian calculus." Gabriele Manfredi's *The Integration of Differential Equations* of 1707 and Charles-René Reynau's *Analyse demonstrée* the next year accorded special prominence to Leibniz.[652]

Newton, who denied seeing the *Acta* review of the *Opticks* until 1711, meanwhile recruited as his champion John Keill, a student of Gregory at Oxford who lectured there on Newtonian science and a Fellow of the Royal Society. In October 1708, Keill submitted for number 317 of the *Philosophical Transactions* a paper entitled "On the Laws of Centripetal Force" that stated that Leibniz's differential calculus simply changed Newton's fluxions and added new notation to conceal this, a virtual, public charge of plagiarism against Leibniz. Keill was merely repeating what Fatio, Wallis, Gregory, Halley, and other English were saying privately. Keill may have written this defense of Newton and accusation of Leibniz before agreeing to serve as Newton's spokesman. The English animus against Leibniz also seems to arise from mounting criticism of Newtonian attraction in *Acta eruditorum*.

Volume 317 of the *Philosophical Transactions,* though printed in 1710, did not reach Leibniz until a year later in Berlin. On March 4 he wrote to Hans Sloane, the secretary of the Royal Society, declaring that Newton knew better than anyone that the charge was untrue and demanding as a remedy a public retraction of this impertinent charge by the upstart (*homo novus*) Keill. The letter was read to the Royal Society on March 23. A week later from the presidential chair Newton gave his account of the invention of calculus. The Royal Society asked Keill to write a letter explaining his position. On May 24 he wrote that he had not meant to imply that Leibniz had known Newton's name for fluxional calculus or notation, but observed that the empirical evidence of the two 1676 letters, which plainly indicated the principles of calculus, would resolve the disagreement. The style of the letter and appeal to the manuscript record suggest that Newton at the least had influenced it. Carefully separating Keill from Newton in his reply dated December 29, Leibniz demanded justice.

In February 1712 the Royal Society met to respond to Leibniz. Newton addressed the body, maintaining that he was the first author of calculus. It was possible that Leibniz had come upon it later. He charged Leibniz with no misconduct. Newton saw Leibniz's last letter as giving the Royal Society cause to establish a tribunal. It appointed a commission, initially including Arbuthnot, Burnet, and Halley and later adding De Moivre and Taylor, to review letters, manuscripts, and extracts pertaining to the case. Operating under the thinnest veneer of impartiality and without hearing any defense of Leibniz, the committee issued within fifty days the volume *Commercium epistolicum D. Johannis Collins, aliorum de analysi promota* containing the documents reviewed in chronological order and its censure of Leibniz. Newton, nearly seventy years old and producing the second edition of the *Principia*, actually wrote it. The *Commercium epistolicum* methodically denigrates the work of Leibniz, adjudges Newton the first inventor, and claims that Keill in no way was "injurious to Mr. Leibniz."[653] The implication is that the second inventor was unimportant. The committee's judgment of Leibniz was harsh, but its assertion that he had not been forthcoming about his early contact with Newton and was the second inventor could not destroy his distinguished reputation in mathematics on the continent.

Upon receiving a copy of the *Commercium epistolicum,* Johann Bernoulli forwarded it to Leibniz with the comment that the Royal Society trial was "hardly a civilized way of doing things." Leibniz asked him to enter into the debate, but Bernoulli declined. Newton had recently praised his work, and he did not want to appear ungrateful. In response to the Royal Society judgment, Leibniz anonymously wrote the flyer *Charta volans*, which his disciple Christian Wolff had printed and circulated in July 1713. Before the *Charta,* Leibniz had thought Newton to be an independent discoverer of calculus. Now he believed that interpretation too generous. The *Charta* held that Newton had arrived at his method of series only in the 1670s and contended that Newton's fluxional calculus imitated Leibniz's differential calculus. *Acta eruditorum* and the *Journal littéraire* of The Hague published these comments. Varignon wrote from Paris supporting Leibniz. When Keill, responding to the *Charta volans*, alleged that Leibniz in computing planetary motion had made errors with second differentials, Wolff thought that remaining silent was a mistake.[654] But Leibniz saw Keill's criticism as unfounded and refused to debate. Wolff then concurred that his Excellency should not respond to an idiot. Leibniz, the *Journal des sçavans* observed, was not satisfied with the Royal Society's decision, so "the public will no doubt receive further information on the subject from him."[655]

The quarrel was far from ended in 1713. Wolff requested that Leibniz prepare his own *Commercium epistolicum* to explain his priority to the learned public. Leibniz wrote "Historia et origo calculi differentialis," which was not published in his lifetime.[656] It gives a detailed account of his invention of differential calculus and charges that his critics had shrewdly waited until the people who knew about the invention directly, such as Huygens, were dead. In 1713 Newton deleted from the *Principia* the part of a scholium in book 2 that listed Leibniz as independently developing calculus. After Leibniz died, the mathematical dispute abated on the continent, but Newton persisted. In 1716 he wrote "Observations upon the Preceding Epistle," but waited until Leibniz died before publishing it.[657] The "Observations" equates Clio, the muse of history, with the goddess of justice in the case presented in the *Commercium*. It pronounces Leibniz guilty of a misdemeanor and refers to Bernoulli, whom Leibniz had called an eminent mathematician supporting him, as a "pretended mathematician" and "knight errant of Leibniz." That rankled Bernoulli, who demanded an apology and urged Newton to put a leash on Keill and Taylor. Newton responded that he had no animosity toward Bernoulli. The English spoke of the "Leipzig rogues" and Bernoulli of the "scurvy English." In an "Account" ("Recensio") appended to an expanded edition of the *Commercium epistolicum* in 1722, Newton criticized Leibniz and Bernoulli for failing to challenge the original version (they in fact considered it a screed not worthy of rebuttal) and turning to squabbles with him over atoms, the nature of space, and attraction.

The calculus priority quarrel split British and continental mathematicians, establishing two distinct traditions that articulated calculus.[658] The British did not adopt the arithmetic and algebraic method of Wallis but primarily followed the authority of Newton, concentrating on seeking geometric foundations, especially after the publication of Colin Maclaurin's *Treatise on Fluxions* in 1742. Most notably in Newton's circle, Brook Taylor (1685–1731) and Abraham De Moivre (1667–1754) advanced fluxional calculus.

Corollary II of theorem III of Taylor's *Methodus incrementorum* of 1715 derives in Newton's dot notation the powerful formula for expanding functions into infinite series, now known as Taylor's theorem. From his revision of Newton's interpolation formula for finite differences and his study of trigonometric functions and uniformly flowing evanescent increments of fluxions, he found in modern notation $f(x+h) = f(x) + f'(x)h/1! + f''(x)h^2/2! + f'''(x)h^3/3! + \ldots + f^{(n)}(x)h^n/n! + R_n$. For φ between 0 and 1, what we call Lagrange's form of the remainder R_n is $f^{(n+1)}(x + \varphi h)h^{(n+1)}/(n+1)!$. But Taylor did not relate the remainder to convergence. That remained for Cauchy. Taylor failed to recognize that among those who had anticipated his series were James Gregory and Newton and that Johann Bernoulli had published a variant of this series in the *Acta* for 1694. Theorem XVII of *Methodus incrementorum* solves a variational isoperimetric problem by applying Jakob Bernoulli's methods.

In a paper in the *Philosophical Transactions* for 1707, De Moivre showed an understanding of, but did not explicitly state, the fundamental theorem of analytic trigonometry: $(\cos\theta \pm i\sin\theta)^n = \cos n\theta \pm i\sin n\theta$, which is now named for him. Here i is $\sqrt{-1}$ and for De Moivre n is a positive integer. In a report dated 1721 that was first published in his *Miscellanea analytica* nine years later, he expresses this theorem as the equivalent of $(\cos\theta \pm i\sin\theta)^{1/n} = \cos\{(2K\pi \pm \theta)/n\} \pm i\sin\{(2Kn \pm \theta)/n\}$. In Chapter 8 of *Analysin infinitorum*, Euler was the first to state the fundamental theorem and generalize it to apply to all real n. Through the *Miscellanea*'s extension of theorems for sectors of circles to those of rectangular hyperbolas, De Moivre came near to the discovery of hyperbolic functions. This requires a clear understanding that $\int \sqrt{a^2 - x^2}\, dx$ gives the area under a circle, while $\int \sqrt{x^2 - a^2}\, dx$ gives the area under a hyperbola. In this case, trigonometric functions can give the area under a circle and logarithm functions the area under the hyperbola.

Stressing geometric foundations, less successful in developing and applying differential equations, and using Newton's dot notation, the British were to fall behind continental Leibnizian scholarship. They did not match the work of the Bernoulli brothers at the start of the century and the massive achievements of Euler, Daniel Bernoulli, and Jean d'Alembert at mid-century.

17.6   Algebra and Analytic Number Theory

Book IV of Leibniz's *Nouveaux essais* declared that while Viète had "substituted letters for numbers to obtain greater generality in algebra," he did not address their order and relations. Leibniz's reintroduction of "the character of numbers" arranged in a way that represents a given situation was to be a great aid for solving equations. Here he was speaking of his creation of determinants and obtaining their properties. As early as 1675, he had begun seeking conditions of consistency in solving a succession of linear and higher-degree equations. By 1678 Leibniz had connected an array of coefficients with general solutions for systems from one linear equation having one unknown up to five equations having four unknowns.[659] In a manuscript of 1684 he had a form of what Euler called Gabriel Cramer's "very beautiful rule" for attaining these solutions and a correct rule of signs. This rule gives the solution of y in the two equations $ax + by = c$ and $dx + ey = f$ as $y = (af - dc)/(ae - db)$.[660] Eliminating x in two equations having one unknown, $a_{10} + a_{11}x = 0$ and $a_{20} + a_{21}x = 0$, gives a second-order determinant

$$\begin{vmatrix} a_{11} & a_{12} \\ a_{21} & a_{22} \end{vmatrix} = 0$$

The first number in the subscript refers to the row and the second to the column.

After many trials, Leibniz's elimination of unknowns using his new simplified symbolic or fictitious numbers and his study of combinatorics gave rise to a satisfactory theory of determinants. By 1696, he had basic rules for interchanging, adding, and subtracting rows or columns of determinants. Although he never formally published on the theory of determinants, he corresponded widely about them with the Bernoullis, Hermann, Huygens, Reynau, Naudé, Theobald Schöttel, Thomas a St. Josepho, and Christian Wolff, and he wrote papers in 1700 and 1710 on representing coefficients in equations with his fictitious numbers.[661] For this research, Eberhard Knobloch adjudges Leibniz, along with Cramer, Laplace, Cauchy, and Cayley, a founder of the modern theory of determinants.

After 1660 British mathematicians debated the relative importance of symbolic algebra and synthetic geometry. In Britain, Oughtred's *Key of Mathematics*, Harriot's *Artis analyticae praxis*, Pell's contributions to the English edition of Johann Rahn's *Teutsche Algebra*, and John Kersey's textbook *Elements of That Mathematical Art Commonly Called Algebra* pioneered early modern symbolic algebra, based on arithmetic and extended to higher-degree equations.[662] These algebraists accepted negative roots. Wallis approved imaginary numbers as well. At a time when specialized disciplinary lines had not yet been clearly drawn, Wallis attempted in *Arithmetica infinitorum* to disentangle algebra from geometry. His *Treatise of Algebra* presents the field as independent, having fundamental computational rules and principles, such as the commutative law of addition and multiplication. Geometers countered. The royalists Hobbes and Barrow argued that geometry is superior to algebra and arithmetic. They found certainty in Euclidean geometry and preferred synthesis over analysis for proofs. Barrow held that the partial basis of geometry in sensory experience gives it an additional strength. This work by Hobbes and Barrow prepared the way for the British preoccupation with geometric foundations into the nineteenth century.

As doubts persisted about the soundness of algebra as an academic subject, John Collins urged resolving them through the preparation of a new textbook that might be read by university mathematicians and literate practitioners. While he encouraged Wallis, Pell, and Kersey in their writings, he urged Newton to compose the desired text. Newton's *Universal Arithmetick* appeared in Latin in 1707. It consists of lectures on the subject that Newton as Lucasian professor had deposited in the library at Cambridge from 1673 to 1684. He had revised these from 1683 with a possible eye to publication. His Lucasian successor William Whiston gathered and published these lectures.

Its pedagogy and pragmatic technical orientation were to make the *Universal Arithmetick* the textbook of choice in algebra in Britain during the eighteenth century and well received abroad, being published in 1732 in Leiden, Milan, and Paris. Newton patiently informs the student how to reduce verbal problems to symbolic algebraic equations depending on the number of conditions, put them into standard Cartesian form, and resolve them. He chooses model practical problems and explains how to solve them step by step. The *Universal Arithmetick* gives formulas, known as "Newton's identities," generalizing the summing of all powers of roots of a polynomial equation. Cardano had discovered that for the equation $x^n + a_1 x^{n-1} + \ldots + a_{n-1} x + a_n = 0$, the sum of its roots is $-a$. Viète had expanded the knowledge of relations between roots and coefficients, and Girard in 1629 showed how to sum the squares, cubes, or fourth powers of roots. Generalizing this to all powers, Newton observes that for $K \leq n$, the two relations $S_K + a_1 S_{K-1} + \ldots + a_K K = 0$ and $S_K + a_1 S_{K-1} + \ldots + a_K S_0 + a_{K+1} S_{-1} + \ldots + a_n S_{K-n} = 0$ hold, as does $S_K + a_1 S_{K-1} + \ldots + a_{n-1} S_{K-n+1} + a_n S_{K-n} = 0$ for $K > n$, where S_q represents the sum of the qth powers of roots. For any integral power, the recursive use of these relations gives sums of powers of roots.

In solving equations, Newton initially recognized a greater domain of numbers than most of his contemporaries. *Universal Arithmetick* accepts negative numbers without reservation. Kersey had defined -5 as a "number less than nothing by 5;" Newton writes of "quantities less than nothing." As a young man he had also considered roots that involve $\sqrt{-1}$. More conservative with age, Newton called such numerical solutions "impossible" and "imaginary."

The *Universal Arithmetick* was not simply an endorsement of algebra. Henry Pemberton's *View of Newton's Philosophy* of 1728 holds that Newton from 1680 was becoming partially disenchanted with Descartes' analytic geometry and drawn more to classical geometry. Newton's study of Books 7 and 8 of the *Mathematical Collection* of Pappus for his Lucasian lectures in the late 1670s seems to have stimulated a change.[663] Supposedly atheistic implications in the Cartesian mechanical philosophy and a desire to repudiate a mechanics requiring action by contact may have contributed to Newton's thought. Notably, he praises Halley's 1706 edition of Apollonius's *De sectione spatii*, which attacks the algebraic methods of Wallis. Against the tediousness of algebra, the *Universal Arithmetick* elects for the simplicity, rigor, and certainty of synthetic geometry. But it only briefly treats foundational issues—and that in the prosaic preface. The book may have served more to confuse than to clarify Newton's shifting views on the foundations of mathematics. Mathematicians and historians have debated these. Richard Westfall interprets Newton's attraction to classical geometry as only a facade behind which stand the original modern methods of his *Principia*.[664] Helena Pycior, who argues that Newton was not seeking an opposite to Wallis's arithmetic and algebraic foundation of mathematics, attributes to him an elliptical foundational model that mixed arithmetic algebra, analytic geometry, geometric analysis, and classical geometry, without fully defining their relative merits.

In the *Acta eruditorum* for 1702, Leibniz expresses hope for the end of the neglect of Diophantine algebra by the Cartesians, who saw little use for it in geometry. A richer harvest of results in integral calculus, he maintains, will depend largely on the arithmetic evolving from a diligent, systematic study of Diophantine algebra.[665] This study was to await Euler. For his part, Leibniz made separate contributions to number theory. In 1674 his exact summing of $\pi/4$ by the famous alternating series known by his name $1 - 1/3 + 1/5 - 1/7 + 1/9 - 1/11 + \ldots = \sum_{n=0}^{\infty}(-1)^n(2n+1)^{-1}$ caused a stir. To evaluate the integral of the quadrant of a circle by long division and term-by-term integration, Leibniz applied the method of Mercator and Wallis for obtaining the quadrature of the hyperbola. He arrived at the result $az - z^3/3a + z^5/5a^3 - z^7/7a^5 + \ldots$.[666] Letting $z = a$ and $a = 1$ reduces the values of its terms to his series for $\pi/4$. At this time, mathematicians were obtaining power series expansions of the inverses of trigonometric functions. Newton had found the series for $\arcsin x$ in 1669 and Gregory the arctangent expansion. Since $\tan(\pi/4) = 1$, $\arctan(1) = \pi/4$. Initially, Leibniz did not know that his series simply replaces x with 1 in Gregory's expansion of $\arctan x$. Although Leibniz considered his series beautiful and stated that "the Lord loves odd numbers," it is not useful for practical computational purposes, since it converges too slowly.[667] It takes 100,000 terms to reach Archimedes's accuracy for π. This prompted Newton to send through Oldenburg a clever variant: $\pi/2\sqrt{2} = 1 + 1/3 - 1/5 - 1/7 + 1/9 + 1/11 - \ldots$.

Throughout the eighteenth century, more rapidly converging series through a transformation of Leibniz's were sought. It was part of the effort to systematize trigonometric functions. The relation $\tan(\pi/6) = 1/\sqrt{3}$ leads to the more efficient formula $\pi = 2\sqrt{3}(1 - 1/3 \cdot 3 + 1/5 \cdot 3^2 - 1/7 \cdot 3^4 + \ldots)$. In 1719 Thomas F. de Lagny (d. 1734) with "incredible labor" summed from this series the value for π to 210 decimal places. The first thirty-five decimal places are

$$\pi = 3.14159265358979323846264338327950288\ldots.\text{[668]}$$

When Euler reproduced Lagny's result in *Analysin infinitorum* of 1748, he or the printer made a mistake in the 113th decimal place. Composite formulas also aided in obtaining series that converge better than Leibniz's. In 1706 John Machin found that $2 \cdot \arctan 1/5 = \arctan\{(2/5)/(1 - 1/25)\} = \arctan 5/12$. This led to $\pi/4 = 4 \cdot \arctan 1/5 - \arctan 1/239$, which is useful for computing in base 10.

Between 1676 and 1680, Leibniz also verified, probably by induction, Fermat's fundamental proposition on aliquot parts. "Given any prime number p and geometric progression $1, a, a^2, \ldots$," so it states, "p divides some number a^{n-1} when n divides $p-1$. If N is a multiple of the smallest n, p divides $a^N - 1$ as well."

In *Novae quadraturae arithmeticae*, published in 1650, the Bolognese mechanist Pietro Mengoli had posed the most prominent challenge problem of this period in number theory: the precise summing of the reciprocals of the square numbers. It is known today as the zeta function of 2: $\zeta(2) = \sum_{n-1}^{\infty} n^{-2} = 1 + 1/2^2 + 1/3^2 + \ldots$. It was probably influenced by Fermat's calculation of primitives, which included the sums $S_m(n) = 1^m + 2^m + 3^m + \ldots$. Wallis's *Arithmetica infinitorum* calculates to three decimal places the sum Mengoli had sought: 1.645. After Leibniz studied this problem in the 1670s, it appeared in Jakob Bernoulli's first dissertation on series in 1689, which made it widely known. Thus, it is called the Basel problem. Bernoulli believed that this series resembles that of triangular reciprocals. Seeking to improve on Wallis's approximation, Bernoulli compared the two series term by term, showing that the sum of the reciprocals of square numbers is less than 2. Since the Basel problem's infinite series converges slowly, an exact summation escaped Leibniz, Jakob and Johann Bernoulli, James Stirling, Abraham De Moivre, and others. In 1728, Daniel Bernoulli computed the sum to be "very nearly 8/5."[669]

To calculate the sums of integral powers, the second section of Jakob Bernoulli's *Ars conjectandi* uses an array that is essentially Pascal's triangle. About 1705 he had noted that

$$1 \;+ 2 \;+ \ldots + n \; = n^2/2 + n/2$$
$$1^2 + 2^2 + \ldots + n^2 = n^3/3 + n^2/2 + n/6$$
$$1^3 + 2^3 + \ldots + n^3 = n^4/4 + n^3/3 + n^2/4 + 0$$
$$1^4 + 2^4 + \ldots + n^4 = n^5/6 + n^4/2 + n^3/3 + 0 - n/30$$
$$\cdots \qquad \cdots$$

From these computations, he derived the general formula for any power k:

$$\sum_{n=1}^{\infty} n^k = n^{k+1}/(k+1) + n^k/2 + (k/2)An^{k-1}$$
$$+ \{k(k-1)(k-2)/(2 \cdot 3 \cdot 4)\}Bn^{k-3}$$
$$+ \{k(k-1)(k-2)(k-3)(k-4)/(2 \cdot 3 \cdot 4 \cdot 5 \cdot 6)\}Cn^{k-5} \ldots .$$

The capital letters are today called Bernoulli numbers.[670] Bernoulli determined that each of these numbers completes unity. Hence, for $\sum n^4 = (1/5)n^5 + (1/2)n^4 + (1/3)n^3 + Cn$, it turns out that $C = 1 - 1/5 - 1/2 - 1/3 = -1/30$.

In December 1735, Euler reported discovering $\zeta(2k) = \{2^{2k-1}/(2k)!\}B_k\pi^{2k}$ and so $\zeta(2) = \sum_{n=1}^{\infty} n^{-2} = \pi^2/2 \cdot 3 = \pi^2/6$.[671] Expending great labor, Euler computed the first twelve Bernoulli numbers. He established the current notation for them, as he computed $B_0 = 1$ and $B_1 = 1/2$. The value of A becomes $B_2 = 1/6$, while $B_4 = -1/30$, $B_6 = 1/42$, $B_8 = 1/30$, $B_{10} = 5/66$, $B_{12} = 691/2730$, $B_{14} = 7/6$, $\ldots$. From the function $M(u)$, having Bernoulli numbers as Taylor coefficients, that is, $M(u) = 1 + \alpha u + \beta u + \gamma u + \ldots = 1 + (B_1/1!)u + (B_2/2!)u + (B_3/3!)u + \ldots$, Euler arrived at the generating function $M(u) = u/(e^u - 1)$. Since this generating function is even, $B_3, B_5, B_7, \ldots = 0$.

The Bernoulli numbers were to be important in analysis, expansions of trigonometric series, and differential topology, as well as number theory, especially in studying Fermat's last theorem. Since π and its powers are irrational, the same holds for the numbers $\zeta(2k)$. Attempts to prove irrational the numbers $\zeta(2k + 1)$, which have no analogous expression to that of Euler for $\zeta(2k)$, seemed intractable until the French mathematician R. Apéry in 1977 devised a procedure that showed $\zeta(3)$ to be so.

17.7   Classical Probability

After 1660 a second generation of scholars and practitioners developed classical probability, changing the theory of risk taking. Close connections continued with issues concerning aleatory contracts, gambling, insurance, and moral certainty. The charges for annuities were treated as the inverse of the pricing of insurance. Political arithmetic, which consists of the gathering and analyzing of data on the economy and population, including mortality charts and shortly afterwards tables on natality, entered university curricula in central and western Europe. For describing these materials the word "statistics," taken from the Italian *statista*, meaning a person who deals with state affairs, came into use. Although sharp fluctuations existed in mortality from plagues, wars, and fires, such as that in London in 1665 and 1666, political arithmetic, which was based on evidence, generally assumed that stable simple patterns must guide mortality statistics. At the same time a theoretical structure for probability began to emerge, primarily in the work of Jakob Bernoulli and Abraham De Moivre.

Among the chief contributors to political arithmetic was John Graunt. In 1662 Graunt wrote for London *Observations Made Upon the Bills of Mortality*. Without fully certifying the accuracy of these bills, he computed the fraction of plague deaths to total burials yearly from 1592. Translated into percentages, these deaths were between 44 and 82 percent of the total. Assuming plague mortality at one-sixth of the population, christenings at one-fiftieth, and two years of migration to offset plague deaths, the third edition of *Observations* in 1665 approximates closely the population of London as 430,000. Statistical studies in France, Holland, and the German states adopted Graunt's methods of organizing data and computing tables.

The Value of Life Annuities in Proportion to Redeemable Annuities, the study of life insurance for the States General of Holland by the grand pensionary Jan de Witt in 1671, was another significant work in political arithmetic. The position of grand pensionary is equivalent to prime minister. Life or fixed annuities were the primary means of raising funding in Holland for paying down the state debt. In 1665 de Witt had reduced from five to four percent the interest rate on life annuities. A second war with the English from 1665 to 1667 drained these resources. De Witt's *Value of Life Annuities* was principally a political tract designed to show that life annuities were being offered at too high a rate of interest. It divides the population from 3 to 83 into four groups: (3–53), (53–63), (63–73), and (73–83), and assumes that the chance of death of the oldest people in each group to be no more than three-halves that of the youngest. From state records, de Witt computes in reverse order the likelihood of death within six months as virtually certain at one, or two-thirds, a half, and a third of the total population. This means that the annuity in the second period is two-thirds the existing value. For almost two hundred years, de Witt's treatise was buried in the state archives.

The findings of Graunt and de Witt are crude step graphs. In *An Estimate of the Degrees of Mortality of Mankind* of 1693, Halley provided the first smooth curve of statistical results. This book is based on tables of births and deaths in Breslau, where records were kept from the sixteenth century. One difficulty is that the total population is unknown. Following Graunt, Halley constructs a life table, giving annual deaths from age 7 to 100. Breaking down the population into four age groups: (9–25), (25–50), (50–70), and (above 70), he considers each life span as part of a coordinate system (l_0, l_d). Halley smooths elements between each age group to reduce chance variation and interpolates between differences in data. He computes Breslau's death rate as roughly 1174/34,000, or 35 per 1,000.[672]

Jakob Bernoulli's *Ars conjectandi*, probably his most original work in mathematics, was long in the making and not yet published at his death. While writing his *Meditations* from 1684 to 1690, he had completed parts and during the last two years of his life he worked intensely on the manuscript. When Bernoulli died, Leibniz urged Jakob Hermann to prepare it for publication. Hermann provided an outline, but no more. In his eulogy for Jakob Bernoulli at the Paris Academy in 1706, Joseph Saurin urged Johann to arrange and supervise the publication of *Ars conjectandi*, but Jakob's widow would not agree. At this time other scholars, notably Pierre Rémond de Montmort (1679–1719), Nicholas (I) Bernoulli, and Abraham De Moivre, were developing mathematical probability and corresponding with one another. Montmort's *Essai d'analyse sur les jeux de hazard (Analytical Essay on Games of Chance)* of 1708 asserts that reports in the *éloges* of Jakob Bernoulli about the *Ars conjectandi* manuscript prompted his attempts to calculate expectations in various games of chance.[673] Before issuing a second edition, Montmort express a desire to see Jakob's results, offering to pay their cost of publication. Nicholas Bernoulli had studied with his uncle Jakob and knew the manuscript well, having used it in writing his master's thesis, "De usu artis conjectandi in jure," the most thorough early study applying probability to legal problems. He and Montmort collaborated in wider studies of probability. Some of their letters appeared in the second edition of Montmort's *Essai* in 1713. Under pressure from Montmort, Nicholas wrote a short preface for the *Ars conjectandi* and it was published in 1713. The Basel printers requested that Nicholas revise the manuscript, but he refused, seeking to keep the book his uncle had written.

The *Ars conjectandi,* which generalized the calculus of chances, in its first part annotates and supplements Huygens' *De ratiociniis in aleae ludo.* Roughly four times Huygens's original text, it offers alternative proofs for basic propositions and extends some, reaching general formulas for obtaining numerical results.[674] In deriving results, Bernoulli often progresses iteratively from the simplest to the most complex cases. In the case of games of chance that are stopped well before the end, his method of reasoning proceeds step by step from what is known. Bernoulli calls this method *synthesis.* His new methods of solution include simultaneous algebraic equations to determine expectations when they depend on previous players, or vice versa. Mathematical reasoning involving recursions or simultaneous equations is analysis. Bernoulli considers the synthetic method more natural. The second part of *Ars conjectandi* comprehensively examines contributions by Schooten, Leibniz, Wallis, and Jean Prestet, but not Pascal, to the theories of permutations and combinations. It gives practically all their results, the Bernoulli numbers derived by use of essentially Pascal's triangle, and by use of induction the first satisfactory proof of the binomial theorem for positive integral n. The third part applies techniques from these theories to solve twenty-four examples of expectations of profit in various popular games of chance, such as cards, coin tossing, and dice. The selection of colored stones from an urn is a game Bernoulli often cites. All these involve conditional probability, the probability of event A given an event B, which is more complex. In drawing a card each from two separate decks, for example, what is the chance that one will be an ace? Symbolically $P(A + B)$, which may represent a case of alternatives, that is, either-or, equals $P(A) + P(B) - P(AB)$. In this instance $P(A) = 4/52$, $P(B) = 4/52$, and $P(AB) = (4/52)^2$. Thus $P(A + B) = 25/169$. The fourth part of *Ars Conjectandi* applies to "civil, ethical, and economic problems" the techniques from the first three sections and introduces Bernoulli's theorem, which the French mathematical physicist Siméon-Denis Poisson later would name the law of large numbers. An appendix contains a letter to a friend on the granting of points in *jeu de paume,* a predecessor of tennis.

The *Ars conjectandi*'s basing of the computation of chance not on symmetry, as in the past, but on a degree of certainty was revolutionary. The concept of degree of certainty seems to have arisen from Bernoulli's worldview. Lorraine Daston and Edith Sylla argue that although unsure of how to reconcile his belief in contingency and human freedom with God's determination of all things and knowledge of the future, Bernoulli held that nature and human life are not inherently statistical, for God knows the future with certainty.[675] Only for a relatively few phenomena can mortals count all possible cases of a phenomenon. But though mortals cannot generally achieve a complete enumeration and absolute certainty, we can approximate ideal probability. An abundance of repeated trials and observations is crucial for establishing which pattern or regularity is more likely. After criticisms from Leibniz, Bernoulli admitted that patterns, such as for deaths from diseases and changes in the weather, can vary over stretches of time. The closest that human beings can approach to certainty is asymptotically. While computation of probability by increasing multiples of numbers of events by induction is useful, it does not assure the discovery of the ideal probability, a true ratio of certainty.

For obtaining the moral certainty of the true probability of an event, Bernoulli proposes what is now called the weak law of large numbers. Moral certainty is defined as having a successful probability of no less than 0.999 and moral impossibility as a probability of no more than 0.001. Absolute certainty is a probability of 1. After keeping his weak law of large numbers secret for twenty years, Bernoulli presents it at the end of the fourth part of *Ars conjectandi*. In examining the difference between actual events and ideal probability, one version of it holds: if the probability of an event is p, if t is the number of times an event successfully happens in n trials, if $\epsilon > 0$ is any arbitrarily small fraction, and if the likelihood of satisfying the inequality $\mid t(n) - p \mid < \epsilon$ is the probability P, then as n grows sufficiently large, P increases to the absolute certainty of 1. Consider applying this law to the case of tossing heads with a coin that is perfectly balanced. It does not mean that the frequency of heads in n tosses will be $n/2$. Bernoulli could demonstrate that the desired outcome $t(n)/n$ approaches $1/2$ as n grows arbitrarily large. Bernoulli takes an indirect approach by holding that the difference between the actual and the ideal, in this case $\mid t(n) - 1/2 \mid$, can be calculated for n approaching infinity and that the possibility that the inequality is greater than ϵ is zero. Bernoulli's weak law of large numbers is the first limit theorem in probability. Drawing on five lemmas from the binomial theorem, he proves it.

At Samur, meanwhile, Abraham De Moivre studied on his own Huygens' *De ratiociniis* before going to Paris, where under Jacques Ozanam he read Euclid and other mathematical classics.[676] Amid the persecution following the revocation of the Edict of Nantes in 1685, De Moivre left France. Not long after arriving in London a year later, he chanced upon Newton's *Principia* and mastered it. In 1692 De Moivre met and became friends with Halley. Three years later Halley communicated to the *Philosophical Transactions* De Moivre's first paper on the subject of Newton's method of fluxions. In 1697 Halley oversaw his election to the Royal Society. No less than the older Newton advised mathematics students: "Go to Mr. De Moivre; he knows these things better than I do." In 1712 the Royal Society appointed him to its commission to resolve the calculus priority quarrel between Leibniz and Newton.

Despite a high reputation in mathematics, De Moivre never obtained the university position that he desired. From 1704 to 1714, he corresponded extensively with Johann Bernoulli. He asked Bernoulli to approach Leibniz about sending a reference letter for him for a professorship at Cambridge, but got no response. Still he proposed Bernoulli for election to the Royal Society in 1712. De Moivre barely made a living as a mathematics tutor, complaining of the wasted time walking from home to home of his pupils, and as an expert on probability in gambling and annuities.

In 1711, two years before the appearance of Bernoulli's *Ars conjectandi*, De Moivre in no. 309 of the *Philosophical Transactions* published "De mensura sortis," a Latin version of his *Doctrine of Chances: or, a Method of Calculating Probability of Events in Play*. Problems posed in Huygens's *De ratiociniis* and Montmort's *Essai d'analyse*, the only two prior systematic treatises on probability, instigated his more detailed book and led to priority quarrels with Montmort. Published in 1718, 1738, and 1756 in expanding editions, the *Doctrine of Chances* includes more than fifty problems. The preface to the first edition expresses a debt to Jakob, Johann, and Nicholas Bernoulli.

De Moivre slowly made fundamental discoveries in probability that he presented in his *Miscellanea analytica*, which he hurriedly published in 1730 to establish priority over the Scottish mathematician James Stirling, and the 1738 edition of *The Doctrine of Chances*. These books improved on the quantification of chance by Huygens, Montmort, and Jakob Bernoulli, especially his weak law of large numbers.

De Moivre first addresses sets of events of equal probability, or the likelihood of $n/2$ for a desired happening in n trials. Like Bernoulli, he estimates odds from a large number of trials, going into the thousands. A mastery of the new calculus allowed De Moivre to obtain better results in probability than conclusions gained from experiment alone. *The Doctrine*'s third problem has a represent the number of chances for success and b the number for failure in $a+b$ trials. For even odds, De Moivre finds that $(a+b)^x = 2b$. The use of logarithms gives $x = \log 2/\{\log(a+b) - \log b\}$. If the odds are not even, $a : b = 1 : m \neq 2$. For this case, $(1 + 1/m) = 2$ and thus $x \log(1 + 1/m) = \log 2$.

The study of chance or probability by employing combinatorial methods produces many computations of probabilities involving coefficients of the binomial $(a+b)^n$, and for large n the computations are very cumbrous. In examining m successes in n trials, Bernoulli had known that as n grows very large the observed proportion m/n lies between limits, but could not compute or prove this. From 1721 De Moivre had investigated the power series expansion of $(1+1)^n$ and four years later Stirling joined him through correspondence in seeking a general method to compute its coefficients. Since factorials increase very rapidly, it is helpful to have an accurate approximation of $n!$. Working with Wallis's infinite product for $\pi/2$, Newton's infinite series for $\ln\{(1+x)/(1-x)\}$, and Bernoulli's formula for summing powers of integers, De Moivre obtained for very large n the factorial approximation that in modern notation is $n! \approx cn^{n+1/2}e^{-n}$, which he presented in a supplement to *Miscellanea analytica*. He recognized that c is a limiting sum of an infinite series, but could not determine it exactly. In his *Methodus differentialis* of 1730, Stirling announced that c is $\sqrt{2\pi}$. Thus, although the equivalent of $n! \approx \sqrt{2\pi n}n^n/e^n$ was first given in 1738 in De Moivre's *Doctrine*, it is called Stirling's formula. De Moivre's approximations could show that observed frequencies lie within certain limits. By using recurring series, he more clearly formulated combinatorial problems and gave a better treatment of these relating to the duration of play.

The general acceptance of Bernoulli's weak law of large numbers largely followed upon two accomplishments of De Moivre in 1733: his discovery of what is called the normal distribution of terms in the binomial expansion $(1+1)^n$ and his introduction of the probability integral $(1/2\pi) \int_{-k}^{k} e^{-x^2/2} dx$ for closely approximating the sums of a large number of such terms. Conceiving as a curve the values in the normal distribution, setting terms of the binomial upright at right angles to a straight line above, and calculating logarithms to obtain their frequencies, he arrived at two inflection points and a maximum. Essentially, he had the normal or bell shaped or Gaussian curve. De Moivre generalized his method to apply to the case $(a+b)^n$, where $a \neq b$, making it possible to achieve required degrees of accuracy with far fewer trials. But he lacked the funding, position, and time to apply his major findings. By the end of the eighteenth century, the investigation of generating functions, which he had pioneered, became central to probability, most significantly to Laplace's theory of errors. For the next two centuries the normal curve, which Gauss was to construct from $\int (1/\sigma\sqrt{2\pi}) e^{-\frac{1}{2}\{(x-\mu)/\sigma\}^2} dx$, would guide advances in probability and statistics.

Endnotes

Preface, pages ix–xii

[1] French philosopher Jean d'Alembert's *Preliminary Discourse* of 1751 divides classical mathematics into three main branches: pure, mixed, and physico-mathematics. Pure mathematics includes arithmetic and geometry; mixed mathematics embraces mechanics, geometric astronomy, optics, acoustics, pneumatics, and the art of conjecturing or the analysis of games of chance. Physico-mathematics applies mathematical calculation to experiment and seeks to deduce physical inferences whose certitude is close to geometric truth. D'Alembert's definitions of algebra, arithmetic, and number are consistent with Newton's identification of number with relationship in *Arithmetica universalis, sive compositione et resolutione liber*, published in 1707 and revised in 1722.

[2] For a criticism of ultraspecialization, see Johan Huizinga, *Men and Ideas: History, the Middle Ages, the Renaissance; Essays*. Trans. by James S. Holmes and Hans van Marle (New York: Meridian Books, 1959).

Prologue, pages xii–xxviii

[3] Konrad Jacobs, *Invitation to Mathematics* (Princeton, NJ: Princeton University Press, 1992), p. 3.

[4] Calculus builds on geometry, trigonometry, arithmetic, and algebra and operates with infinite series. Newton and Leibniz applied the term analysis to it. Today, mathematical analysis refers to the field after Abel, Cauchy, and Weierstrass provided sound foundations.

[5] Before the nineteenth century, the term *mixed* was generally used instead of *applied*.

[6] For the eighteenth century, see Thomas L. Hankins, *Science and the Enlightenment* (Cambridge: Cambridge University Press, 1985), pp. 1–10 and Tore Frängsmyr, J. L. Heilbron, and Robin E. Rider, *The Quantifying Spirit in the Eighteenth Century* (Berkeley: University of California Press, 1990), pp. 1–141.

[7] For literary views, see Ernest H. Coleridge, *The Complete Poetical Works of Samuel Taylor Coleridge* (Oxford: Oxford University Press, 1967), p. 21 and Rebecca Goldstein, *The Mind and Body Problem* (New York: Penguin, 1993), pp. 46–47.

[8] Eugene P. Wigner, "The Unreasonable Effectiveness of Mathematics in the Natural Sciences," *Communications on Pure and Applied Mathematics* 13 (1980): 1–14.

[9] For the symbiosis between business and technology that developed the modern computer, see Martin Campbell-Kelly and William Aspray, *Computer: A History of the Information Machine* (New York: Basic Books, 1997).

[10]See, for example, Subrahmanyan Chandrasekhar, *Truth and Beauty: Aesthetics and Motivations in Science* (Chicago: University of Chicago Press, 1987); Franchino Gaffurio, *The Theory of Music* (New Haven, CN: Yale Univeristy Press, 1993); Dan Pedoe, *Geometry and the Visual Arts* (New York: Dover Reprint, 1983); and Edward Rothstein, *Emblems of the Mind: The Inner Life of Music and Mathematics* (New York: Times Books, 1995).

[11]Jerry P. King, *The Art of Mathematics* (New York: Plenum Press, 1992), p. 25. See also Donald M. Davis, *The Nature and Power of Mathematics* (Princeton, NJ: Princeton University Press, 1993).

[12]Jean Dieudonné, *Mathematics—The Music of Reason* (New York: Springer-Verlag, 1992), p. 19. For more on distinctions between pure and applied mathematics, see G. H. Hardy, *A Mathematician's Apology* (Cambridge: Cambridge University Press, 1993), pp. 121–131.

[13]See Henri Poincaré, *Science and Method*, trans. by G. Bruce Halsted (New York: The Science Press, 1913) and Alfred Adler, "Mathematics and Creativity," *New Yorker*, February 17, 1972: 39–49.

[14]"Connoisseurship: The Penalty of Ahistoricism," *International Journal of Museum Management and Curatorship*, 7 (1988): 261–268; Ivan Gaskell, "History of Images," in Peter Burke, ed., *New Perspectives on Historical Writing* (University Park, PA: Pennsylvania State University Press, 1993), pp. 168–193; and Julien Stock and David Scrase, *The Achievement of a Connoisseur. Philip Pouncy. Italian Old Master Drawings* (Cambridge. MA: Fitzwilliam Musuem, 1985).

[15]See G. H. Hardy, *A Mathematician's Apology* (Cambridge: Cambridge University Press, 1993 printing). In *Emblems of the Mind: The Inner Life of Music and Mathematics* (New York: Random House, 1995), Edward Rothstein returns to this argument.

[16]Subrahmanyan Chandrasekhar, *Truth and Beauty* [fn. 10]. See also James W. McAllister, *Beauty and Revolution in Science* (Ithaca, NY: Cornell University Press, 1996).

[17]See Mortimer Adler, *Six Great Ideas* (New York: Macmillan, 1981).

[18]Jonathan Lear, "The Shrink Is In," *New Republic*, December 25, 1995: 25.

[19]Man-Keung Siu, "Euler and Heuristic Reasoning," in Frank Swetz et al., eds., *Learn from the Masters* (Washington, DC: Mathematical Association of America, 1994), p. 159.

[20]See Nicolas Bourbaki, *Elements of the History of Mathematics* (Berlin: Springer-Verlag, 1994 Edition) and Paul Erdos, *N Is a Number* (Cisery, Video Documentary, 1994).

[21]André Weil, "The History of Mathematics: Why and How," *Collected Papers*, 3 vols. (New York: Springer-Verlag, 1979), 3: 434–442, and Joseph W. Dauben, "Mathematics: An Historian's Perspective," *Philosophy and the History of Science: A Taiwanese Journal* 2, 1, 1993: 1–21.

[22]See, for example, André Weil, "Who Betrayed Euclid?" *Archive for History of Exact Sciences* 15 (1976): 91–93.

[23]Saunders Mac Lane, *Mathematics: Form and Function* (New York: Springer-Verlag, 1986), pp. 446–449.

[24] Saunders Mac Lane, "The Protean Character of Mathematics," in J. Echeverria, A. Ibarra, and T. Mormann, eds., *The Space of Mathematics: Philosophical, Epistmological and Historical Explanations* (Berlin: Walter de Gruyter, 1992), pp. 1–13.

[25] Michael Dummet, "What is Mathematics About?" in Alexander George, ed., *Mathematics and Mind* (Oxford: Oxford University Press, 1994), pp. 11–27.

[26] Bartel L. van der Waerden, "On the Source of My Book *Moderne Algebre*," *Historia Mathematica* 2 (1975): 31–40.

[27] Saunders Mac Lane, "Van der Waerden's *Modern Algebra*," *Notices of the AMS* 44, 3 (1997): 321–322.

[28] See George Polyá, *Induction and Analogy in Mathematics* (Princeton, NJ: Princeton University Press, l954) and Imre Lakatos, *Proofs and Refutations* (Cambridge: Cambridge University Press, 1976).

[29] See Philip Kitcher, *The Nature of Mathematical Knowledge* (New York: Oxford University Press, 1983), pp. 99–100.

[30] Ivars Peterson, *The Mathematical Tourist* (New York: W. H. Freeman and Co., 1988), pp. 164 ff.

[31] Alfred North Whitehead, *An Introduction to Mathematics* (New York: Oxford University Press, 1958), p. 39.

[32] Paul Arthur Schilpp, ed., *The Philosophy of Alfred North Whitehead* (Evanston, IL: Northwestern University, 1941), pp. 308 and 318–319.

[33] See David Hilbert, *The Foundations of Geometry*, second English edition based on the tenth revised and enlarged German edition of Paul Bernays, trans. by Leo Unger (Chicago: Open Court, 1971), and Leo Corry, "David Hilbert and the Axiomatization of Physics," *Archives for History of Exact Sciences* 51 (1997): 83–198.

[34] Michael Detlefson, "Hilbert's Formalism," *Revue Internationale de Philosophie* 47 (1993): 299.

[35] Hermann Weyl, "David Hilbert and His Mathematical Work," *Bulletin of the American Mathematical Society* 50 (1944): 640.

[36] L. E. J. Brouwer, "Intuitionism and Formalism" (1912) in Ronald Calinger, ed., *Classics of Mathematics* (Englewood Cliffs, NJ: Prentice Hall, 1995), p. 735.

[37] See Miriam Franchella, "L. E. J. Brouwer: Towards Intuitionistic Logic," *Historia Mathematica* 22 (1995): 304–322.

[38] David Hilbert, "Mathematical Problems: Lecture Delivered Before the International Congress of Mathematicians at Paris in 1900," trans. by Mary Winston Newson in Ronald Calinger, ed., *Classics of Mathematics* [fn. 36], esp. pp. 698–700.

[39] George Polya, *How to Solve It: A New Aspect of Mathematical Method* (Princeton, NJ: Princeton University Press, 1988) (1st ed., 1945).

[40] Keith Devlin, *Mathematics: The Science of Patterns* (New York: Scientific American, 1994).

[41] An example is the half-century dormancy in the theory of invariants after Hilbert solved its most important problems in the 1890s.

[42] See John L. Casti, *Five Golden Rules* (New York: John Wiley & Sons, 1996) and Wilhelm Magnus, "The Significance of Mathematics: The Mathematicians' Share in the General Human Condition," *American Mathematical Monthly* 104, 3 (1997): 261–270.

[43]See Imre Lakatos, *Proofs and Refutations: The Logic of Mathematical Discovery* (Cambridge: Cambridge University Press, 1976) and Imre Lakatos, *Mathematics, Science, and Epistemology* (Cambridge: Cambridge University Press, 1978), pp. 29–30.

[44]Philip S. Jones, "The Role in the History of Mathematics of Algorithms and Analogies," Frank Swetz et al. eds., *Learn from the Masters* (Washington, DC: Mathematical Association of America, 1994), p. 13.

[45]See George Polyá, *Induction and Analogy in Mathematics* (Princeton, NJ: Princeton University Press, 1954).

[46]Michael Dummett, *Frege: Philosophy of Mathematics* (Cambridge, MA: Harvard University Press, 1991), p. 45, and Ulrich Majer, "Geometry, Intuition, and Experience: From Kant to Husserl," *Erkenntnis* 42 (1995): 261–285.

[47]Stuart Hollingdale, *Makers of Mathemtics* (London: Penguin Books, 1989), p. 341.

[48]Henri Poincaré, *The Foundations of Science* (New York: Science Press, 1913), p. 387; Janet Foline, *Poincaré and the Philosophy of Mathematics* (New York: St. Martin's Press, 1992); and Jacques Hadamard, *The Psychology of Invention in the Mathematical Field* (New York: Dover, 1954).

[49]Max Dehn, "The Mentality of the Mathematician: A Characterization," *The Mathematical Intelligencer* 5, 2 (1983): 18–28. A. C. Crombie's three-volume *Styles of Scientific Thinking in the European Tradition* ... (London: Duckworth, 1994), for example, presents foms of scientific inquiry and argument on an abstract level and only incidentally touches on cultural context.

[50]John G. McEvoy, "Positivism, Whiggism, and the Chemical Revolution," *History of Science* 35, 107 (1997): 1–12.

[51]For methods of historians of science, see Peter Dear, "Cultural History of Science: An Overview with Reflections," *Science, Technology, and Human Values* 20 (1995): 150–170; Peter Galison and David J. Stump, eds., *The Disunity of Science: Boundaries, Context, and Power* (Stanford, CA: Stanford University Press, 1996); and Arnold Thackray, ed., *Constructing Knowledge in the History of Science, Osiris*, vol. 10 (Chicago: University of Chicago Press, 1995).

[52]See J. E. Littlewood, "The Mathematician's Art of Work," *The Rockefeller University Review*, No. 9 (1967).

[53]Detlef Spalt, "*Quo Vadis*—History of Mathematics," *The Mathematical Intelligencer* 3 (1994): 3–5.

[54]Gaston Bachelard, *The New Scientific Spirit* (Boston: Beacon, 1984).

[55]Michel Foucault, "Introduction" to Georges Canguilhem, *On the Normal and Pathological* (Dordrecht: Reidel, 1978), pp. ix–xx.

[56]Paul Forman, "Independence, Not Transcendence, for the Historian of Science," *Isis* 82 (1991): 71–86.

[57]Steven Shapin, "Discipline and Boundary: The History and Sociology of Science as Seen through the Externalism-Internalism Debate," *History of Science* 30 (1992): 333–369. In "Scholarship Epitomized" *Isis* (1991), p. 87, Charles Gillispie refers to this debate as a "passing schizophrenia."

[58]Thomas F. Gieryn, "Distancing Science from Religion in Seventeenth-Century England," *Isis* 79, 299 (1988): 582.

[59] David Rowe, "New Trends and Old Images in the History of Mathematics," in Ronald Calinger, ed., *Vita Mathematica* (Washington, DC: Mathematical Association of America, 1996), pp. 8 ff.

[60] Joan L. Richards, "The History of Mathematics and *L'esprit humain*: A Critical Reappraisal" in Arnold Thackray, ed., *Constructing Knowledge in the History of Science* [fn. 51], pp. 123–124.

[61] Noel M. Swerdlow, "Otto E. Neugebauer (26 May 1899–19 February 1990)," *Proceedings of the American Philosophical Society* 137, 1 (1993): 137–165, see especially pp. 141–142.

[62] Henk J. M. Bos, "The Concept of Construction and the Representation of Curves in Seventeenth Century Mathematics" in *Proceedings of the International Congress of Mathematicians at Berkeley*, ed., Andrew M. Gleason (Providence, RI: American Mathematical Society, 1987) and Michael Mahoney, "Infinitesimals and Transcendent Relations: The Mathematics of Motion in the Late Seventeenth Century," in *Reappraisals of the Scientific Revolution*, eds. David Lindberg and Robert S. Westman (Cambridge: Cambridge University Press, 1988), pp. 461–491.

[63] Ivor Grattan-Guinness, ed., *Companion Encyclopedia of the History and Philosophy of the Mathematical Sciences* (London: Routledge, 1993) and Karen Parshall, *Historia Mathematica* 23, 1, (1996): 1–7.

[64] W. Burkert, *Lore and Science in Ancient Pythagoreanism* trans. by E. L. Milnar, Jr. (Cambridge, MA: Harvard University Press, 1972), pp. 98–103 and 411–412.

[65] Dominic V. O'Meara, *Pythagoras Revived: Mathematics and Philosophy in late Antiquity* (Oxford: Clarendon Press, 1989), p. 37.

[66] S. I. Salem and Alok Kumer, trans. and eds., *Science in the Medieval World: Book of the Categories of Nations* (Austin: University of Texas Press, 1991).

[67] Roshdi Rashed, *The Development of Arabic Mathematics: Between Arithmetic and Algebra*, trans. by A. F. W. Armstrong (Dordrecht: Kluwer Academic Publishers, 1994), pp. 332 ff.

[68] Nicholas Jardine, *The Birth of History and Philosophy of Science: Kepler's "A Defense of Tycho against Ursus" with Essays on Its Provenance and Significance* (Cambridge: Cambridge University Press, 1984), pp. 258–261 and Bruce Eastwood, "Kepler as Historian of Science and Precursors of Copernican Heliocentrism According to *De Revolutionibus*, I, 10," *Proceedings of the American Philosophical Society* 126 (1982): 367–394. In the 1580s Bernardino Baldi had written biographies of mathematicians, depicting mathematics as a component of culture and a field of inquiry. In "The History of Science," *Times Literary Supplement*, 15 February 1985, p. 171, Paul Rose portrays him as the first modern historian of science.

[69] Paul Lawrence Rose, "The History of Science" [fn. 68], p. 171.

[70] Rachel Laudan, "Histories of Science and Their Uses: A Review to 1913," *History of Sciences* 31 (1993): p. 6.

[71] The Johnson Reprint Corporation reissued the 1907 edition in 1965.

[72] Ivo Schneider, "The History of Mathematics: Aims, Results, and Future Prospects," in Sergei S. Demidov, et al., *Amphora: Festschrift für Hans Wussing* (Basel: Birkhäuser, 1992), p. 624 ff. Schneider also covers the monographs of single mathematical fields and the new didactic purposes for the history of mathematics in the nineteenth century.

[73] Arnold Thackray, "Sarton, Science, and History," *Isis* 75, 276 (1984): 9.

[74]See A. C. Crombie, *Augustine to Galileo: The History of Science, A. D. 400–1650* (London: Heinemann, 1959) 2 vols. The second American edition had the title *Medieval and Early Modern Science*. See also William A. Wallace, *Galileo's Logic of Discovery and Proof: The Background, Content, and Use of His Appropriated Treatises on Aristotle's Posterior Analytics* (Dordrecht: Kluwer Academic Publishers, 1992) and, Wallace, *Galileo's Logical Treatises: A Translation, with Notes and Commentary, of His Appropriated Latin Questions on Aristotle's Posterior Analytics* (Dordrecht: Kluwer Academic Publishers, 1992).

[75]See William Clark, "Narratology and the History of Science," *Studies in History and Philosophy of Science* 26 (1995): 1–71.

[76]See his article on Zeno's paradoxes in *Jahrbuch für Philosophes und Phanomeno-logigische Forschung* 5 (1922): 603–628; *Mélanges Alexandre Koyré*, 2 vols. (Paris: Hermann, 1964); and Rob Iliffe, "Theory, Experiment and Society in French and Anglo-Saxon History of Science," *European Review of History* 2 (1995): 65–77.

[77]Frances Amelia Yates, *Giordano Bruno and the Hermetic Tradition* (New York: Random House, 1969) and Lynn Thorndike, *History of Magic and Experimental Science* 8 vols. (New York: Columbia University Press, 1929–1958).

[78]See Gerard Alberts, "On Connecting Socialism and Mathematics: Dirk Struik, Jan Burgers, and Jan Tinbergen," *Historia Mathematica* 21 (1994): 288–305 and Dirk J. Struik, "Some Sociological Problems in the History of Mathematics," in Kostas Gavroglu et al., eds., *Physics, Philosophy, and the Scientific Community* (Dordrecht: Kluwer Academic, 1995), pp. 377–383. For information on Struik's life and the diverse themes in his work, see David E. Rowe," Dirk Struik and His Contributions to the History of Mathematics," *Historia Mathematica* 21 (1994): 245–273.

[79]See his *Science, Technology and Society in Seventeenth-Century England* (originally published in *Osiris* 4, 1938: 360–632) (New York: Howard Fertig, 1988) and Steven Shapin's "Understanding the Merton Thesis," *Isis* 79, 1988: 594–605.

Only since the 1980s has research on ideology and science addressed gender in works that range from specific attempts to reveal political chicanery in scientific claims to thorough social constructions that detail underlying interests, negotiations, and funding potential. Leading scholars in these studies include Donna Haraway, Ludmilla Jordanova, Evelyn Fox Keller, Sally Gregory Kohlstedt, Margaret Rossiter, Cynthia Russet, and Londa Schiebinger. For the mathematical sciences, see Linda Lopez McAlister, ed., *Hypatia's Daughters: Fifteen Hundred Years of Women Philosophers* (Bloomington: Indiana University Press, 1996).

[80]See I. Bernard Cohen, "The Publication of *Science, Technology, and Society*: Circumstances and Consequences," *Isis* 79 (1988): 571–605.

[81]Steven Shapin, "Understanding the Merton Thesis," *Isis* 79, 299 (1988): 597.

[82]For Popper's thought, see Enrico Bellone, "Karl Raimund Popper," *Belfagor*, 50 (1995): 693–716 and Anthony O'Hear, *Karl Popper: Philosophy and Problems* (Cambridge: Cambridge University Press, 1995).

[83]Thomas S. Kuhn, *The Structure of Scientific Revolutions*, 2nd ed., (Chicago: University of Chicago Press, 1970), p. 79.

[84]See, for example, Gary Gutting, ed., *Paradigms and Revolutions: Appraisals and Applications of Thomas Kuhn's Philosophy of Science* (Notre Dame, IN: University of Notre Dame Press, 1980).

[85]Robert S. Westman, "Two Cultures or One? A Second Look at Kuhn's *The Copernican Revolution*," *Isis* 85 (1994): 79–115 and Thomas S. Kuhn, "Possible

Worlds in History of Science," in S Allen, ed., *Possible Worlds in Humanities, Arts and Sciences* (New York: De Gruyter, 1989), pp. 9–32.

[86]Teun Koetsier, *Lakatos' Philosophy of Mathematics: A Historical Approach.* Studies in the History and Philosophy of Mathematics, Vol. 3 (Amsterdam: North-Holland, 1991), pp. 1–3. See also Imre Lakatos, *Proofs and Refutations: The Logic of Mathematical Discovery*, John Worrall and Elie Zahr, eds., (Cambridge: Cambridge University Press, 1991).

[87]Paul K. Feyerabend, *Against Method*, 3rd ed., (New York: Verso, 1993).

[88]Gertrude Himmelfarb, "Revolution in the Library," *The Key Reporter* 62, 3 (1997): 1–5.

[89]Elena Ausejo and Mariano Hormigon, *Paradigms and Mathematics* (Madrid: Siglo XXI de Espana Editores, 1996); Tian Yu Cao, "The Kuhnian Revolution and the Postmodernist Turn in the History of Science," *Physis* 30 (1993): 476–504; Leo Corry, "Kuhnian Issues, Scientific Revolutions, and the History of Mathematics," *Studies in History and Philosophy of Science* 24 (1993): 95–117; and Vasso P. Kindi, "Kuhn's *The Structure of Scientific Revolutions* Revisited," *Zeitschrift für Allgemeine Wissenschaftstheorie* 25 (1994): 75–92.

[90]Michael Crowe, "Ten 'Laws' Concerning Patterns of Change in the History of Mathematics," *Historia Mathematica* 2 (1975): 161–166; Daniel Gillies, ed., *Revolutions in Mathematics* (Oxford: Clarendon Press, 1992); Herbert Mehrtens, "T. S. Kuhn's Theories and Mathematics," *Historia Mathematica* 3 (1976): 297–320; the review of Gillies by David Rowe in *Historia Mathematica* 20 (1993): 320–323; and the Foreword to Clifford Truesdell, *Essays in the History of Mechanics* (Berlin: Springer-Verlag, 1968).

[91]J. E. Montucla, *Histoire des mathématiques* (Paris: Agasse, 1799–1802), 2: 112 and Paolo Mancosu, "Descartes's *Géométrie* and Revolutions in Mathematics," in Daniel Gillies, ed., *Revolutions in Mathematics* [fn. 90], p. 105.

[92]Yvon Belaval, *Leibniz critique de Descartes* (Paris: Gallimard, 1960), pp. 300–301.

[93]Michael S. Mahoney, "The Beginnings of Algebraic Thought in the Seventeenth Century," in S. Gaukroger, ed., *Descartes: Philosophy, Mathematics and Physics* (Sussex: Harvester, 1980), pp. 141–155 and I. Bernard Cohen, *Revolutions in Science* (Cambridge, Mass.: Belknap Press, 1985), pp. 505–507.

[94]Michael J. Crowe, "Ten Misconceptions about Mathematics and Its History," in William Aspray and Philip Kitcher, eds., *History and Philosophy of Modern Mathematics* (Minneapolis: University of Minnesota Press, 1988) *Minnesota Studies in the Philosophy of Science*, Volume 11: 260–277.

[95]See Joseph Dauben, "Conceptual Revolutions and the History of Mathematics," in *Revolutions in Mathematics*, ed., Donald Gillies (Oxford: Clarendon Press, 1992), pp. 49–71.

[96]L. Kruger et al., *The Probabilistic Revolution* (Cambridge, MA: M.I.T. Press, 1988).

[97]Judith V. Grabiner, "Is Mathematical Truth Tine-Dependent," in *New Directions in the Philosophy of Mathematics*, Thomas Tymozko, ed. (Boston: Birkhäuser, 1985), pp. 201–214.

[98]Thomas S. Kuhn, "What Are Scientific Revolutions?" in Lorenz Krüger, Lorraine

J. Daston, and Michael Heidelberger, eds., *The Probabilistic Revolution* (Cambridge, MA: M.I.T. Press, 1987), pp. 7–22, esp. p. 22.

[99]See H. Floris Cohen, *The Scientific Revolution: A Historiographical Inquiry* (Chicago: University of Chicago Press, 1994), esp. pp. 72–73.

[100]David C. Lindberg and Robert S. Westman, eds., *Reappraisals of the Scientific Revolution* [fn. 62], and John A. Schuster, "The Scientific Revolution," in R. C. Olby et al., *Companion to the History of Modern Science* (London: Routledge, 1990), pp. 217–242.

[101]Roy Porter, "The Scientific Revolution: A Spoke in the Wheel," in Roy Porter and Mikuláš Teich, eds., *Revolution in History* (Cambridge: Cambridge University Press, 1986), pp. 290–316 and Steven Shapin, *The Scientific Revolution* (Chicago: University of Chicago Press, 1996). See also Gary Hatfield, "Was the Scientific Revolution Really a Revolution in Science?" in F. Jamil Ragep et al., eds. *Tradition, Transmission, Transformation* (Leiden: Brill, 1996), pp. 489–525.

[102]See Tom Sorell, ed., *The Rise of Modern Philosophy: The Tension Between New and Traditional Philosophies from Machiavelli to Leibniz* (Oxford: Clarendon Press, 1993).

[103]Peter Dear, *Mersenne and the Learning of the Schools* (Ithaca, NY: Cornell University Press, 1988) and Dear, *Discipline & Experience: The Mathematical Way in the Scientific Revolution* (Chicago: University of Chicago Press, 1995) and Paolo Mancosu, *Philosophy of Mathematics and Mathematical Practice in the Seventeenth Century* (Oxford: Oxford University Press, 1996).

[104]Lorraine Daston, *Classical Probability in the Enlightenment*(Princeton, NJ: Princeton University Press, 1988).

[105]Ian Hacking, *The Taming of Chance* (Cambridge: Cambridge University Press, 1990) and Theodore M. Porter, *The Rise of Statistical Thinking, 1820–1900* (Princeton, NJ: Princeton University Press, 1986).

[106]Joella Yoder, "The Best of All Possible Editions and Other Leibniziana," *Isis* 81, 1 (1994): 116–119.

[107]Ronald Calinger, "Leonhard Euler: The First St. Petersburg Years (1727–1741)," *Historia Mathematica* 23 (1996): 159–161 and Derek T. Whiteside, ed., *The Mathematical Papers of Isaac Newton*, 8 vols. (Cambridge: Cambridge University Press, 1967–1981).

[108]See, for example, Cynthia Hay, ed., *Mathematics from Manuscript to Print 1300–1600* (Oxford: Clarendon Press, 1988).

[109]See, for example, Claudia Henrion, *Women in Mathematics: The Addition of Difference* (Bloomington: Indiana University Press, 1997) and Linda Lopez McAlister, ed., *Hypatia's Daughters: Fifteen Hundred Years of Women Philosophers* (Bloomington: Indiana University Press, 1996).

[110]Peter Burke, ed., *New Perspectives on Historical Writing* (University Park, PA: Pennsylvania State University Press, 1993), pp. 1–8.

[111]Michel Serres, ed., *A History of Scientific Thought: Elements of a History of Science* (Oxford: Blackwell Reference, 1995), p. 4.

[112]See Peter Burke, ed., *New Perspectives on Historical Writing* [fn. 110], p. 233.

[113]Roger Chartier, *On the Edge of the Cliff*, trans. by Lydia G. Cochrane (Baltimore: Johns Hopkins University Press, 1996), pp. 51–54. See also Arnold I. Davidson,

Foucault and his Interlocutors (Chicago: University of Chicago Press, 1996); Lois McNay, *Foucault: A Critical Introduction* (New York: Continuum, 1994); and Paul Rainbow, ed., *The Foucault Reader* (New York: Pantheon, 1984).

[114] See Jacques Derrida, *Of Grammatology* (Baltimore, MD: Johns Hopkins University Press, 1976); *Writing and Difference* (Chicago: University of Chicago Press, 1978); *Margins of Philosophy* (Chicago: University of Chicago Press, 1982); and *Archive Fever: A Freudian Impression* (Chicago: University of Chicago Press, 1996) as well as Bruno Latour, "Postmodern? No, Simply Amodern! Steps Toward an Anthropology of Science," *Studies in History and Philosophy of Science* 21 (1990): 145–172. Also note Tian Yu Cao, "The Kuhnian Revolution and the Postmodernist Turn in the History of Science," *Physis* 30 (1993): 492–493.

[115] Clifford Geertz, *The Interpretation of Cultures* (New York: Basic Books, 1973); and "From the Native's Point of View: On the Nature of Anthropological Understanding," in P. Rabinow and W. M. Sullivan, eds., *Interpretive Social Science: A Reader* (Berkeley: University of California Press, 1979), pp. 226–241 and *Local Knowledge* (New York: Basic Books, 1983).

[116] Nicholas Jardine, "A Trial of Galileos," *Isis* 85 (1994): 279–283.

[117] Natalie Zemon Davis, "Toward Mixtures and Margins," *The American Historical Review* 97, 5 (1992): 1409–1427.

[118] See James Tully, ed., *Meaning and Content: Quentin Skinner and His Critics* (Princeton, NJ: Princeton University Press, 1988).

[119] "The Kuhnian Revolution and the Postmodernist Turn in the History of Science," *Physis* 30 (1993): 477–504. See also David A. Hollinger, "Postmodern Theory and *Wissenschaftliche* Practice," *American Historical Review* 96, 3 (19901): 688–698 and Peter Novick, "My Correct Views on Everything," Ibid., pp. 699–703.

[120] Mark Lilla, "The Politics of Jacques Derrida," *New York Review of Books*, 45, 11 (June 25, 1998): 39.

Chapter 1: Origins of Number and Culture, pages 2–16

[121] See Saunders Mac Lane, *Mathematics: Form and Function*, pp. 4–7.

[122] See Tord Hall, *Carl Friedrich Gauss: A Biography* (Cambridge, MA: M.I.T. Press, 1970), pp. 135–153; S. Kleene, ..., and Raymond L. Wilder, *Introduction to the Foundations of Mathematics*, 2nd ed. (New York: John Wiley & Sons, Inc., 1967), pp. 80–165.

[123] See Richard W. Copeland, *How Children Learn Mathematics: Teaching Implications of Piaget's Research* (New York: Macmillan Company, 1970), pp. 32–86; Stanislas Dehaene, *The Number Sense: How the Mind Creates Mathematics* (New York: Oxford University Press, 1998); and Hans Freudenthal, *Mathematics as an Educational Task* (Dordrecht: D. Reidel Publishing Co., 1973), pp. 170–174.

Piaget's once dominant theory of cognitive development, maintaining that children only attain later a stage when they can understand number, has been supplanted. Brain studies by Dehaene and others show that even young babies by differentiating objects can seemingly count up to three.

[124] Emil Grosswald, *Topics from the Theory of Numbers,* 2nd ed. (Basel: Birkhäuser, 1984), p. 13.

[125] To the present, the nature of number and its relationship to the physical world remain difficult and problematic topics.

[126]Lucien Lévy-Bruhl, *How Natives Think*, trans. by Lillian A. Clare (New York: Alfred A. Knopf, 1923), pp. 181–227, esp. p. 192.

[127]V. Gordon Childe, "The Prehistory of Science, Archaeological Documents," in Guy S. Métraux and François Crouzet, eds., *The Evolution of Science* (New York: Mentor Books, 1963), p. 49.

[128]See David R. Harris, *The Archaeology of V. Gordon Childe: Contemporary Perspectives* (Chicago: University of Chicago Press, 1994).

[129]Steven Rose, Review of Stanislas Dehaene, *The Number Sense: How the Mind Creates Mathematics*, *New York Times,* February 8, 1998.

[130]Stanislas Dehaene, *The Number Sense* [fn. 123], Chapter 1.

[131]Raymond L. Wilder, *Evolution of Mathematical Concepts: An Elementary Study* (New York: John Wiley & Sons, Inc., 1968).

[132]R. M. W. Dixon, *The Languages of Australia* (Cambridge: Cambridge University Press, 1980), pp. 107–108. See also John Harris, "Australian Aboriginal and Islander Mathematics," *Australian Aboriginal Studies*, 2 (1987): 29–37.

[133]John N. Crossley, *The Emergence of Number* (Singapore: World Scientific, 1987), pp. 6 ff.; Georges Ifrah, *From One to Zero* (New York: Penguin Books, 1985), p. 1; and O. Koehler, "The Ability of Birds to 'Count'," in James R. Newman, ed., *The World of Mathematics* (New York: Simon and Schuster, 1956), pp. 488–496.

[134]Stansilas Dehaene, *The Number Sense* [fn. 123], pp. 1–2.

[135]This dating uses for referents the Christian calendar.

[136]Noah Edward Fehl, *Science and Culture* (Hong Kong: Chung Chi Publications, 1965), pp. 70–71.

[137]Marcia Asher, *Ethnomathemtics: A Multicultural View of Mathematical Ideas* (Pacific Grove, CA: Brooks/Cole, 1991), pp. 6 ff. and A. Seidenberg, "The Ritual Origin of Counting," *Archive for History of Exact Sciences* 2 (1962–1966): 3.

[138]Epigraphical evidence shows that it was used in Sumeria in 3000 B.C.

[139]Claude Levi-Strauss, *The Savage Mind* (Chicago: University of Chicago Press, 1970), p. 15.

[140]Marcia Ascher, *Ethnomathematics: A Multicultural View of Mathematical Ideas* [fn. 137], pp. 10–11.

[141]Ibid., pp. 4–30; John Crossley, *The Emergence of Number* [fn. 133], pp. 11–13; and Raymond Wilder, *Evolution of Mathematical Concepts* [fn. 131], pp. 40–42.

[142]Levi-Strauss, *The Savage Mind* [fn. 139], p. 142.

[143]Tobias Dantzig, *Number, the Language of Science* (New York: Macmillan, 1943), p. 10.

[144]Claude Levi-Strauss, *Structural Anthropology* (New York: Basic Books, 1963), pp. 67–80.

[145]Ernst Cassirer, *The Philosophy of Symbolic Forms*, 3 vols., (New Haven, CN: Yale University Press, 1953), I: 229.

[146]A. Seidenberg, "The Diffusion of Counting Practices," *University of California Publications in Mathematics*, 31 (1960): 215–300.

[147]James Hurford, *Language and Number* (London: Basil Blackwell, 1987), pp. 61 ff.

[148]Raymond Wilder, "Hereditary Stress as a Cultural Force in Mathematics," in *Historia Mathematica*, 1 (1974): 29–56.

[149] A. Seidenberg, "The Ritual Origin of Counting" [fn. 137], p. 10.

[150] A. Seidenberg, "The Origin of Mathematics," *Archive for History of Exact Sciences* 18 (1978): 323–329 and Bartel van der Waerden, *Geometry and Algebra in Ancient Civilizations* (Berlin: Springer-Verlag, 1983), pp. 22–35 and 66–69.

[151] This dating depends on the ratio of radioactive carbon 14 to normal carbon 12. In living plants, it is roughly equal. After death, the amount of carbon 14 decreases according to the law of radioactive disintegration. The half-life of C-14 is 5,568 years. Van der Waerden improves the C-14 dating by cross-dating it to tree-ring sequences taken from long-lived trees in a region. This modified C-14 dating seems accurate to within 100 years.

Chapter 2: The Dawning of Mathematics in the Ancient Near East, pages 18–55

[152] Again, B. L. van der Waerden's *Geometry and Algebra in Ancient Civilization* (1983) [fn. 150] conjectures a common Neolithic European source that spread to Near Eastern cultures, India, and China, but common problems faced in surveying, architecture, fortifications, and irrigation along with common human characteristics and schooling practices may account for such similarities as Pythagorean triples, decimal counting, and sequences of practical problems.

Taking a global, multicultural approach, ethnomathematics is bringing to light independent origins of mathematical ideas among traditional peoples in sub-Saharan Africa, the Americas, and Pacific Oceania. Ideas from these peoples, however, neither reached the level of ancient Near Eastern contributions nor created a tradition that strongly influenced the development of mathematics.

[153] See Jens Høyrup, *Mathematics and Early State Formation, or the Janus Face of Early Mesoptamian Mathematics: Bureaucratic Tool and Expression of Scribal Professional Autonomy*, Vol. 2 of *Filosofi og Videnskabsteori pa Roskilde Universitets Center* (Roskilde: University Center, 1991).

[154] Noah Fehl, *Science and Culture* [fn. 136], p. 6.

[155] Tom B. Jones, *Ancient Civilization* (Chicago: Rand McNally Publishing Co., 1973), pp. 44–47.

[156] Otto Neugebauer, *A History of Ancient Mathematical Astronomy*, Three Parts (Berlin: Springer-Verlag, 1975), pp. 347–559 and 571–779.

[157] Georges Ifrah, *From One to Zero: A Universal History of Numbers* (New York: Penguin, 1987), p. 51.

[158] Otto Neugebauer, *The Exact Sciences in Antiquity* (New York: Harper & Row, 1962), p. 5.

[159] Otto Neugebauer, "History of Mathematics, Ancient and Medieval," *Encyclopedia Britannica* (1973), vol. II, p. 640.

[160] Computations with Roman numerals are not so difficult as often thought. Roman numerals are numeral-wise additive. Position of parts of the numeral to the left indicate subtraction and, to the right, addition; for example $IV = 5-1$, $VI = 5+1$, and $CIV = 100 + 4$. Thus, the total value of Roman numerals is obtained by subtracting or adding the simple digits within them. Computations with them need not mirror those with Indo-Arabic numerals. Following computational procedures used in finger counting or on the abacus will make for quick addition and multiplication with Roman numerals. See Michael Detlefsen, et al., "Computation with Roman Numerals," in *Archive for History of Exact Sciences* 15, 2 (1976), pp. 141–148.

[161] Joran Friborg, "Methods and Traditions of Babylonian Mathematics," *Historia Mathematica* 8 (1981), pp. 306–315.

[162] Here is an instance of solving a quadratic equation by completing the square. To solve $x^2 + 6x - 12 = 0$, add 12 to both sides of the equation. This gives the result $x^2 + 6x = 12$. To complete the square on the left, add 9 and do the same on the right, so that

$$x^2 + 6x + 9 = 12 + 9 = 21$$
$$(x + 3)^2 = 21$$
$$x + 3 = \pm\sqrt{21}$$
$$x = \pm\sqrt{21} - 3.$$

[163] See Jens Høyrup, *Mathematics and Early State Formation* [fn. 153].

[164] As early as 2000 B.C., the *Gilgamesh Epic* began as oral literature among the Sumerians. A serious and somber myth, it relates the adventures of a hero named Gilgamesh, whose mother is a god, and his friend Enkidu, who symbolizes primeval man. Relating aspects of early civilization, perils of long-distance trade, a nature myth about the change of seasons, and the famous story of the Babylonian flood, the epic teaches about immortality and promotes respect for the dead.

[165] Otto Neugebauer, *The Exact Sciences in Antiquity* [fn. 158], pp. 101–103.

[166] R. J. Gillings, "The Mathematics of Ancient Egypt," in Charles Coulston Gillispie, ed., *Dictionary of Scientific Biography* (New York: Charles Scribner's Sons, 1978), vol. XV: 705.

[167] Marshall Clagett's *Ancient Egyptian Science: A Source Book,* 2 vols. (Philadelphia: American Philosophical Society, 1989), I:137 and 139, was the first to report evidence of a primitive positional numeral system in Egyptian pharaonic record keeping. See also the Aʻhmosè papyrus, problems 48 and 50, for the unit sign, but no sign to distinguish tens.

[168] Richard J. Gillings, *Mathematics in the Time of the Pharaohs* (Cambridge, Mass.: MIT Press, 1972), pp. 45–70.

[169] Problem 24 of the first-century Cairo Papyrus calculates the third side of a right triangle by the Pythagorean theorem. Given sides of 10 and 6, it finds the third side to be the square root of $100 - 36 = 64$ or 8^2.

Because this and other demotic papyri of the Roman era had problems similar to those in Old Babylonian literature, R. A. Parker believes that a transmission of Babylonian knowledge likely occurred to Egypt during the period of Persian rule of Egypt in the sixth and fifth centuries B.C.

[170] John A. Wilson, *The Culture of Ancient Egypt* (Chicago: Phoenix Books, 1951), pp. 54–55.

[171] Ibid., pp. 60–61.

Chapter 3: Beginnings of Theoretical Mathematics in Pre-Socratic Greece, pages 56–95

[172] The convention to render *polis* as city-state has misleading implications. It overlooks the rural population, which comprised the majority of the citizen body, and suggests that the city ruled the surrounding countryside, which is inaccurate.

[173] N. G. L. Hammond, *A History of Greece to 322 B.C.E.* (Oxford: Clarendon Press, 1959), p. 127.

[174] See Aristotle, *Metaphysics* (A3, 983B6) and *De Caelo* (B13, 294a 28).

[175] See Aubrey de Selincourt, *The World of Herodotus* (Boston: Little, Brown and Co., 1962).

[176] Otto Neugebauer, *The Exact Sciences in Antiquity* [fn. 158], p. 142.

[177] B. L. van der Waerden, *Science Awakening* (Groningen: P. Noordhof Ltd., 1954), pp. 87–89.

[178] W. Burkert, *Lore and Science in Ancient Pythagoreanism*, pp. 415–417.

[179] Bless us, divine number, thou who generates gods and men! O holy tetrakys, thou that contain the root and source of the eternally flowing creation! For the divine number begins with the profound pure unity until it comes to the holy four; then it begets the mother of all, the all-comprising, the all-abounding, the first born, the never-swerving, the never-tiring holy ten, the keyholder of all.

[180] In the eighteenth century, Leonhard Euler proved that if a perfect number is even it must have the form stated in Euclid's theorem. The question of whether all perfect numbers are even is still unresolved.

[181]

$$1 + 2 + 4 + 5 + 10 + 11 + 20 + 22 + 44 + 55 + 110 = 284$$
$$1 + 2 + 4 + 71 + 142 = 220.$$

[182] Perhaps the first proof of the Pythagorean theorem was by proportions. The indispensable tool for this is the law of proportions in similar triangles, which was not developed until the late fifth or fourth century B.C. Here is the proof by proportions:

Construct a right-angled triangle ABC and drop a perpendicular CD to AB, and have $c_1 + c_2 = c$. This gives three similar triangles, $\triangle CBD$, $\triangle ABC$, and $\triangle CAD$, because every angle of one of them has its equal in each of the other two. The triangles being similar, it follows that

$$a/c_1 = c/a \qquad\qquad a^2 = cc_1$$
$$b/c_2 = c/b \qquad\qquad b^2 = cc_2$$

and thus

$$a^2 + b^2 = c(c_1 + c_2) = c^2.$$

[183] Wilbur Richard Knorr, *The Ancient Tradition of Geometric Problems* (New York: Dover, 1993), p. 26.

[184] Consider the case in which the radius $AO = 1$ and $AB = \sqrt{2}$ by the Pythagorean theorem. Then $\cap ABC = \pi r^2/2 = \pi/2$, while $\cap \textit{AEB} = \pi r^2/2 = \pi(\sqrt{2}/2)^2/2 = 2\pi/8 = \pi/4$. Subtracting the common area, ABD, from each gives the quadrant $ABO - ADB = \triangle AOB$ and $\cap AEB = ADB + \text{lune } AEBD$. Therefore, $\triangle AOB = \text{lune } AEBD$.

[185] Wilbur Knorr, *The Ancient Tradition of Geometric Problems* [fn. 183], p. 21.

[186] G. Yavis, *Greek Altars* (St. Louis: Saint Louis University Press, 1949), pp. 169–170 and 245.

Chapter 4: Theoretical Mathematics Established in Fourth-Century Greece, pages 96–118

[187]Wilbur R. Knorr, *The Evolution of the Euclidean Elements* (Dordrecht: D. Reidel, 1975), Chaps. 2 and 9.

[188]Jean Itard, *Les livres arithmetiques d'Euclid, Histoire de la Pensée*, X (Paris: Hermann, 1961), see esp. pp. 11–12.

[189]See Philip J. Davis, *Spirals: From Theodorus to Chaos* (Wellesley, MA: A. K. Peters, 1993).

[190]Charles C. Gillispie, ed. *Dictionary of Scientific Biography*, [fn. 166], "Theodorus of Cyrene," XIII: 314–319.

[191]Given two numbers a_0 and a_1, subtract the smaller, a_1, as many times as necessary from a_0 to arrive at zero or a remainder a_2 that is less than a_1. If a_2 turns out to be less than a_1, subtract a_2 from a_1 until a remainder a_3 that is less than a_2 is obtained. If this procedure stops with a_{p+1}, then a_p is the largest common measure. See Euclid VII, 1–3 for numbers and X, 2–4 for homogeneous magnitudes.

[192]Plato uses interchangeably the two Greek words, for which their nearest English equivalents are inadequate. In common speech, ideas or notions exist in people's minds and may be true or false. "Form" admits of many English definitions: Platonic flavor is retained in mathematical "formalism," the mathematical structure of an argument as distinct from its content.

[193]Peter Gay, *Style in History* (New York: McGraw-Hill, 1974), p. 29.

[194]Wilbur R. Knorr, *The Ancient Tradition of Geometric Problems* [fn. 183], pp. 17 ff.

[195]See Harold F. Cherniss, *The Riddle of the Early Academy* (New York: Russell & Russell, 1962, repr. 1981).

[196]D. H. Fowler, *The Mathematics of Plato's Academy: A New Reconstruction* (Oxford: Clarendon Press, 1987), p. 107.

[197]In *Physics, Posterior Analytics, Topics, De Caelo,* and *Meteorologica,* Aristotle presents his physics and cosmology. These ideas dominated the physical sciences until Cartesian science displaced them two millennia later. Aristotle's universe, as I and II of *De Caelo* describe it, is centered on Earth and composed of the four elements of earth, air, fire, and water along with a fifth element (quintessence) in the heavens. In Aristotle's theory of gravity, heavy materials seek the center of Earth and light materials rise toward the heavens. Celestial physics is separate from terrestrial physics, and the moon's orbit is the dividing point. Natural motion in the heavens is circular and celestial bodies are unchanging.

[198]T. L Heath, ed., *The Thirteen Books of Euclid's Elements* 3 vols. (Cambridge: Cambridge University Press, 1908), II: 114.

[199]For the geometric statement of this postulate, see page 145 of Chapter 5.

[200]Stobaeus, *Anthologium,* ed. by C. Wachsmuth, II (Leipzig, 1884), 228.30–33.

[201]Wilbur R. Knorr, *The Ancient Tradition of Geometric Problems* [fn. 183], pp. 62 ff.

Chapter 5: Ancient Mathematical Zenith in the Hellenistic Third Century B.C., I: The Alexandrian Museum and Euclid, pages 119–149

[202]Harold T. Davis, *Alexandria, the Golden City,* vol. I (Evanston, Ill.: Principia Press of Illinois, Inc., 1957), p. 72.

[203] For historical accounts of ancient Alexandria, see H. T. Davis, *Alexandria, The Golden City* [fn. 202], two volumes; J. R. Harris, ed., *The Legacy of Egypt* (Oxford: Clarendon Press, 1971), pp. 323–354; Kenneth Neuer, *City of the Stargazers* (New York: Charles Scribner's Sons, 1972); and Claire Preaux, "Alexandria under the Ptolemies," in Arnold Toynbee, ed., *Cities of Destiny* (New York: McGraw-Hill, 1973).

[204] See Diana Delia, "From Romance to Rhetoric: The Alexandrian Library in Classical and Islamic Traditions," *American Historical Review* 97, 5 (1992): 1449–1468.

[205] Ibid., p. 1461.

[206] Kenneth N. Heuer, *City of the Stargazers* [fn. 203], p. 108.

[207] See Charles Coulston Gillispie, ed., *Dictionary of Scientific Biography*, s. v., "Euclid," IV: 415.

[208] Albert Einstein, "Autobiographical Notes," in Paul Arthur Schlipp, ed., *Albert Einstein: Philosopher-Scientist* (LaSalle, IL: Open Court, 1949), p. 11.

[209] See Ivor Bulmer-Thomas, "Euclid," in Charles Coulston Gillispie, ed., *Dictionary of Scientific Biography* [fn. 166], IV: 419.

[210] See the section on Eudoxus in Chapter 4. The use of this definition to divide all rational numbers into two coextensive classes constitutes the basis for Dedekind cuts.

[211] B. L. van der Waerden, *Geometry and Algebra in Ancient Civilizations* (Berlin: Springer-Verlag, 1983), pp. 73–87.

[212] See Sabetai Unguru, "History of Ancient Mathematics: Some Reflections on the State of the Art," *Isis* 70 (1979): 555–565.

[213] The *Elements* concludes with the construction of the Platonic solids in Book XIII. See George E. Owen, *The Universe of the Mind* (Baltimore, MD: Johns Hopkins Press, 1971), p. 39.

[214] In the nineteenth century, Gauss, Lobachevsky, Bolyai, and Riemann invented non-Euclidean geometries by departing from the parallel postulate and the movement away from Euclid began.

Chapter 6: Ancient Mathematical Zenith in the Hellenistic Third Century B.C., II: Archimedes to Diocles, pages 150–183

[215] Wilbur Knorr, *The Ancient Tradition of Geometric Problems* [fn. 183], pp. 151–209.

[216] For a different version of this anecdote, see E. J. Dijksterhuis, *Archimedes* (New York: Humanities Press, 1957), pp. 15–18.

[217] Ibid., pp. 23–25.

[218] Carthage, or New City, was a colony of the Phoenician city of Tyre. Located on the coast of north Africa near modern Tunis, it had an excellent harbor and a bountiful agriculture from the rich coastal plain. Romans called Carthaginians Phoenicians, or in Latin *Poenia* or *Puni*, hence the adjective Punic.

[219] Book II of *On Floating Bodies* discusses paraboloids of revolution. These are formed by rotating a parabola on its axis.

[220] Notwithstanding his subsequent fame, no bust or picture of Archimedes has survived, although a portrait on a Sicilian coin (of unknown date) is believed to be his. The mosaic of Archimedes before a calculating board with a Roman soldier poised over him is not Hellenistic but Renaissance in origin.

[221]Let *ABCD* be the given circle, *K* the triangle described.

Then, if the circle is not equal to *K*, it must be either greater or less.

I. If possible, let the circle be greater than *K*.

Inscribe a square *ABCD*, bisect the arcs *AB*, *BC*, *CD*, *DA*, then bisect (if necessary) the halves, and so on until the sides of the inscribed polygon whose angular points are the points of division subtend segments whose sum is less than the excess area of the circle over *K*.

Thus the area of the polygon is greater than *K*.

Let *AE* be any side of it, and *ON* the perpendicular on *AE* from the center *O*. Then *ON* is less than the radius of the circle and therefore less than one of the sides about the right angle in *K*. Also the perimeter of the polygon is less than the circumference of the circle, that is, less than the other side about the right angle in *K*. Therefore the area of the polygon is less than λ which is consistent with the hypothesis. Thus the area of the circle is not greater than *K*.

II. If possible, let the circle be less than *K*.

Circumscribe a square, and let two adjacent sides touching the circle in *E*, *H* meet in *T*. Bisect the arc between points of contact and draw the tangents at the points of bisection. Let *A* be the middle point of the arc *EH* and *FAG* the tangent at *A*.

Then the angle *TAG* is a right angle.

Therefore, $TG > GA > GH$.

It follows that the triangle is greater than half the area of *TEAH*.

Similarly, if the arc *AH* be bisected and the tangent at the point of bisection be drawn, it will cut off from the area *GAH* more than one-half.

Thus, by continuing the process, we shall ultimately arrive at a circumscribed polygon such that the spaces intercepted between it and the circle are together less than the excess of *K* over the area of the circle.

Thus the area of the polygon will be less than *K*.

Now, since the perpendicular from *O* on any side of the polygon is equal to the radius of the circle, while the perimeter of the polygon is greater than the circumfernce of the circle, it follows that the area of the polygon is greater than the triangle *K*, which is impossible.

Therefore, the area of the circle is not less than *K*.

Since then the area of the circle is neither greater nor less than *K*, it is equal to it.

[222]Wilbur R. Knorr, *The Ancient Tradition of Geometric Problems* [fn. 183], pp. 156–157.

[223]Other such works may have been composed. Historical accident has turned up this exposition lost for centuries. Does an unknown heuristic of Archytas lurk beneath the hymnody of some MS in a Transylvanian monastery?

[224]With the aid of a heliometer, an improved telescope, and his accurate finding of the distance of a fixed star, Friedrich W. Bessel, director of the observatory in Königsberg, in the 1830s was the first to observe stellar parallax.

[225]Ernst Mach's criticism of these two proofs in the opening pages of *Science of Mechanics* gave rise to an extensive body of literature criticizing or defending them. Among other points, Mach stresses that despite their mathematical deductive form, these proofs rely primarily on experience.

[226]See Plutarch, *Parallel Lives*, Secs. 12–17.

[227] In Archimedes' writings see especially *On the Quadrature of the Parabola* (Propositions 1–3) and *On Conoids* (Propositions 3 and 7–9). Conon and Nicoletes were two other precursors.

[228] Archimedes called the constant p for the parabola "the double of the distance to the axis (of the cone)." See *On Conoids*.

[229] Wilbur R. Knorr, *The Ancient Tradition of Geometric Problems*, p. 157.

[230] Wilbur Richard Knorr, *The Ancient Tradition of Geometric Problems,* pp. 261 ff.

Chapter 7: Mathematics in Roman and Later Antiquity, pages 184–222

[231] For Roman surveying, see O. A. W. Dilke, *The Roman Land Surveyors: An Introduction to the Agrimensores* (Newton Abbot, England: David-Charles, 1971).

[232] Georges Ifrah, *From One to Zero* [fn. 157], pp. 131–150.

[233] Burma P. Williams and Richard S. Williams, "Finger Numbers in the Greco-Roman World and the Early Middle Ages," *Isis* 86 (1995): 587–608.

[234] Charles W. Jones, *Bedae pseudepigraphia: Scientific Writings Falsely Attributed to Bede* (Ithaca, NY: Cornell University Press, 1939), particularly 106–108.

[235] See Lorraine L. Larison, "The Möbius Band in Roman Mosaics," in *American Scientist* (September–October 1973): 544–547.

[236] Otto Neugebauer, *A History of Ancient Mathematical Astronomy* [fn. 156], Part Three: 1071.

[237] Ibid., Part Three: 1061–1062.

[238] Owen Gingerich, *The Eye of Heaven: Ptolemy, Copernicus, Kepler* (New York: American Institute of Physics, 1993), pp. 62–63.

[239] Otto Neugebauer, *A History of Ancient Astronomy* [fn. 156], Part Three: 1061.

[240] Ibid., p. 89.

[241] Ibid., p. 341.

[242] Galen, the son of a geometer, was strongly influenced by an intensive study of geometry in his teens. He later attempted to extend to medicine its abstract rigor and systematic foundation.

[243] See Franz Boll, *Studien über Claudius Ptolemäus* (Leipzig: Teubner, 1894).

[244] Otto Neugebauer, *A History of Ancient Astronomy* [fn. 156], Part Two: 834.

[245] Ibid.

[246] Ibid., p. 836.

[247] Alexander Jones, "The Adaptation of Babylonian Methods in Greek Numerical Astronomy," *Isis* 82 (1991): 440–453.

[248] See Owen Gingerich, *The Eye of Heaven* [fn. 238], pp. 58, 64, and 74 ff.

[249] Ptolemy did not have rules for oblique triangles. Instead he broke them into right triangles.

[250] O. Neugebauer, *Über eine Methode zur Distanzbestimmung Alexandria-Rom bei Heron* Kgl. Danske Vidensk.selsk., Hist.fil.Meddel, 26, 2, 7 (1938 and 1939).

[251] The titles are *On the Construction of Automata, Barulkos, Belopoiica, Catoptrica, Cheirobalistra (On Catapults), Definitiones, Dioptra, Geometrica, Mechanica, Demensuris, Metrica, Pneumatica,* and *Stereometrica.*

[252]See Jens Høyrup, "The Four Sides and the Area: Oblique Light on the Prehistory of Algebra," in Ronald Calinger, ed., *Vita Mathematica* [fn. 59], pp. 45–66.

[253]B. L. van der Waerden, *Geometry and Algebra in Ancient Civilizations* [fn. 211], p. 186.

[254]Thomas Heath, *History of Greek Mathematics*, 2 vols. (Oxford: Clarendon Press, 1921) II: 314–316.

[255]Gerasa may be the city in Palestine referred to in the New Testament as Geradine.

[256]A prominent exception is the Pergamese physician Galen (129/130–199/200). The foremost physician in antiquity, he intensively revised anatomical writings, making the subject the foundation of medicine, but his work suffered from cultural prohibition of human dissection. His corpus of anatomical and physiological studies, including those on the nerves and an ebb and flow movement of the blood, were to dominate Western medicine until the Paduan tradition and William Harvey in the seventeenth century.

[257]But in medicine Galen had known war, famine, plague, intrigue, and corruption.

[258]See the Loeb edition prepared in 1958 by W. R. Paton.

[259]Jean Christianidis, "...: Un traitè perdu de Diophante d'Alexandrie?" *Historia Mathematica* 18, 3 (1991): 229–247.

[260]As translated by Sir Thomas Heath in *Diophantus of Alexandria: A Study in the History of Greek Algebra* (New York: Dover Publications, Inc., 1964), p. 206.

[261]In 1994 Andrew Wiles of Princeton for the most part proved this theorem.

[262]See Harold M. Edwards, *Fermat's Last Theorem: A Genetic Introduction to Algebraic Number Theory*, 4th printing (Berlin: Springer-Verlag, 1993).

[263]See Alexander Jones, *Pappus of Alexandria: Book 7 of the Collection* (Berlin: Springer-Verlag, 1986).

[264]Wilbur Richard Knorr, *Textual Studies in Ancient and Medieval Geometry* (Boston: Birkhäuser, 1989), pp. 386–387.

[265]Maria Dzielska, *Hypatia of Alexandria*, trans. by F. Lyra, *Revealing Antiquity*, Vol. 8 (Cambridge, MA: Harvard University Press, 1995), pp. 74–76.

[266]Ibid., 71–74.

[267]Ibid. And see J. M. Rist, "Hypatia," *Phoenix* 19 (1995): 214–225; and M.E. Waithe, ed., *A History of Women Philosophers* (Dordrecht: M. Nijhoff, 1987), pp. 169–195.

[268]Wilbur Richard Knorr, *Textual Studies in Ancient and Medieval Geometry* [fn. 263], pp. 754–755.

[269]Ibid., p. 754.

[270]See especially pages 109–110.

[271]B. L. van der Waerden, *Science Awakening* [fn. 177], p. 290.

[272]Maria, Dzielska, *Hypatia of Alexandria* [fn. 264], pp. 2–3 and 102.

[273]Proclus, *A Commentary on the First Book of Euclid's Elements*, trans. by Glenn R. Morrow (Princeton, NJ: Princeton University Press, 1970), p. 295.

Chapter 8: Mathematics in Traditional China, pages 223–258

[274]David Pingree, "History of Mathematical Astronomy in India," in Charles Coulston Gillispie, editor in chief, *Dictionary of Scientific Biography* [fn. 166], Vol. XV, Supplement I, p. 533.

[275] Joseph Needham, *Science and Civilization in China* (Cambridge: Cambridge University Press, 1959), 3: 210 ff.

[276] Li Yan and Du Shiran, *Chinese Mathematics: A Concise History* (Oxford: Clarendon Press, 1987) trans. by John N. Crossley and Anthony W. C. Lien, p. 49.

[277] See *The Works of Li Po*, trans. by Shigeyoshi Obata (New York: E. P. Dutton and Co., 1950).

[278] As quoted in Li Yan and Du Shiran, *Chinese Mathematics* [fn. 275], p. 104.

[279] The original problem states: "There are an unknown number of things. Three by three, two remain; five by five, three remain; seven by seven, two remain. How many things?" Given in modern terms, three is a number N, which divided by p_1 has remainder r_1; divided by p_2 has remainder r_2; and divided by p_3, remainder r_3. What is N? In Gauss's modern symbolism, $N \equiv r_1 \pmod{p_1}$ read N is congruent to r_1 modulo p_1, or that N divided by p_1 leaves a remainder of r_1.

[280] Lam Lay-yong, "The Chinese Connection between the Pascal Triangle and the Solution of Numerical Equations of Any Degree," *Historia Mathematica* 7, 4 (1980): 422.

[281] See Frank Swetz, "Enigmas of Chinese Mathematics," in Ronald Calinger, ed., *Vita Mathematica* [fn. 252], pp. 89–90.

[282] George Sarton, *Introduction to the History of Science* (Baltimore, MD: Williams & Wilkins, 1947) III: 700–703.

[283] *Dictionary of Scientific Biography*, [fn. 166], Vol. VIII, "Li Chih."

[284] Ibid.

[285] In 1967 Lam Lay-Yong of the University of Singapore first translated into English *The Method of Computation of Yang Hui* and in 1977 published a critical commentary.

[286] Joseph Needham, *Science and Civilization in China* [fn. 274], 3: 47.

[287] Li Yan and Du Shiran, *Chinese Mathematics* [fn. 275], p. 180.

[288] Ibid., p. 182

[289] Ibid., p. 203.

[290] Ibid., p. 209.

Chapter 9: Indian Mathematics: From Harappan Through Medival Times, pages 259–284

[291] Florian Cajori, *A History of Mathematics* (New York: Macmillan, 1919, rev. 1953), p. 88.

[292] See George Rusby Kaye, *Hindu Mathematics* (Bologna: Zanichelli, 1919).

[293] As stated in George Gheverghese Joseph, *The Crest of the Peacock: Non-Western Roots of Mathematics* (London: I. B. Tauris & Co. Ltd., 1991), p. 229.

[294] See p. 31 below.

[295] George Joseph, *The Crest of the Peacock* [fn. 292], p. 284.

[296] See Lay-Young Lam, *Archive for History of Exact Sciences* 37: 365–392 and 38: 101–108.

[297] See C. T. Rajagopal and N. S. Rangachari, "On an Untapped Source of Medieval Keralese Mathematics," *Archive for History of Exact Sciences* 18 (1978): 89–102; _____, "On Medieval Keralese Mathematics," ibid., 35 (1986): 91–99; and T. A.

Sarasvati, "Development of Mathematical Series in India," *Bulletin of the National Institute of Sciences* 21 (1963): 320–343.

[298]See Donald F. Lach, *Asia in the Making of Europe,* two vols.,(Chicago: University of Chicago Press, 1965).

[299]As quoted in George Joseph, *The Crest of the Peacock* [fn. 292], p. 290.

Chapter 10: Mathematics in the Service of Religion, pages 286–306

[300]In a mid-ninth century exchange, Emperor Michael III equated Latin with a barbarous Scythian language, and Pope Nicholas I retorted that a Roman emperor without command of the Roman tongue was a figure of ridicule.

[301]The last edition of Boethius' *Arithmetica* appeared in Paris in 1521.

[302]William H. Stahl, *Roman Science: Origins, Development and Influence to the Later Middle Ages* (Madison: University of Wisconsin Press, 1962), p. 214.

[303]Bede, *Ecclesiastical History of the English People*, Bertram Colgrave and R. A. B. Mynors, eds. (Oxford: Oxford University Press, reprint, 1991), p. 296.

[304]Jennifer Moreton, "Doubts about the Calendar: Bede and the Eclipse of 664," *Isis* 89 (1998): 51.

[305]See Bede, *De temporum ratione*, ed. by Charles W. Jones in *Bedae Opera de temporibus* (Cambridge, MA: Medieval Academy of America, 1943), pp. 1–111.

[306]Jennifer Moreton, "Doubts about the Calendar," [fn. 303] pp. 64–65.

[307]Finger counting existed in Roman times. Classical writers referred to it, and the lost statue of the god Janus at the gate of ancient Rome formed the number 365 on its fingers. Boethius' classification of numbers as *digiti* (fingers), *articuli* (joints = numbers divisible by ten), and *compositi* reflects the technique. Understanding a symbolic explanation that St. Jerome gives of the gospel parable of the harvest yielding thirty, sixty, and a hundred fold requires a knowledge of finger counting.

[308]This body language seems ornamental, since there was little need in daily life for even four-digit numbers. Largely unnecessary in trade, commerce, or architecture, a four-digit number makes a rare appearance near the end of the old English epic *Beowulf.*

[309]See E. S. Duckett, *Alcuin, Friend of Charlemagne* (New York: Macmillan, 1951).

[310]Gerbert, the first of the Franks elected pope, took the name Sylvester II, remembering Sylvester I, a scholar who had participated in a holy alliance with Constantine.

[311]A. F. Aveni and H. Hartung, "Precision in the Layout of Maya Architecture," *Annals of the New York Academy of Sciences* 385 (1982): 63–80.

[312]A. M. Tozzer, *Landa's Relacíon de las Cosas de Yucatán*, Papers of the Peabody Museum, No. 18 (Cambridge, MA: Harvard University Press, 1941).

[313]See Persis B. Clarkson, "Classic Maya Pictorial Ceramics: A Survey of Content and Theme," in Raymond Sidrys, ed., *Papers on the Economy and Architecture of the Ancient Maya* (Los Angeles: University of California Institute of Archaeology, 1978), Monograph 7, p. 101.

[314]Michael P. Closs, ed., *Native American Mathematics* (Austin: University of Texas Press, 1986), pp. 307–311.

Chapter 11: The Era of Arabic Primacy and a Persian Fluorescence, pages 307–356

[315]The hundreds of thousands of primary source manuscripts produced over seven centuries are now scattered across the globe. Many have yet to be examined and critical editions of crucial documents are scarce. This situation leaves major gaps in the knowledge of different mathematical lines of development and, more generally, supplies only a fragmentary and occasionally unreliable historical record.

[316]The name Gibraltar comes from the Arabic "jabal Tariq," meaning the mountain of Tarik, the Muslim commander who crossed it in 711.

[317]Sabean astronomers claiming a monotheistic "religion of book" were also tolerated. The Sabeans were from Harrān in northern Mesopotamia. It was on the same site as the modern Turkish city of Diyar bakir. The Sabean educated class had a Neoplatonic or Neopythagorean faith. Together with Zoroastrians and Christians, they were early teachers of the Muslims.

[318]Sectarian conflict recurs throughout Islamic history. The two principal sects have been *Sunnis* (followers of "custom" or "path") and *Shi'ites* (partisans of [the Caliph] 'Ali). The original issue was whether succession passed to an *Imam* of blood descent from Muhammad and his nephew and son-in-law 'Ali. *Sunnis* held that succession went to a caliph by election of a candidate from the tribes of the Prophet's Companions. *Shi'ites* have been consistently dissidents with bases in lower, oppressed social strata. Under the Abbāsids they also had an intellectual elite. The nonpartisan egalitarianism of mosque and Meccan pilgrimage has limited the damage of sectarianism.

[319]As quoted in J. L. Berggren, *Episodes in the Mathematics of Medieval Islam* (New York: Springer-Verlag, 1986), p. 30.

[320]*Bayt* is the word for a dwelling. It also refers to a verse containing an independent thought. Several *bayts* make up a *gasidas* or poem. *Dār* also means "abode," but can be more extensive: everything under the crescent constitutes the *Dār al-Islam.*

[321]J. L. Berggren , *Mathematics in Medieval Islam,* p. 5.

[322]Ibid., pp. 3–4.

[323]Yvonne Dold Samplonius, "The *Book of Assumptions* by Thābit ibn Qurra (836–901)," in Joseph Dauben et al., eds., *History of Mathematics: States of the Art* (San Diego, CA: Academic Press, 1996), p. 211.

[324]Roshdi Rashed, *The Development of Arabic Mathematics* (Dordrecht: Kluwer Academic Publishers), p. 206.

[325]From the twelfth century the translated writings, *Verba filiorum,* of the Banū Mūsā were influential in spreading traditions of Euclid and Archimedes in the West. In preparing *Practica Geometriae*, completed in 1220/1221, Leonardo Fibonacci drew ideas and excerpts from it, and Jordanus de Nemore studied it.

[326]Roshdi Rashed, *The Development of Arabic Mathematics* [fn. 323], p. 278.

[327]Seyyed Hossein Nasr, *Science and Civilization in Islam* (New York: New American Library, 1968), p. 153.

[328]As quoted in Berggren, *Mathematics in Medieval Islam* [fn. 318], p. 36.

[329]M. I. Medovoy, "On the Arithmetic Treatise of Abu'l Wafa," *Istoriko-mathematiches-kiissledovanija* 13 (1960), pp. 253–324.

^{330}In one etymology of the word, *jabr* was borrowed from the terminology of the surgeon and meant setting a broken bone or dislocated limb. This tradition persists in modern Spanish, where *algebrista* refers to a bonesetter as well as an algebraist.

^{331}Roshdi Rashed, *The Development of Arabic Mathematics* [fn. 323], p. 264.

^{332}Ibid., p. 8.

^{333}Ibid., p. 265.

^{334}As quoted in Adel Anbouba, "Al-Samaw'al," in Charles Coulston Gillispie, ed., *Dictionary of Scientific Biography* [fn. 273], Vol. XII: 92.

^{335}George Sarton, *Introduction to the History of Science* [fn. 281], vol. 2, p. 176 and Carl Boyer and Uta Merzbach, *A History of Mathematics,* 2nd ed. (New York: John Wiley & Sons, Inc., 1989), p. 277.

^{336}Late medieval Latin translators mistakenly identified half-chords, *jiba,* with a similar Arabic word for bosom and translated it as *sinus,* a bay.

^{337}His correct sexagesimal value to five places is $\sin 30' = 0; 0, 31, 24, 55, 54, 0$. His value for $\sin 30'$ in decimal fractions is 0.0087665373, compared to the more correct 0.0087265355.

^{338}R. Ramsay Wright's English translation of al-Bīrūnī's *Elements of Astrology* was published in London in 1934.

^{339}As quoted in Berggren, *Mathematics of Medieval Islam* [fn. 318], p. 10.

^{340}Roshdi Rashed, *The Development of Arabic Mathematics* [fn. 323], p. 45.

^{341}See P. M. Holt et al., eds., *The Cambridge History of Islam* (Cambridge: Cambridge University Press, 1970), Volume 2, Chapter 13.

^{342}M. Aballagh and E. Djebbar, "Découverte d'un ecrit mathématique d'al-Hassar (XIIIe S.): Le livre I du Kamil," *Historia Mathematica* 14 (1987): 137–158 and Victor Katz, "Combinatorics and Induction in Medieval Hebrew and Islamic Mathematics," in Ronald Calinger, ed., *Vita Mathematica* [fn. 252], pp. 99–107.

^{343}*Risala* . . . , Winter-Araft translation, p. 29.

^{344}J. A. Boyle, ed., *The Cambridge History of Iran* (Cambridge: Cambridge University Press, 1968), Volume 5, p. 289.

^{345}Ibid.

^{346}Roshdi Rashed, *The Development of Arabic Mathematics* [fn. 323], p. 152.

^{347}Ibid., pp. 150–151.

^{348}Ibid., pp. 46 ff. and 152 ff. and Jan P. Hogendijk, "Sharaf al-Din al Tusi on the Number of Positive Roots of Cubic Equations," *Historia Mathematica* 16 (1989): 69–85.

^{349}See George Saliba, "The Original Source of Qutb al-Din al-Shīrāzī's Planetary Model," *Journal for the History of Arabic Science* 3 (1979): 3–18.

^{350}George Saliba, *A History of Arabic Astronomy* (New York: New York University Press, 1994), pp. 197, 203–204, and 259.

^{351}Ibid., pp. 258 ff.

^{352}Willy Hartner, "Copernicus, the Man, the Work, and its History," *Proceedings of the American Philosophical Society* 117 (1973): 413–422 and Noel Swerdlow, "The Derivation and First Draft of Copernicus's Planetary Theory," pp. 423–512.

[353] A. P. Youschkevitch and B. A. Rosenfeld, "Al-Kāshī" in Charles Coulston Gillispie, ed., *Dictionary of Scientific Biography* [fn. 273], Vol. VII: 255 and T. N. Kari-Niazov, "Ulugh Beg," Vol. XIII: 536–537.

[354] J. L. Berggren, *Episodes in the Mathematics of Medieval Islam* [fn. 318], pp. 58 ff.

[355] J. L. Berggren, Ibid., pp. 151–152.

Chapter 12: Recovery and Expansion in Old Europe, 1000–1500, pages 357–394

[356] Several historians recommend revising the traditional periodizations of European history. A recent proposal is to designate as Old Europe the period from 1000 to 1800. Old Europe's formative centuries were the eleventh and twelfth, and its bases were corporate organizations and regional as opposed to national attachments. The year 1500, however, does serve as a useful dividing point for developments in mathematics and the early modern sciences. See Dietrich Gerhard, *Old Europe: A Study in Continuity, 1000–1800* (New York: Academic Press, 1981) and Jerry H. Bentley, "Cross-Cultural Interaction and Periodization in World History," *American Historical Review* 101, 3 (1996): 749–771.

[357] St. Bernard of Clairvaux, mobilizer of the Second Crusade that lasted from 1147 to 1149, judged most of the Crusaders criminals, ravishers, sacrilegious murderers, perjurers, and adulterers. But their departure, he said, relieved the scourge at home and heartened the beleaguered foreigners. Beleaguered residents of Constantinople might concur with the first opinion but take exception to the second.

[358] See R. W. Southern, *Making of the Middle Ages* (New Haven, CN: Yale University Press, 1953) and Jean Leclerq, *The Love of Learning and the Desire for God* (New York: New American Library, 1961). Church construction, mendicant friars, and even spiritual revival obviously require an economic base. See also Robert S. Lopez, *The Commercial Revolution of the Middle Ages, 950–1350* (Englewood Cliffs, NJ: Prentice Hall, 1971) and Lynn White, *Medieval Technology and Social Change* (Oxford: Oxford University Press, 1962).

[359] Richard Lemay challenges the view that this process of translating occurred in states and by committees. See his discussion of Gerard of Cremona in the *Dictionary of Scientific Biography* [fn. 273]. For the counterpoint, see Plato of Tivoli in the same source. A similar two-stage process underlay the translation from Greek to Arabic via Syriac described in chapter 11.

[360] Barnabus B. Hughes, *Jordanus de Nemore: De numeris datis, A Critical Edition and Translation* (Berkeley: University of California Press, 1982), pp. 5–7. In David C. Lindberg, ed., *Science in the Middle Ages* (Chicago: University of Chicago Press, 1978), pp. 160–161, Michael S. Mahoney contests Hughes's claim.

[361] Anthropologists estimate from skeletal remains that life expectancy in thirteenth-century England rose from about twenty-five years in Roman times to about thirty-five. In the fourteenth century, life expectancy apparently again fell below twenty-five.

[362] The Catalan mystic Ramon Lull put forward an equally ambitious but less influential program. Early in the fourteenth century, he proposed an extension of the axiomatic method to all sciences and a systematic examination of all possible combinations of their elements. To conduct that examination, he developed several revolving concentric volvelles that display a symbolic alphabet. In the late seventeenth century, Gottfried Leibniz, who hoped to construct a universal algebra, studied this logic machine attentively.

[363]The popular Adelard II *Elements* had omitted demonstrations as supernumerary. Campanus' edition combines didactic and logical tendencies. These positions help explain why, in place of direct translations from Greek, schoolmen preferred Latin rendering of Arabic texts and why they settled for truncations and popularizations of works of Leonardo of Pisa and Jordanus.

[364]Proposition 18 of *On Spirals* states: "If a straight line is tangent to the extremity of a spiral described in the first revolution, and if from the point of origin of the spiral one erects a perpendicular on the initial line of revolution, the perpendicular will meet the tangent so that the line intercepted between the tangent and the origin of the spiral will be equal to the circumference of the first circle."

[365]See his "From Social into Intellectual Factors: An Aspect of the Unitary Character of Late Medieval Learning" in J. Murdoch and E. Sylla, eds., *The Cultural Context of Medieval Learning* (Dordrecht: D. Reidl, 1995), p. 287. Scholastics did not, however, translate into metric statements the type of statement on which modern physics depends, the theoretical language of proportions.

[366]Recognizing the need for a principle of continuity, thirteenth- and fourteenth-century mathematicians devised through several "postulates of betweenness" less elegant alternatives to Eudoxus. In modern form, they say: for all magnitudes a and b where $a < b$, there exists a third magnitude c, such that $a < c < b$.

[367]See Edward Grant, ed., *A Source Book on Medieval Science* (Cambridge, MA: Harvard University Press, 1974), pp. 150–151.

[368]For a sustained argument that typography is a cultural watershed, see Elizabeth Eisenstein, *The Printing Press as an Agent of Change*, 2 vols. (Cambridge: Cambridge University Press, 1979).

Chapter 13: The First Phase of the Scientific Revolution, ca. 1450–1600, pages 396–446

[369]See Robert S. Westmann, "On Communication and Cultural Change," *Isis* 71 (1980): 474–477 and Anthony T. Grafton, "The Importance of Being Printed," *Journal of Interdisciplinary History* (1980): 265–289.

[370]Traditionally, the European Renaissance is dated from 1350 to about 1550. The High Renaissance is the third and final phase of the Renaissance after 1500.

[371]Among others, Jakob Burckhardt in *Civilization and the Renaissance in Italy* (1860) portrays this era, albeit with some exaggerations.

[372]As quoted in Morris Kline, *Mathematical Thought from Ancient to Modern Times* (New York: Oxford University Press, 1990), p. 224.

[373]See Jeanne Peiffer, "Le style mathématique de Dürer et sa conception de la géométrie," in Joseph Dauben et al., eds., *History of Mathematics: States of the Art* (San Diego, CA: Academic Press, 1996), pp. 49–64.

[374]See Barbara Maria Stafford, *Artful Science: Enlightenment Entertainment and the Eclipse of Visual Education* (Cambridge, MA: MIT Press, 1994).

[375]See Warren van Egmond, *The Commercial Revolution and the Beginnings of Western Mathematics in Renaissance Florence, 1300–1500* (Doctoral Thesis, Indiana University, 1976).

[376]See Allen G. Debus, *Man and Nature in the Renaissance* (Cambridge: Cambridge University Press, 1987).

[377] Herbert Butterfield, *The Origins of Modern Science* (New York: Free Press, 1997), pp. 187–202; David C. Lindberg, "Conceptions of the Scientific Revolution from Bacon to Butterfield: A Preliminary Sketch," in David C. Lindberg and Robert S. Westman, eds., *Reappraisals of the Scientific Revolution* (Cambridge: Cambridge University Press, 1990), pp. 1–27; and Roy S. Porter, "The Scientific Revolution: A Spoke in the Wheel," in Roy Porter and Mikulas Teich, eds., *Revolution in History* (Cambridge: Cambridge University Press, 1986), pp. 290–316.

[378] See H. Floris Cohen, *The Scientific Revolution: A Historiographical Inquiry* (Chicago: University of Chicago Press, 1994) p. xxii.

[379] As translated from *De Revolutionibus* in Thomas Kuhn, *The Copernican Revolution: Planetary Astronomy in the Development of Western Thought* (New York: Vintage Books, 1957), p. 154.

[380] Ibid.

[381] Brian Eslea, *Witch Hunting, Magic and the New Philosophy: An Introduction to the Debates of the Scientific Revolution, 1450–1750* (Atlantic Highlands, NJ: Humanities Press, 1980), pp. 70–78.

[382] Ibid., p. 70.

[383] See V. S. Varadarajan, *Algebra in Ancient and Modern Times* (Providence, RI: American Mathematical Society, 1998).

[384] See Raffaella Franci and Laura Toti Rigatelli, "Maestro Benedetto da Firenze e la Storia dell'Algebra," *Historia Mathematica* 10 (1983): 297–317.

[385] See Stillman Drake and I. E. Drabkin, trans. and intros., *Mechanics in Sixteenth-Century Italy* (Madison: University of Wisconsin Press, 1961), pp. 61–143 and P. L. Rose, *The Italian Renaissance of Mathematics* (Geneva: Libraire Droz, 1975).

[386] *Ibid.*

[387] This autobiography, completed in 1575, was first printed in Paris in 1643. See Jerome Cardan, *The Book of My Life*, trans. by Jean Stoner (New York: Dover Publications, 1929).

[388] Martin Gardner, *Knotted Doughnuts and Other Mathematical Entertainments* (New York: W. H. Freeman and Co., 1986), p. 86.

[389] Michael Sean Mahoney, *The Mathematical Career of Pierre de Fermat 1601–1665*, 2nd ed. (Princeton, NJ: Princeton University Press, 1994), p. 125.

[390] For an intellectual biography of Scaliger, see Anthony Grafton, *Joseph Scaliger: A Study in the History of Classical Scholarship*, Volume I (Oxford: Oxford University Press, 1983).

[391] See G.V. Coyne, S.J., M. A. Hoskin, and O. Pederson, eds., *Gregorian Reform of the Calendar* (Vatican City: Specola Vaticana, 1983).

[392] Jennifer Moreton, "Before Grosseteste: Roger of Hereford and Calendar Reform in Eleventh- and Twelfth-Century England," *Isis* 86 (1995): 562–586.

[393] Ibid., pp. 78–80.

[394] Otto Neugebauer, *A History of Ancient Mathematical Astronomy* [fn. 156], p. 1062.

[395] See John Dee, *The Mathematicall Praeface to the Elements of Euclid of Megara*, intro. by Allen G. Debus (New York: Science History Publications, 1975). Note the normal misidentification of the geometer with the philosopher Euclid of Megara.

[396] Ibid.

[397]Ibid., next to last page.

[398]See Nicholas H. Clulee, *John Dee's Natural Philosophy: Between Science and Religion* (New York: Routledge, 1988), pp. 158 ff.

[399]John Dee, *Mathematicall Praeface* [fn. 394], next to last page.

[400]Peter Dear, *Discipline & Experience: The Mathematical Way in the Scientific Revolution* (Chicago: University of Chicago Press, 1995), pp. 34 ff.

[401]Paolo Mancosu, *Philosophy of Mathematics and Mathematical Practice in the Seventeenth Century* (New York: Oxford University Press, 1996), pp. 178–212.

Chapter 14: Transformation ca. 1600–1660: I: Physico-mathematics, Method, Computational Arithmetic, and Algebra, pages 447–495

[402]Richard S. Westfall, *The Construction of Modern Science: Mechanisms and Mechanics* (New York: Wiley , 1971), p. 1.

[403]Paolo Mancosu, "Aristotelian Logic and Euclidean Mathematics: Seventeenth-Century Developments of the *Quaestio de certitudine*," *Studies in the History and Philosophy of Science* 23 (1992): 241–265.

[404]Derek Thomas Whiteside, "Patterns of Mathematical Thought in the Late Seventeenth Century," *Archive for History of Exact Sciences* 1, 3 (1961): 333.

[405]Wilbur Knorr, "The Method of Indivisibles in Ancient Geometry," in Ronald Calinger, ed., *Vita Mathematica* [fn. 252], pp. 67–87.

[406]See Peter Limm, *The Thirty Years' War* (London: Longman, 1984).

[407]Jean Bérenger, *A History of the Habsburg Empire 1273–1700,* trans. by C. A. Simpson (London: Longman, 1994), pp. 295–296.

[408]William Ashworth, "The Habsburg Circle," in Bruce T. Moran, ed., *Patronage and Institutions* (Rochester, NY: The Boydell Press, 1991), pp. 137–168.

[409]See Charles Avery, *Bernini: Genius of the Baroque* (: Bulfinch Press, 1998).

[410]Lorraine Daston, "The Moral Economy of Science," in Arnold Thackray, ed., *Constructing Knowledge in the History of Science, Osiris,* 10, 1995: 2–24.

[411]Johannes Kepler, *Mysterium Cosmographicum.* trans. by A. M. Duncan (New York: Arabis Books, 1981), p. 23.

[412]See J. V. Field, *Kepler's Geometrical Cosmology* (Philadelphia: American Philosophical Society, 1988).

[413]Allen G. Debus, *The English Paracelsians* (New York: Franklin Watts, 1966), pp. 123–125.

[414]See Arthur Koestler, *The Watershed* (Garden City, New York: Doubleday & Company, 1960).

[415]See Johannes Kepler, *The Harmony of the World,* trans. with Intro. by E. J. Aiton, A. M. Duncan, and J. V. Field. *Memoirs of the American Philosophical Society,* Vol. 209 (Philadelphia: American Philosophical Society, 1997).

[416]Alexandre Koyré, *Études galiléennes,* 3 vols. (Paris: Hermann, 1940), pp. 72–73.

[417]For the influence of the Collegio Romano, see William A. Wallace, *Galileo and His Sources* (Princeton, NJ: Princeton University Press, 1984).

[418]See Stillman Drake, *Galileo Studies: Personality, Tradition, and Revolution* (Ann Arbor: University of Michigan Press, 1970), pp. 19–41.

[419] See Rivka Feldhay, *Galileo and the Church: Political Inquisition or Critical Dialogue?* (Cambridge: Cambridge University Press, 1995); Pietro Redondi, *Galileo Heretic*, trans. by Raymond Rosenthal (Princeton, NJ: Princeton University Press, 1987); and Michael Segre, "Light on the Galileo Case?" *Isis* 88, 3 (1997): 484–504.

[420] See Roger French, *William Harvey's Natural Philosophy* (Cambridge: Cambridge University Press, 1994).

[421] See Kuno Fischer, *Geschichte der neueren Philosophie* 6 vols. (Berlin: 1852–1877).

[422] Michel Serres, ed., *A History of Scientific Thought* [fn.], p. 337. Descartes' appeal to reason alone and his rejection of historical authority Michel Authier dismisses as barbarism.

[423] Renè Descartes, *Discourse on Method and the Meditations* (Hamondsworth, England: Penguin Books, 1968), p. 41.

[424] As quoted in Stephen Gaukroger, *Descartes: An Intellectual Biography* (Oxford: Oxford University Press, 1995), p. 290.

[425] See Bernard Fontenelle, *Conversations on the Plurality of Worlds* (1686), trans. by H. A. Hargreave and intro. by Nina Rattner Gelbart (Berkeley: University of California Press, 1990).

[426] Mario Biagioli, "Galileo's System of Patronage," *History of Science* 28 (1990): 1–62; Mordechai Feingold, *The Mathematicians' Apprenticeship: Science, Universities, and Society in England, 1560–1640* (Cambridge: Cambridge University Press, 1984); Paula Findlen, *Possessing Nature: Museums, Collecting, and Scientific Culture in Early Modern Italy* (Berkeley: University of California Press, 1994); Margaret C. Jacob, *The Newtonians and the English Revolution* (Hassocks, England: Harvester Press, 1976); Alexander Keller, "Mathematics, Mechanics, and the Origins of the Culture of Mechanical Invention," *Minerva* 23 (1985): 348–361; Pamela Long, "Power, Patronage, and the Authorship of *Ars*," *Isis* 88 (1997): 1–41; David Lux, *Patronage and Royal Science in Seventeenth-Century France* (Ithaca, NY: Cornell University Press, 1989); Bruce Moran, ed., *Patronage and Institutions: Science, Technology, and Medicine at the European Court, 1500–1750* (Woodbridge, Suffolk: Boydell, 1991); and Charles Webster, *The Great Instauration* (London: Duckworth, 1973).

[427] William Eamon, "Court, Academy, and Printing House: Patronage and Scientific Careers in Late Renaissance Italy," in Bruce T. Moran, ed., *Patronage and Institutions:* [fn. 425], pp. 25–51.

[428] David S. Lux, "The Reorganization of Science 1450–1700," in Bruce T. Moran, *Patronage and Institutions* [fn. 425], p. 189.

[429] John Murdoch, "The Analytical Character of Medieval Learning: Natural Philosophy without Nature," in L. D. Roberts, ed., *Approaches to Nature in the Middle Ages.* Vol. 16 *Medieval and Renaissance Studies* (Binghampton, NY: Center for Medieval and Renaissance Studies, 1982), pp. 171–213.

[430] William Eamon, "Court, Academy, and Printing House," [fn. 426], pp. 26–27.

[431] Lorraine J. Daston, "Reviews on Artifact and Experiment: The Factual Sensibility," *Isis* 79 (1988): 452–467.

[432] Pamela H. Smith, "Alchemy as a Language of Mediation at the Habsburg Court," *Isis* 85 (1004): 1–25.

[433] Jean Bérenger, *A History of the Habsburg Empire, 1273–1700* (Harlow, Sussex: Longman, 1990), pp. 242–243 and 256–259.

[434]See R. J. W. Evans, *Rudolf II and His World* (Oxford: Clarendon Press, 1973).

[435]David S. Lux, "The Reorganization of Science, 1450–1700," [fn. 427], p. 189.

[436]Stephen Gaukroger, *Descartes: An Intellectual Biography* [fn. 423], p. 322.

[437]See also Eric W. Cochrane, *Tradition and Enlightenment in the Tuscan Academies, 1690–1800* (Chicago: University of Chicago Press, 1961).

[438]Mario Biagioli, "Etiquette, Interdependence, and Sociability in Seventeenth-Century Science," *Critical Inquiry* 22, 3 (Winter 1996): 212–216.

[439]See Margery Purver, *The Royal Society: Concept and Creation* (London: Routledge and Kegan Paul, 1967), pp. 110–112.

[440]Bruce Moran, "Patronage and Institutions" [fn. 426], p. 177.

[441]Florian Cajori, *A History of Mathematical Notation,* 2 Vols. (La Salle, IL: Open Court Publishing Co., 1928), I: 149.

[442]As quoted in Carl Boyer, "Viète's Use of Decimal Fractions," *Mathematical Teacher* 55 (1962): 123–127.

[443]See pp. 480–483.

[444]See Dirk J. Struik, *The Land of Stevin and Huygens,* in *Studies in the History of Modern Science* (Dordrecht: Reidel, 1981), Vol. 7.

[445]Helmuth Gericke und Kurt Vogel, eds., *De Thiende von Simon Stevin* (Frankfurt: Akademische Verlagsgesellschaft, 1965), pp. 45–47.

[446]Martin Gardner, *Knotted Doughnuts and Other Mathematical Entertainments* [fn. 387], pp. 85–86.

[447]Ibid., p. 85.

[448]J. A. Lohne, "Essays on Thomas Harriot," *Archive for History of Exact Sciences* 20, 3/4 (1979): 189–196.

[449]Ibid., pp. 231–233.

[450]Zarko Dadić, "The Early Geometrical Works of Marin Getaldić," in Ronald Calinger, ed., *Vita Mathematica* [fn. 252], p. 116.

Chapter 15: Transformation ca. 1600–1660: II: To the Edge of Modernity, pages 496–554

[451]Michael Sean Mahoney, *The Mathematical Career of Pierre de Fermat 1601–1665* [fn. 388], pp. 75–76.

[452]While at La Fleche in June of 1610, Descartes was one of twenty-four noble students selected to participate in the elaborate funeral procession for the heart of the assassinated French monarch Henry IV.

[453]See Paolo Mancosu, *Philosophy of Mathematics and Mathematical Practice in the Seventeenth Century* [fn. 400].

[454]A. C. Crombie, *Styles of Scientific Thinking in the European Tradition* (London: Duckworth, 1994), II: 898–899.

[455]Stephen Gaukroger, *Descartes: An Intellectual Biography* [fn. 423], pp. 93 and 131.

[456]Ibid., p. 99.

[457]William R. Shea, *The Magic of Numbers and Motion: The Scientific Career of René Descartes* (Canton: MA: Science History Publications, 1991), p. 117.

[458]René Descartes, *La géométrie*, Part III of *Discourse de la methode*, English translation by D. Smith and M. Latham (New York: Dover Reprint, 1954), p. 318.

[459] Stephen Gaukroger, *Descartes: An Intellectual Biography* [fn. 423], p. 323.

[460] Paolo Mancosu, "Descartes's *Géométrie*," in Donald Gillies, ed., *Revolutions in Mathematics* [fn. 90], pp. 100–102.

[461] Ibid.

[462] Michael Sean Mahoney, *The Mathematical Career of Pierre de Fermat 1601–1665* [fn. 388], p. 15.

[463] For more information on Toulouse, see Robert A. Schneider, *Public Life in Toulouse, 1463–1789: From Municipal Republic to Cosmopolitan City* (Ithaca, NY: Cornell University Press, 1989).

[464] Michael Sean Mahoney, *The Mathematical Career of Pierre de Fermat* [fn. 388], p. 24.

[465] See Charles Henry and Paul Tannery, eds., *Oeuvres de Fermat*, 4 vols. (Paris: 1891–1912), I: 91–103. An English translation of his *Isagoge* and *Methodus* appears in D. J. Struik, ed., *A Source Book in Mathematics, 1200–1800* (Cambridge, MA: Harvard University Press, 1969).

[466] Michael Sean Mahoney, *Pierre de Fermat* [fn. 388], p. 418.

[467] Carl B. Boyer, *The History of the Calculus and Its Conceptual Development* (New York: Dover Publications, 1959), p. 164.

[468] Ibid., pp. 170–173.

[469] André Weil, *Number Theory: An Approach through History from Hammurapi to Legendre* (Basel: Birkhäuser, 1984), pp. 104–105.

[470] See J. Field and J. Gray, eds., *The Geometrical Work of Girard Desargues* (New York: Springer-Verlag, 1987).

[471] John Stillwell, *Mathematics and Its History* (Berlin: Springer-Verlag, 1989), p. 86.

[472] Joseph E. Hofmann, *Leibniz in Paris 1673–1676: His Growth to Mathematical Maturity* (Cambridge: Cambridge University Press, 1974), pp. 179–180.

[473] Carl B. Boyer, *The History of the Calculus and Its Conceptual Development* [fn. 466], pp. 102–103.

[474] Ibid., pp. 108–111.

[475] As quoted in Paolo Mancosu, *Philosophy of Mathematics and Mathematical Practice* [fn. 452], p. 48.

[476] Ibid., p. 43.

[477] Ibid., p. 64.

[478] Ibid., p. 119.

[479] Ibid., pp. 119–121.

[480] These include an approximation for an angle trisection.

[481] René Taton, ed., *The Beginnings of Modern Science from 1450 to 1800*, trans. by A. J. Pomerans (New York: Basic Books, 1964), pp. 215–252.

[482] Morris Bishop, *Pascal: The Life of Genius* (New York: Reynal and Hitchcock, 1936), p. 311.

[483] Moritz Cantor, *Geschichte der Mathematik* (Leipzig: Teubner, 1900), II: 753.

[484] Carl Boyer, *The History of the Calculus and Its Conceptual Development* [fn. 466], p. 153.

[485]René Taton, "Blaise Pascal," in Charles Gillispie, ed., *Dictionary of Scientific Biography* [fn. 273], X: 337.

[486]N. Bourbaki, *Éléments d'histoire des mathématiques* (Paris: Hermann, 1969), pp. 238–239.

[487]Morris Bishop, *Pascal: The Life of Genius* [fn. 481], pp. 307–309.

[488]André Weil, *Number Theory* [fn. 468], pp. 32–33.

[489]Stanley J. Bezuska and Margaret J. Kenney, "Even Perfect Numbers (Update)^2" *Mathematics Teacher* 90 (1997): 628–633.

[490]Michael Sean Mahoney, *The Mathematical Career of Pierre de Fermat 1601–1665* [fn. 388], p. 293.

[491]André Weil, *Number Theory* [fn. 468], p. 55.

[492]Ibid., p. 56.

[493]Ibid., p. 294.

[494]Ibid., p. 55.

[495]Michael Sean Mahoney, *The Mathematical Career of Pierre de Fermat 1601–1665* [fn. 388], pp. 338–339.

[496]André Weil, *Number Theory* [fn. 468], p. 75.

[497]Ibid.

[498]Michael Sean Mahoney, *The Mathematical Career of Pierre de Fermat 1601–1665* [fn. 388], p. 348.

[499]Ibid., p. 320.

[500]André Weil, *Number Theory* [fn. 468], pp. 67–69.

[501]Michael Sean Mahoney, *The Mathematical Career of Pierre de Fermat 1601–1665* [fn. 388], p. 320.

[502]Ibid., p. 334.

[503]Ibid., pp. 335–336.

[504]André Weil, *Number Theory* [fn. 468], pp. 50–51.

[505]Ibid., p. 81.

[506]Ibid., p. 342.

[507]Michael Sean Mahoney, *The Mathematical Career of Pierre de Fermat 1601–1665* [fn. 388], p. 345.

[508]Eric Temple Bell, "Wallis on Fermat," *Scripta Mathematica* 15, pp. 162–163.

[509]André Weil, *Number Theory* [fn. 468], p. 97.

[510]Leonard Eugene Dickson, *History of the Theory of Numbers* (New York: G. E. Stechert & Co., 1934), Vol. II, p. 354; Michael Sean Mahoney, *The Mathematical Career of Pierre de Fermat 1601–1665* [fn. 388], p. 328; and André Weil, *Number Theory* [fn. 468], p. 174.

[511]André Weil, *Number Theory* [fn. 468], p. 104.

[512]Ibid., p. 28.

[513]Ronald Calinger, "The Mathematics Seminar at the University of Berlin," in Ronald Calinger, ed., *Vita Mathematica: Historical Research and Integration with Teaching* [fn. 252], pp. 156–157, and Harold M. Edwards, *Fermat's Last Theorem: A Genetic Introduction to Algebraic Number Theory* (Berlin: Springer-Verlag, fourth

printing, 1993), and "The Background of Kummer's Proof of Fermat's Last Theorem for Regular Primes," *Archive for History of Exact Sciences* 14 (1974): pp. 219–236.

[514] Ian Hacking, *The Emergence of Probability* (Cambridge: Cambridge University Press, 1975), pp. 1–15.

[515] Daniel Garber and Sandy Zabell, "On the Emergence of Probability," *Archive for History of Exact Sciences* 21 (1979): 33–53.

[516] Ibid., pp. 42–43.

[517] Morris Bishop, *Pascal: The Life of Genius* [fn. 481], p. 116.

[518] Lorraine Daston, *Classical Probability in the Enlightenment* (Princeton, NJ: Princeton University Press, 1988), pp. x–xx.

[519] Lorraine J. Daston, "The Domestication of Risk: Mathematical Probability and Insurance 1650–1830," in Lorenz Krüger, Lorraine J. Daston, and Michael Heideberger, eds., *The Probabilistic Revolution* (Cambridge, MA: MIT Press, 1987), pp. 237–261.

[520] Lorraine Daston, *Classical Probability in the Enlightenment* [fn. 517], 1–20.

[521] For an English translation of the correspondence, see D. E. Smith, ed., *A Source Book in Mathematics* (New York: Dover Reprint, 1959) and F. N. David, *Games, Gods, and Gambling* (London: Griffin, 1962). In his letter of August 9, Fermat proposes that Carcavi and Pascal edit, amend, and publish his writings but keep his name secret.

[522] Morris Bishop, *Pascal: The Life of Genius* [fn. 481], pp. 114–116.

[523] André Weil, *Number Theory* [fn. 468], p. 44.

[524] Ibid., p. 47.

[525] Anders Hald, *A History of Probability and Statistics and the Applications before 1750* (New York: John Wiley & Sons, 1990), p. 54.

[526] Ronald Calinger, ed., *Classics of Mathematics* (Englewood Cliffs, NJ: Prentice Hall, 1995), pp. 349–353 and A. W. F. Edwards, *Pascal's Arithmetical Triangle* (London: Griffin, 1987).

[527] Anders Hald, *A History of Probability and Statistics and Their Applications before 1750* [fn. 524], p. xx.

[528] Ian Hacking, *The Emergence of Probability* [fn. 513], pp. 64 ff.

[529] In the *Essay,* see ii, 21, 72.

[530] For a translation, see F. N. David, *Games, Gods, and Gambling* [fn. 520].

Chapter 16: The Apex of the Scientific Revolution I : Setting and Laureates, pages 555–597

[531] See Michael Blay, *Reasoning with the Infinite: From the Closed World to the Infinite Universe* (Chicago: University of Chicago Press, 1998).

[532] Bernard le Bovier de Fontenelle, *Histoire de l'Académie royale des sciences en mdcxix et les éloges historiques,* 2 vols. (Amsterdam: Pierre du Coup, 1719–1720), I: 14.

[533] William Doyle, *The Old European Order 1600–1800* (Oxford: Oxford University Press, 1992), p. 242.

[534] See Bruce Boucher, *Italian Baroque Sculpture* (London: Thames and Hudson, 1998) and James Fenton, "How Great Art is Made," *The New York Review,* 45, 7 (April 23, 1998): 22–26.

[535] Pierre Bayle, *Historical and Critical Dictionary: Selections*. Trans. by Richard Popkin (Indianapolis: Bobbs-Merrill Co., Inc., 1965).

[536] Donald Rutherford, *Leibniz and the Rational Order of Nature* (Cambridge: Cambridge University Press, 1995), p. 212, and Catherine Wilson, *Leibniz's Metaphysics: A Historical and Comparative Study* (Princeton, NJ: Princeton University Press, 1989), pp. 181–187.

[537] See John Locke, *An Essay Concerning Human Understanding* (London: Everyman's Library, 1961).

[538] As quoted in Margery Purver, *The Royal Society: Concept and Creation* (Cambridge, MA: M.I.T. Press, 1967), p. 73.

[539] See Peter Remnant and Jonathan Bennett, *G. W. Leibniz: New Essays on Human Understanding* (Cambridge: Cambridge University Press, 1996).

[540] As quoted in Peter Dear, *Discipline and Experience: The Mathematical Way in the Scientific Revolution* (Chicago: University of Chicago Press, 1995), p. 213.

[541] See Isaac Newton, *Principia*, Motte-Cajori transl., (Berkeley: University of California Press, 1966) Book III, pp. 397–399 and I. Bernard Cohen, *The Newtonian Revolution* (Cambridge: Cambridge University Press, 1980), pp. 36–38.

[542] Peter Dear, *Discipline and Experience* [fn. 539], pp. 236–237.

[543] Sir Isaac Newton, *Opticks* (New York: Dover Publications, 1952, based on the fourth edition of 1730), p. 369.

[544] Richard S. Westfall, *The Life of Isaac Newton* (Cambridge: Cambridge University Press, 1993), p. 192.

[545] I. Bernard Cohen, *The Newtonian Revolution* (Cambridge: Cambridge University Press, 1980), pp. 38–39 and 68 ff.

[546] Christa Binder, "Austria and Hungary," in I. Grattan-Guinness, ed., *Companion Encyclopedia of the History and Philosophy of the Mathematical Sciences* (London: Routledge, 1994), vol. 2, p. 1457.

[547] Ibid., p. 1458.

[548] Margery Purver, *The Royal Society: Concept and Creation* (Cambridge, MA: M. I. T. Press, 1967), pp. 128 ff.

[549] Antonia Fraser, *Royal Charles: Charles II and the Restoration* (New York: Alfred A. Knopf, 1979), pp. 194–195.

[550] Alice Stroup, *A Company of Scientists* (Berkeley: University of California Press, 1990), pp. 6 ff.

[551] Roger Hahn, *The Anatomy of a Scientific Institution: The Paris Academy of Sciences, 1666–1803* (Berkeley: University of Californian Press, 1971), pp. 8–14.

[552] Douglas M. Joseph, "Hobbes and Mathematical Method," *Perspectives on Science* 1 (1993): 193–219 and Stephen Shapin, *The Scientific Revolution* [fn. 101], pp. 110–111.

[553] Alice Stroup, *A Company of Scientists* [fn. 548], pp. 47 and 52.

[554] Joella G. Yoder, *Unrolling Time: Christiaan Huygens and the Mathematization of Nature* (Cambridge: Cambridge University Press, 1988), p. 10.

[555] Ibid., p. 9.

[556] Semyon Grigorevich Gindikin, *Tales of Physicists and Mathematicians*, trans. by Alan Shuchat (New York: Springer-Verlag, 1996), p. 74.

[557] See Albert Van Helden, "The Accademia del Cimento and Saturn's Rings," *Physis* 15, 3 (1973): 237–259.

[558] Semyon G. Gindiken, *Tales of Physicists and Mathematicians* [fn. 552], p. 78.

[559] Joella Yoder, *Unrolling Time* [fn. 552a], pp. 155–156.

[560] Semyon G. Gindikin, *Tales of Physicists and Mathematicians* [fn. 552], pp. 90–91.

[561] Fermat had been among the first to achieve rectification, mechanically arriving in 1657 at the perimeter of a triangle equal to a length of the parabola $y/b = 1 - (x/a)^2$. See Joseph E. Hofmann, *Leibniz in Paris* (Cambridge: Cambridge University Press, 1974), pp. 104–105.

[562] See Leopold Defossez, *Les savants du XVIIe siècle et la mesure du temps* (Lausanne: Edition du journal suisse d'horologerie et de bijouterie, 1946) and Bert S. Hall, "The New Leonardo" *Isis* 67 (1976): 463–475 and Ibid., "The Scholastic Pendulum" *Annals of Science* 35 (1978): 441–463

[563] J. E. Hoffmann, *Leibniz in Paris 1672–1676* [fn. 556], pp. 118–120.

[564] The *Philosophical Transactions* 11, No. 129 of 20 (30) November 1676: 749–750.

[565] H. J. M. Bos, "The Influence of Huygens on the Formation of Leibniz's Ideas," in Albert Heinekamp and Dieter Mettler, eds., *Studia Leibnitiana Supplementa: Leibniz a Paris (1672–1676)* (Wiesbaden: Franz Steiner Verlag GMBH, 1978), pp. 59–68.

[566] Alan Hart, "Soul and Monad: Plato and Leibniz," in Ingrid Marchlewitz and Albert Heinekamp, eds., *Leibniz' Auseinandersetzung mit Vorgängern und Zeitgenossen* (Stuttgart: Franz Steiner Verlag, 1990), pp. 42–43.

[567] See Gottfried Wilhelm Leibniz, *Monadology and Other Philosophical Essays.* trans. by Paul Schrecker (Indianapolis: Bobbs-Merrill Co., 1965), pp. 148–163.

[568] See E. A. Fellmann, ed., *G. W. Leibniz: Marginalia in Newtoni Principia Mathematica (1687)* (Paris: J. Vrin, 1973).

[569] Two centuries later, in investigating energetics and seeking absolute constants, Max Planck identified the smallest unit of energy, his quantum of action h.

[570] As quoted in E. J. Aiton, *Leibniz: A Biography* (Bristol: Adam Hilger Ltd., 1985), p. 273.

[571] Sir Isaac Newton, *Principia mathematica* (1687), Andrew Motte and Florian Cajori's translation (Berkeley: University of California Press, 1966) Vol. 1, p. xxxi.

[572] See H. G. Alexander, ed., *The Leibniz-Clarke Correspondence* (Manchester: Manchester University Press, 1956).

[573] As quoted in Richard S. Westfall, *The Life of Isaac Newton* [fn. 542], p. 13.

[574] Ibid., p. 39.

[575] Ibid., p. 162.

[576] Ibid., pp. 438–439.

[577] I. Bernard Cohen, *The Newtonian Revolution* [fn. 543], pp. 105–106.

[578] Richard S. Westfall, *The Life of Isaac Newton* [fn. 542], p. 160.

[579] Derek T. Whiteside, ed., *The Mathematical Papers of Isaac Newton* (Cambridge: Cambridge University Presss, 1967), vol. 1, p. vii.

[580] As quoted in A. R. Hall, *Isaac Newton: Adventurer in Thought* (Oxford: Blackwell, 1992), p. 91.

[581] See Harold L. Klawans, *Newton's Madness (Further Tales of Clinical Neurology* (New York: Harper Perennial Publishing, 1991).

[582] As quoted in Richard S. Westfall, *The Life of Isaac Newton* [fn. 542], p. 213.

[583] Frances Willmoth, ed., *Flamsteed's Stars* (Woodbridge, Suffolk: The Boydell Press, 1998), pp. 31–32.

[584] The twentieth-century rejoinder goes:
'Twas not to be long
For the devil howling "ho!"
Said, let Einstein be
And restored the status quo.

[585] As quoted in Richard S. Westfall, *The Life of Isaac Newton* [fn. 542], p. 309.

Chapter 17: The Apex of the Scientific Revolution II: Calculus to Probability, pages 598–654

[586] C. H. Edwards, *The Historical Development of the Calculus* (New York: Springer-Verlag, 1979), pp. 171–176.

[587] Carl B. Boyer, *The History of the Calculus and its Conceptual Development* [fn. 466], p. 176.

[588] Peter Dear, *Discipline & Expereience* [fn. 539], pp. 222–227.

[589] Robert H. Kargon, "Newton, Barrow, and Hypothetical Physics," *Centaurus* 11 (1965): 46–56.

[590] Joseph E. Hofmann, *Leibniz in Paris 1672–1676* [fn. 556], pp. 43–44.

[591] These volumes were published for the Royal Society by Cambridge University Press in 1959, 1960, 1961, 1967, 1975, 1976, and 1977.

[592] Derek T. Whiteside, "Sources and Strengths of Newton's Early Mathematical Thought," in Robert Palter, ed., *The "Annus Mirabilis" of Sir Isaac Newton, 1666–1966* (Cambridge, MA: M.I.T. Press, 1970), pp. 72–74.

[593] Derek T. Whiteside, ed., *The Mathematical Papers of Isaac Newton*, [fn. 572], Vol. I, p. 7.

[594] Richard S. Westfall, *The Life of Isaac Newton* [fn. 567], p. 31.

[595] D. T. Whiteside, "Newton's Discovery of the General Binomial Theorem," *The Mathematical Gazette* 45 (1961): 175–180.

[596] William Dunham, *Journey Through Genius* (New York: Penguin Books, 1990), pp. 170–171.

[597] Ibid.

[598] See Derek T. Whiteside, ed., *The Mathematical Papers of Isaac Newton* [fn. 572], Vol. I, pp. 400–447.

[599] Ibid., p. 39.

[600] As quoted in Richard S. Westfall, *The Life of Isaac Newton* [fn. 567], p. 81.

[601] Ibid., p. 207.

[602] Eberhard Knobloch, "Zur Vorgeschichte der Determinantentheorie" *Theoria cum Praxi, Studia Leibnitiana Supplementa*, Vol. XXII (Wiesbaden: Franz Steiner Verlag GMBH, 1982), pp. 96–118.

[603] J. M. Child, trans., *The Early Mathematical Manuscripts of Leibniz* (Chicago: Open Court Publishing Company, 1920), p. 11.

[604] Ibid., pp. 11–12.

[605] As quoted in Stuart Hollingdale, *Makers of Mathematics* (London: Penguin Books, 1989), p. 259.

[606] Rupert Hall, "Leibniz and the British Mathematicians 1673–1676," *Leibniz a Paris (1672–1676), Studia Leibnitiana*, Supplement XVII (Wiesbaden: Franz Steiner Verlag GMBH, 1978), p. 133.

[607] E. J. Aiton, *Leibniz: A Biography* [fn. 564], p. 48.

[608] Ibid., p. 49.

[609] Joseph E. Hoffmann, *Leibnis in Paris* [fn. 556], p. 45.

[610] Ibid., p. 49.

[611] Ibid., pp. 74–76.

[612] J. M. Child, trans., *The Early Mathematical Manuscripts of Leibniz* [fn. 591], p. 7.

[613] Ibid., pp. 79–80.

[614] J. E. Hoffmann, *Leibniz in Paris* [fn. 556], pp. 81–82.

[615] J. M. Child, *The Early Mathematical Manuscripts of Leibniz* [fn. 591], pp. 74–76.

[616] Joseph Frederick Scott, *A History of Mathematics from Antiquity to the Beginning of the Nineteenth Century* (London: Taylor and Francis Ltd., 1958), p. 155.

[617] Joseph E. Hofmann, *Leibniz in Paris* [fn. 556], p. 75.

[618] D. T. Whiteside, "Isaac Barrow," in Charles C. Gillispie, ed., *Dictionary of Scientific Biography* [fn. 273] Volume I: 474–475.

[619] René Taton, "L'initiation de Leibniz à la géométrie (1672–1676)" in *Leibniz a Paris (1672–1676), Studia Leibnitiana* (Wiesbaden: Franz Steiner Verlag GMBH, 1978) Supplement XVII, p. 117.

[620] Joseph E. Hoffmann, *Leibniz in Paris* [fn. 556], p. 195.

[621] Ibid., p. 211.

[622] Richard Westfall, *The Life of Isaac Newton* [fn. 567], p. 99.

[623] Eli Maor, *e: The Story of a Number* (Princeton, NJ: Princeton University Press, 1994), p. 90.

[624] Gottfried Leibniz, "Nova methodus" in Ronald Calinger, ed., *Classics of Mathematics* [fn. 525], p. 391.

[625] Gottfried Wilhelm Leibniz, *Mathematische Schriften*, ed. by C. I. Gerhardt (Hildesheim: Olms repr., 1971), vol. 3, Part 1, p. 5.

[626] A. Rupert Hall, *Philosophers at War: the Quarrel between Newton and Leibniz* (Cambridge: Cambridge University Press, 1980), p. 81.

[627] As cited in Carl Boyer, *The History of the Calculus* [fn. 466], p. 215.

[628] A. Rupert Hall, *Philosophers at War* [fn. 614], p. 81.

[629] See Eli Maor, *e: The Story of a Number* [fn. 611].

[630] E. Hairer and G. Wanner, *Analysis by Its History* ( ), p. 135.

[631] The relation that the natural exponential function is the inverse of logarithms is in modern notation $1/y = dx/dy = \lim_{dy \to 0} \log(y + \triangle y)/\triangle y$. Letting $z = 1/y$ gives $z = \lim_{n \to \infty} \log_e(1 + z/n)^n$. The expansion of $(1 + z/n)^n$ using binomial coefficients

gives $(1 + z/n)^n = 1 + z/1! + (1 - 1/n)z^2/2! + (1 - 1/n)(1 - 2/n)z^3/3! + \ldots$. For n arbitrarily large, $e^z = 1 + z/1! + z^2/2! + z^3/3! + \ldots$.

[632] A. Rupert Hall, *Philosophers at War* [fn. 614], p. 83.

[633] Ernst Hairer and Gerhard Wanner, *Analysis by Its History* (New York: Springer-Verlag, 1996), p. 115.

[634] Richard S. Westfall, *The Life of Isaac Newton* [fn. 567], p. 233.

[635] Ernst Hairer and Gerhard Wanner, *Analysis by Its History* [fn. 617a], pp. 136–137.

[636] Herman H. Goldstine, *A History of the Calculus of Variations from the 17th through the 19th Century* (New York: Springer-Verlag, 1980), pp. 58 ff.

[637] E. J. Aiton, *Leibniz: A Biography* [fn. 564], p. 335.

[638] A. Rupert Hall, *Philosophers at War* [fn. 614], pp. 82–83.

[639] Domenico Bertolini Meli, *Equivalence and Priority: Newton versus Leibniz* (Oxford: Clarendon Press, 1993), p. 64. See also M. Horvath, "The Problem of Infinitesimally Small Quantities in the Leibnizian Mathematics," *Studia Leibnitiana, Supplementa* 22 (1982): 149–157 and, "On the Attempts Made by Leibniz to Justify His Calculus," *Studia Leibnitiana* 18 (1986): 60–71.

[640] As quoted in Paolo Mancosu, "The Metaphysics of the Calculus: A Foundational Debate in the Paris Academy of Sciences, 1700–1706," *Historia Mathematica* 16 (1989): 236.

[641] A Rupert Hall, *Philosophers at War* [fn. 614], p. 85.

[642] Ibid., pp. 224–248.

[643] Ibid., pp. 233–234.

[644] Ibid., quoted on p. 87.

[645] As quoted in Richard S. Westfall, *The Life of Isaac Newton* [fn. 567], p. 206 and A. Rupert Hall, *Philosophers at War* [fn. 614], p. 107.

[646] Ibid., p. 207.

[647] Domenico Bertolini Meli, *Equivalence and Priority: Newton versus Leibniz* (Oxford: Clarendon Press, 1993), pp. 7–9.

[648] Ibid., p. 9.

[649] Richard Westfall, *The Life of Isaac Newton* [fn. 567], p. 206.

[650] As quoted in A. Rupert Hall, *Philosophers at War* [fn. 614], p. 95.

[651] Ibid., p. 277.

[652] See Luigi Pepe, "Gabriele Manfredi (1681–1761) et la diffusion du calcul différentiel en Italie," in Albert Heinekamp, ed., *Beiträge zur Wirkungs- und Rezeptiongeschichte von Gotfried Wilehlm Leibniz (Studia Leibnitiana Supplementa*, vol. XXVI) (Stuttgart: Franz Steiner Verlag Wiesbaden GMBH, 1986), pp. 79–87.

[653] Richard S. Westfall, *The Life of Isaac Newton* [fn. 567], p. 285.

[654] E. J. Aiton, *Leibniz: A Biography* [fn. 564], p. 335.

[655] Richard S. Westfall, *The Life of Isaac Newton [fn. 567]*, p. 286.

[656] J. M. Child, trans., *The Early Mathematical Manuscripts of Leibniz* [fn. 591], pp. 22–59.

[657] A Rupert Hall, *Philosophers at War* [fn. 614], p. 234.

[658] See Niccolo Guicciardini, "Three Traditions in the Calculus: Newton, Leibniz, and Lagrange," in I. Grattan-Guinness, ed., *Companion Encyclopedia of the History and Philosophy of the Mathematical Sciences* (London: Routlegde, 1994), pp. 308–317.

[659] Eberhard Knobloch, "Zur Vorgeschichte der Determinantentheorie," in Kurt Müller et al, eds., *Studia Leibnitiana Supplementa: Theoria cum Praxi* (Wiesbaden: Franz Steiner Verlag GMBH, 1982), Vol. XXII, pp. 96–118, esp. p. 99.

[660] E. J. Aiton, *Leibniz: A Biography* [fn. 564], p. 126.

[661] Eberhard Knobloch, "Zur Vorgeschichte der Determinantentheorie" [fn. 644], p. 114.

[662] See Helena M. Pycior, *Symbols, Impossible Numbers, and Geometric Entanglements* (Cambridge: Cambridge University Press, 1997), pp. 40–166.

[663] Ibid., pp. 204–205.

[664] Richard S. Westfall, *Never at Rest: A Biography of Isaac Newton* (Cambridge: Cambridge University Press, 1980), p. 424.

[665] André Weil, *Number Theory* [fn. 468], p. 121.

[666] E. J. Aiton, *Leibniz: A Biography* [fn. 564], p. 52.

[667] Ernst Hairer and Gerhard Wanner, *Analysis by Its History* [fn. 617a], p. 52.

[668] See Leonhard Euler, *Introduction to Analysis of the Infinite* Trans. by John D. Blanton (New York: Springer-Verlag, 1988), Book 1, p. 101.

[669] Ronald Calinger, "Leonhard Euler: The First St. Petersburg Years (1727–1741)," *Historia Mathematica* 23 (1996): 133 and André Weil, *Number Theory* [fn. 468], p. 257.

[670] Jean Dieudonné, *Mathematics: The Music of Reason* (Berlin: Springer-Verlag, 1992), pp. 98–101.

[671] See Ronald Calinger, "Leonhard Euler: The First St. Petersburg Years" [fn. 649], pp. 134–136.

[672] Anders Hald, *A History of Probability and Statistics and Their Applications before 1750* [fn. 524], p. 138.

[673] See pages iii through vi.

[674] The first part consists of pages 2 through 71.

[675] Edith Dudley Sylla, "Jacob Bernoulli on Analysis, Synthesis, and the Law of Large Numbers," in M. Otte and M. Panza, eds., *Analysis and Synthesis in Mathematics* (Amsterdam: Kluwer Academic Publishers, 1997), pp. 79–101, esp. p. 87.

[676] See Ivo Schneider, "Der Mathematiker Abraham de Moivre," *Archive for History of Exact Sciences* 5 (1968–1969): 177–317.

Suggested Further Readings

These suggested further readings for each chapter are not intended to be exhaustive but to indicate selected primary sources and recent useful books and articles available in English.

Origins of Number and Culture

Ascher, Marcia. *Ethnomathematics: A Multicultural View of Mathematical Ideas.* Pacific Groves, CA: Brooks/Cole, 1991.

_____. "Models and Maps from the Marshall Islands: A Case in Ethnomathematics," *Historia Mathematica* 22 (1995): 347–370.

Binford, Lewis R. *In Pursuit of the Past.* New York: Thames Hudson, 1988.

Castledon, Rodney. *The Making of Stonehenge.* London: Routledge, 1993.

Couch, Carl J. "Oral Technologies: A Cornerstone of Ancient Civilizations?" *Sociological Quarterly* 30 (1989): 587–602.

Crossley, John N. *The Emergence of Number.* Singapore: World Scientific, 1987.

Crump, Thomas. *The Anthropology of Numbers.* Cambridge: Cambridge University Press, 1990.

Flegg, Graham, ed. *Numbers Through the Ages.* London: Macmillan, 1989.

Gerdes, Paul. "On Mathematics in the History of Sub-Saharan Africa," *Historia Mathematica* 21 (1994): 345–376. This article contains an extensive bibliography.

Gvozdanovic, Jadranka, ed. *Indo-European Numerals.* Berlin: de Gruyter, 1992.

Hurford, James R. *Language and Number.* London: Basil Blackwell, 1987.

Ifrah, Georges. *From One to Zero: A Universal History of Numbers.* Trans. by Lowell Bair. New York: Penguin, 1987.

Mathews, Jerold. "A Neolithic Oral Tradition for the van der Waerden/Seidenberg Origin of Mathematics," *Archive for History of Exact Sciences*, 34 (1985): 193–220.

Menninger, Karl. *Number Words and Number Symbols: A Cultural History of Numbers.* Cambridge, MA: M. I. T. Press, 1969.

Postgate, J. N. *Early Mesopotamia: Society and Economy at the Dawn of History.* New York: Routledge, 1992.

Schmandt-Besserat, Denise. "Reckoning before Writing," *Archaeology* 32 (1979): 23–31.

Seidenberg, A. "The Origin of Mathematics," *Archive for History of Exact Sciences* 18 (1978): 301–342.

Spencer, Donald D. *Key Dates in Number Theory History from 10,529 B.C. to the Present.* Ormond Beach, FL: Camelot, 1995.

Van der Waerden, Bartel L. *Geometry and Algebra in Ancient Civilizations.* Berlin: Springer-Verlag, 1983.

Zaslavsky, Claudia. *Africa Counts: Number and Pattern in African Culture.* Boston: Prindle, Weber and Schmidt, 1973.

The Dawning of Mathematics in the Ancient Near East

Ancient Mesopotamian and Egyptian Primary Sources

Chace, Arnold B., et al., eds. *The Rhind Mathematical Papyrus.* 2 vols. Reston, Va.: National Council of Teachers of Mathematics, 1967.

Clagett, Marshall, ed. *Egyptian Science: A Source Book.* 2 vols. Philadelphia: American Philosophical Society, 1989.

Gillings, Richard G. *Mathematics in the Time of the Pharaohs.* Cambridge, MA: M.I.T. Press, 1972.

Neugebauer, Otto and Abraham Sachs, eds. *Mathematical Cuneiform Texts.* New Haven, CN: American Oriental Society, 1945.

Parker, Richard A. *Demotic Mathematical Papyri.* Providence, RI: Brown University Press, 1972.

Ancient Mesopotamian Mathematics: Secondary Works

Aaboe, A. "On Period Relations in Babylonian Astronomy," *Centaurus* 10 (1964): 213–231.

Berggren J. L., and B. R. Goldstein, eds. *From Ancient Omens to Statistical Mechanics: Essays on the Exact Sciences Presented to Asger Aaboe.* Copenhagen: University Library, 1987.

Britton, John P. "An Early Function for Eclipse Magnitudes in Babylonian Astronomy," *Centaurus* 31 (1989): 1–52.

Damerow, Peter. *Abstraction and Representation: Essays on the Cultural Evolution of Thinking.* Trans. from the German by Renate Hanauer. Dordrecht: Kluwer Acacdemic, 1996.

Dicks, David. "Pan-Babylonianism Redivivus? Fundamentalism in Ivy League Garb," *DIO* 4, 1 (1994): 4–14.

Høyrup, Jens. "Algebra and Naive Geometry: An Investigation of Some Basic Aspects of Old Babylonian Mathematical Thought," *Altorientische Forschungen* 1990, 17: 27–69 and 262–354.

_____. "Algebra in the Scribal School: Schools in Old Babylonian Algebra?" *Internationale Zeitschrift für Geschichte und Ethik der Naturwissenschaft, Technik und Medizin* 1993, 1: 201–218.

_____. "Changing Trends in the Historiography of Mesopotamian Mathematics: An Insider's View," *History of Science* 34 (1996): 1–32.

_____. *In Measure, Number, and Weight: Studies in Mathematics and Culture.* Albany, NY:State University of New York Press, 1994.

Jones, Alexander. "On Babylonian Astronomy and its Greek Metamorphoses," in F. Jamil Ragep *et al.*, eds., *Tradition, Transmission, Transformation.* Leiden: Brill, 1996.

Leichty, Erle et al., eds. *A Scientific Humanist: Studies in Memory of Abraham Sachs.* Philadelphia: University Museum, 1988.

Mawet, Fr., and Ph. Talon eds. *D'Imhotep à Copernic.* Leuven: Peters, 1992.

Muroi, Kazuo, "Extraction of Cube Roots in Babylonian Mathematics," *Centaurus,* 31 (1989): 181–188.

Nemet-Nejat, Karen Rhea. "Systems for Learning Mathematics in Mesopotamian Scribal Schools," *Journal of Near Eastern Studies* 54 (1995): 241–260.

_____. *Cuneiform Mathematical Texts as a Reflection of Everyday Life in Mesopotamia.* (American Oriental Series) New Haven, Conn.: American Oriental Society, 1993.

Neugebauer, Otto. *The Exact Sciences in Antiquity.* New York: Harper Torchbook Edition, 1962.

_____. *A History of Ancient Mathematical Astronomy.* 3 vols., New York: Springer-Verlag, 1975.

Ritter, James. "Measure for Measure: Mathematics in Egypt and Mesopotamia," in Michel Serres, ed. *A History of Scientific Thought: Elements of a History of Science.* Oxford: Blackwell Reference, 1995, pp. 44–72.

Van der Waerden, Bartel L. "The Date of Invention of Babylonian Planetary Theory," *Archive for History of Exact Sciences,* 5(1968): 70–78.

_____. *Geometry and Algebra in Ancient Civilizations.* Berlin: Springer-Verlag, 1983.

_____. *Science Awakening.* Trans. by Arnold Dresden. Groningen: P. Noordhoff, 1954.

_____. *Science Awakening II: The Birth of Astronomy.* Oxford: Oxford University Press, 1974.

Ancient Egyptian Mathematical Record

Gerdes, Paulus. "Three Alternate Methods of Obtaining the Ancient Egyptian Formula for the Area of a Circle," *Historia Mathematica* 12 (1985): 261–267.

Knorr, Wilbur R. "Techniques of Fractions in Ancient Egypt and Greece," *Historia Mathematica* 9 (1982): 133–171.

Smeur, J. E. M. "On the Value Equivalent to π in Ancient Mathematical Texts: A New Interpretation," *Archive for History of Exact Sciences* 6 (1970): 249–270.

Spalinger, Anthony. "Some Remarks on the Epagomenal Days in Ancient Egypt," *Journal of Near Eastern Studies* 54 (1995): 33–47.

Beginnings of Theoretical Mathematics in Pre-Socratic Greece

Pre-Socratic Greek Mathematics: Primary Sources in Translation

Cohen, Morris R.; and I. E. Drabkin, eds. *A Source Book in Greek Science.* Cambridge, MA: Harvard University Press, 1958.

Heath, Thomas L., ed. *The Thirteen Books of Euclid's Elements.* 2nd ed. 3 vols. Cambridge: Cambridge University Press, 1926 (based on the Heiberg edition, 1883–85).

Morrow, G. R. *Proclus, A Commentary on the First Book of Euclid's Elements.* Princeton, NJ: Princeton University Press, 1970.

Thomas, Ivor, ed. *Greek Mathematical Works.* 2 vols. Cambridge, Mass.: Harvard University Press, 1951–1957.

Pre-Socratic Greek Mathematics: Secondary Works

Bamford, Christopher, ed. *Rediscovering Sacred Science.* Edinburgh: Floris Books, 1994.

Bowen, Alan C., and Bernard R. Goldstein, "Aristarchus, Thales, and Heraclitus on Solar Eclipses: An Astronomical Commentary on P. Oxy. 53.3710 cols. 2.33–3.19," *Physis* 31 (1994): 689–729.

Bunt, Lucas N. H., Phillip S. Jones, and Jack D. Bedient. *The Historical Roots of Elementary Mathematics.* Englewood Cliffs, NJ: Prentice Hall, 1976.

Burkert, W. *Lore and Science in Ancient Pythagoreanism.* Trans. by E. L. Milnar, Jr. Cambridge, MA: Harvard University Press, 1972.

De Vogel, C. J. *Pythagoras and Early Pythagoreanism.* Assen: Van Gorcum & Co., 1966.

Dudley, Underwood. *Numerology, or, What Pythagoras Wrought.* Providence, RI: Mathematical Association of America, 1997.

Goldstein, B. R., and A. C. Bowen. "A New View of Early Greek Geometry," *Isis 74* (1983): 330–340.

Hussey, Edward. *The Presocratics.* Indianapolis, IN: Hackett Publishing, 1995.

Kingsley, Peter. *Ancient Philosophy, Mystery, and Magic: Empedocles and Pythagorean Tradition.* Oxford: Oxford University Press, 1995.

Knorr, Wilbur Richard. *The Ancient Tradition of Geometric Problems.* New York: Dover, 1993.

_____. *The Evolution of the Euclidean Elements.* Dordrecht: Reidel, 1975.

Lloyd, Geoffrey E. R. *Early Greek Science: Thales to Aristotle.* New York: Norton, 1971.

_____. *Methods and Problems in Greek Science.* Cambridge: Cambridge University Press, 1991.

Neugebauer, Otto. *The Exact Sciences in Antiquity.* Princeton, NJ: Princeton University Press, 1952.

Panchenko, Dmitri. "Thales' Prediction of a Solar Eclipse," *Journal of the History of Astronomy* 25 (1994): 275–288.

Rochberg, F., et al., "The Culture of Ancient Science: Some Historical Reflections," *Isis* 83 (1992): 547–607.

Serres, Michael, ed. *Éléments d'histoire des sciences.* Paris: Bordas, 1989.

_____. *Les origines de la géométrie: Tiers livre des fondations.* Paris: Flammarion, 1993.

Van der Waerden, Bartel L. *Geometry and Algebra in Ancient Civilizations*. Berlin: Springer-Verlag, 1983.

Waterfield, Robin. *Before Eureka: The Presocratics and Their Science*. New York: St. Martin's Press, 1989.

Zhmud, Leonid. "All Is Number?: 'Basic Doctrine' of Pythagoreanism Reconsidered," *Phronesis* 34 (1989): 270–292

The Theoretical Mathematics Established in Fourth-Century B.C. Greece

Primary Sources in Translation

Aristotle, *Physics*. Trans. by Robin Waterfield. Oxford: Oxford University Press, 1996.

Sachs, Joe, ed. *Aristotle's Physics: A Guided Study*. New Brunswick, NJ: Rutgers University Press, 1995.

Thomas, Ivor, ed. *Greek Mathematical Works*. 2 vols. Cambridge, MA: Harvard University Press, 1951–1957.

Secondary Works

Algra, Keimpe. *Concepts of Space in Greek Thought*. Leiden: Brill, 1994.

Balaguer, Mark. *Platonism and anti-Platonism in Mathematics*. New York: Oxford University Press, 1998.

Barker, Andrew. "Ptolemy's Pythagoreans, Archytas, and Plato's Conception of Mathematics," *Phronesis* 39 (1994): 113–135.

Bolotin, David. "Continuity and Infinite Divisibility in Aristotle's *Physics*," *Ancient Philosophy* 13 (1993): 323–340.

Fowler, David H. "Could the Greeks Have Used Mathematical Induction? Did They Use It?" *Physis* 31 (1994): 252–265.

_____. *The Mathematics of Plato's Academy: A New Reconstruction*. Oxford: Clarendon Press, 1994.

_____. "The Story of the Discovery of Incommensurability Revisited," in Kostas Gavroglu et al., eds. *Trends in the Historiography of Science*. Dordrecht: Kluwer Academic, 1994, pp. 221–235.

Heath, Thomas. *A Manual of Greek Mathematics*. Oxford: Oxford University Press, 1931; Dover Paperback, 1963.

_____. *Mathematics in Aristotle*. Oxford: Clarendon Press, 1949.

Hetherington, Norriss S. "Plato and Eudoxus: Instrumentalists, Realists, or Prisoners of Themata?" *Studies in the History and Philosophy of Science* 27 (1996): 271–289.

Knorr, Wilbur Richard. *The Ancient Tradition of Geometric Problems*. New York: Dover, 1993.

_____. *The Evolution of the Euclidean Elements*. Dordrecht: D. Reidel, 1975.

_____. "The Practical Element in Ancient Exact Sciences," *Synthese* 81 (1989): 313–328.

Kretzmann, M., ed. *Infinity and Continuity in Ancient and Medieval Thought.* Ithaca, NY: Cornell University Press, 1982.

Lloyd, Geoffrey E. R. *Early Greek Science: Thales to Aristotle.* New York: Norton, 1971.

_____. *Magic, Reason, and Experience.* Cambridge: Cambridge University Press, 1979.

_____. "The *Meno* and the Mysteries of Mathematics," *Phronesis* 37 (1992): 166–183.

_____. "Techniques and Dialectic: Method in Greek and Chinese Mathematics and Medicine," in Jyl Gentzler, ed., *Method in Ancient Philosophy.* Oxford: Clarendon Press, 1998. pp. 351–376.

Pritchard, Paul. *Plato's Philosophy of Mathematics.* Sankt Augustin: Academia Verlag, 1995.

Redell, R. C. "Eudoxan Mathematics and Eudoxan Spheres," *Archive for History of Exact Sciences* 20 (1976): 1–19.

Robins, Ian. "Mathematics and the Conversion of the Mind: *Republic* vii 522c1–531e3," *Ancient Philosophy* 15 (1995): 359–391.

Roochnik, David. "Counting on Number: Plato on the Goodness of *Arithmos*," *American Journal of Philology* 115 (1994): 543–563.

Serres, Michel. *A History of Scientific Thought: Elements of a History of Science.* Oxford: Blackwell Reference, 1995.

_____. *Les origines de la géométrie: Tiers livre des fondations.* Paris: Flammarion, 1993.

Shamsi, F. A. "A Note on Aristotle, *Physics* 239b5–7: What Exactly Was Zeno's Argument of the Arrow?" *Ancient Philosophy* 14 (1994): 51–72.

Smith, R. "The Mathematical Origins of Aristotle's Syllogistic," *Archive for History of Exact Sciences* 19 (1978): 201–209.

Stein, Howard. "Eudoxos and Dedekind: On the Ancient Greek Theory of Ratios and Its Relation to Modern Mathematics," *Synthese* 84 (1990): 163–211.

Taran, Leonardo. "Ideas, Numbers, and Magnitudes: Remarks on Plato, Speusippus, and Aristotle," *Revue de philosophie ancienne* 9 (1991): 199–231.

Taylor, C. C., ed. *From the Beginning to Plato.* London: Routledge, 1997.

Thorup, Anders. "A Pre-Euclidean Theory of Proportions," *Archive for History of Exact Sciences* 45 (1992): 1–16.

White, Michael J. "Concepts of Space in Greek Thought," *Apeiron* 29 (1996): 183–198.

_____. "The Metaphysical Location of Aristotle's *Mathematika*," *Phronesis* 38 (1993): 166–182.

Wilson, Alistair Macintosh. *The Infinite in the Finite.* Oxford: Oxford University Press, 1995.

Wright, M. R. *Cosmology in Antiquity.* London: Routledge, 1995.

Ancient Mathematical Zenith in the Hellenistic Third Century B.C., I: The Alexandrian Museum and Euclid

Primary Sources in Translation

Berggren, J. L., and Robert S. D. Thomas, eds. *Euclid's Phaenomena: A Translation and Study of a Hellenistic Treatise in Spherical Astronomy.* New York: Garland, 1996.

Euclid, *The Data of Euclid.* Trans. by George L. McDowell and Merle A. Sololik. Baltimore, MD: Union Square Press, 1993.

Heath, Thomas L. *The Thirteen Books of Euclid's Elements*, 2nd ed. 3 vols., Cambridge: Cambridge University Press, 1926.

Proclus, *A Commentary on the First Book of Euclid's Elements.* Trans. by Glenn R. Morrow. Princeton, NJ: Princeton University Press, 1992 (1st edition, 1970).

Thomas, Ivor, ed. *Selections Illustrating the History of Greek Mathematics.* 2 vols. Cambridge, MA: Harvard University Press, 1957 (1st edition, 1939–1941).

Secondary Works

Berggren, J. L., and Robert S. D. Thomas, "Mathematical Astronomy in the 4th Century B.C. as Found in Euclid's *Phaenomena,*" *Physis* 29 (1992): 7–33.

Bunt, Lucas H.; Philip S. Jones; and Jack D. Bedient, *The Historical Roots of Elementary Mathematics.* Englewood Cliffs, NJ: Prentice Hall, 1976.

Grattan-Guinness, Ivor. "Numbers, Magnitudes, Ratios, and Proportions in Euclid's *Elements*: How Did He Handle Them?" *Historia Mathematica* 23 (1996): 355–375.

Gray, Jeremy. *Ideas of Space.* Oxford: Oxford University Press, 1989.

Heath, Thomas L. *A History of Greek Mathematics*, 2 vols. Oxford: Clarendon Press, 1921.

Heilbron, John. *Geometry Civilized: History, Culture and Technique.* Cambridge: Cambridge University Press, 1998.

Herz-Fischler, Roger. "Theorem XIV** of the First 'Supplement' of the *Elements,*" *Archives internationale d'histoire des sciences* 38 (1988): 3–66.

Knorr, Wilbur R. *The Ancient Tradition of Geometric Problems.* New York: Dover, 1993.

_____. "When Circles Don't Look Like Circles. An Optical Theorem in Euclid and Pappus," *Archive for History of Exact Sciences* 44 (1992): 287–329.

_____. "The Wrong Text of Euclid: On Heiberg's Text and Its Alternatives" *Centaurus* 38 (1996): 208–276.

Kunitzsch, Paul. " 'The Peacock's Tail': On the Names of Some Theorems in Euclid's *Elements,*" in M. Folkerts and J. P. Hogendijk, eds., *Vestigia mathematica.* Amsterdam: Rodopi, 1992, pp. 205–214.

Levin, Flora R. "Unity in Euclid's 'Sectio canonis'," *Hermes* 118 (1990): 430–443.

Russo, Lucio. *La rivoluzione dimenticata* (The Forgotten Revolution). Milan: Feltrinelli, 1986.

Saito, Ken. "Doubling the Cube: A New Interpretation of Its Significance for Early Greek Geometry," *Historia Mathematica* 22, 2 (1995): 119–138.

Shallit, Jeffrey. "Origins of the Analysis of the Euclidean Algorithm," *Historia Mathematica* 21 (1994): 401–419.

Tobin, Richard. "Ancient Perspective and Euclid's *Optics*," *Journal of the Warburg and Courtauld Institutes* 53 (1990): 14–41.

Van der Waerden, Bartel L. *Geometry and Algebra in Ancient Civilizations*. Berlin: Springer-Verlag, 1983, esp. pp. 73–87.

Hellenistic Science and Alexandria

Delia, Diana. "From Romance to Rhetoric: The Alexandrian Library in Classical and Islamic Traditions," *American Historical Review* 97, 5 (1992): 1449–1468.

Lloyd, Geoffrey E. R. *Greek Science after Aristotle*, New York: Norton, 1973.

_____. *Methods and Problems in Greek Science*. Cambridge: Cambridge University Press, 1991.

Parsons, Edward Alexander. *The Alexandrian Library*. New York: Elsevier: 1952.

Ancient Mathematical Zenith in the Hellenistic Third Century B.C., II: Archimedes to Diocles

Primary Sources in Translation

Apollonius of Perga, *Conics*. 3 vols. Trans. by R. Catesby Taliaferro. Annapolis, MD: Classics of the St. Johns Program, 1964.

Archimedes, *Opera omnia*. Ed. by J. L. Heiberg, 2nd ed. 3 vols. Leipzig: Teubner, 1910–1915 (repr. Stuttgart, 1972).

Dijksterhuis, E. J., ed., *Archimedes*. New York: Humanities Press, 1957.

Thomas, Ivor, ed. *Selections Illustrating the History of Greek Mathematics*. 2 vols. Cambridge, MA: Harvard University Press, 1939–1941, repr. 1957.

Toomer, G. J., ed. and trans. *Apollonius of Perga's Conics: Books V to VII*. New York: Springer-Verlag, 1990.

_____. *Diocles on Burning Mirrors*. New York: Springer-Verlag, 1976.

Secondary Works

Authier, Michel. "Archimedes: The Scientist's Canon," in Michel Serres, ed. *A History of Scientific Thought: Elements of a History of Science*. Oxford: Blackwell Reference, 1995, pp. 124–159.

Dutka, Jacques. "Eratosthenes' Measurement of the Earth Reconsidered," *Archive for History of Exact Sciences* 46, 1 (1993): 55–66.

Hayashi, Eiji. "A Reconstruction of the Proof of Proposition 11 in Archimedes' *Method*: Proofs about the Volume and the Center of Gravity of Any Segment of an Obtuse-Angled Conoid," *Historia scientiarum* 3 (1994): 215–230.

Jones, Alexander. "Studies in the Astronomy of the Roman Period, III: Planetary Epoch Tables," *Centaurus* 40 (1998): 1–41.

Knorr, Wilbur Richard. *The Ancient Tradition of Geometric Problems*. New York: Dover Publications, 1993 (1st ed., 1986).

_____. "Archimedes and the *Elements*: Proposal for a Revised Chronological Ordering of the Archimedean Corpus," *Archive for History of Exact Sciences* 19 (1978): 211–290.

_____. "The Hyperbola Construction in the *Conics*, Book II. Variations on a Theorem of Apollonius," *Centaurus* 25 (1982): 253–291.

_____. "New Readings in Greek Mathematics: Sources, Problems, Publications," *Impact of Science on Society* 40 (1990): 207–218.

_____. *Textual Studies in Ancient and Medieval Geometry*. Boston: Birkhäuser, 1989.

Kouremenos, Theokritos. "Posidonius and Geminus on the Foundations of Mathematics," *Hermes* 122 (1994): 437–450.

Mansfeld, Jaap. *Prolegomena mathematica, from Apollonius of Perga to Late Neoplatonism*. Leiden: Brill, 1998.

Neugebauer, Otto and George Saliba, "On Greek Numerology," *Centaurus* 31 (1989): 189–206.

Simms, D. L. "Archimedes the Engineer," *History of Technology* 17 (1995): 45–111.

Taisbak, Christian Marinus. "Analysis of the So-called "Lemma of Archimedes" for Constructing a Regular Heptagon," *Centaurus* 36 (1993): 191–199.

Wright, M. R. *Cosmology in Antiquity*. New York: Routledge, 1995.

Mathematics in Roman Times, Centering in Alexandria

Primary Sources in Translation

Heath, Thomas L. *Diophantus of Alexandria: A Study in the History of Greek Algebra*. Cambridge: Cambridge University Press, 1910, Dover ed., n. d.

Heiberg, J. L. *Ptolemy's Syntaxis Mathematica (Almagest)*. 2 parts. Leipzig: B. G. Teubner, 1889–1903.

Jones, Alexander, ed. and trans. *Book 7 of Pappus' Collection*. Berlin: Springer-Verlag, 1986.

Jones, Alexander. "Studies in the Astronomy of the Roman Period I: The Standard Lunar Scheme," *Centaurus* 39 (1997): 1–36.

Pederson, Olaf. *A Survey of the Almagest*. Copenhagen: Odense University Press, 1974.

Philiponus. *On Aristotle's Physics 2*. Trans. by A. R. Lacey. Ithaca, NY: Cornell University Press, 1993.

Proclus, *A Commentary on the First Book of Euclid's Elements*. Trans. by Glenn R. Morrow. Princeton: Princeton University Press, 1992 (1st edition, 1970).

Sesiano, J. *Books IV to VII of Diophantus "Arithmetica" in Arabic Translation, Attributed to Qusta ibn Luqa*. New York: Springer-Verlag, 1982.

Secondary Works

Brecher, Kenneth and Michael Feirtag, eds. *Astronomy of the Ancients*. Cambridge, MA: MIT Press, 1979.

Britton, John Phillips. *Models and Precision: The Quality of Ptolemy's Observations and Parameters.* New York: Garland, 1992.

Cameron, Alan. "Isidore of Miletus and Hypatia: On the Editing of Mathematical Texts," *Greek and Roman Byzantium Studies* 135 (1991): 235–254.

Deacon, Michael A. B. "Hypatia and Her Mathematics," *American Mathematical Monthly* 101 (1994): 234–243.

Dunham, William. *Journey Through Genius: The Great Theorems of Mathematics.* New York: Wiley, 1990.

Dzielska, Maria. *Hypatia of Alexandria.* Trans. by F. Lyra. *Revealing Antiquity*, vol. 8. Cambridge, MA: Harvard University Press, 1995.

Glasner, Ruth. "Proclus' Commentary on Euclid's Definitions 1.3 and 1.6," *Hermes* 120 (1992): 320–333.

Grasshof, Gerd. *The History of Ptolemy's Star Catalogue.* New York: Springer-Verlag, 1990.

Høyrup, Jens. "Hero, ps.-Hero, and Near Eastern Practical Geometry: An Investigation of *Metrica, Geometrica,* and Other Treatises." *Antike Naturwissenchaften und ihre Rezeption* 7 (1997): 7–23.

Huby, Pamela, and Gordon Neal, eds., *The Criterion of Truth.* Liverpool: Liverpool University Press, 1989.

Jones, Alexander. *Ptolemy's First Commentator.* (*Transactions of the American Philosophical Society*, vol. 80, part 7). Philadelphia: American Philosophical Society, 1990.

Knorr, Wilbur R. "*Arithmêtikê stoicheiôsis:* On Diophantus and Hero of Alexandria," *Historia Mathematica* 20 (1993): 180–192.

_____, and George Anawati, "Diophantus Redivivus," *Archives internationales d'histoire des sciences* 39 (1989): 345–357.

Krupp, E. C. *Echoes of the Ancient Skies.* Oxford: Oxford University Press, 1994.

McAlister, Linda Lopez, ed. *Hypatia's Daughters: Fifteen Hundred Years of Women Philosophers.* Bloomington: Indiana University Press, 1996.

O'Meara, Dominic V. *Pythagoras Revived: Mathematics and Philosophy in late Antiquity.* Oxford: Clarendon Press, 1989.

Pepin, Jean, and H. D. Saffrey, eds. *Proclus, Lecteur et Interpretre Anciens.* Paris: Centre Nationale de la Recherche Scientifique, 1987.

Pingree, David. "The Teaching of the *Almagest* in Late Antiquity," *Apeiron* 27, 4 (1994): 75–98.

Riley, Mark T. "Ptolemy's Use of His Predecessor's Data," *Transactions of the American Philological Association* 125 (1995): 221–250.

Rist, J. M. "Hypatia," *Phoenix* 19 (1965): 214–225.

Smith, A. Mark. *Ptolemy's Theory of Visual Perception. Transactions of the American Philosophical Society,* vol. 86, part 2. Philadelphia: American Philosophical Society, 1996.

Swerdlow, Noel. "Ptolemy's Theory of the Inferior Planets," *Journal for the History of Astronomy* 20 (1989): 29–60.

Thurston, Hugh. *Early Astronomy.* Berlin: Springer-Verlag, 1994, softcover, 1996.

Toomer, G. J. "The Chord Table of Hipparchus and the Early History of Greek Trigonometry," *Centaurus* 18 (1973): 6–28.

Van Brummelen, Glen. "Lunar and Planetary Interpolation Tables in Ptolemy's *Almagest*," *Journal for the History of Astronomy* 25 (1994): 297–311.

Williams, Burma P. and Richard S. Williams, "Finger Numbers in the Greco-Roman World and the Early Middle Ages," *Isis* 86 (1995): 587–608.

Mathematics in Traditional China

Primary Sources in Translation

Lam, Lay Yong. *A Critical Study of Yang Hui Suan Fa: A Thirteenth–Century Chinese Mathematical Treatise.* Singapore: Singapore University Press, 1977.

Swetz, Frank J. *The Sea Island Mathematical Manual: Surveying and Mathematics in Ancient China.* University Park, PA: Pennsylvania State University Press, 1992.

Secondary Works

Ang, Tian Se. "Chinese Interest in Right-Angle Triangles," *Historia Mathematica* 5 (1978): 253–266.

Chemla, Karine. "Different Concepts of Equations in The Nine Chapters on Mathematical Procedures and the Commentary on It by Liu Hui (3rd Century)," *Historia Scientiarum* 4 (1994): 113–137.

_____. "Similarities Between Chinese and Arabic Mathematical Writings: (1) Root Extraction," *Arabic Science and Philosophy: A Historical Journal* 4 (1994): 207–266.

_____. "Theoretical Aspects of the Chinese Algorithmic Tradition (1st to 3rd Century)," *Historia Scientiarum* 42 (1991): 75–98.

Chen, Cheng-Yih. *Early Chinese Work in Natural Science: A Re-examination of the Physics of Motion, Acoustics, and Astronomy and Scientific Thoughts.* Hong Kong: Hong Kong University Press, 1996.

Crossley, J. N., and A. W. C. Lun. "The Logic of Liu Hui and Euclid as Exemplified in Their Proofs of the Volume of a Pyramid," *Philosophy and the History of Science: A Taiwanese Journal* 3, 1 (1994): 11–27.

Cullen, Christopher. *Astronomy and Mathematics in Ancient China: The Zhou Bi Suanjing.* Cambridge: Cambridge University Press, 1996.

_____. "How Can We Do the Comparative History of Mathematics? Proof in Liu Hui and the *Zhou bi*," *Philosophy and the History of Science: A Taiwanese Journal* 4, 1 (1995): 59–94.

Daiwie, Fu. "Why did Liou Hui Fail to Derive the Volume of a Sphere?" *Historia Mathematica* 18, 3 (1991): 212–238.

Dauben, Joseph W. "The 'Pythagorean Theorem' and Chinese Mathematics...," in Sergei S. Demidov, et al., eds., *Amphora: Festschrift für Hans Wussing.* Basel: Birkhäuser, 1992, pp. 133–155.

Engelfriet, Peter. "The Chinese Euclid and Its European Context," in Catherine Jami and Hubert Delahaye, eds., *L'Europe en Chine.* Paris: College de France, Institut des Hautes Études Chinoise, 1993, pp. 111–135.

Horng, Wann-Sheng. "How Did Liu Hui Perceive the Concept of Infinity: A Revisit," *Historia Scientiarum* 4 (1995): 207–222.

Jami, Catherine. "Scholars and Mathematical Knowledge during the Late Ming and Early Qing," *Historia Scientiarum* 42 (1991): 99–109.

Kangshen, Shen. "Historical Development of the Chinese Remainder Theorem," *Archive for History of Exact Sciences* 38, 4 (1988): 218–315.

Lam, Lay Yong. "*Jiu Zhang Suanshu* (Nine Chapters on the Mathematical Art): An Overview," *Archive for History of Exact Sciences* 47 (1994): 1–51.

Lam, Lay Yong , and Tian Se Ang. *Fleeting Footsteps: Tracing the Conception of Arithmetic and Algebra in Ancient China.* Singapore: World Scientific, 1992.

_____, and _____. "Methods of Solving Linear Equations in Traditional China," *Historia Mathematica* 6 (1989): 107–122.

Li, Yan, and Du Shiran. *Chinese Mathematics: A Concise History.* Oxford: Clarendon Press, 1987.

Libbrecht, Ulrich, *Chinese Mathematics in the Thirteenth Century.* Cambridge, MA: M.I.T. Press, 1973.

Needham, Joseph, and Wang Ling. *Science and Civilization in China.* vol. 3 Cambridge: Cambridge University Press, 1959.

Poor, Robert. "The Circle and the Square: Measure and Ritual in Ancient China," *Monumenta Serica* 43 (1995): 159–210.

Ronan, Colin A. *The Shorter Science and Civilization in China: An Abridgement of Joseph Needham's Original Text,* vol. 5. Cambridge: Cambridge University Press, 1995.

Siu, Man-keung. "Success and Failure of Xu Gwang-Qi: Response to the First Dissemination of European Science in Ming China," *Studies in History, Medicine, & Science* XIV, Nos. 1–2 New Series, 1995/96: 137–179.

So, Jenny F. "Bells of Bronze Age China," *Archaeology* 47, 1 (1994): 42–51.

Swetz, Frank, and Tian Se Ang. "A Brief Chronology Guide to the History of Chinese Mathematics," *Historia Mathematica* 11 (1984): 39–56.

Van der Waerden, Bartel L. *Geometry and Algebra in Ancient Civilizations.* Berlin: Springer-Verlag, 1983.

Volkov, Alexeij. "Calculation of *pi* in Ancient China: From Liu Hui to Zu Chonzhi," *Historia Scientiarum* 4 (1994): 139–157.

Indian Mathematics: From Harappan through Medieval Times

Binford, Lewis R. *In Pursuit of the Past.* New York: Thames & Hudson, 1988.

Cajori, Florian, *A History of Mathematical Notations.* 2 vols. New York: Dover, 1993 (1st ed., 1929–1930).

De Young, Gregg. "Euclidean Geometry in the Mathematical Tradition of Islamic India," *Historia Mathematica* 22 (1995): 138–153.

Hayashi, Takaya, Takanori Kusuba, and Michael Yano. "Indian Value for π Derived from Aryabhata's Value," *Historia scientiarum* 37 (1989): 1–16.

Jha, Parmeshwar. *Aryabhata I and His Contributions to Mathematics.* Patna, India: Bihar Research Society, 1988.

Joseph, George Ghevarghese. *The Crest of the Peacock*. New York: St. Martin's Press, 1991.

Murthy, T. S. Bhanu. *A Modern Introduction to Ancient Indian Mathematics*. New Delhi: Wiley Eastern, 1992.

Pandit, M. D. *Mathematics as Known to the Vedic Samhitas*. Delhi: Sri Satguru Publications, 1993.

Pingree, David. "Bīja Corrections in Indian Astronomy," *Journal for the History of Astronomy* 27 (1996): 161–172.

_____. *Census of the Exact Sciences in Sanskrit*. 5 vols. Philadelphia: American Philosophical Society, 1970–1994.

Rao, S. Balachandra. *Indian Mathematics and Astronomy: Some Landmarks*. Bangalore: Jnana Deep Publications, 1994.

Mathematics in the Service of Religion

Primary Sources in Translation

Bacon, Roger. *Roger Bacon and the Origins of Perspective in the Middle Ages*. Ed. and trans. by David C. Lindberg. Oxford: Oxford University Press, 1996.

Grant, Edward, ed. *A Source Book in Medieval Science*. Cambridge, MA: Harvard University Press, 1974.

Mathematics in Early Medieval Europe

Burge, E. L. *Martianus Capella and the Seven Liberal Arts*. New York: Columbia University Press, 1971.

Butzer, Paul L., and Dietrich Lohrmann, eds. *Science in Western and Eastern Civilization in Carolingian Times*. Basel: Birkhäuser, 1993.

Duckett, E. S. *Alcuin, Friend of Charlemagne*. New York: Macmillan, 1951.

Folkerts, Menso, and J. P. Hogendijk, eds. *Vestigia mathematica*. Amsterdam: Rodopi, 1993.

Fraser, J. T. and Lewis Rowell, eds. *Time and Process: Interdisciplinary Issues*. Madison, CN: International Universities Press, 1993.

Gibson, Craig A. and Francis Newton. "Pandulf of Capua's *De calculatione:* An Illustrated Abacus Treatise and Some Evidence for the Hindu-Arabic Numerals in 11th-Century South Italy," *Mediaeval Studies* 57 ((1995): 293–335.

Grant, Edward. "The Medieval Cosmos: Its Structure and Operation," *Journal of the History of Astronomy* 28 (1997): 147–167.

Jones, Charles W. *Bede, the Schools and the Computus*. ed. by Wesley M. Stevens. Aldershot, Eng.: Variorum, 1994.

Kuehne, Andreas. "A Manuscript Databank for the History of Mathematics in Medieval and Renaissance Europe," *Primary Sources and Original Works* (3–4) 1991: 51–61.

Lindberg, David, ed. *Science in the Middle Ages*. Chicago: University of Chicago Press, 1978.

Molland, A. George. *Mathematics and the Medieval Ancestry of Physics*. Brookefield, VT: Variorum, 1995.

_____. "Roger Bacon's Appropriation of Past Mathematics," in F. Jamil Ragep et al., eds. *Tradition, Transmission, Transformation*. Leiden: Brill, 1996.

Moreton, Jennifer. "Before Grosseteste: Roger of Hereford and Calendar Reform in 11th- and 12th-Century England," *Isis* 86 (1995): 562–586.

Takashi, Ken'ichi. *The Medieval Latin Traditions of Euclid's Catoptrica*. Fukoka: Kyushu University Press, 1992.

The Classical Maya

Ascher, Marcia and Robert Ascher. *Mathematics of the Incas: Code of the Quipu*. New York: Dover Publications, 1997. [A reprint of the 1981 University of Michigan Press edition]

Aveni, Anthony F., Steven J. Morandi, and Polly A. Peterson. "The Maya Number of Time: Intervalic Time Reckoning in the Maya Codices. Part I," *Archaeoastronomy* 20 (1995): 1–28.

Bauer, Brian S. and David S. P. Dearborn. *Astronomy and Empire in the Ancient Andes: The Cultural Origins of Inca Sky Watching*. Austin: University of Texas Press, 1995.

Closs, Michael P., ed. *Native American Mathematics*. Austin: University of Texas Press, 1986.

Sidrys, Raymond, ed. *Papers on the Economy and Architecture of the Ancient Maya*. Monograph 7. Los Angeles: University of California Institute of Archaeology, 1978.

Sullivan, William. *The Secret of the Incas: Myth, Astronomy, and the War Against Time*. New York: Crown, 1996.

The Era of Arabic Primacy and a Persian Fluorescence

Primary Sources in Translation

Al-Bī rūnī, Abu l-Rayhān. *The Determination of the Coordinates of Cities*. Trans. by J. Ali. Beirut: American University of Beirut, 1967.

Al-Khwārizmī, Muhamad ibn Musa. *Le calcul indien*. Paris: Blanchard; Namur: Societé des Études Classiques, 1992.

Al-Sijzi. *Al-Sijzi's Treatise on Geometrical Problem Solving*. Trans. and annotated by Jan P. Hogendijk, including a Persian translation by Mohammed Bagheri. Teheran: Fatemi Publishing Co., 1966.

Avery, Peter and John Heath-Stubbs. *The Ruba'iyat of Omar Khayyam*. Hammondswoth, U.K.: Penguin, 1981.

Berggren, J. L. "Al-Kūhīs 'Filling a Lacunae in Book II of Archimedes' in the Version of Nasir al-Dīn al-Tūsī," *Centaurus* 38 (1996): 140–207.

Ragep, F. J., ed. *Nasīr al-Dīn al-Tūsī's Memoir on Astronomy*. 2 vols. New York: Springer-Verlag, 1992.

Rosen, F., ed. and trans. *The Algebra of Muhammed ben Musa*. Hildesheim: Georg Olms Verlag, repr. 1986.

Saidan, S. *The Arithmetic of al-Uqlīdisī*. Boston: Reidel, 1978.

Secondary Works

Ağargün, Ahmet G. and Colin R. Fletcher. "Al-Farisi and the Fundamental Theorem of Arithmetic," *Historia Mathematica* 21, 2 (1994): 162–174.

Berggren, J. L. *Episodes in the Mathematics of Medieval Islam.* New York: Springer-Verlag, 1986.

Casulleras, Josep, and Julio Samso. *From Baghdad to Barcelona: Studies in the Islamic Exact Sciences in Honour of Prof. Juan Vernet.* Barcelona: University of Barcelona Press, 1996.

Chabás, José, and Bernard R. Goldstein. "Andalusian Astronomy: *al-Zīj al-Muqtabis* of Ibn al-Kammâd," *Archive for History of Exact Sciences* 48 (1994): 1–41.

Al-Daffa, A. A. and J. J. Stroyls. *Studies in the Exact Sciences in Medieval Islam.* New York: John Wiley & Sons, 1984.

Dallal, Ahmad S. "Ibn al-Haytham's Universal Solution for Finding the Direction of the *qibla* by Calculation," *Arabic Sciences and Philosophy: A Historical Journal* 5 (1995): 145–193.

Davidson, H. A. *Alfarabi, Avicenna, and Averroes on Intellect.* Oxford: Oxford University Press, 1992.

De Young, Gregg. "Abu Sahl's Additions to Book II of Euclid's *Elements*," *Zeitschrift für Geschichte der Arabisch-Islamischen Wissenschaften* 7(1991–1992): 73–135.

_____. "Ibn al-Sarī on *ex aequali* Ratios: His Critique of Ibn al-Haytham and His Attempt to Improve the Parallelism Between Books V and VII of Euclid's *Elements*," *Zeitschrift für Geschichte der Arabisch-Islamischen Wissenschaften* 9 (1994): 99–152.

Dold-Samplonius, Yvonne. "Practical Arabic Mathematics: Measuring the muqarnas by al-Kāshī," *Centaurus* 35 (1992): 193–242.

Folkerts, M, and J. P. Hogendijk, eds. *Vestigia mathematica.* Amsterdam: Rodopi, 1993.

Gohlman, W. E. *The Life of Ibn Sina–A Critical Edition.* Albany, NY: State University of New York Press, 1974.

Gray, Jeremy. *Ideas of Space.* 2nd ed. Oxford: Oxford University Press, 1989.

Hogendijk, Jan P. "Al-Mu'taman ibn Hūd, 11th Century King of Saragosa and Brilliant Mathematician" *Historia Mathematica* 22, 1 (1995): 1 - 19.

_____. Four Constructions of Two Mean Proportionals Between Two Given Lines in the Book of Perfection (Istikmál) of al-Mu'taman ibn Hūd," *Journal of the History of Arabic Science* 10 (1994): 13–29.

_____. "The Geometrical Parts of the *Istikmal* of Yusuf al-Mu'taman ibn Hud (11th Century): An Analytic Table of Contents," *Archives internationales d'histoire des sciences* 41 (1991): 207–281.

_____. "Transmission, Transformation, and Originality: The Relation of Arabic to Greek Geometry," in F. Jamil Ragep et al., eds. *Tradition, Transmission, Transformation.* Leiden: Brill, 1996, pp. 31–64.

King, D. A. *Astronomy in the Service of Islam.* Brookfield, Vermont: Ashgate Publishing, 1993.

_____. "Ibn Yūnus' Very Useful Table for Reckoning Time from the Sun," *Archive for History of Exact Sciences* 10 (1973): 342–394.

Kollerstrom, Nick. "The Star Temples of Harran," in Annabella Kitson, ed. *History and Astrology: Clio and Urania Conference.* London: Unwin Paperbacks, 1989, pp. 47–60.

Langermann, Y. Tzvi. "Medieval Hebrew Texts on the Quadrature of the Lune," *Historia Mathematica* 23 (1996): 31–53.

Lévy, Tony. "Hebrew Mathematics in the Middle Ages: An Assessment." in F. Jamil Ragep et al., eds., *Tradition, Transmission, and Transformation.* Leiden: Brill, 1996, pp. 71–88.

Lorch, Richard. *Arabic Mathematical Sciences: Instruments, Texts, Transmission.* Aldershot, England: Variorum, 1995.

_____. "Ptolemy and Maslama on the Transformation of Circles into Circles in Stereographic Projection," *Archive for History of Exact Sciences* 49 (1995): 271–284.

Ozdural Alpay. "Omar Khayyam, Mathematicians, and *Conversazioni* with Artisans," *Journal of the Society of Architectural Historians* 54 (1995): 54–71.

Parshall, Karen. "The Art of Algebra from Al-Khowarizmi to Viète," *History of Science* 26, 2 (1988): 129–164.

Rashed, Roshdi, *The Development of Arabic Mathematics: Between Arithmetic and Algebra.* Trans. by A. F. W. Armstrong. Dordrecht: Kluwer Academic, 1994.

_____, ed. *Encyclopedia of the History of Arabic Sciences.* 3 vols. New York: Routledge, 1996.

Sabra, A. I. *Optics, Astronomy and Logic.* Brookfield, VT: Variorum, 1994.

Saliba, George. *A History of Arabic Astronomy.* New York: New York University Press, 1994.

Sharif, M. M. ed. *A History of Muslim Philosophy.* Wiesbaden: Otto Harrassowitz, 1963.

Smith, John D. "The Remarkable Ibn al-Haytham," *Mathematical Gazette* 76 (1992): 189–198.

Recovery and Expansion in Old Europe, 1000–1500

Selected Primary Sources in Translation

Bradwardine, Thomas. *Geometria speculativa.* Trans. and commentary by George Molland. Stuttgart: Steiner Verlag Wiesbaden, 1989.

Burnett, Charles, and David Pingree, eds., *The Liber Aristotilis of Hugo of Santalia.* London: Warburg Institute, 1997.

Fibonacci, Leonardo Pisano. *The Book of Squares.* Annotated and trans. by L. E. Sigler. Boston: Academic Press, 1987.

Grant, Edward, ed. *A Source Book in Medieval Science.* Cambridge, MA: Harvard University Press, 1974.

Hugh of St. Victor. *Practical Geometry.* Trans. by Frederick A. Homann. Milwaukee, WI: Marquette University Press, 1991.

Jordanus de Nemore. *De numeris datis.* Trans. by Barnabus B. Hughes. Berkeley: University of California Press, 1981.

Oresme, Nicole. *De proportionibus proportionum* and *Ad pauca respicientes.* Ed. by Edward Grant. Madison: University of Wisconsin Press, 1966.

Secondary Works

Aiken, Jane Andrews. "Truth in Images: From the Technical Drawings of Ibn al-Razzazz al-Jazari, Campanus of Novara, and Giovanni de Dondi to the Perspective Projection of Leon Battista Alberti," *Viator* 25 (1994): 325–359.

Berger, Anna Maria Busse. *Mensuration and Proportion Signs: Origins and Evolution.* Oxford: Oxford University Press, 1993.

Clagett, Marshall. *Archimedes in the Middle Ages.* 5 vols. Philadelphia: American Philosophical Society, 1963–1984.

Demidov, Sergei S., et al., eds. *Amphora: Festschrift für Hans Wussing.* Basel: Birkhäuser, 1992.

Flegg, G, C. Hay, and B. Moss, eds. *Nicolas Chuquet, Renaissance Mathematician.* Dordrecht: Reidel, 1985.

Folkerts M., and J. P. Hogendijk, eds. *Vestigia mathematica: Studies in Medieval and Early Modern Mathematics in Honour of H. L. L. Busard.* Amsterdam: Rodopi, 1993.

Gies, J., and F. Gies. *Leonard of Pisa and the New Mathematics of the Middle Ages.* New York: Crowell, 1969.

Grant, Edward. *Planets, Stars, and Orbs: The Medieval Cosmos, 1200–1687.* Cambridge: Cambridge University Press, 1994.

Hay, Cynthia, ed. *Mathematics from Manuscripts in Print, 1300–1600.* Oxford: Clarendon Press, 1988.

Henninger, Mark. *Relations: Medieval Theories, 1250–1325.* Oxford: Clarendon Press, 1989.

Høyrup, Jens. "Jordanus de Nemore, 13th Century Mathematical Innovator," *Archive for History of Exact Sciences* 38 (1988): 307–363.

Knorr, Wilbur Richard. *Textual Studies in Ancient and Medieval Geometry.* Boston: Birkhäuser, 1989.

Lindberg, David C. *The Beginnings of Western Science.* Chicago: University of Chicago Press, 1992.

_____, ed. and trans. *Roger Bacon and the Origins of Perspectiva in the Middle Ages.* Oxford: Oxford University Press, 1996.

_____, ed. *Science in the Middle Ages.* Chicago: University of Chicago Press, 1978.

Molland, George. *Mathematics and the Medieval Ancestry of Physics.* Brookfield, VT: Variorum, 1995.

_____. "The Oresmian Style: Semi-mathematical, Semi-holistic," in Pierre Souffrin and A. Ph. Segonds, eds., *Nicholas Oresme: tradition et innovation chez un intellectuel du XIV siècle.* Paris: Les Belles Lettres, 1988.

_____. "Semiotic Aspects of Medieval Mathematics," in Menso Folkerts, ed., *Mathematische Probleme im Mittelalter. Der Lateinische und Arabische Sprachbereich.* Weisbaden: Harrassowitz, 1996, pp. 1–16.

Murdoch, John E. "The Medieval Euclid: Salient Aspects of the Translations of the *Elements* by Adelard of Bath and Campanus of Novara," *Revue de synthése* 89, 3 (1968): 67–94.

Rusnock, Paul. "Oresme on Ratios of Lesser Inequality," *Archives internationale d'histoire des sciences* 45 (1995): 155–188.

Sylla, Edith D. "The Oxford Calculators and Mathematical Physics,: in Sabetai Unguru, ed. *Physics, Cosmology, and Astronomy, 1300 - 1700.* Dordrecht: Kluwer Academic, 1991, pp. 129 - 161.

———, and M. McVaugh, eds. *Text and Context in Ancient and Medieval Science: Studies on the Occasion of John E. Murdoch's 70th Birthday.* Brill's Studies in Intellectual History, 78. Leiden: Brill, 1997. (This volume contains a bibliography of Murdoch's writings.)

Tanay, Dorit E. "Jean de Meur's Musical Theory and Mathematics in the 14th Century," *Tractrix* 5 (1993): 17–43.

The First Phase of the Scientific Revolution, ca. 1450–1600

Primary Sources in Translation

Bruno, Giordano. *Oeuvres complètes.* 4 vols.Paris: Les Belles Lettres, 1994–1996.

Cardan, Jerome. *The Book of My Life* (1575). Trans. by Jean Stoner. New York: Dover, 1962.

———. *The Great Art.* Ed. and trans. by R. R. Witmer. Cambridge, MA: M.I.T. Press, 1968.

Dee, John. *The Mathematicall Praeface to the Elements of Euclid of Megara.* Preface by Allen G. Debus. New York: Science History Publications, 1975.

Recorde, Robert. *The Grounde of Artes and Whetstone of Witte.* Amsterdam: Theatrum Orbis Terrarum, 1969.

Struik, Dirk J., ed. *A Source Book in Mathematics, 1200–1800.* Cambridge, MA: Harvard University Press, 1969.

Viète, François. *The Analytic Art.* Trans. by R. R. Witmer. Kent, OH: Kent University Press, 1983.

Secondary Works

Alexander, Amir. "The Imperialist Space of Elizabethan Mathematics," *Studies in History and Philosophy of Science* 26 (1995): 559–591.

Anderson, Kirsti. "The Mathematical Treatment of Anamorphoses from Piero della Francesca to Niceron," in Joseph W. Dauben et al., eds., *History of Mathematics: States of the Art.* San Diego, CA: Academic Press, 1996, pp. 3–28.

Baigrie, Brian S., ed. *Picturing Knowledge: Historical and Philosophical Problems Concerning the Use of Art in Science.* Toronto: University of Toronto Press, 1996.

Bennett, J., ed. *The Measurers: A Flemish Image of Mathematics in the Century.* Oxford: Museum of the History of Science, 1995.

Billingsley, Dale B. " 'Authority' in Early Editions of Euclid's *Elements*," *Fifteenth-Century Studies* 20 (1993): 1–14.

Broydrick, Thro. E. "Leonardo da Vinci's Solution to the Problem of the Pinhole Camera," *Archive for History of Exact Sciences* 48 (1994): 343–371.

Crosby, Alfred W. *The Measure of Reality: Quantification and Western Society, 1250–1600.* Cambridge: Cambridge University Press, 1996.

Elkins, James. *The Poetics of Perspective.* Ithaca, NY: Cornell University Press, 1994.

_____. "Renaissance Perspectives," *Journal of the History of Ideas* 53 (1992): 209–230.

Farago, Claire, ed. *Reframing the Renaissance: Visual Culture in Europe and Latin America, 1450–1650.* New Haven, CN: Yale University Press, 1995.

Field, J. V. "Piero della Francesca and the 'Distance Point Method' of Perspective Construction," *Nuncius* 10 (1995): 509–530.

_____, and A. J. L. James, eds. *Renaissance and Revolution: Humanists, Scholars, Craftsmen and Natural Philosophers in Early Modern Europe.* Cambridge: Cambridge University Press, 1993.

Fierz, M. *Girolamo Cardano, 1501–1576.* Boston: Birkhäuser, 1983.

Flegg, G. C. Hall, and B. Moss, eds., *Nicolas Chuquet, Renaissance Mathematician.* Dordrecht: Reidel, 1985.

Folkerts, Menso. "Regiomontanus' Role in the Transmission and Transformation of Greek Mathematics," in F. Jamil Ragep *et al.*, eds. *Tradition, Transmission, Transformation.* Leiden: Brill, 1996, pp. 89–113.

Giusti, Enrico. "Algebra and Geometry in Bombelli and Viète," *Bolletino di Storia delle Scienze Matematiche* 12 (1992): 303–328.

Hay, Cynthia, ed. *Mathematics from Manuscripts to Print, 1300–1600.* Oxford: Clarendon Press, 1988.

Kusukawa, Sachiko. *The Transformation of Natural Philosophy. The Case of Philip Melanchthon.* Cambridge: Cambridge University Press, 1995.

Larvor, Brendan. "History, Methodology and Early Algebra," *Intenational Studies in the Philosophy of Science* 8 (1994): 113–124.

Lattis, James M. *Between Copernicus and Galileo: Christoph Clavius and the Collapse of Ptolemaic Cosmology.* Chicago: University of Chicago Press, 1994.

Macagno, Matilde. "Transformation Geometry in the Manuscripts of Leonardo da Vinci," *Raccolta Vinciana* 26 (1995): 93–134.

Meskens, Ad. "Mathematics Education in Late 16th-century Antwerp," *Annals of Science* 53 (1996): 137–155.

Meli, Domenico Bertolini. "Guidabaldo dal Monte and the Archimedean Revival," *Nuncius* 7, 1 (1992): 3–34.

Moore, Patrick. *The Great Astronomical Revolution, 1543–1687, and the Space Age Epilogue.* Chichester: Albion, 1994.

Neely, Henry M. *The Stars by Clock and Fist.* New York: Viking Press, 1956.

Ore, Oystein. *Cardano, The Gambling Scholar.* Princeton, NJ: Princeton University Press, 1953.

Pesic, Peter. "Secrets, Symbols, and Systems: Parallels between Cryptanalysis and Algebra, 1580–1700," *Isis*, 88 (1997): 674–692.

Pumfrey, Stephen; Rossi, Paolo L.; and Slawinsky, Maurice. *Science, Culture, and Popular Belief in Renaissance Europe.* Manchester: Manchester University Press, 1991.

Rawlins, Dennis. "Tycho 1004-Star Catalog's Completion Was Faked," *DIO*, 2, 1 (1992): 35–50.

_____. *Tycho's Star Catalog: The First Critical Edition. DIO* 3 (1993): 3–106.

Rider, Robin E. "Early Modern Mathematics in Print," in Renato G.Mazzolini, ed. *Non-verbal Communication in Science Prior to 1900.* Florence: Olschki, 1993, pp. 91–113.

Rosen, Edward. *Copernicus and His Successors.* London: Hambledon Press, 1995.

Rosiñska, Grazyna. "A Chapter in the History of Renaissance Mathematics: Negative Numbers and the Formation of the Law of Signs (Ferrara, Italy, ca. 1450)," *Kwartalnik Historii Nauki i Techniki,* 40, 1 (1995): 3–20.

_____. "Decimal Positional Fractions: Their Use for Surveying Purposes (Ferrara, 1442)" *Kwartalnik Historii Nauki i Techniki,* 40, 4 (1995): 17–32.

Rowland, Ingrid D. "Abacus and Humanism," *Renaissance Quarterly* 48 (1995): 695–727.

Sherman, William H. "The Place of Reading in the English Renaissance: John Dee Revisited," in James Raven, Helen Small, and Naomi Tadmor, eds., *The Practice and Representation of Reading in England.* Cambridge: Cambridge University Press, 1996, pp. 62–76.

Siraisi, Nancy G. *The Clock and the Mirror: Girolamo Cardano and Renaissance Medicine.* Princeton, NJ: Princeton University Press, 1997.

Swerdlow, Noel M. "Science and Humanism in the Renaissance: Regiomontanus's Oration on the Dignity and Utility of the Mathematical Sciences," in Paul Horwich, ed., *Thomas Kuhn and the Nature of Science.* Cambridge, MA: MIT Press, 1993.

Swetz, Frank J. *Capitalism and Arithmetic.* La Salle, IL: Open Court, 1987.

Zinner, Ernst. *Regiomontanus: His Life and Work.* Amsterdam: North Holland, 1990.

Transformation of Mathematics, ca. 1600–1660: I

Selected Primary Sources in Translation

Calinger, Ronald, ed. *Classics of Mathematics.* Englewood Cliffs, NJ: Prentice Hall, 1995.

Dijksterhuis, E. J., et al., eds., *The Principal Works of Simon Stevin.* Amsterdam: C. V. Swets & Zeitlinger, 1955–1966. 5 volumes.

Galilei, Galileo. *The Assayer.* Trans. by Stillman Drake, *Discoveries and Opinions of Galileo.* Garden City, NY: Doubleday, 1957.

_____. *Galileo on the World Systems: A New Abridged Translation and Guide.* Trans. with Guide by Maurice A. Finochiaro. Berkeley: University of California Press, 1997.

_____. *Two New Sciences.* Trans. with Intro. by Stillman Drake. Madison: The University of Wisconsin Press, 1974.

Kepler, Johannes. *Epitome of Copernican Astronomy and Harmonies of the World.* Trans. by Charles Glenn Wallis. Amherst, NY: Prometheus Books, 1995.

_____. *New Astronomy.* Trans. by William H. Donahue. Cambridge: Cambridge University Press, 1992.

Girard, A. *A New Discovery in Algebra,* in *The Early Theory of Equations: On Their Nature and Constitution.* Annapolis, MD: Golden Hind Press, 1986.

Secondary Works

Applebaum, Wilbur. "Keplerian Astronomy after Kepler: Researches and Problems," *History of Science* 34 (1996): 451–504.

Biagioli, Mario. "Etiquette, Interdependence, and Sociability in 17th-Century Science," *Critical Inquiry* 22 (1996): 193–238.

_____. *Galileo, Courtier: The Practice of Science in the Culture of Absolutism.* Chicago: University of Chicago Press, 1993.

_____. "Scientific Revolution and Aristocratic Ethos: Federico Cesi and the Accademia dei Lincei," in Carlo Vinti, ed. *Alexandre Koyré: L'avventura intellectuale.* Napoli: Edizioni Scientifiche Italiane, 1994.

Bernstein, Jeremy. "Heaven's Net: The Meeting of John Donne and Johannes Kepler," *American Scholar* 66 (1997): 175–195.

Chapman, Allan. "The Astronomical Work of Thomas Harriot (1560–1621)," *Quarterly Journal of the Royal Astronomical Society* 36 (1995): 97–107.

Clawson, Calvin C. *The Mathematical Traveler.* New York: Plenum, 1994.

Cromwell, Peter R. "Kepler's Work on Polyhedra," *Mathematical Intelligencer* 17, 3 (1995): 23–33.

Dear, Peter. *Discipline & Experience: The Mathematical Way in the Scientific Revolution.* Chicago: University of Chicago Press, 1995.

_____. "The Mathematical Principles of Natural Philosophy: Toward a Heuristic Narrative of the Scientific Revolution," *Configurations* 6 (1998): 173–193.

_____. *Mersenne and the Learning of the Schools.* Ithaca, NY: Cornell University Press, 1993.

De Caro, "Galileo's Mathematical Platonism," in Johannes Czernak, ed., *Philosophie der Mathematik: Akten des 15. Internationalen Wittgenstein-Symposiums.* Vol. 1. Vienna: Hölder-Pichler-Tempsky, 1993, pp. 13–22.

Dooley, Brendan. "The Communications Revolution in Italian Science," *History of Science* 33 (1995): 469–496.

Drake, Stillman. *Galileo at Work: His Scientific Biography.* Repr. of 1978 edition. New York: Dover Publications, 1995.

Edgerton, Jr., Samuel Y. *The Heritage of Giotti's Geometry: Art and Science on the Eve of the Scientific Revolution.* Ithaca, NY: Cornell University Press, 1991.

Feingold, Mordechai. "Decline and Fall: Arabic Science in 17th-Century England," in F. Jamil Ragep, et al., eds., *Tradition, Transmission, Transformation.* Leiden: Brill, 1996, pp. 441–469.

Field, J. V. *Kepler's Geometrical Cosmology.* Chicago: University of Chicago Press, 1988.

_____."The Relation Between Geometry and Algebra: Cardano and Kepler on the Regular Heptagon," in Eckhard Kessler, ed. *Girolamo Cardano: Philosoph, Naturforscher, Arzt.* Harrasowitz Verlag, 1994, pp. 219–242.

Folkerts, M. and J. P. Hogendijk, eds. *Vestigia Mathematica: Studies in Medieval and Early Modern Mathematics in Honour of H. L. L. Busard.* Amsterdam: Rodopi, 1993.

Jesseph, Douglas M. "Hobbes and Mathematical Method," *Perspectives in Science,* 1 (1993): 306–341.

Koyré, Alexandre. *The Astronomical Revolution: Copernicus–Kepler–Borelli.* Trans. by R. E. W. Maddison. Paris: Hermann, 1961, Eng. trans., 1973.

Kozhamthadam, Job. *The Discovery of Kepler's Laws: The Interaction of Science, Philosophy, and Religion.* Notre Dame, IN: University of Notre Dame Press, 1994.

Lohne, J. A. "Essays on Thomas Harriot," *Archive for History of Exact Sciences* 20 (1979): 189–312.

Mathews, Nieves. *Francis Bacon: The History of Character Assassination.* New Haven: Yale University Press, 1996.

Meskens, Ad. "Wine Gauging in Late 16th- and Early 17th-Century Antwerp," *Historia Mathematica* 21 (1994): 121–147.

Moss, Jean Dietz. *Novelties in the Heavens: Rhetoric and Science in the Copernican Controversy.* Chicago: University of Chicago Press, 1993.

Ong, Walter J., *Ramus Method and the Decay of Dialogue.* Cambridge, MA: Harvard University Press, 1958.

Page, Carl. "Symbolic Mathematics and the Intellect Militant: On Modern Philosophy's Revolutionary Spirit," *Journal of the History of Ideas,* 57 (1996): 233–253.

Parkinson, G. H. R., ed. *The Renaissance and 17th-Century Rationalism.* London: Routledge, 1993.

Pitt, Joseph C. *Galileo, Human Knowledge, and the Book of Nature: Method Replaces Metaphysics.* Dordrecht: Kluwer Academic Publishers, 1992.

Ridder-Symeons, Hilde de, ed. *Universities in Early Modern Europe, 1500–1800.* Cambridge: Cambridge University Press, 1996.

Rosen, Edward. *Three Imperial Mathematicians.* New York: Abanis, 1986.

Shank, Michael H. "How Shall We Practice History? The Case of Mario Biagioli's *Galileo Courtier,*" *Early Science and Medicine* 1 (1996): 106–150.

Sharrat, Michael. *Galileo: Decisive Innovator.* Cambridge: Cambridge University Press, 1996.

Stengers, Isabelle. "The Galileo Affair," in Michel Serres, ed., *A History of Scientific Thought: Elements of a History of Science.* Oxford: Blackwell Reference, 1995, pp. 280–314.

Sullivan, Dale L. "Galileo's Apparent Orthodoxy in his *Letter to the Grand Duchess Christina,*" *Rhetorica* 12 (1994): 237–264.

Van Helden, Albert. "Telescopes and Authority from Galileo to Cassini," *Osiris* 9 (1994): 9–29.

Wallace, William A. *Galileo and His Sources.* Princeton, NJ: Princeton University Press, 1984.

_____. *Galileo's Logic of Discovery and Proof: The Background, Content, and Use of his Appropriated Treatises on Aristotle's Posterior Analytics.* Dordrecht: Kluwer Academic Publishers, 1992.

_____. *Galileo's Logical Treatises: A Translation, with Notes and Commentary of his Appropriated Latin Questions on Aristotle's Posterior Analytics.* Dordrecht: Kluwer Academic Publishers, 1992.

Zack, Naomi. *Bachelors of Science: Seventeenth-Century Identity, Then and Now.* Philadelphia: Temple University Press, 1996.

Transformation ca. 1600–1660: II: To the Edge of Modernity

Selected Primary Sources

Descartes, René, *Discourse on Method, Optics, Geometry, and Meteorology.* Trans. by Paul J. Olscamp. Indianapolis, IN: Bobbs-Merrill, 1965.

Field, J. V., and J. J. Gray, eds. *The Geometrical Works of Girard Desargues* New York: Springer-Verlag, 1987.

Pappus, *La collection mathématique.* French translation by Paul Ver Eeecke. Paris: Bruges, 1933.

Pascal, Blaise. *Oeuvres complètes.* Édition présentée, établie et annotée par Michel Le Guern. I. Paris: Gallimard, 1998. Volume one contains his scientific writings.

――――. *Treatise on the Arithmetical Triangle.* Trans. by Richard Scofield. *Great Books of the Western World,* Vol. 33. Chicago: Encyclopedia Britannica, 1952.

Roberval, G. P. de. *Éléments des géométrie.* Paris: Vrin, 1996.

Scott, J. F. *The Scientific Works of René Descartes.* London: Taylor and Francis, 1952; repr. 1976.

Smith, D. E., and M. L. Latham, eds. *The Geometry of René Descartes.* Chicago: Open Court, 1925.

Walker, Evelyn. *A Study of the Traité des indivisibles of Gilles Persone de Roberval.* New York: Columbia University Press, 1932.

Secondary Works

Aczel, Amir D. *Fermat's Last Theorem.* New York: Four Walls Eight Windows, 1996.

Adamson, Donald. *Blaise Pascal: Mathematician, Physicist, and Thinker about God.* New York: St. Martin's Press, 1995.

Andersen, Kirsti. "Cavalieri's Method of Indivisibles," *Archive for History of Exact Sciences* 31 (1985): 291–367.

Ariew Roger, ed. *Descartes versus Gassendi* , an issue of *Perspectives on Science* 3 (1995): 425–581.

Bagley, Paul J. "Descartes, Triangles, and the Existence of God," *Philosophical Forum* 27 (1995): 1–26.

Baron, M. E. *The Origins of Infinitesimal Calculus.* London: Pergamon, 1969.

Bell, David F. "Pascal: Casuistry, Probability, Uncertainty," *Journal of Medieval and Early Modern Studies* 28 (1998): 37–50.

Bird, Alexander. "Squaring the Circle: Hobbes on Philosophy and Geometry," *Journal of the History of Ideas* 57 (1996): 217–231.

Bold, Stephen C. *Pascal Geometer: Discovery and Invention in 17th-Century France.* Geneva: Droz, 1996.

Bos, Henk J. M. "On the Interpretation of Exactness," in Johannes Czermak, ed., *Philosophie der Mathematik: Akten des 15. Internationalen Wittgenstein-Symposiums.* Vol. 1. Vienna: Hölder-Pichler-Tempsky, 1993, pp. 23–44.

Boyer, Carl B. "Early Rectifications of Curves" in *Mélanges Alexandre Koyré.* 2 vols. I:30–39. Paris: Hermann, 1964.

Breger, Herbert. "The Mysteries of Adaesquare: A Vindication of Fermat," *Archive for History of Exact Sciences,* 46 (1994): 193–219.

Brown, James, and Jürgen Mittelstrass, eds. *An Intimate Relation: Studies ... Presented to Robert E. Butts.* Dordrecht: Kluwer, 1989.

Buford, Norman. *Portraits of Thought: Knowledge, Methods, and Styles in Pascal.* Columbus: Ohio State University Press, 1989.

Cole. John R. *Pascal: The Man and His Two Loves.* New York: New York University Press, 1995.

Coolidge, Julian Lowell. *The Mathematics of Great Amateurs.* Oxford: Clarendon Press, 1990.

John Cottingham, ed. *Descartes.* Oxford: Oxford University Press, 1998.

Davidson, Hugh M. *Pascal and the Arts of the Mind.* Cambridge: Cambridge University Press, 1993.

Devlin, Keith. *Goodbye, Descartes.* New York: John Wiley & Sons, 1996.

Edwards, A. W. F. *Pascal's Arithmetical Triangle.* Oxford: Oxford University Press, 1987.

Edwards, Harold. *Fermat's Last Theorem: A Genetic Introduction to Algebraic Number Theory.* New York: Springer-Verlag, 1977.

Gaukroger, Stephen, *Descartes: An Intellectual Biography.* Oxford: Clarendon Press, 1995.

_____, ed. *The Uses of Antiquity: The Scientific Revolution and the Classical Tradition.* London: Kluwer, 1990.

Gillies, Daniel, ed. *Revolutions in Mathematics.* Oxford: Clarendon Press, 1995.

Gingerich, Owen, ed. *The Nature of Scientific Discovery.* Washngton, DC: Smithsonian Institution Press, 1975.

Grosholz, Emily R., *Cartesian Method and the Problem of Reduction.* Oxford: Clarendon Press, 1991.

Hacking, Ian. *The Emergence of Probability.* Cambridge: Cambridge University Press, 1975.

Henry, John. *The Scientific Revolution and the Origins of Modern Science.* New York: St. Martins, 1997.

Hoskin, Michael, ed. *The Cambridge Illustrated History of Astronomy.* Cambridge: Cambridge University Press, 1996.

_____. "Of Analytics and Indivisibles: Hobbes on the Methods of Modern Mathematics," *Revue d'histoire des sciences et de leur applications,* 46 (1993): 153–193.

Jesseph, Douglas M. *Lectures in the History of Mathematics (* History of Mathematics, 7) Providence, RI: American Mathematical Society, 1993.

_____. "On the Representation of Curves in Descartes *Géométrie,*" *Archives for History of Exact Science* 24 (1981): 285–338.

_____. "Recognition and Wonder: Huygens, Tractional Motion, and Some Thoughts on the History of Mathematics," *Tractrix,* 1 (1989): 3–20.

Knobloch, Eberhard. "Harmony and Chaos: Mathematics Serving a Teleological Understanding of the World," *Physis* 32 (1995): 55–89.

Lenoir, Timothy. "Descartes and the Geometrization of Thought: The Methodological Background to Descartes' *Géométrie,*" *Historia Mathematica* 6 (1979): 355–379.

Lindberg, David C. and Robert S. Westman, eds. *Reappraisals of the Scientific Revolution*. Cambridge: Cambridge University Press, 1990.

Lützen, Jesper. "The Relationship Between Pascal's Mathematics and His Philosophy," *Centaurus* 24 (1980): 263–272.

Mahoney, Michael, S. *The Mathematical Career of Pierre de Fermat, 1601–1665*. Princeton, NJ: Princeton University Press, 1973, repr. 1994.

Malet, Antoni. "Gregorie, Descartes, Kepler, and the Law of Refraction," *Archives internationales d'histoire des sciences* 40 (1990): 278–304.

Mancosu, Paolo. "Aristotelian Logic and Euclidean Mathematics: Seventeenth-Century Developments of the *Quaestio de Certitudine Mathematicarum*," *Studies in History and Philosophy of Science* 23 (1992): 241–265.

_____. *Philosophy of Mathematics and Mathematical Practice in the Seventeenth Century*. Oxford: Oxford University Press, 1996.

Molland, Andrew G. "Shifting the Foundations: Descartes' Transformation of Ancient Geometry," *Historia Mathematica* 3 (1976): 21–49.

Otte, Michael and Panza, Marco, eds. *Analysis and synthesis in mathematics: History and philosophy*. Dorecht: Kluwer, 1997.

Ramírez, Santiago and Cohen, Robert S., eds. *Mexican Studies in the History and Philosophy of Science*. Dordrecht: Kluwer Academic, 1995.

Rodis-Lewis, Geneviève. *Descartes: His Life and Thought* Trans. by Jane Marie Todd. Ithaca, NY: Cornell University Press, 1998.

Sepper, Dennis L. *Descartes's Imagination: Proportion, Images, and the Activity of Thinking*. Berkeley: The University of California Press, 1996.

Shapin, Steven. *The Scientific Revolution*. Chicago: University of Chicago Press, 1996.

Shea, William R. *The Magic of Number and Motion: The Scientific Career of René Descartes*. Canton, MA: Science History Publications, 1991.

Singh, Simon. *Fermat's Enigma*. Garden City, NY: Doubleday and Co., 1998.

Smith, Mark. "Galileo's Theory of Indivisibles: Revolution or Compromise," *Journal for the History of Ideas*, 37 (1976): 571–588.

Thackray, Arnold, ed. *Constructing Knowledge in the History of Science. Osiris*, vol. 10. Chicago: University of Chicago Press, 1995.

Weil, André. *Number Theory: An Approach Through History*. Basel: Birkhäuser, 1983.

Weinberg, Steven. "The Revolution That Didn't Happen," *The New York Review of Books*, Vol. XLV, Number 15, October 8, 1998, pp. 48–52.

The Apex of the Scientific Revolution I: Settings and Laureates

Primary Sources

Chandrasekhar, Subrahmanyan. *Newton's Principia for the Common Reader*. Oxford: Oxford University Press, 1995.

Huygens, Christiaan. *Oeuvres complètes de Christiaan Huygens*. 22 vols. The Hague: Martinus Nijhoff, 1888–1950.

_____. *The Pendulum Clock or Geometrical Demonstrations Concerning the Motion of Pendula as Applied to Clocks.* Trans. by Richard J. Blackwell, intro. by H. J. M. Bos. Ames: Iowa State University, 1986.

Leibniz, Gottfried Wilhelm von. *Sämtliche Schriften und Briefe.* (Third Series: Mathematical, Natural Science, and Technical Letters) Heinz-Jürgen Hess, James G. O'Hara, and Herbert Breger, eds., Berlin: Akademie Verlag, 1995.

Newton, Isaac. *The Correspondence of Isaac Newton.* 7 vols. Turnbull, H. W. (vols. 1–3), J. F. Scott (4), and A. Rupert Hall, and Laura Tilling (5–7), eds. Cambridge: Cambridge University Press, 1959–1977.

_____. *Newton's Principia: The Central Argument.* Trans. and Notes by Dana Densmore and William H. Donahue. Santa Fe, NM: Green Lion Press, 1995.

_____. *Texts, Backgrounds, Commentaries.* Ed. by I. Bernard Cohen and Richard S. Westfall. New York: Norton, 1995.

Secondary Works

Achinstein, Peter and Snyder, Laura J., eds. *Scientific Methods: Conceptual and Historical Problems.* Malabar, FL: Krieger, 1994.

Aiton, E. J. *Leibniz: A Biography.* Bristol: Adam Hilger Ltd., 1985.

_____. "Polygons and Parabolas: Some Problems Concerning the Dynamics of Planetary Orbits," *Centaurus* 31 (1989): 207–221.

Aoki, Shinko. "The Moon-Test in Newton's *Principia*: Accuracy of the Inverse-Square Law of Universal Gravitation," *Archive for History of Exact Sciences* 44 2 (1992): 147–190.

Blay, Michael, *Les "Principia" de Newton.* Paris: Presses Universitaires de France, 1995.

Boss, Valentin. *Newton & Russia: The Early Influence, 1698–1796.* Cambridge, MA: Harvard University Press, 1972.

Brackenridge, J. Bruce. *The Key to Newton's Dynamics: The Kepler Problem and the Principia.* Trans. by Mary Ann Rossi. Berkeley: University of California Press, 1995.

Costabel, Pierre. *Leibniz and Dynamics: The Texts of 1692.* Trans. by R. E. W. Maddison. Paris: Hermann, 1973.

De Gandt, François. *Force and Geometry in Newton's Principia.* Trans. by Curtis Wilson. Princeton, NJ: Princeton University Press, 1995.

Densmore, Dana. *Newton's Principia: The Central Argument.* trans. by William H. Donahue. Santa Fe, NM: Green Lion Press, 1995.

Dobbs, Betty J. T. *The Foundations of Newton's Alchemy: The Hunting of the Greene Lyon.* Cambridge: Cambridge University Press, 1974.

Drake, Ellen Tan. *Restless Genius: Robert Hooke and His Earthly Thoughts.* Oxford: Oxford University Press, 1996.

Edwards, Ernest L. *The Story of the Pendulum Clock.* Altrincham: Sherratt, 1977. (This book includes an English translation of Huygens' 1658 *Horologium*.)

Gascoigne, John. "A Reappraisal of the Role of the Universities in the Scientific Revolution," in David C. Lindberg and Robert S. Westman, eds., *Reappraisals of the Scientific Revolution.* Cambridge: Cambridge University Press, 1990, pp. 207–260.

Gowing, Ronald. "Halley, Cotes, and the Nautical Meridian," *Historia Mathematica* 22, 1 (1995): 19–33.

Greenberg, John L. "Isaac Newton and the Problem of the Earth's Shape," *Archive for History of Exact Sciences* 49 (1996): 371–391.

_____. *The Problem of the Earth's Shape from Newton to Clairaut.* Cambridge: Cambridge University Press, 1995.

Hahn, Roger. *The Anatomy of a Scientific Institution: The Paris Academy of Sciences, 1666–1803.* Berkeley: University of California Press, 1971.

Hall, A. Rupert. *Henry More and the Scientific Revolution.* Cambridge: Cambridge University Press, 1996.

_____. *Isaac Newton: Adventurer in Thought.* Cambridge: Cambridge University Press, 1996.

_____. *Philosophers at War: The Quarrel between Newton and Leibniz.* Cambridge: Cambridge University Press, 1980.

_____. "Sir Isaac Newton's Note-book, 1661–65," *Cambridge Historical Journal,* 9 (1948): 239–250.

Harman, P. M., and Alan E. Shapiro, eds. *The Investigation of Difficult Things: Essays on Newton and the History of the Exact Sciences in Honour of D. T. Whiteside.* Cambridge: Cambridge UniversityPress, 1992.

Hill, Katherine. "John Wallis and Isaac Barrow: Tradition and Innovation and the State of Mathematics," *Endeavour* 22 (1998): 117–120.

Iliffe, Robert, " 'Is he like other men?': The Meaning of the *Principia Mathematica,* and the Author as Idol," in Maclean, Gerald, ed.*Culture and Society in the Stuart Restoration: Literature, Drama, History.* Cambridge: Cambridge University Press, 1995.

Keynes, Milo. "The Personality of Isaac Newton," *Notes and Records of the Royal Society of London* 49 (1995): 1–56.

Klawans, Harold L. *Newton's Madness (Further Tales of Clinical Neurology).* New York: Harper Perennial Publishing, 1991.

Koyré, Alexandre. "An Experiment in Measurement," *Proceedings of the American Philosophical Society* 7 (1953): 222–237.

_____. *Newtonian Studies.* Cambridge, MA: Harvard University Press, 1965.

Landes, David S. *Revolution in Time: Clocks and the Making of the Modern World.* Cambridge, MA: Belknap, 1983.

Lindberg, David C., and Robert S. Westman. *Reappraisals of the Scientific Revolution.* Cambridge: Cambridge University Press, 1990.

Lux, David S. and Harold J. Cook. "Closed Circles or Open Networks? Communicating at a Distance During the Scientific Revolution," *History of Science* 36 (1998): 179–211.

Manuel, Frank E. *Isaac Newton, Historian.* Cambridge, MA: Harvard University Press, 1963.

_____. *The Religion of Isaac Newton.* Oxford: Oxford University Press, 1974.

McGuire, J. E. and P. M. Rattansi, "Newton and the 'Pipes of Pan'," *Notes and Records of the Royal Society,* 21 (1966): 108–143.

Mercer, Christia. "Mechanizing Aristotle: Leibniz and Reformed Philosophy," in M. A. Stewart, ed., *Studies in 17th-Century European Philosophy* (Oxford: Clarendon Press, 1997), pp. 117–152.

Palter, Robert, ed. *The "Annus Mirabilis" of Sir Isaac Newton, 1666–1966.* Cambridge, MA: M.I.T. Press, 1970.

Rescher, Nicholas, ed. *Leibnizian Inquiries: A Group of Essays.* Lanham, MD: University Press of America, 1989.

Ross, G. MacDonald. *Leibniz.* Oxford: Oxford University Press, 1984.

Verlet, Loup. " 'F = ma' and the Newtonian Revolution: An Exit from Religion Through Science," *History of Science* 34 (1996): 303–346.

Westfall, Richard S. *The Life of Isaac Newton.* Cambridge: Cambridge University Press, 1993.

_____. *Never at Rest: A Biography of Isaac Newton.* Cambridge: Cambridge University Press, 1980.

_____. "Newton and the Acceleration of Gravity," *Archive for History of Exact Sciences* 35 (1986): 255–272.

Whiteside, Derek T., "Newton's Marvellous Years: 1666 and All That," *Notes and Records of the Royal Society of London* 21 (1966): 32–41.

Wilson, Catherine. *Leibniz's Metaphysics: A Historical and Comparative Study.* Princeton, NJ: Princeton University Press, 1989.

The Apex of the Scientific Revolution II: Calculus to Probability

Primary Sources in Translation

Barrow, Isaac. *The Mathematical Works of Isaac Barrow, D. D.* Ed. by William Whewell. Hildesheim: Georg Olms, 1973.

Bernoulli, Jakob. *Opera.* 2 vols. Geneva, 1744. Reprinted, Brusssels, Culture et Civilisation, 1967.

Bernoulli, Johannes. *Opera omnia.* 4 vols. Lausanne, 1742. Reprinted, Hildesheim: Olms, 1968.

Fontenelle, Bernard le Bovier de. *Élements de la géométrie de l'infini.* Intro. by Michael Blay and Alain Niderst. Paris: Editions Klincksieck, 1995.

Leibniz, Gottfried Wilhelm. *Mathematische Schriften* . Edited by C. I. Gerhardt, 7 vols. Berlin, 1849–63, Reprinted, Hildesheim, Olms, 1971.

_____. *La naissance du calcul differential.* Trans. by Marc Parmentier with a preface by Michael Serres. Paris: Vrin, 1989.

_____. *Oeuvre mathématique, autre que le calcul infinitesimal.* Paris: A.Blanchard, 1989.

Moivre, Abraham De. *The Doctrine of Chances* (London, 1718; 2nd ed., 1738; 3rd ed., 1756); photo. repr. of 2nd ed., London, 1967; photo. repr. of 3rd ed., with a biography by Helen M. Walker, New York, 1967.

Scott, Joseph F. *The Mathematical Work of John Wallis, D. D., F. R. S. (1616–1703).* London: Taylor & Francis, 1938.

Whiteside, Derek T., ed. *The Mathematical Papers of Isaac Newton.* 8 vols. Cambridge: Cambridge University Press, 1967–1981.

Secondary Works

Arthur, Richard T. W. "Newton's Fluxions and Equably Flowing Time," *Studies in the History and Philosophy of Science* 26 (1995): 323–351.

Boyer, Carl. *The History of the Calculus and Its Conceptual Development.* New York: Dover, 1959.

Cohen, I. Bernard. *The Newtonian Revolution.* Cambridge: Cambridge University Press, 1980.

Cook, Alan. *Edmond Halley: Charting the Heavens and the Seas.* Cambridge: Cambridge University Press, 1998.

Eco, Umberto. *History's Best Bad Ideas.* New York: Columbia University Press, 1998.

Erlichson, Herman. "The Visualization of Quadratures in the Mystery of Corollary 3 to Propostion 41 of Newton's *Principia*," *Historia Mathematica* 21, 2 (1994): 148–162.

Feingold, Mordechai, ed. *Before Newton: The Life and Times of Isaac Barrow.* Cambridge: Cambridge University Press, 1990.

_____. "Newton, Leibniz, and Barrow Too," *Isis* 84 (1993): 310–338.

Garrison, James W. "Newton and the Relation of Mathematics to Natural Philosophy," *Journal of the History of Ideas* 48 (1987): 609–627.

Guillen, Michael. *Five Equations That Changed the World: The Power and Poetry of Mathematics.* New York: Hyperion, 1995.

Hairer, Ernst and Gerhard Wanner. *Analysis by Its History.* New York: Springer-Verlag, 1996.

Hill, Katherine. "Neither Ancient nor Modern: Wallis and Barrow on the Composition of Continua. Part I: Mathematical Styles and the Composition of Continua," *Notes and Records of the Royal Society of London* 50 (1996): 165–178.

Hofmann, Joseph E. *Leibniz in Paris, 1672–1676.* Cambridge: Cambridge University Press, 1974.

Hollingdale, Stuart. "Leibniz and the First Publication of the Calculus in 1684," *Bulletin of the Institute of Mathematics and Its Applications* 21 (1985): 88–94.

Kitcher, P. "Fluxions, Limits, and Infinite Littleness: A Study of Newton's Presentation of the Calculus," *Isis* 64 (1973): 33–49.

Knobloch, Eberhard. "The Infinite in Leibniz's Mathematics," in Kostas Gavroglu et al., eds. *Trends in the Historiography of Science.* Dordrecht: Kluwer Academic, 1994, pp. 265–278.

Kracht, Manfred. "E. W. Tschirnhaus: His Role in Early Calculus and His Work and Impact in Algebra," *Historia Mathematica,* 17 (1990): 16–35.

Malet, Antoni. "Barrow, Wallis, and the Remaking of 17th Century Indivisibles," *Centaurus* 39 1997: 67–92.

_____. "James Gregorie on tangents and the "Taylor" Rule for Series Expansions," *Archive for History of Exact Sciences,* 46 (1993): 97–137.

Mancosu, Paolo. "The Metaphysics of the Calculus: A Foundational Debate in the Paris Academy of Sciences, 1700–1706," *Historia Mathematica* 16, 3 (1989): 224–248.

Nahin, Paul J. *An Imaginary Tale: The Story of $\sqrt{-1}$.* Princeton, NJ: Princeton University Press, 1998.

Nastasi, P, and A. Scimone. "Pietro Mengoli and the Six-Square Problem," *Historia Mathematica,* 21 (1994): 10–27.

Otte, Michael. "Two Principles of Leibniz's Philosophy in Relation to the History of Mathematics," *Theoria,* 19 (1993): 113–125.

Paxson, James J. "The Allegory of Temporality and Early Modern Calculus," *Configurations* 4 (1996): 39–66.

Pourciau, Bruce. "The Preliminary Mathematical Lemmas of Newton's *Principia,*" *Archive for History of Exact Sciences,* 52 (1998): 279–295.

Pycior, Helena. *Symbols, Impossible Numbers, and Geometric Entanglements: British Algebra through the Commentaries on Newton's Universal Arithmetic.* Cambridge: Cambridge University Press, 1997.

Ray, Ranjan. "The Discovery of the Series Formulas for $\pi/4$ by Leibniz, Gregory, and Nilakantha," *Mathematics Magazine* 63, 5 (1990): 291–306.

Rickey, V. Frederick. "Isaac Newton: Man, Myth, and Mathematics," *College Mathematical Journal* 19 (1987): 362–89.

Rotman, Brian. "The Teachnology of Mathematical Persuasion," in Timothy Lenoir, ed., *Inscribing Science: Scientific Texts and the Materiality of Communication.* Stanford, CA: Stanford University Press, 1998. pp. 55 – 69.

Ryan, James J. "Leibniz's Binary System and Shao Yang's *Yijing,*" *Philosophy East and West* 46, 1 (1996): 59–90.

Scriba, Christoph J. "The Autobiography of John Wallis, F.R.S.," *Notes and Records of the Royal Society of London* 25 (1970): 17–46.

Swoyer, Chris. "Leibniz's Calculus of Real Addition," *Studia Leibnitiana* 26 (1994): 1–30.

Sylla, Edith Dudley. "Jacob Bernoulli on analysis, synthesis, and the law of large numbers," in Michael Otto and Marco Panza, eds., *Analysis and Synthesis in Mathematics: History and Philosophy.* Dordrecht: Kluwer, 1997. pp. 79–101.

Van Helden, Anne C. and van Gent, Rob. *The Huygens Collection.* Leiden: Museum Boerhaave, 1995.

Van Rijen, Joroen, "Some Misconceptions About Leibniz and the Calculi of 1679," *Studia Leibnitiana* 21 (1989): 195–204.

Vermij, R. H., "Bernard Nieuwentijt and the Leibnizian Calculus," *Studia Leibnitiana,* 21 (1989): 196–204.

Vucinich, Alexander. *Science in Russian Culture: A History to 1860.* Stanford: Stanford University Press, 1963.

Whiteside, Derek T., "Patterns of Mathematical Thought in the Later Seventeenth Century," *Archive for History of Exact Sciences* 1 (1960–62): 179–388.

Whitrow, G. J. "Newton's Role in the History of Mathematics," *Notes and Records of the Royal Society of London* 43 (1989): 71–92.

Yoder, Joella G. *Unrolling Time: Christian Huygens and the Mathematization of Time.* Cambridge: Cambridge University Press, 1988.

Yule, G. Udny. "John Wallis, D.D., F.R.S.," *Notes and Records of the Royal Society of London* 2 (1939): 74–82.

Name Index

Subject Index

Photo Credits

94: (Left, Parthenon) Paul W. Liebhardt, (Right, model of the Acropolis) Royal Ontario Museum. **98**: Corbis. **106**: North Wind Picture Archives. **121**: From Louis Figuier, VIES DES SAVANTS ILLUSTRES (Paris, 1877), "Education of Alexander," p. 135. Dibner Library, Smithsonian Institution. **121**: Smithsonian Institution. **131**: (Left) Line engraving, French, 1584. The Granger Collection, New York. (Right) From Louis Figuier, VIES DES SAVANTS ILLUSTRES (Paris, 1877), "Euclid Presents Geography," p. 250. Dibner Library, Smithsonian Institution. **150**: (Left) Colored French engraving, 1584. The Granger Collection, New York. (Right) From Dibner Library, "Archimedes," Smithsonian Institution. **154**: From Louis Figuier, VIES DES SAVANTS ILLUSTRES (Paris, 1877), "Death of Archimedes," p. 239. Dibner Library, Smithsonian Institution. **176**: From Louis Figuier, VIES DES SAVANTS ILLUSTRES (Paris, 1877), "Apollonius at the Alexandrian Museum," p. 268. Dibner Library, Smithsonian Institution. **198**: Hulton Getty Picture Collection/Tony Stone Images. **221**: From Louis Figuier, VIES DES SAVANTS ILLUSTRES (Paris, 1877), "Death of Hypatia," p. 459. Dibner Library, Smithsonian Institution. **400**: Raphael. "The School of Athens." 1508. Stanze di Raffaello, Vatican Palace, Vatican City. **417**: From Dibner Library, "Niccolo Tartaglia," Smithsonian Institution. **425**: From Hieronymi Canda, "Ni Mediolanensis Medic (Basel, 1554)," Cardano portrait, frontispiece. Dibner Library, Smithsonian Institution. **433**: From Dibner Library, "François Viète," Dibner Library, Smithsonian Institution. **446**: From Dibner Library, "Christopher Clavius," Smithsonian Institution. **453**: Rembrandt, Dutch, 1606–1669. The Anatomy Lesson of Dr. Tulp, 1632. Oil on canvas, 169.5×216.5 cm. The Hague, Mauritshuis. **457**: From Dibner Library, "Johannes Kepler," Smithsonian Institution. **463**: From Giambatista Venturi, ed., "Memorie e lettere ... Galileo Galilei," (Modena, 1818), Galileo (from 1613) frontispiece. Dibner Library, Smithsonian Institution. **484**: From Simon Stevin, "De Beghinsein der Weeghconst (Leiden, 1586)," Wreath of spheres, frontispiece. Dibner Library, Smithsonian Institution. **485**: From Dibner Library, "John Napier," Smithsonian Institution. **499**: From A. Baillet, "La vie de Monsieur Descartes," (Paris, 1691), Dibner Library, Smithsonian Institution. **532**: From Dibner Library, "Blaise Pascal," Smithsonian Institution. **533**: Courtesy of International Business Machines Corporation. Unauthorized use not permitted. **574**: From Christiani Hugeni, OPERA VARIA, vol. 1 (Leiden, 1724), "Christiaan Huygens," frontispiece. Dibner Library, Smithsonian Institution. **583**: From G. G. Leibnitii, "Opera Omnia," vol. 1 (Geneva, 1768), "Leibniz," frontispiece. Dibner Library, Smithsonian Institution. **599**: From Dibner Library, "John Wallis," Smithsonian Institution. **605**: From Dibner Library, "Isaac Barrow," Smithsonian Institution. **636**: From Dibner Library, "Pierre Varignon," Smithsonian Institution.